NEW YORK COTTON EXCHANGE.

CHRONOLOGICAL AND STATISTICAL

HISTORY OF COTTON,

BY

E. J. DONNELL.

"How strange it is that so few attempts have been made to trace the rise and progress of this great branch of industry, the cotton manufacture; to mark the successive steps of its advancement, the solidity of the foundations on which it rests, and the influence which it has already had, and must continue to have, on the number and condition of the people." —MCCULLOCH, in the *Edinburgh Review*.

PUBLISHED BY THE AUTHOR.

NEW YORK:
JAMES SUTTON & CO., PRINTERS, 23 LIBERTY STREET.

1872.

PREFACE.

THIS work has been prepared for the purpose of supplying the cotton trade, and all who are brought into business relations with it, with needed information in a convenient and compact form. No attempt has been made to present anything strikingly novel or altogether new. The facts, which are spread over the succeeding pages, have existed in crude and detached form, but accessible to only a few. In collecting them together, reducing them to shape, and presenting them in a manner admitting of easy reference and comparison, much severe labor has been performed, and there has been no lack of patience and research in its prosecution. Every effort has been made to guard against errors, but among such a mass of figures, perfect accuracy could hardly be expected. In the leading statements, however, confidence may be felt that they are substantially correct.

Important as cotton is, as a staple of commerce, its importance is not more wonderful than the briefness of the period in which it has, with us, reached its pre-eminence. In the course of my researches into the statistics of cotton movements, I find that in only three ports in the United States have figures approximately full been preserved for any considerable period, namely: New York, Charleston, and New Orleans. And yet, in Liverpool, nearly fifty years ago, cotton was regarded by an eminent authority as having then reached the highest pinnacle of its commercial importance.

It must now be apparent to the least reflecting mind, that a new era of development has been opened to the growth and manufacture of cotton; and, to judge what course this development is likely to take, and what progress it will probably make, there can be no better help than in studying closely the history of the growth and manufacture of cotton during the past half century. Fifty years carry us back to the period when Europe began to recover from the effects of the war of the French Revolution; and there investigations and comparisons may properly begin.

In thirty-nine years, before the war of secession, the growth of cotton in the United States increased about one million bales in every decade. Beginning with 500,000 bales in 1822–3, it was eight years before the crop reached a million bales; then nine years to two million bales; then twelve years to three million bales; then eight years to four million bales, 1859–60. We are now back to three million bales, or, where we were twenty years ago. Therefore, when shall we return to four millions? and when shall we reach five millions? or six millions? are questions of great moment. Not less so are the probabilities respecting the increase in the manufacture and consumption of cotton goods, in the near and remote future. In the solution of all these problems so important, not only to the cotton trade but to the commerce and finances of the whole civilized world, this work will furnish effective aid. If I shall be regarded as having been reasonably successful in the difficult task which I have nndertaken, my highest ambition will be gratified.

In conclusion, I desire to express my obligations and return my thanks to many gentlemen connected with the commercial journals of this and other cities, for the valuable assistance which they have afforded me in the preparation of this work.

E. J. D.

New York, September, 1872.

INTRODUCTION.

THE history of cotton in its three departments, agricultural, commercial and manufacturing, is, in some of its aspects, the history of civilization. Its progress seems to have been controlled by the same laws that have governed the progress of the human race. It is a very good illustration of the laws which govern all natural growth, or evolution.

In the remotest records of history we see it just above the surface in India, its roots spreading in every direction. All that dexterity, with such tools as nature furnished, could accomplish, was brought to complete perfection thousands of years ago. Through long ages we see nothing but a naked trunk—no improvement, no change. At last we see branches in every direction. All growth in the individual, in society, in the human race, is by the process of *branching*. It is sometimes termed differentiation; a term which describes the mere phenomena very well, but has the grave fault of being wholly devoid of eteological significance.

It does not require very profound study of the subject to discover that the evolution of the individual is the analogue of the evolution of the race. That is an important fact to know; but the analogy extends much farther. It will be found in the passions and faculties of the human mind.

Take the affections, for instance. All the loves grow out of the one root, self-love. The first necessity is self-preservation—to exist. The first branch from this root is, love for those who first minister to our wants—love of parents, perhaps. The next branch is toward the family with which our name and interests are identified; the next is towards the nation of which we form a part, and is called patriotism; the next is towards the whole race, and is called philanthropy; the next, and last, is universal or Divine love. I give the natural order of succession; each has its proper function in which its action is wholly beneficent; but, when out of place, it is equally pernicious.

The history of the Jews is a well-defined illustration of the action of this law on a large scale. The entire religion and policy of that people, was founded upon the exclusive love of their tribe or race. They believed themselves to be the special favorites of Heaven. This worked admirably in the infancy of the race. It raised them out of the sands of the desert;

and, not only constituted, but fitted them to be the religious teachers of humanity.

In the unfoldment of the race, the time arrived when a new branch—a larger humanity, was demanded. It was the natural outgrowth and demand of the times. The Jews resisted the spirit of the times. Their utter ruin, as a nation, was the result; nor can they ever come into harmonious and happy relations with the times and with humanity, until they recognize, practically, the brotherhood and equality of the whole human family.

There are indications that that day is not far distant. The commercial spirit of the race is sure to lift them above all narrowness or littleness. The growth of the commercial spirit is governed by the same laws as all other growth. In its crude state, it is intensely selfish, but it grows continually toward universal ends.

The human family is *one*, and all its tribes and races are its members, its faculties. The good of each is the good of all. That which we would hoard will corrode and curse us.

There are few things more striking, in the history of human industry, than the fact that the cotton manufacture of India, so perfect in its kind, remained so long stationary. Not only did the Indians themselves remain without any new or improved methods or machinery, but they failed to teach any other people.

It has been said, by a highly philosophical writer, that civilization originated where the paths of two tribes first crossed each other. When a tribe or nation becomes separated from the rest, growth ceases.

In the vast extent of Asia, there was room for each tribe or nation to dwell by itself, developing each its own specialty. When the fullness of time came, tribe after tribe migrated into the peninsula of Europe. Migration with the sun, from east to west, seems always to improve the race.

In Europe, isolation was impossible. Owing to this cause alone, it is probable that permanent stagnation could not exist. The same wisdom that directs man's progress, prepared the continents for his use. Europe is so connected with Asia as to render migration easy. The accidental migration of a single tribe was not sufficient for the purpose. Shut in by the comparatively narrow confines of Europe, these tribes and peoples have been, for more than two thousand years, fighting with and learning from each other. Europe was destined to evolve a new humanity. As between Europe and Asia the centre of activity in the reciprocal action of one upon the other was the shores of the Mediterranean, so in the same action between exclusively European forces, the greatest development is toward central Europe. After America began to exercise a perceptible influence, the centre of activity became western central Europe. Since the opening of our Pacific trade with Asia, the United States began to assume this central position. It is no mere accident that turns the eyes of the whole world toward this continent, with an instinctive faith that it is here all the problems of the future are to be solved. With Europe on the one side, pouring in all the fruits of all past labor and suffering, and Asia on the other from the original fountains of human progress familiarizing us with the simpler instincts of childhood. On the one side, intellect—sight; on the other, instinct—feeling. The positive and negative batteries joined on the American continent, America will be the great reconciler, bringing all forms of humanity together.

It was not possible that manufactures could have advanced beyond the condition in which they existed in India before the eighteenth century. In the order of evolution, the human mind had not got beyond India in that department.

Historians seem to be utterly bewildered by the dense cloud of superstition that settled over Europe during the middle ages. Hardly any attempt is made to assign to the so-called dark ages their true place in the evolution of the race. It seems to me they were a necessary prelude to usher in the intellectual progress of the past four hundred years. Action and reaction—positive and negative—male and female; the same fundamental principle is at the bottom of the ideas expressed by all these terms. The principle is universal. We are always swinging from an extreme in one direction toward an extreme in the opposite direction. In the middle ages, the human mind swung to the greatest extreme in the direction of the supernatural. That movement was necessary to give effective power to the present movement toward the natural. Humboldt seems to intimate that the contest between knowledge and belief commenced early in the Christian era. In one sense, it is a great deal older than that; in another, it is not much older than the twelfth century. The establishment of the inquisition in the early part of the thirteenth century shows that the present movement had then commenced. To the philosophical thinker, the establishment of the inquisition is the best possible evidence of the intense yearning after truth, which was then fermenting in the unconscious instincts of the European mind. The very same motive which originated the inquisition gave birth to free schools, which were established in connection with the monasteries in the middle ages. The free school system is yet young, but is the greatest power in modern society, while the inquisition is dead. All the motive forces of the mind have this double action, a knowledge of which would prevent much lamentation and regret over the misfortunes of humanity. In the childhood of the race (*even now it is so among children*), this dual action was mistaken for a conflict of opposing and independent forces. The Greek mythology, which is a faithful poetical record of this mistake, has not yet lost its influence on the human mind. Childhood never loses its fascination; it will continue to charm the oldest and wisest, even when its true position in the order of evolution is fully understood.

In the order of intellectual evolution, the idea of beauty precedes the idea of use. That is the reason why art so early reached perfection.

Dress, as ornament to the person, is the first of the fine arts. Dress was first used as ornament to the person; no other idea was attached to it. That was the germ from which sprang the arts and architecture of Greece. Use is higher than beauty, or, rather, it is the highest form of beauty. Art is prophecy of something better. Painting and sculpture were revived in Italy three hundred years before the *Renaissance* and general revival of learning in the sixteenth century. The movement in that century began with the discoveries of Copernicus, in astronomy, and the establishment of English colonies in North America. The mental horizon began to expand in all directions; yet the light, to most eyes, was very uncertain. In 1552, books of astronomy and geometry were destroyed in England as magical. In 1579 was published George Buchanan's celebrated treatise on the principles of government, in which he inculcates the doctrine that governments exist for the sake of the governed. As late as 104 years afterward, the work was burnt at Oxford, along with the works of Milton; nor did it cease to be generally condemned by the governing classes, in both Church and State, until about the time of the American Revolution.

Until this development in government was, in some degree, realized, no great and general expansion of industrial art was possible. It required two centuries for the doctrine, as to the responsibility and duty of governments to the governed, to be generally recognized and admitted. It will probably require as long for them to learn how this duty and responsibility can be most intelligently discharged.

The success of the American Revolution derived its main importance from the fact that it marked the progress of the human mind into a new and higher plane of development than it had before reached. The permanent, peaceful establishment of the Constitution of the United States was the first successful attempt in the history of the race to establish authority on a large scale, either in Church or State, upon a foundation purely and simply rational. The deep significance and far-reaching importance of such a fact are beyond all calculation. No matter what irregularities may take place in the practical working of our government, so long as the fact remains, it will be the landmark of a new departure. Reason may be defined as the blended *aroma* of all the faculties and powers of the human mind. All that we know as growth, or unfoldment, is in the direction of the supreme dominion of reason in everything.

In the fifteenth century, the intellectual movement was made manifest in many directions, by useful inventions, such as watches, the art of printing, painting in oil colors, delf-ware, and the manufacture of glass; by geographical discoveries, such as America by Columbus, the coast of Guinea, and the Cape de Verd Islands by the Portuguese. Algebra first began to be taught in Venice in 1495. In these movements we discover the germ of all that has been done since.

It may be here remarked that in all intellectual movements of the race one of its earliest manifestations is in astronomical research. Naturally the grand mystery of the visible heavens first rivits the attention of the roused and hungry intellect. Astronomical observations began at Babylon 2234 B. C., a register of which was transmitted by Callisthenes to Aristotle for 1903 years, to the capture of that city by Alexander, 331 B. C. As I before remarked, the sixteenth century commenced with grand colonization movements, and still grander astronomical discoveries. Geographical discoveries were pushed forward with great vigor. People began to doubt everything that was not capable of demonstration. When men doubt they must reason, because the soul hungers for the positive; therefore doubt is essential to growth. This fermentation in the European mind produced an immense augmentation of activity. It also caused great suffering; but this is the order of nature. The power of the human mind is in proportion to its capacity for suffering.

The seventeenth century showed intellectual development far exceeding any previous century in the history of the race. It seemed as if the human mind had, for the first time, been freed from bondage, and disported itself in its new-found liberty. In poetry, philosophy, and science, the seventeenth century has names that will, probably for a long time to come, be ranked the very first in each department. There were Shakespeare, Milton, Dryden, Moliere, Corneille, and Racine, Newton, Kepler, Galileo, and Descartes, besides a host of others.

There is no better evidence of growing power than the boldness with which individual sovereignty was asserted and maintained. This attitude of the human mind was an indispensable preliminary to success in those inventions through which the forces of nature have been harnessed to the car of human industry. In the order of evolution, there is an inseparable connection between this mastery over nature and a consciousness of the possession of sovereign attributes. Both are the legitimate result of a high degree of development. The one is self-knowledge, the other a knowledge of external nature.

All true growth is from within outward, because man possesses, in the constitution of his own mind, all the elements of infinite progress. This is because he is the offspring of the infinite. Even in the earliest infancy of the race, when man manifested little, if anything, more than blind instinct, we can read his future. He grew toward the truth as a plant toward the sunlight. The earliest instincts of the mind express themselves in the form of religious sentiment. Instinct is feeling, intellect is sight. Whatever instinct has felt, intellect will yet see and understand.

From the earliest records of the race, this idea that man is the offspring of the Infinite—of the Deity has been struggling up toward human consciousness. This has appeared, sometimes in a refined form, as when the Germanic tribes called the Deity the Allfather; at others, in a gross form, as when Apollo is represented as overshadowing the mother of Pythagoras, and countless other stories of carnal intercourse between the gods and the daughters of men. Every one of these myths has its root in the blind, but yet infallible instinct that we are the children of Deity—Sons of God.

In one case in history this truth seems to have revealed itself with an intensity so dazzling as to have bewildered the whole civilized world. This seems to me to have been the central idea that animated the founder of Christianity. It is the key that will unlock many mysteries. In one form or another it ran through everything He said. Of course, He was misunderstood, and still is. There are still vague ideas floating through men's minds of the fatherhood of God, but at the same time we are positively taught that we are the creation, not the offspring, of Deity.

This idea that man is not the creation, but the offspring of the Deity, is intensely revolutionary. It goes to the root of the matter. If it were possible for the human mind to comprehend it fully, without first growing up to it, it would pluck society up by the roots; but Divine wisdom has so ordered that men shall often realize a truth in practice before it rises fully into their consciousness—we feel a truth before we see it.

All that is called modern progress is the fruit of this truth, which is thus gradually preparing the way for its full revelation. It will be seen that during the sixteenth and seventeenth centuries, the human mind began to approach this great truth from the side of the intellect. "The whole movement was an appeal to the individual reason and conscience." What are the individual reason and conscience, was a question that was sure to suggest itself to reflecting minds; nor did the question remain entirely unanswered. Several writers in the sixteenth century claimed for the individual conscience the absolute right of final appeal; but it was left for Descartes, in the early part of the seventeenth century, to build around it a consistent compact system of philosophy. The whole truth Descartes, probably, did not see; his discoveries were not, therefore, the less valuable, but more so. He does not infer the individual power and its absolute right from its Divine origin, but he proves its possession of that power and this right by a system of reasoning at once lucid, comprehensive and convincing. "The thought of each man is the last element to which analysis can carry us; it is the supreme judge of every doubt; it is the starting point for all wisdom." Such is the essence of his system.

This is the same truth, in a different form, that was announced by Jesus more than sixteen hundred years before. It is because the individual soul is a child of Deity, that it possesses this attribute of sovereignty. It is only in the light of the intellectual progress of the present age—*perhaps, future ages*—that the true grandeur of the great truth that inspired and consumed the Man of Nazareth can be understood. He found none who could understand Him; the ages alone can interprete Him.

Not only is this the true key to what is called Christianity, it is the key to the future as well as the past. It is the keystone, without which, no system of philosophy, no matter how ingeniously elaborated, can stand. All social science, all statesmanship, must have this truth for their foundation and guide, consciously or unconsciously, or they will fail. All modern progress is in this direction—the dignity of the individual and the brotherhood of the race. These ideas began to produce fruit in the eighteenth century, which commenced with the war of the Spanish succession and ended amidst the wars of the French Revolution, with liberty organized and established on the American continent.

About the middle of the eighteenth century began those discoveries and inventions which have transformed civilized society, so that men now live and see and learn as much in one year as formerly in a lifetime. This is the vision seen by the Hebrew seer when he wrote, "In the latter days men will run to and fro, and knowledge shall be increased." That migration, which I before stated was a condition of improvement, is now all but universal, facilitated by modern improvements in traveling. There is now a universal commerce of ideas, which is the best part—the essence of all commerce.

I have already tried to illustrate the manner in which the affections unfold themselves in the progress of the race. It was fit that the law of universal love, the natural complement of the brotherhood of the race, should have been announced to the world, at the time that nearly all the nations in the then known world had been welded together under one empire; forced into a recognition of mutual dependence and equality. It was then a religious sentiment. It was also fit that in the eighteenth century, when the human intellect demanded an explanation for everything, and the laws of commerce were rapidly becoming the sole guide for the intercourse of men and nations, that it should be announced that the law of love is the true philosophy of trade.

In 1776 was published Adam Smith's "Inquiry into the Nature and Sources of the Wealth of Nations." The progress of economical science during the past hundred years, though only a part of the general movement, is by no means the least important. De Quincy affirms that no progress has been made in political economy since 1817; and interested or superficial politicians and writers, try to confuse public opinion by asserting that there is no such science. It is true that since 1817, or about that time, the progress made has been less, and a great deal less striking, than during the previous fifty years. Still there has been much practical, and some abstract progress all the time. Even Mr. H. C. Cary, notwithstanding the extraordinary use he has made of his acquirements in advocating the most extreme protective system, has made very material additions to our knowledge of political economy, and successfully refuted some theories that had been generally accepted.

The great and leading object of Mr. Smith's work is to show that man possesses, in the constitution of his own nature and in the circumstances of his external situation, ample provision to insure the progressive augmentation of the national wealth; and he demonstrates that the most effectual means of advancing a people to greatness is, to allow every man, as long as he observes the rules of justice, to pursue his own interest in his own way, and to bring his industry and his capital into the freest competition with those of his fellow-citizens. He shows that all legislative restraints on the perfect freedom of industry and exchanges are, in reality, subversive of the great purpose which they are really, or ostensibly, intended to promote.

With regard to the short-sighted selfishness, whose maxim seems to have been that other men's loss is our gain, he expresses himself in a tone of honest indignation. "In this," he says, "the sneaking arts of underling tradesmen are erected into political maxims for the conduct of a great empire. By such maxims as these, nations have been taught that their interests consisted in beggaring all their neighbors. Each nation has been made to look with an invidious eye upon the prosperity of all the nations with which it trades, and to consider their gain as its own loss. Commerce, which ought naturally to be among nations as among individuals, a bond of union and friendship, has become the most fertile source of discord and animosity." "The violence and injustice of the rulers of mankind is an ancient evil, for which, perhaps, the nature of human affairs can scarce admit of a remedy. But the mean rapacity, the monopolizing spirit of merchants and manufacturers, who neither are, nor ought to be, the rulers of mankind, though it cannot perhaps be corrected, may very easily be prevented from disturbing the tranquility of anybody but themselves."

What is this but the law of love becoming the true philosophy of trade?

The above passage shows the largeness, but also marks the limitations, of Mr. Smith's genius. Merchants are not always to remain the mean, short-sighted, selfish, degraded creatures they once were. When the true interests of commerce are understood by the merchant, it enlarges and liberalizes the mind as nothing else can.

About the time Mr. Smith's book was published, cotton began to attract attention, in connection with the newly-invented machinery for spinning and weaving. The importation of cotton cloth from India was the immediate cause of the efforts made to improve existing methods of manufacturing. It was evident that competition with India, by the old methods, was impossible. At first an attempt was made to exclude India calicoes by act of Parliament, but that was impossible. Happily, the time was ripe for great inventions. Europe had so far evolved a new and advanced humanity, that all that was required was to feel the necessity for action. It is an immense benefit to the whole human race to have this free competition. It is a violation of the laws of nature to enact laws to place imbecility on a par with natural force. As to protecting the people of England against the people of Hindoostan, when the true sources of power are understood, it is supremely ludicrous.

The true way to protect people is to cultivate them, and make them strong to protect themselves. Whatever government can do in this direction is legitimate. Rightly considered, this is the whole duty of government. There is everywhere an instinct that prompts men to expect and demand help from government which is the depository of the collective forces of society. Assuredly this feeling, which is universal, should not be disregarded. The demand must be answered, either in an intelligent and effective way, or in a way that will corrupt both the people and the government. There is but one way in which government can effectually aid the people without doing more injury than good, viz.: by making them more competent to take care of themselves; in other words, by educating them. But it should be very different from what is now called education. In the first place, it should be in accordance with the laws and order of nature; in the next place, it should be universal. No plant in the garden of society should be permitted to be without cultivation.

What are the laws and order of nature? The object of true education is to strengthen every faculty. Knowledge of principles is the nutrition of the mind. The process of assimilating truth is precisely analagous to that of chylification in the animal system. A dogma is to the mind what an indigestible substance is to the digestive functions. Not only does it afford no nutrition, but it weakens and deranges the functions.

Nature teaches by example. Everything should be taught in the school, as nearly as possible, in the same way as it is used in practical life. The universe is the natural educator of man. The school should be a miniature universe. No question here as to the comparative value of classical and scientific education. All departments hold equal dignity, and the individual follows his attractions. Practical science, practical mechanics, practical agriculture are all taught in the laboratory, the workshop, and the field. This would be a university indeed—something worthy of the name. I do not know whether the art of war should find a place in such a system ; but it is more than probable that, if such a system were general among the nations, war would soon become obsolete.

Such a system as this would at once put the sceptre into the hands of labor. Such things as bounties, monopolies, or any other kind of protection, would be remembered as the most childish of follies.

Diversification of industry is one of the necessary results of growth, as the natural phenomena of growth is branching. To bring about this much-desired result, we should promote growth. Social growth is but the aggregate of individual growth. There is no end at which the protectionist *professes* to aim that can be reached in any other way so well as this.

I do not believe, with some free traders, that governments should do nothing. It is one of the uses of social organism, and by no means the least important, that the aggregate, collective powers of society should be used for the promotion of its own welfare. The great danger to be feared from government interference is that it may weaken, instead of strengthening the people. Any kind of assistance that may be rendered, with the single exception of assistance to grow, that is, culture, education, will surely tend to make the people less self-reliant, and, consequently, less strong.

Communism, socialism, internationalism, they have, of late, become terrible words. They strike the public ear of Europe like a fire-bell in the night. Let us not try to persuade ourselves that our Republican form of government, and our abundant and cheap land, will always save us from the responsibility that now hangs about the neck of Europe. Neither the selfish expedients of accumulated wealth, nor the ignorant denunciations of accumulated superstition, will be able to dispose of these questions that are now demanding a solution.

The influence of real education on the human mind is very little understood. Knowledge of principles, each one of which is the key to a new world, is to the mind what wholesome, stimulating food is to the body ; while what is commonly called learning, instead of strengthening the mind, enervates it. The former produces health, strength, sanity ; the latter, intoxication and disease. The mind becomes strong in proportion as it comes near to causes. Technical education is indispensable. I would not underestimate it ; but it is certainly more important to have something worth expressing than to know how to express it.

Compared with the object to be accomplished, the largest amount at present expended for educational purposes is utterly insignificant. Why should not the mechanical arts and agriculture be taught by experiment, as well as chemistry ; and why should not chemistry be taught in a system of experiments, vastly more extensive and enterprising than at present? We profess to understand something about chemistry, yet in every trial for poisoning, every chemist contradicts every other. In fact, a large part of what is called science is the merest empiricism. Let the sciences be taught in their practical application to the duties of life. I have some doubts as to whether they should ever be taught in any other way, but I suppose a place must be allowed for students, whose vocation is the closet, for it is a fundamental principle of the system that all forms of humanity, and every peculiarity of taste or attraction, shall have ample provision for its full development.

Humanity is one, and each individual an essential atom. In the earliest ages, when the masses of mankind had not yet emerged from what might be called animalism, there were individuals of great genius, who are still regarded as standards of excellence. These were a beneficent provision of nature, intended to teach mankind their own possibilities. From a similar provision of nature, we sometimes find in an individual a single faculty developed to a degree of perfection that, in our present stage of progress, seems nothing less than miraculous. Everybody has either seen or heard of persons wholly uneducated, who could solve any problem in mathematics, or answer any question in arithmetic instantly, with scarce a perceptible lapse of time. So far as I know, it has never yet been suspected by anybody that this degree of perfection is possible, and will yet be attained by every faculty of the intellect; yet such is certainly the truth.

If we will only try, however inadequately, to realize what this means: the noble, the divine being man will be, when all his faculties act with the rapidity and precision of a sunbeam, penetrating not only all the arcana of nature, but the spiritual universe also, we may then begin to perceive what is really meant by education, culture—whatever contributes to the growth and unfoldment of the human mind.

I am not disposed to magnify the importance of cotton in the progress of civilization. All the elements of progress are in the constitution of the human mind; all mineral, vegetable, and animal nature seem to have been a preparation for man. In few of the productions of nature is this preparation more marked than in cotton. Less indispensable than iron, it is hardly less useful or less extensively used. No single article gives remunerative employment to a larger number of persons. The large addition which is made to its value between the hands of the producer and the back of the wearer, constitutes the financial life-blood of whole communities and governments, without which, so far as we can see, they could hardly exist.

The United States and Europe manufacture, at the present time, about seven million bales of cotton per annum, averaging not far from four hundred pounds each. For this cotton, the producers receive about four hundred million dollars, gold value. When this cotton is manufactured, and ultimately sold to the consumers in all parts of the world, it has risen in market value to probably sixfold its original cost, leaving to the merchants, ship-owners, manufacturers, and tax receivers not less than two thousand millions per annum as remuneration for their capital and labor.

Nearly all the inventions for spinning and weaving by power originated in Great Britain. The French revolution, and the disturbed condition of the continent until after the battle of Waterloo, retarded the advance of manufactures there. At the same time, England used every effort in her power to prevent a knowledge of her inventions from reaching other countries. Practically, the English people had, during their whole contest with France, a monopoly of these inventions, and, consequently, a monopoly of certain kinds of manufactures. This was especially true of cotton, which did more than anything else to sustain her financial system under the tremendous strain to which it was then exposed. It was to her industrial progress that England owed her success in that great struggle. To what did she owe her industrial progress? Undoubtedly to the genius of her people. The philosophical works of Lord Bacon mark the drift of the English mind when it first began to manifest extraordinary vigor. Bacon was the natural forerunner of Newton. Philosophy first, science afterward. It is common to give Bacon credit for the great intellectual development that followed the publication of his works. In one sense, this is just; in another, it is not.

Such men as Bacon follow quite as much as they lead. Great men are the product of the nation and the time. They are, in the highest sense, representative men. No great discovery is wholly the act of one man. Bacon, like all men of great genius, was exquisitely receptive. He felt the spirit of the genius of his countrymen before it had yet assumed a body. He incarnated it, if I may use the expression. The inductive method was then due in the order of evolution. The first attempt at reasoning in the childhood of the race, or the individual, takes a synthetic form. I before stated that the first budding of a faculty is prophetic of its ultimate future. So it is even with the reasoning faculties. When the human mind attains to a knowledge of fundamental laws, it will resume its native instinct, and sweep the whole domain of nature with a single synthesis. Childhood, in the absence of a knowledge of principles, takes principles for granted. It adopts theories which become, in some sense, matters of faith, and reasons from them. How many thousands of years the human family reasoned in this way, there is no record to show. Nine in ten still continue the same method, but it is a great deal to be able to say that there is ten per cent. of the whole who demand an examination of the premises, as well as the deductions. Until about the commencement of the seventeenth century, all departments of human knowledge were filled with dogmas, that is, things taken for granted. The English mind, as represented by Lord Bacon, began to question and ask for proof. That single fact placed the English people in the vanguard of the race for the time being. It was the first ripe fruit of the swing of the human mind from the supernatural toward the natural. All the scientific discoveries and mechanical inventions, of which the eighteenth century was so prolific, were the natural fruit of this movement.

I would not underestimate the importance of men of great genius to the world. I would only avoid that kind of hero-worship which exaggerates the importance of one man that it may pour contempt upon the mass of mankind. Communities are responsible for the crimes that are committed by their members. This is manifest in the fact that crimes have always the color and bias of the popular sentiment. In a community where animal pride, in the form of personal force and courage, are esteemed above the other qualities, crimes of personal violence prevail most. In communities where the possession of wealth is unduly esteemed, crimes against property prevail most. If a community is thus responsible for the crimes of its members, it has a right to be credited with the achievements of its worthies.

Men of great parts are always being born into the world, but it is the drift of the times and the popular sentiment that shape their work. The people of central western Europe, after the discovery of America, led the advance guard of progress. I have before explained what I suppose to be one of the natural causes of this. After America was discovered, Spain advanced to the position of the leading power of Europe. Why did she not retain it? I stated that the shores of the Mediterranean were the centre of activity in the action and reaction of European and Asiatic influences. The truth is, that southern Europe is not yet distinctly European, but is largely Asiatic. Eastern and northeastern Europe are also tinctured with the same quality. It was the evident purpose of nature that Europe should evolve a new humanity, different from Asia; as it is the purpose that America shall evolve a new humanity, different from Europe. I suppose that the reason why Spain lost her position was because she was not in the current of progress, but seemingly in one of its eddies.

We are now on the eve of great changes, which will probably increase, more rapidly than ever before, the production and consumption of cotton. The progress of European influences westward has already penetrated Asia with a new force. Japan is already melting at our touch. Strange that contact with Europe from the west has never had any such influence upon any Asiatic people. It would seem as if the European was not, and could not be, prepared to exercise such potent influence on the Asiatic mind until it first becomes Americanized. Within a short time, certainly before the end of this century, the vast populations of eastern Asia will be opened up to trade with this country, even to the most interior hamlet. The railroad and the telegraph will penetrate Japan and China in every direction. There will be a large emigration of Asiatics to the United States, and of Americans to Asia, each supplying the other with what it most needs. The dense population of China, with their wants increased by European civilization, will require nearly all their land for the production of food. There will, consequently, be an immense increase in the production of cotton in this country. It is not improbable that, before the end of this century, the cotton crop of this country may reach ten million bales, and at least half of it be manufactured here. I suppose the time cannot be very far distant when the commercial intercourse between different nations will be as free and unrestricted as it is now between the States of this Union. Every enlightened mind, every merchant especially, should labor to hasten the coming of that day.

The times are almost ripe for this consummation of the prophetic instincts of the religious sentiment. Everywhere men feel that they are brothers, and that all barriers between nations are in some sort sacrilege. The pulpits preach universal love as a sentiment, but they should understand that it could not be a true sentiment if it were not also true philosophy. There is no sacrifice in the law of love. It is all pure gain. The more good we do, the more we are benefited.

A knowledge of this will lift commerce out of the mire and place it in the atmosphere of the purest ethics.

The swindling, the deception, and all the desperate expedients of speculators are merely a reflex of the policy pursued by nations towards each other. The policy of international jealousy and hatred cannot last much longer. If the churches were to proclaim a crusade against all custom-houses, they would be doing more for the human race than they can accomplish in any other way at present. It is the most imperative demand of the age.

EARLY HISTORY OF COTTON.

CHAPTER I.

THE object of this work is to supply a convenient book of reference for all who are in any way interested in the cotton trade. To facilitate such reference, the facts and statistics will be arranged chronologically. Of course all who deal in an article of world-wide use, liable to great changes in value, would be glad to learn something of the causes which so often influence the market. I know of no better way to obtain that knowledge, than through a carefully arranged statement of the facts connected with the trade in past time.

History furnishes no means of ascertaining when, or by what progressive stages of discovery and invention, cotton was first utilized to human use. The arts of spinning and weaving are probably as old as agriculture. The Egyptians ascribed their origin to Isis. According to Pliny, Semiramis was believed to have been the inventress of weaving. The Peruvians ascribed them to Manco Capac, their first sovereign. These traditions point to their extreme antiquity. It is certain that they have been found among almost all the nations of the old and new worlds, where anything like a social organization existed.

The first mention of cotton by any European writer is by Herodotus, called the father of history, about 450 B.C. Even then the manufacture of cotton cloth in India seems to have been as perfect as at any later period.

Wool was principally used for weaving in Palestine, Asia Minor, Greece, Italy and Spain; hemp in the northern countries of Europe; flax in Egypt, and silk in China. In like manner, cotton was always characteristic of India. What ancient Egypt was in the culture and manufacture of flax, India was in cotton. The "fine linen of Egypt" was not more celebrated on the shores of the Mediterranean, than the beautiful soft cotton fabrics of India. The great perfection attained in this manufacture in India is scarcely credible.

Tavernier, a merchant, who traveled in the middle of the seventeenth century, says "The white calicuts (calicoes, or rather muslins, so called from the great commercial city of Calicut, whence the Portuguese and Dutch first brought them) are woven in several places in Mogulistan and Bengal, and are carried to Rioxary and Baroche to be whitened, because of the large meadows and plenty of lemons that grow thereabouts; for they are never so white as they should be until they are dipped in lemon water. Some calicuts are made so fine you can hardly feel them in your hand, and the thread, when spun, is scarce discernible." The same writer says, "There is made at Laconge (in the province of Malwa) a sort of calicut so fine that when a man puts it on, *his skin will appear as plainly through it, as if he was quite naked;* but the merchants are not permitted to transport it, for the governor is obliged to send it all to the Great Mogul's seraglio and the principal lords of the court, to make the sultanesses and noblemen's wives shifts and garments for the hot weather; and the king and the lords take great pleasure to behold them in these shifts, and see them dance with nothing else upon them." Of the turbans of the Mohammedan Indians, Tavernier says "The rich have them of so fine cloth that twenty-five or thirty ells of it put into a turban will not weigh four ounces." The Decca muslins were designated, in Oriental phraseology, "Webs of woven wind," and nobody will dispute the poetic truthfulness of the name.

There is satisfactory evidence that the Greeks made use of muslins, or cotton cloths of some kind, which were brought from India at least 200 B.C. It is probable that the expedition of Alexander the Great (330 B.C.) first introduced cotton goods into Europe as articles of traffic.

About 60, B.C., we find them used in Rome for awnings and tent covers, but there is no evidence that the finer fabrics of cotton were ever in much demand in the Roman Empire. Of silk there is frequent mention; to cotton as a curious product of the East there are several allusions, but to cotton cloths, as articles of apparel among the Roman people, there is no allusion whatever. Cottons are among the imports of the Empire, taxed by a law under Justinian, from which it is evident they were in use for some purpose. Indeed it would be strange if it were otherwise, for before the Christian era India had begun to supply cottons to Persia, parts of Arabia, Abyssinia and all the eastern parts of Africa. For centuries the Phenicians had traded in these fabrics, and Egypt was certainly well acquainted with their use.

There is good reason to believe that the art of manufacturing cotton had existed in Arabia before the present era, and that cotton constituted a considerable part of the clothing of the people.

That the Roman empire carried on an extensive trade with India is attested by the well known statement that this trade drained the empire every year of a large amount of the precious metals. One hundred and twenty ships sailed annually from the Arabian Gulf, from Oceles at its mouth, across the great ocean to the coast of Malabar. They returned with the eastern monsoons, ladened with the products of India, clearing from the general mart Musiris.

It is just possible that the semi-transparent robes with which the Roman ladies clothed, or rather exposed their beauties, in the decline of the empire, were India muslins.

There is no record of any cotton being manufactured in Europe before the tenth century ; and then it was only by the Mohammedans in Spain.

Though the Arabs seem to have learned something of cotton culture from India long before the Christian era, China, whose intercourse with India was probably as old as the pyramids of Egypt, did not learn the arts of cotton culture and manufacture until the thirteenth century, when they were introduced into that empire by the dynasty of the Mongol Tartars. Yet the court had long held in high estimation the cotton garments which had been presented to them by foreign ambassadors. As early as 510 A.D., the Emperor Ou-ti had a robe of cotton. Toward the end of the seventh century it is known that the cotton plant had long been cultivated in the gardens *for its flowers*. There was great opposition to the new article.

It is amusing to observe how like the objections of the Chinese were to those raised by the English, in the eighteenth century, to the importation of India goods into their country ; also to the objections to the importation of machine-manufactured cottons into this country in the beginning of this century. Even now a distinguished editor and writer, on what he calls "political economy," assigns his conversion to the restrictive or protective policy to the distress produced in New England households by the importation of goods made by the then lately invented methods and machinery at such low prices and of such superior attractiveness, that the domestic manufacture by hand (almost exactly the methods of India and Egypt) was completely prostrated.

I should have mentioned that long prior to the tenth century, a manufacture of indigenous cotton had existed in southern parts of Italy ; there was also something of the sort in the Crimea, but it was very trifling in extent, crude and wholly domestic ; not at all an article of commerce. The rise of Mohammedanism and the conquests of the Saracens, were by far the most important events in the Middle Ages.

They opened an active commerce from the Straits of Gibralter to Bagdad and fartherest India. The Arabs had preserved some of the learning of the Alexandrian schools. They brought into Europe the figures of arithmetic and Euclid's works on geometry. After the conquest of Constantinople the Turks introduced the cotton culture into Macedonia.

The manuscript of Marco Polo's travels was first circulated in 1298 at Genoa. At this time a considerable trade was carried on by Venetian and Syrian merchants in India muslins. Marco Polo was confidentially employed in the service of the Tartar conqueror of China and returned in the year 1295, after having visited a great many countries in Asia. He makes no mention of any cotton goods in China. The cotton culture was then in its infancy in that empire.

In consequence of the dearth of provisions about seventy years ago, an imperial mandate was issued, to convert to the cultivation of corn a considerable portion of the land then appropriated to the cotton plant; since which the Chinese have imported a large quantity of cotton from the east coast of India, from Calcutta, and even from Bombay. The quantity of cotton produced in China is enormous, some writers estimate it equal to twelve millions of our bales. The lower orders are all clothed in cottons, and the higher classes in silks. In cold weather they do not change the character of their clothing, but increase the quantity, adding garment upon garment.

It is supposed that Mohammedans first introduced the use of the bow-string in opening up the fibres of the cotton, from the circumstance that the bow-string operation is never executed by Hindoos. but by Mohammedans. The Hindoos maintain their ancient superiority in all the finer fabrics. This is attributed to the greater delicacy and susceptibility of their organization.

The bow-string was once used in this country, and gave rise to the term "Bowed Georgia," still used in Liverpool.

The rollers used in India for separating the seed from the lint, are still used in this country for long staple cotton. It is thought that the best is that worked by the treadle, in the same way it was worked in India, as described by Nearchus, one of Alexander's officers, 325 B.C.

During the thirteenth, fourteenth and fifteenth centuries, the cotton manufacture continued to flourish in Spain. Its chief mart was Barcelona, in the neighborhood of which the plant is still found growing wild.

During the same period, and probably much earlier, the cotton manufacture was very extensively established all over the southern shores of the Mediterranean. Indeed it may be taken for granted that wherever the Mohammedans obtained a foothold this industry was established, from the Atlantic Ocean to the River Euphratus.

Humaine, a small African city in the Mediterranean, frequented in the fifteenth century by the Venetians, is spoken of with high commendation on account of its eminence in this manufacture.

An Italian writer says of them: "The inhabitants were a noble, civilized race of men, and almost all engaged in the production of cotton and cotton cloth."

At Amon, a place five days journey from Damascus, it is said that a very great quantity of cotton was grown.

According to Odoardo Barbosa, of Lisbon, who made a voyage to southern Africa in 1516, the Caffres then wore cotton dresses. At Cefala, he says, the Moors grow a large quantity of fine cotton, and use it in white cloth, being unable to dye it on account of the want of coloring stuffs.

Cotton cloth, woven on the coast of Guinea, was imported into London in the year 1590.

Travelers, who have penetrated into the interior of Africa, concur in showing that cotton is indigenous to that continent, and that it is spun and woven into cloth, which is used for raiment by the inhabitants of every class and latitude. It is probable that a much larger surface of Africa is suited to the cotton culture than of either Asia or America.

Lord Palmerston predicted that Africa will yet supply Europe with cotton.

The time is drawing near when Africa will be in the line of Colonization. The currents of commercial exchanges between Oceanica and Europe, and between South America and Europe, cannot much longer be prevented from overflowing the African continent. The movement from Egypt is not likely to accomplish much; the Cape of Good Hope is the key to Africa's future.

In 1110, A.D., the revival of learning at Cambridge, England, is reported, and the statement is also made, that paper made of cotton is commonly used in writing. The art of making paper from cotton came into Europe from Arabia, where it was first known, though the Chinese had long made paper from refuse silk. The

Mohammedans, in Spain, subsequently discovered that linen was superior to cotton for that purpose.

In 1498 Vasco De Gama sailed to the East Indies around the Cape of Good Hope, thus opening to Western Europe the India trade, which for more than three hundred years had made Venice the envy and admiration of Europe. In the same year Americus Vespucius discovered North America, six years after Columbus made his discovery of the New World.

After the voyage of Vasco De Gama, the Portuguese made large importations of muslins, and other cotton goods into Europe, but did not attempt to establish any cotton manufacture in their own country. When the Dutch, sometime afterwards, succeeded in depriving the Portuguese of their eastern colonies, they not only extended the traffic in cotton goods, but, towards the end of the six teenth century, began to fabricate them at home.

The earliest notice of cotton, as an article of English trade, is about the end of the fifteenth century. It was naturally included in the trade of the Mediterranean, and was carried by the ships of the Italian cities wherever they sailed.

Early in the sixteenth century English commerce began to expand. Though Italy had some knowledge of cotton manufacture three hundred years before it was known in Western Europe, yet in the year 1870 there were only 500,000 spindles in the whole of Italy.

Columbus found cotton in use among the natives of Hispaniola, but only in the most primitive forms. Cortez found the manufacture in a much more advanced condition in Mexico. The Spanish historian of Mexico informs us that "The Mexicans made large webs, and as delicate and fine as those of Holland. They wove their cloths of different figures and colors, representing different animals and flowers. Of feathers interwoven with cotton, they made mantles and bed-curtains, carpets, gowns, and other things, not less soft than beautiful. With cotton also they interwove the finest hair of the belly of rabbits and hares—after having spun it into thread; of this they made most beautiful cloths, and in particular winter waistcoats for their lords."

It is said that in 1536 the cotton plant was found growing in some of the country drained by the Mississippi, and in Texas.

In 1589 the stocking frame, one of the most complex and ingenious machines then known, was invented. In 1530 the spinning-wheel was invented by Jurgen of Brunswick.

In 1563 the slave trade was actively carried on by England. This may not seem to be a very conclusive indication of progress, yet in view of its connection with cotton culture in this country, it is well worthy of record, as it was undoubtedly a part of the great movement and expansion of industrial activity which began with the geographical discoveries of the fifteenth and sixteenth centuries.

Nearly all the cotton in the world is produced by the colored races. It is a tropical plant, and the tropical races cultivate it, though it seems destined to supply the whole race with a large part of their clothing. The trade is yet in its infancy. Even in this country, where cotton culture is most successful, the methods of culture are most primitive. The time is not distant when four times as much cotton will be produced to the acre as is now produced, and of quality much superior to the cotton of the present time. Cotton culture in this country is in about the same stage of progress that marked the condition of agriculture in England one hundred years ago. Since the emancipation of the slaves a new impetus has been given to improvement, but still the business is so profitable that it is done very carelessly. "Necessity is the mother of invention."

Every other textile material has some special merit; cotton alone is suited to all climates, conditions, and races. It is the only article that can be produced in such abundance, and so cheaply, that the demand never can for any long time exceed the supply.

The rapid progress of railroads at the present time in all parts of the world—in Europe, Asia, and America—stimulating activity, facilitating exchanges, increasing wealth and multiplying the wants of hundreds of millions of people, will, for a long time to come, keep the price of cotton far above the cost of production, but the supply is merely a question of price. There is no other textile material that can be grown so profitably on such an extensive area of the earth's surface.

Baine's "History of the Cotton Manufacture" says: "No mention has yet been found of the cotton manufacture in England earlier than 1641. In 1631 calicoes were first imported into England from India; and at once attracted attention and excited emulation. It was not until forty-five years afterwards that calico printing commenced in London."

CHAPTER II.

COTTON IN NORTH AMERICA.

In "Historical Collections of South Carolina," by B. R. Carroll, I find several allusions to cotton as an article of culture in that colony.

Some colonists from Barbadoes, who settled on the Cape Feare River in 1664, brought with them cotton seed, which they cultivated for domestic purposes.

In a description of the Province of Carolina, by Samuel Wilson, addressed to the Earl of Craven, in 1682, it is stated that "cotton of the Smyrna and Cypress sort grows well, and good plenty of the seed is sent thither."

In "Rivers' Historical Sketches of South Carolina," p. 343–4, I find the following passages:

"In the instructions given by the proprietors to Mr. West, the first Governor of South Carolina, we find the following:

"'Mr. West, God sending you to Barbadoes, you are then to furnish yourself with cotton seed, indigo seed, ginger roots. * * Your cotton and indigo is to be planted where it may be sheltered from ye north-west winde, for they are both apt to blast.'

"On page 351 'West was instructed to receive the products of the country in payment of rents at the following valuations: ginger, scaled, at 2d. per lb.; scraped ginger, at 3d. per lb.; indigo, at 3s. per lb.; silke, at 10s. per lb.; cotton, at 3½d. per lb.'"

About that time great efforts were being made to establish indigo culture in the Carolinas. Indigo was also introduced into Louisiana by the French in 1718, and within ten years became an object of export.

About 1740, when rice became reduced in price, the seed of the East India indigo plant, which had been for many years extensively cultivated in the West Indies, was sent, along with cotton, ginger, lucerne, etc., from Antigua by Mr. Lucas, the governor of the island. Previous to the war of the Revolution, indigo held the position among the products of South Carolina afterwards occupied by cotton. It was hardly less important in Georgia.

Miss Lucas, the daughter of the Governor of Antigua, and the mother of General Charles Cotesworth Pinckney, was, at the age of eighteen, in charge of a plantation in South Carolina. In her journal, 1739 and 1741, she speaks of the pains she had taken to bring cotton and indigo to perfection. The first export of cotton was from Savannah. An exportation of seven bags, valued at £3. 11s. 5d. per bag, was made from Charleston, between November, 1747, and November, 1748.

Peter Purry, in his description of Carolina in 1731, says: "Flax and cotton thrive admirably, and hemp grows thirteen to fourteen feet in height; but, as few people know how to order it, there is very little cultivated." In 1742 a French planter of enterprise, M. Dubreuil, invented a cotton-gin for separating the fibre from the seed, which greatly stimulated the culture of cotton in that colony. The separation of the seed had previously been effected by picking it with the fingers, at the rate of one pound a day. This operation, as the evening task of the family, black and white, long continued to be the practice in the cotton region, until increased production called for mechanical appliances. About the commencement of the controversy that led to the Revolution, the roller-gin was introduced. It is still used for long staple cotton, and was the best contrivance until the invention of the saw-gin, by Whitney, in 1793. This introduced a new era in the cotton trade; not less important than the splendid inventions of Watt and Arkwright

It was not until after the Revolution that cotton culture was prosecuted with a view to export. Even so late as 1784 an American ship, which exported eight bags of cotton into Liverpool, was seized on the ground that so much cotton could not be produced in the United States. The first regular exportation of cotton from Charleston was in 1785, when one bag arrived at Liverpool, per *Diana*, to John and Isaac Teasdale & Co. In the same year, twelve bags from Philadelphia and one from New York were received at that port. During the next five years the receipts of American cotton were respectively 6, 109, 389, 842, and 81 bags, estimated at 150 pounds each, or 1440 bags, weighing 216,150 pounds in six years.

In the meantime the necessities of the Seven Years' War had taught the people to raise the material for the greater part of their own clothing, and laid the foundation for the cotton culture, which has since grown to such gigantic proportions. Though the inventions of Hargreaves, Arkwright, and others, in England date earlier than the Revolution, they had not at that time assumed sufficient importance to stimulate production.

The *green seed*, or short staple cotton, was the kind principally cultivated before the Revolution.

The *black seed*, or Sea Island cotton, was introduced into Georgia from the Bahamas about the year 1786, and in 1788 the first attempt was made in South Carolina by Mrs. Kinsey Burden, of St. Paul's Parish, whose husband, having introduced the roller-gin in that State, had, nearly ten years before, clothed his slaves in cotton cloth. The first successful crop is said, by Mr. Seabrook, to have been grown by Mr. Elliott on Hilton Head, near Beaufort, in 1790, with 5½ bushels of seed, purchased in Charleston at 14s. per bushel. The price then varied from tenpence to two or three shillings per pound. Though in 1786 Mr. Madison said at the Annapolis Convention that "there was no reason to doubt that the United States would one day become a great cotton producing country;" yet in the treaty negotiated by Mr. Jay in 1792, it was stipulated by the 12th Article "that no cotton should be imported from America." The Senate of the United States of course refused to ratify this article. This was only eighty years ago, yet it reads like a record of the Middle Ages.

In the year following the treaty, Whitney's invention of the saw-gin gave an immediate impetus to the cotton culture, and caused great excitement all over the Southern States. The inventor was robbed of the fruits of his genius, but the benefit to the country and to the world was beyond all calculation. The very first year in which the saw-gin was used, South Carolina exported cotton to value of $1,109,653. The growth of the whole country in that year (1795) was estimated at eight millions, and the exports six millions. In 1801 the product had risen to about forty-eight millions and the exports to twenty millions pounds. In 1798 indigo entirely yielded to cotton as an article of commerce, though it was still cultivated for domestic purposes as late as 1850. The writer has seen the small farmers bring it to the stores to exchange for necessaries. The old women called it "Spanish float," at least this was the name they had for the best quality of indigo.

The efforts made to introduce the silk culture into Georgia, during the colonial times, were of course an entire failure, notwithstanding liberal expenditure by the mother country.

The time will come when the United States will be one of the greatest silk and wine producing countries in the world. Leaving out California, which seems to be unequaled for the variety and fertility of its productions, there are large districts in the older States admirably suited to the production of both wine and silk. The country near the dividing line of North and South Carolina is wonderfully adapted to the cultivation of the grape. I have been informed by experts, who are well acquainted with all the wine producing countries in the old world, that it has no equal in Europe.

The qualities required to subdue the forest and reclaim the swamp are wholly unsuited to the successful cultivation of silk and the grape. Negro labor was exactly suited to the cultivation of cotton; the Anglo-Saxon, bold, ingenious, enterprising and avaricious of power, and of wealth as the means of power, was exactly suited to direct the labor of the negro.

Negro slavery served its time, and served it as nothing else could, but it could not be of long duration in the present age of the world. Slavery being founded upon force, was, philosophically speaking, a state of war. It cultivated the virtues, but also the vices that are promoted by war. This was shown in our late civil war. The Southern army fought as well, and even better in their first campaign than in their last. The admiration of the Southern people for the ancient orders of chivalry, their tilts and tournaments was not a mere affectation, as many suppose; it was the natural outgrowth of their social system. The tendency of modern progress is to unfit men for war, by depriving them of a taste for it. The combative faculty is neither wasted nor lost, it is only turned into a higher sphere of action. The people of the Northern States, and especially of New England, have probably less of the purely warlike propensity than any other people in existence. The English have, owing to the predominance of the spirit and habits of commercial enterprise, a strong tendency in the same direction, but it is greatly impeded by their aristocratic institutions. Subordination and command are the life-blood of the war-making spirit.

Of course these two antagonistic systems of society could not long exist under the same government. As in all such conflicts, between the old and the new, but one result could be final.

Negro labor is still necessary to the cultivation of cotton, and the problem of the future is still unsolved. If white labor can take the place of free labor in the cultivation of cotton the negro will disappear, dying out; if not, it is to be hoped the negro will be preserved as a permanent inhabitant of the cotton region. Under slavery the blacks would undoubtedly have increased at least as rapidly as the whites, immigration excepted, but I doubt if this was the intention of Providence; that is, of the laws of Nature. I do not think Nature ever intended that an inferior race should occupy ground to the exclusion of a superior race, which the latter could occupy with more advantage to the world.

Under the new order of free negro labor, it seems to me probable that the white race will, for a long time to come, apply themselves to manufactures of cotton. There is no country in the world possessing so many advantages for the successful prosecution of industry in this direction as the Southern cotton growing States. White labor for the cotton factory can there be obtained much lower than at the North—as low, probably, as in England or the continent of Europe. Northern Georgia and North Alabama are admirably suited for manufacturing enterprise. Water power is abundant, and coal can be obtained much cheaper than in any part of the Northern States. The climate is unsurpassed. Labor is abundant.

Labor-saving machinery was always popular, and eagerly sought after in this country, though it was quite otherwise in Europe.

In 1766 a society in New York announced through its secretary, Benjamin Kissard, among other premiums for the encouragement of home manufactures, £10 for the first three stocking looms of iron set up in that year, and £5 for the next three, and £15 for the first stocking loom made in the province in that year.

An extract from a letter, dated at Baltimore, January 11, 1772, appeared in the Pennsylvania *Gazette*, on the 30th, in which the writer says: "We learn that a person who has been for many years a master in several large manufactories for linen, cotton, and calico printing, likewise cutting and stamping of the copper plates for the same, intends, sometime this month, to leave England for America with six journeymen and all the machinery for carrying on the said business, previous to which, and unknown to the English manufacturers, he has shipped sundry machines, some of which will spin ten, and others from twenty to one hundred threads at one time, with the assistance of one hand to each machine. These machines are not allowed at home, and so inveterate are the common people against them that they burn and destroy not only these, but the houses, also, where they are found. The Americans being able to purchase cotton to more advantage than Europeans, a manufactory of this kind will doubtless be properly encouraged by the well-wishers to America."

The above will be better understood when it is known that the British Government prohibited the exportation of machinery used in manufacturing. This prohibition did not at first apply to shipments to the colonies, because, at first, the colonies were content to be entirely dependent upon the mother country; but as soon as they began to manufacture articles which they had previously bought from England, the most stringent laws were enacted to prevent it. In 1774, by the 14 Geo. III., c. 71, it was enacted that if any person exports any such tools or utensils as are commonly used in the cotton or linen manufactures, or other goods wherein cotton or linen are used (excepting wool cards) to North America, or any parts of such tools or utensils, he shall not only forfeit the same, but £200. The collection, or having in possession such implements, made them liable to seizure, and the possessor to arrest.

The Revolutionary War and the patriotic abstinence of the American people, during the controversy that preceded it, from purchasing English manufactures, laid the foundation of the cotton manufacture in this country. That it would have grown naturally in this soil, and rapidly too, is certain, but the measures taken by the mother country to prevent it, undoubtedly pushed it into premature growth.

Arkwright erected his first spinning-frame in 1769. The first spinning-jenny probably seen in America was exhibited at Philadelphia, early in the year 1775. This machine was made by Mr. Christopher Tully. It was on the plan of Hargreave's, and was probably made in England. In this year a manufactory was established in Philadelphia for cotton, flax and wool. It was the first joint-stock company formed for such a purpose, and the first to attempt the cotton manufacture in this country. This factory was supplied with native cotton during the Revolution, at two shillings per pound. The utter dependence of the colonies upon the mother country previous to the Revolution is shown by the destitution of the people and the armies during that period. It was the ragged condition of the American soldiers and officers that gave rise to the term *sans culotte.* Baron Steuben's aids on one occasion invited a number of young officers to dine at their quarters, torn clothes being an indispensable requisite of admission. The baron loved to speak of his ragged guests as his *san culottes*, little dreaming that the name which honored the followers of Washington would afterwards be assumed by the satellites of Marat and Robespierre. Even when the army was clothed it was principally in English cloth, bought in Holland, and sent to America.

In 1780 an association was formed in Worcester, Mass., for spinning and weaving cotton, and a subscription raised to procure a "jenny" for that purpose. On the 30th of April it was announced in the *Spy* that "on Tuesday last the first piece of corduroy made in the manufactory, in this town, was taken from the loom." After the close of the war, great efforts were made to obtain models of the then lately-invented machines for spinning, weaving, and finishing cotton and linen goods. A set of complete brass models of Arkwright's machinery was made and packed in England by the agent of Tench Coxe, of Philadelphia, in 1786, but was seized on the eve of its shipment, and the object defeated.

The invention of the power-loom by Cartwright in 1774; of the mule "jenny," in 1775, by Compton, which soon superseded the machine of Hargreave's; the several improvements of Arkwright and others in carding, drawing and roving, were crowned by the adaptation, in 1783, of the steam-engine of Watt, to the spinning and carding of cotton in Manchester. Cylinder printing was invented by Bell in 1785, and the use of acid in bleaching was introduced at Glasgow, by Watt, in 1786, and at Manchester in 1788. The use of acid for the same purpose was very old in India.

These improvements gave a great impetus to manufactures, attracting attention and exciting emulation everywhere. The first "jenny" manufactured in this country was in Massachusetts, by two brothers, Robert and Alexander Barr, from Scotland, under the patronage of the Hon. Hugh Orr, and by the aid of a grant from the State.

The American Revolution ended, the material for a great industrial nation was all here, but it was in the form of scattered fragments. Jealousies began to arise between the States, with regard to trade and taxation. There may have been, and no doubt was, an instinctive feeling that free trade among the states would, in the end, be mutually profitable, but as to freedom of trade as a principle, very little was known about it. There was a very general sentiment that freedom was best, but unless by very few, perhaps not a dozen in all, was it accepted as a principle, but it was these few (so short-sighted are the wisest of mankind) who opposed the acceptance of the present Federal Constitution, on the ground that it endangered liberty, by establishing a "central despotism."

That Jefferson erred in opposing the Constitution, is no more than he himself acknowledged afterwards, though his admitted error was afterwards made the animating principle of a great party. Perfect freedom of intercourse between the people of all the States was then, and is now absolutely indispensable to the prosperity of this country. No price is too high to pay for it, and I doubt if there could be any abuse of power in the Federal Government that this freedom of intercourse and trade would not sooner or later remove. This could only be obtained under one government. The world has not yet arrived at that stage of progress when States will consent to make intercourse with each other perfectly free and equal. It is because the Federal Government is essential to freedom that it is successful and has triumphed over all its enemies; when it ceases to be necessary to freedom, or becomes antagonistic to it, it will fall without a blow.

Cotton manufacturing, in this country, is now as secure as freedom; and its growth and extension will be commensurate with the growth and extension of freedom.

The time is coming, and cannot be very distant, when we will have the same freedom of intercourse with the whole world that we now have with each other. Then the markets of the world will be open to our manufacturers, in which to obtain the raw materials used in manufacturing—as machinery, chemicals, dye-stuffs, or the necessaries of life; and in which to dispose of the products of our industry to the best advantage.

What the manufacturing interests of the United States need, and the cotton interest especially, is complete exemption from Government intermeddling. If it ever needed government aid, that time is past; it is no longer an infant, but is prepared to enter the field with any and all competitors. The cotton manufacturers do not ask the country to sacrifice its mercantile marine, or any other legitimate interest for their benefit. They have learned that the prosperity of all other legitimate enterprises will add to their own prosperity. They would be only too glad to have their own ships to carry their products to the ends of the earth. For nearly two years the cotton manufacturing power of this country has been increasing at the rate of 20 per cent. per annum, notwithstanding the fact that the cost of their buildings and machinery has been nearly doubled by unwise legislation. They know that the increase should and would be much greater than it is, if the present system of legislation did not practically exclude them from all but the home market.

The extent to which labor-saving machinery is yet destined to relieve man from the drudgery of labor is not yet dreamed of. This is the country which, above all others, must lead in the redemption of the race from toil. We cramp and stultify ourselves by attempting to exclude competition. We fight as an enemy that which we should treat as our dearest friend. But I am now treading on ground which should be more fitly occupied in the introduction. I will now introduce the reader to the chronological and statistical history of the cotton trade, after which I may or may not, make some reflections for the benefit of dealers in the great staple.

CHAPTER III.

CHRONOLOGICAL AND STATISTICAL HISTORY OF COTTON.

1519.

Magellan, while circumnavigating the globe in this year, found the Brazillians using "this vegetable down" (cotton) in making their beds.

Cotton fabrics were sent by Cortes from Mexico to Spain this year as presents to the Emperor Charles V.

Cotton was cultivated and manufactured as early as this year by different nations on the coast of Guinea.

1536.

De Vica, it is stated, found the cotton plant in Texas and Louîsiana as early as this. (*See* year 1621.)

1560.

Cotton was an article of importation from Antwerp into England this year. (*See* year 1569.)

1563.

Cæzar Frederick, a merchant of Venice, while traveling in India during this year, mentions the extensive cotton trade between St. Thorne and Pequ. He says the trade is "in *bombast* (cotton) cloth of every sort, painted, which is a rare thing, because this kind of clothes show as they were gilded with divers colours, and the more they be washed the livelier the colours will show; and there is made such account of this kind of cloth that a small bale of it will cost 1,000 or 2,000 duckets."

1569.

Gaspar Campion this year published, in England, "A Disçourse of the Trade to Chio," in which he says: "There is cotton, wooll, etc., and also coarse wooll to make beds." (*See* year 1560.)

1582.

Abul Fazel celebrates the town of Sinnugan, or Soonergong, in India, for the manufacture of a beautiful cotton cloth named *cassas*.

1583.

Mr. Ralph Fitch, an English traveller, in this year visited "Sinnergau, a towne sixe leagues from Senapore, where there is the best and finest cloth made of cotton that is in all India." (*See* year 1582.)

1589.

The stocking frame, though a complex and ingenious machine, was invented this year by a Mr. William Lee, of Woodborough in Nottinghamshire, England, who, from want of patronage in that country, took his machine to France and established the stocking manufacture at Rouen under the patronage of Henry IV., upon whose death Lee got into difficulties and died at Paris in great poverty. (*See* years 1768 and 1787.)

1590.

Camden, in speaking of Manchester, England, in this year, says: "This town excels the towns immediately around it in handsomeness, populousness, *woollen manufacture*, market place, church, and college; but did much more excel them in the last age, as well as by the glory of its *woollen* cloths, which they call *Manchester cottons*." (*See* years 1313, 1322, 1519, and 1538.)

Macphersen, in his "Annals of Commerce," states that cotton cloths were imported into London this year from the Bight of Benin.

1601.

A list of foreign goods imported by the "English Society of Merchants and Adventurers" in this year from Holland and Germany, mentions cotton fabrics—"fustian"—said to have been manufactured at Nuremburgh.

1621.

This year is generally regarded as the birth year of cotton culture in the United States. It had previously been found growing in a wild state, however, in various portions of the South, more particularly in the country bordering upon the Mississippi (Meschachebe) and its many tributaries. A volume entitled "Purchas's Pilgrims" thus records the fact: "Cotton seeds were first planted as an experiment in 1621, and their plentiful coming up was, at that early day, a subject of interest in America and England." A tract called "A Declaration of the State of Virginia," published in London in 1620, mentions *cotton wool* as one of the commodities of that "collony." A list of articles "to be had in the Virginia collony" in 1621 mentions "cotton wool 8d. per pound" as among the number. The cotton thus introduced was probably from seed from the West Indies or the Levant, and its cultivation was for a long time limited to such qualities only as were needed for domestic use, as the cost of hand cleaning or separation of the seed by hand exceeded the commercial value of all cotton so cleaned.

1631.

Calicoes were first brought into Great Britain from Calicut, India, this year. (*See* year 1772.)

1641.

Roberts, in his "Treasure of Traffic," published this year in England, says: "The town of Manchester buys cotton wool from London that comes from Cyprus and Smyrna, and works the same into Fustians, vermillions and dimities."

Baines' "History of the Cotton Manufacture" (London, 1835) says: "No mention has yet been found of the cotton manufacture earlier than the year 1641." In this year it had become well established at Manchester, England. (*See* years 1313, 1322, 1520, 1538, 1590, and 1712.)

1664.

Pepys, in his diary, under date of February 27, this year, says: "Sir Martin Noel told us the dispute between him, as framer of the additional duty, and the East India Company, whether calico be *linen* or no, which he says it is, having been ever returned so. They say it is made of *cotton wool* and *grows upon trees*."

1666.

Carroll's "Historical Collections of South Carolina" mentions the growth of the cotton plant.

1676.

Calico printing commenced in London, England, in a very imperfect state. (*See* years 1690, 1712, 1750, and 1830.)

1678.

The Dutch and English East India Companies had, during the seventeenth century, imported Indian muslins, chintzes and calicoes in such quantities into Great Britain, and they were so cheap and popular that those interested in ancient woolen manufactures indulged in a loud demonstration against their further importation, maintaining that the advent of cottons was ruining the woolen interest. A pamphlet was issued this year in London, entitled "The Ancient Trades Decayed and Repaired Again," in which the author laments the interference of cotton with woolen fabrics. (*See* years 1696 and 1708.)

A loom moved by water power—"a new engine to make linen cloth without the help of an artificer"—is described in the Philosophical Transactions of the Royal Society of England for this year.

1690.

A print-ground was established on the banks of the Thames at Richmond, England, by a Frenchman. (*See* years 1676, and 1712.)

1696.

A pamphlet was this year published in England, entitled "The Naked Truth, in an Essay upon Trade," bewailing the introduction of cotton fabrics, saying they were "*becoming the general wear in England.*" (*See* years 1678, and 1708.)

1697.

Cotton imported into Great Britain, 1,976,359 lbs. Value of all kinds British cotton goods, £5,915.

1700.

So great had become the dissatisfaction with, and opposition to, the introduction of cotton fabrics into Great Britain during the seventeenth century (*see* years 1678, 1696, 1708 and 1728), that in this year Acts were passed which prohibited the introduction of printed calicoes for domestic use, either as apparel or furniture, under a penalty of £200 on the wearer and seller. But it did not prevent the continued use of cotton goods, quantities of which were smuggled into the country.

Population of Lancashire, England, 166,200. (*See* 1750, 1801 and 1831.)

About 1,000,000 lbs. of cotton used in Great Britain, requiring the services of 25,000 persons. (*See* year 1800).

1701.

Cotton imported into Great Britain, 1,985,856 lbs. (*See* year 1800 and nett increase). Value of all kinds British cotton goods exported, £23,253.

1702.

Average amount of cotton imported into England, 1,170,881 lbs.

1703.

Average amount of cotton imported into England, 1,170,881 lbs.

1704.

Average amount of cotton imported into England, 1,170,881 lbs.

1705.

Average amount of cotton imported into England, 1,170,881 lbs.

1708.

The *Weekly Review* of Daniel De Foe, January 31st, deplores the growing popularity, of late years and previous to the passage of the Prohibition Act (*see* year 1700), of cotton goods, thus: "Above half of the woolen manufacture was entirely lost, half of the people scattered and ruined, and all this by the interference of the East India trade." (*See* years 1696 and 1708).

1710.

Cotton imported into Great Britain, 715,008 lbs. Value all kinds British cotton goods exported, £5,698.

1712.

Cotton manufacture had become sufficiently extensive (*see* year 1641) in England to lead parliament to impose an excise duty of 3d. per square yard on calicoes printed, stained, painted or dyed. (*See* years 1676, 1690 and 1714).

1714.

The excise duty of 3d. per square yard "on calicoes printed, stained, painted or dyed," was raised to 6d. per square yard. (*See* year 1712).

1715.

Col. Hugh Orr, afterward instrumental in first introducing cotton machinery into this country (United States), was born at Lochwinnock, Scotland on January 2d, this year. (*See* years 1740, 1786 and 1798).

1720.

Cotton imported into Great Britain, 1,972,805 lbs. Value, all kinds British cotton goods exported, £16,200.

An Act was passed prohibiting altogether "the *use* or *wear* in Great Britain, in any garment or apparel whatsoever, of *any printed, painted, stained, or dyed calico*, under the penalty of forfeiting to the informer the sum of £5." By the same Act, the use of printed or dyed calicoes "in or about any bed, chair, cushion, window curtain, or any other sort of household stuff or furniture," was forbidden under a penalty of £20, and the same penalty attached to the *seller* of the article. (*See* years 1736 and 1774.)

About 2,200,000 lbs. of cotton used in Great Britain.

1721.

An Act of British Parliament this year prohibited the use or wear of printed calico, whether printed in England or elsewhere.

1728.

A volume published this year in England, called "A Plan of the English Commerce," speaks of the still prevalent "evil" of a consumption of Indian cotton manufactures (*see* years 1678, 1696, 1700, and 1708), and ascribed it to the will of the ladies, or, to use the author's words, to their "passion for their fashion."

1730.

Cotton imported into Great Britain, 1,545,472 lbs. Value, all kinds British cotton goods exported, £13,524.

John Wyatt, then living at a village near Lichfield, England, first conceived the project of spinning by rollers, and prepared to carry it into effect. (*See* years 1733, 1738, 1766, and 1769.)

1732.

Richard Arkwright (*see* years 1761, 1764, 1782, 1786, 1769, and 1792) was born at Preston, England, on the 23d of December of this year.

1733.

Cotton seed brought into Carolina by Mr. Peter Purry, who settled a colony of Swiss near Purrysville this year.

In this year John Wyatt (*see* years 1730, 1738, and 1769), "by a model of about two feet square"—to use the language of his son in a letter to another son, dated November 15th, 1817—"in a small building near Sutton Coldfield (England) without a single witness to the performance, was spun the first thread of cotton ever produced without the intervention of the human fingers. The wool had been carded in the common way, and *was passed between two cylinders, from whence the bobbin drew it by means of the twrit.*"

The earliest patent granted in Great Britain for any important improvement in manufacturing was that to John Kay for the fly shuttle, May 26th, this year.

1734.

Cotton was planted in Georgia from seed sent to the trustees by Philip Miller, of Chelsea, England.

1736.

The cotton plant was known on the Eastern shores of Maryland, lat. 39° N.

The prohibition to use mixed goods containing cotton in the dyed or printed state (*see* years 1720 and 1774), as it struck at the existence of the then rising cotton manufacture of England, was repealed this year.

Long before our Southern States took up its regular culture, cotton was raised on the Eastern shore of Maryland, lower counties of Delaware, and at other places in the Middle States. As early as this year, and for some time after, it was chiefly regarded, however, as an ornamental plant and confined to gardens; but it soon came to be appreciated for its useful qualities, and was brought under regular cultivation.

1738.

Mr. John Kay, a native of Bury, in Lancashire, England, at this time residing at Colchester, "suggested a mode of throwing the shuttle whereby the man could make nearly twice as much cloth as he could before." This invention was first applied to woolen manufactures, and not till the year 1760 was it much used among cotton weavers. The invention of Kay and his son, Robert, (*see* year 1760) was opposed by the operatives, who found they would lose their employment, and the elder Kay was so persecuted that he left his native country and afterward resided in Paris.

On the twenty-fourth day of June, this year, a patent was granted by George II. to Lewis Paul, a partner of one John Wyatt, of Birmingham, England, for *spinning by rollers*. Wyatt is supposed to have been the inventor, although the patent appears in the name of Paul, the latter having means and the former none. This is the same process which was, at a later period (*see* year 1764) brought out by Richard Arkright, and to whom credit of the invention has been generally awarded. There can be no dispute, however, as to the fact that a patent was issued in 1738 to Lewis Paul, which set forth the same claims as those made by Arkwright thirty years later. (*See* years 1741 and 1769.)

1739.

The deposition of Samuel Auspourguer, a Swiss, who had been living in Georgia, was taken for the use of the Georgia grant, in London, in the controversy about the introduction of slaves, which had been disapproved by Oglethorpe and others of the company, and opposed by the Highlanders (Scotch) and Galtzburgers, who had been settled in Georgia. Auspourguer said: "The climate of Georgia is very healthy, the climate and soil is very fit for raising silk, wine, and *cotton*, by this deponent's own experience, who has planted it there, grows very well in Georgia." A specimen of this cotton Auspourguer brought over with him and produced before the trustees. "All these produces," the deponent said, "can be raised by white persons without the use of negroes."

An article in the London *Daily Advertiser*, September 5th, this year, says: "The manufacture of cotton, mixed and plain, is arrived at so great perfection within these twenty years, that we not only make enough for our own consumption, but supply our colonies, and many of the nations of Europe. (*See* year 1641.)

1740.

Colonel Hugh Orr (*see* years 1715, 1786, and 1798) came to America and settled at Bridgewater, Mass., June 17th, this year.

1741.

A sample of Georgia cotton was taken to England.

Cotton imported into Great Britain 1,645,031 lbs. Value all kinds British cotton goods exported, £20,709.

Baines' "History of the Cotton Manufacture," p. 125, says: "I have before me the hanks of cotton yarn spun about 1741, and wrapped in a piece of paper, on which is written the following, in the handwriting of Mr. Wyatt: "The inclosed yarn, spun by the spinning engine (without hands) about the year 1741. The movement was at that time turned by two (or more) asses, walking round an axis in a large warehouse near the mill in the Upper Priory, in Birmingham. It owed the condition it was then in to the superintendency of John Wyatt. The above was wrote June 3d, 1756.'" (*See* years 1738 and 1769.)

1742.

In Louisiana, M. Dubreuil, a French planter, invented a machine for separating the seed from the fibre. It is to be inferred that the culture of this plant had become somewhat extensive. The machine was probably only an adjustment of rollers, like the contrivance of Cribs, which was the best machine for cleaning cotton until the invention of the saw-gin by Whitney. Previous to these instruments, the fibre was detached from the seed by the picking of the fingers. The bowstring in its use, intermediate between the fingers and the rollers, and used for beating up as well as cleaning the cotton, was borrowed from India, and having been first introduced in Georgia, gave use to the term "Bowed Georgia." The term is still used in Liverpool, though not a pound of "Bowed Georgia" has been in the market for 50 years.

1743.

Cotton imported into Great Britain, 1,132,288 lbs. ; exported, 40,870 lbs. ; home consumption, 1,091,481 lbs.

1744.

Cotton imported into Great Britain, 1,882,873 lbs. ; exported, 182,765 lbs. ; home consumption, 1,700,108 lbs.

1745.

Cotton imported intc Great Britain, 1,469,523 lbs. ; exported, 73,172 lbs. ; home consumption, 1,369,351 lbs.

1746.

Cotton imported into Great Britain, 2,264,808 lbs. ; exported, 73,279 lbs. ; home consumption, 2,191,529 lbs.

1747.

Cotton imported into Great Britain, 2,224,869 lbs. ; exported, 29,438 lbs. ; home consumption, 2,195,431 lbs.

1748.

Among the exports in this year from Charleston, S. C., were recorded "seven bags of cotton wool," valued at £3 11s. 5d. per bag! (*See* year 1754). Some writers have expressed a doubt if this cotton was of American growth, but as the culture had commenced in Carolina at least fifteen years before, there is no good reason to doubt it.

Cotton imported into Great Britain, 4,852,966 lbs.; exported, 291,717 lbs. ; home consumption, 4,561,249 lbs.

On the 30th of August, this year, Lewis Paul, of Birmingham, "gentleman," procured a patent for two carding machines, one a flat and the other a cylindrical arrangement. (*See* years 1760, 1772, and 1773.)

1749.

Cotton imported into Great Britain, 1,658,365 lbs. ; exported, 330,998 lbs. ; home consumption, 1,327,367 lbs.

1750.

Lancashire, England, the chief seat of the cotton trade, had a population of only 297,400. (*See* year 1831.)

About this year it was computed that 50,000 pieces of linen and cotton goods were annually printed in England. (See years 1676 and 1690.)

1751.

Cotton imported into Great Britain, 2,976,610 lbs. Value, all kinds of British cotton goods exported, £45,986.

1753.

A small shipment of cotton wool was made from Charleston, S. C., to England. (*See* year 1748.)

A liberal citizen of Delaware offered premiums for the promotion of industry, among them one of £4 for the most and best cotton off an acre.

1758.

Lewis Paul took out a new patent June 29, this year, for a *spinning machine* in England. (*See* years 1730, 1733, and 1738.) In this patent Paul is thus described: "Lewis Paul, of Kensington, Gravel Pitts, in the County of Middlesex, Esquire."

1760.

Up to this year the machines used in the manufacture of cotton goods in England were nearly as primitive as those of India. Mr. Robert Kay, of Bury, son of Mr. John Kay, (*see* year 1738) invented the *drop box*, "by means of which the weaver can at pleasure use any of those shuttles, each containing a different colored weft, without the trouble of taking them from and replacing them in the latter."

Richard Arkwright this year (*see* years 1732, 1761, and 1769) established himself as a barber in Bolton, England.

The carding machine of Lewis Paul (*see* years 1748, 1772, and 1773) was introduced into Lancashire, England, by a gentleman by the name of Morris, in the neighborhood of Wigan.

During this year it is estimated that not more than about 43,000 persons were supported by the whole cotton manufacture of England. (*See* year 1835.)

A considerable share of the calico printing business was transferred during this year from London to Lancashire in consequence of the cheaper accommodation for carrying on the work and the lower wages of the workmen.

1761.

Richard Arkwright was married this year to a lady of Leigh. (*See* years 1732, 1774, 1782, and 1769.)

A spinning wheel was invented by a Mr. John Webb, in this year, in England, also one by a Mr. Thomas Perrin. (*See* year 1765.)

1763.

Glasgow, Scotland, had a population of but 28,300. (*See* year 1831.)

1764.

An English mechanic, James Hargreaves, invented the "Spinning Jenny"—the name supposed to have been derived from "gin," a contraction of the word "engine." A number of young people were one day assembled at play in Hargreaves house, during the hour generally allotted for dinner, and the wheel at which he or some of his family were spinning, was by accident overturned. The thread still remained in the hand of the spinner, and as the arms and periphery of the wheel were prevented by the framing from any contact with the floor, the velocity it had acquired still gave motion to the spindle, which continued to revolve as before. Hargreaves surveyed this with mingled curiosity and attention. He expressed his surprise in exclamations which were long afterward remembered by those who heard them, and continued again and again to turn round the wheel as it lay on the floor, with much interest (which was at that time mistaken for mere indolence)! It is not, therefore, improbable that he derived from this circumstance the first idea of that machine, which paved the way for subsequent improvements. Hargreaves' first jenny was a very rude machine, *made entirely with a pocket knife*, and the clasp by which the thread was drawn out was *the stalk of a briar split in two!*

The first "jenny" had eight spindles set in a frame, and made to spin eight threads at once. The number was afterwards increased to eighty spindles. Envy on the part of his companions drove Hargreaves to Nottingham, where he erected a mill for the spinning of yarns by his machines.

About this period another mechanic, Richard Arkwright, came to Nottingham with an improvement upon the "Jenny" of Hargreaves: a combination of rollers, "which drew out the *slives* or rolls as they came from the carding machine, elongating and strengthening the fibres." Arkwright, in connection with other parties, soon after built a mill, in which the machinery was run by water, and the yarn produced thereby was called the "water twist." A great impulse was given to cotton manufacture by the ingenuity of Hargreaves and Arkwright. (*See* year 1782.)

Cotton imported into Great Britain, 3,870,392 lbs. Value all kinds British cotton goods exported, £200,354.

A horizontal spinning wheel was invented in England this year by a Mr. William Harrison.

Calico printing (*see* years 1676, 1690, and 1750) was introduced into Lancashire, England, by the Messrs. Clayton, of Bamber Bridge, near Preston, who began the business on a small scale. They were succeeded by Robert Peel. (*See* year 1773.)

About 3,900,000 lbs. of cotton used in Great Britain.

Eight *bags* of cotton imported into Liverpool from the United States.

According to Richard Grant, the "spinning jenny" was invented by Thomas Highs. (*See* year 1770.)

1765.

A spinning wheel was invented in England this year by a Mr. Perrin. (*See* year 1761.)

A "weaving factory," probably filled with "swivel-looms"—an invention of M. Vanconson during this century—was erected at Manchester, England, by a Mr. Gartside, but no advantage was realized, as a man was required to tend each loom.

1766.

John Wyatt, the inventor of the first machine for *spinning by rollers* (*see* years 1730, 1733, 1738, 1741, and 1769) died.

A spinning wheel was invented in England this year by a Mr. Ganat. (*See* year 1767.)

Official value of British cottons exported, £220,759.

Postlethwayt estimates the whole value of cotton goods *manufactured* in Great Britain this year at £600,000.

An Act passed in the British Parliament this year exempted cotton wool from duty, on importation into, or exportation from, any British colony, and on importation into Great Britain in British-built ships. In foreign ships it was subject to a duty.

In the correspondence of the Earl of Chatham, Vol. II., p. 420, it is stated in a note that, "in 1766, cotton, as an article of commerce, was scarcely known in Great Britain."

1767.

In this year the annual value of cotton manufactures of Great Britain was estimated at £600,000. (*See* year 1787 and nett increase). The goods, however, were a compound of linen warp and cotton weft.

Thomas Highs (or Hays), a reed maker of Leigh, claims (in 1785) that this year he made *rollers* for the purpose of *spinning cotton*, that he met Arkwright after he had taken out his (Arkwright's) patent for the water frame (see year 1769), and reproached him with having got his (Highs') invention, which Arkwright did not deny. (*See* years 1730, 1733, 1738, 1758, 1769, and 1782). Highs also asserted, as proof of his claim to the invention, that he hired a clockmaker at Warrington, named Kay, to make him a model of his machine, but never produced the model, nor did any one else ever see it save Kay, who having been afterwards in the employ of Arkwright, but was discharged therefrom, saying nothing of the model he made for Highs until after his discharge; and when Arkwright was endeavoring to establish his claim to the invention, which was disputed by Highs, Kay said, in his evidence, that Arkwright induced him, during this year, to make a model of Highs' machine, and took it away with him.

North Providence, R. I., was incorporated this year. (*See* year 1840.)

1768.

James Hargreaves retired to Nottingham and went into partnership with one Thomas James, and the two erected a small mill. (*See* years 1730, 1738, 1767, and 1769.)

A frame-work knitter of Nottingham, England, named Hammond, while looking at the lace on his wife's cap, thought he could make a similar article by means of his stocking frame. He tried, and was, on the first attempt, partially successful. (*See* years 1589, 1787, 1809, 1831, and 1835.)

Samuel Slater (*see* year 1789) was born June 9th, this year, near Belper, in Derbyshire, England.

1769.

Richard Arkwright patented his machine for *spinning by rollers*. (*Sec* years 1738 and 1741.) The specification, which was enrolled July 15, although the patent is dated July 3, says. "he had by great study and long application invented a new piece of machinery, never before found out, practised or used, for the making of weft, or yarn, from cotton, flax, and wool." Up to this time, Arkwright had struggled fiercely with poverty in perfecting his machine. The Messrs. Wright, bankers, of Nottingham, had aided him somewhat, but deserted him because his invention was not made remunerative soon enough. Mr. Samuel Need, of the same place, and Mr. Jedediah Strutt (the patentee of the stocking frame) thereupon entered into partnership with Arkwright, and from this time on, his prosperity steadily increased. (*See* year 1792.)

James Watt took out his patent "for lessening the consumption of steam and fuel in fire engines." (*See* years 1781, 1782, and 1784.)

1770.

Three shipments, amounting in all to ten bales, were made from Charleston, S. C., to Liverpool. There were also shipped to the same place three bales from New York, four from Virginia and Maryland, and three barrels full from North Carolina.

James Hargreaves obtained a patent for his spinning jenny this year, in England. (*See* years 1730, 1738, 1764, 1767, 1768, and 1769.)

In this year the land in the township of Miller, fourteen miles from Lancashire, England, was occupied by between fifty and sixty farmers; rents did not exceed 10s. per statute acre, and out of these fifty or sixty farmers there were but six or seven who raised their rents directly from the produce of their farms; all the rest got their rent partly in some branch of trade, such as spinning and weaving woolen, linen, or cotton.

Liverpool, England, had a population of but 34,050; Blackburn had but 5,000. (*See* year 1831.)

Imported into Liverpool from the United States, three bales from New York, four *bags* from Virginia, and three *barrels* from North Carolina.

The planters in our Southern States began turning their attention more particularly to the cultivation of cotton as an article of commerce.

At Pittsfield, Berkshire County, Mass., during this year, Valentine Rathbun erected a fulling-mill—"an old-fashioned, double action crank mill, driven by a three foot open bucket water wheel," requiring a strong head of water. Rathbun charged from forty to fifty cents per yard for fulling and finishing cloth.

Attempts were made to introduce the manufacture of cotton goods this year into Ireland, which were successful, but on a very limited scale. (*See* year 1854.)

1771.

Richard Arkwright's first mill was built at Cumford, England. (*See* years 1760, 1764, 1767, 1769, 1773, and 1775.)

Average amount of cotton imported into England, 4,764,589 lbs.

1772.

Thomas Highs (*see* year 1767) received a present of two hundred guineas from the manufacturers of Manchester, England, for a very ingenious invention of a double jenny, which was publicly exhibited in the Exchange. He afterward constructed spinning machines at Nottingham, Kidderminster, and in Ireland.

One of the first improvements made in the carding machine (*see* years 1748 and 1760) was invented this year by John Lees, a Quaker, of Manchester, England, and consisted in the fixing of a perpetual revolving cloth called the feeder, in which a given weight of cotton was spread, and by which it was conveyed to the cylinder. (*See* year 1773.)

Average amount of cotton imported into England, 4,764,589 lbs.

The manufacture of calicoes was begun this year in Lancashire, England. (*See* year 1631.)

1773.

About this year a very ingenious carding contrivance was invented (claimed by both Richard Arkwright and James Hargreaves, with preponderance of proof in favor of Arkwright), being "a plate of metal, finely toothed at the edge like a comb, which being worked by a crank in a perpendicular direction, with slight but frequent strokes on the teeth of the card, stripped off the cotton in a continuous flimsy fleece. The fleece as it came off was contracted and drawn through a funnel at a little distance in front of the cylinder, and was thus reduced into a roll or slive, which, after passing betwixt two rollers, and being compressed into a firm, flat riband, fell into a deep can, when it coiled up into a continuous length, till the can was filled." (*See* years 1748, 1760, and 1772.)

Average amount of cotton imported into England 4,764,589 lbs.

Robert Peel, son of Sir Robert Peel, Bart., quitted his father's concern in Lancashire, England, and established a partnership with his uncle, Mr. Haworth, and his afterward father-in-law, Mr. William Yates, at Bury, where cotton spinning and printing were carried on successfully for many years. Sir Robert, with his other sons and another Mr. Yates, established the print works at Church, and also had large works at Burnley, Salley Abbey, and Foxhill-bank, and spinning mills at Altham, and afterward at Burton-upon-Trent, in Staffordshire. The history of the two houses, the Peels of Bury and the Peels of Church, is the history of spinning weaving and printing in Lancashire for many years.

Bolton, England, had a population of but 5,339. (*See* year 1831.)

1774.

A law was passed in England sanctioning the manufacture of cotton goods, such as had heretofore been prohibited under heavy penalties (*see* years 1720 and 1736), and rendering English calicoes subject to a duty of 3d. per square yard on being printed; and providing that each piece be stamped *British Manufactory*, and that all persons exposing such goods to sale without the mark (unless for exportation) should forfeit the stuff and £50 for every piece, and that persons *importing* such goods should lose them, and forfeit £10 for each piece. It was death to counterfeit the stamp, or to sell the goods knowing them to have counterfeit stamps thereon. Cotton velvet, velverets, and fustians were not affected by this Act.

Average amount of cotton imported into England, 4,764,589 lbs.

Oxymuriatic acid was discovered by Scheele, the Swedish philosopher. (*See* year 1785.)

Manchester, England, had a population of 41,032. (*See* year 1831.)

Thomas Wood, of England, invented what was called a perpetual or endless carding by nailing the cards on the cylinder spirally instead of longitudinally. (*See* year 1776.)

Act of Parliament in Great Britain to prohibit the exportation of machinery.

1775.

The Assembly of the Province of Virginia, on the 27th of March, this year, in view of the changing relations with Great Britain, adopted a plan for the encouragement of arts and manufactures, including resolutions of non-importation; and "that all persons having proper land ought to cultivate and raise a quantity of hemp, flax, and *cotton*, not only for the use of his own family, but to spare to others on moderate terms." The planting of cotton had been recommended in the previous January by the first Provisional Congress held in South Carolina.

In the five years ending with this year, the average import of cotton wool into Great Britain did not exceed 4,764,589 lbs. per year, only four times as much as the average import at the beginning of the century.

Machinery was still very imperfect in England, especially in the preparation of the cotton for the spinning frame

Richard Arkwright took out a second patent on the 16th of December, this year, in England, "for a series of machines, comprising the carding, drawing, and roving machines, all used in preparing silk, cotton, flax, and wool for spinning." (*See* years 1738, 1741, and 1769.)

Ashton, England, had a population of but 5,097. (*See* year 1831.)

About 4,800,000 lbs. of cotton used in Great Britain.

The first spinning jenny probably seen in America was exhibited at Philadelphia early in this year, and was manufactured by Christopher Tully. The "Society for the Improvement of Arts, Manufactures, and Commerce," in England, repeatedly offered a premium of £100 sterling for a machine of this plan, but never had any presented to them which answered the purpose—"notwithstanding which," says an old record, "a very large one has been erected at Nottingham (England), which performs to great advantage, but no person as a speculatist is allowed to see it." (This referred, undoubtedly, to Arkwright's experimental machine. Tully's machine was, in all probability, manufactured in England.

The "United Company of Philadelphia, for promoting American Manufactures," was formed this year, previous to February 22d, at Philadelphia.

Nathaniel Niles, of Norwich, Conn., set up at that place a manufactory of iron wire for the making of cotton cards, which he continued throughout the Revolution.

Jeremiah Wilkinson, of Cumberland, R. I., about this time engaged in the manufacturing of hand cards.

1776.

The year which saw the "Declaration of Independence," also first witnessed cotton culture in New Jersey, in the county of Cape May.

Average amount of cotton imported into England, 6,766,613 lbs.

Deacon Barber erected a fulling-mill "on an improved plan," at Pittsfield, Mass, this year. Jacob Ensign and others soon followed, and fulling-mills became numerous.

Thomas Wood, of England, obtained a patent for his "perpetual or endless carding." (*See* year 1774.)

Samuel Wetherell, Jr., had a cotton factory, including dye-house, fulling-mill, etc., in South Alley, between Market and Arch, and Fifth and Sixth streets, Philadelphia.

1777.

Average amount of cotton imported into England, 6,766,613 lbs.

James Wallace, stocking weaver from abroad, petitioned the Assembly of Connecticut for a loan of £100 to erect stocking looms and a machine to spin the materials. He possessed a thorough knowledge of the manufacture of silk, cotton, and worsted stockings, which he claimed he could make as cheap as any imported. But his petition was not allowed.

Oliver Evans, of Philadelphia, Pa., then a young man of about twenty-one, having been engaged in the making of card teeth by hand, as then practiced, invented a very efficient machine for manufacturing them, it is said, at the rate of 1,500 per minute.

1778.

James Hargreaves, inventor of the spinning jenny, died at Nottingham, April 22, of this year, some authors say in obscurity and great distress. (*See* years 1764, 1767, 1768, 1769, and 1770.)

Average amount of cotton imported into England, 6,766,613 lbs.

Benjamin Hanks, of Windham, Conn. (afterward the inventor of the ingenious self-winding clock), sought, from the Assembly a premium for making stockings in looms.

1779.

Samuel Crompton, of Bolton, aged 21, invented a machine, combining the results of Arkwright's "roller spinner" and Hargreave's "spinning jenny" (*see* year 1764), which was named the "mule jenny," the spindles being attached to a carriage or mule, which ran out about five feet on wheels, stretching and twisting the roving into thread at the same time, and as the carriage ran back, the threads were wound on the spindles. The first machines carried but 20 to 30 spindles, but were enlarged to carry 2,200, all operating at once and superintended by one person.

Mobs rose in England to "put down" Hargreave's spinning jenny machines, and demolished all jennies, which ran but twenty spindles, all carding engines, water frames, and every machine turned by water or horses. The twenty spindle jennies were admitted to be useful, but all others were destroyed. (*See* years 1730, 1738, 1767, 1768, 1769, 1779, and 1789.)

Average amount of cotton imported into England, 6,766,613 lbs.

During this year the English Parliament imposed additional duties upon printed calicoes (*see* years 1712, 1714, 1720, and 1774), which, together with further duties imposed (*see* year 1782), increased the duty five per cent., making on the whole fifteen per cent.

Patent granted in England to Robert Peele for carding, roving, and spinning, February 18th.

Mule spinning invented by Samuel Crompton, of England—not patented.

1780.

Official value British cotton exported, £355,060.

Average amount of cotton imported into England in five years, 6,666,613 lbs.

Prices of various kinds of cotton in England this year were as follows: Berbice, 2s. 1d.; Demerara, 1s. 11d. to 2s. 1d.; Surinam, 2s.; Cayenne, 2s.; St. Domingo, 1s. 10d.; Tobago, 1s. 9d.; Jamaica, 1s. 7d.

An Act passed this year by the British Parliament allowed the importation of cotton in foreign ships at a duty of 1¼d. per lb., and 5 per cent. additional—the produce to be devoted to "the encouragement of the growth of cotton in His Majesty's (George III.) Leeward Islands, and for encouraging the importation thereof into Great Britain."

An attempt was made this year to manufacture muslin, both at Lancashire, England, and Glasgow, Scotland, with weft spun by the jenny, but it failed owing to the coarseness of the yarn. (*See* years 1783, 1785, and 1787.)

Preston, England, had a population of but 6,000. (*See* year 1831.)

An association was formed in Worcester, Mass., for spinning and weaving cotton, and a subscription was raised to procure a jenny for that purpose. On the 30th of April this year, the *Spy* announced "that on Tuesday the first piece of corduroy made in the manufactory of this town was taken from the loom."

The British Parliament enacted that "any person who took or put on board or caused to be brought to any place in order to be put on board any vessel with a view to exportation, any machine, engine, tool, press, paper, utensil or implement, or any part thereof, which now is, or hereafter may be, used in the woollen, cotton, linen, or silk manufactures of this kingdom, or goods wherein wool, cotton, linen, or silk are used, or any model or plan thereof, etc., etc., should forfeit every such machine and the goods taken therewith and £200, and suffer imprisonment for twelve months.

There were at this time twenty frame factories in England, the property of Sir Richard Arkwright, or of persons who had paid him for permission to use his machinery. (*See* year 1790.)

1781.

Richard Arkwright, of England, brought an action against one Colonel Mordaunt for the invasion of his patent. (*See* year 1769.) He also instituted eight other similiar actions in the same year.

Cotton imported into England, 5,198,788 pounds.

James Watt secured further patents on his steam engine inventions. (*See* years 1769, 1782, and 1784.)

Brazilian cotton was first imported into England this year, in a very dirty state.

Muslins were first made in England this year.

Raw material consumed in cotton manufactories in England, 6,000,000 lbs. (*See* year 1787.)

Great Britain commenced re-exporting a portion of her imports of cotton to the continent.

1782.

Richard Arkwright (*see* year 1764), now had nearly 5,000 persons employed in his mills at Nottingham, and was rapidly acquiring a fortune from the result of his inventions. Arkwright this year presented his "case"—his claim for important inventions, etc.—to Parliament, and as much as acknowledged that the principle upon which his spinning machines worked were discovered by others before his patents were taken out. (*See* years 1730, 1733, 1738, 1758, and 1769.) He says: "About forty or fifty years ago, one Paul, and others, of London, *invented an engine for spinning cotton, and obtained a patent for such invention.*"

Cotton imported into England, 11,828,039 lbs.

James Watt secured further patents on his steam engine inventions. *See* years 1769, 1781, and 1784.) Up to this time engines had been used for little else than pumping water out of mines, but in this year it was adapted to the production of rotative motion and the working of machinery.

During this year the English Parliament imposed additional duties upon printed calicoes (*see* years 1712, 1714, 1720, and 1774) which, together with duties imposed in (*see*) the year 1799, increased the duty 5 per cent., making, on the whole, 15 per cent.

An Act of the British Parliament passed this year made the destruction of cotton, woolen, silk, and linen goods, or any tools, or utensils used in spinning, preparing, or weaving such goods, in England, a capital felony. This law was meant to check the riotous attacks on machinery. (*See* years 1779 and 1789.)

Paisley had a population of but 17,700. (*See* year 1831.)

The exportation and the attempt to put on board for that purpose, "any blocks, engines, tools, or utensils used in, or which are proper for the preparing or finishing of the calico, cotton, muslin, or linen printing manufactures, or any part thereof," was prohibited in England under a penalty of £500.

The whole produce of the cotton manufacture in Great Britain did not exceed £2,000,000.

About 23,000,000 pounds of cotton used in Great Britain, requiring the service of 60,000 persons, 143 cotton factories, 550 mule jennies, 50,000 mule spindles, 20,070 hand jennies, and 1,600,000 jenny spindles.

1783.

Cotton imported into England, 9,735,663 pounds.

Arkwright & Simpson, of Shude-hill, Manchester, England, erected an atmospheric engine in their cotton mill this year. (*See* 1789 and 1790.)

An Act passed this year in the British Parliament reduced the heavy duties on muslins, calicoes, and nankeen cloths, to 18 per cent. *ad valorem*, with a drawback of 10 per cent.; another Act gave bounties on the exportation of British printed cottons as follows:

Under the value of 5d. per yard (before printing) ½d. per yard. Of the value of 5d. per yard, and under 6d. per yard, 1d. per yard. Of the value of 6d. per yard, and under 8d. per yard, 1½d per yard, besides the drawback on the excise duty. Another Act gave the manufacturers of cotton and flax a drawback of the excise duties on hard and soft soap, amounting to ¾d. per lb. weight, and on starch amounting to 1½d. per lb.

During this year there were above a thousand looms set up in Glasgow, Scotland, for the manufacture of muslins. (*See* year 1780.)

During this year Mr. Jefferson, in a letter to a Mr. Digges, said that "in general it is impossible for manufactories to succeed in America from the high price of labor," and that it was "not the policy of the government of this country to give aid to works of any kind." (*See* years 1785 and 1786.)

Cylinder printing invented and patented by Thomas Bell, at Glasgow, Scotland. (*See* year 1785.)

1784.

About fourteen bales of American cotton were shipped to Liverpool, of which eight bales were seized as improperly entered, on the ground that so much cotton could not have been produced in the United States; and this was more than 150 years after the first importation to England of cotton grown in the same country.

Cotton imported into England, 11,482,083 lbs.

James Watt secured further patents on his steam engine inventions. (*See* years 1768, 1781, and 1782.)

By an Act passed this year in the British Parliament—to impose new taxes to repair the finances of that country injured by the war with America—a new duty of 1d. per yard was laid on all cottons and mixed goods if bleached or printed, which were under 3s. per yard in value, and 2d. on all above that value, in addition to the former duties of 3d. per yard; (*see* years 1774, 1779, and 1782) and 15 per cent. additional was charged on the new duties as well as the old. This Act also compelled bleachers, printers, and dyers to take out licenses, for which the sum of £2 was paid annually. (*See* year 1785.)

A machine for binding and cutting card teeth was invented by Mr. Chittenden, of New Haven, Conn., this year, capable of making 87,000 per hour.

A German was fined £500 in England for persuading cotton operatives to go to Germany. A native of Amiens succeeded, the same year, in importing into France the first machine for spinning cotton.

Cotton imported into Great Britain estimated at 11,000,000 lbs.

New Jersey had in this year forty-one fulling mills for household woolens, but no woolen factories.

1785.

This year Thomas Highs (*see* years 1767 and 1772) first *publicly* laid claim to his invention of *spinning cotton by rollers*.

During a trial this year in England, concerning the validity of Richard Arkwright's patents (*see* years 1767 and 1775), Mr. Bearcroft, the counsel opposed to Arkwright, stated that "30,000 people were employed in the establishments set up *in defiance of the patents*, and that nearly £300,000 had been expended in the buildings and machinery of these establishments."

The factory system in England takes its rise from about this period, the work to a great extent having hitherto been done in the homes of the workmen. (*See* year 1719.)

Richard Arkwright's patents were set aside, and the benefit of his inventions was thrown open to the public. (*See* year 1786.)

Cotton imported into England, 18,400,384 lbs.

Hands employed in cotton manufactures in Great Britain, 80,000. (*See* years 1787 and 1831.)

The first steam engine ever applied to the propelling of cotton manufactures was erected this year by Boulton and (James) Watt in the works of the Messrs. Robinsons, of Popplewick, in Nottinghamshire, England.

The Rev. Dr. Edmund Cartwright, of Hollander House, Kent, this year, invented a power loom, which was the parent of all now in use, although, according to the inventor's own words, it was a most cumbersome affair. "The warp was placed perpendicularly, the reed fell with a weight of at least half a hundred weight, and the springs which threw the shuttle were strong enough to have thrown a congreve rocket. In short, it required the strength of two powerful men to work the machine at a slow rate and only for a short time. This was patented on the 4th of April this year. (*See* years 1787, 1801, 1809, 1812, 1813, 1820, and 1829.)

Oxymuriatic acid (*see* year 1774) was first applied to the bleaching of cotton goods.

A Scotchman, by the name of Bell, had invented a machine for cylinder printing—it had hitherto been done by blocks, either wooden or copper, and was a very tedious operation, and the machine was first successfully applied about this year at Morney, near Preston, in Lancashire, England, by the house of Linsey, Hargreaves, Hall & Co. (*See* year 1788.)

The imposition of the Act of (*see* year) 1784 excited so much alarm and discontent throughout the cotton manufacturing districts of England and Scotland, that petitions were sent to the House of Commons, and memorials to the Lords of the Treasury, representing that these new duties would crush the rising manufacture and render the English altogether unable to compete with Indian goods. The manufacturers were heard by counsel at the bar of the House during this year; much evidence was given, and a short bill was brought in repealing all the new duties imposed by the bill of (*see* year) 1784, on the linen and cotton manufactures. The repeal was celebrated as a jubilee in the manufacturing sections. During this year, however, a considerable addition was made to the former duties on cotton, linen, and mixed goods, which, on the average, more than doubled the duties existing previously to (*see* year) 1784, but they only applied to *printed* goods, not to goods which were merely *bleached.*

Weft and warp were produced sufficiently fine for muslins (*see* years 1780 and 1787.)

Pullicat handkerchiefs first made this year in Glasgow, Scotland.

Thomas Somers, of Baltimore, Md., having been brought up a

cotton manufacturer, visited England and brought away with him descriptions and models of machines for spinning cotton. The Assembly of Massachusetts, on the 2d of March of this year, ordered £20 to be deposited to encourage Somers in the trial of manufacturing one of the machines. It was afterwards known as the "States' Model," and was an imperfect form of Arkwright's machine.

The "Society for the Promotion of Agriculture" was chartered, this year, in South Carolina, and among other matters, offered premiums or medals for the best mode of destroying the caterpillar, which infested the cotton plant, and for a practical method of discharging stains from cotton and rendering it perfectly white.

In his "Notes on Virginia," within this year, Mr. Jefferson opposed the establishment of manufactures, believing that the people would be more happy, virtuous, and prosperous as an agricultural people than they could be with the vices and evils of manufacturing towns in their midst. His views afterward underwent a change. (*See* year 1786.) He even himself became a manufacturer in a small way in his household, and employed two spinning jennies, a carding machine, and a loom with a flying shuttle, by which he made more than two thousand yards of cloth, which his family and servants required yearly.

1786.

This year, a convention was held at Annapolis, Maryland, to consider what means could be resorted to for the purpose of remedying the embarrassment of the country, then so much exhausted in its finances. The late President Madison, a member of this convention from Virginia, then expressed it as his opinion, that, from the results of cotton raising in Talbot County, Maryland, and numerous other proofs furnished in Virginia, there was no reason to doubt "that the United States would one day become a great cotton producing country."

First Sea Island cotton was produced in Georgia, the seed having been obtained from the Bahamas.

The State of Massachusetts made a grant of £200 lawful money" (six tickets in the State lottery, in which there were no blanks), for the encouragement of two mechanics from Scotland named Robert and Alexander Barr, brothers, who made the first machines in the United States for carding, roving, and spinning, while in the employ of a Mr. Orr, of East Bridgewater. (*See* years 1787, 1788, and 1790.)

It does not appear that the machinery at East Bridgewater was used to any extent for manufacturing purposes, but rather for models and to diffuse information upon the subject, and the State Legislature had provided in their resolve "that public notice be given for three weeks successively in Adam's and Nurse's newspaper, that said machines may be seen and examined at the house of the Hon. Hugh Orr (*see* years 1715, 1740, and 1798), in Bridgewater, and that the manner of working them will be explained." There seems to be no doubt that this machinery was the first built or introduced into this country for the manufacture of cotton, which included Arkwright's roller spinning and other patent improvements.

In South Carolina, the family manufactures in interior parts of the State furnished a sufficient supply of substantial middling and coarse cotton, woolen, and linen goods. It was the same in Georgia. In North Carolina they were nearly as attentive to domestic manufactures as in Virginia, and some good cotton stuffs were made. In Connecticut the household manufactures were such as to furnish a surplus sold out of the State. In Massachusetts the importation of foreign manufactures was less by one half than it was twenty years before, although the population had greatly increased, and considerable quantities of home-made articles were shipped out of the State. In Rhode Island and New Hampshire the same progress had been made. The number of regular factories in Rhode Island was great in proportion to the population.

Mr. Jefferson, in a letter addressed to M. de Warville, August 15, of this year, said: "The four southernmost States make a great deal of cotton. Their poor are almost entirely clothed in it in winter and summer. In winter they wear shirts of it, and outer clothing of cotton and wool mixed. In summer their shirts are linen, but the outer clothing cotton. The dress of the women is almost entirely of cotton, manufactured by themselves, except the richer class, and even many of these wear a great deal of homespun cotton. It is as well manufactured as the calicoes of Europe. (*See* year 1785.)"

A complete set of brass models of Arkwright's machine was made this year, and packed in England by the agent of Mr. Tench Coxe, of Philadelphia, Pa., but was seized on the evening before it was to have been shipped and its object defeated.

Ohl Buell, an ingenious merchant of Killingsworth, Conn., visited England this year, ostensibly to purchase copper, but in reality to obtain a knowledge of the various kinds of machinery used in the manufacture of cloth.

In Lancaster, Pa., then the largest inland town in the United States, there were in this year about 700 families, of whom 234 were manufacturers, in which were included 25 weavers of woolen, linen and cotton cloth, 3 stocking weavers, and 4 dyers.

On the 25th of October this year, Richard Cranch, of the Massachusetts Senate, and Mr. Clarke and Mr. Bowdoin, of the House, were appointed "to view any new invented machines that are making within this Commonwealth for the purpose of manufacturing sheep's and cotton wool, and report what measures are proper for the Legislature to take to encourage the same." (*See* year 1787.)

Legislature of Massachusetts made a grant to Robert and Alexander Barr to aid them in building cotton spinning machinery.

Tench Coxe, of Philadelphia, Pa., in his Report, fourteen years afterward—1810—said: "In 1786 I became acquainted with the fact that labor saving machinery was considerable in Great Britain. It was understood that it was applicable at that time only to the carding and spinning of cotton, which was then constantly imported from foreign countries, apparently to the amount of our whole consumption."

At this period, one third of the English consumption of cotton was brought from the British West Indies, one third from the foreign West Indies, one quarter from Brazil, and the remainder from the Levant.

Richard Arkwright was appointed high sheriff of Derbyshire, England, and received the honor of knighthood. (*See* years 1732, 1761, 1764, 1767, 1769, 1782, 1785, and 1792.)

Cotton imported into England, 19,900,000 lbs. from the following sources: British West Indies, 5,800,000 lbs; French and Spanish colonies, 5,500,000 lbs.; Dutch colonies, 1,600,000 lbs.; Portuguese colonies, 2,000,000 lbs.; Smyrna and Turkey, 5,000,000 lbs.

A small quantity of cotton, of the best quality then known, was received from the Isle of Bourbon by way of Ostend and sold at from 7s. 6d. to 10s. per lb.

Sir Charles Wilkins brought to England in this year a specimen of Decca muslin from India.

1787.

At Beverly, Mass., a company was formed for the purpose of carding, roving and spinning cotton by machinery invented by the Barr Brothers. The legislature made a grant of £500 to assist the design. (*See* years 1786, 1788 and 1790.)

Import of American cotton into Liverpool was 16,350 lbs.

Cotton imported into England from all quarters, 23,250 268 lbs.

Official value British cottons exported, £1,101,457.

The purposes for which cotton were used in Great Britain during this year are thus stated: calicoes and muslins, 11,600,000 lbs.; fustians, 6,000,000 lbs.; mixtures with silk and linen, 2,000,000 lbs.; hosiery, 1,500,000 lbs.; *candle-wicks*, 1,500,000 lbs., (*see* year 1701)—the latter article alone consuming nearly as much in 1787 as the whole importation in 1701).

Estimated annual value of cotton manufactures in Great Britain, £3,304,371. (*See* year 1767 and note increase.)

Hands employed in cotton manufactures in Great Britain, this year, numbered 162,000. (*See* years 1785 and 1831.)

Number of cotton mills in Great Britain this year, 143, as follows: Lancashire, 41; Derbyshire, 22; Nottinghamshire, 17; Yorkshire, 11; Cheshire, 8; Staffordshire, 7; Westmoreland, 5; Berkshire, 2; rest of England, 6; Flintshire, 3; Pembrokeshire, 1; Lanarkshire, 4; Renfrewshire, 4; Perthshire, 3; Edinburghshire, 2; rest of Scotland, 6; Isle of Man, 1.

The Rev. Edmund Cartwright, on the first of August this year, took out his last loom patent, a great improvement on his first invention. (*See* years 1785, 1801, 1809 and 1812.)

In the consolidation of the customs in this year, all former duties were repealed, and cotton, linen, or mixed goods of every kind were subjected to a duty of 3½*d.* per square yard, when printed or dyed, and the whole duty was returned by drawback on the exportation of the goods. At the same time, foreign calicoes and muslins were charged with a duty of 7d. per square yard when printed or dyed in Great Britain. (*See* year 1831.)

An Act was passed, this year, in England, to encourage the art of designing original patterns for printing on calicoes, muslins and linens, vesting in the proprietors the sole right of vending the goods printed with original patterns for two months after the day of publishing them, afterward enlarged to three months. Also, an Act allowing importation of cotton from British plantations duty free, and of cotton not from British plantations at a duty of 1*d.* per lb. in foreign ships, free in British ships.

500,000 pieces of muslin were manufactured in Great Britain during this year. (*See* years 1780 and 1785.)

It was estimated that during this year, 1,500,000 lbs. of cotton were consumed in the manufacture of hosiery. (*See* years 1589, 1768 and 1809.)

The average exports of British manufactures to the United States, for several years preceding this year, notwithstanding a great increase in the population of the States, were nearly half a million dollars less than the average of several years preceding the war with England.

During this year, Samuel Loomis, of Colchester, Conn., announced that he was "prepared to introduce a new epoch in the manufacture of wool, cotton, etc., upon a newly constructed plan."

The descriptions of cotton imported into Great Britain during this year appears to have been as follows:

From the British West Indies,	6,800,000 lbs.
" " French and Spanish Colonies,	6,000,000 "
" " Dutch,	1,700,000 "
" " Portuguese,	2,500,000 "
" " Isle of Bourbon, by Ostend,	100,000 "
" Smyrna and Turkey,	5,700,000 "
	32,800,000 lbs.

The number of cotton mills in Great Britain at this time, as near as could be ascertained, was 143, the cost of which was estimated at £715,000; there were in operation 550 mules, and 20,700 jennies (*see* year 1767), containing, together with the water frames, 1,951,000 spindles, the cost of which, and of the auxiliary machinery, together with that of the buildings, is stated to have been at least £285,000, making a total investment of £1,000,000. They were supposed to give employment to 26,000 men, 31,000 women, and 53,000 children, and in the subsequent stages of manufacture the number of persons employed were supposed to be 133,000 men, 59,000 women, and 48,000 children, making a total of 159,000 men, 90,000 women, and 100,000 children—350,000 persons in all. The quantity of raw material consumed exceeded 22,000,000 lbs. (*See* year 1781.)

Daniel Anthony, of Providence, R. I., had a spinning-jenny of twenty-eight spindles, built on the model of the Beverly, Mass., machines.

The Board of Managers of the "Pennsylvania Society for the Encouragement of Manufactures and the Useful Arts," in November of this year, offered "a gold medal of the value of twenty dollars for the most useful engine or machine, to be moved by water, fire, or otherwise, by which the ordinary labor of hands in manufacturing cotton, wool, flax, or hemp, should be better saved than by any then in use in this State."

Not less than 500,000 pieces of muslin were made at Bolton, Glasgow and Paisley, in this year, with yarn of British production. (*See* year 1780.)

"The Society for the Encouragement of the Useful Arts" was formed at Philadelphia, Pa., Aug. 9th.

On the 8th of March, this year, "Richard Cranch was appointed by the Massachusetts Senate, with such as the House should join, to examine the machines, which are now nearly completed." (*See* year 1786.)

Cotton machinery first introduced into France this year.

Grant made by Massachusetts Legislature to Thomas Somers, of Baltimore, to aid him in completing cotton spinning machinery.

First cotton factory built in the United States, at Beverly, Mass. (*See* year 1791.)

A factory was commenced this year at Beverly, Mass., expressly for the manufacture of cotton goods (*see* year 1791), with such machinery as could then be procured. The act for the incorporation of this company contained the following provision: "That all goods which may be manufactured by said corporation, shall have a label of lead affixed to one end thereof, which shall have the same impression with the seal of said corporation; and that if any person shall knowingly use a like seal or label, with that used by said corporation, by annexing the same to any cotton, or cotton and linen goods not manufactured by said corporation, with a view of vending or disposing thereof as the proper manufactures of said corporation, every person so offending shall forfeit and pay treble the value of such goods, to be sued for and recovered for the use of said corporation, by action of debt, in any court of record proper to try the same.

General Washington visited this factory on Friday, October 30th, 1789.

1788.

The East India Company, stimulated by the representations of the English manufacturers, commenced operations this year for improving the quality and increasing the quantity of cotton exported into Great Britain. (*See* years 1793 and 1800.)

First exportation of Georgia Sea Island cotton, made by Alexander Bissell of St. Simon's Island.

A "home-spun cloth" company was incorporated at Providence, R. I., with machinery modeled after English plans and those used by Mr. Orr, at Bridgewater, Mass. (*see* year 1786), and by the Beverly, Mass., company. (*See* year 1790.)

Import of American cotton into Liverpool was 58,500 lbs.

Upland cotton was worth, in Liverpool, 2s. 2d. per lb.

Cotton imported into England from all quarters, 20,467,436 lbs.

The large concern of Linsey, Hargreaves, Hall & Co., of Lancashire, England, disastrously failed, causing a severe shock to the industry of that section. (*See* year 1785.)

A pamphlet was published this year in England entitled, "An Important Crisis in the Calico and Muslin Manufactures of this Country Explained," the purport of which was to warn the nation of the bad consequences which would result from the rivalry of the East India cotton goods, which then began to be poured into the market in increased quantities, and at diminished prices.

Richard Teake, in a letter dated Savannah, Ga., December 11th, this year, to Tench Coxe, of Philadelphia, says: "I have been this year an adventurer, and the first that has attempted on a large scale in the article of cotton. Several here, as well as in Carolina, have followed me and tried the experiment. I shall raise about 5,000 pounds in the seed from about eight acres of land, and the next year I expect to plant from fifty to one hundred acres. The lands in the southern part of this State are admirably adapted to the raising of this commodity. The climate is so mild, so far to the South, scarce any winter is felt, and—another grand advantage—whites can be employed. The labor is not severe attending it, not more than raising Indian corn."

Giles, Richards & Co., of Boston, Mass., began the manufacture of cotton cords with newly invented machinery.

Joseph Alexander and James McKevins, weavers from Scotland, who understood the use of the fly shuttle, came to Providence, R. I., to weave corduroy. McKevins went to East Greenwich, Conn., but Alexander stopped at Providence, and a loom was built and put in operation in the Market House, with the first fly shuttle ever used in Providence, and probably in America.

The Legislature of Pennsylvania granted £100 in October of this year to John Hague, for introducing a machine for carding cotton.

The first loom in Philadelphia was built and worked on the 12th of April, this year.

About this year Daniel Anthony, Andrew Doctor, and Lewis Peck, all of Providence, R. I., entered into an agreement to make what was then called *home spun cloth*. The idea, at first, was to spin by hand and make jeans with linen warps and cotton filling;

but hearing that Mr. Orr, of Bridgewater, Mass., (*see* 1786 and ante) had imported some models of machinery from England, for the purpose of spinning cotton, it was agreed that Daniel Anthony should go to Bridgewater, and get a draft of the model of said machine. A draft was taken, but laid aside, and they built a jenny with twenty-eight spindles, which was first set up in a private house, and afterward in a chamber in the Market-house at Providence, where it was operated.

Joshua Lindley, of Providence, R. I., was this year engaged to build a carding machine for carding cotton, after a draft obtained from the Beverly, Mass., mill, which, after some delay, was finished, although very imperfect. It consisted of eight heads of four spindles each, and was operated by a crank turned by hand.

"Spinning jennies" were put in operation in Providence, R. I., and Philadelphia, Pa.

Bleaching by oxymuriatic acid practically introduced (*See* year 1785) at Manchester, England.

1789.

In November of this year Samuel Slater, a young Englishman from the Derbyshire mills, aged 21, with an experience of seven years in the mills of England, arrived in New York (having sailed from London on the 13th day of September) for the avowed purpose of introducing the Arkwright and Hargreaves processes (*see* years 1764 and 1782) into the manufacture of cotton in this country. (*See* year 1790.)

Import of American cotton into Liverpool was 127,500 lbs.

Cotton imported into England from all quarters, 32,576,023 lbs.

First steam engine used for cotton spinning in Manchester, England, erected this year by Boulton & Watt for the mill of Mr. Drinkwater. (*See* years, 1783 and 1790.)

The Act of the British Parliament (*see* years 1782) for the protection of machinery, etc., was extended into Scotland this year.

The parish of Oldham, England, had a population of but 13,916. (*See* year 1831.)

On the 29th of March, this year, the Legislature of Pennsylvania passed an "Act to encourage and protect the manufacturers of the State." This Act, which was limited to two years, prohibited under certain penalties, the exportation of manufacturing machines, the scarcity of which was the great obstacle to such undertakings. This Act is stated by Mr. Carey, the editor of the *American Museum*, to have been prompted by the fact that in the year previous (1787) two carding and spinning machines in the possession of a citizen of Philadelphia, and calculated to save the labor of one hundred and twenty persons, were purchased by the agency of a British artizan, packed as common merchandise, and shipped to Liverpool.

A quantity of cotton seed is stated to have been purchased and burned this year in Virginia, in order to prevent, if possible, the extent of cotton manufactures in America and their injurious effects upon the importation of Manchester goods.

In October of this year, a reward of £100 was given John Hague, of Alexandria, Va., for a carding machine, completed for the "Pennsylvania Society." (*See* year 1775.)

Samuel Slater, who came from England to this country, was employed at New York city, where he said they had in operation one carding engine and two spinning jennies at the close of the year.

Thomas Hubbard and Christopher Leffingwell, of Norwich, Conn., who had erected eight stocking-looms, asked for themselves and their apprentices, an exemption from poll-taxes, which was granted by the Lower but refused by the Upper House.

About this year a large manufactory of sail duck was established in Frog Lane, in Boston, Mass.—a two-story building, one hundred and eighty feet long, being erected for the purpose. The company was duly incorporated by the General Court, and encouraged by a bounty upon its manufacture. Careless workmanship was punished by a system of fines, which went into a common fund for the relief of sick members, and the goods if unsaleable were to be made good. The product of the establishment was said to be the best ever seen in America, and sold lower than imported sail cloth. The ship *Massachusetts*, of about eight hundred tons, had her sails and cordage wholly of Boston manufacture. (*See* year 1792.)

The account of Andrew Dexter, with Messrs. Almy & Brown, of Providence, R. I., in May of this year, shows a charge "for completing a spinning jenny—£24 4s. 10d."

First cotton spinning commenced by Messrs. Almy & Brown, of Providence, R. I., about the 11th of June, this year, between which time and the close of the year they made of corduroys, royal ribs, denims, cottonets, jeans, fustians, etc., 189 pieces, containing 4,566 yards, which sold from 1s. 8d. to 4s. per yard.

John Heusen, a Revolutionary soldier was, in March of this year, granted a loan of £200 from the State of Pennsylvania by the Legislature, to enable him "to enlarge and carry on the business of calico printing and bleaching within this State." His factory was near Richmond, or where Dyottville now is.

A sail cloth factory was commenced in Haverhill, Mass., this year, which did not finally succeed, although for some years it did moderately well. Factories of the same kind were likewise established at Salem and Nantucket, Mass., Exeter, N. H., and Newport, R. I., about this year. Those at Salem and Nantucket, Mass., became flourishing concerns.

A Mr. Austin, of Glasgow, Scotland, invented a power-loom, which was not, however, perfected for operation until some years after. (*See* year 1798.)

The origin of the celebrated "Sea Island" cotton is thus related in a letter by one Patrick Walsh to Dr. Meare: "I had settled in Kingston, Jamaica, some years ago, when finding my friend, Frank Leavet, with his family and all his negroes, in a distressed situation, he applied to me for advice as to what steps he should take, having no employment for his slaves. I advised him to go to Georgia and settle on some of the islands, and plant provisions until something better turned up. I sent him a large quantity of various seeds of Jamaica; and Mr. Moss and Colonel Brown requested me to get some of the Pernambuco cotton seed, of which I sent him three large sacks, of which he made no use but by accident. In a letter to me during the year 1789 he said: 'Being in want of the sacks for gathering in my provisions, I shook their contents on the dunghill, and it happening to be a very wet season, in the Spring multitudes of plants covered the place. These I drew out and transplanted them into two acres of ground, and was highly gratified to find an abundant crop. This encouraged me to plant more. I used all my strength in cleaning and planting, and have succeeded beyond my most sanguine expectations.' "

Ædanus Burke, in a debate on the tariff, on the 16th of April this year, to induce the House to lay a considerable duty on hemp and cotton, said: "The staple products of South Carolina and Georgia were hardly worth cultivation on account of their fall in price. The lands were certainly well adapted to the growth of hemp, and cotton was likewise in contemplation among them, and if good seed could be procured, he hoped might succeed."

1790.

Hon. Whitmarsh Seabrooks' statements as to exports of "cotton wool" from Charleston (*see* years 1748, 1754, 1770, and 1784) are flatly contradicted by McCulloch, who claims that *no* cotton whatever was exported from any portion of the United States previous to the year 1790.

First successful crop of cotton in South Carolina was raised this year on "Hilton Head" Island by William Elliott. The success of Elliott caused many to engage in cotton culture, and many of the largest fortunes in that State were thus realized. (*See* years 1805, 1806, and 1816 for prices of this quality as compared with others.)

A company at Beverly, Mass., (*see* years 1787 and 1788) having expended £4,000 in erecting machinery for carding, spinning, and roving, obtained a grant of £1,000 from the legislature and introduced the manufacture of cotton goods, but with indifferent success, as their machinery was very imperfect.

Samuel Slater (*see* year 1789), in January of this year was hired by Messrs. Almy & Brown, of Providence, R. I., to manufacture and perfect the spinning machines of Richard Arkwright, (*see* year 1764 and 1782), and in December the first Arkwright machinery was put into operation—three cards, roving and drawing, and a seventy-two spindle frame, worked by a water wheel (*see* year 1764.)

Import of American cotton into Liverpool was 14,000 lbs.

Upland cotton was worth in Liverpool only 10d. per lb. This may account for the small shipments of American cotton this year. It was probably poorer in staple, and certainly less clean than the upland of the present day.

Thomas Highs, the inventor (*See* year 1767), was disabled by a stroke of palsy. (*See* year 1803.)

Mr. William Kelly, of Lanark Mills, England, was the first to turn the "mule" by water power (*See* year 1792) this year, and Mr. Wright, a machinist, of Manchester, constructed a *double mule*. (*See* year 1793.)

Cotton imported into England from all quarters, 31,447,605 lbs.

Official value British cotton goods exported, £1,662,369.

Not until this year did Richard Arkwright (*See* year 1732) adopt the use of the steam engine (see year 1769) in his mills, when one of Boulton & Watt's engines was put up in his mill at Nottingham, England. (*See* years 1783, 1789, and 1792.)

About this year the Messrs. Grimshaw, of Gorton, England, under a license from Dr. Cartwright (*See* years 1785, 1787, 1801, 1809, and 1812) erected a weaving factory at Knott's Mills, Manchester, and attempted to improve the power-loom. They spent much money, did not succeed, their factory was burned down by a mob, and they abandoned the undertaking.

A correspondent of the *American Museum*, writing from Charleston, S. C., in July of this year, states that a gentleman well acquainted with the cotton manufacture had already completed and in operation on the high hills of the Santee, near Statesburg, ginning, carding, and other machines, driven by water, and also spinning machines with eighty-four spindles each, with every necessary article for manufacturing cotton.

Twenty-two ships arrived in American ports, from St. Petersburg, laden with cordage, tickings, drillings, diapers, etc.

Philadelphia had a total population of 43,000, of whom but 2,200 were what might be properly denominated manufacturers.

In twenty families, rich and poor, taken indiscriminately, in Virginia, during this year, among a total of 301 persons of both colors, there were made of fine table linen, sheeting, shirting, etc., 1,907 yards. The finer qualities of cloth were worth sixty cents per yard, and the coarser forty-two cents. The total value of this industry was $1,670; the highest value made in one family was $267, and the lowest $21.50. There was but one family in the twenty that did not manufacture.

About this year, a Mr. Grimshaw, of Manchester, England, under a license from Mr. Cartwright (*see* year 1787), erected a weaving factory which was to have contained five hundred looms, but, after a small part of the machinery had been set going, the mill was destroyed by fire and was not rebuilt.

The sale of spinning-wheel irons from one shop in Philadelphia, this year, amounted to 1,500 sets, an increase of twenty-nine per cent. over the previous year.

The average product of the spinner of yarn No. 40, was but little more than a hank per spindle, per day, at this period. (*See* years 1812 and 1830.) The price of spinning No. 100 was 4*s*. per pound.

The manufactory of the "Pennsylvania Society" (*see* year 1775) was burned on the night of the 29th of March, this year, at the southwest corner of Market and Ninth Streets, Philadelphia, Pa., and evidence appearing that it was fired by design, a reward was offered by the State for the arrest of the incendiary.

From this year the progress of cotton manufactures in Ireland became considerable. (*See* years 1770 and 1854.)

About 31,000,000 lbs, of cotton used in Great Britain.

A person who had been employed in the Beverly, Mass., factory (*see* year 1789) was this year engaged to go to Norwich, Conn., to put in operation some cotton machinery, which was understood to be similar to that used in Beverly. This machinery was not built in this country, but was supposed to have been imported by some means, from England. The parties engaged in the business at Norwich were a Mr. Huntington, Dr. Lathrop, and others. (This Dr. Lathrop was the same in whose druggists' store Benedict Arnold is said to have been employed, before the Revolutionary War.)

There were, at this time. one hundred and fifty cotton factories in England and Wales. (*See* year 1780.)

Samuel Brazier, of Worcester, Mass., advertised for sale, in the *Spy*, "jeans, corduroys, 'Federal Rib,' and *cottons*."

The managers of the Beverly, Mass., manufactory, in a memorial to the General Court, in June of this year, stated that they had encountered more obstacles and difficulties than they had anticipated, especially in the purchase and construction of machinery. They had expended nearly £4,000.

The account of Andrew Dexter, of Providence, R. I., with Almy & Brown, shows a charge for a jenny and a carding and spinning machine, "completed at the joint and equal expense of Lewis Peck and Andrew Dexter." A machine for calendering cotton goods was also charged in the books of the same firm, in March of this year, which was put up in Moses Brown's barn, and worked by a horse!

The first sheetings, shirtings, checks and ginghams made in America, were made this year.

At East Greenwich, Conn., a German named Herman Vandausen commenced the calico printing business. He cut his own blocks, and printed India cottons, and the coarse cottons woven in families, for the people generally. But it was found cheaper to import than to print, and the business was given up.

Total number of slaves in the United States, this year, 697,897. (*See* years 1830 and 1850.)

1791.

Cotton crop of the United States was 2,000,000 lbs, of which three-fourths was grown in South Carolina, and one-fourth in Georgia. Export, 189,500 lbs; worth twenty-six cents average.

A cotton mill was erected, this year, at Providence, R. I., hitherto commonly supposed to have been the first one in America. (*See* years, 1787, 1807, 1810 and 1831.)

Previous to this year, Great Britain obtained her supplies of cotton from the West Indies, South America, and the countries around the eastern parts of the Mediterranean.

From the 1st of January to the 15th of October of this year, the firm of Almy & Brown, of Providence, R. I., with the new machinery of Samuel Slater (*see* years 1768 and 1789), made of corduroys, royal ribs, denims, cottonets, jeans, fustians, etc., 326 pieces of 7,823 yards. (*See* year 1789.)

William Pollard, of Philadelphia, obtained a patent for cotton spinning, December 30th, of this year, and erected the first water frame put in motion in Pennsylvania.

Cotton exported from the United States into Great Britain, 189,316 lbs. (*See* years 1800, 1810, 1821, 1831 and 1841.)

William Pollard, of Philadelphia, Pa., on the 20th of December, of this year, patented a machine for spinning cotton, which is said to have been the first water frame erected there.

Cotton machinery "of all kinds" were, at this period, manufactured at Philadelphia, Pa., Hartford, Conn., and Providence, R. I.

At Philadelphia, Pa., "John Butler, Cotton Machine Maker and Plane Maker," carried on business at No. 111 North Third Street, and a Mr. Felix Crawford made flying shuttles at No. 364 South Second Street.

"A Society for the Establishment of Useful Manufactures," was, under the patronage of the Secretary of the Treasury, chartered by the Legislature of New Jersey, in November of this year, with a large capital in shares of $400 each and extensive privileges, to carry on all kinds of manufactures at the falls of the Passaic. Although not immediately successful, the enterprise was the foundation of the city of Paterson, N. J., now the seat of numerous cotton manufactories.

The Legislature of New Jersey, November 22d, this year, incorporated "The Society for the Establishment of Useful Manufactures," and the company was organized at New Brunswick during the same month, with a capital of about $200,000, and obtained extensive rights in the Great Falls of the Passaic at Paterson, N. J. (*See* years 1792, 1793 and 1794.)

Samuel Slater (*see* year 1799) began spinning in the machines he had erected for Almy & Brown at Pawtucket, R. I. The machine for preparing the cotton for spinning was a very imperfect affair. "The cotton was laid on by hand, taking up a handful and pulling it apart with both hands and shifting it all into the right hand, to get the staple of the cotton straight, and fix the handful so as to hold it firm, and then applying it to the surface of the breaker, moving the hand horizontally across the card, to and fro, until the cotton was fully prepared." This description will serve to show the rude state of the Arkwright machinery, as introduced by Slater, at this period.

On the 19th of April, this year, Moses Brown, of Almy & Brown, Providence, R. I., addressed a letter to Moses Brown, of Beverly, Mass.—"To be Communicated to the Proprietors of the Beverly Manufactory"—in which he said: "I have for some time thought of addressing the Beverly manufacturers on the subject of an application to Congress for some encouragement to the cotton manufacture, by an additional duty on the cotton goods imported, and the applying such duty as a bounty, partly for *raising and saving cotton in the Southern States, of a quality and clearness suitable to be wrought by machines*,"—(when Slater first began to spin he used Cayenne and Surinam cotton)—"and partly as a bounty on cotton goods of the kind manufactured in the United States."

1792.

Yarns were spun with the "mule," in Manchester, of the fineness of 278 hanks to the pound, of 840 yards each.

Sir Richard Arkwright (*see* year 1732), the noted inventor (*see* years 1761, 1764, 1767, 1769 and 1785), died at his house at Cremford, England, on the 3d of August, this year, aged sixty.

Mr. Kelly, formerly of Lanark Mills, England, made a self-acting mule (*see* year 1790), and took out a patent for it in the summer of this year. (*See* years 1825 and 1830.)

In Glasgow, Scotland, the first steam engine for cotton spinning was set up for the Messrs. Scott & Stevenson, during this year. (*See* years 1783, 1789 and 1790.)

Cotton exported from the United States, 138,328 lbs.

The Frog Lane Factory of Boston, Mass. (*see* year 1789), produced about 2,000 yards of duck weekly, and employed four hundred hands. Its annual production, for a number of years after, was between two and three thousand bolts, of forty yards each, worth $13 per bolt.

Barrow stated, at the time of his visit to China, in this year, that the manufacture of cotton fabrics was stationary, owing to the want of proper encouragement on the part of the government, and to the rigid adherence of the people to ancient usages.

Mr. Jonathan Pollard, of Manchester, England, succeeded in spinning yarn, upon the mule (*see* year 1775), of the fineness of 278 hanks to the pound, from cotton wool grown by Mr. Robley in the island of Tobago. The yarn was sold at twenty guineas per pound to the muslin manufacturers of Glasgow.

Cotton exported from the United States was only 138,328 lbs.

In May, of this year, "The Society for the Establishment of Useful Manufactures" (*see* years 1791, 1793 and 1794) selected the site for their operations at Paterson, N. J., and on the "Fourth of July" made appropriations for building factories, machine shops, and print-works, and for the extensive use of water power from the Passaic Falls.

1793.

This year was made memorable forever in the cotton trade by Eli Whitney's invention of the saw-gin. Whitney was a native of Massachusetts who moved to Georgia, where he was employed as a teacher in the family of General Green, of Revolutionary memory. He had constructed several ingenious articles for Mrs. Green, and during a conversation about the great value of cotton to Georgia, if anything could be invented to separate the lint from the seed rapidly and cheaply, Mrs. Green remarked that if anybody could invent such a machine, Mr. Whitney could. Thereupon she urged the young man to make the attempt. He was entirely without mechanical assistance, or even the ordinary aids which almost any village in Georgia, or well-supplied plantation, would now furnish in abundance. He had only the rudest tools, and had to make his own wire with no better aid than the country blacksmith. As usual, genius overcame all difficulties, and the cotton gin is the result. (*See* year 1825.)

Messrs. Almy, Brown & Slater, of Providence, R. I. (*see* year 1790), built a spinning mill at Pawtucket, commencing with seventy-two spindles, afterwards enlarging it, and it is still known as "the old factory."

Mr. Kennedy, the author of the "Memoir of Crompton" (*see* year 1779), made a considerable improvement in the whole work of the "mule," this year, in England, which accelerated the movement of the machine.

Cotton exported from United States, 487,600 lbs.

In a "Report of the Select Committee of the Court of Directors of the East India Company upon the subject of Cotton Manufacture," in England, made this year, it is said: "Every shop offers British muslins (*see* year 1780) for sale, equal in appearance, and of more elegant patterns than those of India, for one-fourth, or perhaps more than one-third, less in price."

The factory of G. Richards, Amos Whittemore and Mark Richards (at Boston, Mass.), turned out 12,000 dozen cards annually.

The Byfield, Mass., cotton factory was established this year, but for several years manufactured woolens only.

"The Society for the Establishment of Useful Manufactures" (*see* years 1791, 1792 and 1794) confided the construction of their canals at Paterson, N. J., to Major L'Enfant, a French engineer (the same who was originally employed by General Washington to survey and lay out the city of Washington, but who had some difficulty with the commissioners before the business was completed), whose gigantic schemes were far beyond the pecuniary means of the company, and during this year the business was taken from the hands of L'Enfant and put under charge of Peter Colt, then comptroller of the State of Connecticut.

East India cotton imported into Great Britain, 729,643 lbs. (*See* years 1788 and 1800.)

1794.

A power-loom was invented by a Mr. Bell, of Glasgow, Scotland, which was abandoned. (*See* years 1785, 1787, 1796, 1801 and 1809.)

Cotton exported from the United states, 1,601,700 lbs.

Peter Colt, who had taken charge of the construction of the canals, etc., of "The Society for the Establishment of Useful Manufactures," at Paterson, N. J. (*see* years 1791, 1792 and 1793), completed the water-courses and built a factory in which they began spinning cotton yarn this year.

A cotton mill was erected in the west part of New Haven, Conn., this year, by John R. Livingston and David Dickson, of New York. These persons had, previously to this, a small mill not far from Hurlgate (or "Hellgate"), on the New York side, the machinery of which was moved to the New Haven mill. (*See* year 1807.)

There were but ten houses in Paterson, N. J., at this period. (*See* year 1827.)

Eli Whitney's cotton-gin patented March 14th.

1795.

Import of cotton into the United States, this year, was 4,107,000 lbs., and the export was 6,276,300 lbs., including a quantity of foreign cotton.

Georgia cotton of good quality was offered in New York at 1*s.* 6*d.* per lb.

A cotton mill—the second one in the United States—was erected in Rhode Island, this year. (*See* year 1791.)

1796.

On the 6th of June, of this year, a Mr. Robert Miller of Glasgow, Scotland, took out a patent for a power loom. (*See* years 1785, 1787, 1801 and 1809.)

During this year the quantity of British calicoes and muslins, which paid the print duty, was 28,621,797 yards. (*See* year 1829.)

The following table shows the gross produce of the excise duty in England and Scotland, on printed calicoes and muslins, during this year, minus the drawback:

IN ENGLAND.

Foreign calicoes and muslins,	rate 7d. ;	yards, 1,750,270 ;	Amt. £ 51,490
British " " "	" 3½d ;	" 24,363,240 ;	" 355,297

IN SCOTLAND.

Foreign calicoes and muslins,	rate 7d. ;	yards, 141,403 ;	Amt. £11,124
British " " "	" 3½d. ;	" 4,258,557 ;	" 62,103

Cotton exported from the United States, 6,106,729 lbs., including a quantity of foreign cotton.

Amos Whittemore, of Boston, Mass., took out a patent for an improved loom.

1797.

The Scutching machine was invented by a Mr. Snodgrass, of Glasgow, Scotland. (*See* year 1808.)

Cotton exported from United States, 3,788,429.

1798.

Several of the hands employed in the various mills at Pawtucket, R. I., which had sprung up after the founding of the "old factory" at that place by Messrs. Almy, Brown & Slater (*see* year 1793) started a factory at Cumberland, R. I.

Mr. Tennant, of Glasgow, Scotland, took out a patent for a bleacher, consisting of a saturated liquid of chloride of lime. (*See* years 1797, 1799 and 1802.)

Cotton exported from United States, 9,360,005 lbs.

First importation of cotton from the East Indies, 4,637 bales of about 350 lbs. into Great Britain.

Austin's power loom (*see* year 1789) was put in operation at Mr. Monteith's mill, near Glasgow, Scotland, but with what success does not appear.

Cotton machinery first introduced into Switzerland this year.

Col. Hugh Orr (*see* years 1715, 1740 and 1780) died at Bridgewater, Mass., December 6th, of this year, aged ninety-three years.

Cotton mill built by Samuel Slater (*see* years 1789, 1790. 1791 and 1805) and associates at Pawtucket, Mass.

First cotton mill and machinery in Switzerland built and operated.

No duty upon raw cotton imported into Great Britain up to this year. Tariff passed House of Parliament, this year, imposing the following duties:

On cotton imported by the East India Company,	£4 per cwt. *ad val.*	
" " " " " British colonies or plantations,	8s. 9d. per 100 lbs	
" " " " " Turkey and the United States,	6s. 6d.	"
" " " " " other parts,	12s. 6d.	"

This tariff lasted till, and inclusive of, the year 1800. (*See* year 1801.)

1799.

Mr. Tennant, of Glasgow, Scotland, took out a patent for impregnating slacked lime in a dry state with chloride of lime, to be used as a bleacher. (*See* years 1797, 1798 and 1802).

Cotton exported from United States, 9,532,263 lbs.

The second cotton mill built by Samuel Slater (*see* years 1798 and 1801), and the first one erected on the Arkwright system in Massachusetts (?) (*see* year 1787) was the "White Mill," begun this year, on the east side of the Pawtucket river, in what was then the town of Rehoboth, Slater entering into partnership with Oziel Wilkinson, Timothy Green and William Wilkinson.

Cotton machinery first introduced into Saxony this year.

1800.

Consumed in the United States, this year, 500 bales of 300 lbs. each (*See* years 1810 and 1815.)

The ravages of the cotton worm were first noticed. (*See* years 1804, 1825 and 1846.)

Official value British cottons exported, £5,406,501.

Cotton imported into England from all quarters, £56,010,730. (*See* year 1701, and note increase.)

The following table shows the gross produce of the excise duty in England and Scotland on printed calicoes and muslins during this year, minus the drawback:

IN ENGLAND.

Foreign calicoes and muslins,	rate 7d.;	yards, 1,577,536;	Amt. £ 46,011
British " " "	" 3½d.;	" 28,692,790;	" 418,436

IN SCOTLAND.

Foreign calicoes and muslins,	rate	7d.;	yards,	78,868;	Amt. £	2,300
British " " "	"	3½d.;	"	4,176,939;	"	60,913

Cotton exported into Great Britain from the United States, 17,789,803 lbs.

Upwards of 51,000,000 lbs. of cotton used in Great Britain. (*See* year 1700.)

East India cotton imported into Great Britain, 6,629,822 lbs. (*See* years 1788 and 1973.)

1801.

Cotton crop of the United States was 48,000,000 lbs., of which were contributed by South Carolina, 20,000,000 lbs., Georgia, 10,-000,000 lbs., Virginia, 5,000,000 lbs., North Carolina, 4,000,000 lbs., Tennessee, 1,000,000 lbs.; exports, 20,000,000 lbs. Up to this time, tables of exports of cotton at the Custom House did not distinguish home-grown from foreign cotton.

An Act was passed, this year, in the British Parliament, prolonging the loom patents of Rev. Dr. Cartwright. (*See* years 1785, 1787, 1809 and 1812.)

Mr. John Monteith, this year, adopted the loom patent of Robert Miller (*see* year 1796), and fitted up a mill at Pollackshaus, Glasgow, with two hundred looms, but it was several years before the business was made to pay.

Cotton exported from United States, 20,911,201 lbs.

Population of Lancashire, England, 762,565 (*see* years 1700, 1750 and 1831); of Wigan, 20,774. (*See* year 1831.)

During this year the cotton manufacturers of Ireland were protected by duties of sixty-eight per cent. *ad valorem* on gray and white cottons imported, and of forty-six per cent. on prints, and the quantity of raw cotton imported was only 1,575,789 lbs. (*See* years 1816, 1817 and 1825.)

Mr. Benjamin S. Wolcott, who was employed by Samuel Slater in building his first mill (*see* year 1791), having acquired considerable knowledge of the construction of machinery, united this year with Rufus and Elisha Waterman, and erected a factory at Cumberland, R. I. The machinery was afterward removed to Central Falls, a short distance from Pawtucket, and a new company was formed, with the addition of Mr. Stephen Jenks.

About 54,000,000 lbs. of cotton used in Great Britain.

Mr. Monteith, of Pollakshaus, Glasgow, fitted up 200 of Dr. Cartwright's power looms this year. (*See* year 1785.)

The first cotton mill in the vicinity of Boston, Mass., and the first in that State, after that built by Samuel Slater, at Rehoboth (*see* year 1799), was a small establishment on Bass River, in Beverly, was put in operation in the fall of this year, with six water-frames of seventy-two spindles each. The machinery was built at Paterson, N. J. (*see* years 1791, 1792, 1793 and 1794), by a man named Clark, who went to Beverly to put it into operation. It was unsuccessful on account of the insufficiency of water-power and other causes.

All duties upon raw cotton imported into Great Britain were taken off. (*See* year 1798.)

1802.

On the 2d of January, this year, William Radcliffe, of the Messrs. Radcliffe & Ress, cotton manufacturers, of Stockport, England, "shut himself up in his mill" for the purpose of trying experiments to overcome the hitherto great difficulty with the power-loom, viz.: "that it was necessary to stop the machine frequently in order to dress the warp as it unrolled from the jenny; which operation required a man to be employed for each loom, so that there was no saving of expense." Radcliffe had for one of his assistants an ingenious but dissipated young man, named Thomas Johnson, and his own perseverance and judgment, together with young Johnson's genius, at length produced the dressing machine (*see* years 1803 and 1804). When the patents were taken out, Johnson received by deed, the sum of £50 for his services, and was continued in their employment.

Tannant's first patent for a bleacher (*see* years 1797 and 1798) was set aside. But his second patent (*see* year 1799) was not contested.

About this year an important improvement was made in England, in the construction of blocks for printing calicoes, by sinking copper wire into the wood so as to stand out about an eighth of an inch and receive the coloring, afterward printing it upon the cloth. (*See* years 1758, 1805 and 18(8.)

Sir Robert Peel, of Bury, England, was the first, in this year, to print calicoes by the system known as *resist work*. (*See* years 1676, 1690, 1785, 1805 and 1808.) It consisted in printing various mordants on those parts of the cloth intended to be colored, and a paste or *resist* on such as were intended to remain white. It was discovered by a commercial traveler named Grouse, who sold the process for the sum of £5!

Cotton exported from United States, 27,501,075 lbs.

Bandana handkerchiefs and Bandana cloths, for garments, were first made in this year, at Glasgow, Scotland, by Mr. Henry Monteith.

At the instance of Sir Robert Peel, a law was passed in England, this year, prohibiting the employment of apprentices for more than twelve hours a day. (*See* years 1819, 1831, 1832 and 1833.)

Water-mill built at Beverly, Mass. (*see* year 1787), with Arkwright machinery.

Tariff passed British House of Parliament, this year, imposing the following duties:

On cotton imported by the East India Company,	£4 10s. 0d.	per cwt. *ad val.*
" " " " British colonies and plantations,	10s. 6d.	per 100 lbs.
" " " " Turkey and United States,	7s. 10d.	"
" " " " other parts,	15s. 0d.	"

1803.

Thomas Highs, the inventor (*see* years 1776 and 1790) died December 13th, this year, aged eighty-four years.

Radcliffe & Ress, of Stockport, England, took out patents in the name of Thomas Johnson, for improvements in looms, and for the new mode of warping and dressing. (*See* years 1802 and 1804.)

A patent for another power loom was taken out, this year, by Mr. H. Horrocks, cotton manufacturer, of Stockport, England, which he further improved and took out subsequent patents. (*See* years 1805 and 1813.)

Cotton exported from United States, 41,105,623 lbs.

A Mr. Toad, of Bolton, England, took out a patent for a power loom, this year.

First cotton mill erected in New Hampshire, at New Ipswich. (*See* years 1804 ad 1808.)

Duty in Great Britain, on cotton imported by the East India Company, from Turkey, the United States, or any British colony or plantation, 10s. 6d. per 100 lbs. From other parts, 25s. 0d. per 100 lbs.

1804.

Remarkable ravages by the cotton worm occurred. (*See* years 1800, 1825 and 1846).

Radcliffe and Ress of Stockport, England, took out patents in the name of Thomas Johnson, for improvements in looms, and for the new mode of warping and dressing. (*See* years 1802 and 1803.)

Cotton exported from United States, 38,118,041 lbs.

The first cotton mill in New Hampshire, (*see* year 1803), was put in operation this year at New Ipswich, by a man named Robbins, a former workman of Samuel Slater's. The original proprietors were Ephraim Hartwell, Charles Barrett and Benjamin Champney. (*See* years 1805 and 1807). It ran 500 spindles.

1805.

The cotton raised at Hilton Head, South Carolina, (*see* year 1790), by William Elliott brought higher prices than any other kinds, (*see* years 1806 and 1816), with one exception. Kinsey Burden, of St. John's, Colleton District, South Carolina, made a most careful selection of seeds, and with rigid care in cultivation produced cotton worth 25 cents per lb. more than that of any of his competitors. (*See* years 1826, 1827 and 1828.)

Cotton export of the United States for this year was 38.400,000 lbs., of which 8,787,659 was Sea Island.

Mr. H. Horrocks of Stockport, England (*see* years 1803 and 1813), improved his former loom patent and took out another.

The union of the two systems of calico printing—cylindrical and surface—(*see* years 1785, 1802 and 1808), were this year combined in one machine, the invention of Mr. James Burton, an engineer in the factory of Robert Peel & Co., of Church, England.

A large weaving factory was erected at Catrine in Ayrshire, by Messrs. James Finlay & Co., to be carried on in conjunction with their extensive spinning work in that place.

Thomas Johnson and James Kay, received a patent in England for improvements in looms, among them a "revolving *temple.*" (*See* years 1816, 1825, 1850 and 1855).

A man named Johnson, of Preston, England, took out a patent for a power loom this year, in which the warp, instead of being in a horizontal was in a perpendicular position. An attempt of the same kind was also made this year at Dorchester, England.

The "Yellow Mill" (*see* year 1799), the second cotton factory erected on the east side of the river and in the village of Pawtucket, R. I. (town of Rehoboth), was built this year by "Eliphalet Slack, Oliver Starkweather, Eleazer Tyler, 2d; Elijah Ingraham, and others."

The New Hampshire Legislature granted the proprietors of the first cotton mill in the State (*see* year 1804) an exemption from taxes for five years.

Duty in Great Britain on Cotton imported by the East India Company; from Turkey, the United States, or any British Colony or Plantation, 16s. 10½d. per 100 lbs. From other parts, 25s. 3¾d. per 100 lbs.

1806.

The South Carolina cotton of Elliott (*see* years 1790, 1805 and 1816) realized 30 cents per lb., while other qualities brought but 22 cents.

Samuel Slater, the first man to erect the Arkwright machinery (*see* year 1764) in America (*see* year 1789), was joined by his brother John Slater, from England, and the prosperous village of Slatersville, R. I., was quickly founded, and still continues to prosper.

Mexican cotton seeds introduced into Mississippi by Walter Burling, of Natchez, and supposed to have improved the character of cotton then grown.

Mr. Peter Marsland of Stockport, England, a spinner, took out a patent for a power loom with a double crank, but from its complexity it was not adopted by anybody but himself. Superior cloth, however, was made by it.

Cotton exported from United States, 37,491,282 lbs.

General Humphrey built his mill at Derby, Conn. (afterwards Humphreysville), for both cotton and woolen manufacture.

About 58,000,000 lbs. of cotton used in Great Britain.

A power-loom was built this year at Exeter, N. H., by T. M. Mussey, which, as an experiment, would perform all the operations of weaving, but was not a success as a labor saving machine.

Richard Guest, the historian, says a factory for steam-looms was built at Manchester, England, this year. (*See* year 1814).

Cotton mill built at Pomfret, Conn.

A mill for weaving with power-looms was built at Manchester, England, this year, which may be considered the date of the *successful* commencement of power-loom weaving.

American cotton in Liverpool worth from 15d. to 24d. per lb. (*See* years 1820 and 1830.

1807.

Eli Whitney, the inventor of the cotton-gin, brought suit in Savannah, Ga., to sustain the validity of his patent.

Cotton exported from the United States, 66,212,737 lbs.

There were but 15 cotton mills in operation in America at this time, producing about 300,000 lbs. of yarn per year, (*see* year 1791).

Upland cotton exported from the United States, 55,018,448 lbs.

Mr. Peter Marsland, of Stockport, England, this year obtained a patent for an improvement upon the power loom (*see* year 1787), by means of a double crank which gave the lathe a quick blow to the cloth on coming in contact with it, and by that means rendered it more stout and even.

A machine of most ingenious contrivance for performing the operation of tambouring (for tamboured muslins) was invented this year by Mr. John Duncan of Glasgow, Scotland, and a patent taken out. Each machine contained about forty tambouring needles, and was superintended by one person, who pieced the thread when it broke. The machine now came into general use.

The cotton mill of Messrs. Livingston and Dickson (*see* year 1794), at New Haven, Conn., was converted into a woolen mill, and afterward into a paper mill.

The second cotton mill in New Hampshire was built on the Souhegan, by Seth Nason, Jesse Holton and Samuel Batchelder. It ran 500 spindles. (*See* years 1804 and 1808.)

February 27th, this year, an exemption from taxes for five years was granted by the Massachusetts Legislature for a cotton mill, erected at Watertown by Seth Bemis and Jeduthan Fuller.

June 20th, a cotton factory was incorporated at Fitchburg, Mass.

Mr. Zachariah Allen estimated the number of spindles in operation in the United States, at about 4,000.

Cotton mill built at Smithfield, R. I., by John Slater, a brother of Samuel Slater. (*See* years 1789, 1790, 1791 and 1805).

Cotton mill built at Watertown, Massachusetts.

1808.

Snodgrass's scutching machine (*see* year 1797), was introduced into England by Mr. James Kennedy.

Mr. Joseph Lockett, engraver for calico printers in Manchester England, introduced the system of transferring the engravings for printing from a small steel cylinder to the larger copper ones which came in direct contact with the goods, thus saving immense labor and much time. (*See* years 1676 and 1690).

Cotton exported from United States, 12,064,366 lbs.—the year of American embargo on Foreign trade.

Messrs. James Finlay & Co., (of Catrine in Ayrshire, *see* year 1805), erected a weaving factory at Doune, connected with the spinning works there, the two containing 462 looms.

The brick cotton factory at Beverly, Mass., was burned. (*See* year 1841).

Second mill built at New Ipswich, New Hampshire. (*See* years 1803 and 1804).

The Globe factory, with a capital of $80,000, was established under Dr. Redman Coxe, of Philadelphia, Pa.

The first cotton mill in Oneida County, New York, four miles west of Utica, was erected this year by Benjamin S. Wolcott, Jr. (*see* year 1801), assisted by his father.

Cotton Manufacture established in Boston, Mass.

March 12th, the "Norfolk Cotton Manufactory" at Dedham, Mass., was incorporated.

The New Hampshire Legislature granted the proprietors of the second cotton mill in the State (*see* year 1807) an exemption from taxes for five years.

In December, this year, the New Hampshire Legislature, by a general law, granted exemption from taxes for five years to all who should erect works for the manufacture of cotton.

The first Peterborough and Exeter, N. H., cotton manufactories were incorporated this year.

1809.

The inventor of the power-loom, Rev. Dr. Samuel Cartwright, of England, obtained from Parliament this year a grant of £10,000 as a reward for his ingenuity. (*See* years 1785, 1787, 1801 and 1812).

Mr. H. Horrocks of Stockport, England, made still further improvements in his former loom patents (*see* years 1803 and 1805), and took out another.

Cotton exported from United States, 53,210,225 lbs.

The application of the stocking frame to the making of lace (*see* year 1768), was perfected and patented this year by Mr. John Heathcoat, M. P. for Tiverton, England. (*See* years 1787, 1823 and 1835.)

There were seventeen cotton mills in operation within the town of Providence, R. I., and its vicinity, working 14,296 spindles. (*See* years 1791, 1810 and 1812).

A second cotton manufactory was incorporated at Peterborough, N. H. (*See* year 1808).

The first cotton mill in the State of Maine, at this time comprised in Massachusetts, was erected this year, at Brunswick, and soon after another at Gardiner.

Richard Guest, the historian, says a factory for steam-looms was built at West Houghton, England, this year. (*See* year 1814).

A power-loom, capable of weaving about twenty yards of coarse cloth per day, was in operation this year at Dedham, Mass.

The "old Loughborough" lace machine of Heathcoat and Lacy was patented in England this year, and formed a new epoch in the lace trade. The machine was a complicated one, requiring sixty motions for the formation of a mesh.

Duty in Great Britain on cotton imported, from whatever country, 16s. 11d. per 100 lbs.

1810.

Consumed in home manufactures in the United States this year, 10,000 bales of 300 lbs. each (*see* years 1800 and 1815).

The art of giving the fine red color, known as Turkey or Adrianople red, to cloth was unknown till this year, when it was first practiced by M. Daniel Koechlin. of Mulhausen, in Alsace. (*See* year 1811).

Cotton exported from United States, 93,874,201 lbs.

French cotton manufactures consumed 25,000,000 lbs. of cotton.

Number of cotton mills in America, 102. (*See* years 1791, 1807 and 1831).

The "throstle frame" was introduced about this year.

Cotton manufactories were incorporated at Milford. Swanzey, Cornish, Pembroke and Amoskeag Falls, N. H.

The number of cotton mills erected in the United States at this time was eighty-seven, sixty-two of which were in operation, working 31,000 spindles. (*See* years 1787 and 1791.)

The statistics of Tench Coxe, from the census of this year, give for the State of Rhode Island, cotton factories twenty-eight, and spindles 21,178. (*See* years 1791, 1809 and 1812.)

Number of cotton factories in New Hampshire, twelve, of which eight were in the county of Hillsborough. In other States, as follows:

Massachusetts, 54; Vermont, 1; Rhode Island, 28; Connecticut, 14; New York, 26; New Jersey, 4; Pennsylvania, 64; Delaware, 3; Maryland, 11; Ohio, 2; Kentucky, 15; Tennessee, 4.

There were none in any other State.

Cotton exported from the United States into Great Britain, 93,900,000 lbs. (*See* years 1790 and 1791.)

1811

January 1st, an association was formed for building a cotton mill at Dorchester, Mass., with two thousand spindles and incorporated June 13th with a capital of $60,000.

Spindles in operation in the United States, estimated at 80,000. (*See* year 1791.)

About 90,000,000 lbs. of cotton used in Great Britain.

Cotton manufactories were incorporated at Walpole, Hillsborough, Meredith and (the third—*see* years 1808 and 1809) at Peterborough, N. H.

M. Daniel Koechlin, of Mulhausen, in Alsace (*see* year 1810), added an improvement to his discovery of the art of dyeing cloth Turkey red. It consisted in printing upon Turkey red, or any dyed color, some powerful acid, and then immersing the cloth in a solution of chloride of lime. Neither of these agents singly or alone affects the color, but those parts which have received the acid, on being plunged into chloride of lime, are speedily deprived of their dye and made white by the acid of the liberated chlorine. (*See* years 1813 and 1816.)

Cotton export 62,186,081 lbs.

Cotton mill incorporated at Dorchester, Mass.

1812.

Up to this year all the mills put in operation in the United States were designed simply for spinning, carding and roving. England had a power-loom (*see* years 1785 and 1787), the invention, of a clergyman—Rev. Dr. Cartwright, but the method of its construction was unknown in this country. Mr. Francis C. Lowell, of Boston, Mass., who had made a voyage to England, and visited Scotland, returned from abroad this year, and he, in conjunction with Mr. Patrick T. Jackson, a relative, undertook the invention of a loom, and in the autumn had completed a satisfactory model and decided upon erecting a suitable mill to operate it. (*See* year 1813.)

Samuel Crompton of England, the inventor of the "mule jenny," (*see* years 1779 and 1793), who had never had his invention patented was granted £5,000 by Parliament. (*See* year 1827.)

Cotton exported, 28,952,544 lbs.—War with England.

Number of mule spindles in Great Britain estimated at between 4,000,000 and 5,000,000.

The average product of the spinner of yarn No. 40, was two hanks per spindle per day. (*See* years 1790 and 1830.)

At this time two thirds of the steam engines in Great Britain were employed in the cotton manufacture.

A survey of all the manufacturing districts of Great Britain this year showed the number of spindles at work to be between 4,000,000 and 5,000,000, on the principle of Crompton's mule jenny.

There were said to be within thirty miles of Providence, R. I., at this period, thirty-three cotton factories running 30,660 spindles, (*see* years 1791, 1809 and 1810), and in Massachusetts 20 factories, running 17,370 spindles.

All cotton mills built in the United States, previous to this year, were modeled after the plan first introduced by Samuel Slater (*see* year 1791), with but very little modification. There were said to have been nearly forty mills in Rhode Island at this time, with about 30,000 spindles, and thirty in Massachusetts, within thirty miles of Providence, R. I., with about 18,000 spindles.

War with England. This raised the price of goods to such extravagant rates that cotton goods, such as had been previously imported from England at from seventeen to twenty cents per yard, were sold by the package at seventy-five cents. This stimulated the erection of cotton factories in this country to a wonderful degree. (*See* year 1815.)

Fifteen cotton mills in operation in New Hampshire, averaging not more than 500 spindles each.

1813.

The first mill in America (if not in the world) where a power-loom was used, and spinning and weaving were carried on under the same roof, was erected at Waltham, Mass., by Mr. Francis C. Lowell, Mr. Patrick T. Jackson and Mr. Paul Moody (*see* year 1812). It ran about 1,700 spindles. This factory is still in operation.

At this period there were in operation, so far as known, in all quarters of England and Scotland but one hundred dressing machines (*see* years 1802, 1803 and 1804), and 2,400 power-looms. (*See* years 1785, 1820 and 1829.)

Mr. James Thomson, of Primrose, near Clitheroe, England, took out a patent for the process of dyeing, discovered by M. Daniel Koechlin, of Mulhausen, in Alsace. (*See* years 1810, 1811 and 1816.)

Cotton exported from United States, 19,399,911 lbs.—War with England.

February 23d. an Act was passed in the Massachusetts Legislature to corporate "The Boston Manufacturing Company," better known as the "Waltham Company," for the purpose of manufacturing cotton, woolen or linen goods. The Act authorized them to conduct their business at "Boston in the County of Suffolk," or within fifteen miles thereof, or at any other place or places, not exceeding four. The Company had a capital of $400,000.

Mr. Metcalf, an American, was sent from England to the East Indies with improved cotton machinery.

1814.

British Calicoes, etc., printed at an average duty of 5s. per piece..5,192,228 pieces.
Duty on same received by Government......................£1,228,057
Calicoes, etc., exported—average drawback of 5s. per piece......3,324,160 pieces.
Drawback paid by Government on same.....................£831,040
Calicoes, etc., taken for home consumption at average duty of 5s. per piece......................................1,868,068 pieces.
Net amount duty received by Government for same.............£467,017

Cotton exported from United States, 17,806,479 lbs.—War with England.

Attempts were begun this year to introduce the Bourbon cotton seed into Guzerat.

Good authority (*The London Quarterly Review—see* year 1825) avers that at this time not a power-loom was in use in Manchester, England.

Power-looms in operation at Waltham, Mass., the first in the United States.

1815.

British calicoes, etc., printed at an average duty of 5s. per piece..5,326,656 pieces.
Duty on same received by Government.....................£1,331,664
Calicoes etc., exported—average drawback of 5s. per piece......3,813,000 pieces.
Drawback paid by Government on same....................£953,250
Calicoes, etc., taken for home consumption at average duty of 5s. per piece......................................1,513,652 pieces.
Net amount of duty received by Government for same..........£378,413

Cotton exported from United States, 82,998,747 lbs.

Up to this year it had been thought that the cotton wool of India, from the shortness of its staple, could not be spun with advantage upon the machinery then in use, but it was about this time discovered that by mixing it with the longer stapled cottons of other countries it could be brought into a state fit for the mule and spinning frames.

William Gilmon, patentee of the crank-loom (*see* years 1816 and 1817) came to Boston, Mass., from the British Provinces in September of this year. He located at Smithfield, Mass., as a machinist, where he began paying rent October 21st.

A list of cotton factories in New England at this time estimates:

	Mills	Spindles
Rhode Island..............	99 mills —	68,142 spindles.
Massachusetts..............	52 "	39,46 "
Connecticut................	13 "	11,700 "
Total..............	165 mills—	119,310 spindles.

A Report of the Committee on Manufactures, to Congress, in this year, gives the following particulars of the cotton manufacture in the United States.

Capital employed	$40,000,000
Males employed over seventeen years old	10,000
" " under " " "	24,000
Women and girls employed	66,000
Wages of 100,000 averaging $1.50 per week	$15,000,000
Cotton manufactured, pounds	27,000,000
" " yards	81,000,000
Cost, averaging 30 cents per yard	$20,300,000

The recommencement of the importation of goods at the close of the war with England, in this year, and the sudden reduction of prices consequent thereon (*see* year 1812) was very destructive to manufacturing operations; business was prostrated and many establishments that had been built at extravagant rates became almost if not quite worthless. For the purpose of protecting this interest, which was supposed to have some claim upon the country on account of the aid afforded during the war, a tariff was passed. (*See* year 1816.)

From the commencement of the French Revolution to the restoration of peace after the battle of Waterloo, I find few records and no reliable statistics of the cotton trade. During the first fifteen years of this century, the number of patents issued in England for improvements in machinery used in manufacturing textile fabrics averaged only two per annum; during the next fifteen years they averaged over four per annum; the next fifteen over twelve per annum; and for the sixty years ending with 1860, about twenty-four per annum.

It is possible that the European wars may have retarded the progress of peaceful industry, and turned inventive genius in other directions, but, on the whole, I suppose the progress was all that could reasonably be expected.

Until 1816 we are almost wholly dependent upon government records for commercial statistics. It was not until 1825 that the movements of the cotton trade were regularly recorded and preserved.

CHAPTER IV.

1816.

This year South Carolina cotton (Sea Island) brought 47 cents when other kinds brought 27. The gradual appreciation of this cotton is shown by the fact, that in 1806 it brought 30 cents to 22 cents for other kinds. The business of home manufacture had increased since 1800 to 90,000 bales of 300 lbs. each (*see* year 1810); 81,000,000 yards cotton cloth were manufactured, costing $24,000,000; 100,000 operatives were employed and $40,000,000 capital was invested. (*See* year 1800.)

Mr. James Thomson, of Primrose, near Clitheroe, England, took out a patent for a useful improvement on his former patent for dyeing (*see* year 1813) which consisted in combining with the acid some mordant, or metallic oxide, capable, after the dyed color was removed, of having imparted to it some other color. (*See* years 1810 and 1811.) This laid the foundation of a series of processes in which the chromic acid and its combinations have since been employed with such great success.

British calicoes, etc., printed at an average duty of 5s. per piece..4,511,244 pieces.
Calicoes, etc., exported—average drawback of 5s. per piece.....2,878,704 pieces.
Calicoes, etc., taken for home consumption at an average duty of
5s. per piece..1,632,540 pieces.

Cotton exported from United States, 81,747,116 lbs.

Duties on gray and white cottons and prints imported into Ireland had been reduced to 10 per cent *ad valorem*. (*See* year 1801.)

A "protective" tariff was passed this year by Congress. It was supported by the South on the ground of encouraging the manufacture of our own cotton, but met with much and decided opposition at the North, where navigation and commerce were the favorite pursuits.

Until this time the operation of cotton factories had been confined to the production of yarn, which was woven upon the hand-loom.

William Gilmore, the patentee of the crank-looms, (*see* years 1815 and 1817) was employed in the early part of this year to build twelve power-looms, and also machinery for warping and dressing, from the plans he had brought with him from the British Provinces, which he accomplished. For the sum of "$10 lawful money" he allowed another firm than the one for whom he built the above looms, to use his patterns for the building of twelve others, which were built and put in operation quite as soon as his own.

"Self-acting temples" were invented and patented January 7th, this year, by Ira Draper, of Weston, Massachusetts. (*See* years 1805, 1825, 1850 and 1855.)

The first lace machine—a traverse warp—was taken to France this year, and set up at Douay by Corbit, Black & Cutts for M. Thornassin.

The price semi-weekly at New York and the course of Exchange on London, from January 1st, to October 1st, 1816.

1816.	Price of New Orleans	Price of Upland.	Exchange on London.	1816.	Price of New Orleans	Price of Upland.	Exchange on London.
Jan. 2 . . .	31@33	28@30		May 24 . . .	31@35	31@33	108½
" 5 . . .	31@33	28@30	107	" 28 . . .	31@35	31@33	
" 9 . . .	31@33	28@30		" 31 . . .	31@35	31@33	108½@¾
" 12 . . .	31@33	28@30	107	June 4 . . .	31@35	31@33	
" 16 . . .	30@32	26@29		" 7 . . .	31@35	31½@33	108½@¾
" 19 . . .	30@32	26@29	107	" 11 . . .	31@35	31½@33½	
" 23 . . .	30@32	26@29		" 14 . . .	31@35	31½@33½	108½
" 26 . . .	30@32	26@29	107	" 18 . . .	31@35	31½@33½	
" 30 . . .	30@32	26@29		" 21 . . .	30@33	28½@32	108
Feb. 2 . . .	30@32	26@29	107	" 25 . . .	30@32	27@31	
" 6 . . .	30@32	26@29		" 28 . . .	30@32	27@31	106½@107
" 9 . . .	30@32	26@29	107	July 2 . . .	30@32	27@31	
" 13 . . .	30@32	26@29		" 6 . . .	32@33	27@32	105½
" 16 . . .	30@32	26@29	107	" 9 . . .	28@33	27@32	
" 20 . . .	30@32	26@29		" 12 . . .	28@33	30@32	104
" 23 . . .	30@32	26@29	107	" 16 . . .	30@33	30@32	
" 27 . . .	30@32	26@29		" 19 . . .	30@33	30@32	104
Mch. 1 . . .	30@32	26@29	107	" 23 . . .	30@33	29@32	
" 5 . . .	30@32	26@29		" 26 . . .	30@32	26@31	104½
" 8 . . .	30@32	25@27	108	" 30 . . .	3 @32	27@32	
" 12 . . .	31@32	27@29		Aug. 2 . . .	30@32	26@31	104½@105
" 15 . .	31@32	27@30	108½@109	" 6 . . .	27@30	26@29	
" 19 . . .	31@32	27@30		" 9 . . .	27@30	26@29	104@104½
" 22 . . .	30@32	29@30	108½@109	" 13 . . .	26@29	24@28	
" 26 . . .	30@32	29@30		" 16 . . .	27@30	25@28	104@104½
" 29 . . .	29@31	28@29	109½	" 20 . . .	27@30	2 @28	
Apr. 2 . . .	29@31	28@29½		" 23 . . .	29@30	25@29	105
" 5 . . .	30@32	28@29½	109½	" 27 . . .	29@30	25@29	
" 9 . . .	30@32	28@29½		" 30 . . .	29@31	25@30	106
" 12 . . .	30@31	28@29	109½	Sept. 3 . . .	29@31	25@30	
" 16 . . .	30@31	28@29		" 6 . . .	30@32	25@30	106½
" 19 . . .	30@31	27½@30	109½	" 10 . . .	30@32	25@30	
" 23 . . .	30@33	28@31		" 13 . . .	30@32	25@30	106½
" 26 . . .	30@33	29@32	109½@110	" 17 . . .	30@32	25@30	
" 30 . .	30@33	29@32		" 20 . . .	30@32	25@30	106
May 3 . . .	33@34	31@33	109½	" 24 . . .	. . @ . .	25@30	
" 7 . . .	33@34	31@33		" 27 . . .	. . @ . .	25@28	106
" 10 . . .	33@35	31@33	108@108½	Oct. 1 . . .	. . @ . .	25@28	
" 14 . . .	33@35	32@33					
" 17 . . .	33@35	32@33	108@108½	Average for the 9 months	31 27	28.96	
" 21 . . .	33@35	32@33					

Weekly Quotations in the Charleston Market. Year 1816.

Week	Sea Island.	Maine and Santee.	Short Staple.	Week	Sea Island.	Maine and Santee.	Short Staple.	Week	Sea Island.	Maine and Santee.	Short Staple.	Week	Sea Island	Maine and Santee.	Short Staple.
1	50	..	26	14	43	..	27	27	55	..	32	40	43	40	27
2	45	..	25	15	44	..	27	28	53	..	31	41	43	40	27
3	45	..	26	16	45	..	28	29	53	..	32	42	43	40	26
4	45	.	27	17	45	..	28	30	53	..	32	43	43	40	26
5	45	..	28	18	45	..	28	31	53	..	32	44	43	40	26
6	43	..	27½	19	46	..	28	32	50	..	32	45	41	40	25
7	43	..	27½	20	50	..	30	33	50	..	31	46	40	38	23
8	43	..	28	21	50	..	30	34	48	45	30	47	38	35	23
9	43	..	28	22	50	..	31¼	35	48	45	30	48	38	35	24
10	43	..	28	23	55	..	32	36	45	42	25	49	37	35	25
11	40	..	26½	24	55	..	32	37	45	42	28	50	37	35	25½
12	41	..	26½	25	55	..	32	38	44	40	25	51	39	35	25½
13	43	..	27	26	55	..	31	39	43	40	27	52	39	35	25½

This was the first year after the close of the Napoleonic wars, and is as far back as I am able to find any record of the market; even this is meagre, but valuable as showing how slight the fluctuations were during the first year of peace.

During the previous twenty years it was the cotton trade, moved by the machinery of Watt, Arkwright, and others, that sustained the tremendous strain upon British finances caused by the policy which rendered it necessary for England to be constantly either subsidising half the nations on the continent, or fighting them all. That was but a foretaste of what cotton has since accomplished; and but as a drop in the bucket to what it is destined to accomplish.

1817.

The "fly-frame," for preparing rovings for the middle and coarser numbers of both warp and weft (*see* years 1764, 1782, 1789 and 1790) came into use.

British calicoes, etc., printed at an average of 5s. per piece......4,695,264 pieces.
Calicoes, etc., exported—average drawback of 5s. per piece. ..3,282,216 pieces.
Calicoes, etc., taken for home consumption at an average duty of
5s. per piece...1,413,048 pieces.

Cotton exported from United States, 85,649,328 lbs.

Import of raw cotton into Ireland had increased to 3,286,429 lbs. (*See* years 1801 and 1816.)

Value of cotton goods manufactured in France, from 200,000,000 to 300,000,000 francs.

There were in Glasgow, Scotland, or belonging to it, at this time, fifteen weaving factories containing 2,275 looms. (*See* year 1832.)

The number of power looms in Lancashire, England, this year was estimated at 2,000, one-half of which only were said to be in employment.

The manufacturers of cotton goods in New England, to manifest their gratitude for the services rendered them by Mr. William Gilmore (*see* years 1815 and 1816), raised a fund of $1,500 and presented the same to him during this year, when his crank-looms were first put in operation in this country.

Webster, Clark and Bonniton, of England, took a thirty-six inch (straight bolt) lace machine to Calais.

The price semi-weekly at New York and the course of Exchange on London, for the Crop Year ending October 1, 1817.

1816.	Price of New Orleans	Price of Upland.	Exchange on London.	1817.	Price of New Orleans	Price of Upland.	Exchange on London.
Oct. 1..	......	25@28	106	Apr. 4..	30@31	26½@28	
" 4..	......	25@28		" 8..	30@31	26½@28	102½
" 8..	......	25@28	106	" 11..	30@33	26½@30	
" 11..	......	25@28		" 15..	30@33	26½@28	102¼
" 15..	......	25@28	106	" 18..	30@33	26½@28	
" 18..	......	25@31		" 22..	30@33	26½@28	102¼
" 22..	......	25@31	106	" 25..	31@33	26½@28½	
" 25..	30@31	22@28		" 29..	31@34	27@28½	102½
" 29..	30@31	22@28	106@106½	May. 2..	31@32	27@28½	
Nov. 1..	30@31	22@28		" 6..	31@32	28@29	102½
" 5..	30@31	22@29	106½@107	" 9..	31@33	28@31	
" 8..	30@31	22@29		" 13..	31@33	28@31	101¼
" 12..	30@31	22@29	107	" 16..	31@34	28@32	
" 15..	30@31	22@28		" 20..	32@34	28@32	101¼
" 19..	30@31	22@28	106	" 23..	32@34	26@32	
" 22.	30@31	25@26		" 27..	32@34	26@32	101¼
" 26 .	30@31	25@26	101¾	" 30..	31@34	26@30	
" 29..	......	26@27		June 3..	31@34	26@30	101¼
Dec. 3..	......	26@27	103@103½	" 6..	31@34	26@32	
" 6..	......	26@26½		" 10..	31@34	26@32	101
" 10..	......	26@26½	103@103½	" 13..	32@34	26@32	
" 13..	..@29	25½@26½		" 17..	32@34	26@32	101
" 17..	.@29	25½@26½	103	" 20..	32@33½	26@31	
" 20..	28@29	25½@26½		" 24..	32@33½	26@31	101½
" 24..	28@29	25½@26½	103¼	" 27..	32@33	26@31	
" 27..	28@29	25½@26½		July 1..	32@33	26@31	101½
" 31..	28@29	25½@26½	par	" 4..	32@33	26@31	
1817.				" 8..	32@33	26@31	101½
Jan. 3..	28@29	25½@26½		" 11..	32@33½	26@32	
" 7..	28@29	25½@26½	par	" 15..	32@33½	26@32	101½
" 10..	28@29	25½@26½		" 18..	32@34	26@32	
" 14..	28@29	25½@26½	par	" 22..	32@34	26@32	101½
" 17..	28@29	25@26		" 25..	32@34	26@32	
" 21..	27@29	25@26	par	" 29..	32@34	26@32	101¾
" 24..	28@29	26@27		Aug. 1..	32@34	26@32	
" 28..	28@29	26@27	par	" 5..	31@33	26@31	100½
" 31..	28@29	25@26		" 8..	31@33	26@31	
Feb. 4..	28@29	25@26	par@1 pm	" 12..	28@33	25@30	par
" 7..	28@29	25@26		" 15..	28@33	26@30	
" 11..	28@29	25@26	par@1 pm	" 19..	28@33	26@30	par
" 14..	28@29	25@26		" 22..	28@33	26@30	
" 18..	28@29	25@26	101@101½	" 26..	28@33	26@30	101@101½
" 21..	28@29	25@26		" 29..	28@33	26@30	
" 25..	28@29	25@26	101@101½	Sept. 2..	28@33	26@30	101½@102
" 28 .	28@29	26@27		" 5..	28@33	26@30	
Mch. 4..	28@29	26@27	101½@102	" 9..	28@33	26@30	102
" 7..	28@30	26½@27		" 12..	28@33	26@30	
" 11..	28@30	26½@27	101½@102	" 16..	28@33	26@30	103
" 14..	28@30	26½@27		" 19..	28@35	26@32	
" 18..	28@30	26½@27	101½@102	" 23..	28@35	26@32	103
" 21..	28@30	26½@27		" 26..	28@35	26@32	
" 25..	28@30	26½@27	102	" 30..	30@35	26@32	102½
" 28..	30@31	26½@27					
Apr. 1..	30@31	26½@27	102½	Average for the year.	30.70	27.25	

Weekly Quotations in the Charleston Market. Year 1817.

Week	Sea Island.	Maine and Santee.	Short Staple.	Week	Sea Island.	Maine and Santee.	Short Staple.	Week	Sea Island.	Maine and Santee.	Short Staple.	Week	Sea Island.	Maine and Santee.	Short Staple.
1	39	35	26	14	42	38	29	27	45	43	30	40	45	40	30
2	39	35	25½	15	42	38	28½	28	45	43	30	41	45	42	30
3	39	35	25½	16	43	40	28	29	45	43	30	42	45	42	30
4	39	35	25½	17	43	40	28	30	45	43	29	43	45	43	31¼
5	40	35	26	18	45	40	28	31	45	43	29	44	45	43	31¼
6	39	35	26	19	45	40	29	32	45	43	29	45	45	45	31
7	39	35	26	20	45	43	30	33	45	43	30	46	47	45	32
8	39	35	26½	21	46	44	30	34	45	43	30	47	50	45	33½
9	40	36	26½	22	46	44	30	35	45	43	30	48	50	45	33½
10	39	35	26½	23	46	44	30	36	45	40	30	49	51	45	33½
11	39	35	26½	24	46	44	30	37	45	40	30	50	51	47	33½
12	40	37	27½	25	45	43	30	38	45	40	30	51	56	50	34
13	42	38	29	26	45	43	30	39	45	40	30	52	58	..	35

1818.

British calicoes, etc., printed at an average duty of 5s. per piece . . 6,282,544 pieces.
Calicoes, etc., exported—average drawback of 5s. per piece . . . . . 4,317,508 pieces.
Calicoes, etc., taken for home consumption at an average duty of 5s. per piece . . . . 1,965,036 pieces.

Cotton exported from United States, 92,471,178 lbs.

Only 2,000 power-looms in operation in Manchester, England, at this period.

Richard Guest, the historian, says there were in Manchester, England, and vicinity, this year, fourteen cotton factories containing about 2,000 power looms. (*See* year 1814.)

The price semi-weekly at New York and the course of Exchange on London, for the Crop Year ending October 1, 1818.

1817.	Price of New Orleans	Price of Upland.	Exchange on London.	1818.	Price of New Orleans	Price of Upland.	Exchange on London.
Oct. 3..	30@35	26@32		April 7..	31@33	29@32	101¼
" 7..	30@35	26@32	103	" 10..	31@34½	29@33	
" 10..	30@34	26@32		" 14..	31@34½	29@33	101¼
" 14..	30@34	26@32	102	" 17..	31@34	29@33	
" 17..	31@35	28@33		" 21..	31@34	29@33	99@100
" 21..	31@35	28@33	102	" 24..	31@34½	31@33	
" 24..	34@35	28@33		" 28..	31@34½	31@33	99@100
" 28..	34@35	28@33	101½	May 1..	32½@34½	32@34	
" 31..	34@35	28@34		" 5..	32½@34½	32@34	99@100
Nov 4..	34@35	28@34	101½	" 8..	32½@34½	32@34½	
" 7..	32@35	28@35		" 12..	32½@34½	32@34½	par.
" 11..	32@35	28@35	101½	" 15..	32½@34½	32@34½	
" 15..	32@36	29@35		" 19..	32½@34½	32@34½	par.
" 18..	32@36	29@35	101½	" 22..	32½@35	32@34½	
" 21..	32@35	29@35		" 26..	32½@35	32@34½	99½@100
" 25..	32@35	29@35	101½	" 29..	33½@34½	32@34	
" 28..	32@36	29@35		June 2..	33½@34½	32@34	par.
Dec. 2..	32@36	29@35	102	" 5..	33@35	32@33½	
" 5..	32@37½	29@35½		" 9..	33@35	32@33½	par.
" 9..	32@37½	29@35½	102¼@102½	" 12..	33@34½	32@33½	
" 12..	32@36½	29@35½		" 16..	33@34	32@33	par.
" 16..	32@36½	29@35½	102@102¼	" 19..	33@34	32@33½	
" 19..	32@36½	29@35½		" 23..	33@34	32@33½	par.
" 23..	32@36½	29@35½	102@102¼	" 26..	33@34	32@33½	
" 27..	32@36	29@35		" 30..	33@34	32@33½	par.
" 30..	32@36	29@35	101¾@102	July 3..	32@33	30@33	
1818.				" 7..	32@33	30@33	100@100½
Jan. 3..	32@35	29@34		" 10..	32@34	30@33½	
" 6..	32@35	29@34	101¾@102	" 14..	32@34	30@33½	par.
" 9..	32@35	29@34		" 17..	32@34	30@33½	
" 13..	32@35	29@34	101¾@102	" 21..	32@34	30@33½	par.
" 16..	32@35	29@34		" 24..	31@34	30@33½	
" 20..	32@35	29@34	101¾@102	" 28..	31@34	30@33½	par.
" 23..	32@35	29@34		" 31..	31@34	30@33½	
" 27..	32@35	29@34	102	Aug. 4..	31@34	30@33½	par.
" 30..	32@35	29@34		" 7..	32@33	30@33½	
Feb. 3..	32@35	29@34	102@102¼	" 11..	32@33	30@33½	par.
" 6..	32@35	29@34		" 14..	33@34½	31@34	
" 10..	32@35	29@34	102	" 18..	33@34½	31@34	par.
" 13..	33@35	29@33½		" 21..	33@34½	31@34	
" 17..	33@35	29@33½	102½@102¾	" 25..	33@34½	31@34	par.
" 20..	33@35	29@33½		" 28..	33@34½	30@33½	
" 24..	33@35	29@33½	101@101½	Sept. 1..	33@34½	30@33½	par.
" 27..	32@34	29@32		" 4..	33@35	31@34½	
Mch. 3..	32@34	29@32	par ½ prem	" 8..	33@35	31@34½	par.
" 6..	31@33	29@32		" 11..	33@35	31@34½	
" 10..	31@33	29@32	par.	" 15..	33@35	31@34½	par.
" 13..	31@33	29@32		" 18..	33@35	31@34½	
" 17..	31@33	29@32	par.	" 22..	33@35	31@34½	par.
" 20..	31@33	29@32		" 25..	33@34	30@34	
" 24..	31@33	29@32	par@101	" 29..	33@34	30@34	100¾
" 27..	31@33	29@32					
" 31..	31@33	29@32	par@101	Average for the year.	33.33	31.70	
April 3..	31@33	20@32					

Weekly Quotations in the Charleston Market. Year 1818.

Week	Sea Island.	Maine and Santee.	Short Staple.	Week	Sea Island.	Maine and Santee.	Short Staple.	Week	Sea Island.	Maine and Santee.	Short Staple.	Week	Sea Island.	Maine and Santee.	Short Staple.
1	58	....	35	14	.60	52	31	27	75	60	33½	40	65	....	32
2	58	....	35	15	65	55	32½	28	75	60	33	41	60	....	32
3	58	....	34½	16	70	55	28	29	75	....	33	42	60	....	31½
4	56	....	34	17	70	60	33	30	75	....	33	43	60	....	32
5	56	....	34	18	72	60	34	31	73	....	33	44	60	....	32
6	56	....	33	19	72	60	33	32	73	....	33	45	60	....	31
7	56	....	32½	20	72	60	33	33	73	....	33	46	60	....	30
8	56	....	33	21	72	60	32	34	70	....	33	47	60	....	30
9	56	50	33	22	75	58	33	35	70	....	33	48	55	....	28
10	56	50	33	23	75	58	33	36	65	....	33	49	55	50	27
11	58	50	32	24	75	60	33	37	65	....	32	50	54	50	25
12	58	50	31	25	75	60	34	38	65	....	32	51	53	48	26
13	60	50	31	26	75	60	33½	39	65	....	32	52	55	50	27

1819.

British calicoes, etc., printed at an average duty of 5s. per piece..5,938,572 pieces.
Calicoes, etc., exported—average drawback of 5s. per piece.....3,519,868 pieces.
Calicoes, etc., taken for home consumption at an average duty of
5s. per piece.................................2,418,704 pieces.

Cotton exported from United States, 87,997,045 lbs.

United States cotton crop 303,589 bales of 300 lbs.

Sir Robert Peel this year obtained the passage of an Act in England, extending the prohibition of his Act of (*see* year) 1802 to the labor of all children under sixteen years of age, and making it illegal to employ any children under nine years of age in cotton factories. (*See* years 1831, 1832 and 1833.)

During this year, Kirk Boot, Esq., a wealthy Boston merchant, explored the wilds as a hunter, where the City of Lowell—"The Manchester of America"—now stands, and, discovering its resources, in company with others purchased the land and water privileges, under the name of "*The Proprietors of the Locks and Canals on Merrimac River.*"

Duty in Great Britain on cotton imported from any British Colony or Plantation in America, and imported directly from thence, 6s. 3d. per 100 lbs. From other parts, 8s. 7d. per 100 lbs.

The price semi-weekly at New York and the course of Exchange on London, for the Crop Year ending October 1, 1819.

1818.	Price of New Orleans	Price of Upland.	Exchange on London.	1819.	Price of New Orleans	Price of Upland.	Exchange on London.
Oct. 2..	33@34	30@34	100¾@101	Apr. 6..	20@21	20@20½	
" 6..	33@34	30@34		" 9..	20@21	19@20	99@100
" 9..	33@34	30@33½	100¾@101	" 13..	20@21	19@20	
" 13..	33@34	30@33½		" 16..	21@22½	20@22	100@101
" 16..	32@34	30@34	100¾@101	" 20..	21@22½	20@22	
" 23..	32@35	29@34½		" 23..	21@22½	20@22	100@101
" 27..	32@35	29@34½	100¾@101	" 27..	19@20	18@19	
" 30..	32@35	29@34½		" 30..	18@19	15½@17	101@102
Nov. 3..	32@35	29@34½	100	May 4..	17@18	15½@17	
" 7..	32@34	29@34		" 7..	16@18	13½@17	102½
" 10..	32@34	29@34	2½@3 disc.	" 11..	16@18	13½@17	
" 13..	32@34	29@33½		" 14..	17@19	13@17½	102¼@102½
" 17..	32@34	29@33½	1½@2 disc.	" 18..	17@18	13@17	
" 20..	30@33	29@33		" 21..	17@18	13@17	102¼@102½
" 24..	30@33	29@33	2@2½ disc.	" 25..	17@18	13@17½	
" 27..	30@33	29@33		" 28..	16@18	13@17	102
Dec. 1..	30@33	29@33		June 1..	16@18	13@17	
" 4..	31@32	25@31	2@2½ disc.	" 4..	16@18	13@16½	101½
" 8..	31@32	25@31		" 8..	16@18	13@16½	
" 11..	30@..	24@29	2@2½ disc.	" 11..	16@18	13@16½	101¼
" 15..	30@..	24@29		" 15..	16@18	15@17	
" 18..	29@..	24@27	2@2½ disc.	" 18..	16@18	15@17	100
" 22..	29@..	24@27		" 22..	16@18	15@17	
" 25..	29@..	24@27½	2@2½ disc.	" 25..	16@18	15@17	100
" 29..	29@..	24@27½		" 29..	16@18	15@17	
1819.				July 2..	16@18	15@17	100
Jan. 1..	29@..	24@27½	2@2½ disc.	" 7..	16@18	15@17	
" 5..	29@..	24@27½		" 9..	16@18	15@17	100
" 8..	26@27	24@26	2@2½ disc.	" 13..	16@18	15@17	
" 12..	25½@26½	24@25		" 16..	16@18	15@17	100
" 15..	25½@27	24@26	2@2½ disc.	" 20..	16@18	15@17	
" 19..	25½@27	24@26		" 23..	16@18	14@17	100
" 22..	27@28	26½@27¼	2@2½ disc.	" 27..	16@18	14@17	
" 26..	25@26½	24@26		" 30..	16@18	14@17	100
" 29..	26½@27½	25½@26½	1½@2 disc.	Aug. 3..	16@18	14@17	
Feb. 2..	26½@27½	25½@26½		" 6..	16@18	14@17	100
" 5..	26½@27	25½@26	1@1½ disc.	" 10..	16@18	14@17	
" 9..	26½@27	25½@26		" 13..	16@18	14@17	100
" 12..	25@26½	25@26	1 disc.	" 17..	16@18	14@17	
" 16..	25@26½	25@26		" 20..	16@18	14@17	100
" 19..	25½@26½	25@26	1 disc.	" 24..	16@18	14@17	
" 23..	25½@26½	25@26		" 27..	16@18	14@17	100
" 26..	25@26	24@25½	1 disc.	" 31..	16@18	14@17	
Mch. 2..	25@26	24@25½		Sept. 3..	17@18½	14@17½	100
" 5..	25@26	24@25½	1 disc.	" 7..	17@18½	14@17½	
" 9..	25@26	24@25½		" 10..	17@18½	14@17½	101@102
" 12..	26@27½	25½@26½	1 disc.	" 14..	17@18½	14@17½	
" 16..	26@27½	25½@26½		" 17..	17@18½	14@17½	102@103
" 19..	25@26	23½@25	1 disc.	" 21..	17@18½	14@17½	
" 23..	24@25	23@24		" 24..	18½@21	17½@20½	102@103
" 26..	23@24	21@22	1 disc.	" 28..	18½@21	17½@20½	
" 30..	23@24	21@22					
April 2..	22@23	20@21½	1 disc.	Average for the year.	23.38	21.86	

Weekly Quotations in the Charleston Market. Year 1819.

Week	Sea Island.	Maine and Santee.	Short Staple.	Week	Sea Island.	Maine and Santee.	Short Staple.	Week	Sea Island.	Maine and Santee.	Short Staple.	Week	Sea Island.	Maine and Santee.	Short Staple.
1	55	50	27	14	50	40	21	27	40	35	18	40	40	35	18
2	55	50	27	15	50	40	20	28	40	35	17	41	40	35	18
3	55	50	26	16	45	38	18	29	40	35	17	42	40	35	18½
4	55	50	26	17	45	38	18	30	40	35	18	43	40	35	18½
5	54	48	25½	18	45	38	18	31	40	35	18	44	40	35	18½
6	53	48	25½	19	45	38	16	32	40	35	16½	45	40	35	18
7	53	45	25	20	37½	33	16	33	40	35	16½	46	40	35	17
8	53	45	25	21	37½	33	16	34	40	35	16	47	42	37	17
9	53	45	25	22	37½	33	16	35	40	35	16	48	43	37	17
10	55	45	25	23	37½	33	15	36	40	35	16	49	40	35	16
11	55	45	25	24	37½	33	15	37	40	35	16	50	40	35	15
12	55	45	25	25	37½	33	16½	38	40	35	16	51	39	35	16
13	50	40	21	26	40	35	17½	39	40	35	18	52	37½	34	16½

It was early this year that the general depression began to affect the price of cotton. I suppose this general depression may have had some connection with Sir Robert Peel's Act passed during this year fixing the time of return to specie payments in 1823, but this is doubtful, as the premium on gold was only 4½ per cent.—and two years earlier was only 2½ per cent. The Bank of England actually resumed in 1821, though not legally bound to do so. It may have been, and probably was the case, that the resumption of specie payments in England caused a demand for the precious metals, that may have contracted the financial resources of other countries, and thereby caused general depression, but it is just as likely that the contraction of enterprise, confidence and credits may have facilitated the resumption of cash payments.

1820.

In operation in England 12,150 power-looms, and in Scotland 2,000. (*See* years 1785, 1813 and 1829.)

British calicoes, etc., printed at an average duty of 5s. per piece . . 5,456,196 pieces.
Calicoes, etc., exported—average drawback of 5s. per piece . . . . . 3,727,820 pieces.
Calicoes, etc., taken for home consumption at an average duty of
5s. per piece . . . . 1,728,340 pieces.

Cotton exported from United States, 127,860,152 lbs.

United States cotton crop, 369,800 bales of 300 lbs.

Imports of cotton into Great Britain from foreign countries, 120,527,826 lbs., as follows: United States, 89,999,174; Brazil, 29,198,155; Turkey and Egypt, 285,350; miscellaneous, 2,045,147. From British possessions, 30,144,829 lbs., as follows: East Indies and Mauritius 23,125,825; British West Indies, the growth of, 6,219,625, foreign, 617,191; miscellaneous, 182,188. Total imports, 151,672,655 lbs; Exports, 6,024,038; Home consumption, 152,829,633 lbs.

The price semi-weekly at New York and course of Exchange on London for the Crop Year ending October 1, 1820.

1819.	Price of New Orleans	Price of Upland.	Exchange on London.	1820.	Price of New Orleans	Price of Upland.	Exchange on London.
Oct 1..	18½@21	17½@20½	102	April 4..	16@18	14@17	
" 5..	18½@21	17½@20½		" 7..	16@18	14@16	100
" 8..	18½@21	17½@20½	102¼	" 10..	16@18	14@16	
" 12..	18½@21	17½@20½		" 14..	16@18	14@16	100
" 16..	18½@21	17½@20½	102¼	" 18..	16@18	14@16	
" 19..	18½@21	17½@20½		" 21..	16@17	14@17	100
" 22..	18½@21	17½@20	102¼	" 25..	16@17	14@17	
" 26..	18½@21	17½@20		" 28..	16@17	14@16	100
" 29..	18½@21	17½@20	102¼	May 2..	16@17	14@16	
Nov. 2..	18½@21	17½@20		" 6..	16@17	14@16	1 disc.
" 5..	18½@20	16½@19	102½	" 9..	16@18	14@17	
" 9..	18½@20	16½@19		" 12..	16@19	14@18	1 disc.
" 12..	18½@21	16½@19	102½	" 16..	16@19	14@18	
" 16..	18½@21	16½@19		" 19..	16@19	14@18	1 disc.
" 19..	18½@21	16½@19	102½	" 23..	16@19	14@18	
" 23..	18½@21	16½@19		" 26..	16@18	14@18	1 disc.
" 26..	18½@21	16½@19	102½	" 30..	16@18	14@18	
" 30..	18½@21	16½@19		June 2..	16@18	14@18	99@100
Dec. 3..	17@19	14@18½	102	" 6..	16@18	14@18	
" 7..	17@19	14@18½		" 9..	17@18	15@19	99@100
" 10..	18@20	14@18	102¾@103¼	" 13..	17@18	15@19	
" 14..	18@20	14@18		" 16..	19@20	15@19	100@100½
" 17..	18@20	14@18	102½	" 20..	19@20	15@19	
" 21..	18@20	14@17½		" 23..	19@20	15@20	101
" 24..	18@19	14@17½	102½	" 27..	19@20	15@20	
" 28..	18@19	14@17½		" 30..	20@21½	17@20½	101½@102
" 31..	18@19	14@17½	102	July 4..	20@21½	17@20½	
1820				" 7..	20@21½	17@21	101½@102
Jan. 4..	18@19	14@17½		" 11..	20@21½	17@21	
" 7..	18@19	14@17½	102	" 14..	20@21½	17@21	101½@101¾
" 11..	18@19	14@17½		" 18..	19@21	17@21	
" 14..	18@19	14@17½	100	" 21..	19@21	17@21	101½@101¾
" 18..	18@19	14@17½		" 25..	20@21½	17@21	
" 21..	18@19	14@17½	100	" 28..	20@21½	17@20½	101½@101¾
" 25..	18@19	14@17½		Aug. 1..	20@21½	17@20½	
" 28..	18@19	14@17½	100	" 4..	20@21½	17@20½	101½@101¾
Feb. 1..	18@19	14@17½		" 8..	20@21½	17@20½	
" 4..	18@19	14@17½	100	" 11..	20@21½	17@20½	101½@101¾
" 8..	17@19	14@17		" 15..	20@21½	17@20½	
" 11..	17@19	14@17	100	" 18..	20@21½	17@20½	101¼@101½
" 15..	17@19	14@17		" 22..	20@21½	17@20½	
" 18..	17@18½	14@17	100	" 25..	20@22	17@21	101½@102
" 22..	17@18½	14@17		" 29..	20@22	17@21	
" 25..	17@18½	14@17	100	Sept. 1..	20@22	17@21	101½@102
" 29..	17@18½	14@17		" 5..	20@22	17@21	
March 3..	17@18½	14@17	100	" 8..	20@22	17@21	102½@103
" 7..	17@18½	14@17		" 12..	20@22	17@21	
" 10..	17@18½	14@17	100	" 15..	20@22	17@21	102½@103
" 14..	17@18½	14@17		" 19..	20@22	17@21	
" 17..	16@18	14@16	100	" 22..	20@21	15@20	102
" 21..	16@18	14@16		" 26..	20@21	15@20	
" 24..	16@18	14@16	100	" 29..	20@21	15@20	101½@102
" 28..	16@18	14@16					
" 31..	16@18	14@17	100	Average for the year.	18.70	16.92	

Weekly Quotations in the Charleston Market. Year 1820.

Week	Sea Island.	Maine and Santee.	Short Staple.	Week	Sea Island.	Maine and Santee.	Short Staple.	Week	Sea Island.	Maine and Santee.	Short Staple.	Week	Sea Island.	Maine and Santee.	Short Staple.
1	37½	34	16½	14	30	26	16	27	37½	32	20	40	34	30	20
2	37½	34	16½	15	30	25	15	28	37½	32	20	41	33	28	19
3	38	33	17	16	32	26	17	29	37½	31	20	42	33	28	18½
4	38	33	17	17	33	28	17½	30	37½	31	20	43	33	28	16½
5	37	33	16½	18	33	28	17½	31	37½	31	20	44	33	28	16½
6	38	31	16	19	33	28	17½	32	37½	31	20	45	30	26	16
7	38	31	16	20	35	30	17½	33	37½	31	20	46	30	25	15½
8	38	31	16	21	35	30	17½	34	35	30	19	47	28	24	15½
9	37	30	16½	22	35	30	17½	35	35	30	19	48	28	24	15½
10	35	30	16½	23	35	29	18	36	35	30	19	49	28	24	16
11	35	30	16½	24	35	29	18	37	35	30	19	50	28	24	16½
12	30	28	15	25	35	29	20	38	34	30	19	51	31	25	16
13	30	26	16	26	35	29	20	39	34	30	19	52	30	25	16½

In this year Mehemet Ali, Pasha of Egypt, seeing the climate of his country in everyway adapted for the cultivation of the cotton plant on a much larger scale than it had hitherto been carried, and knowing the rising importance of it as an article of commerce and manufacture, commenced active measures for the further production of the same. (*See* year 1823.)

1821.

Cotton culture introduced into Egypt, which has since yielded fruitfully, England drawing a good portion of its supplies from thence. In quality the Egyptian product is second only to the Sea Island of the United States.

British calicoes, etc., printed at an average duty of 5s. per piece. . 7,005,484 pieces.
Calicoes, etc., exported—average drawback of 5s. per piece. . . . . . 4,333,664 pieces.
Calicoes, etc., taken for home consumption at an average duty of 5s. per piece. . . . . 2,671,820 pieces.

Cotton exported from United States, 124,893,405 lbs., 11,344,066 lbs. being Sea Island: total value, $20,157,484.

United States cotton crop, 539,033 bales of 300 lbs. each.

Cotton cultivation on anything like a large scale, first began in Upper Egypt during this year, the crop being sixty bags. (*See* year 1824.)

Imports of cotton into Great Britain from foreign countries 116,367,579 lbs., as follows: United States, 93,470,745; Brazil, 19,535,786; Turkey and Egypt, 856,868; miscellaneous, 2,504,180. From British Possessions, 16,169,041 lbs., as follows: East Indies and Mauritius, 8,827,107; British West Indies, the growth of,

5,854,944, foreign, 1,284,036; miscellaneous, 302,954. Total imports, 132,536,620 lbs; Exports, 14,589,497 lbs· Home consumption, 137,401,549 lbs.

Richard Guest, the historian, says there were in Manchester, England, and vicinity, this year, thirty-two cotton factories containing 5,732 power-looms.

The price semi-weekly at New York and course of Exchange on London, for the Crop Year ending October 1, 1821.

1820.	Price of New Orleans	Price of Upland.	Exchange on London.	1821.	Price of New Orleans	Price of Upland.	Exchange on London.
Oct. 3..	20@21	15@20	101½@102	April 6..	15@17	11@14	107½
“ 6..	19@20	15@18		“ 10..	15@17	11@14½	
“ 10..	19@20	14½@18	102@102¼	“ 13..	15@17	11@14½	107¼
“ 13..	19@20	14½@18		“ 17..	15@17	11@14½	
“ 17..	19@20	14½@18	101¾@102¼	“ 20..	15@17	11@14½	107¼
“ 20..	17@19	14@17		“ 24..	15@17	11@14½	
“ 24..	17@18	12@16	101¾@102¼	“ 27..	15@17	11@14½	107½
“ 31..	17@18	12@16		May 1..	15@17	11@14½	
Nov 3..	17@18	12@16	102½@103	“ 4..	15@17	11@14½	107¼@107¾
“ 7..	17@18	12@16		“ 8..	15@17	11@14½	
“ 10..	17@18	12@16	102½@103	“ 11..	15@17	11@14½	107¼@107¾
“ 14..	16@16¾	12@16		“ 15..	15@17	11@14½	
“ 17..	16@16¾	12@16	102½@103	“ 18..	15@17	11@15	107½@108
“ 21..	16@16¾	12@16		“ 22..	15@17	11@15	
“ 24..	16@18¾	12@17½	103½	“ 25..	15@17	11@16	109¾
“ 28..	16@18¾	12@17½		“ 29..	15@17	11@16	
Dec. 1..	16@18¾	12@17	104@104½	June 1..	17@18	13½@16½	110
“ 5..	16@18¾	12@17		“ 5..	17@18	13½@16½	
“ 8..	16@18¾	12@17	104@104¼	“ 8..	17@18	13½@16½	110
“ 12..	16@18¾	12@17		“ 12..	17@18	13½@16½	
“ 15..	16@17	12@16	103@103½	“ 15..	17@18	13½@16½	109
“ 19..	16@17	12@16		“ 19..	17@18	13½@16½	
“ 22..	16@17	12@16	103@103½	“ 22..	17@18	13½@16½	109
“ 26..	16@17	12@16		“ 26..	17@18	13½@16½	
“ 29..	16@17	12@16	103@103¼	“ 29..	17@18	13½@16½	109
1821.				July 3..	17@18	13½@16½	
Jan. 2..	16@17	12@16		“ 6..	17@19	13½@18	109
“ 5..	16@17	12@16	103@103¼	“ 10..	17@19	13½@18	
“ 9..	16@17	12@16		“ 13..	17@19	13½@18	108¾@109
“ 12..	16@17	12@16	103@103¼	“ 17..	17@19	13½@18	
“ 16..	16@17	12@16		“ 20..	17@19	13½@18	108¾@109
“ 19..	16@18	12@16	103@103½	“ 24..	17@19	13½@18	
“ 23..	16@18	12@16		“ 27..	17@19	13½@18	108¾@109
“ 26..	16@18	12@16	103½@104	“ 31..	17@19	13½@18	
“ 30..	16@18	12@16					
Feb. 2..	16@18	12@16	103½@104	Aug. 3..	17@19	13½@18	108¾@109
“ 6..	16@18	12@16		“ 7..	17@19	13½@18	
“ 9..	16@18	12@16	104@104½	“ 10..	17@19	13½@18	108¾@109
“ 13..	16@18	12@16		“ 14..	17@19	13½@18	
“ 16..	16@17	12@16	104½@105	“ 17..	17@19	13½@18	109
“ 20..	16@17	12@16		“ 21..	17@19	13½@18	
“ 23..	16½@17½	11@16	105	“ 24..	17@19	13½@18	109
“ 27..	16½@17½	11@16		“ 28..	17@19	13½@18	
Mch. 2..	16½@17½	11@16	105	“ 31..	17@19	13½@18	109
“ 6..	16½@17½	11@16		Sept. 4..	17@19	13½@18	
“ 9..	16½@17½	11@16	105	“ 7..	17@19	13½@18	109
“ 13..	15@16½	11@14		“ 11..	17@19	13½@18	
“ 16..	15@16½	11@14	106@106½	“ 14..	17@19	12@17	108¾@109
“ 20..	15@16½	11@14		“ 18..	17@19	12@17	
“ 23..	15@16½	11@14	106¾@107	“ 21..	17@19	12@17	108¼@109
“ 27..	15@16½	11@14		“ 25..	17@19	12@17	
“ 30..	15@16½	11@14	107@107½	“ 28..	17@19	12@17	108¼@109
April 3..	15@16½	11@14		Average for the year.	17.14	14.32	

COTTON AT LIVERPOOL. YEAR 1821.

Week Ending.	Receipts.						Sales.				Prices.			Actual Export.	Consumpt'n.
	American.	E. I.	Egypt.	Brazil.	Other.	Total.	Consumpt'n.	Speculation.	Export.	Total.	Mid. Up.	Mid. Orl	Dohl.		
Jan. 6	1,146					1,146	6,778	5,020		11,798	8	8¼	6¾		6,778
" 13							5,466			5,466	7½	8	6¾		12,244
" 20	10,940			6,827	1,017	18,785	4,320			4,320	7½	8	6¾		16,564
" 27	5,878			2,937	2,802	11,617	8,201			8,201	7	7¾	6¾		24,765
Feb. 3	10,859				1,551	12,410	7,805	600	200	8,605	7	8	6¾	200	32,570
" 10	7,730				128	7,858	5,797		200	5,997	7	8	6¾	200	38,367
" 17							6,281			6,281	8	8½	6¾		44,648
" 24	1,436					1,436	5,986	2,000		7,986	7¾	8½	6¾		50,634
Mch. 3	1,240					1,240	8,500	500		9,000	7¼	8½	6¼		59,134
" 10	No Circ.														
" 17	11,077	402		2,644	1,049	15,172	8,240	300	300	8,840	8	8½	6¾	300	67,374
" 24	5,867			805	1,247	7,919	15,178	3,500	300	18,978	8¼	8¾	6¼	300	82,552
" 31	8,546			1,286	2,914	19,746	11,077	2,800		13,877	8¾	9	7		93,629
April 7	16,525			1,996	989	19,510	2,096		100	2,196	9	9	7	100	95,725
" 14	6,054			2,287	2,558	10,899	4,300		300	4,600	8¾	9	7	300	100,025
" 21	2,314				1,311	3,625	8,970	2,000		10,970	9	9	7		108,995
" 28	1,657					1,657	8,219	400	200	8,819	8½	9	7	200	117,214
May 5	829					829	5,277			5,277	8½	9	7		122,491
" 12	17,501			4,474	2,400	24,375	2,859	4,500		7,359	8¼	9	7		125,350
" 19	2,907	202			369	3,478	4,800			4,800	8¾	9	7		130,150
" 26	3,241				240	3,481	4,714	400		5,114	8½	9	7		134,864
June 2	3,241				1,446	4,687	6,175	400	100	6,675	8½	9	7	100	141,039
" 9	9,013			2,562	2,671	14,246	7,534			7,534	8½	9	7		148,573
" 16	10,746			617	1,613	12,976	5,143	1,000		6,143	8½	9	7		153,716
" 23	4,132			1,023	1,881	7,036	13,696	3,000		16,696	8½	9	7		107,412
" 30	954					954	6,358		300	6,658	8½	8¾	7	300	173,770
July 7	1,089				871	1,960	8,855			8,855	8½	8¾	7		182,625
" 14	No Circ.														
" 21	11,450			2,503		13,963	3,082			3,082	9¼	9½	7		185,707
" 28	18,098	700		7,474	2,069	28,341	10,717	1,000		11,717	9¼	10½	7		196,424
Aug. 4	2,250			1,003	330	3,583	8,999			8,999	9½	10½	7		205,423
" 11	7,170			1,189	617	8,976	4,970		100	5,070	9¼	10½	7	100	210,393
" 18	3,997			7,812	308	12,117	5,045			5,045	9	10	7		215,438
" 25	3,608				1,912	5,520	5,362	200		5,562	8¾	8¾	7	200	220,800

Sept. 1	No Circ.														
" 8	No Circ.														
" 15	6,694			2,055	2,273	11,022	13,551	2,000		15,551	8¾	8¾	7		234,351
" 22	5,605			1,816	1,839	9,260	13,007	1,000		14,007	8¾	8¾	7		247,358
" 29	4,629			3,371	1,936	9,936	6,685		150	6,835	8¾	8¾	6¾	150	254,043
Oct. 6	8,081			1,651	716	3,175	4,092			4,092	8¾	8¾	6¾		258,135
" 13	1,490			1,068	4,228	6,786	6,164		300	6,464	8¾	8¾	6¾	300	264,299
" 20	4,055			879		4,934	5,990			5,990	8	8½	6¾		270,289
" 27	1,476			1,182	168	2,826	5,443		300	5,743	8	8½	6¾	300	275,732
Nov. 3	3,189	1,969		1,923	1,924	9,005	5,699		250	5,949	7¼	8½	6¾	250	281,431
" 10	5,778			297	1,594	7,669	9,635			9,635	7¼	8½	6¾		291,066
" 17	6,406				3,891	10,297	10,663			10,663	7¼	8½	6½		301,729
" 24	903					903	11,124			11,124	7¾	8½	6½		312,853
Dec. 1	397			5,265	367	6,029	9,337	500		9,837	7¼	8½	6½		322,190
" 8	5,294					5,294	5,140			5,140	7¾	8½	6½		327,330
" 15	2,303			43		2,346	6,951			6,951	7¾	8½	6½		334,281
" 22	2,769			1,596	1,917	6,282	5,776	800		6,576	7¾	8½	6½		340,057
" 29	5,866			1,475	537	7,878	6,401	1,400		7,801	7¾	8½	6½		346,458
Average prices & total sales, receipts & stocks.	240,257	3,273		70,060	54,673	376,673	346,458	33,320	3,100	382,878	$8\frac{143}{625}$	$8\frac{51}{64}$	$6\frac{13}{16}$	3,300	7,217,875

Weekly Quotations in the Charleston Market. Year 1821.

Week	Sea Island.	Maine and Santee.	Short Staple.	Week	Sea Island.	Maine and Santee.	Short Staple.	Week	Sea Island.	Maine and Santee.	Short Staple.	Week	Sea Island.	Maine and Santee.	Short Staple.
1	30	25	16	14	28	22	14½	27	30	25	17½	40	30	25	16¾
2	30	25	16	15	28	22	14½	28	30	25	17	41	30	25	16¾
3	30	25	16	16	28	22	14½	29	30	25	17	42	30	25	16¾
4	28	24	16	17	28	22	15	30	30	25	17	43	30	25	16¾
5	30	24	16	18	26	21	15	31	30	25	17	44	30	25	16¾
6	30	24	16	19	28	21	15	32	30	25	17	45	30	25	17
7	30	24	16	20	26	21	15	33	30	25	17	46	30	25	18
8	30	24	16	21	26	22	15	34	30	25	16½	47	28	25	18
9	30	24	14	22	26	24	15½	35	30	25	16½	48	28	23	18
10	30	24	14	23	30	25	15¼	36	30	25	16½	49	28	23	18
11	30	24	14½	24	30	25	16	37	30	25	16½	50	28	23	18
12	30	24	14½	25	30	25	16	38	30	25	16¾	51	28	23	17½
13	30	22	15	26	30	25	17½	39	30	25	16¾	52	30	23	18

This is the first year of which I was able to obtain any regular Liverpool quotations. It will be seen that that market had not entirely recovered from the decline commenced about the end of 1818, and it will hereafter be seen that this was to be a permanent reduction in value, to which the abnormal advance in 1825 was merely an accidental exception.

1822.

The first cotton mill was erected in Lowell, Mass. (*See* year 1852).

The bronze color so extensively used in common prints was first produced in this year by Messrs. Hartman of Munster, Ireland, from solutions of manganese.

British calicoes, etc., printed at an average duty of 5s. per piece.	6,730,880	pieces.
Calicoes, etc., exported—average drawback of 5s. per piece....	4,730,228	"
Calicoes, etc., taken for home consumption at an average duty of 5s. per piece..	2,000,580	"

Cotton exported from United States, 144,675,095 lbs.; 11,250,-635 lbs. being Sea Island; total value $24,035,058.

United States cotton crop, 588,139 bales of 300 lbs.

Cotton crop in upper Egypt, 50,000 bags. (*See* year 1821.)

Imports of cotton into Great Britain, from foreign countries, 127,666,532 lbs., as follows: United States 101,031,766; Brazil 24,705,206; Turkey and Egypt 395,077; miscellaneous 1,534,483; 1,534,484; from British possessions 15,171,096 lbs., as follows: East Indies and Mauritius 4,554,225; British West Indies—the growth of—9,031,904; foreign 1,263,210; miscellaneous 321,757.

Total imports, 142,837,628 lbs.; Exports, 18,267,776 lbs.; Home consumption, 143,428,127 lbs.

Among the improvements in cotton machinery that have originated in this country, one of the most important is the combination of the train of three bevel wheels, to regulate the variable velocity requisite for winding the slender filaments of cotton on the bobbin of the roving frame, which was originally applied and put in operation this year by Mr. Aza Arnold, a native of Rhode Island. (*See* years 1823, 1825 and 1826).

COTTON AT LIVERPOOL. YEAR 1822.

Week Ending.	Receipts.						Sales.				Prices.			Actual Export.	Consumptn.
	American.	E. I.	Egypt.	Brazil.	Other.	Total.	Consumpt'n.	Speculation.	Export	Total.	Mid. Up	Mid. Orl	Dhol.		
Jan. 5	2,009					2,009	5,065			5,065	7¼	8½	6½		5,065
" 12	332					332	7,835			7,835	7¾	8½	6½		12,900
" 19	9,241			3,421		12,662	7,015			7,015	7½	8½	6½		19,915
" 26	3,839			4,052	3,334	11,225	7,410			7,410	7¼	8½	6½		27,325
Feb. 2	2,525			2,424		4,949	6,298			6,298	7½	8½	6½		33,623
" 9	3,630			988		4,618	7,066			7,066	7½	8½	6½		40,689
" 16	8,849			2,438	4,533	15,820	11,313			11,313	7½	8½	6½		52,002
" 23	6,438			4,995	4,654	16,087	9,546			9,546	7½	8½	6½		61,548
Mch. 2	4,784			2,733		7,517	10,974	600		11,574	7½	8½	6½		72,522
" 9	8,783				1,841	10,624	15,186			15,186	7½	8½	6½		87,708
" 16	8,527			1,000	2,983	12,510	6,426			6,426	7½	8½	6½		94,134
" 23	1,164	201		1,301	320	2,986	8,155			8,155	7½	8¼	6½		102,289
" 30	6,070				5,906	11,976	9,380			9,380	7½	8¼	6½		111,669
Apl. 6	1,221				535	1,756	7,242			7,242	7½	8¼	6½		118,911
" 13	2,506				140	2,646	6,684		200	6,884	7½	8¼	6½	200	125,595
" 20	3,628				672	4,300	5,403			5,403	7¾	8¼	6¼		130,988
" 27	12,743			4,645	639	18,027	9,837			9,837	7¾	8¼	6¼		140,835
May 4	2,714				800	3,514	8,597		600	9,197	7¾	8¼	6¼	600	149,432
" 11	854					854	8,502			8,502	7¾	8¼	6¼		157,934
" 18	3,870				758	4,628	6,608			6,608	7¾	8¼	6¼		164,542
" 25	15,730			1,963	7,451	25,144	4,066		950	5,016	7¾	8¼	6¼	950	168,608
June 1	8,374			7,485	4,906	20,765	4,175			4,175	7¾	8¼	6½		172,783
" 8	2,701				280	2,981	5,695			5,695	7¾	8¼	6½		178,478
" 15	2,597					2,597	11,000			11,000	7¼	8	6½		189,478
" 22					618	618	8,006			8,006	7¼	10¾	6½		197,484
" 29	20,043			2,853	4,561	27,457	7,339			7,339	7¼	7½	6¼		204,823
July 6	22,204			5,022	2,393	29,619	5,072			5,072	7	7½	6¼		209,895
" 13	7,690				778	8,468	10,658	2,500		13,158	7	7½	6¼		220,553
" 20	1,991				3,210	5,201	15,610	800		16,410	7	7½	6¼		236,163
" 27	20,965				3,597	24,562	6,832			6,832	7	7½	6¼		242,995
Aug. 3	8,772			2,309	1,135	12,216	10,288			10,288	7	7½	6¼		253,283
" 10	744				620	1,364	8,703			8,703	7	7½	6		261,986
" 17	8,448				5,869	14,317	6,886			6,886	6½	7	6		268,872
" 24	6,090				145	6,235	5,431			5,431	6½	6¾	6		274,303

Aug. 31	3,796			4,334	8,157	16,287	13,304			13,304	5¾	6	5½		287,607
Sept. 7	4,932			2,174	1,224	8,330	11,295			11,295	5¾	6	5½		298,902
" 14	1,591			1,182	1,002	3,775	11,617	2,200	1,000	14,817	5¾	6	5½	1,000	310,519
" 21	885				114	999	11,713	3,000	1,000	15,713	5¾	6	5½	1,000	322 232
" 28					5	5	7,353			7,353	5¾	6	5½		329,585
Oct. 5	2,899			2,739	449	6,087	7,640			7,640	5¾	6	5½		337,225
" 12	14,910			3,401	4,483	22,794	22,044			22,044	6	6½	5½		359,269
" 19	2,841				1,427	4,268	8,910	2,500		11,410	6¼	6¾	5½		368,179
" 26	4,591			1,632	2,135	8,358	14,922	5,000		19,922	6¾	7	5½		383,101
Nov. 2	2,277				173	2,450	6,378			6,378	6¾	7	5½		389,479
" 9	1,667			2,267	4,631	8,565	6,926	1,500		8,426	6½	7	5½		396,405
" 16	3,587	1,352			2,008	6,947	3,472			3,472	6½	7	5½		399,877
" 23	1,672				1,382	3,054	3,829			3,829	6½	7	5½		403,706
" 30	313	60		1,748	97	2,218	4,628			4,628	6½	7	5½		408,334
Dec. 7	1,084				2,957	4,041	7,767			7,767	6¼	7	5½		416,101
" 14	1,252				1,500	2,752	8,490			8,490	6	7	5½		424,591
" 21	3,614				914	4,528	8,431			8,431	6	7	5½		433,022
" 28	2,845				144	2,989	13,456		200	13,656	6¼	7	5½	200	446,478
Average prices & total sales, receipts & stocks.	274,832	1,613		67,106	95.380	439,031	446,468	18,100	3,950	468,518	6$\frac{95}{100}$	7$\frac{63}{100}$	6.087	3,950	8,586$\frac{11}{100}$

The price semi-weekly at New York and Course of Exchange on London, for the Crop Year ending October 1, 1822.

1821.	Price of New Orleans	Price of Upland.	Exchange on London.
Oct. 2..	17@19	12@17	108¼@109
" 5..	17@19	12@17	
" 9..	17@19	12@17	109
" 12..	17@19	12@17	
" 16..	17@19	12@17	109¾
" 19..	17@19	12@17	
" 23..	17@19	12@17	109¾
" 26..	17@19	12@17	
" 30..	17@19	12@17	109¾
Nov. 2..	17@19	12@17	
" 6..	17@19	12@17	109¾
" 9..	17@19	12@17	
" 13..	17@19	12@17	109½@110
" 16..	17@19	12@17½	
" 20..	17@19	12@17½	110
" 23..	17@19	12@17½	
" 27..	17@19	12@17½	110
" 30..	17@19	12@17½	
Dec. 4..	17@19	12@17½	110½@111
" 7..	17@19	12@17½	
" 11..	17@19	12@17½	112½@113
" 14..	17@19	12@17½	
" 18..	17@19	14@18	112@112½
" 21..	17@19	14@18	
" 25..	17@19	14@18	111½
" 28..	17@19	14@17½	
1822.			
Jan. 1..	17@19	14@17½	111½@112
" 4..	17@19	14@17½	
" 8..	17@19	14@17½	111½@112
" 11..	19@21	12@17½	
" 15..	19@21	12@17½	111@111½
" 18..	19@21	12@17½	
" 22..	19@21	12@17½	112½
" 25..	19@21	12@17½	
" 29..	19@21	12@17½	112½
Feb. 1..	19@21	12@17½	
" 5..	19@21	12@17½	112
" 8..	19@21	12@17½	
" 12..	19@21	12@17½	113½@114
" 15..	19@21	12@17½	
" 19..	19@21	12@17½	113½@114
" 22..	19@21	12@17½	
" 26..	19@21	12@17½	113½@114
Mch. 1..	19½@22	12@17½	
" 5..	19@21	12@17½	113½@114
" 8..	19@21	12@17½	
" 12..	19@21	12@17½	113@113½
" 15..	19@21	12@17½	
" 19..	19@21	12@17½	112@112½
" 22..	19@21	12@17½	
" 26..	20@21	12@17½	112@112½
" 29..	18@21	12@17½	
April 2..	17½@20	12@17½	112¾

1822.	Price of New Orleans	Price of Upland.	Exchange on London.
April 5..	17½@20	12@17½	
" 9..	17½@20	12@17½	112@112¾
" 12..	17½@20	12@17½	
" 16..	17½@20	12@17½	113@113¼
" 19..	17½@20	12@17½	
" 23..	17@20	12@18	113@113¼
" 26..	17@20	12@18	
" 30..	17@20	12@18	113@113¼
May 3..	17@20	12@18	
" 7..	17@19	12@18	100
" 10..	17@19	12@18	
" 14..	17@19	12@18	100
" 17..	17@19	12@16	
" 21..	17@19	12@17	100
" 24..	17@19	12@17	
" 28..	17@19	12@17	109½@110
" 31..	17@19	12@17	
June 3..	16@18	11@17	109@109¼
" 7..	16@18	11@17	
" 11..	16@18	11@17	108@108½
" 14..	16@18	11@17	
" 18..	16@18	11@17	108@108½
" 21..	16@18	11@16	
" 25..	16@18	11@17	108@108½
" 28..	16@18	11@17	
July 2..	15@18	11@15	108@108½
" 6..	15@18	11@15	
" 9..	15@17	11@16	108@108½
" 12..	14@17	10@16	
" 16..	14@17	10@16	109
" 19..	14@17	10@16	
" 23..	14@17	11@15	110
" 26..	14@17	11@15	
" 30..	14@17	11@15	110
Aug. 2..	14@17	11@15	
" 6..	14@17	11@15	110
" 9..	14@17	10½@15	
" 13..	14@17	10½@15	110
" 16..	12@16	10@14	
" 20..	12@16	10@14	110¼
" 24..	12@16	10@14	
" 27..	12@16	10@14	110¼
" 30..	12@16	10@14	
Sept. 3..	12@16	10@14	110¼
" 6..	12@16	10@14	
" 10..	12@16	10@15	110½@111
" 13..	12@16	10@15	
" 17..	12@16	10@15	110½@111
" 20..	12@16	10@15	
" 24..	14@16	10@15	112½@113
" 27..	14@16	10@15	
Oct. 1..	14@16	10@15	112½
Average for the year,	17@59	14@32	

1823.

United States crop 509,600 bales of 300 lbs.

Machinery was so far perfected at this period that a steam-loom weaver, about fifteen years of age, attending two looms, could weave seven pieces 9-8ths shirting per week (each 24 yards long, containing 100 shoots of weft an inch, the reed of the cloth being a 44 Bolton count and the warp and weft 40 hanks to the lb). *Two* pieces per week for a hand-weaver was considered good work. (*See* years 1826 and 1833.)

British calicoes, etc., printed at an average duty of 5s. per piece.	7,247,676	pieces.
Calicoes, etc., exported—average drawback of 5s. per piece....	4,587,004	"
Calicoes, etc., taken for home consumption at an average duty of 5s. per piece..................................	2 660,672	"

Cotton exports from United States, 173,723,270 lbs., 12,136,688 lbs. being Sea Island; total value $20,445,520.

About this year, long stapled cotton of an excellent quality—superior to any other kind except Sea Island—began to be imported into England from Egypt.

Cotton crop in upper Egypt 120,000 bags. (*See* year 1821.)

Egyptian cotton was introduced into Great Britain for the first time this year, when Mehemet Ali exported 5,623 bales.

Mr. John Heathcoat, M. P.'s patent (*see* year 1809) for making lace expired.

Imports of cotton into Great Britain from foreign countries, 169,370,073 lbs., as follows: United States, 142,532,112; Brazil, 23,414,641; Turkey and Egypt 1,334,547; miscellaneous, 1,988,-773: from British Possessions, 22,032,430 lbs.. as follows: East Indies and Mauritius 14,839,117; British West Indies—the growth of—5,719,610, foreign, 1,315,183; miscellaneous 158,520. Total Imports, 191,402,503 lbs.; Exports, 9,318,402 lbs.: Home consumption, 186,311,070 lbs.

Richard Guest, the historian, says there were this year in England 10,000 steam-looms in operation.

United States export of cotton to France, 25,000,000 lbs. (*See* year 1829).

Aza Arnold received a patent January 21st, this year, for his bevel-wheel combination. (*See* years 1822, 1825 and 1826).

COTTON AT LIVERPOOL. YEAR 1823.

Week Ending.	Receipts.						Sales.				Prices.			Actual Export.	Consumpt'n.
	American.	E. I.	Egypt.	Brazil.	Other.	Total.	Consumpt'n.	Speculation.	Export	Total.	Mid. Up	Mid Orl	Dhol		
Jan. 4...	....	......	......	....	....	....	5,708	....	......	5,708	6¼	7	5¼	....	5,708
" 11...	3,418	......	......	....	....	3,418	4,792	....	......	4,792	6¼	7	5¼	....	10,500
" 18...	1,941	1,000	......	1,574	1,223	5,738	10,519	5,000	......	15,519	6¼	7	5¼	....	21,019
" 25...	2,435	......	......	275	353	3,063	7,296	....	......	7,296	6¼	7	5¼	....	28,315
Feb. 1...	9,094	......	......	1,090	230	10,414	4,989	....	......	4,989	6¼	7	5½	....	33,304
" 8...	999	......	......	1,114	....	2,113	13,4[illegible]	....	......	13,422	6¼	7	5¼	....	46,726
" 15...	19,292	......	......	2,261	2,577	24,130	11,297	5,000	......	16,297	6½	7	5¼	....	58,023
" 22...	7,522	......	......	1,188	294	9,004	20,222	1,500	500	22,222	6½	7	5¼	500	78,245
Mch. 1...	7,036	......	......	4,194	....	11,230	8,901	....	......	8,901	6½	7	5¼	....	87,146
" 8...	9,845	1,851	......	1,660	1,560	14,916	9,706	....	......	9,706	6½	7	5¼	....	96,852
" 15...	17,417	......	......	2,277	....	19,694	9,750	800	300	10,850	6½	7	5¼	300	106,602
" 22...	6,217	1,289	......	2,256	1,174	10,936	14,927	3,000	......	17,927	6½	7	5¼	....	121,529
" 29...	11,222	......	......	....	1,709	12,931	7,118	....	......	7,118	6½	7	5¼	....	128,647
April 5...	14,501	......	......	3,807	170	18,478	7,788	....	......	7,788	6½	7	5¼	....	136,435
" 12...	4,031	......	......	....	44	4,075	11,092	1,500	1,000	13,592	6½	7	5¼	1,000	147,527
" 19...	14,624	......	......	3,317	2,384	20,325	10,153	1,000	......	11,153	6½	7	5¼	....	157,680
" 26...	11,895	......	......	....	3,043	14,938	7,577	600	......	8,177	6½	7	5¼	....	165,257
May 3...	2,825	......	......	....	....	2,825	12,240	1,000	......	13,240	6½	7	5¼	....	177,497
" 10...	14,787	......	......	....	3,185	17,972	11,022	1,500	......	12,522	6½	7	5¼	....	188,519
" 17...	14,819	......	......	2,768	1,017	18,604	6,525	2,000	......	8,525	6½	7	5¼	....	195,044
" 24...	6,440	......	......	1,195	....	7,635	12,536	12,000	......	24,536	6½	7	5¼	....	207,580
" 31...	3,566	......	......	1,615	660	5,841	11,760	6,700	......	18,460	6½	7	5¼	....	219,340
June 7...	23,340	......	......	3,815	49	27,204	10,335	7,200	......	17,535	6½	7	5¼	....	229,675
" 14...	16,820	1,519	......	2,231	1,412	21,982	16,089	10,000	......	26,089	7	7½	5¼	....	245,764
" 21...	4,591	......	......	....	114	4,705	8,220	1,300	......	9,520	7	7½	5¼	....	253,984
" 28...	11,610	......	......	3,091	460	15,161	11,315	12,000	......	23,315	7¼	7½	5¼	....	265,299
July 4...	8,386	......	......	....	1,571	9,957	12,758	9,000	......	21 758	7½	7¾	6	....	278,057
" 11...	21,589	......	......	3,081	140	24,810	17,604	21,000	......	38,604	8½	8¾	6½	....	295,661
" 18...	26,603	......	......	915	4,122	31,640	8,016	2,000	......	10,016	8½	8¾	6½	....	303,677
" 25...	5,588	......	......	544	1,393	7,525	7,175	3,000	......	10,175	8½	8¾	6½	....	310,852
Aug. 1...	4,697	......	......	3,304	2,690	10,691	6,854	5,000	......	11,854	8½	8¾	6½	....	317,706
" 8...	7,159	......	......	2,247	284	9,690	12,033	....	......	12,033	8½	8¾	6½	....	329,739
" 15...	1,059	......	......	981	292	2,332	3,069	1,500	......	4,569	8½	8¾	6½	....	332,808
" 22...	15,210	......	......	2,325	1,156	18,691	2,980	....	......	2,980	8½	8¾	6½	....	335,788

April 29...					493	493	4,034	2,000		6,034	8	8½	6½		339,822
Sept 5...	10,148			2,858	3,024	16,030	6,110	4,000		10,110	8	8½	6½		345,932
" 12...	1,752				308	2,060	24,105	5,000		29,105	8½	8½	6½		370,037
" 19...	4,470			1,654	755	6,879	3,322	3,000		6,322	8½	8½	6½		373,359
" 26...	4,978			2,693	1,300	8,971	7,241	2,000		9,241	8½	8½	6½		380,600
Oct. 3...	100			855		955	5,809			5,809	8½	8½	6½		386,409
" 10...	4,442			4,577	260	9,279	7,072			7,072	8½	8½	6½		393,481
" 17...	2,387	1,355			1,153	4,895	3,414			3,414	7¾	8½	6½		396,895
" 24...	50			608		658	6,456	1,200		7,656	7¾	8½	6½		403,351
" 31...	1,856			3,046	287	5,189	5,819			5,819	7¾	8½	6½		409,170
Nov. 7...	1,982			2,135	1,653	5,770	5,530			5,530	7½	8	6½		414,700
" 14...	4,826			2,947	316	8,089	6,686	1,000		7,686	7¼	8	6		421,386
" 21...	978				2,244	3,222	11,694	2,000		13,694	7¼	8	6		433,080
" 28...	6,759			10,369	1,216	18,344	5,041			5,041	7¼	8	6		438,121
Dec. 5...	6,361			1,207	2,960	10,528	7,850			7,850	6¾	7½	6		445,971
" 12...	5,319			5,736	1,308	12,363	12,184			12,184	6¾	7½	6		458,155
" 19...	3,916	6,670			5,368	15,954	17,115	7,000		24,115	7	7½	6		475,270
" 26...	12			2,788	2,846	5,646	10,168	5,000		15,168	7¼	8	6		485,438
Average prices & total sales, receipts & stocks.	390,914	13,684	......	94,598	58,797	557,993	485,438	145,800	1,800	633,038	7.21	7.69	5.80	1.800	9.335.23

The price semi-weekly at New York, rates of Freight to Liverpool and Course of Exchange on London, for the Crop Year ending October 1, 1823.

1822.	Price of New Orleans.	Price of Middlings.	Exchange on London.	Freight to Liverpool.
Oct. 4..	14@16	10@15	112½	½@¾d.
" 8..	14@16	10@15		
" 11..	14@16	10@14½	112½	
" 15..	14@16	10@14½		
" 18..	14@16	10@14½	112¾	
" 22..	13@15	9@14		
" 25..	13@15½	9@14½	112½@113	
" 29..	13@15½	9@14½		
Nov. 1..	13@15½	9@14½	112½@113	⅜@½d.
" 5..	13@15½	9@14		
" 8..	13@15½	9@14	112½@113	
" 12..	13@15½	9@14		
" 15..	13@15½	9@14	112¾	
" 19..	13@15½	9@14		
" 22..	13@15½	9@14	112¾	
" 26..	13@15½	9@14		
" 29..	13@15½	9@14	112¾	
Dec. 3..	13@15½	10@11		⅜@½d.
" 7..	13@15½	10@11	112¾	
" 10..	13@15½	10@11		
" 13..	13@15½	10@11	113	
" 17..	13@15½	10@11		
" 20..	13@15½	10@11	112⅜@112⅝	
" 24..	13@15½	10@11		
" 27..	13@15½	10@11	112⅜@112⅝	
" 31..	11@14	9@11		
1823.				
Jan. 3..	11@14½	9@11	112@112½	⅜@½d.
" 7..	11@14½	9@11		
" 10..	11@14½	9@11	112@112½	
" 14..	17@...	9@12½		
" 17..	17@...	9@12½	112@112½	
" 21..	17@...	9@12½		
" 24..	16½@..	9@12½	112@112¼	
" 28..	16½@..	9@12½		
" 31..	16½@..	9@12½	111@111¼	
Feb. 4..	16½@..	9@12		½d.
" 7..	16½@..	9@12	110¾	
" 11..	16½@..	9@12		
" 14..	16½@..	9@12	109@109½	
" 18..	14@15	9@12		
" 21..	14@15	9@12	109@109½	
" 25..	14@15	9@12		
" 28..	14@15	9@12	109@109½	
Mch. 4..	14@15	9@12		½d
" 7..	14@15	9@12	110@110½	
" 11..	14@15	9@12		
" 14..	14@..	9@11½	107½@107¾	
" 18..	14@..	9@11½		
" 21..	14@..	9@11½	106½@107	
" 25..	14@..	9@11½		
" 28..	14@..	9@11½	105½@106½	½d.
April 1..	12½@14	9@12		
" 4..	12½@14	9@12	103½@104	

1823.	Price of New Orleans.	Price of Middlings.	Exchange on London.	Freight to Liverpool.
April 8..	12½@14	9@12		
" 11..	12½@14	9@12	105@105½	
" 15..	12½@14	9@12		
" 18..	12½@14	9@12	105@105½	
" 22..	12½@14	9@12		
" 25..	12½@14	9@12	105@105½	
" 29..	12½@14	9@12		
May 2..	12½@14	9@12	104	1d.
" 6..	12½@14	9@12		
" 9 .	12½@14	9@12	103½	
" 13..	12½@14	9@12		
" 16..	12½@14	9@12	104@104½	
" 20..	12½@14	9@12		
" 23..	11½@14	9@12	104@104½	
" 27..	11½@14	9@12		
" 30..	11½@14	9@12	104@104½	
June 3..	11½@14	9@12		1d.
" 6..	11½@14	9@12	104½@105	
" 10..	11½@14	9@12		
" 13..	11½@14	9@12	106½@107	
" 17..	11½@14	9@12		
" 20..	11½@14	9@12	106½@107	
" 24..	11½@14	9@12		
" 27..	16½@17	9@15	106@106¼	
July 1..	16½@17	9@15		1d.
" 4.	16½@17	9@15	106@106¼	
" 8..	16½@17	9@15		
" 11..	16½@17	9@15	105¾@106	
" 15..	16½@17	9@15		
" 18..	16½@17	9@15	106¼@106¾	
" 22..	16½@17	9@15		
" 25..	16½@17	9@15	106¼@106¾	
" 29..	14@18	11½@16		
Aug. 1..	14@18	11½@16	106¼@106¾	½d.
" 5..	14@18	11½@16		
" 8..	14@18	11½@16	107½	
" 12..	16@18	10½@16		
" 15..	16@18	10½@16	107½	
" 19..	16@18	10½@16		
" 22..	16@18	10½@16	107¼	
" 26..	16@18	10½@16		
" 29..	16@18	10½@16	107	
Sept. 2..	16@18	10½@16		½d.
" 5..	16@18	10½@16	106½	
" 9..	16@18	10@17		
" 12..	16@18	10@17	106½	
" 16..	16@18	12@17		
" 19..	16@18	12@17	106¾@107	
" 23..	17@19	12@17		
" 26..	17@19	12@17	106¾@107	
" 30..	17@19	12@17		
Average for year.	14.46	11.40		

Weekly Quotations in the Charleston Market. Year 1823.

Week	Sea Island.	Maine and Santee.	Short Staple.	Week	Sea Island.	Maine and Santee.	Short Staple.	Week	Sea Island.	Maine and Santee	Short Staple.	Week	Sea Island.	Maine and Santee.	Short Staple.
1	30	20	12½	14	25	18	11½	27	32	20	14½	40	30	20	17
2	30	21	12½	15	25	18	11½	28	32	20	14½	41	30	20	17
3	30	21	12½	16	25	18	12	29	30	20	14½	42	30	20	16
4	30	21	12½	17	25	18	12	30	30	20	15	43	30	24	16½
5	30	21	12	18	25	18	12	31	33	20	15	44	30	24	17
6	25	21	12	19	25	18	12½	32	33	20	15	45	30	24	17
7	25	21	12	20	25	19	12½	33	33	20	15	46	30	24	17
8	25	21	12	21	25	19	12½	34	33	20	14½	47	30	24	16
9	25	21	12	22	25	19	12	35	33	20	14½	48	28	23	16
10	25	21	12	23	30	19	13	36	33	20	15	49	28	22	15½
11	25	19	12	24	30	19	13	37	30	20	16	50	28	22	15
12	25	19	12	25	30	19	13½	38	30	20	17	51	28	22	14½
13	25	18	11½	26	32	20	14½	39	30	20	17	52	28	22	14½

This year "the house of Cropper, Benson & Co. of Liverpool drew up a manifest, in which by a variety of calculations it strove to show by logical conclusions and reckonings, that the production of cotton had its limit, and that in consequence of the abolition of the slave trade, and the annual decrease of the colored population, as well as by the natural restrictions which northern latitudes place upon cotton growing, the necessary and approaching consequence must be that the importation would become daily less, and obtainable only at very high prices, and that from this would result—so at least they believed—that the consumption would far exceed the production, and make the cost of cotton immensely higher. This manifest was circulated with a certain pomp all over the cotton manufacturing districts in England and the United States."

Vincent Nolte's Reminiscences, p. 291–2.

I copy the above because the arguments are so much like what we have often heard in the past ten years; and also because it is a faint indication of the direction in which some men's thoughts were turning at that time. Cropper, Benson & Co., were naturally men of great enterprise and, *sometimes*, had a kind of unconscious *instinct of the future*, as shown in their early investments in the Liverpool and Manchester Railway. The commercial world began this year, if not earlier, to tire of the inaction and hum-drum of legitimate trade. Cropper, Benson & Co. were like other people, only more sensitive to the great currents of thought, and hence, in a small way, leaders. To be sure the arguments and logic of their manifest look rather awkward and bungling at this day; but this was all natural.

McCulloch says "the price of corn, which had been very much depressed in 1821 and 1822, rallied in 1823 ; and this circumstance contributed, along with others peculiar to that period, to promote an extraordinary rage for speculation. The issues of the country banks being in consequence far too much extended, the currency became redundant in the Autumn of 1824."

Here we see one of the *immediate* causes of the cotton speculation and the financial crash among the country banks in 1825.

1824.

Average price for United States middling uplands in Liverpoll, 8½d.

British calicoes, etc., printed at an average duty of 5s. per piece.	8,167,872 pieces.
Duty on same received by Government........................	£2,040,718
Calicoes, etc., exported—average drawback of 5s. per piece....	5,527,864 pieces.
Drawback paid hy Government on same......................	£1,381,941
Calicoes, etc., taken for home consumption at an average duty of 5s. per piece........	2,635,108 pieces
Net amount of duty received by Government for same..........	£658,677

Cotton exported from United States, 142,469,663 lbs., 9,525,722 lbs being Sea Island ; total value $21,947,401.

Cotton crop in upper Egypt 140,000 bags. (*See* year 1821).

Egyptian cotton imported into Great Britain, 38,022 bags.

Imports of cotton into Great Britain from foreign countries, 126,035,391 lbs., as follows: United States, 92,187,662; Brazil, 24,849,552; Turkey and Egypt, 7,719,368; miscellaneous, 1,278,-720; from British Possessions, 23,344,820 lbs., as follows; East Indies and Mauritius, 16,420,005; British West Indies—the growth of—5,006,002, foreign, 1,263,304; miscellaneous, 655,509. Total imports, 149,830,122 lbs; Exports, 13,299,505 lbs; Home consumption, 141,038,743 lbs.

George Danforth of Massachusetts invented the tube frame, or "Taunton Speeder," as it was sometimes called, having been first built and brought into use in Taunton, Mass., and it was patented September 2d of this year. (*See* year 1825).

A 7 qr. bobbin-net machine was easily sold this year for £600. (*See* year 1835.)

COTTON CROP OF THE UNITED STATES.

Statement and Total Amount of the Cotton Crop of the United States, for the year ending 1st October, 1824.

		Bales.	Total.
NEW ORLEANS.			
Amount of Exports, Foreign and Coastwise		143,943	
Remaining on hand		1,501	
		145,444	
DEDUCT			
On hand 1st October, 1823	2,869		
Received from Mobile	13,094		
do Florida	3,000		
		18,963	
Total at New Orleans, which includes Louisiana, Mississippi, Tennessee and the North of Alabama			126,481
FLORIDA.			
Export to New Orleans		3,000	
do New York		1,039	
do other ports		461	
			4,500
ALABAMA.			
Export from Mobile and Blakely (the crop from the north part of the State included in the export from New Orleans)			44,924
GEORGIA.			
Amount of Exports, Foreign and Coastwise		152,208	
Remaining on hand		527	
			152,735
SOUTH CAROLINA.			
Amount of Exports Foreign and Coastwise		154,518	
Deduct amount received from Savannah and included in the Georgia Exports	4,000		
Received from North Carolina through Cheraw and Wilmington	16,000		
		20,000	134,518
NORTH CAROLINA			
No positive data could be obtained in regard to the crop in this State; but, the estimate is believed to be substantially correct at			46,000
Total amount of crop in the United States			509,158

The following is ascertained to be the deficiency, compared with the previous year ending 1st October, 1823.

		Bales.	Total.
NEW ORLEANS.			
Amount of Exports for the year ending, 1st October, 1823	171,431		
Remaining on hand	2,869		
		174,300	
Export for the year ending, 1st October 1824	143,943		
Remaining on hand	1,501		
	145,444		
Deduct difference in the receipts from Mobile, more than the previous year	9,131		
		136,313	
			37,987
ALABAMA.			
At Mobile and Blakely ascertained			4,137
GEORGIA.			
Deficiency in Uplands		16,534	
do Sea Islands		1,478	
			18,012
SOUTH CAROLINA.			
Deficiency in Uplands		6,280	
do Sea Islands		2,115	
			8,395
NORTH CAROLINA.			
Deficiency estimated at			5,000
			73,531
Deduct for increase in Florida			1,500
Total deficiency			72,031
Consisting of Uplands		68,438	
do Sea Islands		3,593	
			72,031

COTTON AT LIVERPOOL. YEAR 1824.

Week Ending.	Receipts.						Sales.				Prices,			Actual Export.	Consumpt'n.
	American.	E. I.	Egypt.	Brazil.	Other.	Total.	Consumpt'n.	Speculation.	Export	Total.	Mid. Up	Mid. Orl	Dhol.		
Jan. 2	7,639			2,316	520	10,475	5,369			5,369	7¼	8	6		5,369
" 9	8,403			1,240		9 643	3,458			3,458	7¼	8	6		8,827
" 16	2.785			681		3,466	9,529			9,529	7	8	6		18,356
" 23	2,713			988	1,598	5,299	10,688			10.688	7	8	6		29,044
" 30	11,227			1,423	3,600	16,250	15,366			15,366	7	8	6		44,410
Feb. 6	6,870			1,877	2,501	11.254	12,450	1.000		13,450	7¼	8	6		56,860
" 13	2.759		498	1,948	503	5,708	14,110	3,000		17,110	7¼	8	6		70,970
" 20	5,080			3,576	880	9,536	9,840			9,840	7¼	8	6		80,810
" 27	2,033					2,033	13,427			13,427	7¼	8	6		94,237
Mch. 5	2,920		630			3,550	5,843			5,843	7¼	8	6		100,080
" 12	18,268		6.879	4,663		29,810	9,838			9,838	7¼	8	6		109,918
" 19	11,224		314	2,480	1,755	15,773	12,798	9,000		21,798	7½	8	6		122,716
" 26	7,129			2,361	1,260	10,750	15,647	2,000		17,647	7½	8	6		138 363
Apr. 2	1,697			2,571	606	4 874	10,499			10,499	7¾	8	6		148,862
" 9	7,914				2,005	9,919	10,322			10,322	7¾	8	6		159,184
" 16	6,621	1,872			184	8,677	11,922			11,922	7¾	8	6		171.106
" 23	5,783		420		766	6,969	17,241			17,241	8	8½	6		188,347
" 30	22.089	145			2,655	24.889	12,509			12,509	8	8½	6		200,856
May. 7	4,245			2,247	84	6,576	8,416	1,000		9,416	8	8½	6		209,272
" 14	4,058			3,080	1,352	8,485	8,102		800	8,902	8	8½	6	800	217,374
" 21	9,667			1,464	224	11,355	6,909			6,909	8	8½	6		224,283
" 28	3,312			1,210	1,750	6,272	6,197			6,197	8	8½	6		230,480
June 4	1,415	2.271		1,923	25	5.634	9,904	500	800	11,204	8	8½	6	800	240.384
" 11	9,776	2,836		3,873	830	17.315	11,943	700	500	13,143	8	8½	6	500	252.327
" 18	4.637				1,167	5,804	10,553	1,500	500	12,553	8	8½	6	500	262,880
" 25	19,396	1,733		4.099	7,035	32,263	6,009			6,009	8	8½	6		268,889
July 2	7,916			2 155	1,543	11,614	8,644			8,644	7¾	8½	6		277.533
" 9	3,374			2,565	2,042	7.981	15,079	1,000		16,079	7¾	8½	6		292,612
" 16	6,601		1,716	1,315	547	10,179	7,589			7,589	7½	8½	5½		300,201
" 23	5,125		3,010	1,260		9.395	7,476			7,476	7½	8½	5½		307,677
" 30	675				2,831	3,506	10,975	2,000		12,975	7½	8½	5½		318,652
Aug. 6	5,470				286	5,756	11.331	2,000	1,000	14,331	7½	8½	5½	1,000	329,983
" 13	6,264			1,817	667	8,748	7,300			7,300	7½	8½	5½		337,283
" 20	2,038	1,000		956	311	4,305	7,221			7,221	7½	8½	5½		344,504
" 27	3,373	1,843			2,011	7,227	9,905			9,905	7½	8½	5½		354,409

Sept. 3...	618			1,757	498	2,873	5,908			5,908	7½	8½	5½		360,317
" 10...	2,495			7,214	722	10,431	7,185			7,185	7½	8½	5½		367,502
" 17...	4,153			3,548	1,728	9,429	9,771			9,771	7¼	8	5½		377.273
" 24...	2,595		1,636	4,780	2,184	11,195	11,345			11,345	7¼	8	5½		388,618
Oct. 1...	2,652		4,375		1,847	8,874	8,622			8,622	7¼	7¾	5½		397,240
" 8...	1,257				582	1,839	10,601			10,601	7¼	7¾	5½		407,841
" 15...	686		941		370	1,997	19,827			19,827	7⅜	7¾	5½		427,668
" 22...	2,177	2,463	1 560	3.570	528	10.298	9,480	1,000		10,480	7⅜	7¾	5½		437,148
" 29...	2.053		724	3,887	471	7,135	25,478			25,478	7¾	8	5½		462,626
Nov. 5...	2,685		474	1,386		4,545	16,965			16.965	7⅛	8¼	5½		479,591
" 12...	1,621		3,170	867	2,753	8 411	7,513			7,513	7¾	8¼	5½		487 104
" 19...	240			2,069	390	2,699	13,234	13,300		26,534	8	8⅛	5½		500,338
" 26...	23			493	1,229	1,745	6,238	1,800		8,038	8	8⅛	5½		506,576
Dec. 3...	201			4,603	103	4,907	12,303	2,000		14,303	8¼	8½	5½		518.879
" 10...	438					438	11,680	1,000		12,680	8¼	8½	5½		530,559
" 17...			845			845	15,373	9,500		24,873	8⅞	9	5¾		545,932
" 24...	1.608		481	1,720	1,275	5,084	8,804			8,804	9	9	5¾		554,736
" * 31...	2,620		1.597	4,358	388	8,963	1,904			1,904	9	9	5¾		556, 640
Average prices & total sales, receipts & stocks.	265,413	13,863	28,170	94,460	55,722	457,628	556,640	46,600	3,600	606,840	7.66	8.27	5.77	3,600	1,050,264

* American and other stocks, 121,000.

The price semi-weekly at New York, rates of Freight to Liverpool and Course of Exchange on London, for the Crop Year ending October 1, 1824

1823.	Price of New Orleans.	Price of Upland.	Exchange on London.	Freights to Liverpool.
Oct. 3	17@19	12@17	106¾@107¼	½d.
" 7	17@19	12@17		
" 10	17@19	12@17	108	
" 14	17@19	12@17		
" 17	15@17	12@16	107¾@108	
" 21	15@17	12@16		
" 24	17½@...	14@17	107½@108	
" 28	17@18½	14@18		
" 31	17@19	14@18	107½@108	
Nov. 4	17@19	14@18		½d.
" 7	17@19	14@18	106½@107½	
" 10	17@19	14@18		
" 14	16@18½	13@18	106¾@107	
" 18	14½@18½	13@17		
" 21	14½@18	13@16½	107@107¼	
" 25	14½@18	13@16		
" 28	14½@18	13@16	107½	
Dec. 2	15@17	13@16		½@⅝d.
" 5	15@17	13@16	108	
" 9	15@17	12½@16		
" 12	15@17	12½@16	108	
" 16	15@17	12½@16		
" 19	15@17	12½@16	108	
" 23	15@19	11½@16		
" 26	15@18½	13@16	107¾@108	
" 30	15@18½	13@16		
1824.				
Jan. 2	15@18½	13@16	107¾@108	½@⅝d.
" 6	15@18	13½@15½		
" 9	15@18	13½@15½	108	
" 13	15@18	13½@15		
" 16	15@17	13½@15	107¾@108	
" 20	15@17	13½@15		
" 23	15@17	13½@15	107¾	
" 27	15@17	13½@15		
" 30	15@17	13½@15	107¾	
Feb. 3	15@17	13½@15		
" 6	15@17	13½@15	107¾	½d.
" 10	15@17	13½@15		
" 13	15@17	13½@15	108	
" 17	15@17	13½@15		
" 20	15@17	13½@15	108	
" 24	15@17	13½@15		
" 27	15@17	13½@15	108¼	
Mch. 2	15@17	13½@15		⅜d.
" 5	15@17	13½@15	108½	
" 9	15@17	13½@15		
" 12	15@17	13½@15	108¾@109	
" 16	15@17	13½@15		
" 19	15@17	13½@15	109	
" 23	15@17	13½@15		
" 26	15@17	13½@15	109	
" 30	15@17	13½@15		
April 2	15@17	13½@15	109	⅜d.

1824.	Price of New Orleans.	Price of Upland.	Exchange on London.	Freights to Liverpool.
April 6	15@17	13½@15		
" 9	15@17	13½@15	109	
" 13	16@17	14@15		
" 16	15½@17	14@15	109	
" 20	15½@17	14@15		
" 23	15½@17½	14@15	108¾	
" 27	16@18½	14½@15½		
" 30	16@18½	14½@16	108¾	
May 4	16@18½	14½@16		⅜@½d.
" 7	16@18½	14½@16	109	
" 11	16@18½	15@16		
" 14	16@18½	15@16	109@109¼	
" 18	16@18½	15@16		
" 21	16@18½	15@16½	109¾@110	
" 25	16@18½	15@16½		
" 28	16@18½	15@16½	110	
June 1	16@18½	15@16½		⅜@½d.
" 4	16@18½	15@16½	110	
" 8	16@18½	15@16½		
" 11	16@18½	15@16½	109¼@109½	
" 15	16@18½	15@16		
" 18	16@18½	15@16	109	
" 22	16@18½	15@16		
" 25	16@18½	15@16	108½	
" 29	16@18½	15@16		
July 2	16@18½	15@16	109	⅜@½d.
" 6	16@18½	15@16		
" 9	16@18½	15@16	109	
" 13	16@18	15@16		
" 16	16@18	15@16	108¾@109	
" 20	16@18	15@16		
" 23	16@18	15@16	108¾	
" 27	16@18	15@16		
" 30	16@17½	14½@15¾	108½	
Aug. 3	16@17½	14@15		⅜@½d.
" 6	16@17½	14@15	108¼	
" 10	16@17½	14@15		
" 13	16@17½	14@15	108¼	
" 17	16@17½	14@15		
" 20	16@17½	14@15	108½	
" 24	16@17½	14@15		
" 27	16@17½	14@15	109	
" 31	16@17½	14@15		
Sept. 3	16@17½	14@15	109@109¼	⅜@½d.
" 7	16@17½	14@15		
" 10	14@17	13@15	109¼	
" 14	14@17	13@15		
" 17	14@17	13@15	109¼@109½	
" 21	14@17	13@15		
" 24	14@17	13@15	109½@109¾	
" 28	14@17	13@15		
Average for year.	16.64	14.75		

Weekly Quotations in the Charleston Market. Year 1824.

Week	Sea Island.	Maine and Santee.	Short Staple.	Week	Sea Island.	Maine and Santee.	Short Staple.	Week	Sea Island.	Maine and Santee.	Short Staple.	Week	Sea Island.	Maine and Santee.	Short Staple.
1	30	22	14½	14	26	22	14½	27	28	23	16	40	26	21	14½
2	30	22	14½	15	26	22	15	28	28	23	15½	41	26	21	14
3	26	19	14	16	26	22	15	29	26	23	15½	42	26	21	14½
4	26	22	14	17	26	22	15½	30	26	23	15¼	43	26	23	14
5	26	22	14	18	26	22	16	31	26	23	15	44	26	23	14
6	26	22	14	19	28	22	16	32	26	23	14	45	26	22	14
7	26	22	14	20	28	23	16	33	26	23	14	46	26	22	14½
8	26	22	14	21	28	23	16	34	26	23	14	47	26	22	14½
9	25	21	14½	22	28	23	16	35	26	23	14	48	26	22	14½
10	25	21	14½	23	30	23	16½	36	26	23	14	49	26	24	15
11	25	21	14½	24	30	23	16½	37	26	23	14	50	28	24	15
12	25	21	14½	25	30	23	16½	38	26	21	14	51	28	24	15½
13	25	21	14½	26	30	23	16½	39	26	21	14	52	28	24	15½

1825.

Previous to this year there were no full and reliable statistics, either commercial or official, of the cotton production and trade.

The cotton worm committed extensive ravages this year. (*See* years 1800, 1804 and 1846.)

Consumption reduced. Cotton costing twenty-five cents in the United States, was sold in Liverpool, after a long holding, so as to return but six cents per lb.

Number of cotton spindles in the United States estimated at 800,000, using 100,000 bales per annum.

Mr. Roberts, of Sharp, Roberts & Co., of Manchester, England, took out a patent for a self-acting mule. (*See* years 1792 and 1830.)

Mr. Henry Houldsworth, cotton spinner, of Manchester, England, this year invented a combination of wheels for attaching the movements of the spindle and bobbin together, so as to regulate the speed of the latter in proportion to that of the former.

Mr. Dyer, of Manchester, England, introduced the "tube-frame" from America, obtaining the first patent this year. (*See* year 1829 and 1833.)

British calicoes, etc., printed at an average duty of 5s. per piece . . 8,140,876 pieces.
Calicoes, etc., exported—average drawback of 5s. per piece . . . . 6,662,368 pieces.
Calicoes, etc., taken for home consumption at an average duty of 5s. per piece . . . . 1,478,508 pieces.

Cotton exported from United States, 176,439,907 lbs., 9,655,278 lbs. being Sea Island; total value, $36,346,649.

Egyptian cotton imported into Great Britain, 111,023 bags. (*See* year 1823.)

The importations of cotton goods into England from other sources than the East Indies were inconsiderable up to this year.

Importation of raw cotton into Ireland this year, 6,768,453 lbs. (*See* years 1801, 1816 and 1817.)

Imports of cotton into Great Britain from foreign countries, 199,272,665 lbs., as follows: United States, 139,908,699; Brazil, 33,180,491; Turkey and Egypt, 18,938,246; miscellaneous, 7,245,-229; from British Possessions, 28,732,626 lbs., as follows: East Indies and Mauritius, 20,294,262; British West Indies, the growth of, 7,413,764, foreign 780,184; miscellaneous, 244,416. Total imports, 228,005,291 lbs. Exports, 18,004,953 lbs; Home Consumption, 202,546,869 lbs.

Eli Whitney, the inventor of the cotton gin (*see* year 1793), died this year on the 8th of January, and is buried in the cemetery of New Haven, Connecticut. His tomb is after the model of Scipio's at Rome, and bears the following inscription:

ELI WHITNEY,

THE INVENTOR OF THE COTTON GIN.

OF USEFUL SCIENCE AND ARTS THE EFFICIENT

PATRON AND IMPROVER.

ON THE SOCIAL RELATIONS OF LIFE A MODEL

OF EXCELLENCE.

WHILE PRIVATE AFFECTION WEEPS AT

HIS TOMB, HIS COUNTRY HONORS

HIS MEMORY.

BORN DECEMBER EIGHTH, **1765.**

DIED JANUARY EIGHTH, **1825.**

The *London Quarterly Review* contained an article this year which said: "At this moment there are upwards of 30,000 looms worked by steam engines" in Manchester, England. (*See* year 1814.)

A model of Arnold's bevel-wheel combination (*see* years 1822, 1823 and 1826) was taken to Manchester, England.

Danforth's tube frame, or "Taunton Speeder," was patented in England by a Mr. Dyer, of Manchester. (*See* year 1824.)

The "self-acting temples" of Ira Draper (*see* year 1816) were first introduced at Waltham, Mass., this year. (*See* years 1805, 1850 and 1855.)

Factory Island, Maine, was purchased by a company, principally from Boston, Mass., for the purpose of erecting a cotton factory, the whole cost being $110,000. (*See* years 1826, 1829 and 1830.)

Self-acting mule, patented by Richard Roberts, March 29th, in England.

The immediate or ostensible cause of the great speculation in cotton during this season was the delay in shipments from this country. On the 1st January, 1825, it was found that the stock of American cotton in Liverpool was reduced to about 100,000 bales, whereas the trade had counted on twice that quantity. Without steam or telegraph to convey information, Liverpool was in the dark, and at once became alarmed for her supply. This excited the speculators in this country. The speculation in cotton was only a part of the general movement, which extended to nearly all departments of business.

Vincent Nolte, in his "Fifty Years in Both Hemispheres," gives some account of the movement and crisis in the Liverpool cotton market, in his usual spicy style. He says, in substance, Liverpool led and this country followed. The price in Liverpool advanced 110 per cent; in this country 85 per cent. The Liverpool cotton holders, led by Messrs. Cropper, Benson & Co, Rathbone, Hodgson & Co., in union with the brokers Cooke &Comer, tacitly united in upholding the price. The first result of this was that buyers of cotton and cotton goods held off. In the meantime the stock increased, and another crop commenced growing. The whole month of May passed over without a single important sale having taken place.

In June, J. & A. Dunstown, of Glasgow, received in Liverpool from New Orleans 5,000 bales which they resolved to sell at best price. The *confederation* implored them to keep up the price, which was 16½d. for Georgia; but in vain. The 5,000 bales were sold at from 2½ to 2¾d. below quotations. This burst the bubble.

Mr. Nolte is never accurate when he is personal. There was no confederation, nor anything like it. The fact is, there was a general infatuation. Probably if money had been as easy with others as with Rathbone, Hodgson & Co. and Cropper, Benson & Co., the price might have been sustained much longer; and certainly it would not have declined so low on the reaction. The elements of the cotton speculation in the spring of 1825 were all either formed or forming in 1823 and 1824. They were sure to combine, either with or without cause. The unexpected reduction in stock was sufficient for the purpose.

COTTON CROP OF THE UNITED STATES.

Statement and Total Amount of the Cotton Crop of the United States, for the Year ending 1st October, 1825.

	Bales.	Bales.	Total.	Same period last year.
NEW ORLEANS.				
Received from—				
Louisiana and Mississippi		124,630		
Tennessee and North Alabama		68,895		
Mobile		7.615		
Across the Lake		4,998		
Arkansas		403		
Florida		226		
Missouri		26		
On hand 1st October, 1824		1,501		
Deduct—		208,294		
Quantity received from Mobile	7,615			
" " Florida	226			
		7,841		
Total at New Orleans which includes Louisiana, Mississippi, Tennessee and North Alabama, and of which there remained on hand 1st October, 3,737 bales			200,453	126,481
FLORIDA.				
The whole crop of West Florida is estimated at 6@7000 bales of which there have been received at—				
New York		2,521		
New Orleans		226		
Philadelphia, about		253		
			3,000	4,500
As there was no shipment to any foreign port during the year, the remainder must have reached New Orleans, and Mobile without being specified, and is included in the crop of those places.				
ALABAMA.				
Exports from Mobile, Foreign and Coastwise	55,919			
On hand at Mobile	37			
		55,956		
Export from Blakely		2,840		
			58,796	44,924
The crop of the North part of the State included in the Export from New Orleans.				
GEORGIA.				
Export Foreign and Coastwise—				
Uplands		129,926		
Sea Islands		7,769		
On hand—probably		305		
			138,000	152,735
SOUTH CAROLINA.				
Export Foreign and Coastwise—				
Uplands	141,074			
Sea Islands	18,253			
On hand—about	673	160,000		
Deduct—				
Amount received from Georgia and included in the Exports from that State	39,500			

Cotton Crop of the United States, continued.

		Bales.	Bales.	Total.	Same period last year.
Received from North Carolina through—					
Wilmington	1,000				
Cherau	12,000				
Camden	8,000				
Columbia	2,500				
		23,500	63,000	97,000	134,518
These items are below the estimates of well informed merchants residing in those places.					
NORTH CAROLINA AND VIRGINIA.					
Received at New York from—					
Wilmington		12,298			
Newbern		5.362			
Washington		2,676			
Plymouth		1,554			
Edenton		1,023			
Murfreesboro'		662			
Swansboro'		306			
			23,881		
Petersburg and Richmond		10,921			
Norfolk and Fredericksburg		431			
			11,352		
Total received at New York		....	35,233		
Received at Cherau, Camden and Columbia and deducted from the South Carolina exports		....	22,500		
Shipped from Wilmington to Charlestown		1,000			
To Europe and elsewhere		2,000			
			3,000		
Shipped from Petersburg to Europe and elsewhere (New York excepted) estimated at 11 to 14,000—say		....	11,267		
				72,000	46,000
Total Crop of the United States		....	....	569,249	509,158
Crop of last year		....	....	509,158	
Increase		....	....	60,091	
Amount received at New York for the year ending 1st October, 1825		....	....	175,629	
Deduct for what was received from the West Indies, and South America, and 13 bales from Calcutta		....	....	629	
Leaves in round numbers		....	....	175,000	
Or nearly one-third of the whole crop of the United States.					

COTTON AT LIVERPOOL. YEAR 1825.

Week Ending.	Receipts.						Sales.				Prices.			Actual Export.	Consumpt'n.
	American.	E. I.	Egypt.	Brazil.	Other.	Total.	Consumpt'n.	Speculation.	Export.	Total.	Mid. Up.	Mid. Orl	Dhol.		
Jan. 7...	2,629		2,401	1,688		6,718	12,789	18,000		30,789	9	10	6		12,789
" 14...	949			1,628	50	2,627	7,915	1,000		8,915	9	10	6		20,704
" 21...	3,363			3,880	869	8,112	5,155			5,155	9	10	6		25,859
" 28...	4,123		1,140			5,263	5,194			5,194	8½	10	6		31,053
Feb. 4...	5,403			2,399	1,170	8,912	15,366			15,366	8½	10	6		46,419
" 11...	6,418		2,000	1,895	510	10,823	12,537	5,000		17,537	9	10½	6½		58 956
" 18...	7,264			5,061		12,325	22,908	15,000		37,908	10	11½	7		81,864
" 25...	1,281		512		1,126	2,919	13,599	26,000		39,599	11½	12½	8		95,463
Mch. 4...	13,826		2 431		978	17,235	14,906	28,000		42,906	11½	12½	8		110,369
" 11...	6,737	1,400	9.963	1,454	49	19,603	17,107			17,107	12	13	8		127,476
" 18...	1,938		1,620		30	3,588	22,869	7,000		29,869	12	13½	8		150,345
" 25...	7,981			701	901	9,583	10,741	24,000		34,741	12½	13½	8½		161,086
" 31...	4,949				1,643	6,592	11,735	31.000		42,735	13¾	14½	9½		172,811
April 8...	5,848		2,400	2,556	67	10,871	12,088	20,000		32,088	14	14½	9½		184,899
" 15...	14,710	2,270	1,324	9,586	2,292	30,182	31,944			31,944	14½	15	10		216,843
" 22...	16,672			1,456	2,719	20,847	85,920	15,000		100,920	16¾	17½	12		302,763
" 29...	6,938	3,168		5.000	4,314	19,360	14,704	16,000		30,704	16¾	16¾	12		317,467
May 6...	15,624	1,575		7,618	3,092	27,909	9,711			9,711	15½	17	12		327,178
" 13...	3,828		1,075	1,930	1,370	8,203	14,341			14,341	15½	17	11½		341,519
" 20...	13,404			3,460	520	17,384	4,611	3,000		7,611	15½	17	11½		346,130
" 27...	15,320		4,017		4,219	23.556	4,400	9,500		13,900	15½	17	11½		350,530
June 3...	7,182			2,451	1,613	11,246	1,920			1,920	15½	17	11½		352,450
" 10...	12,681			9,017	3,257	24,955	1,952			1,952	15	16½	11		354,402
" 17...	5,498		1,355	920	1,866	9.639	4,489			4,489	12	13½	9		358,891
" 24...	9,294		4,203		1,067	14,564	4,680	10,500		15,180	12½	13	9		363,571
July 1...	18,069		2,226	7,648	5,790	33,723	17,954			17,954	12	12½	9		381,525
" 8...	9,546				1,040	10,586	2,241			2,241	10½	11	8		383,766
" 15...	25,616			4,107	7,219	36,942	4,417	5,500		9,917	10½	11½	8		388,183
" 22...	9,385		1.380	3,625	633	15,023	5,173	5,000		10,173	9½	11	8		393,356
" 29...	4,400		5,079	1,483	1,119	12,081	11,503	2,000		13,503	8	8¾	7		404,859
Aug. 5...	15,896		350	6,650	1,362	24,258	7,393	5,000		12,393	8	8¾	7		412,252
" 12...	16,146			5,674	4,637	26,457	13,941	6,000		19,941	8	8¾	7		426,193
" 19...	10,210			2,819	1,403	14,432	8,964	5,000		13,964	8	8¾	7		435,157
" 26...	5,336				922	6,258	9,755		2,000	11,755	7½	8¾	7	2,000	444,912

Sept. 2	9,664	......	3,736	10,968	1,946	26,314	9,707	......	......	9,707	7¼	8½	6½	......	454.619
" 9	3,305	......	......	......	4,111	7,416	7,782	5,000	2,000	14,782	6½	8	6	2,000	462.401
" 16	16,949	......	......	3,019	1,396	21,364	12,513	7,000	3,000	22,513	7	8	6	......	474,914
" 23	9,772	......	3 551	4,026	1,766	19,115	7,458	2,000	2,500	11,958	7	8	6	2,500	482,372
" 30	12,208	......	1,106	2,979	2,909	19,202	8.411	......	2,000	10,411	6⅜	8	6	2,000	490,783
Oct. 7	4,959	2,178	......	6,486	834	14,457	8,136	2,000	4,000	14,136	6¾	8	5¾	4,000	498,919
" 14	8,861	......	......	4,467	2,819	16,147	17,299	2,000	2,500	21,799	7	8	5¾	2,500	516,218
" 21	1,695	......	......	......	1,326	3,021	10,891	10,000	1,500	22,391	7	8½	5¾	1,500	527,109
" 28	......	......	......	......	195	195	11,866	7,000	1,500	20,366	7	8½	5¾	1,500	538,975
Nov. 4	3,161	......	......	4,306	865	8,332	8,229	2,500	500	11,229	7	8½	5¾	500	547,204
" 11	1,309	......	......	1,849	1,141	4,299	8,891	......	1,200	10,091	7	8½	5¾	1,200	556.095
" 18	238	......	......	......	100	338	11,994	2,000	1,500	15,494	7	8½	5¾	1,500	568,089
" 25	418	......	1,151	508	125	2,202	9,250	2,000	1,000	12,250	6¾	8	5¾	1,000	577,339
Dec 2	2,538	......	6,595	1,060	1,388	11,581	9,290	......	2,000	11,290	6⅜	8	5¾	2,000	586,629
" 9	322	......	6.488	279	1,362	8,451	8.453	1,500	......	9,953	6¾	8	5¾	......	595,082
" 16	1,223	......	1,616	2,451	77	5,367	6,897	......	......	6,897	6¾	8	5¾	......	601,979
" 23	1,442	2.109	617	......	631	4,799	10,010	1,500	......	11,510	6½	8	5½	......	611,989
" * 30	5,253	2,420		3,012	1,322	12,007	10,117	......	......	10,117	6½	8	5½	......	622,106
Average prices & total sales, receipts & stocks.	419,490	15,060	71.486	140,057	82,820	728,913	622,106	302,000	27,200	951.136	10 . $\frac{1}{16}$	11 : 15	7 : 63	24.200	11,963:63

* American and other Stocks. 311,000.

The semi-weekly price at New York, Weekly Sales, Receipts, and Exports, rates of Freight to Liverpool and Course of Exchange on London, for the Crop Year ending October 1, 1825.

1824.	Price of New Orleans	Price of Upland.	Sales for week ending	Receipts for week ending	Exports for week ending	Exchange on London.	Freight to Liverpool
October 1..	15@17½	13½@15				110¼@110½	⅜@½d.
" 5..	15@17½	13½@15	700	350			
" 8..	15@17½	14@16				111½	
" 12..	15@17½	14@16	1,100	652	1,695		
" 15..	15@17½	14@16				110¾	
" 19..	15@17½	14@16	300	264	388		
" 22..	15@17½	14@16				110½	
" 26..	15@17½	14@16	350	381	192		
" 29..	15@17½	14@16				109¾	
Novem. 2..	15@17½	14@15½	1,200	631	414		⅜@½d.
" 5..	15@17½	14@15½				109¼@109½	
" 9..	15@17½	13½@15½	400	2,027	954		
" 12..	15@17½	13½@15½				109½	
" 16..	15@17½	13½@15½	1,000	935	127		
" 19..	15@17½	13½@15½				109½	
" 23..	15@17½	13½@15½	700	732	766		
" 26..	15@17½	13@15½				109½	
" 30..	15@17½	13@15½	1,700	3,288	371		
Decem. 3..	15@17½	13@15½				109½	⅜@½d.
" 7..	15@17½	13@15	650	676	709		
" 10..	15@17½	12½@15				109¼@109½	
" 14..	15@17½	12½@15	2,400	5,649	1,785		
" 17..	15@17	12@15				109½	
" 21..	15@17	12@15	3,100	1,928	888		
" 24..	15@17	12@15				109¼@109½	
" 28..	15@17	12½@15	2,000	1,286	661		
1825.							
January 1..	15½@18	12½@15				109¼@109½	½d.
" 3..	15½@18	12½@15	1,200	2,327			
" 7..	16@18	13@15½				110	
" 11..	16@18	13@15½	7,000	4,688	884		
" 14..	16@18	13@15½				110½@111	
" 18..	16@18	13@15½	3,000	3,019	1,184		
" 21..	16@18	13½@16				110¼@110½	
" 25..	16@18	13½@16	4,500	4,704	2,863		
" 28..	16@18	13½@16				109¾	
February 1..	18@20	15½@19	4,600	3,277	1,760		½@⅜d.
" 4..	18@20	15½@19				109	
" 8..	18@20	15½@19	2,600	7,916	2,030		
" 11..	18@20	16@19				108¾	
" 15..	18@21	16@19	10,000	3,850	1,976		
" 18..	18@21	16@19				109	
" 22..	18@21	16@19	5,000	7,822	2,375		
" 25..	18@21	16@19				109	
March 1..	18@21	16@19	4,000	3,092	3,395		⅝@¾d.
" 4..	18@21	16@19				109	
" 8..	18@21	16@19	2,100	7,848	2,127		
" 11..	18@21	16@18½				108¾@109	
" 15..	18@21	16@18½	6,100	12,530	2,679		
" 18..	18@21	16@18½				109@109¼	
" 22..	18@21	16@18½	12,500	5,603	3,301		
" 25..	20@23	17½@20				109@109¼	
" 29..	20@23	17½@21	19,000	1,830	1,134		
April 1..	21@24	17½@22				109¼@109½	¾d.
" 5..	23@25	18½@23	24,000	11,405	3,686		
" 8..	25@30	22@26				109½	
" 12..	25@30	22@26	20,000	4,626	5,853		
" 15..	25@30	22@26				109¼@109½	
" 19..	25@30	22@26	2,500	4,724	2,036		

Year 1825—*Continued.*

1825.		Price of New Orleans	Price of Upland.	Sales for week ending	Receipts for week ending	Exports for week ending	Exchange on London.	Freight to Liverpool
April	22..	25@30	22@26				109	
"	26..	25@30	22@26	14,000	6,089	5,394		
"	29..	27@30	23@27				108¾	
May	3..	27@30	23@27	15,000	5,437	1,640		¾d.
"	6..	27@30	23@27				108	
"	10..	27@30	23@27	4,500	4,690	880		
"	13..	26@30	22@26½				106¾	
"	17..	26@30	23@27	13,000	1,532	2,356		
"	20..	27@31	24½@30				106½@106¾	
"	24..	27@30	23½@27½	8,000	6,130	3,688		
"	27..	25@30	23@27				105½	
"	31..	25@30	23@27	2,000	6,138	8,375		
June	3..	25@30	23@27				104¾@105	¾@⅞d.
"	7..	25@30	23@27	3,700	5,315	8,273		
"	10..	25@30	22@26				105@105¼	
"	14..	25@30	22@26	1,100	3,766	8,199		
"	17..	25@29	21½@25				105@105¼	
"	21..	23@27	20@25	2,000	7,013	6,554		
"	24..	23@27	20@25				105½	
"	28..	23@27	20@25	2,700	4,067	8,224		
July	1..	23@27	20½@25				105¾@106	⅝@¾d.
"	5..	23@27	20½@25	2,000	2,294	9,756		
"	8..	23@26	20@24				105½	
"	12..	23@26	20@24	1,500	1,750	6,417		
"	15..	23@26	20@24				105	
"	19..	23@26	20@24	1,600	1,504	9,675		
"	22..	23@26	19@23				104¾	
"	26..	23@26	19@23	800	2,406	4,632		
"	29..	23@26	19@23				104¾	
August	2..	22@25	18@22	600	2,333	7,476		⅝@¾d.
"	5..	22@25	18@22				104¾@105	
"	9..	22@25	18@22	500	1,452	173		
"	12..	22@25	17@20				105	
"	16..	22@25	17@20	400	1,909	5,038		
"	19..	22@25	17@20				106	
"	23..	20@25	16@20	800	1,179	1,348		
"	26..	20@23	16@18				107½@108	
"	30..	20@23	16@18	1,000	249	3,733		
Septem.	2..	20@23	16@18				107½@107¾	½d.
"	6..	20@23	16@18	1,100	336	799		
"	9..	20@23	16@18				107¾	
"	13..	20@23	16@18	300	86	2,285		
"	16..	20@23	15@18				109½@110	
"	20..	20@23	15@18	600	211	1,837		
"	23..	18@22	14@18				110@111	
"	27 .	18@22	14@18	900	519	772		
"	30..	18@22	13½@17½				110½@111	
Average price and total receipts and exports.		21.82	18.59	221,800	174,465	153,757		

Weekly Quotations in the Charleston Market. *Year* 1825.

Week	Sea Island.	Maine and Santee.	Short Staple.	Week	Sea Island.	Maine and Santee.	Short Staple.	Week	Sea Island.	Maine and Santee.	Short Staple.	Week	Sea Island.	Maine and Santee.	Short Staple.
1	23	24	15	14	50	38	22	27	75	60	24	40	65	40	16½
2	32	26	15½	15	60	40	23	28	75	60	24	41	65	40	16
3	32	26	15½	16	75	50	30	29	75	60	24	42	65	40	15
4	40	29	15½	17	80	55	30	30	75	60	24	43	65	40	13½
5	40	29	15½	18	80	55	28	31	65	40	23	44	65	40	13½
6	40	28	15½	19	80	55	28	32	65	40	20	45	50	30	13¾
7	40	28	18½	20	85	55	30	33	65	40	20	46	50	30	14
8	50	30	19	21	87½	60	30	34	65	40	20	47	50	30	14½
9	50	30	19	22	87½	65	32	35	65	40	20	48	50	30	14
10	50	30	19	23	87¼	65	30	36	65	40	18	49	50	30	14
11	50	31	18½	24	.87½	65	30	37	65	40	18	50	50	30	14½
12	50	31	18½	25	87½	65	29	38	65	40	17	51	55	30	14½
13	50	35	20	26	85	60	26	39	65	40	17	52	55	32	14½

Fluctuations in price are, to a great extent, the result of expansion and contraction. This is always the case when the movement is general.

Trade is always under the influence of one or the other movement.

Contraction always follows great wars. This process followed the battle of Waterloo and continued about seven years. The expansion in 1823, 1824 and 1825 was in exact proportion to the previous contraction.

The cotton speculation commenced in December, 1824, and the revulsion set in during June, 1825.

1826.

Kinsey Burden (*see* years 1805, 1827 and 1828), of South Carolina, raised a crop of sixty bags Sea Island, which brought $1.10 per lb.

Calico printing first introduced in the United States, in Dover, N. H., and Taunton, Mass., under the supervision of Mr. John D. Prince, who came from Manchester. (*See* year 1855.)

A steam-loom weaver, about 15 years of age, attending two looms, could now (*see* year 1823 for contract) weave twelve pieces 9-8 shirtings per week (*see* year 1823); some could weave fifteen pieces. (*See* year 1833.)

British calicoes, etc., printed at an average duty of 5s. per piece. . 6,098,656 pieces.
Calicoes, etc., exported—average drawback of 5s. per piece. . . . . . 4,082,684 pieces.
Calicoes, etc., taken for home consumption at an average duty of
5s. per piece. . . . . 2,015,972 pieces.

Value of foreign cotton goods imported into England, £110,365.

Imports of cotton into Great Britain from foreign countries, 151,516,848 lbs., as follows: United States, 130,858,203; Brazil, 9,871,092; Turkey and Egypt, 10,032,400; miscellaneous, 755,153. From British Possessions, 26,090,553 lbs., as follows: East Indies and Mauritius, 21,187,900; British West Indies, the growth of, 4,510,302, foreign, 240,768; miscellaneous, 151,583. Total imports, 177,607,401 lbs; Exports, 24,474,920 lbs.; Home consumption, 162,889,012 lbs.

French cotton manufactures had advanced since 1812 at a rate of 310 per cent.

Henry Holdsworth, Jr., of England, secured a patent in that country for what he called a "*differential or equation box*," a combination of the same kind and for the same purpose as Arnold's patent bevel-wheel combination (*see* years 1822, 1823 and 1825), and the idea was undoubtedly taken from a model of Arnold's machine. (*See* year 1825.) Dr. Ure, in his "Cotton Manufacture," says of this machine: "It may be considered the most ingeniously combined apparatus in the whole range of productive industry." Mr. Arnold was hardly known, even by the public of his own country, as the inventor of this machine, and acquired neither wealth or fame by an improvement which has been of immense advantage to manufacturers both at home and abroad.

The Factory Island (Me.) mill was erected this year, 210 feet in length, 47 feet broad and seven stories high, calculated to operate 12,000 spindles and 300 looms.

With the national financial collapse of this year came that of the "bobbin-net prosperity," as it had been miscalled. Many lost their means, and fell into hopeless poverty; others died, or went into self-imposed exile. (Since 1809 the increase in this line of the cotton trade had deluded many into rash speculations and the mania came to be known as the "bobbin-net fever.") Another "mart" (*see* year 1819) proposed this year, did not come into operation. (*See* year 1828.)

COTTON CROP OF THE UNITED STATES.

Statement and Total Amount of the Cotton Crop of the United States, for the Year ending 1*st October,* 1826.

	Bales.	Bales.	Total.	Same period 1825.
NEW ORLEANS.				
Received from—				
Louisiana and Mississippi	....	143,124	....	124,630
Tennessee and North Alabama	....	96,574	....	68,895
Mobile	....	2,685	....	7,615
Across the Lake	....	7,512	....	4,998
Arkansas	....	1,002	....	403
Florida	....	1,076	....	226
Missouri	....	10	....	26
On hand 1st October, 1825	....	3,737	....	1,501
Deduct—		255,720		208,294
Received from Mobile	2,685			
Received from Florida	1,076			
Which is included in the crops of those places		3,761		7,841
			251,959	200,453
FLORIDA.				
Received at—				
New Orleans	....	1,076		
New York	....	1,141		
On hand at Pensacola	....	600		
			2,817	3,000
The whole crop of West Florida is estimated at 7 to 8,000 bales. The remainder, therefore, must have found its way to the neighboring places, and is included in their exports.				
ALABAMA.				
Exports from Mobile—				
To foreign ports	38,495			
Coastwise	34,857			
On hand at Mobile	423			
		73,775		
Export from Blakely	....	426		
			74,201	58,796
The crop of the north part of the State is included in the export from New Orleans.				
GEORGIA.				
Exports, Foreign and Coastwise—				
Uplands	....	184,238		
Sea Islands	....	6,354		
On hand in Savannah and Augusta—not known	....	....	....	....
			190,592	138,000
It is understood that a considerable quantity remains in the interior in the hands of the planters.				
SOUTH CAROLINA.				
Exports, Foreign and Coastwise—				
Uplands	164,543			
Sea Islands	12,647			
On hand		177,190		
Deduct—				
Amount received from Georgia, and included in the exports of that State	44,412			

Statement and Total Amount of the Cotton Crop of the United States, for the Year ending 1*st October,* 1826—*Continued.*

	Bales.	Bales.	Total.	Same period 1825.
Received from North Carolina, through—				
Wilmington......about...	800			
Cheraw...... " ...	10,000			
Camden and Columbia...... " ...	10,000			
		65,212	111,978	97,000
No positive data could be obtained in regard to the quantities from North Carolina, but the estimates are believed to be within bounds.				
NORTH CAROLINA AND VIRGINIA.				
Received at New York, from—				
Wilmington......	9,562			
Newbern......	5,107			
Washington......	3,578			
Plymouth......	2,578			
Edenton......	2,086			
Murfresborough......	775			
Swansborough......	416			
Windsor......	532			
		24,634		
Received at Cheraw, Camden and Columbia, and deducted from the South Carolina export......	....	20,000		
Shipped from Wilmington—to Charleston......	800			
do do to Europe and elsewhere..	2,000			
		2,800		
Export from Petersburgh—				
To foreign ports......	21,000			
Coastwise......	14,446			
On hand in Petersburgh......	2,100			
		37,546		
Export from Richmond, Norfolk, &c......	....	3,500		
			88,480	72,000
Total crop of the United States......	....	....	720,027	569,249
Crop of last year......	....	....	569,249	
Increase......	....	....	150,778	

The very great transition from the high prices of 1825, to the comparatively low rates of 1826, it is presumed, tended to keep the article back; and the quantity withheld from the market, in the interior, and in the hands of the planters, in some of the cotton-growing States, was probably much larger than it was last year—but as we have no satisfactory data on which to found an estimate, readers are left to their own conclusions.

The price semi-weekly at New York, Weekly Sales, Receipts and Exports, Rates of Freight to Liverpool and Course of Exchange on London, for the Crop Year ending October 1, 1826.

1825.	Price of New Orleans	Price of Upland.	Sales for week ending	Receipts week ending	Exports week ending	Exchange on London.	Freight to Liverpool.
October 4..	18@22	13½@17½	700	53	934	110½@111	⅜@½d.
" 7..	18@21	13½@16½					
" 11..	17@21	13 @16	600	45	112	109½@110	
" 14..	17@20	13 @16					
" 18..	17@20	12½@16	500	361	1,134	109¼@109½	
" 21..	17@20	12½@16					
" 25..	17@20	12½@16	600	1,085	1,140	109¼@....	
" 28..	..@..	11@15					
Novem. 1..	..@..	11@15	1,200	1,220	1,448	109½@....	⅜@½d.
" 4..	..@..	11@15					
" 8..	..@..	11@15	800	804	574	109½@....	
" 11..	..@..	..@15					
" 15..	..@..	15@..	500	653	271	109@109¼	
" 18..	..@..	15@16					
" 22..	..@..	15@16	1,300	1,341	637	109@....	
" 26..	..@..	15@16					
" 29..	..@..	15@16	1,000	879	526	108@108½	
Decem. 2..	..@..	15@16					⅜@½d.
" 6..	..@..	15@16	500	653	934	108@108½	
" 10..	18@..	15@16					
" 13..	18@..	14½@15	1,500	3,284	645	108@108¼	
" 16..	18@..	14½@15					
" 20..	18@..	14@14¾	1,900	2,427	1,212	108½@....	
" 23..	18@..	14@14½					
" 27..	18@..	14@14½	1,200	2,225	687	108@108½	
" 30..	17@18	13½@14½					
1826.							
January 3..	17@18	13½@14½	900	539	2,792	108@108½	⅜@½d.
" 6..	17@18.	13½@14½					
" 10..	17@18	13½@14½	1,200	2,047	1,351	108@....	
" 13..	17@..	12½@13½					
" 17..	17@..	12½@13½	2,300	2,723	1,645	108@108¼	
" 20..	17@18	13@13½					
" 24..	16@18	13@14	1,600	1,345	1,626	108@108½	
" 27..	16@18	13½@14					
" 31..	15@17	13½@14	1,400	2,158	1,831	108¼@....	
February 3..	15@17	13½@14					½@⅝d.
" 7..	15@17	13½@14	1,300	2 317	1,083	108@108¼	
" 10..	15@17	13½@14					
" 14..	15@17	13@14	1,900	3,060	2.965	108@108½	
" 17..	14@16	12½@13½					
" 21..	14@16	12½@13½	1,600	710	842	108@108½	
" 24..	14@16	12¼@13					
" 28..	14@16	12@13	1,200	1,983	1,129	108@108½	
March 3..	14@16	12@13					..@½d.
" 7..	14@16	12@13	600	234	344	108@108½	
" 10..	14@16	12@13					
" 14..	14@16	12@13	2,100	3,222	1,919	108@108½	
" 17..	14@15	12@13					
" 21..	14@15	12@13	700	1,475	1,282	107¾@108	
" 24..	13½@15	11¼@12½					
" 28..	13½@15	11@12½	1,000	7,039	386	107½@107¾	
" 31..	13½@15	11@12					
April 4..	12½@15	10½@11½	3,000	5,408	1,049	107½@....	½@⅜d.
" 7..	12½@15	10½@11½					
" 11..	12½@15	10½@11½	4,200	3,951	673	109@109½	
" 14..	12½@15	10½@12					
" 18..	12½@15	10½@12	4,300	6,578	1,412	109¾@ ...	
" 21..	12½@15	10½@12					
" 25..	12½@15	10½@12	2,500	5,246	1,269	109¾@110	
" 28..	12@15	10@11½					

New York Statement for Year 1826—*Continued.*

1826.		Price of New Orleans	Price of Upland.	Sales for week ending	Receipts week ending	Exports week ending	Exchange on London.	Freight to Liverpool.
May	2..	12@15	10@11½	1,600	3,787	6,339	109½@110	⅝@¾d.
"	5..	12@15	10@11½					
"	9..	12@15	10@11½	2,400	4,511	5,234	109½@109¾	
"	12..	12@15	10@11½					
"	16..	12@14	10@11	2,100	2,230	1,691	109@....	
"	19..	11@14	10@11					
"	23..	11@14	9½@10½	1,800	7,444	1,729	108¾@109	
"	26..	11@14	9½@10½					
"	30..	11@13	9½@10½	3,800	5,766	5,618	109¼@109½	
June	2..	11@13	9½@10½					⅝d.@.
"	6..	11@13	9½@10½	1,900	2,844	3,176	109@109½	
"	9..	10½@12½	9@10½					
"	13..	10½@12½	9@10½	2,600	2,054	3,379	109@109½	
"	16..	10½@12½	9½@11					
"	20..	10½@12½	9½@11	2,400	3,346	1,335	109½@111	
"	23..	10½@13	10@11					
"	27..	10½@13	10@11	2,900	3,671	927	109½@111	
"	30..	10½@14	10@11					
July	4..	10½@14	10@11½	1,800	5,490	1,372	109½@111	½@⅝d.
"	7..	11½@14	10@12					
"	11..	11½@14	10@12	3,600	5.486	1,637	110@111	
"	14..	11½@14	10@12					
"	18..	11½@14	10@12	1,100	2,397	2,857	110@111	
"	21..	11@14	10@12					
"	25..	11@14	10@12	1,900	3,165	3,266	110@111	
"	28..	11@14	10@12					
August	1..	11@14	10@12	1,800	2,498	6,979	110@110½	½d.@.
"	4..	11@14	10@12					
"	8..	11@14	10@11½	3,400	2,057	3,767	110@110½	
"	11..	11@14	10@11½					
"	15..	11@14	10@11½	1,600	2,701	2,388	110@110½	
"	18..	11@14	10@11½					
"	22..	11@14	10@11½	1,000	640	2,378	110@110½	
"	25..	11@14	9½@11					
"	29..	11@14	9½@11	1,600	3,409	2,119	110¾@112	
Septem.	1..	11@14	9½@11					½@⅝d.
"	5..	11@14	9½@11	1,500	1,419	2,515	111½@113	
"	8..	11@14	9½@11					
"	12..	11@14	9½@11	2,300	165	1,916	112@113	
"	15..	11@14	9½@11					
"	19..	11@14	9½@11	1,300	897	1,824	112@113	
"	22..	11@14	9½@11					
"	26..	11@14	9½@11	2,000	1,320	951	112¼@112½	
"	29..	11@14	9½@11					
Average prices and total sales, receipts and exports.		14.41	12.19	90,500	132,362	97,249		

COTTON AT LIVERPOOL. YEAR 1826.

Week Ending.	Receipts.						Sales.				Prices.			Actual Export.	Consumpt'n.
	American.	E. I.	Egypt.	Brazil.	Other.	Total.	Consumpt'n.	Speculation.	Export	Total.	Mid. Up	Mid Orl	Dhol.		
Jan. 6...	718	......	......	....	111	829	7,681	....	......	7,681	6¼	7½	5½	....	7,681
" 13...	66	......	......	....	144	210	7,974	....	......	7,974	6	7½	5½	....	15,655
" 20...	7,942	......	620	....	1,435	9,997	6,100	1,500	......	7,600	6	7	5½	....	21,755
" 27...	4,982	300	115	1,208	50	6,655	10,818	2,500	700	14,018	5½	6½	5½	700	32,573
Feb. 3...	5,014	......	2,068	2,959	25	10,066	11,115	3,000	500	14,615	5¾	7	5	500	43,688
" 10...	4,583	......	2,226	....	1,030	7,839	6,068	7,000	500	13,568	5½	7	5	500	49,756
" 17...	7,773	......	1,574	2,092	58	11,497	6,006	1,000	1,000	8,006	5½	7	5	1,000	55,762
" 24...	10,661	......	781	3,002	912	15,356	8,954	1,000	800	10,754	5½	7	4¾	800	64,656
Mch. 3...	9,293	2,110	......	....	160	11,563	8,128	2,000	2,000	12,128	5¾	7	4¾	2,000	72.784
" 10...	4,070	......	4,477	707	....	9,254	11,325	3,000	1,000	15.325	5¾	7	4¾	1,000	84,109
" 17...	....	......	......	....	64	64	10,368	2,500	1,000	13,868	5¾	7	4¾	1,000	94,477
" 24...	3,902	......	......	....	....	3,962	6,028	....	......	6,028	5¾	7	4¾	....	100,505
" 31...	7,752	......	895	1,274	216	10,137	7,214	1,000	......	8,214	5¾	6¾	4¼	....	107,719
April 7...	11,331	......	......	....	39	11,370	7,180	2,000	......	9,180	5¾	6½	4½	....	114,899
" 14...	12,256	2,820	......	2,825	1,239	19,140	6,735	1,500	700	8,935	5¾	6½	4½	700	121,634
" 21...	11,266	......	......	....	3,334	14,600	5,961	800	1,500	8,261	5¾	6¼	4½	1,500	127,595
" 28...	14,287	......	2,344	2,685	103	19,419	4,999	1,000	1,000	6,999	5¼	6	4½	1,000	132,594
May 5...	4,446	......	......	....	....	4,446	8,362	2,500	500	11,362	5½	6	4½	500	140,956
" 12...	6,013	......	......	....	613	6,626	8,540	2,000	......	10,540	5½	6	4½	....	149,496
" 19...	10,282	......	......	1,552	44	11,878	13,923	4,000	......	17,923	5½	6	4½	....	163,419
" 26...	12,254	......	......	....	1,715	13,969	11,463	4,000	......	15,463	5¾	6¼	4½	....	174,382
June 2...	15,093	......	......	368	....	15,537	9,761	3,000	......	12,761	6	6¼	4½	....	184,643
" 9...	7,052	1,403	2,322	2,744	228	13,749	5,328	....	......	5,328	6	6¼	4½	....	189,971
" 16...	18,002	......	......	....	421	18,423	7,335	....	......	7,335	6	6¼	4½	....	197,306
" 23...	9,381	2,950	......	1,372	427	14,130	8,987	....	......	8,987	6	6¼	4½	....	206,293
" 30...	1,007	......	......	....	....	1,007	7,225	500	400	8,125	6	6¼	4½	400	213.518
July 7...	19,785	......	......	....	2,401	22,186	6,333	....	1,500	7,833	5¾	6¼	4½	1,500	219,851
" 14...	12,053	......	2,350	....	327	14,730	6,766	....	......	6,766	5½	6¼	4½	....	226,617
" 21...	25,975	......	......	....	323	26,298	5,212	....	1,000	6,212	5½	6	4½	1,000	231,829
" 28...	6,833	......	4,158	....	72	11,063	10,071	2,000	1,000	13,071	5½	6	4½	1,000	241,900
Aug. 4...	4,365	......	6,005	1,913	....	12,283	13,732	....	500	14,232	5½	6	4½	500	255,632
" 11...	2,614	......	......	....	384	2,998	9,514	1,000	800	11,314	5½	6	4½	800	265,146
" 18...	23,632	......	......	....	2,381	26,013	12,379	1,500	700	14,579	5½	6	4½	700	277.525
" 25...	6,723	......	......	2,160	2,013	10,896	9,843	2,000	1,000	12,843	5½	6	4½	1,000	287,368

Sept. 1	7,034					7,034	14,094	1,000	1,000	16,094	5½	6	4½	1,000	301,462
" 8	2.425				296	2,721	20,621	2,000		22.621	5½	6	4½		322,083
" 15	18,769			3,338	3,484	25,591	11,326	1,000		12,326	6	6¼	4½		333,409
" 22							11 552	3,000	600	15,152	6	6¼	4½	600	344,961
" 29	3,052			2,965		6,017	15,257	1,000	500	25,757	6¼	6½	4¾	500	360,218
Oct. 6	5,373				580	5,952	9,214	2,000		11.214	6¼	6½	4¾		369,432
" 13	1,580	710	1,600		1,560	5,450	7,608			7,608	6¼	6½	4¾		377.040
" 20	3,156			641	215	4,012	10,273		1,000	11,273	6¼	6½	4¾	1,000	387,313
" 27	3,875				852	4,727	7,651		650	8,301	6⅛	6½	4¾	650	394,964
Nov. 3	450				1,590	2,040	7,456			7,456	6⅛	6½	4¾		402,420
" 10	622					622	6,653		500	7,153	6⅛	6½	4¾	500	409,073
" 17	3,186	641			918	4,745	8.620	500		9,120	6⅛	6½	4¾		417,693
" 24	526					526	10,496	500		10,996	6⅛	6½	4¾		428,189
Dec. 1	1,549				1,031	2,580	11,379			11,379	6⅛	6¼	4¾		439,568
" 8	569		4,486	772		5.827	17,499			17,499	6¼	6½	4¾		457,067
" 15	1,360				2,828	4,188	19,382		600	19,982	6⅜	6¾	5	600	474,449
" 22	3,691				1,659	5,350	10,033			10,033	6⅜	6¾	5		484,482
" 29	4,420	639	750	1,188	28	7,025	11,106	1,000		12,106	6¾	7	5		495,588
Average prices & total sales, receipts & stocks.	371,143	11,573	36,767	35,765	35,366	490,614	495,588	64,300	22,950	582,838	5.85	6.48	4.71	22,950	* 9,530.53

* American and other stocks, 237,800.

LIVERPOOL STATEMENT FOR 1826.

UNITED STATES, 1825–1826.

Stock, October 1, 1825	55,000	Export	598,000
Crop, 1825–'26	710,000	Consumption	100,000
		Stock, October 1, 1826	67,000
Bags	765,000	Bags	765,000

	Liverpool.	Gt. Britain.	France.	Continent.	Tot. Europe.
United States	126,900	131,500	9,583	8,410	149,493
Brazil	96,200	108,740	4,532	800	114,072
West Indies	8,400	12,630	3,855	4,332	20,817
East Indies	16,500	80,450	688	8,043	89,181
Egypt	63,000	82,680	16,648	4,947	104,275
Stock	311,000	416,000	35,306	26,532	477,838
IMPORT.					
United States	369,337	393,914	216,320	39,722	649,956
Brazil	53,128	56,402	17,586	17,553	91,541
West Indies	11,810	17,519	11,440	15,522	44,481
East Indies	13,914	64,662	1,892	37,154	103,708
Egypt	39,343	47,682	72,936	23,381	143,999
Consumption	487,532	58,0179	320,174	133,332	1,033,685
Stock supply	789,532	996,179	355,480	159,864	1,511,523
	U. States.	Brazil & W.I.	East India.	Egypt.	
Export deducted from Import	23,900	26,800	41,000	2,000	93,700
Total supply (bags)					1,417,823

CONSUMPTION.

Tot. Europe.	Continent.	France.	Gt. Britain.	Liverpool.
605,968	41,056	182,813	382,099	362,872
108,001	16,717	16,137	75,147	73,913
42,202	14,689	10,908	16,605	12,129
70,270	37,693	2,580	29,997	19,379
144,066	19,272	68,563	56,231	50,239
970,507	129,427	281,001	560,079	518,532
447,316	30,437	74,479	342,400	238,000
			93,700	42,000
1,417,823	159,864	355,480	996,179	798,532

The high price existing when the crop of 1825 was planted caused an extension of acreage, and a largely increased crop. It will hereafter be seen that this is always the case, though many Southern people say that they always plant all they can, the price making little or no difference. Owing to the low price prevailing during the season of 1825–6, planters held back a portion of their crop, so that the actual increase over the previous crop was probably not far from 40 per cent. This was equal to the increase in the production of 1870.

The course of the market shows that speculation was entirely suspended, and the market was consequently guided by legitimate influences alone. It is always so after a great speculative movement. It may be counted upon with perfect safety.

1827.

Kinsey Burden (*see* years 1805, 1826 and 1828), of South Carolina, got $1.25 per lb. for his crop.

Samuel Crompton, inventor of the "mule jinny" (*see* years 1779, 1793 and 1812), died on the 26th of January, this year, in his cottage, King street, Bolton, England.

British calicoes, etc., printed at an average duty of 5s. per piece . . 8,089,028 pieces.
Calicoes, etc., exported—average drawback of 5s. per piece . . 5,440,272 pieces.
Calicoes, etc., taken for home consumption at an average duty of 5s. per piece . . . . 2,648,756 pieces.

Value of foreign cotton goods imported into England, £115,026.

Imports of cotton into Great Britain from foreign countries, 243,992,426 lbs., as follows: United States, 216,924,812; Brazil, 20,716,162; Turkey and Egypt, 5,071,519; miscellaneous, 1,279,873. From British Possessions, 28,456,483 lbs., as follows: East Indies and Mauritius, 20,984,916; British West Indies—the growth of—6,227,172; foreign, 938,709; miscellaneous, 305,686. Total imports, 272,448,909 lbs.; Exports, 18,134,170 lbs.; Home consumption, 249,804,396 lbs.

Paterson, N. J. (*see* year 1794), had 6,236 inhabitants; 1,046 heads of families; 7 houses for public worship; 17 schools; a philosophical society; 15 cotton factories, operating 24,000 spindles; 2 canvas factories, operating 1,644 spindles, and employing 1,453 persons; 6,000 bales of cotton were consumed annually, and 1,630,000 lbs. of cotton yarn, manufacturing 3,354,500 yards of cotton cloth.

COTTON CROP OF THE UNITED STATES.

Statement and Total Amount of the Cotton Crop of the United States, for the Year ending 30*th September,* 1827.

	Bales.	Bales.	Total.	Same period 1826.
NEW ORLEANS.				
Received from—				
Louisiana and Mississippi		170,295		143,124
Tennessee and North Alabama		152,166		96,574
Across the Lake		9,627		7,512
Mobile		2,613		2,685
Arkansas		1,739		1,002
Florida		1,481		1,076
Illinois		10		....
Missouri		3		10
On hand 1st October, 1826		3,030		3,737
Deduct—		340,964		255,720
Received from Mobile	2,613			
" Florida	1,481			
Included in the crops of those places		4,094		3,761
FLORIDA.			336,870	251,959
Received at—				
New Orleans		1,481		
New York		1,736		
Philadelphia		256		
Boston		190		
Shipped to France		500		
			4,163	2,817
The whole crop of Florida is estimated at about 10,000 bales, including 500 bales of Sea Island. The remainder must have found its way to the neighboring places, and is included in their exports.				
ALABAMA.				
Exported from Mobile—				
To Foreign Ports		42,247		
Coastwise		47,050		
On hand at Mobile		410		
Export from Blakeley		....		
			69,707	74,201
The crop of the North part of the State is included in the receipts at New Orleans.				
GEORGIA.				
Export, Foreign and Coastwise—				
Uplands		219,254		
Sea Islands		14,666		
			233,920	190,592
SOUTH CAROLINA.				
Export, Foreign and Coastwise—				
Uplands	199,175			
Sea Islands	31,828			
Received at New York from Georgetown	3,307			
		234,310		
Deduct—				
Received from Georgia, and included in the exports from Savannah	29,000			
Received from North Carolina through—				
Wilmington	500			
Cheraw—about	9,000			
Camden "	11,000			
Columbia "	5,000			
		54,500		
			179,810	111,978

Statement and Total Amount of the Cotton Crop of the United States, for the Year ending 30th September, 1827.—Continued.

	Bales.	Bales.	Total.	Same period 1826.
NORTH CAROLINA AND VIRGINIA.				
Received at New York from—				
Wilmington	19,502			
Newbern	8,591			
Washington	5,376			
Plymouth	2,562			
Edenton	2,176			
Murfreesboro'	1,564			
Windsor	757			
Swansboro'	460			
Elizabeth City	146			
Ocracock	308			
		41,442		
Received at Philadelphia from North Carolina	....	1,223		
Received at Boston from North Carolina	....	219		
Received at Cheraw, Camden, and Columbia, and deducted from the South Carolina export	....	25,000		
Shipped from Wilmington to Charleston	....	500		
Shipped from Wilmington to Europe	....	2,200		
Export from Petersburgh—				
To Foreign ports	20,633			
Coastwise	16,094			
On hand in Petersburgh	500			
		37,227		
Export from Richmond, Norfolk, &c.	....	5,000		
			112,811	88,480
Total crop of the United States	....	....	957,281	720,027
Crop of last year	....	....	720,027	
Increase	....	....	237,254	Bales.

Export to Foreign Ports, from 1st October, 1826, to 30th September, 1827.

FROM	To Great Britain.	To France.	Other parts of Europe.	Total.
New Orleans	193,539	60,101	9,279	262,919
Florida	....	500	....	500
Alabama	35,690	5,717	840	42,247
Georgia	100,524	19,034	4,220	123,778
South Carolina	138,186	32,600	13,927	184,713
North Carolina	2,200	....	....	2,200
Virginia—about	17,000	....	5,000	22,000
New York—whole export 198,441 bales—of which, it is estimated, there went to Great Britain	148,000	....	....	198,441
Philadelphia	10,000	....	....	15,000
Boston	1,000	....	....	2,000
Total	646,139	117,952	33,266	853,798

Growth.

The increasing cultivation of this all important staple may be seen by reference to our Annual Statement for three years past.

Total crop of 1824–5, 569,249 bales.
do 1825–6, 720,027 do.
do 1826–7, 957,281 do.

Consumption.

The quantity manufactured in the United States has been variously estimated at 100 to 150,000 bales per annum. It is matter of regret that no positive data can be found on which to rest these estimates.

If we take the total crop of the United States, for the year ending 30th September last	957,281 bales.
Deduct therefrom the export to foreign ports for the same period	853,798 do.
And assume that the stock on hand at the close of each year was the same —it will result that the consumption, for the year ending 30th September last, was	103,483 do.

LIVERPOOL STATEMENT FOR 1827

UNITED STATES, 1826–27.

Stock 1st October, 1826	67,000	Export	853,000
Crop 1826–27	937,000	Consumption	103,000
		Stock 1st Oct., 1827	48,000
Bags	1,004,000	Bags	1,004,000

CONSUMPTION.

Tot. Europe.	Continent.	France.	Gt. Britain.	Liverpool.
708,930	39,112	167,578	502,240	466,178
132,893	20,627	20,363	91,909	90,819
61,896	20,102	13,089	28,705	20,516
62, 24	31,150	384	30,990	9,691
163,557	25,286	78,279	59,992	43,059
1,129,800	136,277	279,693	713,830	630,263
562,360	24,757	85,403	452,200	342,700
....	....	....	70,000	21,000
1,692,160	161,034	365,096	1,236,030	993,963

	Liverpool.	Gt. Britain.	France	Continent.	Tot. Europe.
United States	113,400	118,715	43,090	7,076	168,881
Brazil	56,800	64,655	5,981	1,636	72,272
West Indies	6,680	10,790	4,387	5,165	20,342
East Indies	9,920	74,665	....	7,504	82,169
Egypt	51,200	73,575	21,021	9,056	103,652
Stock	238,000	342,400	**74,479**	**30,437**	**447,316**
IMPORT.					
United States	592,163	648,075	168,819	40,496	857,390
Brazil	115,704	118,648	21,297	20,077	160,022
West Indies	20,485	29,812	14,049	17,551	61,412
East Indies	13,656	73,728	384	33,496	107,608
Egypt	13,955	23,367	86,068	18,977	128,412
Consumption	755,963	893,630	290,617	130,597	1,314,844
Stock supply	993,963	1,236,030	365,096	**161,034**	1,762,160
	U. States.	**Braz. & W.I.**	**East India.**	**Egypt.**	
Export deducted from Imports	8,279	14,726	**44,471**	2,524	**70,000**
Total supply of bags	....	....	....	....	**1,692,160**

The Price semi-weekly at New York, weekly Sales, Receipts and Exports, Rates of Freight to Liverpool and Course of Exchange on London, for the Crop Year ending October 1, 1827.

1826.	Price of New Orleans	Price of Upland.	Sales for week.	Receipts week.	Exports for week.	Exchange on London.	Freight to Liverpool.
October 3..	11@14	9½@11	3,000	988	....	112	½@⅝d.
" 6..	11@14	9½@11					
" 10..	11@14	9½@11	700	1,718	4,550	112@112¼	
" 13..	11@14	9½@11					
" 17..	11½@14	10@11	1,000	324	1,032	111¾@112	
" 20..	11½@14	10@11					
" 24..	11½@14	10@11½	2,500	2,360	2,688	111½@112	
" 27..	11½@14	10@11½					
" 31..	11½@14	10@11½	1,700	2,341	835	111@111¾	
Novem. 3..	11½@14	10@11½					⅜@½d.
" 7..	12@14	10½@11¾	2,800	1,596	1,163	111¼@111½	
" 10..	12@14	10½@11¾					
" 14..	12@14	10½@11¾	2,500	2,536	4,053	111¼@111½	
" 17..	12@14	10½@11¾					
" 21..	12@14	10½@11¾	2,000	4,089	1,789	111½@111¾	
" 24..	12@14	10½@11¾					
" 28..	12@14	10½@11¾	1,400	1,198	3,661	111½@111¾	
Decem. 1..	12@14	10½@11¾					⅜@½d.
" 5..	12@14	10½@11¾	2,500	6,623	3,739	111½@111¾	
" 8..	12@14	10½@11¾					
" 12..	11½@14	10@11	1,500	6,386	3,058	111½@112	
" 15..	11@14	10@11					
" 19..	11@14	10@11	3,500	7,091	1,905	111½@112	
" 22..	11@13½	9¾@10½					
" 26..	11@13½	9¾@10½	2,600	5,979	2,433	111¾@112	
" 29..	11@13	9¾@10½					
1827.							
January 2..	11@13	9¾@10½	3,700	1,066	6,046	111¾@112	½@⅝d.
" 5..	11@13	9¾@10½					
" 9..	10½@ 3	9½@10½	2,000	3,649	4,358	111¾@112	
" 12..	11@13	10@10¾					
" 16.	11@13	10@10¾	3,200	7,254	4,578	111	
" 19..	10½@13	9½@11					
" 23..	10½@13	9½@11	3,500	2,241	2,833	110¾@111¼	
" 26..	10½@13	9¾@10¾					
" 30..	10½@13	9¾@10¾	1,800	10,184	3,334	110¾	
February 2..	10½@13	9¾@10¾					⅝@¾d.
" 6..	10½@13	9¾@10¾	2,100	3,725	4,261	110@110¼	
" 9..	10½@13	9¾@10¾					
" 13..	10½@13	9¾@10¾	2,600	5,765	1,163	110	
" 16..	10½@13	9¾@10½					
" 20..	10½@13	9¾@10½	3,300	6,895	4,420	110	
" 23..	10½@13	9¾@10½					
" 27..	10½@13	9¾@10½	3,000	9,132	6,660	110@110¼	
March 2..	10¼@12½	9½@10½					½@¾d.
" 6..	10¼@12½	9½@10½	8,000	4,684	4,691	110	
" 9..	10¼@12½	9½@10½					
" 13..	10¼@12½	9½@10½	4,200	9,383	6,118	110@110¼	
" 16..	10¼@12½	9¼@10½					
" 20..	10¼@12½	9¼@10½	3,200	7,498	3,605	110½@110¾	
" 23..	10@12½	8¾@10					
" 27..	10@12½	8¾@10	6,000	12,380	8,686	110½@110¾	
" 30..	10@12½	8¾@10					
April 3..	10@12½	8¾@10	6,600	6,880	6,359	110@110½	¾@⅞d.
" 6..	10@12½	8¾@10					
" 10..	10@12	8¾@10	4,200	7,010	7,750	110@110¼	
" 13..	10@12	8¾@10					
" 17..	10@12	8¾@10	4,000	9,598	6,624	110	
" 20..	10@12	8¾@10					

New York Statement for Year 1827—*Continued.*

1827.		Price of New Orleans	Price of Upland.	Sales for week.	Receipts for week.	Exports for week.	Exchange on London.	Freight to Liverpool.
April	24..	10@11½	8¾@10	4,300	5,999	6,415	110	
"	27..	10@11½	9@10					
May	1..	10@11½	9@10	3,300	3,971	4,338	110	¾@⅞d.
"	4..	9½@11½	9@10					
"	8..	9½@11½	9@10	4,000	6,221	8,597	110½	
"	11..	10@12	9@10					
"	15	10@12½	9@10	5,400	5,472	3,118	110¾@111	
"	18..	10@12½	9@10					
"	22..	10@12½	9@10	5,500	6,038	2,050	110¾@111	
"	25..	10@12½	9¼@10½					
"	29..	10@12½	9¼@10½	3,000	4,867	6,359	110¾	
June	1..	10@12½	9¼@10½					½@⅝d.
"	5..	10@12½	9¼@10½	1,700	1,561	4,537	110¾	
"	8..	10@12½	9¼@10½					
"	12..	10@12½	9¼@10½	3,600	4,338	3,051	110½	
"	15..	10@12½	9¼@10½					
"	19..	10@12½	9¼@10½	3,800	6,407	5,317	110@110¼	
"	22..	10@12½	9½@11					
"	26..	10@12½	9½@11	2,100	2,040	3,946	110	
"	29..	10@12½	9½@11					
July	3..	10@12½	9½@11	4,000	1,859	5,174	110	½@⅝d.
"	6..	10@12½	9½@11					
"	10..	10@12½	9½@11	2,100	9,328	512	109¾@110	
"	13..	10@13	9½@11					
"	17..	10@13	9½@11	2,200	2,407	1,399	109½@109¾	
"	20..	10@13	9½@11					
"	24..	10@13	9½@11	2,500	2,466	3,326	109½@109¾	
"	27..	10@13	9½@11					
"	31..	10@13	9½@11	4,000	2,231	5,771	109¾@110	
August	3..	10¼@13	10@11½					⅜d.
"	7..	10¼@13	10@11½	3,300	2,219	1,942	109¾@110	
"	10..	10¼@13	10@11½					
"	14..	10¼@13	10@11½	1,200	2,310	3,941	110@110½	
"	17..	10¼@13	10@11½					
"	21..	10¼@13	10@11½	1,100	1,130	4,099	110@110½	
"	24..	10¼@13	10@11½					
"	28..	10¼@13	10@11½	2,200	1,226	1,664	111	
"	31..	10¼@13	10@11½					
Septem.	4..	10¼@13	10@11½	1,600	400	2,520	111	⅜@½d.
"	7..	10¼@13	10@11½					
"	11..	10¼@13	10@11½	2,500	453	1,295	111	
"	14..	10¼@13	10@11½					
"	18..	10¼@13	10@11½	1,800	2,418	1,347	111@111½	
"	21..	10¼@13	10@11½					
"	25..	10¼@13	10@11½	1,500	3,366	3,285	111	
"	28..	10¼@13	10@11½					
Average price and total sales, receipts and exports.		11.75	9.29	155,800	219,300	196,395		

COTTON AT LIVERPOOL. YEAR 1827.

Week Ending.	Receipts.						Sales.				Prices.			Actual Export.	Consumpt'n.
	American.	E. I.	Egypt.	Brazil.	Other.	Total.	Consumpt'n.	Speculation.	Export	Total.	Mid. Up	Mid. Orl	Dhol.		
Jan 5	No Circ.	......	......	......	......	......	......	......	......	......	......	......	......	......	......
" 12	8,224	......	......	1,265	602	10,091	5,392	......	......	5,392	$6\frac{3}{4}$	$6\frac{3}{4}$	5	......	5,392
" 19	4,419	......	......	......	381	4,800	11,754	......	......	11,754	$6\frac{3}{4}$	$6\frac{3}{4}$	5	......	17,146
" 26	2,093	......	......	1,204	290	3,587	10,644	......	......	10,644	$6\frac{5}{8}$	$6\frac{3}{4}$	5	......	27,790
Feb. 2	7,012	......	......	2,377	1,000	10,380	10,661	......	......	10,661	$6\frac{7}{8}$	7	5	......	38,451
" 9	70	......	......	......	......	70	12,940	......	......	12,940	$6\frac{5}{8}$	7	5	......	51,391
" 16	442	......	......	......	......	442	8,505	......	1,000	9,505	$6\frac{5}{8}$	7	5	1,000	59,896
" 23	5,687	......	......	......	223	5,910	9,743	......	......	9,743	$6\frac{5}{8}$	7	5	......	69,639
Mch. 2	9,456	......	......	......	......	9,456	8,857	......	......	8,857	$6\frac{5}{8}$	7	5	......	78,496
" 9	21,460	656	......	3,299	240	25,655	13,055	......	......	13,055	$6\frac{5}{8}$	7	5	......	91,551
" 16	29,687	......	......	5,773	369	35,829	13,761	......	......	13,761	$6\frac{1}{4}$	$6\frac{7}{8}$	5	......	105,312
" 23	12,761	......	120	170	538	13,589	12,550	......	......	12,550	$6\frac{1}{4}$	$6\frac{3}{8}$	5	......	117,862
" 30	11,626	1,100	602	3,670	902	17,900	12,029	1,000	......	13,029	6	$6\frac{3}{8}$	5	......	129,891
Apl. 6	17,197	1,778	......	1,030	1,341	21,346	13,241	......	......	13,241	6	$6\frac{3}{8}$	5	......	143,132
" 13	14,507	......	2,020	......	750	17,277	13,478	......	......	13,478	$5\frac{7}{8}$	$6\frac{1}{8}$	$4\frac{3}{4}$	......	156,610
" 20	6,244	2,655	1,275	630	433	11,227	19,770	1,500	1,000	22,270	$5\frac{7}{8}$	$6\frac{1}{4}$	$4\frac{1}{2}$	1,000	176,380
" 27	14,932	......	1,719	......	12	16,763	8,618	......	......	8,618	$5\frac{7}{8}$	$6\frac{1}{4}$	$4\frac{1}{2}$	......	184,998
May 4	8,651	......	1,370	2,113	11	12,145	17,072	2,000	......	19,072	$5\frac{7}{8}$	$6\frac{1}{4}$	$4\frac{1}{2}$	......	202,070
" 11	24,301	......	......	6,380	195	30,876	11,328	2,000	......	13,328	$5\frac{7}{8}$	$6\frac{1}{4}$	$4\frac{1}{4}$	......	213,398
" 18	30,650	905	......	1,000	374	32,929	15,222	......	......	15,222	$5\frac{7}{8}$	$6\frac{1}{4}$	$4\frac{1}{4}$	......	228,620
" 25	50,711	450	......	8,721	611	60,493	10,214	3,000	1,000	14,214	$5\frac{7}{8}$	$6\frac{1}{4}$	$4\frac{1}{4}$	1,000	238,834
June 1	15,525	......	......	2,078	248	17,851	12,426	4,000	......	16,426	$5\frac{3}{4}$	$6\frac{1}{4}$	$4\frac{1}{4}$	......	251,260
" 8	13,059	......	......	4,129	1,060	18,248	19,716	4,500	700	24,916	$5\frac{3}{4}$	$6\frac{1}{4}$	$4\frac{1}{4}$	700	270,976
" 15	2,792	......	......	1,300	......	4,092	12,243	......	......	12,243	$5\frac{3}{4}$	$6\frac{1}{4}$	$4\frac{1}{4}$	......	283,219
" 22	18,647	2,663	3,779	2,130	1,608	28,827	9,863	......	700	10,563	$5\frac{3}{4}$	$6\frac{1}{4}$	$4\frac{1}{4}$	700	293,082
" 29	5,454	......	1,639	399	......	7,492	9,071	......	......	9,071	$5\frac{3}{4}$	$6\frac{1}{4}$	$4\frac{1}{4}$	......	302,153
July 6	30,295	......	......	......	......	30,295	7,920	......	500	8,420	$5\frac{3}{4}$	$6\frac{1}{4}$	$4\frac{1}{4}$	500	310,073
" 13	13,185	......	......	......	403	13,588	23,847	3,000	1,000	27,847	$5\frac{3}{4}$	$6\frac{3}{8}$	$4\frac{1}{4}$	1,000	333,920
" 20	21,673	......	......	2,626	459	24,758	15,291	......	......	15,291	$5\frac{3}{4}$	$6\frac{3}{8}$	$4\frac{1}{4}$	......	349,211
" 27	10,948	......	......	2,734	40	13,722	7,200	......	......	7,200	$5\frac{3}{4}$	$6\frac{3}{8}$	$4\frac{1}{4}$	......	356,411
Aug. 3	42,047	80	......	1,597	1,889	45,613	8,278	......	......	8,278	$5\frac{3}{4}$	$6\frac{3}{8}$	$4\frac{1}{4}$	......	364,689
" 10	6,760	......	......	......	379	7,139	9,979	2,000	......	11,979	$5\frac{3}{4}$	$6\frac{3}{8}$	$4\frac{1}{4}$	......	374,668
" 17	24,010	......	......	3,316	234	27,560	10,260	4,500	......	14,760	$5\frac{3}{4}$	$6\frac{3}{8}$	$4\frac{1}{4}$	......	384,928
" 24	2,672	......	......	......	46	2,718	6,859	1,000	......	7,859	$5\frac{3}{4}$	$6\frac{3}{8}$	$4\frac{1}{4}$	......	391,787

Aug. 31...	6,020	......	......	....	....	6,020	9,127	1,500	......	10,627	5½	6¼	4¼	....	400,914
Sept 7...	4,228	......	......	1,000	1,214	6,442	7,796	1,500	300	9,596	5½	6¼	4¼	300	408,710
" 14...	38,042	......	......	5,402	2,240	45,684	7,276	5,000	600	12,876	5½	6¼	4¼	600	415,986
" 21...	4.507	......	......	....	1,002	5,509	9,019	2,000	......	11,019	5½	6¼	4	....	425,005
" 28...	2,127	......	......	....	1,545	3,672	8.787	....	......	8,787	5½	6¼	4	....	433,792
Oct. 5...	4,593	......	......	4,886	2,171	11,650	8,417	500	......	8,917	5½	6¼	4	....	442,209
" 12...	1,785	2,115	......	1,876	274	6,050	11,976	....	......	11,976	5½	6¼	4	....	454,185
" 19...	4,564	......	......	2,750	816	8,130	8,610	....	......	8,610	5½	6¼	4	....	462,795
" 26...	1,405	......	......	....	....	1,405	11,545	....	......	11,545	5¼	6¼	4	....	474,340
Nov. 2...	1,634	......	......	1,400	121	3,155	9,412	....	......	9,412	5¼	6¼	4	....	483,752
" 9...	2,216	......	......	218	262	2,696	12,380	2,500	500	15,380	5	6¼	4	500	496,132
" 16...	6,247	......	......	3,166	1,807	11,220	15,466	2,500	......	17,966	5	6¼	3⅞	....	511,598
" 23...	7,299	......	......	3,751	699	11,749	16,091	7,000	......	23.091	5	6¼	3⅞	....	527,689
" 30...	2,513	500	......	....	542	3,555	11,098	5,000	......	16,098	5	6¼	3⅞	....	538,787
Dec. 7...	1,583	......	......	5,552	542	7,677	15,144	....	......	15,144	5	6⅛	3⅞	....	553,931
" 14...	1,918	......	......	1,660	523	4,101	8,958	5,000	500	14,458	5	6⅛	3⅞	505	562,889
" 21...	1,436	......	......	3,771	4,194	9,401	6,638	500	......	7,138	4⅞	6	3⅞	....	569,527
" 28...	No Circ.	......	......	....	....	....	....	....	......	....	.....	.....	.....	....	....
Average prices & total sales, receipts & stocks.	579,134	12,902	12,524	93,373	30,590	728,523	569,527	60,500	7,800	637,827	5.79	6.395	4.585	7,800	1,119,054

The growth of 1826, was swollen by the cotton held back from the previous crop and hence appears larger than it really was. The planting of 1826, was probably influenced by the belief that the depression in price would be temporary. In the crop of 1827, the natural influence of a low price was manifest in a large falling off.

1828.

Kinsey Burden (*see* years 1805, 1826 and 1827), of S. C., sold two bags of cotton at two dollars per lb., the highest price ever touched, so far as known, in the market.

British calicoes, etc., printed at an average duty of 5s. per piece. . 8,395,848 pieces.
Calicoes, etc., exported—average drawback of 5s. per piece . . . . 5,769,828 pieces.
Calicoes, etc., taken for home consumption at an average duty of 5s. per piece . . . . 2,631,020 pieces.

Value of foreign cotton goods imported into England, £68,528.

Imports of cotton into Great Britain from foreign countries, 189,401,567 lbs., as follows: United States, 151,752,289; Brazil, 29,143,279; Turkey and Egypt, 6,926,288; miscellaneous, 1,579,-711; from British Possessions, 38,359,075 lbs., as follows: East Indies and Mauritius, 32,247,187; British West Indies, 5,893,800; miscellaneous, 218,088. Total imports, 227,760,642 lbs. Exports, 17,396,776 lbs.; Home Consumption, 208,987,744 lbs.

The "Cap Spinner"—or "Danforth Spinner"—was invented and patented on September 2d. of this year, by Charles Danforth, of Paterson, N. J. (*See* year 1830.)

The use of leather belts, instead of iron gearing for transmitting motion to the main shafting of mills, was introduced by Mr. Paul Moody, at Lowell, Mass., this year.

Another, and an unsuccessful effort was made, this year, to establish a "mart," by the bobbin-net manufacturers. In its place, however, a committee of owners of machines and their delegates was constituted, to regulate and control working hours of the machines already in the trade, and to dissuade from constructing more.

COTTON CROP OF THE UNITED STATES.

Statement and Total Amount of the Cotton Crop of the United States, for the Year ending September 30, 1828.

	Bales.	Bales.	Total.	Same period 1827.
NEW ORLEANS.				
Received from—				
Louisiana and Mississippi	....	190,966		170,295
Tennessee and North Alabama	....	92,438		152,166
Across the Lake	....	8,017		9,627
Mobile	....	3,484		2,613
Arkansas	....	1,201		1,739
Florida	....	1,543		1,481
Illinois	....	326		10
Missouri	....	67		3
On hand 1st October, 1827	....	11,171		3,030
		309,213		340,964
Deduct—				
Received from Mobile	3,484			
" Florida	1,543			
Included in the crops of those places	....	5,027		4,094
FLORIDA.			304,186	336,870
Received at—				
New Orleans	....	1,543		
New York	....	1,062		
Philadelphia	....	920		
Baltimore	....	415		
			3,940	4,163
ALABAMA.				
Exports from Mobile—				
To Foreign Ports	....	37,209		
Coastwise	....	34,300		
On hand at Mobile	....	54		
			71,563	89,707
GEORGIA.				
Export Foreign and Coastwise—				
From Savannah—Uplands	141,128			
do Sea Islands	11,648			
		152,776		
From Darien to Great Britain	912			
do New York	61			
		973		
			153,749	233,920
SOUTH CAROLINA.				
Export Foreign and Coastwise—				
From Charleston—Uplands	124,887			
do Sea Islands	22,750			
		147,637		
From Georgetown to New York	....	1,949		
		149,586		
Deduct—				
Received from Georgia and included in the export from Savannah	19,353			
Received from North Carolina through—				
Wilmington	1,500			
Cheraw—about	11,000			
Camden do	6,000			
Columbia do	2,000			
		39,853		
			109,733	179,810

Statement and Total Amount of the Cotton Crop of the United States, for the Year ending September 30, 1828—*Continued.*

	Bales.	Bales.	Total.	Same period last year.
NORTH CAROLINA AND VIRGINIA.				
Received at New York from—				
Wilmington	11,301			
North Counties	11,005			
		22,306		
At Philadelphia from North Carolina	....	2,412		
At Boston from North Carolina	....	31		
At Baltimore from North Carolina	....	17		
At Cheraw, Camden and Columbia and deducted from the South Carolina exports	....	19 000		
Shipped from Wilmington to Charleston	....	1,500		
Shipped from Wilmington to Great Britain	....	1,750		
Export from Petersburg—				
To Foreign Ports	10,406			
Coastwise	15,000			
		25,406		
Export from Richmond, Norfolk, &c	....	5,000		
			77,422	112,811
Total Crop of the United States	....	....	720,593	957,281
				720,593
Deficiency compared with last year	....	....	....	236,688

Export to Foreign Ports, from October 1, 1827, *to September* 30, 1828.

FROM	To Great Britain.	France.	Other parts of Europe.	Total.
New Orleans	142,546	66,425	9,594	218,565
Alabama	30,042	5,321	1,846	37,209
Georgia	65,245	9,946	545	75,736
South Carolina	96,145	22,929	3,194	122,268
North Carolina	1,750	....	....	1,750
Virginia	8,000	4,500	1,100	13,600
New York	74,515	37,198	9,158	120,871
Philadelphia	5,000	1,500	505	7,005
Boston	1 500	700	331	2,531
Other Ports	....	....	....	465
Grand total	424,743	148,519	26,273	600,000
Total last year	....	....	....	853,798
Deficiency	....	....	....	253,798

Growth.

Total crop of 1824–5, 569,259 bales.
do 1825–6, 720,027 do.
do 1826–7, 957,281 do.
do 1827–8, 720,593 do.

Consumption.

To estimate the quantity manufactured in the United States, we take the total crop of the year past	720,593
Deduct therefrom the export to foreign ports	600,000
And assuming that the stock on hand, was the same as at the close of the preceding year—it results that the quantity manufactured the year past, ending 30th September, was	120,593
Estimated consumption last year	103,483
Increase	17,110

Note.—The quantity on hand in the Northern ports, is believed to be considerably larger than at the same period last year. The quantity set down for increased consumption will, therefore, be diminished in proportion. It may also be noted that no account is taken of any cotton manufactured in the *Cotton growing States*, either in the statement of the crop or in the quantity consumed.

The Price semi-weekly at New York, Weekly Sales, Receipts, and Exports, Rates of Freight to Liverpool and Course of Exchange on London, for the Crop Year ending October 1, 1828.

1827.	Price of New Orleans	Price of Upland.	Sales for week.	Receipts for week.	Exports for week.	Exchange on London.	Freight to Liverpool.
October 2..	10¼@13	10@12	2,200	379	1,927	111@111¼	¼@½d.
" 5..	10¼@13	10@12					
" 9..	10¼@13	10@12	1,800	2,140	861	111@111¼	
" 12..	10¼@13	10@12					
" 16..	10¼@13	10@12	2,200	3,640	2,840	111@111¼	
" 19..	10¼@13	10@12					
" 23..	10¼@13	10@12	1,700	692	2,113	111@111¼	
" 26..	10¼@13	10@12					
" 30..	10¼@13	10@12	2,800	2,092	2,450	111	
Novem. 2..	10¼@13	10@12					¼@½d.
" 6..	10@13	9½@11	1,300	2,695	3,521	111@111¼	
" 9..	10@13½	9@11½					
" 13..	10@13½	9@11	1,000	3,517	734	111¼	
" 16..	10@13½	9@11					
" 20..	10@13½	9@11	1,900	475	1,715	111¼	
" 23..	9¾@13½	8¾@11					
" 27..	9¾@13½	8¾@11	2,200	1,602	1,280	111¼	
" 30..	9¾@13½	8¾@11					
Decem. 4..	9¾@13½	8¾@10½	1,200	4,410	5,381	111¼	½d.
" 7..	9¾@13	8¾@10½					
" 11..	9¾@13	8½@10½	1,500	2,903	311	111@111¼	
" 14..	9¾@13	8¼@10½					
" 18..	9¾@13	8¼@10½	1,100	1,224	1,292	111@111¼	
" 21..	9¾@13	8¼@10½					
" 25..	9¾@13	8¼@10½	1,300	405	2,301	111@111¼	
" 28..	9¾@13	8¼@10½					
1828.							
January 1..	9¾@13	8¼@10½	1,000	3,753	4,579	110½@111	⅜@½d.
" 4..	9¾@13	8¼@10½					
" 8..	9¾@13	8¼@10½	1,900	3,144	1,225	110¼@110¾	
" 11..	9¾@13	8¼@10½					
" 15..	9¾@13	8¼@10½	500	726	....	110¼@110¾	
" 18..	9¾@13	8¼@10½					
" 22..	9¾@13	8¼@10½	2,300	921	3,692	110¼@110½	
" 25..	10@12½	9@10½					
" 29..	10@12½	9@10½	1,200	2,589	....	110¼@110½	
February 1..	10@12½	9@10½					¼@⅜d.
" 5..	10@12½	9@10½	2,100	3,621	2,589	110¼@110½	
" 8..	10@12½	9@10½					
" 12..	10@12½	9@10½	1,200	3,843	811	110¼@110½	
" 15..	10@12½	9@10½					
" 19..	10@12½	9@10½	1,800	737	900	110¼@110½	
" 22..	10@12½	9@10½					
" 26..	10@12½	9@10½	1,100	2,984	1,515	110¾@111	
" 29..	10@12½	9@10½					
March 4..	10@12½	9@10½	1,400	4,662	938	111@111¼	¼@⅜d.
" 7..	10@12½	9@10½					
" 11..	10@12½	8½@10½	1,700	5,765	928	110½@110¾	
" 14..	10@12½	8½@10½					
" 18..	10@12½	8½@10½	2,400	1,740	140	110½@111	
" 21..	10@12½	8½@10½					
" 25..	10@12½	8½@10½	2,300	7,282	1,388	110¾@111	
" 28..	10@12½	8¾@10½					
April 1..	10@12½	8¾@10½	2,400	7,657	4,834	110¾@111	⅜@½d.
" 4..	10@12½	8¾@10½					
" 8..	10@12½	8¾@10½	1,700	4,553	1,603	110¾@111	
" 11..	10@12½	8¾@10½					
" 15..	10@12½	8¾@10½	1,800	2,430	380	111@111½	
" 18..	10@12½	8¾@10½					
" 22..	10@12½	8¾@10½	3,700	3,757	2,989	110½@111	

New York Statement for Year 1828—*Continued.*

1828.		Price of New Orleans	Price of Upland.	Sales for week.	Receipts for week.	Exports for week.	Exchange on London.	Freight to Liverpool.
April	25..	10@12½	8¾@10½					
"	29..	10@12½	8¾@10½	4,000	7,917	2,775	110½	
May	2..	10@12½	9@10½					¾@½d.
"	6..	10@12½	9@10½	2,400	5,682	1,769	110½	
"	9..	10@13	9@11					
"	13..	10½@13	9½@11½	5,000	3,374	2,103	110½	
"	16..	10½@13½	10@12					
"	20..	10½@13½	10@12	3,700	830	3,138	110½@111	
"	23..	11@14	10@13					
"	27..	11@14	10@13	4,200	9,403	4,324	110¾@111	
"	30..	11@14	10@13					
June	3..	11@14	10@13	800	2,021	3,998	110¾@111	⅜d.
"	6..	11@14	10@13					
"	10..	11@14	10@13	1,600	5,843	1,689	110¾@111	
"	13..	11@14	10@13					
"	17..	11@14	10@13	2,000	3,501	1,019	110¾@111	
"	20..	11@14	10@13					
"	24..	11@14	10@13	1,000	2,073	1,443	110¾	
"	27..	11@14	10@13					
July	1..	11@14	10@13	1,100	8,243	2,574	110¾	¼d.
"	4..	11@14	10@13					
"	8..	11@14	10@13	1,500	5,964	....	110@110¼	
"	11..	11@14	10@13					
"	15..	11@14	10@13	800	2,626	973	110	
"	18..	11@14	10@13					
"	22..	11@13½	10@12½	1,600	3,396	2,905	109½	
"	25..	11@13½	10@12½					
"	29..	10½@13½	9¾@12½	1,700	2,972	1,085	109@109¼	
August	1..	10½@13½	9¾@12½					⅜d.
"	5..	10½@13½	9¾@12	2,100	517	4,046	109@109½	
"	8..	10½@13	9½@11½					
"	12..	10½@13	9½@11½	2,600	1,151	3,809	109¾@110	
"	15..	10½@13	9½@11½					
"	19..	10½@13	9½@11½	2,100	1,258	2,911	110	
"	22..	10½@13	9½@11½					
"	26..	10½@13	9½@11½	2,200	678	5,695	110	
"	29..	10½@13	9½@11½					
Septem.	2..	10½@13	9½@11½	1,800	18	7,822	110½	⅜d,
"	5..	10½@13	9½@11½					
"	9..	10½@13	9½@11½	2,500	816	1,708	110½@111	
"	12..	10½@13	9½@11½					
"	16..	10½@13	9½@11½	2,100	462	1,669	110½@111	
"	19..	10½@13	9½@11½					
"	23..	10½@13	9½@11½	4,400	124	3,199	110½@111	
"	26 .	10½@13	9½@11½					
"	30..	10½@13	9@11½	1,700	1,457	4,434	110½@111	
Averago price and total sales, receipts and exports.		11.68	10.32	105,600	156,664	120,325		

COTTON AT LIVERPOOL. YEAR 1828.

Week Ending.	Receipts.						Sales.				Prices.			Actual Export.	Consumpt'n.
	American.	E. I.	Egypt.	Brazil.	Other.	Total.	Consumpt'n.	Speculation.	Export.	Total.	Mid Up	Mid. Orl	Dhol.		
Jan. 4...	833	......	......	7,313	....	8,146	9,015	5,000	.....	14,015	5	6	3⅞	....	9,015
" 11...	3,689	......	......	556	118	4,363	11,816	7,500	.....	19.316	5	6	3⅞	....	20,831
" 18...	846	......	......	1,801	....	2.647	13,950	....	.....	13 950	5⅜	6	3⅞	....	34,781
" 25...	11.272	......	......	....	1,475	12,747	8,913	1,000	.....	9,913	5⅜	6	3⅞	....	43,694
Feb. 1...	10,368	......	......	3,269	2,056	15,693	9,183	....	.....	9,183	5⅜	6	3⅞	....	52,877
" 8...	6,244	......	......	2,358	519	9,121	12 363	....	.....	12,363	5¼	5⅞	3⅞	....	65,240
" 15...	1,354	......	......	1,422	103	2,879	11,469	....	.....	11,462	5¼	5⅞	3⅞	....	76,702
" 22...	11,043	2,212	740	....	....	13,995	12 930	1,500	.....	14.430	5¼	5⅞	3⅞	....	89,632
" 29...	6,428	1,691	......	1,566	....	9.685	12,929	....	.....	12,929	5¼	5⅞	3⅞	....	102,561
Mch. 7...	7,074	......	......	2,921	508	10,503	9,780	....	.....	9,780	5¼	5⅞	3⅞	....	112,341
" 14...	4,306	......	......	....	3,330	7,636	14,515	....	.....	14,515	5¼	5⅞	3⅞	....	126,856
" 21...	11,947	......	......	....	4,600	16.547	15,751	2,700	.....	18,451	5¼	6	3½	....	142,607
" 28...	6,622	......	......	2,077	528	9,227	12,150	5,500	.....	17,650	5¼	6	3½	....	154,757
April 3...	9,298	......	......	....	1,439	10,737	12.535	1,500	.....	14,035	5¼	6	3½	....	167,292
" 11...	11,359	......	2,460	1,882	1,112	16.813	18,757	2,000	.....	20,757	5⅜	6	3½	....	186,049
" 18...	19,791	2,714	1,840	2,494	697	27.536	15,315	7,000	.....	22,315	5½	6⅛	3½	....	201,364
" 25...	6,438	325	......	864	241	7,868	14.168	8,000	.....	18,168	5½	6⅛	3½	....	215,532
May 2...	9,452	......	......	1,372	1,628	12,452	7,339	2,000	.....	9,339	5½	6⅛	3½	....	222,871
" 9...	2,294	......	......	1,003	4,125	7,422	14,069	4,000	.....	18,069	5½	6⅛	3½	....	236,940
" 16...	15,069	......	......	3,987	1,907	20.963	14,748	3,000	.....	17,748	5⅞	6⅛	3½	....	251,688
" 23...	1,975	......	......	....	151	2.126	16.760	7,000	.....	23,760	6	6¼	3½	....	268,448
" 30...	12,709	......	......	3,116	133	15,958	10,771	6,000	.....	16,771	6⅛	6⅜	3¾	....	279,219
June 6...	14,021	2,065	......	5,726	110	21,922	7,462	2,000	.....	9,462	6⅛	6⅜	3¾	....	286.681
" 13...	16,945	......	2,200	2,979	110	22,234	10,016	....	.....	10,016	6⅛	6⅜	4	....	266,697
" 20...	2,670	......	2,675	....	34	5,379	8,194	....	.....	8,194	6⅛	6⅜	3⅞	....	304,891
" 27...	38,217	......	......	5,327	871	44,415	5,870	....	.....	5,870	6⅛	6⅜	3⅞	....	310,761
July 4...	16,344	......	......	1,475	339	18,158	8,110	....	.....	8,110	6⅛	6⅜	3⅞	....	318,871
" 11...	10,040	......	900	3.815	1,455	16,210	15,969	3,000	1,200	20,169	6¼	6⅜	3⅞	....	334,840
" 18...	5,248	1,881	......	4,171	1,989	13,289	13,580	3,000	.....	16,580	6¼	6⅜	3⅞	....	348,420
" 25...	21,179	......	......	950	298	22,427	11,834	....	1,000	12,834	6¼	6⅜	3⅞	1,200	360,254
Aug. 1...	3.427	......	......	5,425	352	9,204	11,320	....	.....	11,320	6¼	6⅜	3⅞	....	371,570
" 8...	15,979	......	2,811	....	1,701	20,491	22,520	....	.....	22,520	6¼	6⅜	3⅞	1,000	394,094
" 15...	7,955	......	......	1,511	....	9,466	9,763	....	.....	9,763	6¼	6⅜	3⅞	....	403,857
" 22...	5,283	......	1,980	1,356	1,474	10,093	9,270	....	.....	9,270	6⅛	6⅜	3⅞	....	413,127

Aug. 29...	2,455	......	......	4,960	59	7,474	13,474	....	......	13,464	6⅛	6⅜	3⅞	....	426,601
Sept. 5...	1,997	......	......	1,120	28	3,145	16,733	....	......	16,733	6	6⅜	3¾	....	443,334
" 12...	3,399	......	......	827	2,082	6,308	8,393	1,000	600	9,993	6	6⅜	3¾	....	451,727
" 19...	2,568	......	980	1,606	172	5.326	15,126	....	......	15,126	6	6⅜	3¾	600	466,853
" 26...	3,585	......	1,250	6,788	118	11,741	15,999	....	......	15,999	6	6⅜	3¾	....	482,852
Oct. 3...	9,452	1.693	1,800	5,589	995	19,529	12.270	....	200	12,470	6	6⅜	3¾	....	495,122
" 10...	2,589	......	......	....	839	3,428	20.030	2,500	......	22,530	6⅛	6⅜	3¾	200	515,152
" 17...	2,124	......	......	3,206	674	6,004	23,160	8.000	......	31,160	6⅜	6⅝	3¾	....	538.312
" 24...	1,526	......	......	....	142	1,668	7,565	2,000	......	9,565	6⅜	6⅝	3¾	....	545,877
" 31...	6,454	......	......	7,410	22	13,886	12,042	4,500	......	16,542	6⅜	6⅝	3¾	....	557,919
Nov. 7...	2,558	644	......	2,261	2,784	8,247	9,807	1,100	......	10,907	6⅜	6⅝	3¾	....	567,726
" 14...	....	......	......	....	531	531	6,329	1,800	500	8,629	6¼	6½	3¾	....	574,055
" 21...	7,853	1,851	......	2,666	265	12.635	6,550	1,300	500	8,350	6¼	6½	3¾	500	580,066
" 28...	7.307	......	......	4,792	4,444	16.543	7,464	1,000	......	8,464	6⅛	6⅜	3¾	....	588,069
Dec. 5...	1,107	......	......	5,399	....	6,506	7,300	1,200	......	8,500	6⅛	6⅜	3¾	....	595,369
" 12...	8,144	......	......	1,251	569	9,964	7,830	....	......	7,830	6⅛	6⅜	3½	....	603,199
" 19...	4,250	......	......	1,329	244	5,823	13,507	....	......	13,507	6⅛		3½	....	616,706
" 26...	8,168	......	......	1,646	96	9,910	10,250	....	......	10,250	6⅛	6⅜	3½	....	626,956
Average prices & total sales, receipts & stocks.	403,255	15,076	19,636	121,586	49,367	608,920	626,956	96,100	4,000	727,056	5.84	6¼	3¾	4,000	1,205,684

LIVERPOOL STATEMENT FOR 1828.

UNITED STATES—1827–1828.

Stock, 1st October, 1827	48,000	Export	600,000
Crop, 1827–28	712,000	Consumption	120,000
		Stock, 1st October, 1828	40,000
Bags	760,000	Bags	760,000

	Liverpool.	Gt. Britain.	France.	Continent.	Tot. Europe.
United States	230,250	249,950	44,331	8,460	302,741
Brazil	71,000	79,200	6,915	1,086	87,201
West Indies	6,650	11,700	5,347	2,614	19,661
East Indies	12,700	74,600	....	9,850	84,450
Egypt	22,100	36,750	28,810	2,747	68,307
Stock	342,700	452,200	85,403	24,757	562,360
IMPORT.					
United States	414,248	444,938	142,230	42,960	630,128
Brazil	162,465	165,038	16,020	7,808	188,866
West Indies	13,375	23,565	15,322	18,422	57,309
East Indies	16,645	84,700	....	39,180	123,880
Egypt	24,667	31,103	32,560	34,114	97,777
Consumption	631,400	749,344	206,132	142,484	1,097,960
Stock supply	974,100	1,201,544	291,535	167,241	1,660,320
	U. States.	Brazil & W.I	East India.	Egypt.	
Export deducted from import	20,300	6,080	38,050	....	64,430
Total supply, bags	....	....	....	....	1,595,890

CONSUMPTION.

Tot. Europe.	Continent.	France.	Gt. Britain.	Liverpool.
707 728	43,040	162,881	501,807	471,768
166,602	8,094	20,190	138,318	134,945
55,841	15,839	18,412	21,590	14,075
74,992	37,422	....	37,570	14,095
98,644	28,441	38,240	31,963	22,447
1,103,807	132,836	239,723	731,248	657,330
492,083	34,405	51,812	405,866	295,470
			64,430	21,300
1,595,890	167,241	291,535	1,201,544	974,100

1829.

Mr. Dyer, of Manchester, England, obtained a patent in Great Britain for an improvement in the "tube frame" introduced from America by him. (*See* years 1825 and 1833).

In operation in England 45,500 power looms, and in Scotland 10,000. (*See* years 1785, 1813 and 1820.)

During this year the quantity of British calicoes and muslins, which paid the print duty, was 128,340,004 yards. (*See* year 1796).

British calicoes, etc., printed at an average duty of 5s. per piece.	7,768,072	pieces.
Calicoes, etc., exported—average drawback of 5s. per piece....	5,562,136	"
Calicoes, etc., taken for home consumption at an average duty of 5s. per piece..	2,105,936	"

Cotton exports from United States, 264,847,186 lbs., 12,833,307 lbs. being Sea Island; total value $26,574,311.

Value of foreign cotton goods imported into England £60,770.

Imports of cotton into Great Britain from foreign countries, 193,122,967 lbs., as follows: United States, 157,187,396; Brazil, 28,878,386; Turkey and Egypt, 5,986,385; miscellaneous, 1,070,800: from British Possessions, 29,644,444 lbs., as follows: East Indies and Mauritius 24,908,399; British West Indies, 4,640,414; miscellaneous, 95,631. Total Imports, 222,767,411 lbs.; Exports, 30,285,115 lbs.: Home consumption, 204,097,037 lbs.

A very cheap machine for making roving, called the "Eclipse Speeder," capable of any rapid operation, was patented April 18th, this year, by Gilbert Brewster of Poughkeepsie, N. Y. (*See* year 1835).

Five hundred hands employed about the establishment of the Factory Island, Maine, Mill. (*See* years 1825, 1826 and 1830.)

COTTON CROP OF THE UNITED STATES.

Statement and Total Amount of the Cotton Crop of the United States, for the Year ending September 30, 1829.

	Bales.	Bales.	Total.	Same period 1828.
NEW ORLEANS.				
Export—				
To Foreign ports	....	226,932		218,565
Coastwise	....	41,017		86,283
Stock on hand 1st October, 1829	....	5,557		4,365
		273,506		309,213
Deduct—				
Received from Mobile	6,350			
do Florida	2,907			
		9,257		5,027
			264,249	304,186
FLORIDA.				
Received at—				
New Orleans	....	2,907		
Mobile	....	371		
New York	....	721		
Baltimore	....	147		
			4,146	3,940
ALABAMA.				
Export from Mobile—				
To Foreign ports	....	48,534		
Coastwise	....	31,314		
On hand at Mobile	....	481		
		80,329		
Deduct received from Florida	....	371		
			79,958	71,563
GEORGIA.				
Export from Savannah—				
To Foreign ports—Uplands	128,233			
do Sea Islands	13,729			
		141,962		
Coastwise	....	103,375		
		245,337		
From Darien—				
To New York	....	128		
On hand in Savannah	2,300			
do Augusta	1,401			
		3,701		
			249,166	153,749
SOUTH CAROLINA.				
Export from Charleston—				
To Foreign ports—Uplands	161,531			
do Sea Islands	23,047			
		184,578		
Coastwise	....	29,450		
		214,028		
From Georgetown to New York	....	3,649		
On hand in Charleston	....	4,323		
		222,000		
Deduct—				
Received from Savannah and included in export from that place	23,635			
Received at North Carolina through—				
Wilmington	90			
Cheraw, about	15,000			
Camden, "	10,000			
Columbia "	5,000			
		53,725		
			168,275	109,733

Statement and Total Amount of the Cotton Crop of the United States, for the Year ending September 30, 1829—*Continued.*

	Bales.	Bales.	Total.	Same period 1828.
NORTH CAROLINA AND VIRGINIA.				
Received at New York from—				
Wilmington	11,172			
North Counties	22,205			
		33,377		
At Philadelphia from North Carolina	....	2,731		
At Boston	....	197		
At Cheraw, Camden and Columbia	....	30,000		
Shipped from Wilmington to Charleston	....	90		
do do Europe	....	4,210		
On hand at Fayetteville	....	1,000		
		71,605		
Export from Petersburg and Richmond—				
To Foreign ports	19,516			
Coastwise	12,000			
On hand in Petersburgh	1,500			
		33,016		
			104,621	77,422
Total crop of the United States	....	....	870.415	720,593
			720,593	
Increase compared with last year	....	....	149,822	

Export to Foreign Ports, from 1st *October*, 1828, *to* 30*th September*, 1829.

FROM	Great Britain.	France.	Other parts of Europe.	Total.
New Orleans	130,514	78,370	18,043	226,932
Alabama	37,818	7,051	3,665	48,534
Georgia	118,404	21,988	1,570	141,962
South Carolina	138,086	33,847	12,645	184,578
North Carolina	3,100	1,110	....	4.210
Virginia	13,302	5,803	411	19,516
New York	53,575	36,059	28,138	117,772
Philadelphia	2,928	477	596	4,001
Boston	243	116	925	1,284
Other Ports	31	....	180	211
Grand total	498,001	184,821	66,178	749,000
Total last year	424,743	148,519	26,738	600,000
Increase	73.258	36,302	39,440	149,000

Growth.

Total crop of	1824–5,	569,259	bales.
do	1825–6,	720,027	do.
do	1826–7,	957,281	do.
do	1827–8,	727,593	do.
do	1828–9,	870,415	do.

Consumption.

To estimate the quantity manufactured in the United States, we take the total crop for the year past		870,415
Deduct therefrom the export to foreign ports	749,000	
Stocks on hand in the Southern ports	16,562	
		765,652
And assume that the quantity at the Northern ports was the same as last year, it results that the consumption was		164,853
Estimated consumption for 1827–8		120,593
do do 1826–7		103,483

Note.—The quantity on hand in the Northern ports at the close of last year, is believed to have been much greater than the present. It is probable, therefore, that the actual manufacture has varied but little.

LIVERPOOL STATEMENT. FOR 1829.

UNITED STATES—1828–1829.			
Stock, 1st October, 1828............	40.000	Export...........	749,000
Crop, 1828–29.....	853,000	Consumption......	119,000
		Stock, 1st October, 1829............	30,000
Bags.............	898,000	Bags.............	898.000

	Liverpool.	Gt. Britain.	France.	Continent.	Tot. Europe.
United States...............	157,590	172,781	23,680	8,380	204,841
Brazil......................	95.510	100,910	2,745	800	104,455
West Indies.................	5,950	12,655	2,257	5,197	20,109
East Indies.................	12,100	83,680	....	11,608	95,288
Egypt...................	24,320	35,840	23,130	8,420	67,390
Stock	295,470	405,866	51,812	34,405	492,083

CONSUMPTION.					IMPORT.					
Tot. Europe.	Continent.	France.	Gt. Britain.	Liverpool.		Liverpool.	Gt. Britain.	France.	Continent.	Tot. Europe.
758,205	66,553	186,433	505,219	471,830	....United States..............	429,640	460,782	185,508	72,463	718,753
205,557	25.088	27.006	153,463	125,639	....Brazil.....................	157,329	159,013	26,168	26,302	211,483
58,513	20,129	12,045	26,339	17,643	....West Indies................	14,663	20.637	12,605	20,161	53,403
100,348	64,759	....	35,589	17,907	....East Indies................	16,587	80,402	....	64,428	144,830
93,232	29,519	39,226	24,447	19,878	....Egypt......................	22,458	24,739	17,949	23,524	66,212
1,215,855	206,048	264,750	745,057	679,897	....Consumption	640,677	745,573	242,230	206,878	1,194,681
353,909	35,235	29,292	289,382	203,250	Stock. Supply,	936,147	1,151,439	294,042	241,283	1,686,764
						U. States.	Brazil & W.I.	East India.	Egypt.	
....	....	....	117,000	53,000	Export deducted from import.....	21,000	28,500	54,500	13,000	117,000
1,596,764	241,283	294,042	1,151,439	936,147	Total supply, bags..........	....	....	....	....	1,569,764

The Price semi-weekly at New York, Weekly Sales, Receipts and Exports, Rates of Freight to Liverpool and Course of Exchange on London, for the Crop Year ending October 1, 1829.

1828.	Price of New Orleans	Price of Upland.	Sales for week.	Receipts for week.	Exports for week.	Exchange on London.	Freight to Liverpool.
October 3..	10½@13	9@11½				110½@111	¼@⅜d.
" 7..	10½@13	9@11½	2,100	1,771	734		
" 10..	10½@13	9@11½				110½@111	
" 14..	10½@13	9@11½	1,900	1,101	2,298		
" 17..	10½@13	9@11½				110½@111	
" 21..	10½@13	9@11½	2,200	1,498	5,325		
" 24..	10½@13	9@11½				110½@111	
" 28..	10½@13	9@11½	2,100	1,961	2,366		
" 31..	10¼@13	9¼@11				110½@111	
Novem. 4..	10¼@13	9¼@11	1,900	2,766	3,364		¼@⅜d.
" 7..	10¼@13	9¼@11				110½	
" 11..	10¼@13	9¼@11	2,900	2,346	2,390		
" 14..	10¼@13	9¼@11				110@110½	
" 18..	10¼@13	9¼@11	2,800	953	2,669		
" 21..	10¼@13	9½@11				110	
" 25..	10¼@13	9½@11	1,000	2,585	3,599		
" 28..	10¼@13	9½@11				109¾@110	
Decem. 2..	10¼@13	9½@11½	1,900	3,813	1,432		½d.
" 5..	10¼@13	9½@11½				109¾@110	
" 9..	10¼@13	9½@11½	2,400	4,143	2,520		
" 12..	10½@13	9¾@11½				109¾@110	
" 16..	10½@13	9¾@11½	2,500	5,572	2,240		
" 19..	10½@13	9¾@11½				109½@110	
" 23..	10½@13	9¾@11½	2,000	3,222	2,785		
" 27..	10½@13	9¾@11				108½@108¾	
" 30..	10½@13	9¾@11	1,100	3,299	851		
1829.							
January 2..	10½@13	9¾@11				108¼	½@⅝d.
" 6..	10½@13	9½@11	700	3,296	4,375		
" 9..	10½@13	9½@11				107½	
" 13..	10½@13	9¼@10½	2,100	4,912	632		
" 16..	10½@12½	9¼@10½				107½	
" 20..	10@12½	9@10¼	1,700	6,312	2,160		
" 23..	10@12	9@10				107¾@108	
" 27..	10@12	9@10	3,300	3,839	1,564		
" 30..	10@12	9@10				108½	
February 3..	10@12	9@10	3,000	1,653	5,598		½@⅝d.
" 6..	10@12	9@10				108¾@109	
" 10..	10@12	9@10	1,100	4,814	1,133		
" 13..	10@12	9@10				108¾@109	
" 17..	10@12	8¾@10	2,900	1,086	2,492		
" 20..	10@12	8¾@10				108¼@108½	
" 24..	10@12	8¾@10	1,600	1,407	1,678		
" 27..	10@12	8½@10				108½	
March 3..	10@12	8½@10	1,700	3,540	3,203		½@⅝d.
" 6..	10@12	8¾@10				108½	
" 10..	10@12	8¾@10	1,000	8,540	1,323		
" 13..	10@12	8½@10				108¼	
" 17..	10@12	8½@10	1,300	4,142	1,747		
" 20..	10@12	8½@ 0				108¼	
" 24..	10@12	8½@10	1,600	2,861	3,650		
" 27..	10@12	8½@10				108¼	
" 31..	10@12	8½@10½	2,900	918	1,839		
April 3..	10@12	8¾@10½				108	½@⅝d.
" 7..	10@12	8¾@10½	1,500	5,531	2,287		
" 10..	10@12	8¾@10½				108¼	
" 14..	10@12	8¾@10½	1,000	2,002	1,948		
" 17..	10@12	8¾@10½				108½@109	
" 21..	10@12	8¾@10½	1,600	2,167	3,748		
" 24..	10@12	8¾@10½				109¾@110	

New York Statement for Year 1829—*Continued.*

1829.		Price of New Orleans	Price of Upland.	Sales for week.	Receipts for week.	Exports for week.	Exchange on London.	Freight to Liverpool.
April	28..	10@12	8¾@10½	1,700	2,108	1,505		
May	1..	10@12	8¾@10½				109¾@110	⅜d.
"	5..	10@12	8¾@10½	2,000	1,967	1,372		
"	8..	10@12	8¾@10½				109¾@110	
"	12..	10@12	8¾@10½	1,300	3,030	2,419		
"	15..	10@12	8¾@10½				109½	
"	19..	10@12	8¾@10½	2,100	2,900	3,389		
"	22..	10@12	8½@10½				109½	
"	26..	10@12	8¾@10½	500	2,725	4,293		
"	29..	10@12	8¾@10½				109@109½	
June	2..	10@12	8¾@10½	1,000	2,307	1,157		¼d
"	5..	10@12	8¾@10½				108¾@109	
"	9..	10@12	8¾@10½	900	2,255	600		
"	12..	10@12	8¾@10½				108¾	
"	16..	10@12	8½@10½	700	1,790	743		
"	19..	10@12	8½@10½				108½	
"	23..	10@12	8½@10½	1,500	5,913	1,539		
"	26..	9½@12	8½@10½				108½	
"	30..	9½@12	8¼@10½	1,300	1,748	710		
July	3..	9½@12	8¼@10½				108¼@108½	¼d.
"	7..	9½@12	8¼@10½	1,400	4,063	1,298		
"	10..	9½@12	8¼@10½				109	
"	14..	9½@12	8¼@10½	1,800	6,860	2,380		
"	17..	9½@12	8¼@10½				109¼@109½	
"	21..	9½@12	8¼@10½	1,400	1,257	1,413		
"	24..	9½@12	8¼@10½				109¼@109½	
"	28..	9½@12	8¼@10½	1,500	2,534	1,095		
"	31..	9½@12	8@10½				109½	
August	4..	9½@12	8@10½	1,000	1,780	1,604		¼@⅜d.
"	7..	9½@12	8@10½				109½	
"	11..	9½@12	8@10½	1,700	1,886	1,832		
"	14:	9½@12	8@10½				109½	
"	18..	9½@12	8@10½	1,500	4,277	555		
"	21..	9½@12	8@10½				109¼@109½	
"	25..	9½@12	8@10½	2,200	562	2,204		
"	28..	9½@12	8@10½				109¼@109½	
Septem.	1..	9½@12	8@10½	2,100	54	4,581		¼@⅜d.
"	4..	9½@12	8@10½				109½@109¾	
"	8..	9½@12	8@10½	2,900	2,118	645		
"	11..	9½@12	8@10½				109½@109¾	
"	15..	9½@12	8@10½	1,500	1,130	4,636		
"	18..	9½@12	8@10½				109¾	
"	22..	9½@12½	8@10½	1,300	334	2,029		
"	25..	9½@12½	8@10½				109¾	
"	29..	9½@12½	8@10¾	1,400	1,156	2,644		
Average price and total sales, receipts and exports.		10.65	9.88	90,500	146,404	116,590		

COTTON AT LIVERPOOL. YEAR 1829.

Week Ending.	Receipts.						Sales.				Prices.			Actual Export.	Consumpt'n
	American.	E. I.	Egypt.	Brazil.	Other.	Total.	Consumpt'n.	Speculation.	Export	To'al.	Mid. Up	Mid. Orl	Dhol.		
Jan. 2	2,769			3,062	60	5,891	10,021			10,021	$6\frac{1}{8}$	$6\frac{3}{8}$	$3\frac{1}{2}$		10,021
" 9	2,234			350		2,594	8,787			8,787	$5\frac{7}{8}$	$6\frac{1}{4}$	$3\frac{1}{2}$		18,808
" 16	4,579			1,405	2,807	8,791	8,556			8,556	$5\frac{3}{4}$	$6\frac{1}{4}$	$3\frac{1}{4}$		27,364
" 23	9,387			1,261	2,251	12,890	6,768			6,768	$5\frac{1}{2}$	6	$3\frac{1}{4}$		34,132
" 30	10,339			2,238		12,577	14,313	4,000		18,313	$5\frac{5}{8}$	6	$3\frac{1}{4}$		48,445
Feb. 6	30,909			9,472	624	41,005	13,634		700	14,334	$5\frac{5}{8}$	6	$3\frac{1}{4}$	700	62,079
" 13	16,977			3,479	3,474	23,930	11,790			11,790	$5\frac{5}{8}$	6	$3\frac{1}{4}$		73,869
" 20	11,906			922	1,897	14,725	7,600			7,600	$5\frac{1}{2}$	$5\frac{7}{8}$	$3\frac{1}{4}$		81,469
" 27	8,484	1,324		3,570	1,328	14,698	11,804			11,804	$5\frac{1}{2}$	$5\frac{7}{8}$	$3\frac{1}{4}$		93,273
Mch. 6	15,521				1,854	17,375	11,755			11,755	$5\frac{1}{4}$	$5\frac{1}{2}$	$3\frac{1}{4}$		105,028
" 13	3,817					3,817	14,210			14,210	$5\frac{1}{4}$	$5\frac{1}{2}$	$3\frac{1}{4}$		119,238
" 20	1,677	916			449	3,042	7,746	2,000		9,746	$5\frac{1}{4}$	$5\frac{1}{2}$	$3\frac{1}{4}$		126,984
" 27	2,748				685	3,433	8,530			8,530	$5\frac{1}{4}$	$5\frac{1}{2}$	$3\frac{1}{4}$		135,514
Apr. 3	3,385			1,127		4,512	8,197			8,197	$5\frac{1}{4}$	$5\frac{1}{2}$	$3\frac{1}{4}$		143,711
" 10	33,490			5,129	992	39,611	10,711	600		11,311	$5\frac{1}{4}$	$5\frac{1}{2}$	$3\frac{1}{4}$		154,422
" 16	38,954	4,185		7,225	436	50,800	10,470			10,470	$5\frac{1}{4}$	$5\frac{1}{2}$	$3\frac{1}{4}$		164,892
" 24	11,218	6,144	1,380		2,332	21,074	17,347		1,500	18,847	$5\frac{1}{4}$	$5\frac{1}{2}$	$3\frac{1}{4}$	1,500	182,239
May 1	5,152					5,152	13,585	2,000		15,585	$5\frac{1}{8}$	$5\frac{1}{2}$	$3\frac{1}{4}$		195,824
" 8	11,452			3,613	878	15,943	11,440	500		11,940	$5\frac{1}{8}$	$5\frac{3}{8}$	$3\frac{1}{4}$		207,264
" 15	11,026	2,109		7,686	444	21,265	14,635	1,500	200	16,335	$5\frac{1}{8}$	$5\frac{3}{8}$	3	200	221,899
" 22	5,360			2,644	376	8,380	15,490	1,000	900	17,390	$5\frac{1}{8}$	$5\frac{3}{8}$	3	900	237,389
" 29	13,043			1,199	1,504	15,746	17,351	3,000		20,351	$5\frac{1}{4}$	$5\frac{1}{2}$	3		254,740
June 5	6,663			1,138	300	8,101	8,944			8,944	$5\frac{1}{4}$	$5\frac{1}{2}$	3		263,684
" 12	2,114			2,407	6	4,527	13,384			13,384	$5\frac{1}{4}$	$5\frac{1}{2}$	3		277,068
" 19	9,542		6,740	11,064	978	28,324	13,797			13,797	$5\frac{1}{4}$	$5\frac{1}{2}$	3		290,865
" 26	3,537			1,588		4,625	15,610			15,610	$5\frac{1}{4}$	$5\frac{1}{2}$	3		306,475
July 3	9,060			1,000	2,053	12,113	11,994			11,994	$5\frac{1}{4}$	$5\frac{1}{2}$	3		318,469
" 10	34,523			11,634	227	46,384	10,300			10,300	$5\frac{1}{4}$	$5\frac{1}{2}$	3		328,769
" 17	13,126			2,833	601	16,560	13,580			13,580	$5\frac{1}{8}$	$5\frac{3}{8}$	3		342,349
" 24	8,931		7,125	1,462	1,563	19,081	10,942			10,942	$5\frac{1}{8}$	$5\frac{3}{8}$	3		353,291
" 31	4,341			2,333	133	6,707	14,139			14,139	$5\frac{1}{8}$	$5\frac{3}{8}$	3		367,430
Aug. 7	22,012			2,753	4,570	29,335	8,880		500	9,380	$5\frac{1}{8}$	$5\frac{3}{8}$	3	500	376,310
" 14	1,630	1,296			699	3,625	10,914			10,914	$5\frac{1}{8}$	$5\frac{1}{4}$	3		387,224
" 21	3,264			2,455	336	6,055	15,097			15,097	5	$5\frac{1}{8}$	3		402,321
" 28	10,250			1,824	2	12,076	11,852	1,000	1,000	11,852	5	$5\frac{1}{8}$	3	1,000	414,173

Sept. 4	3,265	......	1,900	......	6	5,171	16,460	......	3,500	19,960	5	$5\frac{1}{8}$	3	3,500	430,633
" 11	18,643	......	......	4.375	1,697	24,715	13.841	......	4,000	17,841	5	$5\frac{1}{8}$	3	4,000	444,474
" 18	1,110	......	......	......	......	1,110	11,190	1,000	2,800	14,990	5	$5\frac{1}{8}$	3	2,800	455,664
" 25	9,448	......	......	3,408	683	13,539	13,500	1,500	1,000	16,000	5	$5\frac{1}{8}$	3	1,000	469,164
Oct. 2	938	......	......	4,755	244	5,937	18,860	5,500	1,000	25,360	$5\frac{1}{8}$	$5\frac{1}{4}$	$2\frac{7}{8}$	1,000	488,024
" 9	2,325	......	2,254	1,161	103	5,843	17,886	2,500	500	20,886	$5\frac{1}{4}$	$5\frac{3}{8}$	$2\frac{7}{8}$	500	505,910
" 16	552	......	......	2,233	662	3,447	16,264	......	2,500	18,764	$5\frac{1}{4}$	$5\frac{3}{8}$	$2\frac{7}{8}$	2,500	522,174
" 23	1,583	1,479	......	3,601	743	7,406	18,844	......	300	19,144	$5\frac{1}{4}$	$5\frac{3}{8}$	$2\frac{7}{8}$	300	541,018
" 30	731	......	......	1,101	......	1,832	18,584	......	700	19,284	$5\frac{1}{4}$	$5\frac{3}{8}$	$2\frac{7}{8}$	700	559,602
Nov. 6	358	......	......	......	......	358	12,730	500	......	13,230	$5\frac{1}{4}$	$5\frac{1}{2}$	$2\frac{7}{8}$	......	572,332
" 13	561	......	......	3,540	1,050	5.151	21,525	3,000	......	24,525	$5\frac{3}{8}$	$5\frac{1}{2}$	3	......	593,857
" 20	......	......	......	......	......	......	22,270	6,000	......	28,270	$5\frac{5}{8}$	$5\frac{3}{4}$	3	......	616,127
" 27	367	......	......	......	3	370	13,340	1,000	......	14,340	$5\frac{5}{8}$	$5\frac{3}{4}$	3	......	629,467
Dec. 4	1,380	......	2,860	1,531	......	5,771	8,437	......	......	8,437	$5\frac{5}{8}$	$5\frac{3}{4}$	3	......	637,904
" 11	3,446	......	......	5,427	36	8,909	5,276	......	......	5,276	$5\frac{1}{2}$	$5\frac{1}{2}$	3	......	643,180
" 18	2,924	......	......	1,302	502	4,728	14,694	......	......	14,694	$5\frac{1}{2}$	$5\frac{3}{8}$	3	......	657.874
" 24	1,158	......	......	......	65	1,223	15,120	1,000	......	16,120	$5\frac{3}{4}$	$5\frac{7}{8}$	3	......	672,994
" 31	......	......	......	......	......	......	15,321	......	......	15,321	6	$6\frac{1}{8}$	3	......	688,315
Average prices & total sales, receipts & stocks.	322,109	17,453	22,259	128,707	39,645	530,173	688,315	37,100	21,100	746,515	5.32	5.545	3.085	21,100	1,298,694

* Total American and other stocks, 203,250.

10

1830.

Mr. William Strutt, F.R.S., of Derby, son of Jedediah Strutt, the partner of Richard Arkwright (*see* years 1732, 1767, 1790 and 1792), died on the 29th of December this year.

Mr. Roberts, of Sharp, Roberts & Co., of Manchester, England, took out a patent for an improvement upon his self-acting mule. (*See* years 1792 and 1825.)

British calicoes, etc., printed at an average duty of 5s. per piece..8,596,952 pieces.
Calicoes, etc., exported—average drawback of 5s. per piece......6,315,440 pieces.
Calicoes, etc., taken for home consumption at an average duty of 5s. per piece...........................2,281,512 pieces.

An extremely well informed calico printer of England, this year made the following calculation of the number of individuals employed in the printing trade and in the manufacturing of the cloth printed:

"The duty is, in round numbers, £2,000,000, which is equal to 8,000,000 pieces of prints."

	s.	d.
Average price of printing cloth per piece........................	7	0
Deduct value of raw material............................	2	6
	4	6
Deduct for profits of machinery, etc...........................	1	0
Supposed amount paid in wages on each piece....................	3	6

8,000,000 pieces of cloth—wages for spinning and weaving at 3s. 6d..	£1,400,000
Average of wages for printing do. at 2s. 6d........................	1,000,000
	£2,400,000

Or equal to £46,154 of wages paid weekly for labor in spinning, weaving and printing, the average of which is about 8s. per head; divide £46,154 by 8s. and there results 115,385 individuals employed in spinning, weaving and printing; and it may be said that as many more are dependent upon them—thus giving 230,770 individuals employed in and dependent upon the printing trade." (*See* year 1676.)

Value of foreign cotton goods imported into England £42,277.

Imports of cotton into Great Britain from foreign countries, 248,018,963 lbs., as follows: United States, 210,885,358; Brazil, 33,092,072; Turkey and Egypt, 3,401,710; from British Possessions, 15,942,489 lbs., as follows; East Indies and Mauritius, 12,483,217; British West Indies, 3,429,247; miscellaneous, 30,025.

Total imports, 263,961,452 lbs.; Exports, 8,534,976 lbs.; Home consumption, 269,616,640 lbs.

The average product of the spinner of yarn No. 40, was two and three quarter hanks per spindle per day. (*See* years 1790 and 1812.)

Computed number of pounds of yarn and twist made in Great Britain, 223,000,000; 64,000,000 lbs. of which were imported. Also 442,000,000 yards of cotton goods exported—value of all, £19,000,000.

A patent was taken out in England, in the name of John Hutchinson, of Liverpool, this year, for the "Cap" or "Danforth" Spinner (*see* year 1828), when it went into extensive use throughout Europe.

Total number of slaves in the United States this year, 2,009,000; 983,000 of whom were in the cotton States alone. (*See* years 1790 and 1850.)

The Mill on Factory Island, Maine (*see* years 1825, 1826 and 1829), which had been in course of erection for five years, was completed early this year, but was no sooner finished than it was burned to the foundation and the company lost all their stock. The wreck of the mill, with all other property, was sold at a very low price to another Boston Company, who obtained a charter under the name of "The York Manufacturing Company of Saco." (*See* year 1833.)

American cotton in Liverpool, worth from 7d. to 9d. per pound. (*See* year 1806.)

COTTON CROP OF THE UNITED STATES.

Statement and Total Amount of the Cotton Crop of the United States, for the Year ending 30th September, 1830.

	Bales.	Bales.	Total.	Same period 1829.
NEW ORLEANS.				
Export—				
To Foreign Ports	....	295,774	....	226,932
Coastwise	....	56,115	....	41,017
Stock on hand 1st October, 1830	....	9,505	....	5,557
		361,394	....	273,506
Deduct—				
Stock on hand 1st October, 1829	5,557			
Received from Mobile	6,095			
" Florida	3,521			
		15,173	....	13,192
		346,221		
Add—				
Burnt	7,250			
Loss by damage	553			
		7,803		
			354,024	260,314
FLORIDA.				
Received at—				
New Orleans	....	3,521		
Baltimore	....	77		
Philadelphia	....	686		
New York	....	1,433		
Boston	....	70		
			5,787	4,146
ALABAMA.				
Export from Mobile—				
To Foreign Ports	....	61,323		
Coastwise	....	41,702		
Stock on hand at Mobile	....	140		
		10,165		
Deduct—				
Stock on hand 1st October, 1829	....	481		
			102,684	79,904
GEORGIA.				
Export from Savannah—				
To Foreign Ports—Uplands	139,238			
" Sea Islands	9,579			
		148,817		
Coastwise	....	98,845		
		247,662		
From Darien—				
To Liverpool	2,422			
To New York	634			
		3,056		
		250,718		
On hand in Savannah	1,600			
" Augusta	4,500			
		6,100		
		256,818		
Deduct—				
Stock on hand 1st October, 1829	....	3,701		
			253,117	246,000

Cotton Crop of the United States—Continued.

	Bales.	Bales.	Total.	Same period 1830.
SOUTH CAROLINA.				
Export from Charleston—				
To Foreign Ports—Uplands	165,636			
" Sea Islands	16,536			
		182,172		
Coastwise	....	27,256		
		209,428		
From Georgetown—To New York	....	3,451		
On hand in Charleston	....	3,906		
		216,785		
Deduct—				
On hand 1st October, 1829	4,323			
Received from Savannah, and included in the Export from that place	23,591			
		27,914		
			188,871	195,365
NORTH CAROLINA.				
Export—				
To Europe	....	3,324		
New York	....	31,581		
Philadelphia	....	1,218		
Boston	....	732		
Baltimore	....	7		
On hand	....	1,000		
Deduct—		37,862		
Stock on hand 1st October, 1829	....	1,000		
			36,862	40,515
VIRGINIA.				
Export—				
To Foreign Ports	....	28,753		
Coastwise	....	8,000		
On hand	....	247		
Deduct—		37,000		
Stock on hand 1st October, 1829	....	1,500		
			35,500	31,500
Total crop of the United States	....	....	976,845	857,744
			857,744	
Increase compared with last year	....	....	119,101	

Export to Foreign Ports, from October 1, 1829, *to September* 30, 1830.

FROM	To Great Britain.	To France.	Other parts of Europe.	Total.
New Orleans	196,892	93,446	5,436	295,774
Alabama	43,165	9,972	8,186	61,323
Georgia	117,939	27,278	3,600	148,817
South Carolina	134,820	36,119	11,233	182,172
North Carolina	2,349	975	....	3,324
Virginia	22,274	6,309	170	28,753
New York	73,596	26,642	12,844	113,082
Philadelphia	2,256	50	56	2,362
Boston	....	....	687	687
Other Ports, (Darien)	2,422	....	....	2,422
Grand total	595,713	200,791	42,212	838,716
Total last year	498,001	184,821	66,178	749,000
Increase	97,712	15,970	....	89,716
Decrease	....	....	23,966	

Growth.

Total crop of 1824–5, 560,000 bales.
" 1825–6, 710,000 "
" 1826–7, 937,000 "
" 1827–8, 712,000 "
" 1828–9, 857,744 "
" 1829–30, 976,845 "

Consumption.

To estimate the quantity manufactured in the United States, take the growth of the year past	976,845 bales.	
Add—Stocks on hand in the Southern ports at the commencement of the year, (October 1, 1829)	16,562	
		993,407 bales.
Deduct therefrom—The Export to Foreign ports	838,716	
Stocks on hand in the Southern ports at the close of the year, (October 1, 1830)	20,898	
Destroyed at New Orleans	7,803	
	867,417	
Less—Amount of Foreign cotton included in the export from N.Y.	522	
		866,895

Assuming that the quantity on hand in the Northern ports on the 1st Oct., was the same as last year, it results that the consumption was	126,512
Consumption of 1828–9	104,853
" 1827–8	120,593
" 1826–7	103,483

Note.—In regard to the consumption of cotton in 1827–8, it may be remarked, that the quantity on hand in the manufacturing districts at the close of that year was much greater than usual, which will account for the *apparent* falling off in the succeeding year. We remark, also, that the consumption as above does not include any cotton manufactured in the cotton-growing States.

LIVERPOOL STATEMENT FOR 1830.

UNITED STATES, 1829–1830.			
Stock, October 1, 1829	30,000	Export	846,000
Crop, 1829–'30	977,000	Consumption	126,000
		Stock, Oct. 1, 1830	35,000
Bags	1,007,000	Bags	1,007,000

	Liverpool.	Gt. Britain.	France.	Continent.	Tot. Europe.
United States	99,400	107,344	22,755	14,290	144,389
Brazil	74,500	78,960	1,907	2,014	82,881
West Indies	2,970	5,953	2,817	5,229	13,999
East Indies	9,480	73,993	....	11,277	85,270
Egypt	16,900	23,132	1,813	2,425	27,370
Stock	203,250	289,382	29,292	35,235	353,909

CONSUMPTION.					IMPORT.					
Tot. Europe.	Continent.	France.	Gt. Britain.	Liverpool.		Liverpool.	Gt. Britain.	France.	Continent.	Tot. Europe.
753,389	43,298	171,033	539,058	503,731	United States	571,848	618,754	191,450	41,507	851,711
232,472	15,527	34,601	182,344	178,899	Brazil	189,754	192,814	38,651	17,562	249,027
43,464	18,700	11,700	13,064	8,159	West Indies	7,289	10,925	12,748	19,669	43,342
78,369	32,140	....	46,229	17,767	East Indies	14,087	34,619	....	24,100	58,719
82,842	23,521	34,541	24,780	19,962	Egypt	11,062	14,839	40,994	27,360	83,193
1,190,536	133,186	251,875	805,475	728,518	Consumption	794,040	871,951	283,843	130,198	1,285,992
414,365	32,247	61,260	320,858	258,000	Stock. Supply,	997,290	1,161,333	313,135	165,433	1,639,901
						U. States.	Brazil & W.I.	East India.	Egypt.	
....	....	....	35,000	10,772	Export deducted from Import	7,800	4,130	22,000	1,070	35,000
1,604,901	165,433	313,135	1,161,333	997,290	Total supply (bags)	....	....	....	....	1,604,901

The Price semi-weekly at New York, Weekly Sales, Receipts and Exports, Rates of Freight to Liverpool and Course of Exchange on London, for the Crop Year ending October 1, 1830.

1829.	Price of New Orleans	Price of Upland.	Sales for week.	Receipts for week.	Exports for week.	Exchange on London.	Freight to Liverpool.
October 2..	9½@12½	8@10¾				109¾	¼@⅜d.
" 6..	9½@12½	8@10¾	1,300	1,099	2,298		
" 9..	9½@12½	8@10¾				109¾	
" 13..	9½@12½	8@10¾	1,600	124	706		
" 16..	9½@12½	8@10¾				109¾	
" 20..	9½@12½	8@10¼	600	159	1,856		
" 23..	9½@12½	8@10¾				109¾	
" 27..	9½@12½	8@10¾	2,000	1,855	2,054		
" 30..	9@12½	8@10¾				109¾	
Novem. 3..	9@12½	8@10¾	1,400	44	1,391		¼@⅜d.
" 6..	9@12½	8@10¾				109¾@110	
" 10..	9@12½	8@10¾	3,200	2,713	112		
" 13..	9@12½	8@10¾				109¾	
" 17..	9½@12½	8½@10¾	2 800	1,938	2,298		
" 20..	9½@12½	8½@10¾				109¾	
" 24..	9½@12½	8½@10¾	2,200	2,857	2,000		
" 27..	9½@12½	8½@11				109¾	
Decem. 1..	9½@12½	8½@10¾	1,400	1,408	2,840		¼d.
" 4..	9½@12½	8¼@10¾				109¾	
" 8..	9½@12½	8¼@10½	1,600	2,667	....		
" 11..	9½@12½	8¼@10½				109½@109¾	
" 15..	10@12½	8¼@10½	1,100	2,339	1,047		
" 18..	10@12½	8¼@10½				109½@109¾	
" 22..	10@12½	8¼@10½	1,700	4,649	1,247		
" 25..	10@12½	8¼@10½				109½@109¾	
" 29..	10@12½	8¼@10½	2,300	6,651	1,543		
1830.							
January 1..	10@12½	9@10¾				109½	⅜d.
" 5..	10@12½	9@10¾	1,400	2,214	1,476		
" 8..	10@12½	9@10¾				109¼@109½	
" 12..	10@12½	9@10¾	800	5,823	2,404		
" 15..	10@12½	8¾@10½				109	
" 19..	10@12½	8½@10½	2,500	3,033	632		
" 22..	10@12½	8½@10½				108¾@109	
" 26..	10@12½	8½@10½	4,400	7,850	867		
" 29..	10@12½	8½@10½				108½@108¾	
February 2..	10@12½	8½@10½	3,200	2,560	3,153		½@9-16d.
" 5..	10@12½	8½@10½				108½@108¾	
" 9..	10¼@12½	8½@10½	2,400	3,852	59		
" 12..	10¼@12½	8¾@10½				108½	
" 16..	10¼@12½	8¾@10½	5,300	3,621	2,244		
" 19..	10½@12½	9@10¾				108¼@108½	
" 23..	10@12½	9@10¾	4,200	7,328	3,637		
" 26..	10@12½	9@10¾				108@108¼	
March 2..	10@12½	9@10¾	3,900	3,724	4,928		½@⅝d.
" 5..	10@12½	9@10¾				108@108¼	
" 9..	10@12½	9@10¾	1,600	2,792	163		
" 12..	10@12½	9@10¾				108@108¼	
" 16..	10@12½	9@10¾	1,200	7,348	3,852		
" 19..	10@12½	9@10¾				108	
" 23..	10@12½	8¾@10¾	1,800	5,555	2,433		
" 26..	10@12½	8¾@10¾				108	
" 30..	10@12½	8¾@10¾	1,600	4,671	1,189		
April 2..	10@12½	8¾@10¾				108	½@9-16d
" 6..	10@12½	8¾@11	4,700	4,540	2,749		
" 9..	10@12½	8¾@11				108	
" 13..	10@12½	8¾@11	3,800	2,117	1,671		
" 16..	10½@12½	9@11				108	
" 20..	10½@12½	9@11	5,300	2,161	2,644		
" 23..	10½@12½	9@11				107½@107¾	

New York Statement for Year 1830—*Continued.*

1830.	Price of New Orleans	Price of Upland.	Sales for week.	Receipts for week.	Exports for week.	Exchange on London.	Freight to Liverpool.
April 27..	10½@12½	9@11	3,000	9,899	1,090		
“ 30..	10½@12½	9@11				107½	
May 4..	10½@12½	9@11	1,600	4,258	5,209		½@⅝d.
“ 7..	10½@12½	9@11				107	
“ 11..	10½@12½	9@11	3,000	3,721	2,397		
“ 14..	10½@12½	9@11				107	
“ 18..	10½@12½	9@11	2,900	2,479	5,963		
“ 21..	10½@12½	9@11				107	
“ 25..	10½@12½	9@11	2,100	2,596	1,480		
“ 28..	10½@12½	9@11				107	
June 1..	10½@12½	9@11	1,900	3,344	7,141		½@⅝d
“ 4..	10½@12½	9@11				107¼	
“ 8..	10½@12½	9@11	4,200	5,890	554		
“ 11..	10½@12½	9@11				107	
“ 15..	10½@12½	9@11	3,400	2,319	4,604		
“ 18..	10½@12½	9@11				106¾@107	
“ 22..	10½@12½	9@11	4,900	2,999	3,092		
“ 25..	10½@12½	9½@11½				106½@106¾	
“ 29..	10½@12½	9½@11½	2,100	1,219	6,297		
July 2..	10½@12½	9¾@11½				106@106¼	⅜@½d.
“ 6..	10½@12½	9¾@11½	1,900	1,570	2,833		
“ 9..	10½@13	9¾@11½				106	
“ 13..	10½@13	9¾@11½	800	4,030	3,076		
“ 16..	10½@13	9¾@11½				105½@106	
“ 20..	10½@13	9¾@11½	1,000	2,643	1,862		
“ 23..	10½@13	9¾@11½				106	
“ 27..	10½@13	9¾@11½	700	1,117	1.170		
“ 30..	10½@13	9½@11½				106@106¼	
August 3..	10½@13	9½@11½	2,300	493	4,197		⅜@½d.
“ 6..	10½@13	10@12				106¼@106½	
“ 10..	10½@13½	10@12	3,400	1,177	1,751		
“ 13..	10½@13½	10@12				106½	
“ 17..	11@13½	10@12	1,500	512	2,739		
“ 20..	11@12½	10@12½				106½	
“ 24..	11@13½	10@12½	1,800	1,015	2,492		
“ 27..	11@13½	10@12½				106½	
“ 31..	11@13½	10@12½	900	....	195		
Septem. 3..	11@13½	10@12½				106	¼@⅜d.
“ 7..	11@13½	10@12½	1,100	2,149	1,478		
“ 10..	11@13½	10@12½				106	
“ 14..	11@13½	10@12½	2,000	2,048	236		
“ 17..	11½@13½	10½@12½				106	
“ 21..	11½@13½	10½@12½	1,800	234	728		
“ 24..	11½@13½	10½@12½				106	
“ 28..	11½@13½	10½@12½	900	541	1,318		
Average prices and total sales, receipts and exports.	11.44	10.04	120,500	153,945	115,395		

COTTON AT LIVERPOOL. YEAR 1830.

Week Ending.	RECEIPTS.						SALES.				PRICES.			Actual Export.	Consumpt'n
	American.	E. I.	Egypt.	Brazil.	Other.	Total.	Consumpt'n.	Speculation.	Export	Total.	Mid. Up	Mid. Orl	Dhol.		
Jan. 8...	1,305	......	......	....	702	2,007	5,158	500	......	5,658	6	6⅛	3	....	5,158
" 15...	307	......	......	2,101	....	2,408	19,095	8,200	......	27,295	6¼	6½	3¼	....	24,253
" 22...	2,452	......	......	....	13	2,465	9,677	....	......	9,677	6¼	6½	3¼	....	33,930
" 29...	13,501	......	......	1,655	....	15,156	7,692	1,000	......	8,692	6⅛	6½	3½	....	41,622
Feb. 5...	6,934	......	......	3,201	186	10,321	12,315	....	......	12,315	6	6¼	3½	....	53,937
" 12...	23,573	......	......	9,988	447	34,008	11,740	....	......	11,740	6	6¼	3½	....	65,677
" 19...	14,968	......	......	5,785	824	21,577	19,650	....	300	19,950	6	6¼	3½	300	85,327
" 26...	2,583	......	......	1,789	....	4,372	26,165	3,000	......	29,165	6⅛	6¼	3½	....	111,492
Mch. 5...	17,393	......	......	1,403	454	19,250	18,535	1,000	......	19,535	6¼	6½	3½	....	130,027
" 12...	11,913	......	......	3,534	369	15,816	16,871	3,800	......	20,671	6¼	6½	3½	....	146,898
" 19...	25,620	......	......	....	2,925	28,545	18,709	2,300	......	21,009	6⅜	6⅝	4	....	165,607
" 26...	20,218	1,758	......	4,153	15	26,144	22,214	2,000	......	24,214	6⅜	6¾	4¼	....	187,821
April 2...	5,283	......	......	....	....	5,283	15,430	1,500	......	16,930	6¾	6¾	4¼	....	203,251
" 8...	819	......	......	1,500	5	2,324	16,357	1,900	......	18,257	6⅞	6⅞	4⅜	....	219,608
" 16...	13,898	......	......	6,438	343	20,679	13,599	....	......	13,599	6⅞	6⅞	4⅜	....	233,207
" 23...	23,247	6,534	3,270	5,820	305	39,176	4,552	....	......	4,552	6¾	6⅞	4¼	....	237,759
" 30...	16,915	......	......	7,680	3,920	28,515	5,440	....	......	5,440	6½	6½	4	....	243,199
May 7...	10,062	......	......	....	623	10,685	18,114	1,000	......	19,114	6⅝	6¾	4	....	261,313
" 14...	4,623	......	1,509	1,078	3	7,213	14,745	1,000	......	15,745	6⅝	6¾	4	....	276,058
" 21...	38,287	......	......	....	138	38,425	9,110	1,000	......	10,110	6½	6½	4	....	285,168
" 28...	7,993	......	......	7,666	34	15,693	15,496	....	......	15,496	6⅜	6⅜	4	....	300,664
June 4...	16,795	......	......	....	2,240	19,035	12,021	....	......	12,021	6⅜	6⅜	4	....	312,685
" 11...	20,409	......	......	....	....	20,409	11,636	1,000	......	12,636	6⅜	6⅜	4	....	324,321
" 18...	19,927	......	......	2,785	1,174	23,886	20,999	1,500	......	22,499	6⅜	6½	4	....	345,320
" 25...	5,580	......	......	3,487	46	9.113	26,980	3,500	......	3,480	6½	6⅝	4	....	372,300
July 2...	41,114	......	......	5,076	590	46,780	17,300	3,000	......	20,300	6½	6⅝	4	....	389,610
" 9...	29,372	......	......	2,751	231	32,354	20,009	6,000	......	26,009	6¾	6⅞	4½	....	409,609
" 16...	12,122	......	......	1,105	16	13,243	28,382	6,000	......	34,382	6¾	6⅞	4½	....	437,991
" 23...	19,829	......	......	5,924	1,360	27,113	14,173	1,000	......	15,173	6¾	6⅞	4½	....	452,164
" 30...	6,313	......	......	....	55	6,368	21,910	5,000	......	26,910	6⅞	7	4½	....	474,074
Aug. 6...	21,722	......	......	6,017	167	27,906	13,522	1,000	......	14,522	6⅞	7	4½	....	487,596
" 13...	5,478	......	......	....	1,310	6,788	16,963	....	......	16,963	6⅞	7	4½	....	504,559
" 20...	9,758	......	......	3,434	219	13,411	6,850	400	......	7,250	6⅞	7	4½	....	511,409
" 27...	18,255	......	......	....	2,290	20,545	15,115	500	......	15,615	6⅞	7	4½	....	526,524

Sept. 3	10,058			4,992	250	15,300	10,160			10,160	6⅞	7	4½		436,684
" 10	2,434					2,434	7,070	2,000		9,070	6⅝	6⅞	4½		543.754
" 17	13,097			5,821	418	19,336	5,040			5,040	6½	6¾	4½		548,794
" 24	4,274				2,536	6,810	6,850	1,000		7,850	6½	6¾	4½		555,644
Oct. 1	4,566		2,346	9,297	46	16,255	23,440	1,000		24,440	6½	6¾	4½		579,084
" 8	1,893	575		1,865	460	4,793	6,230	2,000	220	8,450	6½	6¾	4½	220	585,314
" 15	878				104	982	11,189			11,189	6⅜	6¾	4½		596,503
" 22	289		1,720	4,015	50	6,074	7,992			7,992	6¼	6⅝	4½		604,495
" 29	3,051			3,401	1,447	7,899	6,961			6,961	6¼	6⅝	4½		611,456
Nov. 5	3,308	1,856		807		5,971	9,375			9,375	6¼	6½	4½		620,831
" 12	2,373	1,993		8,619	1,872	14,857	8,082			8,082	6¼	6⅜	4½		628,913
" 19	1,488				385	1,873	20,163	2,000		22,163	6¼	6½	4½		649,076
" 26	1,070			9,403	2,088	12,561	10,933			10,933	6¼	6½	4½		660 009
Dec. 3							17,486			17,486	6⅜	6½	4½		677,495
" 10			2,178			2,178	9,743			9,743	6⅜	6½	4½		687.238
" 17	13,807	300		12,246	6,466	32,819	8,023			8,023	6⅛	6½	4½		695,261
" 23	2,162				1,812	3,974	5,550			5,550	6	6½	4½		700,811
" 31	17,492	200		6,389		24,081	8,450			8,450	6	6½	4½		709,261
Average prices & total sales, receipts & stocks.	570,808	12,276	11,023	161,225	38,538	793,870	709,261	65,100	520	774,881	6.44	6.62	4.16	520	1,364,000

Total American and other stocks. 1,034,500.

1831.

Lancashire, England, had a population of 1,336,854. (*See* years 1700, 1750 and 1801.)

One hundred and seventeen "natives," of Calcutta, "of high respectability," petitioned the "Right Honorable the Lords of His Majesty's Privy Council for Trade" of England to be admitted to the privilege of British subjects—(the fabrics of Great Britain being consumed in Bengal without any duties being levied thereon to protect the native fabrics)—and that the cotton and silk fabrics of Bengal, be allowed to be used in Great Britain "free of duty," or at the same rate which might be charged on British fabrics consumed in Bengal. This request was not complied with.

The extra duty of 3½d. per yard on printed cottons imported into Great Britain, was taken off this year.

Excise duty on English prints was repealed.

Hands employed in cotton manufactures in Great Britain, 833,000. (*See* years 1785 and 1787.)

The duties fixed on cotton goods in (*see* year) 1787, were entirely remitted in England and relieved the trade. (*See* years 1779, 1782 and 1784.)

Cotton exported from United States, 270,979,784 lbs., 8,312,762 lbs., being Sea Island; total value, $25,289,492.

Value of foreign cotton goods imported into England, £35,180.

Manchester, England, had a population of 270,961. (*See* year 1774.) Liverpool had 165,175. (*See* year 1770.) Glasgow had 202,426. (See year 1763.) Paisley had 57,466. (*See* year 1782.) Preston had 33,172. (*See* year 1780.) Blackburn had 27,091. (*See* year 1770.) Bolton had 43,396. (*See* year 1773.) Wigan had 20,774. (*See* year 1801.) Ashton had 33,597. (*See* year 1775.) Parish of Oldham had 50,513. (*See* year 1789.)

Sir John Hobhouse brought a bill in the English House of Commons, to shorten the term of labor for young persons under eighteen years of age in all factories, to eleven and a-half hours per day, but was defeated in his object. The bill passed, but fixed the term of labor at twelve hours, and was confined in its operation to the cotton mills. (*See* years 1802, 1819, 1832 and 1833.)

Imports of cotton into Great Britain from foreign countries, 259,808,104 lbs., as follows: United States, 219,333,628; Brazil, 31,695,761; Turkey and Egypt, 8,081,024; miscellaneous, 697,691. From British Possessions, 28,866,749 lbs., as follows: East Indies and Mauritius, 25,805,153; British West Indies—the growth of—2,228,927; foreign, 172,758; miscellaneous, 659,911. Total imports,

288,674,853 lbs.; Exports, 22,308,555 lbs.; Home consumption, 273,249,653 lbs.

Number of cotton mills in the United States, 795. (*See* years 1791, 1807 and 1810.)

Cotton exported from the United States, 773,000 bales; to Great Britain, 619,000, to France, 127,000, and to other parts, 27,000 bales.

The number of lace frames in Great Britain this year, was 4,501, 3,501 of which were hand machines. They worked up 1,600,000 lbs. annually of Sea Island cotton, which were spun into 1,000,000 pounds of yarn, valued at £500,000, and gave employment to fifty-five spinning factories at Manchester, containing 860,000 spindles.

A spinning frame called the "ring spinner," was invented this year by John Sharp, of Providence, R. I.

The first cotton mill erected at Amoskeag Falls in Manchester, N. H.

The practice of "frame breaking"—the destruction of lace-making machinery—was inaugurated this year, by hand-weavers, in the three hosiery Midland counties of England. (*See* year 1846.)

The following table sets forth the state of the cotton manufacture in twelve of the United States this year. The figures are official:

STATES.	Capital.	Number of Spindles.	Yards of cloth produced yearly.	Pounds of cloth produced yearly.	Pounds of cotton consumed yearly.
Maine	$765,000	6,500	1,750,000	525,000	588,500
New Hampshire	5,300,000	113,776	29,060,500	7,255,060	7,845,000
Vermont	295,500	12,392	2,238,400	574,500	760,000
Massachusetts	12,891,000	339,777	79,231,000	21,301,062	24,871,981
Rhode Island	6,262,340	235,753	37,121,681	9,271,481	10,414,578
Connecticut	2,825,000	115,528	20,055,500	5,612,000	6,777,209
New York	3,669,500	157,316	21,010,920	5,297,713	7,661,670
New Jersey	2,027,644	62,979	5,133,776	1,877,418	5,832,204
Pennsylvania	3,758,500	120,810	21,332,467	4,207,192	7,111,174
Delaware	384,500	24,806	5,203,746	1,201,500	1,535,000
Maryland	2,144,000	47,222	7,649,000	2,224,000	3,008,000
Virginia	290,000	9,844	675,000	168,000	1,152,000
Total	40,612,984	1,246,703	230,461,990	59,514,926	77,457,316

The following table sets forth the total number of mills, etc., in the above twelve States this year:

Number of Mills	801
" Looms	33,433
" Males employed	18,590
" Females "	38,927
" Children "	4,091

COTTON CROP OF THE UNITED STATES.

Statement and Total Amount of the Cotton Crop of the United States, for the Year ending September 30, 1831.

	Bales.	Bales.	Total.	Same period 1830.
NEW ORLEANS.				
Export—				
To Foreign Ports	....	289,598		295,774
Coastwise	....	135,086		56,115
Stock on hand 1st October, 1831	....	13,697		9,505
		438,381		361,394
Deduct—				
Stock on hand 1st October, 1830	9,505			
Received from Mobile	367			
" Florida	2,024			
		11,896		15,173
Add—			426,485	346,221
Burnt and damaged	....	....	....	7,803
			426,485	354,024
FLORIDA.				
Received at—				
New Orleans	....	2,024		
Baltimore	....	330		
New York	....	9,246		
Providence	....	994		
Boston	....	479		
			13,073	5,787
ALABAMA.				
Exported from Mobile—				
To Foreign Ports	....	71,839		
Coastwise	....	40,626		
Stock on hand at Mobile	....	861		
Deduct—		113,326		
Stock on hand 1st October, 1830	....	140		
			113,186	102,684
GEORGIA.				
Export from Savannah—				
To Foreign Ports—Uplands	116,097			
" Sea Islands	7,582			
		123,679		
Coastwise	....	94,590		
		218,269		
From Darien—				
To Liverpool	343			
New York	2,985			
Providence	1,058			
		4,386		
		222,655		
On hand in Savannah	3,947			
" Augusta	10,000			
		13,947		
		236,602		
Deduct stock on hand 1st October, 1830	......	6,100		
			230,502	253,117

Statement and Total Amount of the Cotton Crop of the United States, for the Year ending September 30, 1831.—*Continued.*

	Bales.	Bales.	Total.	Same period 1830.
SOUTH CAROLINA.				
Export from Charleston—				
To foreign ports—Uplands	148,127			
" Sea Islands	18,597			
		166,724		
Coastwise		38,574		
		205,298		
From Georgetown to New York	6,521			
" Providence	23			
		6,544		
Destroyed part of Isaac Hicks' cargo		70		
On hand in Charleston		8,551		
Deduct—		220,463		
On hand 1st October, 1830	3,906			
Received from Savannah and included in the export from that place	31,391			
		35,297		
			185,166	188,871
NORTH CAROLINA.				
Export—				
To Europe		2,532		
New York		29,893		
Philadelphia		1,599		
Providence		124		
Boston		2,492		
On hand		900		
Deduct—		37,540		
Stock on hand 1st October, 1830		1,000		
			36,540	36,862
VIRGINIA.				
Export—				
To Foreign Ports		20,642		
Coastwise		8,500		
On hand		5,000		
Deduct—		34,142		
Stock on hand 1st October, 1830		247		
			33,895	35,500
Total crop of the United States			1,038,847	976,845
			976,845	
Increase compared with last year			62,002	

Export to Foreign Ports, from 1st October, 1830, to 30th September, 1831.

FROM	To Great Britain.	To France.	Other parts of Europe.	Total.
New Orleans (bales)	223,374	60,913	5,311	289,598
Alabama	63,490	7,423	926	71,839
Georgia (Savannah)	111,297	10,874	1,508	123,679
South Carolina	151,265	11,181	4,278	166,724
North Carolina	1,882	650	....	2,532
Virginia	18,323	1,726	593	20,642
New York	45,906	33,716	10,202	89,824
Philadelphia	2,678	46	619	3,343
Boston	100	500	3,100	3,700
Providence, Baltimore and Darien	403	....	499	902
Grand total	618,718	127,029	27,036	772,783
Total last year	595,713	200,791	42,212	838,716
Increase	23,005			
Decrease	....	73,762	15,176	65,933

Growth.

Total crop of 1824–5, 560,000 bales.
" 1825–6, 710,000 "
" 1826–7, 937,000 "
" 1827–8, 712,000 "
" 1828–9, 857,744 "
" 1829–30, 976,845 "
" 1830–1, 1,038,847 "

Consumption.

To estimate the quantity manufactured in the United States, we take the growth of the year		1,038,847 bales.
Add—Stocks on hand at the commencement of the year October 1, 1830, in the Southern ports	20,898	
In the Northern ports	13,997	
		34,895
		1,073,742
Deduct therefrom the export to Foreign ports	772,783	
Stocks on hand at the close of the year October 1, 1831.		
In the Southern ports	42,956	
In the Northern ports	76,467	
		119;423
		892,206
Less Foreign cotton included in the export		606
		891,600
Quantity consumed and in the hands of manufacturers, 1830–1, (bales)		182,142
Consumption of 1829–30		126,512
" 1828–9		118,853
" 1827–8		120,593
" 1826–7		103,483

Note.—The quantity of cotton in the hands of manufacturers, is known to be much larger than it was last year and will probably reduce the actual consumption to about 150,000 bales.

LIVERPOOL STATEMENT FOR 1831.

UNITED STATES, 1830–1831.

Stock 1st October, 1830	35,000	Export	772,000
Crop 1830–31	1,038,000	Consumption	182,000
		Stock 1st Oct., 1831	119,000
Bags	1,073,000	Bags	1,073,000

11

						Liverpool.	Gt. Britain.	France.	Continent.	Tot. Europe.
					United States	160,800	179,240	43,172	12,499	234,911
					Brazil	81,300	85,500	5,957	4,049	95,506
					West Indies	2,100	3,614	3,865	6,198	13,677
					East Indies	5,800	40,383	....	3,237	43,620
					Egypt	8,000	12,121	8,266	6,264	26,651
					Stock	258,000	320,858	61,260	32,247	414,365
CONSUMPTION.										
Tot. Europe.	Continent.	France.	Gt. Britain.	Liverpool.	IMPORT.					
834,190	41,503	174,079	618,608	572,399	United States	560,389	608,768	152,868	36,160	797,796
218,498	21,872	17,822	178,804	175,954	Brazil	164,344	170,234	14,164	19,600	203,998
32,941	13,760	7,468	11,713	8,097	West Indies	7,947	12,859	5,287	12,598	30,744
79,625	39,898	....	39,727	19,861	East Indies	33,601	76,654	....	40,500	117,164
133,739	56,339	44,764	32,636	29,809	Egypt	27,089	36,675	47,472	60,022	144,169
1,298,993	173,372	244,133	881,488	806,120	Consumption	793,370	905,190	219,791	168,880	1,293,861
339,483	27,755	36,918	274,810	212,350	Stock. Supply,	1,051,370	1,226,048	281,051	201,127	1,708,226
						U. States.	Braz. & W.I.	East India.	Egypt.	
....	....	....	69,750	32,900	Export deducted from Import	15,212	15,218	38,970	350	69,750
1,638,476	201,127	281,051	1,226,048	1,051,370	Total supply of bags	....	....	....	....	1,638,476

COTTON AT LIVERPOOL. YEAR 1831.

Week Ending.	Receipts.						Sales.				Stocks.			Prices.			Actual Export	Con-sumption.
	Americ'n.	E. I.	Egypt.	Brazil.	Other.	Total.	Con-sumption.	Specu-lation.	Export.	Total.	Amer'n.	Other.	Total.	Mid. Up.	Mid. Orl.	Dhol		
Jan. 7	1,494					1,494	23,047			23,047				6	$6\frac{3}{8}$	$4\frac{1}{2}$		23,047
" 14	5,440				143	5,583	11,826			11,826				$5\frac{7}{8}$	$6\frac{1}{8}$	$4\frac{1}{2}$		34,873
" 21	3,497	2,687				6,184	27,577			27,577				$5\frac{7}{8}$	$6\frac{1}{8}$	$4\frac{1}{2}$		62,450
" 28	4,857					4,857	18,000	1,300	800	20,100				$5\frac{7}{8}$	$6\frac{1}{8}$	$4\frac{1}{2}$	800	80,450
Feb. 4	28,703			14,655	886	44,244	10,601			10,601	137,000	85,300	222,300	$5\frac{3}{4}$	6	$4\frac{1}{2}$		91,051
" 11	8,949			4,884	2,993	16,826	14,137	1,000		15,137				$5\frac{5}{8}$	$5\frac{7}{8}$	$4\frac{1}{2}$		105,188
" 18	12,521			6,334	712	19,567	24,258	500		24,758				$5\frac{3}{4}$	6	$4\frac{1}{2}$		129,446
" 25	11,859			2,100		13,959	16,120	1,500		17,620				$5\frac{3}{8}$	$5\frac{7}{8}$	$4\frac{1}{2}$		145,566
Mch. 4	22,210		1,070	1,756	491	25,527	17,620			17,620	143,000	90,000	233,000	$5\frac{5}{8}$	$5\frac{7}{8}$	$4\frac{1}{2}$		163,186
" 11	13,789				2,071	15,860	15,411			15,411				$5\frac{5}{8}$	$5\frac{7}{8}$	$4\frac{1}{2}$		178,597
" 18	9,753	1,862		4,155	1,609	17,379	16,308			16,308				$5\frac{5}{8}$	$5\frac{7}{8}$	$4\frac{1}{2}$		194,905
" 25	5,068				82	5,150	15,960			15.960				$5\frac{1}{2}$	$5\frac{7}{8}$	$4\frac{1}{2}$		210,865
" 31	9,535		2,741			12,276	21,092	4,000		25,092	118,000	84,000	202,000	$5\frac{3}{4}$	6	$4\frac{1}{2}$		231,957
April 8	2,610		2,298			4,908	12,388			12,388				$5\frac{3}{4}$	6	4		244,345
" 15	23,109	4,685		8,904	899	37,597	10,753			10,753				$5\frac{5}{8}$	6	4		255,098
" 22				120	30	150	16,903			16,903				$5\frac{1}{2}$	$5\frac{7}{8}$	$3\frac{7}{8}$		272,001
" 29	886	2,422		1,610	608	5,526	10,500			10,500				$5\frac{1}{2}$	$5\frac{7}{8}$	$3\frac{7}{8}$		282,501
May 6	29,259	103		2,813	1,141	33,316	9,341	1,000		10,341	125,500	87,000	212,500	$5\frac{1}{2}$	$5\frac{7}{8}$	$3\frac{7}{8}$		291,842
" 13	3,079	300		570	394	4,343	15,467		2,600	18,067				$5\frac{3}{8}$	$5\frac{3}{4}$	$3\frac{7}{8}$	2,600	307,309
" 20	4,169		2,220	882		7,271	15,972		1,250	17,222				$5\frac{3}{8}$	$5\frac{3}{4}$	$3\frac{7}{8}$	1,250	323,281
" 27	10,349		3,011	2,725	22	16,107	10,643		2,000	12,643				$5\frac{3}{8}$	$5\frac{3}{4}$	$3\frac{7}{8}$	2,000	333,924
June 3	18,261			6,225	572	25,058	10,912		700	11,612	138,000	86,000	224,000	$5\frac{3}{8}$	$5\frac{5}{8}$	$3\frac{7}{8}$	700	344,836
" 10	35,739			10,497	3,597	49,833	14,703		2,800	17,503				$5\frac{3}{8}$	$5\frac{5}{8}$	$3\frac{7}{8}$	2,800	359,539
" 17	41,520	1,965	1,622	9,164	2,627	56,898	11,430		2,900	14,330				$5\frac{1}{4}$	$5\frac{1}{2}$	$3\frac{3}{4}$	2,900	370,969
" 24	8,218			899	1,380	10,497	15,750		1,000	16,750				$5\frac{1}{4}$	$5\frac{1}{2}$	$3\frac{3}{4}$	1,000	386,719
July 1	11,581					11,581	13,972	1,500	220	15,692	190,000	99,000	289,000	$5\frac{1}{4}$	$5\frac{1}{2}$	$3\frac{3}{4}$	220	400,691
" 8	7,882		2,010		52	9,944	18,877			18,877				$5\frac{1}{4}$	$5\frac{1}{2}$	$3\frac{3}{4}$		419,268
" 15	17,625	400		4,894	1,476	24,395	12,859			12,859				$5\frac{1}{4}$	$5\frac{1}{2}$	$3\frac{3}{4}$		432,127
" 22	17,003	867				17,870	13,157	1,000		14,157				$5\frac{1}{8}$	$5\frac{1}{2}$	$3\frac{3}{4}$		445,284
" 29	9,926					9,926	20,070	2,000		23,070				$5\frac{1}{4}$	$5\frac{1}{2}$	$3\frac{3}{4}$		466,354
Aug. 5	7,568				1,964	9,532	11,860			11,860	193,000	92,000	285,000	$5\frac{1}{4}$	$5\frac{1}{2}$	$3\frac{3}{4}$		478,214
" 12	33,321	1,740	3,814	8,733	155	47,763	8,212			8,212				$5\frac{1}{8}$	$5\frac{1}{2}$	$3\frac{3}{4}$		486,426
" 19	5,796				104	5,900	10,020			10,020				$5\frac{1}{8}$	$5\frac{1}{2}$	$3\frac{3}{4}$		496,446
" 26	19,549			1,373	101	21,023	17,621	500		18,121				$5\frac{1}{8}$	$5\frac{1}{2}$	$3\frac{3}{4}$		514,067

Sept. 2...	23,517	3,642	......	7,788	5,101	40,048	23,703	......	1,300	25,003	231,000	108,000	339,000	5⅛	5½	3¾	1,300	537,370
" 9...	7,989	1,297	......	2,689	252	12,227	11,821	......	350	12,171	......	......	......	5⅛	5½	3¾	350	549,191
" 16...	2,823	......	1,425	3,519	......	7,767	22,702	2,000	800	25,502	......	......	......	5⅛	5½	3¾	800	571,893
" 23...	7,327	......	2,166	1,370	190	11,053	16,040	3,000	1,500	20,540	......	......	......	5⅛	5½	3¾	1,500	587,933
" 30...	5,903	......	......	......	4	5,907	18,500	2,000	2,500	23,000	194,500	96,500	291,000	5⅛	5½	3¾	2,500	606,433
Oct. 7...	7,415	1,325	......	1,687	......	10,427	12,948	1,500	2,000	16,448	......	......	......	5¼	5⅝	3¾	2,000	619,381
" 14...	2,275	1,965	1,113	4,548	......	9,901	8,620	300	1,000	9,920	......	......	......	5¼	5⅝	3¾	1,000	628,001
" 21...	5,519	1,592	......	......	148	7,259	17,538	2,000	1,800	21,338	......	......	......	5¼	5⅞	3¾	1,800	645,539
" 28...	2,724	......	1,045	3,153	......	6,922	18,600	2,500	900	22,000	......	......	......	5¼	5⅞	3¾	900	664,139
Nov. 4...	4,618	104	......	1,563	45	6,330	9,629	......	400	10,029	174,000	93,000	267,000	5¼	5⅞	3¾	400	673,768
" 11...	554	......	......	......	1,319	1,873	8,239	......	......	8,239	......	......	......	5¼	5⅞	3¾	......	682,007
" 18...	......	......	......	4,671	24	4,695	8,290	......	......	8,290	......	......	......	5⅛	5⅞	3¾	......	690,297
" 25...	5,179	......	......	4,741	770	10,690	9,123	......	300	9,423	......	......	......	5⅛	5¾	3¾	300	699,420
Dec. 2...	2,870	4,254	......	2,884	25	10,033	17,243	1,500	......	18,743	149,500	102,500	252,000	5⅛	5¾	3¾	......	716,663
" 9...	1,916	2,091	484	3,049	110	7,650	22,272	4,000	......	26,272	......	......	......	5⅛	5¾	3¾	......	738,935
" 16...	7,130	......	......	......	414	7,544	16,703	2,000	......	18,703	......	......	......	5⅛	5¾	3½	......	755,638
" 23...	8,716	......	......	1,360	......	10,076	12,090	......	500	12,590	......	......	......	5⅛	5¾	3½	500	767,728
" 30...	5,081	300	......	......	......	5,381	12,920	500	300	13,720	......	......	......	5¼	5¾	3½	300	780,648
Average prices & total sales, receipts & stocks.	560,181	33,601	25,019	138,312	32,511	789,624	780,648	35,600	27,920	844,168				5.38	5.72	3.95	27,920	1,501,246

The semi-weekly Price and Weekly Sales and Receipts at New York, Weekly Exports from New York and Rates of Freight to Liverpool 1st of each month, for the Crop Year ending October 1, 1831.

1830.	Price of New Orleans	Price of Upland.	Sales for week.	Receipts for week.	EXPORTS FOR THE WEEK. To Great Britain.	To France.	North of Europe.	Other Fo'n Ports	Total Exports.	Rates of Freight to Liverpool, 1st each month.	GENERAL REMARKS.
October 1..	12@14¼	11@13	1,700	224	....	....	....	....	....	¼d.	
" 5..	12@14	11@13									
" 8..	12@14	11@13¼	800	1,879	....	....	....	....	....		
" 12..	12@14	11@13									
" 15..	12@14	11@13	700	2,081	....	....	....	....	....		
" 19..	12@14	11@13									
" 22..	12@14	11@13	1,200	600	*1,341	25	200	2	1,568		*These figures include the shipments of the previous weeks, no statement having been completed *ad interim*.
" 26..	12@14	11@13									
" 29..	12@14	11@13	1,000	1,608	769	....	60	....	829		
Novem 2..	12@14	11@13								⅜d.	
" 5..	12@13½	10½@12½	1,000	1,819	96	8	....	....	104		
" 9..	12@13½	10½@12½									
" 12..	12@13½	10@12½	2,100	1,117	1,320	....	41	....	1,361		
" 16..	12@13½	10@12½									
" 19..	11½@13½	9½@12	1,300	1,025	558	....	....	....	558		
" 23..	11½@13½	9½@12									
" 29..	11½@13½	9½@11½	800	2,247	425	....	....	165	590		
Decem. 3..	11½@13½	9½@11½								⅜@7-16d	December 24.—Latest European advices October 23, and buyers for export wait further accounts.
" 7..	11½@13½	9½@11½	2,300	4,757	266	....	....	75	341		
" 10..	11½@13	9½@11½									
" 14..	11½@13	9½@11½	1,900	2,043	497	....	....	....	497		
" 17..	11¼@13	9¼@11½									On the 28th December, 1830, advices were received from Europe per packet ships to 1st, noting a change in the British Ministry, blockade of Antwerp, Ghent and the ports of the Netherlands, and the probability of a general European war.
" 21..	11¼@13	9¼@11	1,900	2,570	1,590	913	....	114	2,617		
" 24..	11¼@12½	9¼@11									
" 28..	11¼@12½	9¼@11	1,300	686	399	679	....	72	1,150		
1831.											
January 1..	11¼@12½	9¼@10¾								⅜@7-16d	
" 4..	11¼@12½	3¼@10¾	1,800	3,329	327	354	50	....	731		
" 7..	11¼@12½	9¼@10¾									
" 11..	11½@12½	9¼@10¾	1,100	2,532	385	....	....	....	385		
" 14.	11@12	9¼@10¾									
" 18..	11@12	9@10¾	600	290	1,103	....	....	....	1,103		
" 21..	11@12	9@10½									
" 25..	11@12	9@10½	1,100	1,057	141	172	....	....	313		

January 28..	11@12	9@10½									
February 1..	11@12	9@10½	1,350	2,010	1,278	264	....	....	1,542	½d.	
" 4..	11@12	9@10½									
" 8..	11@12	9@10½	800	4,552	658	....	100	121	879		
" 11..	10½@11½	8¾@10½									
" 15..	10½@11½	8¾@10½	700	3,490	989	....	15	354	1,358		European accounts per packet at hand February 17, up to January 1 ; markets abroad unsettled by the state of political affairs and the usual consumption greatly checked ; prices off here ¼ cent.
" 18..	10@11½	8½@10½									
" 22..	10@11½	8½@10½	1,470	8,179	338	154	150	....	642		
" 25..	10@11½	8½@10									
March 1..	10@11½	8½@10	3,230	3,681	684	....	233	12	929	½d.	
" 4..	10@11½	8½@10									
" 8..	10@11½	8½@10	3,900	5,331	958	....	356	....	1,314		
" 11..	10@11½	8½@10									The unfavorable accounts counteracted by an increased demand from spinners and prices advanced again.
" 15..	10@11½	8½@10	3,100	4,041	919	913	....	169	2,001		
" 18..	10@11¾	8½@10¼									
" 22..	10@12	8½@10¼	3,150	5,398	848	....	68	....	916		
" 25..	10@12	8½@10½									
" 29..	10@12	8½@10½	2,800	5,900	445	840	176	308	1,769		
April 1..	10@12	8½@10½								½d.	
" 5..	10@12	8½@10½	960	4,613	1,025	645	135	....	1,805		
" 8..	10@12	8½@10½									
" 12..	10@12½	8½@10½	2,200	4,183	1,384	....	....	....	1,384		
" 15..	10@12½	8½@10½									Advices at hand April 15, from Liverpool, to March 19, announcing the addition by Parliament of ⅝d. per lb. on the previous duty.
" 19..	10@12½	8½@10½	1,400	2,728	558	502	96	....	1,156		
" 22..	10@12½	8¼@10½									
" 26..	10@12¼	8¼@10½	1,090	8,083	263	913	48	156	1,380		
" 29..	10@12¼	8¼@10½									
May 3..	10@12¼	8¼@10½	3,550	2,108	195	....	125	....	320	7-16@½d	
" 6..	10@12¼	8¼@10½									
" 10..	10@12¼	8¼@10½	2,050	11,479	212	944	100	....	1,256		
" 13..	10@12¼	8¼@10½									
" 17	10@12¼	8¼@10½	1,450	3,168	1,266	669	265	56	2,256		
" 20..	10@12¼	8¼@10½									
" 24..	10@12¼	8¼@10½	2,850	8,716	977	778	....	79	1,834		
" 27..	10@12¼	8¼@10½									The course of EXCHANGE for the crop year, was thus :
" 30..	10@12¼	8¼@10½	1,700	7,848	519	387	....	48	954		The opening price on October 1, 1830, for 60 days bills in London, was 6¼ per cent premium, on the 19th, advanced to 7@7¼, on November 23, the United States Bank drew at 6¾@7 ; this was followed by a further decline on December 28, to 6½.
June 3..	10@12¼	8@10								½@⅝d.	
" 7..	10@12¼	8@10	1,800	6,651	1,742	764	....	....	2,506		
" 10..	10@12¼	8@10									
" 14..	10@12	8@10	4,400	2,075	1,211	1,141	....	13	2,365		
" 17..	10@12	8@10									
" 21..	10@12	8@10	4,900	10,318	1,398	214	....	....	1,612		
" 24..	10@12	8@10									

New York Statement for 1831.—*Concluded.*

1831.	Price of New Orleans	Price of Upland.	Sales for week.	Receipts for week.	EXPORTS FOR THE WEEK.					Rates of Freight to Liverpool, 1st each month.	GENERAL REMARKS.
					To Great Britain.	To France.	North of Europe.	Other Fo'n Ports	Total Exports.		
June 28..	10@12½	8@10	1,900	8,516	2,054	1,020	....	243	3,317		February 22, 1831, bills declined to 6; market began to rise on March 18, when the quotations were 7@7½; this was followed by successive advances through April until 9½@9¾ was touched. May 13, prices down to 8½@8¾, and on 24th, to 7¼@7½; on June 1, up to 8, and continued to rise from thence to the close of the crop year, when 10¾@11 was the quotation.
July 1..	10@12½	8@10								½@11-16	
" 5..	10@12½	8@10	1,200	6,290	531	2,423	490	150	3,594		
" 8..	10@12½	8@10									
" 12..	10@12½	8@10	1,600	5,992	714	....	401	....	1,115		
" 15..	10@12½	8@10									
" 19..	10@12½	8@10	3,300	2,882	1,624	1,067	307	....	2,998		
" 23..	10@12½	8@10									
" 26..	10@12½	8@10	3,550	7,458	1,874	3,720	356	80	6,030		
" 29..	10@12½	8@10									
August 2..	10@12½	8@10	1,950	3,357	1,998	....	269	....	2,267	½@11-16	
" 5..	10@12½	7½@10									
" 9..	10@12½	7½@10	2,600	1,425	2,757	1,366	325	....	4,448		
" 12..	10@12½	7½@10									
" 16..	10@12½	7½@10	800	5,273	948	1,203	294	316	2,761		
" 19..	10@12½	7¼@10									
" 23..	10@12½	7½@10	4,500	2,250	256	362	102	192	912		
" 26..	10@12	7½@10									
" 30..	10@12	7½@10	2,950	2,702	991	2,120	11	....	3,122		
Septem. 2..	10@12	7½@10								⅜@½d.	
" 6..	10@12	7½@10	2,150	632	1,326	1,128	44	26	2,524		September 10, accounts received of violent storms of wind and rain in the United States cotton belt, with much damage to the crop, but aside from holders being firm, there is not much change in this market.
" 9..	10@12	7½@10									
" 13..	10@12	7½@10	3,500	1,151	221	2,930	262	364	3,777		
" 16..	10@12	7½@10									
" 20..	10@12	7½@10	3,500	923	1,291	1,983	261	1,016	4,551		
" 23..	10@12	7½@10									
" 27..	10@12	7½@10	3,600	686	98	1,589	35	....	1,722		
" 30..	10@12	7½@10									
October 4..	10@12	7½@10	3,900	1,917	2,139	1,526	272	424	4,361	⅜@7-16d	
Average price and total sales, receipts and exports.	11.51	9.71	112,450	195,870	45,906	33,716	5,647	4,555	89,824		

1832.

Cotton export from United States, 322,215,122 lbs.; 8,743,373 lbs. being Sea Island; total value, $31,742,682.

Quantity of cotton yarn spun in England, 222,596,907 lbs.; number of spindles used, 7,949,208; capital invested, £6,955,557.

During this year one cotton establishment alone in Ireland, near Dublin, sent upwards of 100,000 pieces of prints to Manchester and London. (*See* years 1801, 1816, 1817 and 1825.)

Mr Sadler attempted, in the English House of Commons, to reduce working hours in cotton mills to ten hours per day, but without success. (*See* years 1802, 1819, 1831 and 1833.)

Imports of cotton into Great Britain from foreign countries, 249,578,251 lbs., as follows: United States, 219,756,753; Brazil, 20,109,560; Turkey and Egypt, 9,113,890; miscellaneous, 598,048; from British Possessions, 37,254,274 lbs., as follows: East Indies and Mauritius, 35,178,625 · British West Indies—the growth of—1,708,764; foreign 331,664; miscellaneous, 35,221. Total imports, 286,832,525 lbs. Exports, 18,027,940 lbs.; Home Consumption, 259,412,463 lbs.

According to the report of a committee appointed by Congress, this year, there were in twelve States the following cotton mills, etc.:

Number of Mills		795
" Spindles		1,246,503
" Looms		33,506
" Males employed	18,539	
" Females "	38,927	
Total "	57,466	

There were in Glasgow, Scotland, or belonging to it, at this time, sixty-three weaving factories, containing 14,127 looms. (*See* year 1817.)

80,000 power looms in Lancashire, England.

The "stop-motion" in the drawing frame, was designed by Samuel Batchelder, and brought into use in Saco, Maine, this year by its designer, who is also author of a book of 108 pages on cotton manufacture, published by Little, Brown & Co., of Boston, Mass. No patent was taken for it in this country, its inventor not fully appreciating its importance until it had been put in use by other parties. It was, however, patented in England by H. Houldsworth, from which the inventor derived some profit, and no machinery is now built without this improvement.

COTTON CROP OF THE UNITED STATES.

Statement and Total Amount of the Cotton Crop of the United States, for the Year ending September 30, 1832.

	Bales.	Bales.	Total.	Same period 1831.
NEW ORLEANS.				
Export—				
To Foreign Ports	....	291,678		289,598
Coastwise	....	64,728		135,086
Stock on hand 1st October, 1832	....	7,088		13,697
			363,494	438,381
Deduct—				
Stock on hand 1st October, 1831	....	13,697		
Received from Mobile	....	17,663		
" Florida	....	9,499		
			40,859	11,896
			322,635	126,485
FLORIDA.				
Received at—				
New Orleans	....	9,499		
Baltimore	....	138		
Philadelphia	....	2		
New York	....	12,469		
Providence	....	485		
Boston	....	58		
			22,651	13,073
ALABAMA.				
Exports from Mobile—				
To Foreign Ports	....	72,922		
Coastwise	....	53,773		
Stock in Mobile 1st October, 1832	....	87		
Deduct—		126,782		
Stock on hand 1st October, 1831	....	861		
			125,921	113,186
GEORGIA.				
Export from Savannah—				
To Foreign Ports—Uplands	161,248			
" Sea Island	9,964			
		171,212		
Coastwise	....	112,721		
		283,933		
From Darien—				
To New York	2,617			
Providence	564			
Stock in Savannah 1st October, 1832	1,770			
" Augusta "	1,500			
		6,451		
Deduct—		290,384		
Stock on hand 1st October, 1831	....	13,947		
			276,437	230,502
SOUTH CAROLINA.				
Export from Charleston—				
To Foreign Ports—Uplands	165,687			
" Sea Island	16,941			
		182,628		
Coastwise	....	36,648		
		219,276		

Statement and Total Amount of the Cotton Crop of the United States, for the Year ending September 30, 1832—*Concluded.*

	Bales.	Bales.	Total.	Same period 1831.
From Georgetown—				
To New York	6,578			
Stock in Charleston 1st October, 1832	2,987			
		9,565		
Deduct—		228,841		
On hand 1st October, 1831	8,551			
Received from Savannah, and included in the exports from that place	46,418			
		54,969		
			173,872	185,166
NORTH CAROLINA.				
Export—				
To Philadelphia	....	655		
New York	....	28,026		
Providence	....	13		
Boston	....	467		
Stock on hand 1st October, 1832	....	200		
Deduct—		29,361		
On hand 1st October, 1831	....	900		
			28,461	36,540
VIRGINIA.				
Export—				
To Foreign Ports	....	31,915		
Coastwise	....	9,500		
Stock on hand 1st October, 1832	....	1,085		
Deduct—		42,500		
Stock on hand 1st October, 1831	....	5,000		
			37,500	33,895
Total Crop of the United States	....	....	987,477	1,038,847
Crop of last year	....	....	1,038,847	
Decrease	....	....	51,370	

Export to Foreign Ports, from October 1, 1831, *to September* 30, 1832.

FROM	To Great Britain.	France.	Other parts of Europe.	Total.
New Orleans ... Bales	203,365	78,138	10,175	291,678
Alabama	54,707	18,114	101	72,922
Georgia	141,768	26,743	2,701	171,212
South Carolina	138,683	35,901	8,044	182,628
Virginia	23,375	4,752	3,788	31,915
Philadelphia	2,991	893	88	3,977
Baltimore	1,940	....	60	2,000
New York	70,919	42,063	19,414	132,396
Boston	400	600	2,000	3,000
Grand total	638,148	207,209	46,371	891,728
Total last year	618,718	127,029	27,036	772,783
Increase	19,430	80,180	19,335	118,945

Growth.

Total crop of	1824–5,	560,000	bales.
"	1825–6,	710,000	"
"	1826–7,	937,000	"
"	1827–8,	712,000	"
"	1828–9,	857,744	"
"	1829–30,	976,845	"
"	1830–31,	1,038,848	"
"	1831–32,	987,477	"

Consumption.

To estimate the quantity manufactured in the United States, we take the growth of the year			987,477 bales.	
Add—Stocks on hand at the commencement of the year, 1st October, 1831.—In the Southern ports		42,956		
" Northern "		76,467		
			119,423	
				1,106,900
Deduct therefrom—The export to Foreign ports		891,728		
Stocks on hand at the close of the year, 1st Oct., 1832.—				
In the Southern ports	14,717			
" Northern "	26,882			
		41,599		
			933,327	
Less—Foreign cotton included in the export			227	
				933,100

Quantity consumed, and in the hands of the manufacturers,	1831–2	173,800
" " "	1830–1	182,142
" " "	1829–30	126,512
" " "	1828–9	118,853
" " "	1827–8	120,593
" " "	1826–7	103,483

Note.—It will be perceived that the product, or quantity received from the planters, during the year ending 1st Oct., 1832, was 51,370 bales less than that of the preceding year, while the actual quantity, applicable to the purposes of the year, was 33,158 bales more.—Thus :

Growth of the year ending 1st October, 1832		987,477 bales.
Add, stocks on hand at the commencement		119,423
		1,106,900
Growth of the year ending 1st October, 1831	1,038,847	
Stocks on hand at the commencement	34,895	
		1,073,742
Difference		33,158

It may be noticed also that there is an apparent falling off in the home consumption, compared with the preceding year, while the probability is that the consumption has been increased. The difference is found in the fact, that, at the close of the last year the manufacturers had an unusually large stock on hand, while at the close of the present, they had very little.

LIVERPOOL STATEMENT FOR 1832.

UNITED STATES—1831-1832.

Stock, 1st October, 1831	119,000	Export	891,000
Crop, 1831–32	987,000	Consumption	174,000
		Stock, 1st October, 1832	41,000
Bags	1,106,000	Bags	1,106,000

1832	Liverpool.	Gt. Britain.	France.	Continent.	Tot. Europe.
United States	135,790	154,180	21,961	7,156	183,297
Brazil	58,290	62,140	2,299	1,777	66,216
West Indies	1,950	4,340	1,684	5,036	11,060
East Indies	11,040	38,340	....	3,839	42,179
Egypt	5,280	15,810	10,974	9,947	36,731
Stock	212,350	274,810	36,918	27,755	339,483

CONSUMPTION.					IMPORT.					
Tot. Europe.	Continent.	France.	Gt. Britain.	Liverpool.		Liverpool.	Gt. Britain.	France.	Continent.	Tot. Europe.
873,536	46,910	196,380	630,246	581,126	United States	584,006	628,858	196,439	45,666	870,963
159,290	9,911	14,393	134,986	132,507	Brazil	111,607	113,871	13,359	9,831	137,061
31,024	18,461	3,000	9,563	5,652	West Indies	6,342	8,497	4,336	16,515	29,348
100,064	41,866	....	58,198	27,887	East Indies	45,007	109,285	....	40,000	149,285
172,362	78,632	48,946	44,784	30,950	Egypt	32,470	41,174	39,800	79,491	160,465
1,336,276	195,780	262,719	877,777	778,122	Consumption	779,432	901,685	253,934	191,503	1,347,122
296,729	23,478	28,133	245,118	197,960	Stock. Supply,	991,782	1,176,495	290,852	219,258	1,686,605
						U. States.	Brazil & W.I.	East India.	Egypt.	
....	....	....	53,600	15,700	Export deducted from import	14,800	3,750	34,050	1,000	53,600
1,633,005	219,258	290,852	1,176,495	991,782	Total supply, bags	....	....	....	....	1,633,005

COTTON AT LIVERPOOL. YEAR 1832.

Week Ending.	Receipts.						Sales.				Stocks.			Prices.			Actual Export	Con-sumption.
	Americ'n.	E. I.	Egypt.	Brazil.	Other.	Total.	Con-sumption.	Specu-lation.	Export	To'al.	Amer'n.	Other.	Total.	Mid. Up.	Mid. Orl.	Dhol		
Jan. 6...	5,291	110	......	8,974	......	14,375	23,770	2,000	500	26,270	123,000	78,600	201,500	5 3/8	5 3/4	3 1/2	500	23,770
" 13...	5,465	2,080	......	......	2,028	9,573	15,989	1,500	700	18,189	120,000	76,000	196,000	5 3/8	5 3/4	3 1/2	700	39,759
" 20...	4,427	......	1,616	3,912	74	10,029	10,285	1,000	200	11,485	115,000	78,500	193,500	5 1/4	5 3/4	3 1/2	200	50,044
" 27...	5,426	......	......	......	1,731	7,157	11,670	2,000	1,000	14,670	112,500	77,500	190,000	5 3/8	5 3/8	3 1/2	1,000	61,714
Feb. 3...	8,321	......	......	......	790	9,111	18,715	6,000	500	25,215	107,000	71,000	178,000	5 1/2	5 7/8	3 1/2	500	80,429
" 10...	5,732	......	......	2,120	23	7,875	18,172	6,000	300	24,472	101,500	66,000	167,500	5 5/8	5 7/8	3 7/8	300	98,601
" 17...	3,060	......	......	......	1,660	4,720	13,200	3,200	......	16,400	94,500	63,500	158,000	5 5/8	5 7/8	3 7/8	......	111,801
" 24...	16,006	797	......	3,262	209	20,274	10,133	......	......	10,133	102,500	65,000	167,500	5 5/8	5 7/8	3 7/8	......	121,934
Mch. 2...	7,313	......	......	2,495	2,452	12,260	21,181	2,500	400	24,081	94,500	61,500	156,000	5 7/8	6	4	400	143,115
" 9...	23,077	1,415	3,387	......	671	28,550	14,505	3,500	500	18,505	107,000	61,500	168,500	5 7/8	6	4 1/8	500	157,620
" 16...	12,658	......	......	2,371	......	15,029	21,170	4,800	......	25,970	103,000	59,500	162,500	6	6 1/8	4 1/4	......	178,790
" 23...	11,429	280	2,696	2,772	258	17,435	14,967	1,500	200	16,667	103,500	60,500	164,000	6 1/8	6 1/4	4 1/4	200	193,757
" 30...	2,590	......	685	......	......	3,275	12,808	1,500	800	15,108	96,000	56,500	152,500	6 1/8	6 1/4	4 1/4	800	206,565
Apr. 6...	26,158	......	854	3,836	279	31,127	7,981	......	......	7,981	118,500	56,500	175,000	6 1/8	6 1/4	4 1/4	......	214,546
" 13...	10,867	199	......	......	......	11,066	8,853	300	300	9,453	123,000	54,000	177,000	6 1/8	6 1/4	4 1/4	300	223,399
" 19...	10,433	2,580	......	2,092	369	15,474	10,066	......	300	10,366	125,000	66,000	181,000	6 1/8	6 1/4	4 1/4	300	233,465
" 27...	25,949	2,668	2,568	5,165	110	36,460	9,905	......	460	10,365	141,500	64,000	205,500	6 1/8	6 1/4	4 1/4	460	243,370
May 4...	2,368	......	......	4,571	......	6,939	13,703	......	1,000	14,703	131,500	64,000	195,500	6 1/8	6 1/4	4	1,000	257,073
" 11...	11,048	3,112	5,213	3,915	338	23,626	10,876	......	......	10,876	133,000	74,000	207,000	6	6 1/8	4	......	267,949
" 18...	20,703	1,751	2,548	......	205	25,207	13,050	......	700	13,750	141,500	75,000	216,500	6	6 1/4	4	700	280,999
" 25...	22,728	......	......	2,070	1,254	26,052	19,291	......	500	19,791	149,500	71,500	221,000	6	6 1/4	4	500	300,290
June 1...	28,448	......	2,862	6,079	998	38,387	12,774	......	300	13,074	167,500	77,500	245,000	6	6 1/4	4	300	313,064
" 8...	7,833	......	......	......	......	7,833	15,322	......	......	15,322	163,000	74,000	237,000	6	6 1/8	4	300	328,386
" 15...	19,394	......	2,088	1,991	321	23,794	12,670	600	800	14,070	172,000	74,500	246,500	6	6 1/8	4	800	341,056
" 22...	14,983	115	......	3,199	277	18,574	21,800	3,300	500	25,600	168,000	73,500	241,500	6 1/8	6 1/4	4	500	362,856
" 29...	23,645	......	......	5,107	129	28,881	18,530	4,500	200	23,230	178,500	71,500	250,000	6 1/8	6 1/4	4	200	381,386
July 6...	531	......	......	......	......	531	17,907	5,500	200	23,607	167,500	63,500	231,000	6 3/8	6 1/2	4 1/8	200	399,293
" 13...	45,954	......	560	......	51	46,565	17,161	800	......	17,961	200,500	59,500	260,000	6 3/8	6 1/2	4 1/8	......	416,454
" 20...	17,775	......	......	1,545	174	19,494	8,529	......	......	8,529	210,500	59,000	269,500	6 1/4	6 3/8	4 1/8	......	424,983
" 27...	......	......	2,247	1,831	......	4,078	17,250	1,000	700	18,950	196,000	59,000	255,000	6 1/4	6 3/8	4 1/8	700	442,233
Aug. 3...	......	......	......	......	......	......	13,140	......	400	13,540	185,000	57,000	242,000	6 1/4	6 3/8	4 1/8	400	455,373
" 10...	54,025	1,984	2,036	7,840	304	66,189	13,019	......	300	13,319	229,500	66,500	295,000	6 1/4	6 3/8	4 1/8	300	468,392
" 17...	20,864	......	......	......	69	20,933	16,610	1,000	......	17,610	235,500	62,000	297,500	6 1/4	6 3/8	4 1/8	......	485,002
" 24...	21,439	......	1,614	......	761	23,814	12,555	......	......	12,555	247,000	61,000	308,000	6 1/4	6 3/8	4 1/8	......	497,557

Aug. 31...	6,411	......	......	......	6	6,417	20,620	......	......	20,620	236,500	57,000	293.500	6¼	6⅜	4⅛	......	528,177
Sept. 7...	5,933	......	......	......	......	5,933	22,330	3,000	500	25,830	222,500	54,000	276,500	6¼	6½	4⅜	500	550,507
" 14...	4,829	......	......	5,368	76	10,273	20,065	14,500	......	34,565	212,500	55,000	267,500	6½	6⅝	4½	......	570,572
" 21...	3,974	......	......	......	805	4,779	17,370	10,000	......	27,370	202,500	52,500	255,000	6⅞	7	4¾	......	587,942
" 28...	4,583	......	......	......	68	4,651	13,650	5,000	......	18,650	195,500	51,000	246,500	7	7	5⅛	......	601,592
Oct. 5...	7,137	1,703	......	2,524	522	11,886	10,330	......	......	10,330	195,000	53,000	248,000	7	7	5⅛	......	612,922
" 12...	5,793	3,075	......	4.060	209	13,137	9,592	1,000	......	10.592	1[illegible]3,000	68,000	251,000	7	7	5⅛	......	621,514
" 19...	7,861	985	1,227	2,092	40	12,205	13.856	1,000	......	14,856	189,000	60,500	249,500	6⅞	7	5⅛	......	635,370
" 26...	2,197	......	......	......	33	2,230	7,430	......	......	7,430	183,500	61,000	244,500	6¾	6⅞	5	......	642,800
Nov. 2...	5,203	3,992	......	879	115	10,189	8,540	......	......	8,540	182,500	65,000	247,500	6⅝	6¾	4¾	......	651,340
" 9...	541	......	......	559	6	1,106	10,119	2,500	......	12,619	174,500	54,500	239,000	6⅝	6¾	4¾	......	661,459
" 16...	1,694	......	......	4,522	180	6,396	8,760	......	......	8,760	169,000	68,500	237,500	6⅝	6¾	4¾	......	670,219
" 23...	2,475	......	......	......	170	2,645	14,792	600	......	15,392	159,500	66,000	225,500	6⅝	6¾	4⅝	......	685,011
" 30...	2.703	......	......	......	1	2,704	9,310	500	......	9,810	155,500	63,500	219,000	6⅝	6¾	4½	......	694,321
Dec. 7...	2,047	7.696	......	6,842	48	16,633	10,750	......	......	10,750	148,500	76,500	224,000	6⅝	6¾	4½	......	705,071
" 14...	12,953	1,675	......	......	171	14,799	10,135	......	......	10,135	153,000	76,000	229,000	6⅝	6¾	4½	......	715,206
" 21...	3,771	1,465	......	1.178	92	6,506	13,710	......	......	13,710	145,500	75,000	220,500	6⅝	6¾	4¼	......	728,916
" 28...	5,670	2,096	......	1,589	72	9,427	14,480	......	......	14,480	141,000	75,000	216,000	6⅝	6¾	4¼	......	743,396
Average prices & total sales, receipts & stocks.	581,695	39.778	32,196	104,760	18,849	777,278	743,396	90,600	12,260	844,956				6.22	6.37	4.23	12,560	1,429,607

The semi-weekly Price and Weekly Sales and Receipts, at New York, Weekly Exports from New York and Rates of Freight to Liverpool 1st *of each Month, for the Crop Year ending October* 1, 1832.

1831.	Price of New Orleans	Price of Upland.	Sales for week.	Receipts for week.	Exports for Week. To Great Britain.	To France.	North of Europe.	Other Fo'n Ports	Total Exports.	Rates of Freight to Liverpool.	General Remarks.
October 7 ..	10@12	7½@10								$\frac{5}{8}$@7-16d	
" 11 ..	10@12	7½@10	1,600	675	....	....	....	....	....		* These figures embrace the shipments of the previous two weeks, no statement having been compiled *ad interim.*
" 14 ..	10@12	7½@10									
" 18 ..	10@12	7½@10	1,700	1,085	....	....	....	....	....		
" 21 ..	10@12	7½@10									
" 25 ..	10@12	7½@10	900	1,377	* 4,726	6,154	487	360	11,727		
" 28 ..	10@12	7½@10									
Novem. 1 ..	10@12	7½@10	2,500	1,137	1,126	1,808	....	....	2,934	$\frac{3}{8}$@½d.	
" 4 ..	10@12	7@10									
" 8 ..	10@12	7@10	2,000	1,752	977	179	....	....	1,156		
" 11 ..	10@11½	7@10									1831.
" 15 ..	10@11½	7@10	2,200	1,560	1,983	855	154	10	3,002		
" 18 ..	10@11½	7@10									On October 14, was received 26 bales New Orleans, the first of the new crop.
" 22 ..	10@11½	7@10	2,100	2,534	579	346	446	....	1,371		
" 25 ..	10@11½	7@10									
" 29 ..	10@11½	7@10	4,050	1,892	1,057	761	....	79	1,897		
Decem. 2 ..	10@11½	7@10								$\frac{3}{8}$@½d.	October 19, dates from Liverpool received to September 8, advising the enforcement of the new rate of duty of $\frac{5}{8}$d.
" 6 ..	10@11½	7@10	2,550	1,254	3,036	2,058	60	....	5,154		
" 9 ..	10@11½	7@10									
" 13 ..	10@11½	7@10	2,600	1,432	1,578	376	....	92	2,046		
" 16 ..	10@11½	7@10									
" 20 ..	10@11½	7@10	2,000	1,052	749	2,498	....	178	3,425		
" 23 ..	10@11½	7@10									
" 27 ..	10@11½	7@10	1,200	3,510	1,065	986	221	728	3,000		
" 30 ..	10@11½	7@10									
1832.											1832.
January 3 ..	10@11½	7@10	630	1,079	1,473	2,093	....	....	3,566	$\frac{3}{8}$@½d.	
" 6 ..	10@11½	7@10									Liverpool advices were received on February 16, noting an advance there of ½d. with large sales, which imparted much buoyancy to this market.
" 10 ..	10@11½	7@10	1,600	2,854	....	....	....	647	647		
" 13 ..	10@11½	7@10									
" 17 ..	10@11½	7@10	1,000	2,004	366	....	235	54	655		
" 20 ..	10@11½	7@10									
" 24 ..	10@11½	7½@10	2,900	694	2,500	511	....	35	3,046		

January 27..	10@12	7½@10									
" 31..	10@12	7½@10	1,000	2,003	1,108	119	126	270	1,623		
February 3..	10@12	7½@10								⅜@½d.	
" 7..	10@12¼	7½@10¼	2,900	4,890	1,404	163	307	....	1,874		
" 10..	10@12¼	7½@10¼									
" 14..	10@12¼	7½@10¼	1,300	4,566	782	....	....	188	970		
" 17..	10@12¼	7½@10¼									
" 21..	10@12¼	7½@10¼	3,500	1,565	809	....	....	830	1,639		
" 24..	10@12¼	7½@10¾									
" 28..	10@12¼	7½@10¾	1,750	3,859	828	211	185	....	1,224		
March 2..	10@12¼	7½@10¾								7-16@½d	
" 6..	10@12½	7½@10¾	1,100	5,600	1,446	706	270	....	2,422		March 8, advices at hand from England to February 6, noting a further advance of ⅛@¼d. ; upland being quoted 5½@7d. ; market here advanced, with large business.
" 9..	10@13	7½@11									
" 13..	10@13	8@11	1,810	7,189	2,535	....	....	273	2,808		
" 16..	10@13	8@11									
" 20..	10@13	8@11	3,100	3,519	1,161	632	127	85	2,005		
" 23..	10@13	8@11									
" 27..	10½@13	8@11¼	1,550	6,917	314	383	180	90	967		About July 1, the Asiatic cholera appeared in a violent form, and mercantile operations almost wholly suspended. 30 to 40,000 persons left the city and the stagnation continued until last of August, when with an abatement in the epidemic, citizens returned and trade resumed its usual course.
" 30..	10½@13	8½@11¼									
April 3..	10½@13	8½@11¼	1,700	5,455	1,302	....	....	....	1,302	⅜@½d	
" 6..	10½@13	8½@11¼									
" 10..	10½@13	8½@11¼	6,900	6,257	2,016	840	205	....	3,061		
" 13..	10½@13	8½@11¼									
" 17..	10½@13	8½@11¼	2,850	3,792	601	549	91	....	1,241		
" 20..	10½@13	8½@11¼									
" 24..	10½@13	8½@11¼	4,500	9,897	871	764	301	....	1,936		
" 27..	10½@13	8½@11¼									
May 1..	10½@13	8½@11¼	1,300	8,411	977	....	323	....	1,300	7-16@9/16	December 26, accounts were received from London, advising the blockade of the Scheldt by the English and French fleets.
" 4..	10½@13	8½@11¼									
" 8..	10½@13	8½@11¼	1,500	5,270	....	....	....	....	....		
" 11..	10½@13	8½@11									
" 15..	10½@13	8½@11	2,050	2,863	5,942	1,242	577	400	8,161		
" 18..	10½@13	8½@11									
" 22..	10½@13	8½@11	1,100	4,950	2,553	237	40	....	2,830		
" 25..	10½@13	8½@11									
" 29..	10½@13	8½@11	4,400	5,148	2,605	2,120	628	....	5,353		
June 1..	10½@13	8½@11								⅜@9-16d	
" 5..	10½@13	8½@11	2,100	4,326	1,905	3,258	198	....	5,361		
" 8..	10½@13	8½@11									
" 12..	10½@13	8½@11	2,900	4,444	1,919	813	1,058	....	3,790		
" 15..	10½@13	8½@11									
" 19..	10½@12½	8½@11	1,250	3,794	2,577	2,291	89	....	4,957		
" 22..	10½@12½	8½@11									

The semi-weekly Price and Weekly Sales and Receipts at New York, Weekly Exports from New York and Rates of Freight to Liverpool 1st of each Month, for the Crop Year ending October 1, 1832—Concluded.

1832.		Price of New Orleans	Price of Upland.	Sales for week.	Receipts for week.	EXPORTS FOR WEEK.					Rates of Freight to Liverpool.	GENERAL REMARKS.
						To Great Britain.	To France.	North of Europe.	Other Fo'n Ports	Total Exports.		
June	26..	10½@12½	8½@11	2,450	989	2,203	2,558	821	850	6,432		The course of Exchange for the crop year, was thus: On October 1, 1831, the quotation was for 60 days bills on London, 10¾@11 per cent. premium. Prices fell off in November to 9½@10; early in February, 1832, rose again to 10¾@11; later in the month declined to 9@9½; then advanced in March to 9½@10; in April to 10¼@10½; in May fell off to 9@10; June 19, up again to 10¼@10½; on the 29th declined to 9¼; July 5, to 8½@9, 10th to 8@8½, and continued to decline until June 30, when 6½@7 was the ruling rate; in August the price went up to 8@8½; in September, it fell off to 7¾@8, advancing again in October to 8@8¼.
"	29..	10½@12½	8½@11									
July	3..	10½@12½	8½@11	2,300	657	932	....	767	....	1,699	¼@7-16d	
"	6..	10½@12½	8½@11									
"	10..	10½@12½	8½@11	800	1,732	1,426	1,274	1,204	....	3,904		
"	13..	10½@12½	8½@11									
"	17..	10½@12½	8½@11	1,340	1,462	1,467	1,706	71	....	3,244		
"	20..	10½@12½	8½@11									
"	24..	10½@12½	8½@11	....	....	....	....	....	....	....		
"	27..	10½@12½	8½@11									
"	30..	10½@12½	8½@11	1,150	3,413	2,373	784	1,824	1,781	6,762		
August	4..	10½@12½	8½@11								¼@⅜d.	
"	7..	10½@12½	8½@11	1,500	1,362	200	1,011	141	....	1,352		
"	10..	10½@12½	8½@11									
"	14..	10½@12½	8½@11	2,700	1,233	313	232	107	....	652		
"	17..	10½@13	8½@11¼									
"	21..	10½@13	8½@11¼	2,250	451	721	....	137	....	858		
"	24..	10½@13	8½@11¼									
"	28..	10½@13	8½@11½	1,650	164	1,469	383	267	281	2,400		
"	31..	10½@13	9¼@11½									
Septem.	4..	10½@13	9¼@11½	1.600	54	2,984	61	234	....	3,279	¼d.	
"	7..	10½@13	9¼@11¾									
"	11..	10½@13	9½@12	1,370	281	....	....	60	....	60		
"	14..	10½@13	9½@12									
"	18..	10½@13	9½@12	1,750	379	93	54	....	....	147		
"	21..	10½@13	9½@12									
"	25..	10½@13	9½@12	2,500	317	682	482	33	....	1,197		
"	28 .	10½@13	9½@12									
October	2..	10½@13	9½@12	1,500	868	1,237	575	120	89	2,021	¼d.	
Average price and total sales, receipts and exports.		11.34	9.38	106,950	143,968	70,970	42,073	12,094	7,320	132,457		

1833.

British Parliament passed an act putting an end to the commercial character of the East India Company, abolishing the oppressive monopolies which it had hitherto exercised.

Cotton imported into Great Britain, 303,726,199 lbs. ; home consumption, 293,682,976 lbs. ; value of British cotton goods exported—real or declared—£18,486,400.

Not less than 1,000 "tube frames," were in operation in Great Britain, each capable of working 1,000 lbs. of cotton per week. (*See* years 1825 and 1829.)

Evidence was given this year, before the Commons' Committee on Manufactures, Commerce, etc., in England, that "in the whole of Scotland, there were 14,970 steam-looms." In England alone there were estimated to be 85,000 at this period. (*See* years 1785, 1787, 1813, 1820 and 1829.)

The hand loom weavers in the United Kingdom during this year were estimated at from 200,000 to 250,000.

A steam-loom weaver, from 15 to 20 years of age, could now (*see* years 1823 and 1826 for contract), assisted by a girl about 12 years of age, attending four looms, weave *eighteen* pieces of 9-8 shirting (*see* year 1823) per week, and some could weave *twenty* pieces. (*See* year 1785.)

Cotton exports from United States this year, were valued at $36,191,102.

Exported from England, 468,602 dozen pairs of cotton stockings, estimated as worth £257,931. (*See* years 1589, 1768, 1787, 1809, 1823 and 1835.)

Exported from England, 1,187,601 lbs. of sewing cotton.

Estimated number of power-looms in operation in England and Scotland, from 80,000 to 90,000. (*See* year 1785.)

Lord Ashley renewed the attempt of Mr. Sadler (*see* year 1832), in the English House of Commons, to reduce the hours of labor in cotton mills to 10 hours per day, but failed. (*See* years 1802, 1819 and 1831.)

Imports of cotton into Great Britain, from foreign countries, 268,953,949 lbs., as follows: United States, 237,506,758 ; Brazil, 28,463,821 ; Turkey and Egypt, 987,262 ; miscellaneous, 1,696,108 ; from British possessions, 35,002,888 lbs., as follows: East Indies and Mauritius, 32,755,164 ; British West Indies—the growth of—1,653,166 ; foreign, 431,696 ; miscellaneous, 162,862. Total imports, 303,656,837 lbs. ; exports, 17,363,882 lbs. ; home consumption, 293,682,976 lbs.

"The York Manufacturing Company of Saco," Maine, completed their building, on the spot formerly occupied by another company whose immense mill was destroyed by fire. (*See* year 1830.) The building was the same size as the former one, but was only four stories and an attic in height.

This year, embroidering and finishing of lace employed in Europe, about 55,000 hands, their wages amounting to £550,000. (*See* years 1831, 1836 and 1856.)

COTTON CROP OF THE UNITED STATES.

Statement and Total Amount of the Cotton Crop of the United States, for the Year ending 30th September, 1833.

	Bales.	Bales.	Total.	Same period 1832.
NEW ORLEANS.				
Export—				
To Foreign Ports	....	312,997	....	291,678
Coastwise	....	94,223	....	64,728
Stock on hand 1st October, 1833	....	9,657	....	7,088
			416,877	363,494
Deduct—				
Stock on hand 1st October, 1832	....	7,078		
Received from Mobile	....	1,383		
" Florida	....	6,278		
			14,749	40,859
Add—			402,128	
Burnt	....	....	1,315	
			403,443	322,635
FLORIDA.				
Received at—				
New Orleans	....	6,278		
Philadelphia	....	18		
New York	....	17,070		
Providence	....	176		
Boston	....	99		
			23,641	22,651
ALABAMA.				
Export from Mobile—				
To Foreign Ports	....	101,916		
Coastwise	....	27,398		
Stock in Mobile, 1st October, 1833	....	139		
Deduct—		129,453		
Stock on hand 1st October, 1832	....	87		
			129,366	125,921
GEORGIA.				
Export from Savannah—				
To Foreign Ports—Uplands	146,010			
" Sea Islands	12,168			
		158,178		
Coastwise	....	94,281		
		252,459		
From Darien—				
To Liverpool	3,076			
To New York	2,698			
To Providence	40			
		5,814		
Stock in Savannah, 1st October, 1833	5,522			
" Augusta, "	10,500			
		16,022		
Deduct—		274,295		
Stock on hand 1st October, 1832	....	3,270		
			271,025	276,437

Statement and Total Amount of the Cotton Crop of the United States, for the Year ending September 30, 1833—*Concluded.*

	Bales.	Bales.	Total.	Same period 1832.
SOUTH CAROLINA.				
Export from Charleston—				
To Foreign Ports—Uplands	143,166			
" Sea Island	21,787			
		164,953		
Coastwise	....	34,027		
		198,980		
From Georgetown—				
To New York	8,884			
Stock in Charleston, 1st October, 1833	4,364			
		13,248		
		212,228		
Deduct—				
Stock in Charleston, 1st October, 1832	2,987			
Received from Savannah, and included in the exports from that place	27,365			
		30,352		
			181,876	173,872
NORTH CAROLINA.				
Export—				
To Foreign Ports	....	715		
Philadelphia	....	595		
New York	....	27,613		
Providence	....	9		
Boston	....	724		
Stock on hand, 1st October, 1833	....	1,000		
		30,458		
Deduct—				
Stock on hand 1st October, 1832	....	200		
			30,258	28,461
VIRGINIA.				
Export—				
To Foreign Ports	....	20,064		
Coastwise	....	11.350		
Stock on hand, 1st October, 1833	....	500		
		31,914		
Deduct—				
Stock on hand 1st October, 1832	....	1,085		
			30,829	37,500
Total crop of the United States	....	....	1,070,438	987,477
Crop of last year	....	....	987,477	
Increase compared with last year	....	....	82,961	

Export to Foreign Ports, from October 1, 1832, *to September* 30, 1833.

FROM		To Great Britain.	To France.	Other parts of Europe.	Total.
New Orleans	Bales	225,667	82,302	5,028	312,997
Alabama		77,201	22,343	2,372	101,916
Georgia		134,975	25,584	695	161,254
South Carolina		114,912	37,795	12,246	164,953
North Carolina		....	517	....	517
Virginia		14,236	4,806	1,022	20,064
Baltimore		750	25	....	775
Philadelphia		4,970	117	....	5,087
New York		55,979	33,948	6,568	96,495
Boston		1,455	80	1,862	3,397
Grand total		630,145	207,517	29,793	867,455
Total last year		638,148	207,209	46,371	891,728
Increase		....	308	....	....
Decrease		8,003	....	16,578	24,273

Growth.

Total crop of	1824–5,	560,000	bales.
"	1825–6,	710,000	"
"	1826–7,	937,000	"
"	1827–8,	712,000	"
"	1828–9,	857,744	"
"	1829–30,	976,845	"
"	1830–31,	1,038,848	"
"	1831–32,	987,477	"
"	1832–33,	1,070,438	"

Consumption.

To estimate the quantity manufactured in the United States, we take the growth of the year.................................1,070,438 bales.

Add—Stocks on hand at the commencement of the year, (October 1, 1832.)—In the Southern ports.................. 14,717

" Northern ports.................. 26,882

41,599

1,112,037

Deduct therefrom—The Export to Foreign ports............... 867,455

Stocks on hand at the close of the year, (October 1, 1833.)—

In the Southern ports............ 31,682

" Northern ports............ 18,774

50,456

917,911

Less—Foreign cotton, included in the export........................ 286

917,625

Quantity consumed, and in the hands of	manufacturers,		1832–3	194,412
"	"	"	1831–2	173,800
"	"	"	1830–1	182,142
"	"	"	1829–30	126,512
"	"	"	1828–9	118,853
"	"	"	1827–8	120,593
"	"	"	1826–7	103,483

Note.—The quantity taken for home manufacture, as shown by the above statement, does not include any cotton spun in the cotton-growing States. We have no means of ascertaining the quantity taken for domestic use in the States south and west of the Potomac, and if we had, we are not aware of any practical use that could be made of the information. Our statement furnishes the sum total of all the cotton *brought to market* from year to year, and whether the quantity used on the plantations, or in their immmediate vicinity, be more or less, it can have little or no bearing on the commerce in it.

COTTON AT LIVERPOOL. YEAR 1833.

Week Ending.		Receipts.						Sales.				Stocks.			Prices.			Actual Export	Average Consumption.
		Americ'n.	E. I.	Egypt.	Brazil.	Other.	Total.	Consumption.	Speculation.	Export.	Total.	Amer'n.	Other.	Total.	Mid. Up.	Mid. Orl.	Dhol		
Jan.	4	6,382					6,382	20,334			20,334	121,000	62,000	184,000	6¾	6⅞	4¼		20,334
"	11	3,664				51	3,715	34,284	2,000		36,284	101,000	55,000	156,000	6⅞	7	4½		54,618
"	18	9,588	2,030		280	40	11,938	18,785	3,000		21,785	93,500	46,500	14,000	6⅞	7	4½		73,403
"	25	4,767				4,096	8,863	13,500	5,000		18,500	85,500	46,000	131,500	6⅞	7	4½		86,903
Feb.	1	15,493			2,168	733	18,394	11,030	800		11,830	92,000	45,500	137,500	6⅞	7	4½		97,933
"	8	29,585			1,663	260	31,508	13,084			13,084	111,000	44,000	155,000	6⅞	7	4½		111,017
"	15	17,560			4,447	30	22,037	12,520	800		13,320	117,500	45,000	162,500	6⅞	7	4½		123,537
"	22	10,476				3	10,479	14,212			14,212	118,000	39,000	157,000	6⅞	6⅞	4¼		137,749
Mch.	1	12,435				1,656	14,091	20,164	900		21,064	115,500	35,000	150,500	6⅞	6⅞	4¼		157,913
"	8	16,741				1,966	18,707	20,077	4,300		24,377	116,000	33,000	149,000	6⅞	6⅞	4¼		177,990
"	15	14,821				534	15,355	17,505	950	1,100	19,555	113,000	30,000	143,000	6⅞	7	4¼	1,100	195,495
"	22	4,257				2	4,259	14,160			14,160	105,000	27,500	132,500	6⅞	7	4¼		209,655
"	29	12,168			2,103	622	14,893	8,778	400		9,178	110,000	28,500	138,500	6⅞	7	4¼		218,433
April	4	30,447	5,105		4,558	25	40,135	9,360			9,360	131,500	36,000	167,500	6⅞	7	4¼		227,793
"	12	52,165	2.142		5,611	1,419	61,337	18,401	3,000	1,000	22,401	167,000	40,000	207,000	6⅞	7	4¼	1,000	246,194
"	19	3,041	1,840			151	5,032	14,147		2,000	16,147	155,500	38,500	193,000	6¾	6⅞	4¼	2,000	260,341
"	26	7,401	1,905			404	9,710	15,562			15,562	151,500	35,000	186,500	6¾	6⅞	4¼		275,903
May	3	50,319	3,003		13,021	368	66,711	12,380		800	13,180	189,000	48,500	237,500	6¾	6⅞	4¼	800	288,283
"	10	6,878			1,056	457	8,391	11,000		3,000	14,000	185,000	45,500	230,500	6¾	6⅞	4¼	3,000	299,283
"	17	19,367				5,843	25,210	11,219	2,900	700	13,819	195,000	48,000	243,000	6¾	6⅞	4¼	700	310,502
"	24	10,162	1,115	2,450	2,083	230	16,040	13,533	850	2,000	16,383	194,000	59,000	243,000	6⅞	6⅞	4¼	2,000	324,035
"	31	5,304			5,052	30	10,386	21,100	6,900	2,100	30,100	181,500	46,500	228,000	7	7⅛	4¾	2,100	345,135
June	7	37,070				1,061	38,131	31,855	15,400		47,255	193,500	38,500	231,000	7½	7½	5¼		376,990
"	14	30,013				430	30,443	28,008	4,500	300	32,808	201,000	31,000	232,000	7½	7½	5¼	300	404,998
"	21	15,451	1,252		8,030	88	24,821	18,150	8,300	400	26,800	198,000	36,500	234,500	7¾	7¾	5½	400	423,148
"	28	24,733			2,482	4,064	31,279	38,530	32,000	800	71.330	192,000	34,000	226,000	8¾	8¾	5¾	800	461,678
July	5	4,842			92		4,934	33,420	19,650	200	53,270	184,000	31,000	215,000	8¾	8¾	6	200	495,098
"	12	6,177				5,009	11,186	17,311	19,150		36,461	179,000	30,500	209,500	9	9	6		512,409
"	19	14,850				2,314	17,164	17,723	13,600		30,923	180,500	29,000	209,500	9⅛	9¼	6½		529,732
"	26	27,289			3,942	168	31,399	17,846	29,200		47,046	193,500	30,500	223,000	9⅞	10	7		547,578
Aug.	2	4,143			1,562		5,705	12,569	6,500	1,000	20,069	190,000	30,000	220,000	9⅞	10	7	1,000	560,147
"	9	2,003				9	2,012	9,410	5,400		14,810	184.000	29,500	213,500	9¾	9⅞	7		569,557
"	16	3,218				1,162	4,380	9,000	16,500		25,500	182,500	28,000	210,500	9⅞	10	7		578.557
"	23	7,802	2,031		2,834	562	13,229	21,940	22,600		44,540	173,500	28,000	201,500	10½	10⅝	7		600,497

Aug. 30...	2,878	.....	.....	2,918	117	5,913	9,265	6,000	.....	15,265	168,500	29,400	197,900	$10\frac{3}{8}$	$10\frac{1}{2}$	7	.....	609,762
Sept. 6...	4,675	2 022	.....	1,990	12	8,699	9,420	3,000	.....	12,420	165,000	32,200	197,200	$10\frac{1}{8}$	$10\frac{3}{8}$	7	.....	619,182
" 13...	6,500	2,827	.....	4,396	194	13,917	2,890	1,500	.....	4,390	169,500	39,000	208,500	$9\frac{7}{8}$	$10\frac{1}{8}$	$6\frac{3}{4}$	.....	622,072
" 20...	1,879	2,673	.....	3,640	1,227	9,419	6,320	400	.....	6,720	166,000	45,500	211,500	$9\frac{1}{2}$	$9\frac{5}{8}$	$6\frac{1}{2}$	.....	628,392
" 27...	3 882	1,878	.....	.....	1,614	7,374	3,873	.....	.....	3,873	167,500	48,000	215,500	$8\frac{7}{8}$	9	$6\frac{1}{2}$	.....	632,265
Oct. 4...	3,674	.....	.....	3,217	1,599	8,490	8,610	5,300	.....	13,910	165,000	50,000	215,000	$9\frac{1}{4}$	$9\frac{1}{4}$	$6\frac{1}{2}$	.....	640,875
" 11...	1,900	6,907	.....	1,729	34	10,570	4,120	300	.....	4,420	164,000	57,500	221,500	$8\frac{7}{8}$	$8\frac{7}{8}$	$6\frac{1}{8}$	.....	644,995
" 18...	4,190	2,842	.....	11,412	67	18,513	3,820	.....	.....	3,820	164,500	72,500	236,000	$8\frac{1}{2}$	$8\frac{5}{8}$	6	.....	648,815
" 25...	4,249	.....	.....	.....	498	4,747	10,060	1,400	.....	11,460	162,500	67,500	229,000	$8\frac{1}{2}$	$8\frac{5}{8}$	$5\frac{1}{2}$	.....	658.875
Nov. 1...	1,794	2,404	.....	4,477	.....	8,675	9,030	1,200	.....	10,230	156,500	68,000	224,500	$8\frac{1}{8}$	$8\frac{3}{8}$	$4\frac{3}{4}$	.....	667,905
" 8...	9,006	.....	.....	6,935	671	16,612	12,947	700	.....	13,647	156,000	69,500	225.500	$7\frac{3}{8}$	$7\frac{1}{2}$	$4\frac{3}{4}$	.....	680,852
" 15...	4,512	4,380	.....	6,797	51	15,740	20,920	1,900	900	23,720	144,500	75,000	219,500	$7\frac{3}{8}$	$7\frac{1}{2}$	$4\frac{1}{2}$	900	701.772
" 22...	3,237	1,052	.....	4,580	291	9,160	11,578	.....	300	11,878	138,500	77,000	215,500	$6\frac{7}{8}$	7	$4\frac{1}{2}$	300	713.350
" 29...	7,575	2,034	.....	1,932	239	11,780	11,434	3,900	3,830	19,164	137,000	77,500	214,500	7	$7\frac{1}{8}$	$4\frac{1}{2}$	3,830	724,784
Dec. 6...	8,727	.....	.....	.....	672	9,399	24,880	4,300	100	29,280	128,000	69,000	197,000	$7\frac{1}{4}$	$7\frac{1}{4}$	$4\frac{3}{4}$	100	749.664
" 13...	11,265	1,840	.....	5,137	.....	18,242	21,020	6,500	4,500	32,020	122,500	78,500	191,000	$7\frac{3}{8}$	$7\frac{1}{2}$	5	4,500	770,684
" 20...	2,982	1,112	.....	3,564	127	7,785	32.475	2,900	.....	35,375	164,500	60,500	165,000	$7\frac{3}{8}$	$7\frac{5}{8}$	$5\frac{1}{4}$	.....	803,159
" 27...	8.492	.....	.....	.....	2,329	10,821	27,039	4,300	.....	31,339	97,000	35,500	152,500	$7\frac{7}{8}$	8	$5\frac{1}{4}$	.....	830,198
Average prices & total sales, receipts & stocks.	612,031	52,694	2,450	123,688	43,329	834,192	830,198	168250	25,030	1,023,478				7.87	7.97	5.22	25,030	1,596,534

The semi-weekly Price and Weekly Sales and Receipts at New York, Weekly Exports from New York and Rates of Freight to Liverpool 1st of each month, for the Crop Year ending October 1, 1833.

1832.	Price of New Orleans	Price of Upland.	Sales for week.	Receipts for week.	EXPORTS FOR THE WEEK.					Rates of Freight to Liverpool.	GENERAL REMARKS.
					To Great Britain.	To France.	North of Europe.	Other Fo'n Ports	Total Exports.		
October 5..	10½@13	9½@12								¼d.	
" 9..	10½@13	9½@12	3,300	712	....	....	....	....	....		
" 12..	10½@13½	10@12									
" 16..	10½@13½	10@12	3,200	758	....	....	....	....	....		
" 19..	10½@13½	10@12½									
" 23..	10½@13	10@12½	2,000	587	....	....	....	....	....		
" 26..	11@13	10½@12½									
" 30..	11@13	10½@12½	2,400	27	....	....	....	....	....		
Novem. 2..	11@13	10½@12½								¼d.	
" 6..	11@13	10½@12½	750	1,617	....	....	....	....	....		
" 9..	11@13	10½@12½									
" 13..	11@13½	10½@12½	1,100	4,456	....	....	....	....	....		*The figures for exports given December 4, embrace the shipments of the previous eight weeks, no statement being compiled in the mean time.
" 16..	11½@13¾	10½@12½									
" 20..	11½@13¾	10½@12½	2,150	2,122	....	....	....	....	....		
" 23..	11½@14	10½@13									
" 27..	11½@14	10½@13	1,450	3,899	....	....	....	....	....		
" 30..	11½@14	10½@12½									
Decem. 4..	11½@14	10½@12½	2,450	3,624	*6,221	3,857	310	....	10,388	¼d.	The prominent feature in the market during the year was, great firmness, much speculation, and, with the exception of one or two brief periods, a steady advance in price; the crop year closing with the large difference in favor of sellers, of 5@5½ cents per lb.
" 7..	11½@13½	10½@12½									
" 11..	11½@13½	10½@12½	2,300	3,023	....	....	....	....	....		
" 14..	11½@13½	10½@12½									
" 18..	11½@13¼	10½@12	1,500	2,970	1,193	390	....	....	1,583		
" 21..	11@13¼	10½@12									
" 26..	11@13¼	10½@12	800	1,762	1,121	144	....	....	1,265		
" 28..	11@13	10¼@11½									
1833.											
January 2..	11@13	10@11½	1,650	6,054	587	193	....	....	780	3-16@¼d	
" 4..	11@13	10@11½									
" 8..	11@13	10@11½	2,200	4,032	925	643	....	....	1,568		
" 11..	11@13	10@11½									
" 15..	11@13	10@11½	2,500	4,429	296	190	....	....	486		
" 18..	11@13	10@11½									
" 22..	11@13	10@11½	1,500	3,173	1,160	1,072	40	....	2,272		

January 25..	11@13	9¾@11½								
" 29..	11@13	9¾@11½	1,000	4,849	745	156	....	30	931	
February 1..	11@13	9¾@11½								⅜@½d.
" 5..	11@13	9¾@11	1,450	1,266	1,815	1,402	85	....	3,302	
" 8..	11@13	9¾@11								
" 12..	11@13	9¾@11	4,350	5,898	1,171	425	318	....	1,914	
" 15..	11@13	9¾@11								
" 19..	11@13	9¾@11¼	3,000	4,208	51	....	225	....	276	
" 22..	11@13	10@11¾								
" 26..	11@13	10@11¾	5,200	1,367	947	....	158	....	1,105	7-16@9/16
March 1..	11@13	10@11¾								
" 5..	11@13	10@11¾	1,500	1,795	1,346	1,453	....	42	2,841	
" 8..	11@13½	10@11¾								
" 12..	11@13½	10@11¾	2,200	5,713	747	898	....	....	1,645	
" 15..	11@13½	10@12								
" 19..	11@13½	10@12½	3,000	4,214	2,592	....	428	....	3,020	
" 22..	11@13½	10@12½								
" 26..	11@13½	10@12½	2,900	6,335	395	852	380	....	1,627	
" 29..	11@13½	10@12½								
April 2..	11@13½	10@12½	3,550	3,899	1,461	783	140	....	2,384	7-16@9/16
" 5..	11@13½	10@13								
" 9..	11@13½	10@13	2,800	6,660	812	698	504	....	2,014	
" 12..	11@13½	10@13								
" 16..	11@13½	10@13	7,050	5,756	1,986	507	426	....	2,919	
" 19..	11@13¾	10@12¼								
" 23..	11@13¾	10@12¼	3,500	4,025	699	97	199	....	995	
" 26..	11@14	10¼@12½								
" 30..	11¼@14	10½@12½	6,100	4,115	1,740	664	554	....	2,958	
May 3..	11½@14½	10¾@12¾								7-16@9/16
" 7..	11½@14½	10¾@12¾	2,200	2,473	389	888	263	....	1,540	
" 10..	11½@14½	10¾@12¾								
" 14..	12@14½	11@13	4,000	5,941	1,189	1,240	420	38	2,887	
" 17..	12@15	11@13								
" 21..	12@15	11@13	2,300	3,043	2,239	326	....	....	2,565	
" 24..	12@15	11@13								
" 28..	12@15	11@13	900	4,505	1,104	238	....	....	1,342	
" 31..	12@15	11@13								
June 4..	12@15	11@13	1,200	3,301	310	943	26	....	1,279	¼d.
" 7..	12@15	11@13½								
" 11..	12@15	11@13½	1,700	3,042	682	....	537	....	1,219	
" 14..	12@15	11@13½								
" 18..	12@15	11@13½	1,900	7,538	168	847	150	....	1,165	

On the 23d February, 1833, advices were received from Liverpool per packet ship New York, up to January 16, noting the destruction by fire, at Liverpool, of 10 to 12,000 bales cotton; which, with the increased consumption of the Kingdom, made the stock fully 100,000 bales smaller than for the same time the previous year; this gave a great impetus to the business; and from this time, until the end of September, the course of prices was, for the most part, rapidly upward.

On the 2d of July, advices at hand from Liverpool per ship Silas Richards, to May 24, announcing a reduction of duty of ¼d. and with a stock further reduced by another fire, the market here became very active, and advanced 1 cent per lb.; these favorable accounts were followed by others, in August and September, and hence the appreciation in values, as noted in the quotations annexed.

New York Statement for Year 1833—*Concluded.*

1833.		Price of New Orleans	Price of Upland.	Sales for week.	Receipts for week.	EXPORTS FOR THE WEEK					Rates of Freight to Liverpool.	GENERAL REMARKS.
						To Great Britain.	To France.	North of Europe.	Other Fo'nPorts	Total Exports.		
June	21..	12@15	11@13½									
"	25..	12@15	11@13½	1,500	3,378	91	529	80	....	700		
"	28..	12@15	11@13½									
July	2..	12@15	11@13½	7,900	3,771	785	1,764	178	....	2,727	¼d.	
"	5..	13@15½	11@13½									
"	9..	14@16½	13@15	3,500	5,897	....	....	100	....	100		Prices of 60 days bills on London, for this crop year, were as follows: The opening price in October, was 8 per cent. premium; in November there was an advance to 9@1½ premium; in December the price fell off to 8@8¼; 1st January, to 7½@8; later in the month, 8@8½, was the quotation; in February the price declined to 7½@7¾; advancing on the 1st of March to 8@8½; declining again on the 15th to 7½@8; prices steady until May, when they rose to 8½@9; in June the quotation was $4.80@$4.83 per £; in July down to $4.78@$4.80; steady until September, when they touched $4.75@$4.77 per £.
"	12..	14@17	13@16									
"	16..	14@17	13@16	4,500	2,036	537	804	147	....	1,488		
"	19..	14@17	13@16									
"	23..	14@18	14@17	3,900	1,405	997	1,014	194	....	2,205		
"	26..	16¼@17½	15@17									
"	30..	16¼@17½	15@17	2,300	2,546	....	....	....	....	....		
August	2..	15@17½	14@17								¼d.	
"	6..	15@17½	14@17	600	1,373	1,176	399	....	....	1,575		
"	9..	15@17½	14@17									
"	13..	15@17½	14@16½	3,500	714	1,166	236	....	......	1,402		
"	16..	15@17½	14@16½									
"	20..	15@17½	14@16	5,700	387	1,060	....	29	....	1,089		
"	23..	15½@18½	14@17									
"	27..	15½@18½	14½@17	5,400	837	786	708	146	......	1,640		
"	30..	15½@18½	14½@17									
Septem.	3..	15½@18½	14½@17	3,900	1,100	3,258	1,693	127	....	5,078	¼@⅜d,	
"	6..	15½@18½	14½@17									
"	10..	15½@18½	14½@17	3,000	2,875	1,731	2,353	....	....	4,084		
"	13..	15½@18½	14½@17									
"	17..	15½@18	14½@17	2,600	5,889	1,447	2.332	....	....	3,779		
"	20..	15½@18	14½@17									
"	24..	15½@18½	14½@17	4,550	596	4,448	954	100	......	5,502		
"	27..	16@18½	15@17									
October	1..	16@18½	15@17	2,800	1,619	4,376	2,256	174	....	6,806	⅜@7-16d	
Average prices and total sales, receipts and exports.		13.46	12.32	148,700	167,611	55,949	33,948	6,438	110	96,445		

LIVERPOOL STATEMENT FOR 1833.

UNITED STATES, 1832–1833.

Stock, Oct., '32..	41,000	Export	867,000
Crop, " " ..	1,070,000	Consumption....	194,000
		Stock, Oct., '33..	50,000
Bags............	1,111,000	Bags	1,111,000

Stock 1st Jan. in............	Liverpool.	Gt. Britain.	France.	Continent.	Tot. Europe.
United States..........Bags.	129,170	137,992	18,898	5,994	162,884
Brazil	35,990	37,625	897	2,205	40,727
West Indies................	2,440	2,924	981	2,815	6,720
East Indies	23,560	55,377	......	1,500	56,877
Egypt and Levant...........	6,800	11,200	1,730	9,922	22,852
Bags.	197,960	245,118	22,506	22,436	290,060

CONSUMPTION, 1833. / IMPORT, 1833.

Tot. Europe.	Continent.	France.	Gt. Britain.	Liverpool.	IMPORT, 1833.	Liverpool.	Gt. Britain.	France.	Continent.	Tot. Europe.
895,970	43,824	198,409	653,737	618,377	....United States,..........Bags.	619,987	656,735	217,276	42,612	891,273
181,480	11,814	24,351	145,315	144,624	....Brazil	162,414	164,190	26,795	11,124	194,859
37,371	16,248	7,849	13,274	10,054	....West Indies................	10,034	14,940	9,196	17,175	37,762
110,253	35,573	......	74,680	46,726	....East Indies	49,256	94,683	......	37,145	100,878
111,141	55,274	43,778	12,089	8,269	....Egypt and Levant...........	2,169	2,569	52,366	56,761	110,976
1,336,215	162,733	274,387	899,095	528,050	Bags,	843,860	933,117	305,633	164,817	1,335,748
....	1,819	2.000	64,000	33,000	Export.					
289,593	22.701	51,752	215.140	180,770	Stock, Dec....Stock above, "	197,960	245,118	22.506	22,436	290,060
1,625,808	187,253	328,139	1,178,23	1,041,820	Supply, "	1,041,820	1,178,235	328,139	187,253	1,625,808

1834.

In March of this year the patentees of Roberts' Self-Acting Mule, Manchester, England (*see* years 1825 and 1830), had manufactured 500 of the machines, containing upwards of 200,000 spindles.

By an alteration in the tariff, made by royal ordinance on the 8th of July, this year, in France, cotton yarns of the high numbers—those above No. 142 French, which answers to No. 189 English—were admitted into France on payment of a duty of 7 francs per killogramme, or about 2s. 7¾d. per lb., which was a duty of from 27 to 33 per cent. *ad valorem* on the qualities chiefly used.

The Nottingham, England, bobbin-net trade had for some time cast an anxious eye upon the increase of French bobbin-net machinery, and this year prayed the Board of Trade to "endeavor strenuously for commercial reciprocity with our neighbors," and tried hard, but unsuccessfully, to prevent the export of Nottingham machinery. (*See* year 1841.)

COTTON CROP OF THE UNITED STATES.

Statement and Total Amount of the Cotton Crop of the United States, for the Year ending September 30, 1834.

	Bales.	Bales.	Total.	Same period 1833.
NEW ORLEANS.				
Export—				
To Foreign Ports	401.548			
Coastwise	60,705			
Burnt, &c	1 500			
Stock on hand 1st October, 1834	8,756			
		472,509		
Deduct—				
Stock on hand 1st October, 1833	7,406			
Received from Mobile	5,063			
" Florida	5,321			
		17,790		
			454,719	403,443
FLORIDA.				
Export—				
To Foreign Ports	....	2,900		
New Orleans	....	5,321		
Baltimore	....	262		
Philadelphia	....	127		
New York	....	27,709		
Providence	....	376		
Boston	....	43		
			36,738	23,641
ALABAMA.				
Exported from Mobile—				
To Foreign Ports	117,140			
Coastwise	32,566			
Stock in Mobile 1st October, 1834	411			
		150,117		
Deduct—				
Stock on hand 1st October, 1833	....	139		
			149,978	129,366
GEORGIA.				
Export from Savannah—				
To Foreign Ports—Uplands	155,158			
" Sea Island	9,055			
	164,213			
Coastwise	96,233			
		260,446		
From Darien—				
To Foreign Ports	2,500			
New York	5,873			
Providence	232			
		8,605		
Stock in Savannah 1st October, 1834	2,844			
" Augusta and Hambro' 1st October, 1834	2,782			
		5,626		
		274,677		
Deduct—				
Stock on hand 1st October, 1833	....	16,022		
			258,655	271,025

Statement and Total Amount of the Cotton Crop of the United States, for the Year ending September 30, 1834.—*Concluded.*

	Bales.	Bales.	Total.	Same period 1833.
SOUTH CAROLINA.				
Export from Charleston—				
To foreign ports—Uplands	195,269			
" Sea Island	17,149			
	212,418			
Coastwise	42,463			
		254,881		
From Georgetown—				
To New York	....	9,905		
Stock in Charleston 1st October, 1834	....	3,806		
Deduct—		268,592		
Stock in Charleston 1st October, 1833	4,364			
Received from Savannah and included in the exports from that place	36,869			
		41,233		
			227,359	181,876
NORTH CAROLINA.				
Export—				
To Foreign Ports	2,817			
Baltimore	100			
Philadelphia	355			
New York	29,894			
Providence	100			
Boston	704			
Stock on hand 1st October, 1834	250			
		34,220		
Deduct—				
Stock on hand 1st October, 1833	....	1,000		
			33,220	30,258
VIRGINIA.				
Export—				
To Foreign Ports	31,045			
Coastwise	7,550			
Taken by Manufacturers	6,000			
Stock on hand 1st October, 1834	630			
		45,225		
Deduct—				
Stock on hand 1st October, 1833	....	500		
			44,725	30,829
Total crop of the United States	....	....	1,205,394	1,070,438
Crop of last year	....	....	1,070,438	
Increase	....	....	134,956	

Export to Foreign Ports, from 1st October, 1833, to 30th September, 1834.

FROM	To Great Britain.	To France.	To North Europe.	Other Foreign Ports.	Total.
New Orleans (bales)	289,169	101,253	9,742	1,384	401,548
Alabama	96,180	15,256	944	4,760	117,140
Florida	2,900	....	....	....	2,900
Georgia (Savannah and Darien)	146,979	17,638	544	1,552	166,713
South Carolina	142.194	46,433	21,554	2,237	212,418
North Carolina	2,817	....	....	....	2,817
Virginia	19,749	9,144	1,764	388	31,045
Baltimore	300	....	150	....	450
Philadelphia	500	724	522	....	1,746
New York	54,568	25,876	7,655	895	88,994
Boston	935	100	405	740	2,180
Grand total	756,291	216,424	43,280	11,956	1,027.951
Total last year	630,145	207,517	23,860	5,933	867.455
Increase	126,146	8,907	19,420	6,023	160,496

Growth.

Total crop of 1824–5, 560,000 bales.
" 1825–6, 710,000 "
" 1826–7, 937,000 "
" 1827–8, 712,000 "
" 1828–9, 857,744 "
" 1829–30, 976,845 "
" 1830–1, 1,038,848 "
" 1831–2, 987,477 "
" 1832–3, 1,070,438 "
" 1833–4, 1,205,394 "

Consumption.

To estimate the quantity manufactured in the United States, we take the growth of the year		1,205,394 bales.
Add—Stocks on hand at the commencement of the year October 1, 1833, in the Southern ports	29,431	
In the Northern ports	18,774	
		48,205
		1,253,599
Deduct therefrom the export to Foreign ports	1,027,951	
Stocks on hand at the close of the year October 1, 1834.		
In the Southern ports	19,283	
In the Northern ports	10,334	
		29,617
		1,057,568
Less Foreign cotton included in the export		382
		1,057,186

Quantity consumed and in the hands of manufacturers, 1833–4, (bales)	196,413
Consumption, etc., of 1832–3	194,412
" 1831–2	173,800
" 1830–1	182,142

Consumption, etc., of	1829-30	126,512
"	1828-9	118,853
"	1827-8	120,593
"	1826-7	103,483

Note.—The manufacture of Cotton having very much increased, within the last two years, in the vicinity of Petersburgh and Richmond, our correspondent in Petersburgh, after investigating the matter, advises us that about 6,000 bales have been taken during the past year for the supply of those mills; we have accordingly included that quantity in our statement of the crop. It may be remarked, also, that the quantity of *new* Cotton which reached the Southern ports previous to the first of October, last year, was unusually large—while the quantity brought to market this year has been unusually small.

LIVERPOOL STATEMENT FOR 1834.

UNITED STATES, 1833–1834.

Stock 1st October, 1833	50,000	Export	1,029,000
Crop	1,205,000	Consumption	197,000
		Stock 1st Oct., 1834	29,000
Bags	1,255,000	Bags	1,255,000

13

	Liverpool.	Gt. Britain.	France.	Continent.	Tot. Europe.
United States .... Bags	110,580	117,640	35,765	4,782	158,187
Brazil	47,280	49,250	3,341	1,515	54,106
West Indies	1,920	2,860	2,328	1,923	7,111
East Indies	20,890	44,430	....	3,072	47,502
Egypt and Levant	100	960	10,318	11,409	22,687
Bags	180,770	215,140	51,752	22,701	289,593

CONSUMPTION. Tot. Europe.	Continent.	France.	Gt. Britain.	Liverpool.	IMPORT.	Liverpool.	Gt. Britain.	France.	Continent.	Tot. Europe.
1,006,716	76,899	234,212	695,605	644,395	United States	670,755	731,335	218,761	81,309	996,005
167,943	12,451	20,330	135,162	133,132	Brazil	100,372	103,532	19,128	11,479	128,289
39,518	15,712	8,608	15,198	11,888	West Indies	15,218	18,378	7,846	18,183	39,945
89,081	36,957	....	52,124	32,164	East Indies	46,384	88,124	....	37,397	93,921
88,779	43,942	38,502	6,335	5,265	Egypt and Levant	6,155	7,125	28,572	38,521	70,068
1,392,037	185,961	301,652	904,424	826,844	Bags	838,884	948,494	274,307	186,889	1,328,228
....	7,812	....	73,650	47,500	Export.					
225,784	15,817	24,407	185,560	145,310	Stock Dec. — Stock above (bags)	180,770	215,140	51,752	22,701	289,593
1,617,821	209,590	326,059	1,163,634	1,019,654	Total supply, bags	1,019,654	1,163,634	326,059	209,590	1,617,821

COTTON AT LIVERPOOL. YEAR 1834.

Week Ending.	RECEIPTS.						SALES.				STOCKS.			PRICES.			Actual Export	Con-sumption.
	Americ'n.	E. I.	Egypt.	Brazil.	Other.	Total.	Con-sumption.	Specu-lation.	Export	Total.	Amer'n.	Other.	Total.	Mid. Up.	Mid. Orl.	Dhol		
Jan. 3...	4,420	......	......	3,061	45	7,526	2,467	......	......	2,467	113 000	72,500	185,500	7⅞	8	5¼	......	2,467
" 10...	12,542	......	......	4,183	......	16,725	9,488	2,500	......	11,988	117,500	74,500	192,000	7⅞	8	5¼	......	11,955
" 17...	13,273	......	......	2,164	......	15,437	9,480	......	......	9,480	123,000	75,000	198,000	7⅝	7¾	5	......	21,435
" 24...	15,869	2,484	......	871	......	19,224	12,777	1,750	1,200	15,727	129,500	73,000	202,500	7½	7¾	5	1,200	34,212
" 31...	10,684	......	......	2,088	215	12,987	17,506	1,200	200	18,906	125,500	70,500	196,000	7½	7¾	5	200	51,718
Feb. 7...	16,274	......	......	1,878	1,028	19,180	13,708	500	1,000	15,208	131,500	79,500	200,500	7½	7¾	5	1,000	65,426
" 14...	14,201	2,441	......	130	228	17,000	15,043	......	2,100	17,143	133,500	66,000	199,500	7½	7½	5	2,100	80,469
" 21...	1,551	......	......	......	1,421	2,972	21,350	10,500	700	32,550	118,500	71,500	180,000	7⅝	7⅝	5	700	101,819
" 28...	16,024	1,773	......	......	1,372	19,169	19,603	6,000	150	25,753	121,000	57,000	178,000	7⅞	7⅞	5⅜	150	121,422
Mch. 7...	11,873	......	......	......	234	12,107	20,797	8,100	1,020	29,917	114,500	53,000	167,500	8	8	5½	1,020	142,219
" 14...	26,297	......	......	2,206	380	28,883	8,160	600	700	9,460	132,500	54,500	186,000	8	8	5¼	700	150,379
" 21...	9,700	......	......	......	953	10,653	17,490	2,700	1,000	21,190	126,000	51,000	177,000	8	8	5¼	1,000	167,869
" 28...	15,046	2,224	......	2,181	316	19,767	16,600	2,600	400	19,600	127,500	52,000	179,500	8	8	5	400	184,469
Apr. 4...	13,600	3,661	......	7,277	86	24,624	14,280	......	1,450	15,730	129,500	59,000	188,500	8	8	5	1,450	198,749
" 11...	5,472	......	......	2,127	34	7,633	17,770	......	750	18,520	120,500	57,000	177,500	8	8	5¼	750	216,519
" 18...	2,275	......	......	......	......	2,275	17,451	......	1,850	19,301	109,000	51,500	160,500	8	8	5	1,850	233,970
" 25...	6,034	1,224	......	554	......	7,812	9,800	......	1,400	11,200	106,500	51,500	157,000	8	8	5	1,400	243,770
May 2...	14,619	1,019	......	2,025	475	18,138	17,600	......	1,800	19,400	107,000	47,500	154,500	8	8	5	1,800	261,370
" 9...	73,282	3,002	......	3,736	485	80,505	18,200	500	2,300	21,000	166,000	47,000	213,000	8	8	5¼	2,300	279,570
" 16...	22,492	......	......	......	1,479	23,971	36,600	9,000	1,600	47,200	158,000	39,000	197,000	8	8	5½	1,600	316,170
" 23...	24,452	1,234	......	5,331	147	31.164	16,592	5,300	1,500	23,392	159,000	39,500	198,500	8	8	5¾	1,500	332,762
" 30...	3,539	......	......	......	......	3,539	17,390	......	480	17,870	147,000	36,500	183,500	8	8	5¾	480	350,152
June 6...	28,611	......	......	3,138	670	32,419	28,540	3,000	1,500	33,040	149,000	36,500	185,500	8	8	5¾	1,500	378,692
" 13...	35,915	4,371	......	3,326	200	43,812	15,160	2,200	1,950	19,313	169,500	41,500	211,000	8	8	5¾	1,950	393,852
" 20...	29,225	......	......	1,205	603	31.036	20.153	300	705	21,158	180,000	40 000	220,000	7⅞	8	5¾	705	414,005
" 27...	31,854	1,494	......	6,654	89	40,091	16,784	......	4,300	21,084	195,000	42,500	237,500	7¾	7⅞	5¾	4,300	430,789
July 4...	5,064	......	......	......	602	5,666	9,015	1,250	1,810	12,075	189,000	40,000	229,000	7¾	7⅞	5¾	1,810	439,804
" 11...	31,429	......	......	2,072	208	33,709	12,600	900	2,300	15,800	208,500	38,500	247,000	7⅝	7¾	5¾	2,300	452 404
" 18...	27,775	......	......	1,598	344	29,717	17.800	2,500	1,700	22,500	217,500	37,600	255,100	7¾	7⅞	5¾	1,700	470,204
" 25...	9.539	......	1,262	3,857	4	14,662	40,030	1.900	1,700	43,630	198,500	36,200	234,700	8	8⅛	5¾	1,700	510,234
Aug. 1...	10,046	......	......	621	......	10,667	17,740	6,200	1,200	25,140	193,000	42,700	225,700	8⅛	8¼	5¾	1,200	527,974
" 8...	29,020	......	......	......	2,177	31,197	10,700	1.000	550	12,250	214,500	31,000	245,500	8⅛	8¼	5¾	550	538,674
" 15...	13,474	......	......	......	2,145	15,619	11,675	1,200	1,050	13,925	218,500	30,000	284,500	8⅛	8¼	6	1,050	550,349
" 22...	5,777	5,365	......	1,946	688	13,777	8,402	1,000	1,500	10,902	215,000	35,000	250,000	8	8⅛	6	1,500	558,751

Aug. 29...	14,826	......	......	4,112	112	19,050	10,461	......	1,000	11,461	220,000	37,500	257,500	7⅞	8	6	1,000	569,212
Sept. 5...	10,634	......	......	......	365	10,999	10,290	800	1,100	12,190	222,500	33,500	258,000	7¾	8	6	1,100	579,502
" 12...	1,317	......	......	......	1,246	2,563	16,649	3,500	2,500	22,649	208.500	32,500	241,000	7¾	8	6	2,500	596,151
" 19...	1,397	......	624	1,089	893	4,003	17,043	2,700	800	20,543	196 500	30,500	227,000	7¾	8	6	800	613,194
" 26...	216	......	......	3,301	1,990	5,507	22,581	9,150	560	32,291	179,000	30,000	209,000	8	8¼	6	560	635,775
Oct. 3...	2,777	2,198	......	410	2,142	7,527	15,700	9,300	500	25,500	169,000	31,500	200,500	8	8¼	6	500	651,475
" 10...	2.019	......	......	3,876	441	6,336	11,676	4,412	200	16,288	162,000	31,000	193,000	8	8¼	6	200	663,151
" 17...	4,776	......	......	......	......	4,776	21,860	14,100	800	36,760	148,500	26,000	174,500	8¼	8⅜	6¼	800	685,011
" 24...	2,658	......	......	3,654	75	6,387	19,700	16,200	100	36,000	134,000	26,500	160,500	8½	8⅜	6½	100	704,711
" 31...	......	......	......	2,655	......	2,655	9,700	2,700	100	12,500	126,000	26,000	152,000	8½	8⅝	6½	100	714,411
Nov. 7...	1,662	......	......	5,215	290	7,167	16,300	20.600	200	37,100	114,000	35,000	149,000	8⅞	8⅞	6½	200	730,711
" 14...	542	......	......	......	......	542	12,665	7,300	125	20,090	104,000	32,500	136,500	8⅞	8⅛	6½	125	743,376
" 21...	343	2,306	......	......	......	2,649	14,602	17,220	......	31,822	95,500	29,200	124,700	9⅛	9⅛	6½	......	757,978
" 28...	1,080	......	......	......	1,308	2,388	9,906	19,300	......	29,206	91,000	26,000	117,500	9½	9½	6¾	......	767,884
Dec. 5...	5,711	4,802	......	6,057	686	17,256	4,828	8,000	......	12,828	93,000	37,500	130,500	9½	9½	6¾	......	772,712
" 12...	18.205	5,119	......	......	492	23,816	4,500	4,500	......	9,000	107.500	41,500	148,000	9¼	9¼	6¾	......	777,212
" 19...	2,995	2,498	......	......	1,092	6,585	6,095	1,700	160	7,955	106,000	43,500	149,500	8⅞	8⅞	6¾	160	783,307
" 26...	1,797	......	......	......	1,443	3,240	6,563	500	......	7,063	102,500	42,500	145,000	8¾	8¾	6⅝	......	789,870
Average prices & total sales, receipts & stocks.	664,023	47,216	1,886	94,598	29,203	836,926	789,870	238332	50,010	1,078,212				8.10	8.18	5¾	50,010	151,898

The semi-weekly Price and Weekly Sales and Receipts at New York, Weekly Exports from New York and Rates of Freight to Liverpool 1st of each month, for the Crop Year ending October 1, 1834.

1833.	Price of New Orleans	Price of Upland.	Sales for week.	Receipts for week.	RECEIPTS FOR THE WEEK. To Great Britain.	To France.	North of Europe.	Other Fo'n Ports	Total Exports.	Rates of Freight to Liverpool.
October 4..	16½@18½	15@17								⅜@7-16d
" 8..	16@18½	15@17	700	1,562	....	....	....	....	....	
" 11..	16@18½	14½@18								
" 15..	16@18½	14½@18	1,900	2,902	....	....	....	....	....	
" 18..	16@18	14@18								
" 22..	16@18	14@17½	550	2,529	*5,343	2,022	760	....	8,125	
" 25..	15@18	13@16½								
" 29..	15@18	13@16½	1,900	2,764	1,760	2,371	118	45	4,294	
Novem. 1..	15@18	13@16½								⅜@7-16d
" 5..	15@18	13@16½	1,600	2,001	1,308	447	56	....	1,811	
" 8..	14½@18	13@16								
" 12..	14½@18	13@16	1,700	2,594	305	1,268	....	....	1,573	
" 15..	14½@18	13@16								
" 19..	14½@18	13@16	2,800	4,695	280	31	474	....	785	
" 22..	14½@17½	13@15¾								
" 26..	14½@17	13½@15	1,700	4,637	154	180	....	....	334	
" 29..	14@16½	13@15								
Decem. 3..	14@16½	12½@14½	1,510	4,046	680	304	....	....	984	⅜@7-16d
" 6..	14@16	12@14								
" 10..	14@15½	12@13½	2,000	3,586	....	....	51	....	51	
" 13..	13½@14½	11½@12½								
" 17..	13½@14½	11@12½	3,500	905	755	....	....	....	755	
" 20..	13½@14½	11@12½								
" 24..	13½@14½	11@12½	1,100	2,854	1,704	327	....	....	2,031	
" 27..	13½@14½	11@12								
" 31..	13½@14½	11@12	1,300	9,436	690	....	....	....	690	
1834.										
January 3..	12@14	10@12								⅜@½d.
" 7..	12@14	10@12	1,700	688	2,326	223	....	....	2,549	
" 10..	12@14	10@12								
" 14..	12@13½	10@12	1,800	2,101	1,856	704	291	....	2,851	
" 17.	12@13½	10@12								
" 21..	11½@13	10@11	1,200	954	1,203	192	50	....	1,445	

GENERAL REMARKS.

*These figures for exports include the two previous weeks, no statements having been compiled in the meantime.

This crop year was chiefly remarkable for the monitary revulsion and panic that set in early in 1834, when business was in a great measure brought to a stand, the depression in cotton circles, being also assisted by unfavorable accounts from abroad that came to hand during the closing months of '33, and the early ones of '34, and by the hesitancy caused by the long passages that were made at that time, by the packet ships. On February 9, '34, the latest European advices were 72 days old.

January 24..	$11\frac{1}{2}$@13	10@11								
" 28..	$11\frac{1}{2}$@13	10@11	700	1,293	601	....	....	....	601	
" 31..	$11\frac{1}{2}$@13	10@11								
February 4..	$11\frac{1}{2}$@13	10@11	750	2,649	1,384	93	354	....	1,831	$\frac{1}{4}$d.
" 7..	11@13	$9\frac{1}{2}$@11								
" 11..	11@13	10@$11\frac{1}{2}$	1,600	3,709	472	54	478	....	1,004	
" 14..	11@13	10@12								
" 18..	11@13	10@12	1,750	5,055	544	332	....	....	876	
" 21..	11@13	10@12								
" 25..	11@13	10@12	2,150	1,435	813	197	115	....	1,125	
" 28..	11@13	10@12								
March 4..	11@13	10@12	1,000	1,219	3,407	218	122	....	3,747	7-16@$\frac{9}{16}$
" 7..	11@13	$10\frac{1}{2}$@$12\frac{1}{2}$								
" 11..	11@13	$10\frac{1}{2}$@$12\frac{1}{2}$	1,500	2,958	1,356	....	....	100	1,456	
" 14..	11@13	$10\frac{1}{2}$@$12\frac{1}{2}$								
" 18..	11@13	$10\frac{1}{2}$@$12\frac{1}{2}$	1,500	3,127	550	....	....	....	550	
" 21..	11@13	$10\frac{1}{2}$@$12\frac{1}{2}$								
" 25..	$11\frac{1}{2}$@$13\frac{1}{2}$	$10\frac{1}{2}$@$12\frac{1}{2}$	1,900	2,011	1,020	677	229	....	1,926	
" 28..	$11\frac{1}{2}$@$13\frac{1}{2}$	$10\frac{1}{2}$@$12\frac{1}{4}$								
April 1..	$11\frac{1}{2}$@$13\frac{1}{2}$	$10\frac{1}{2}$@$12\frac{1}{2}$	2,100	6,625	469	692	....	....	1,161	$\frac{3}{8}$@$\frac{1}{2}$d.
" 4..	$11\frac{1}{2}$@14	11@13								
" 8..	$11\frac{1}{2}$@14	11@13	1,500	2,371	420	560	....	....	980	
" 11..	$11\frac{1}{2}$@14	11@$13\frac{1}{2}$								
" 15..	$11\frac{1}{2}$@14	11@$13\frac{1}{2}$	1,100	2,428	587	2,537	142	....	3,266	
" 18..	$11\frac{1}{2}$@15	$11\frac{1}{2}$@14								
" 22..	$11\frac{1}{2}$@15	$11\frac{1}{2}$@14	1,200	6,257	2,371	1,107	235	....	3,713	
" 25..	$11\frac{1}{2}$@15	$11\frac{1}{2}$@14								
" 29..	$11\frac{1}{2}$@15	$11\frac{1}{2}$@14	1,150	4,177	520	603	73	....	1,196	
May 2..	$11\frac{1}{2}$@15	$11\frac{1}{2}$@14								5-16@$\frac{1}{2}$d
" 6..	$11\frac{1}{2}$@15	11@14	2,250	3,630	368	1,159	....	....	1,527	
" 9..	$11\frac{1}{2}$@15	11@14								
" 13..	$11\frac{1}{2}$@15	11@14	1,400	4,755	1,472	1,173	175	....	2,820	
" 16	$11\frac{1}{2}$@15	11@14								
" 20..	$11\frac{1}{2}$@15	11@14	1,900	4,023	926	680	135	....	1,741	
" 23..	11@15	$10\frac{1}{2}$@14								
" 27..	11@15	10@14	2,200	4,502	217	642	26	36	921	
" 30..	11@$14\frac{1}{2}$	10@$13\frac{1}{2}$								
June 3..	11@$14\frac{1}{2}$	10@$13\frac{1}{2}$	3,100	175	1,143	1,112	475	31	2,761	5-16@$\frac{7}{16}$
" 6..	11@$14\frac{1}{2}$	10@$13\frac{1}{2}$								
" 10..	11@$14\frac{1}{2}$	10@$13\frac{1}{2}$	2,400	9,608	....	....	304	....	304	
" 15..	11@$14\frac{1}{2}$	10@$13\frac{1}{2}$								
" 17..	11@$14\frac{1}{2}$	10@$13\frac{1}{2}$	2,800	5,336	1,535	147	122	....	1,804	

The financial crisis referred to, covered the months of January, February, March and April ; in May, confidence began to be measurably restored, and from thence to the close of the year, business resumed its customary channels.

The Commercial Historian of that period, says, "confidence is daily diminished by the uncertainty of the future, in regard to merchandise ; holders are afraid to sell, and purchasers are afraid to buy. Negotiable paper is disposed of for cash with much difficulty, at the very heavy loss of $1\frac{1}{2}$@3 per cent. per month. The occurrence of numerous bank and mercantile failures, destroys all inclination to let goods pass from the control of owners, and our city is overspread with gloom."

The state of affairs outlined above, caused a very unsettled feeling in exchange, and the fluctuations were wide and frequent, as noted below.

New York Statement for 1834.—*Concluded.*

1834.		Price of New Orleans	Price of Upland.	Sales for week.	Receipts for week.	EXPORTS FOR THE WEEK.					Rates of Freight to Liverpool.
						To Great Britain.	To France.	North of Europe.	Other Fo'n Ports	Total Exports.	
June	20..	11@14½	10½@14								
"	24..	11@14½	10½@14	2,700	4,466	2,809	424	646	....	3,879	
"	27..	11½@14½	11@14								
July	1..	11½@14½	11@14	1,700	4,155	3,636	1,056	463	96	5,251	¼d.
"	4..	12@15	12@14½								
"	8..	12@15	12@14½	1,450	2,857	....	365	41	....	406	
"	11..	12@15½	12@15								
"	15..	12@15½	12@15	1,850	1,195	....	....	....	....	....	
"	18..	12@15½	12@15								
"	22..	12@15½	12@15	1,650	1,189	....	....	....	....	....	
"	26..	12½@16	12½@15½								
"	29..	12½@16	12½@15½	2,600	1,817	600	177	366	....	1,143	¼d.
August	1..	12½@16	12½@15½								
"	5..	12½@16	12½@15½	1,000	2,281	177	73	280	....	530	
"	8..	12½@16	12½@15½								
"	12..	12½@16	12½@15½	1,050	2,299	94	100	300	....	494	
"	15..	12½@16	12½@15½								
"	19..	12½@16	12½@15½	950	3,176	353	....	31	....	384	
"	22..	12½@16	12½@15½								
"	26..	12½@16	12½@15½	1,950	1,199	5	307	92	....	404	
"	29..	12½@16	12½@15½								¼d.
Septem.	2..	12½@16	12½@15½	1,100	790	1,201	1,518	....	....	2,719	
"	5..	12½@16	12½@15½								
"	9..	12½@16	12½@15½	1,600	3,239	....	....	22	....	22	
"	12..	12½@16	12½@15½								
"	16..	12½@16	12½@15½	2,800	1,812	272	639	....	....	911	
"	19..	12½@16	12½@15½								
"	23..	12½@16	12½@15½	1,900	567	622	708	57	239	1,626	
"	26..	12½@16	12½@15½								
"	30..	12½@16	12½@15½	1,400	1,118	1,254	965	....	298	2,517	¼@⅜d.
Average price and total sales, receipts and exports.		13.79	12.90	88,610	157,757	49,562	26,384	7,083	845	83,874	

GENERAL REMARKS.

Exchange.

On the 1st October, 1833, the quotation for 60 days bills on London, was 7@7½ per cent. premium ; on the 29th declined to 6½@7 ; in November to 5@5¾ ; in December to 3@4, the downward tendency continuing until the succeeding April.

In January, 1834, the price receded from 2@2½ to ½ per cent premium, and on the 24th to par ; in February to 1@2 per cent. below par, and bills generally declined except those drawn against consignments ; in early March, the quotation was par to 1 per cent. discount, later in the month par to 1 per cent. premium ; in April prices began to rise until they touched on the 22d 5 per cent premium ; subsequently there was a downward turn, and in May 2@2½ per cent. premium was the quotation ; on June 3d 1½@2 per cent. premium, 17th ½@1, 24th 1@1½. At the opening of July, prices began to advance, and continued to rise until the close of the crop year ; in July the range was from 2½ up to 4½ per cent. premium ; in August 4½ up to 7, and in September the quotation was steady at 7¼@7½ per cent. premium.

1835.

The actual workers in cotton factories this year in Great Britain, were 220,134; number of power looms in operation, 109,626; obtainable power of these looms, if all fully employed throughout the year, 700,000,000 yards; number of hand-looms *estimated* at from 200,000 to 250,000. Cotton weavers used this year, for the mere purpose of dressing the warp threads before weaving, 650,000 bushels of flour. The factories were 1,304 in number. Total number of persons supported by the cotton manufactures *estimated* at 1,500,000. (*See* year 1760.)

Cotton imported into Great Britain, 364,000,000 lbs.

"Brewster's Eclipse Speeder" (*see* year 1829), was introduced in Manchester, England, this year, and built by Sharp, Roberts & Co., and known as the "Eclipse Roving Frame."

The plate speeder was introduced into Manchester this year from America, by Mr. Neil Snodgrass.

"Edward Baines, Jr., Esq.," published a "History of the Cotton Manufacture in Great Britain," an illustrated volume of 532 pages and an appendix, dedicated to the Right Hon. C. Poulett Thomson, M.P.

The consumption of cotton in the manufacture of hosiery, was estimated to be 4,584,000 lbs., of the value of £153,000. (*See* years 1589, 1768, 1787 and 1805.)

Computed number of pounds of cotton yarn and twist spun in Great Britain this year, 281,000,000. (*See* year 1840.) Exports of cotton piece goods, 558,000,000 yards.

Another mill was built on Factory Island, Saco, Maine. (*See* years 1830, 1833 and 1837.)

Samuel Batchelder patented in the United States, his steam-drying cylinder, this year. (*See* years 1832 and 1839.)

Bobbin-net machines—7 qr.—"were cast out of top story windows at the cry of 'old rags and twist machines to sell,'" in England. (*See* year 1824.)

The following table sets forth the state of the cotton manufacture in New York at this period. The figures are from Williams's Annual Register for this year:

Number of Mills	112
" Spindles	157,316
Pounds of cotton consumed yearly	7,961,670
" yarn sold "	1,867,790
Yards of cotton manufactured yearly	21,010,920
Number of hands employed	12,954
Capital invested	$3,669,500

COTTON CROP OF THE UNITED STATES.

Statement and Total Amount of the Cotton Crop of the United States, for the Year ending September 30, 1835.

	Bales.	Bales.	Total.	Same period 1834.
NEW ORLEANS.				
Export—				
To Foreign Ports	412,281			
Coastwise	122,484			
Burnt, &c	515			
Stock on hand 1st October, 1835	4,842			
		540,122		
Deduct—				
Stock on hand 1st October, 1834	8,756			
Received from Mobile	17,456			
" Florida	2,764			
		28,976		
			511,146	454,719
FLORIDA.				
Export—				
To Foreign Ports	....	9,156		
New Orleans	....	2,764		
Mobile	....	155		
Charleston	....	2,316		
New York	....	33,984		
Savannah	....	135		
Boston	....	873		
Pittsburg	....	996		
Burnt and sunk on board steamboats	....	1,406		
Stock on hand 1st October, 1835	....	300		
			52,085	36,738
ALABAMA.				
Exports from Mobile—				
To Foreign Ports	115,437			
Coastwise	82,333			
Stock in Mobile 1st October, 1835	488			
		198,258		
Deduct—				
Stock on hand 1st October, 1834	411			
Received from Florida	155			
		566		
			197,692	149,978
GEORGIA.				
Export from Savannah—				
To Foreign Ports—Uplands	135,519			
" Sea Island	8,362			
	143,881			
Coastwise	71,374			
		215,255		
From Darien—				
To Foreign Ports	2,245			
New York	6,501			
		8,746		
Stock in Savannah 1st October, 1835	1,027			
" Augusta and Hambro', 1st October. 1835	3,403			
		4,430		
		228,431		
Deduct—				
Stock on hand 1st October, 1834	5,626			
Received from Florida	135			
		5,761		
			222,670	258,655

Statement and Total Amount of the Cotton Crop of the United States, for the Year ending September 30, 1835—*Concluded.*

	Bales.	Bales.	Total.	Same period 1834.
SOUTH CAROLINA.				
Export from Charleston—				
To Foreign Ports—Uplands	145,242			
" Sea Island	15,171			
	160,413			
Coastwise	41,268			
		201,681		
From Georgetown—				
To New York	....	13,927		
Stock in Charleston 1st October, 1835	....	4,766		
" Georgetown, "	....	300		
		220,674		
Deduct—				
Stock in Charleston, 1st October, 1834	3,806			
Received from Savannah, and included in the exports from that place	11,386			
Received from Florida	2,316			
		17,508		
			203,166	227,359
NORTH CAROLINA.				
Export—				
To Foreign Ports	351			
Baltimore	284			
Philadelphia	316			
New York	33,053			
Boston	445			
Stock on hand 1st October, 1835	200			
		34,649		
Deduct—				
Stock on hand 1st October, 1834	....	250		
			34,399	33,220
VIRGINIA.				
Export—				
To Foreign Ports	19,897			
Coastwise	8,000			
Taken by Manufacturers	5,400			
Stock on hand 1st October, 1835	503			
		33,800		
Deduct—				
Stock on hand 1st October, 1834	....	630		
			33,170	44,725
Total Crop of the United States	....	....	1,254,328	1,205,394
Crop of last year	....	....	1,205,394	
Increase	....	....	48,934	

Export to Foreign Ports, from October 1, 1834, *to September* 30, 1835.

FROM	To Great Britain.	To France.	To North Europe.	Other Foreign Ports.	Total.
New Orleans..............Bales	259,123	141,872	4,368	6,918	412,281
Alabama	88,739	24,236	1,051	1,411	115,437
Florida	9,156				9,156
Georgia (Savannah and Darien)	135,728	10,223	175		146,126
South Carolina	108,149	29,948	14,974	7,342	160,413
North Carolina	351				351
Virginia	17,910	1,477	510		19,897
Baltimore					
Philadelphia	2,100	2,667		10	4,777
New York	98.630	41,259	6,001	1,414	147,304
Boston	2,832	788	2,724	1,413	7,757
Grand total	722,718	252,470	29,803	18,508	1,023,499
Total last year	756,291	216,424	43,280	11,956	1,027,951
Increase		36,046		6,552	
Decrease	33,573		13,477		4,452

Growth.

Total crop of	1824–5,	560,000 bales.
"	1825–6,	710,000 "
"	1826–7,	937,000 "
"	1827–8,	712,000 "
"	1828–9,	857,744 "
"	1829–30,	976,845 "
"	1830–31,	1,038,848 "
"	1831–32,	987,477 "
"	1832–33,	1,070,438 "
"	1833–34,	1,205,394 "
"	1834–35,	1,254,328 "

Consumption.

To estimate the quantity manufactured in the United States, we take the growth of the year			1,254,328 bales.	
Add—Stocks on hand at the commencement of the year, 1st October, 1834.—In the Southern ports	19,283			
" Northern "	10,334			
		29,617		
				1,283,945
Deduct therefrom—The export to Foreign ports			1,023,499	
Stocks on hand at the close of the year, 1st Oct., 1835.— In the Southern ports	15,829			
" Northern "	25,794			
			41,623	
Burnt and lost at New Orleans	515			
" Apalachicola	1,406			
" New York	500			
		2,421		
			1,067,543	
Less—Foreign cotton included in the export from New York			486	
				1,067,057

Quantity consumed,	and in the hands	of manufacturers,	1834–5	....	216,888
"	"	"	1833–4	....	196,413
"	"	"	1832–3	....	194,412
"	"	"	1831–2	....	173,800
"	"	"	1830–1	....	182,142
"	"	"	1829–30	....	126,512
"	"	"	1828–9	....	118,853
"	"	"	1827–8	....	120,593
"	"	"	1826–7	....	103,483

Note.—In our estimate of the quantity taken for consumption, it will be perceived that we do not include any cotton manufactured in the States south and west of Virginia, nor any in that State, except in the vicinity of Petersburgh and Richmond.

The injury done to the crops in the Atlantic States towards the autumn of 1834 led to various opinions as to the extent of the whole crop, and it was asserted with great confidence, as late as January last, that the amount could not exceed 1,150,000 bales. The event shows, that although there was a deficiency in the Atlantic States of about 70,000 bales, compared with the previous year—on the other hand, there was a gain from the Gulf of Mexico of about 120,000, making a net increase over the previous year in the total, of about 49,000, and exceeding by that amount any crop ever before produced.

The semi-weekly Price and Weekly Sales and Receipts, at New York, Weekly Exports from New York and Rates of Freight to Liverpool 1st of each Month, for the Crop Year ending October 1, 1835.

1834.	Price of New Orleans	Price of Upland.	Sales for week.	Receipts for week.	EXPORTS FOR WEEK. To Great Britain.	To France.	North of Europe.	Other Fo'n Ports	Total Exports.	Rates of Freight to Liverpool.
October 3..	12½@16	12½@15½								¼@⅜d.
" 7..	12½@16	12½@15½	1,400	1,744	1,749	482	....	....	2,231	
" 10..	12½@16	12½@15½								
" 14..	12½@16	12½@15½	1,000	1,421	....	....	....	....	....	
" 17..	12½@16	12½@15½								
" 21..	12½@16	12½@15½	1,400	4,688	899	1,685	....	....	2,584	
" 24..	14 @17	12½@16								
" 28..	14 @17	12½@16	2,900	5,650	696	....	118	61	875	
" 31..	14 @17½	13 @16½								
Novem. 4..	14 @17½	13 @16½	2,600	497	385	1,260	....	32	1,677	¼@5-16d
" 7..	14½@18	13½@17								
" 11..	14½@18	13½@17	1,900	1,836	....	....	57	....	57	
" 14..	15½@19	14½@18								
" 18..	16½@20	15 @19	2,800	6,700	1,217	549	....	....	1,766	
" 21..	16½@20	15 @19								
" 25..	16½@20	15 @19	2,600	7,558	785	....	....	....	785	
" 28..	16½@19½	15 @18								
Decem. 2..	16½@19½	15 @18	1,900	7,946	883	....	....	....	883	5-16@⅜d
" 5..	15½@19	15 @17½								
" 9..	15½@19	15 @17	4,000	6,489	....	....	....	....	....	
" 12..	16 @19	15 @17								
" 16..	16 @19	15 @17	5,500	5,525	416	324	....	....	740	
" 19..	15½@18½	15 @17								
" 23..	15½@18½	15 @17	5,000	5,432	1,964	913	186	....	3,063	
" 26..	16 @19	15½@17½								
" 30..	16½@19	16 @18	3,400	2,415	4,980	625	30	....	5,635	
1835.										¼@⅜d.
January 2..	16½@19	15½@18							3,108	
" 6..	16½@18½	15½@17½	1,100	7,685	3,108	....	....	....		
" 9..	16 @18½	15½@17½							2,447	
" 13..	16 @18½	15½@17½	1,400	260	1,420	1,027	....	....		
" 16..	16 @18½	15 @17								
" 20..	16 @18½	15 @17	2,700	26,321	2,823	641	....	....	3,464	

GENERAL REMARKS.

The fluctuations in prices of Cotton during this crop year were considerable—the difference between the lowest range at the opening of the season and the highest touched in August, 1835, were 5 @ 6 cents per lb. The most noteworthy events of the year were the accounts received here, early in November, of injury to that portion of the crop still in the fields, and a consequent yield predicted of less than the previous year. These advices excited the market at the time, and, being supplemented by favorable foreign accounts, caused an advance of 1¾ cents per lb. Subsequently it was found that the short crop reports were greatly exaggerated, and prices at once fell back 1½ cents, notwithstanding advices from abroad were still favorable for the staple.

At the close of the year 1834, and at the beginning of 1835, the markets of the country were excited and unsettled by a message from the President, and Congressional action thereon, relative to the enforcement of the French Treaty of 1831. Accounts were received from France, on February 21, stating

"	23..	15 @18	15 @17								
"	27..	15½@18½	15 @17	5,500	8,384	302	150	87	....	539	
"	30..	15½@18½	15 @17								
February	3..	15½@18½	15 @17	5,000	1,897	2,028	1,836	105	....	3,969	7-16@9-16
"	6..	16 @18½	15 @17								
"	10..	16 @18½	15 @17	3,800	2,489	1,407	....	140	....	1,547	
"	13..	16 @18½	15 @17								
"	17..	16 @18½	15½@17	1,600	7,576	4,633	1,107	....	46	5,786	
"	20..	16 @18½	15½@17								
"	24..	16 @18½	15½@17	2,500	6,247	2,168	968	....	....	3,136	
"	27..	16 @18½	15½@17								
March	3..	16½@19	15½@17½	2,400	2,041	1,215	1,952	715	....	3,882	7-16@9-16
"	6..	16½@19	15 @17½								
"	10..	16½@19	15 @17½	2,000	1,624	....	....	145	....	145	
"	13..	16½@19	15 @17½								
"	17..	16½@19	15½@17½	6,100	7,622	1	854	....	....	855	
"	20..	16½@19½	15½@18								
"	24..	16½@20	15½@18	9,300	6,760	948	764	....	....	1,712	
"	27..	16½@20	15½@18								
"	31..	16½@20	15½@18	2,000	2,829	1,725	1,059	....	....	2,784	
April	3..	16½@20	15½@18								5-16@⅜d
"	7..	16½@20	16 @18½	3,700	5,821	619	738	254	....	1,611	
"	10..	17 @20	16 @18½								
"	14..	17 @20	16 @18½	3,500	8,977	1,586	....	....	....	1,586	
"	17..	17½@21	16 @19								
"	21..	17½@21	16 @19	2,950	7,433	1,678	1,432	....	....	3,110	
"	24..	17½@21	16 @19								
"	28..	17½@21	16 @19	1,659	4,151	625	829	260	....	1,714	
May	1..	17½@21	16 @19								¼@⅜d.
"	5..	17½@21	16 @19	3,440	7,302	295	876	....	....	1,171	
"	9..	17½@21	16 @19								
"	12..	18 @21	16½@19½	8,600	4,790	642	1,388	50	....	2,080	
"	15..	18 @21	16½@19½								
"	19..	18 @21	16½@19½	2,500	5,764	2,265	....	....	....	2,265	
"	22..	18 @21	16½@19½								
"	26..	18 @21	17 @19½	4,700	7,344	877	3,385	....	....	4,262	
"	29..	18 @21	17 @19½								
June	2..	18 @21	17 @19½	5,100	9,446	2,088	1,870	199	282	4,439	¼@5-16d
"	5..	18 @21	17 @19½								
"	9..	18 @21	17 @19½	2,600	7,157	4,985	1,008	50	....	6,043	
"	12..	18 @21	17 @19½								
"	16..	18 @21	17 @20	3,600	3,600	3,541	2,240	53	....	5,834	
"	19..	18 @21	17 @20								

that the French Minister had been recalled, which caused the underwriters to double their premiums on European risks. Stocks fell 10 @12 per cent, and all French goods, notably Brandies, Wines, etc., advanced largely. Later, however, the *Sully*, from Havre, brought more pacific advices, and, with an abatement in the belligerent attitude of Congress, mercantile circles became reassured, and business was resumed, Cotton becoming quite buoyant.

On April 1 the latest European advices were to February 11, and accounts were awaited with anxiety. On the 7th accounts came to hand one month later, and, being favorable, the market became active at better prices. Early in September advices were received from New Orleans, to August 18, noting severe injury to the crop by heavy rains, and the assurance that it would be fully a month later than usual. These reports had no effect upon the market, however, being neutralized by the receipt of unfavorable foreign accounts, and the market receded, the crop year closing at prices about 4 cents below the highest point touched

The semi-weekly Price and Weekly Sales and Receipts at New York, Weekly Exports from New York and Rates of Freight to Liverpool 1st of each Month, for the Crop Year ending October 1, 1835—Concluded.

1835.		Price of New Orleans	Price of Upland.	Sales for week.	Receipts for week.	EXPORTS FOR WEEK.					Rates of Freight to Liverpool.
						To Great Britain.	To France.	North of Europe.	Other Fo'n Ports	Total Exports.	
June	23..	18 @21	17 @20	4,050	6,966	4,404	751	....	....	5,155	
"	26..	18 @21	17 @20								
"	30..	18 @22	17 @20	3,600	4,066	2,313	2,702	59	....	5,074	¼d.
July	3..	18 @22	17 @20								
"	7..	18 @22	17 @20	1,800	3,677	4,388	851	185	....	5,424	
"	10..	18 @22	17 @20								
"	14..	18 @22	17 @20	2,600	8,452	8,714	3,730	403	224	13,071	
"	17..	18 @22	17½@19½								
"	21..	18 @22	17½@19½	2,000	1,086	....	....	....	....	....	
"	24..	18 @22	17½@19½								
"	28..	18 @22	17½@19½	2,300	3,412	....	....	....	....	....	
"	31..	18 @22	17½@19½								
August	4..	18 @22	17½@19½	2,500	674	5,364	668	83	....	6,115	⅜@5-16d
"	7..	18 @22	17½@19½								
"	11..	18 @22	17½@19½	1,300	2,399	1,976	1,028	634	....	3,638	
"	14..	18 @22	17½@19½								
"	18..	18 @22	17½@19½	900	2,241	3,047	728	339	....	4,114	
"	21..	18 @22	17½@19½								
"	25..	18 @22	17 @19½	700	1,071	2,639	582	392	....	3,613	
"	28..	18 @22	17 @19½								
Septem.	1..	18 @22	17 @19½	900	702	2,664	46	161	....	2,871	¼@5-16d
"	4..	18 @22	17 @19½								
"	8..	18 @22	17 @19	3,800	375	492	....	191	138	821	
"	11..	18 @22	16½@19								
"	15..	18 @22	16½@19	1,800	1,035	1,969	334	200	....	2,503	
"	18..	18 @22	16½@19								
"	22..	17½@21	16 @19	800	1,154	2,191	162	770	....	3,123	
"	25 .	17 @20	15½@19								
"	29..	16½@20	15½@19	1,400	356	5,268	197	135	....	5,600	⅜@7-16d
Average price and total sales, receipts and exports.		18.11	17.45	154,590	249,087	100,382	41,741	6,001	783	148,907	

GENERAL REMARKS.

Exchange.

Foreign bills were comparatively steady throughout this crop year, opening in October at 7¼ @ 7½ per cent. premium, gradually receding through November to 6 @ 6¼, rising in December to 6¼ @ 6½, and continued to advance in early January until 7¼ @ 7½ was touched; then fell back to 6 @ 6½. In February the price advanced again to 7@7½; then declined to 6½@6¾ In March the price went up until the quotation stood, on the 20th, at 8½ @ 9; then receded to 8¼ @ 8¾. Through April the quotation was steady at 8¾@9; May, 9@9¾; June, 9@9½. In July it advanced from 9¼ up to 10; in August, fell off to 9@9¼; in September, to 8¾@9, closing, October 1 at 9@9¼ premium.

LIVERPOOL STATEMENT FOR 1835.

UNITED STATES, 1834–1835.			
Stock, October 1, 1834	29,000	Export	1,024,000
Crop	1,254,000	Consumption	217,000
		Stock, Oct. 1, 1835	42,000
Bags	1,283,000	Bags	1,283,000

	Liverpool.	Gt. Britain.	France.	Continent.	Tot. Europe.
United States .... Bags.	105,940	117,970	20,314	9,192	147,476
Brazil	9,720	11,770	2,139	543	14,452
West Indies	4,860	5,390	1,566	582	7,538
East Indies	23,510	48,830	....	3,512	52,342
Egypt and Levant	1,280	1,600	388	1,988	3,976
Bags	145,310	185,560	24,407	15,817	225,784

CONSUMPTION.					IMPORT.					
Tot. Europe.	Continent.	France.	Gt. Britain.	Liverpool.		Liverpool.	Gt. Britain.	France.	Continent.	Tot. Europe.
1,036,848	80,807	225,903	730,138	679,917	United States .... Bags.	707,317	763,238	225,498	85,696	1,032,632
150,683	7,888	22,405	120,390	118,212	Brazil	142,312	143,580	28,527	8,086	179,043
55,273	13,896	18,433	22,944	21,047	West Indies	21,277	24,014	19,093	18,490	59,297
104,660	46,887	....	57,773	40,599	East Indies	63,409	118,433	....	50,108	119,591
105,799	40,676	41,995	23,128	18,499	Egypt and Levant	35,699	41,958	51,307	58,391	140,956
1,453,263	190,154	308,736	954,373	878,274	Bags	970,014	1,091,223	324,425	220,771	1,531,519
....	12,500	....	92,400	52,350	Export.					
304,040	33,934	40,096	230,010	184,700	Stock, Dec. .... Stock above,	145,310	185,560	24,407	15,817	225,784
1,757,303	236,588	348,832	1,276,783	1,115,324	Total supply, bags,	1,115,324	1,276,783	348,832	236,588	1,757,303

COTTON AT LIVERPOOL. YEAR 1835.

Week Ending.	Receipts.						Sales.				Stocks.			Prices.			Actual Export	Consumption.
	Americ'n.	E. I.	Egypt.	Brazil.	Other.	Total.	Consumption.	Speculation.	Export.	Total.	Amer'n.	Other.	Total.	Mid. Up.	Mid. Orl.	Dhol		
Jan. 9	14,808				3,146	17,954	17,128	1,500		18,628	136,000	46,000	182,000	8¾	8⅞	6½		17,128
" 16	6,174	5,524		2,055	1,556	15,309	26,480	500	250	22,230	120,500	50,500	170,000	8¾	8⅞	6½	250	43,608
" 23	24,074				3,021	27,095	34,126	1,500	200	35,826	118,500	40,500	159,000	9	9	6⅝	200	77,734
" 30	21,608			4,205		25,813	14,580	1,000	400	15,980	129,000	39,000	168,000	9	9	6⅝	400	92,314
Feb. 6	4,671				1,377	6,048	15,260	8,500		23,760	99,500	37,500	136,000	8⅞	9	6⅝		107,574
" 13	47,191	1,923		11,111	470	60,695	13,125	1,100		14,225	136,000	46,500	182,500	8¾	8⅞	6½		120,699
" 20	15,764				8	15,772	19,880	2,500	850	23,230	127,500	34,000	161,500	9	9	6⅝	850	140,579
" 27	13,483	3,695		2,333	481	19,992	14,000	1,100	40	15,140	125,500	37,500	163,000	9	9⅛	6⅝	40	154,579
Mch. 6	19,364				7	19,371	13,270	2,500	700	16,470	131,500	36,000	167,500	9	9⅛	6⅝	700	167,849
" 13	18,688			3,478		22,166	19,650	2,820	700	23,170	134,500	33,500	167,000	9	9⅛	6⅝	700	187,499
" 20	10,254				2,603	12,857	20,900	5,850	2,800	29,550	126,500	29,000	155,500	9⅛	9¼	6⅝	2,800	208,399
" 27	8,463				902	9,365	21,940	10,700	2,650	35,290	116,500	22,500	184,500	9½	9½	6⅝	2,650	230,339
April 3	16,703				445	17,148	10,077	1,150	1,250	12,477	124,500	20,200	144,700	9⅝	9⅝	6⅝	1,250	240,416
" 10	38,323	5,244		6,889	594	51,050	16,730	7,800	1,650	26,180	149,000	28,500	177,500	9⅞	9⅞	6⅝	1,650	257,146
" 16	8,426				735	9,161	20,510	12,350	2,300	35,160	138,000	25,500	163,500	10⅛	10⅛	7¼	2,300	277,656
" 24	21,720		700	3,338		25,758	18,810	11,000	3,760	33,570	143,500	22,500	166,000	10⅛	10⅛	7	3,760	296,466
May 1	3,723				1,539	5,262	12,030	2,500	1,550	16,080	134,500	20,500	155,000	10⅛	10⅛	7	1,550	308,496
" 8	34,143	300			1,247	35,690	12,850	9,200	2,900	24,950	147,500	17,500	164,000	10⅛	10⅛	7	2,900	321,346
" 15	18,865	250	1,000	3,284	216	23,615	14,960	3,000	2,050	20,010	151,000	20,000	171,000	10¼	10¼	7	2,050	336,306
" 22	22,425			3,804	553	26,782	18,660	3,600	1,650	23,910	156,000	20,000	176,000	10¼	10¼	7	1,650	354,966
" 29	12,985	3,847		2,013	377	19,222	10,188	1,750	800	12,738	159,000	23,500	182,500	10⅛	10⅛	7	800	365,154
June 5	13,573	1,337			320	15,230	8,830	2,200	1,300	12,330	164,000	23,500	187,500	10½	10½	7	1,300	373,984
" 12	6,235				3,657	9,892	7,650	1,500	800	9,950	162,500	26,000	188,500	10⅛	10⅛	7	800	381,634
" 19	17,615	1,897		4,810	398	24,720	12,550	1,500	1,990	16,040	167,500	30,000	197,500	10⅛	10⅛	7	1,990	394,184
" 26	16,508				1,603	18,111	15,687	1,500	1,050	18,237	169,000	30,000	198,000	10⅛	10⅛	7	1,050	409,871
July 3	24,318				844	25,162	11,421	500	1,180	13,101	182,500	28,000	210,500	10⅛	10⅛	7	1,180	421,292
" 10	19,688		2,698	6,936	2,254	31,576	10,220	550	750	11,520	192,500	38,500	230,000	10	10⅛	7	750	431,512
" 17	13,316			3,733	217	17,266	15,420	450	300	16,170	192,500	39,000	231,500	10	10⅛	7	300	446,932
" 24	14,453				305	14,758	7,909	200	200	8,309	197,000	37,500	234,500	9⅞	10	7	200	454,841
" 31	6,233			2,010	88	8,331	8,909	500	1,450	10,859	194,500	46,500	231,000	9⅞	10	7	1,450	463,750
Aug. 7	34,097			2,706	2,119	38,922	19,870	4,000	500	24,370	211,500	42,000	253,500	9⅞	10	7	500	483,620
" 14	8,447			4,379	605	13,431	14,880	750	450	16,080	207,000	42,000	249,000	9⅞	10	6⅞	450	498,500
" 21	19,210		256		3,427	22,893	6,370		840	7,210	219,500	43,500	263,000	9⅝	9⅝	6⅞	840	504,870
" 28	16,622	1,378	1,150	2,555	1,509	23,214	8,420	1,000	350	9,770	228,500	49,500	277,000	9¼	9¼	6⅞	350	513,290

Sept. 4...	7,036	......	782	......	466	8,284	15,646	1,500	1,850	18,996	222,500	47,000	269.500	$9\frac{1}{8}$	$9\frac{1}{8}$	$6\frac{3}{4}$	1,850	528,936
" 11...	17,709	......	3,490	2,304	4,514	28,017	9,720	200	1,600	11,520	233,000	54,500	287,500	9	$8\frac{3}{4}$	$6\frac{3}{4}$	1,600	538,656
" 18...	6,019	1,585	592	7,283	1,316	16,795	10,536	500	1,600	12,636	229,000	33,500	292,500	$8\frac{7}{8}$	$8\frac{3}{4}$	$6\frac{1}{2}$	1,600	549,192
" 25...	4,901	10,813	1,050	2,371	1,989	21,114	16,440	1,500	1,700	19,640	216,500	75,000	291,500	$8\frac{3}{4}$	$8\frac{5}{8}$	$6\frac{1}{2}$	1,700	565,632
Oct. 2...	693	......	2,210	......	1,169	4,072	15,000	......	1,800	16,800	203,000	76,000	279,000	$8\frac{3}{8}$	$8\frac{3}{8}$	$6\frac{1}{4}$	1,800	580,632
" 9...	....	......	1,902	......	285	2,187	18,385	2,500	2,000	22,885	185,500	75,000	260,500	$8\frac{1}{8}$	$8\frac{1}{8}$	$6\frac{1}{4}$	2,000	599,017
" 16...	4,823	......	......	5,450	782	11,055	17,488	6,600	2,000	26,088	176,000	76,500	252,500	$8\frac{1}{8}$	$8\frac{1}{8}$	$6\frac{1}{4}$	2,000	616,505
" 23...	14,160	1,839	......	5,370	423	21,792	16,270	700	2,400	19,370	175,500	80,000	255,500	8	8	$6\frac{1}{4}$	2,400	632,775
" 30...	1,531	1,014	5,030	2,387	510	10,472	13,950	1,000	1,100	16,050	165,000	85,500	250,500	$7\frac{3}{4}$	$7\frac{3}{4}$	$6\frac{1}{4}$	1,100	646,725
Nov. 6...	....	1,994	......	650	......	2,644	20,710	4,600	1,950	27,260	146,500	83,500	230,000	$7\frac{7}{8}$	$7\frac{3}{4}$	$6\frac{1}{4}$	1,950	667,435
" 13...	786	......	2,445	4,609	773	8,613	19,260	4,100	900	24,260	132,000	86,000	218,000	8	8	$6\frac{1}{8}$	900	686,695
" 20...	5,066	......	1,700	......	15	6,781	22,890	1,550	960	25,400	119,500	82,500	201,000	8	8	$6\frac{1}{8}$	960	709,585
" 27...	10,399	325	......	5,828	619	17,171	9,730	500	200	10,430	123,500	84,500	208,000	$7\frac{7}{8}$	$7\frac{7}{8}$	$6\frac{1}{8}$	200	719,315
Dec. 4...	12,194	7,401	......	650	223	20,468	14,470	2,100	500	17,070	123,000	90,500	213,500	$7\frac{3}{8}$	$7\frac{3}{8}$	$6\frac{1}{8}$	500	733,785
" 11...	3,976	3,708	......	4,005	301	11,990	19,587	1,250	1,630	22,467	109,500	95,000	204,500	$7\frac{5}{8}$	$7\frac{5}{8}$	6	1,630	753,372
" 18...	18,896	......	......	5,625	777	25,298	19,130	800	1,290	21,220	113,500	96,000	209,500	$7\frac{5}{8}$	$7\frac{5}{8}$	6	1,290	772,502
" 24...	....	476	1,250	......	......	1,726	19,740	3,500	700	23,940	97,000	93,500	190,500	$7\frac{5}{8}$	$7\frac{5}{8}$	6	700	792,242
" 31...	....	......	......	......	......	....	22,461	5,600	......	28,061	......	......	......	....	....	....	......	814,703
Average prices & total sales, receipts & stocks.	700,359	54,560	26,255	106,071	50,785	938,030	814,703	145070	61,840	1,021,683				9.13	9.15	6.64	61,840	1,566,736

1836.

This year, embroidering and finishing of lace employed in Europe but 35,000 hands, their wages amounting to but £350,000. (*See* years 1831 and 1833.)

COTTON CROP OF THE UNITED STATES.

Statement and Total Amount of the Cotton Crop of the United States, for the Year ending September 30, 1836.

	Bales.	Bales.	Total.	Same period 1835.
NEW ORLEANS.				
Export—				
To Foreign Ports	400,470			
Coastwise	92,535			
Burnt, &c	2,030			
Stock on hand 1st October, 1836	3,702			
Deduct—		503,737		
Stock on hand 1st October, 1835	4,842			
Received from Mobile	17,366			
" Florida	6,882			
		29,090	474,647	511,146
NATCHEZ.				
Export—				
To Liverpool	....	4,841		
New York	....	248		
Burnt there, 8th December	....	1,800		
			6,889	
FLORIDA.				
Export—				
To Foreign Ports	20,047			
New Orleans	6,882			
Mobile	875			
Charleston	2,128			
New York	43,778			
Providence	1,682			
Boston	2,813			
Philadelphia	508			
Burnt on board brig William Osborn	300			
Stock on hand 1st October, 1836	1,049			
Deduct—		80,062		
Stock on hand 1st October, 1835		300		
			79,762	52,085
ALABAMA.				
Export from Mobile—				
To Foreign Ports	151,434			
Coastwise	86,580			
Stock in Mobile, 1st October, 1836	64			
Deduct—		238,078		
Stock on hand 1st October, 1835	488			
Received from Florida	875			
		1,363		
			236,715	197,692
GEORGIA.				
Export from Savannah—				
To Foreign Ports—Uplands	169,102			
" Sea Island	7,983			
	177,085			
Coastwise	68,102			
		245,187		

Statement and Total Amount of the Cotton Crop of the United States, for the Year ending September 30, 1836—Concluded.

	Bales.	Bales.	Total.	Same period 1835.
From Darien—				
To Foreign Ports	5,745			
New York	7,356			
Charleston	10,177			
		23,278		
Stock in Savannah, 1st October, 1836	2,096			
" Augusta and Hambro', 1st Oct., 1836	3,390			
Burnt on Savannah River	600			
		6,086		
		274,551		
Deduct—				
Stock on hand 1st October, 1835, in Sav. and Aug.	....	4,430		
			270,121	222,670
SOUTH CAROLINA.				
Export from Charleston—				
To Foreign Ports—Uplands	181,328			
" Sea Island	15,168			
Coastwise	47,831			
		244,327		
From Georgetown—				
To New York	....	13,930		
Stock in Charleston, 1st October, 1836	....	2,997		
		261,254		
Deduct—				
Stock in Charleston, 1st October, 1835	4,766			
Received from Savannah and Darien	22,672			
Received from Florida	2,128			
Received from Key West	451			
		30,017	231,237	203,166
NORTH CAROLINA.				
Export—				
To Foreign Ports	762			
Baltimore	30			
Philadelphia	2,622			
New York	28,043			
Boston	400			
Stock on hand, 1st October, 1836	400			
		32,257		
Deduct—				
Stock on hand 1st October, 1835	....	200		
			32,057	34,398
VIRGINIA.				
Export—				
To Foreign Ports	14,871			
Coastwise	4,900			
Taken by Manufacturers	8,000			
Stock on hand, 1st October, 1836	1,929			
		29,700		
Deduct—				
Stock on hand 1st October, 1835	....	503		
			29,197	33,170
Received at Philadelphia from Tennessee	....	....	100	
Total crop of the United States	....	....	1,360,725	1,254,328
Crop of last year	....	....	1,254,328	
Increase compared with last year	....	....	106,397	

Export to Foreign Ports, from October 1, 1835, *to September* 30, 1836.

FROM		To Great Britain.	To France.	To North of Europe.	Other Fn. Ports.	Total.
New Orleans	Bales	236,526	133,881	17,989	12,074	400,470
Natchez		4,841	....	....	....	4,841
Alabama		125,858	21,661	3,192	723	151,434
Florida		18,789	1,258	....	....	20,047
Georgia (Savannah and Darien)		166,434	15,923	150	323	182,830
South Carolina		112,437	57,913	19,974	6,172	196,496
North Carolina		762	....	....	....	762
Virginia		10,345	3,160	1,366	....	14,871
Baltimore		....	24	....	....	24
Philadelphia		3,948	130	228	327	4,633
New York		89,230	32,238	10,030	3,225	134,723
Boston		1,978	....	1,960	1,534	5,472
Grand total		771,148	266,188	54,889	24,378	1,116,603
Total last year		722,718	252,470	29,803	18,508	1,023,499
Increase		48,430	13,718	25,086	5,870	93,104

Growth.

Total crop of	1824–5,	560,000	bales.
"	1825–6,	710,000	"
"	1826–7,	937,000	"
"	1827–8,	712,000	"
"	1828–9,	857,744	"
"	1829–30,	976,845	"
"	1830–1,	1,038,848	"
"	1831–2,	987,477	"
"	1832–3,	1,070,438	"
"	1833–4,	1,205,394	"
"	1834–5,	1,254,328	"
"	1835–6,	1,360,725	"

Consumption.

To estimate the quantity manufactured in the United States, we take the growth of the year			1,360,725 bales.
Add—Stocks on hand at the commencement of the year (October 1, 1835)—In the Southern ports	15,829		
" Northern ports	25,794		
		41,623	
			1,402,348
Deduct therefrom—The Export to Foreign ports		1,116,603	
Stocks on hand at the close of the year (October 1, 1836)—			
In the Southern ports	20,627		
" Northern ports	22,714		
		43,341	
Burnt and Lost at New Orleans	2,030		
" Apalachicola	300		
" New York, 16th Dec.	1,500		
" Natchez, 8th Dec.	1,800		
" Savannah River	600		
		6,230	
		1,166,174	
Less—Foreign cotton, included in the export		559	
			1,165,615

Quantity consumed, and in the hands of manufacturers,	1835-6	236,733
" " "	1834-5	216,888
" " "	1833-4	196,413
" " "	1832-3	194,412
" " "	1831-2	173,800
" " "	1830-1	182,142
" " "	1829-30	126,512
" " "	1828-9	118,853
" " "	1827-8	120,593
" " "	1826-7	103,483

New Crop Received.

AT	1835.			1836.		
	Date of first receipt.	No. Bales.	Bales rec'd before 1st October.	Date of first receipt.	No. Bales.	Bales rec'd before 1st October.
New Orleans	....	....	3,412	18th Aug.	1	7,239
Apalachicola	1st Sept.	1	22	....	....	....
Mobile	....	....	....	2d Sept.	2	433
Savannah	....	....	....	17th Aug.	2	2,000
Charleston	13th Aug.	1	....	....	....	....
Augusta	....	....	....	....	....	....
Petersburg	22d Sept.	1	....	20th Sept.	1	30

Note.—Our estimate of the quantity taken for consumption does not include any Cotton manufactured in the States south and west of Virginia, nor any in that State, except in the vicinity of Petersburg and Richmond.

*** It has been objected to our statements that the Cotton received from Texas does not properly belong to the crop of the United States. As we are inclined to make our account literally what it purports to be, it is our intention hereafter to deduct it: the quantity received during the past year was 3,564 bales.

The semi-weekly Price and Weekly Sales and Receipts at New York, Weekly Exports from New York and Rates of Freight to Liverpool 1st of each month, for the Crop Year ending October 1, 1836.

1835.	Price of New Orleans	Price of Upland.	Sales for week.	Receipts for week.	EXPORTS FOR THE WEEK. To Great Britain.	To France.	North of Europe.	Other Fo'n Ports	Total Exports.	Rates of Freight to Liverpool.	General Remarks.
October 2..	16@20	15@18½								⅜@½d.	
" 6..	16@19	14@18	1,800	1,641	....	....	....	....	....		Prices of cotton did not vary as much this crop year, as in many others, the extreme fluctuations being from 3½@4¾ cents per lb. The most prominent event that occurred was, what is known in the annals of the city, as "the great fire;" this calamity befell the community the night of December 16, 1835, when about 50 acres of the densest business part of New York was desolated, and some 500 buildings, mostly stores and warehouses, were utterly destroyed, bringing ruin to most of the local Fire Insurance Companies then in existence. A considerable quantity of cotton was destroyed, say 1,500 to 2,000 bales, but the fire occurring at the period of the season when stocks are usually light, the misfortune was not as great as it would have been earlier, or later, in the year. Business was of course largely interrupted, but in an almost incredibly short time, the former activity was resumed, and all vestiges of this great conflagration disappeared.
" 9..	16@19	14@18									
" 13..	16@19	14@18	2,000	702	3,251	728	....	10	3,989		
" 16..	16@19	14@18									
" 20..	16@19	14@18	1,100	213	1,620	471	102	67	2,260		
" 23..	16@19	14@18									
" 27..	16@18	14@18	600	535	1,787	....	165	....	1,952		
" 30..	16@18	14@18									
Novem. 3..	16@18	14@18	1,000	304	3,282	66	68	....	3,416	5-16@⅜d	
" 6..	15@19	13@18									
" 10..	15@19	13@18	1,450	5,312	3,191	415	199	16	3,821		
" 13..	15@19	13@18									
" 17..	15@19	13@17	2,500	3,191	387	582	227	....	1,196		
" 20..	15@18	13@17									
" 24..	15@18	13@17	1,550	3,386	3,142	1,186	....	5	4,333		
" 27..	15@18	14@16¾									
Decem. 1..	15@18	14@16¾	1,850	2,238	640	1,151	99	....	1,890	⅜@½d.	
" 4..	15@18	14@16¾									
" 8..	15@18	14@16¾	1,500	4,031	772	807	....	156	1,735		
" 11..	15@18	14@16¾									
" 15..	15@18	14@16¾	1,300	4,450	470	780	....	100	1,350		
" 18..	15@18	14@16¾									
" 22..	15@18	14@16½	1,100	3,569	233	1,278	....	14	1,525		
" 25..	15@18	14@16½									
" 29..	15@18	14@16½	1,900	6,174	396	685	....	....	1,081		
1836.											
January 1..	15@18½	14@16								⅜@½d.	
" 5..	15@18½	14@16	2,450	4,913	526	....	....	....	526		
" 8..	15@18½	14@16									
" 12..	15@18	14@15½	1,800	2,396	690	423	....	....	1,113		
" 15..	15@17½	13½@15½									
" 19..	15@17½	13½@15½	2,450	2,286	....	780	....	87	867		

January 22..	15@17½	14@16								
" 26..	15@17½	14@16	3,500	8,306	1,641	1,283	....	....	2,924	
" 29..	16@18	14@16½								
February 2..	16@18	14@16½	2,400	1,756	485	668	100	....	1,253	⅜@½d.
" 5..	16@18	14@16¼								
" 9..	16@18¼	14@16½	3,050	1,898	1,020	879	....	103	2,002	
" 12..	16@18¼	14½@17								
" 16..	16½@18½	14½@17	2,150	6,219	290	403	398	....	1,091	
" 19..	16½@18½	15@17½								
" 23..	16½@19	15@18	3,900	3,376	896	213	....	....	1,109	
" 26..	16½@19	15@18¼								
March 1..	17@20	15½@19	5,500	6,609	52	358	....	136	546	¼@⅜d.
" 4..	17@20	15½@19								
" 8..	17@20	15½@19	3,400	4,131	95	1,367	....	....	1,462	
" 11..	17½@21	16@19½								
" 15..	17½@21	16@19½	4,200	11,730	330	723	174	....	1,227	
" 18..	18@21	16@19½								
" 22..	18@21	16@19½	2,700	4,732	....	543	190	....	733	
" 25..	18@21	16@19½								
" 29..	18@21	16@19½	3,600	5,831	608	147	....	....	755	
April 1..	18@21	16@19½								⅜@7-16d
" 5..	18@21	16@19½	2,800	3,279	773	498	....	....	1,271	
" 8..	18@21	16@19½								
" 12..	18@21	16@19½	2,200	14,315	507	742	156	....	1,405	
" 15..	18@21	16@19								
" 19..	18@21	16@19	1,300	8,405	....	....	116	731	847	
" 22..	17½@20	16@18½								
" 26..	17@20	16@18½	6,000	7,546	1,009	48	1,090	....	2,147	
" 29..	16@20	16@18½								
May 3..	16@20	16@19	5,000	4,546	4,390	1,890	220	17	6,517	⅜@½d.
" 6..	16@20	16@19								
" 10..	16@20	16@19	3,600	14,486	1,292	742	396	....	2,430	
" 13..	16@20	16@19								
" 17..	16@20	16@19	2,400	4,242	6,263	884	351	....	7,498	
" 20..	16@20	16@19								
" 24..	16@20	16@19	1,900	12,878	2,918	1,476	640	....	5,034	
" 27..	16@20	16@19								
" 31..	16@20	16@19	2,300	2,695	4,307	606	574	204	5,691	5-16@⅜d
June 3..	16@20	16@19								
" 7..	16@20	16@19	2,900	650	1,376	998	80	54	2,508	
" 10..	16@20	15½@19								
" 14..	16@20	15½@19	2,300	13,839	5,528	585	426	....	6,539	

New York Statement for Year 1836—Concluded.

1836.	Price of New Orleans	Price of Upland.	Sales for week.	Receipts for week.	EXPORTS FOR THE WEEK					Rates of Freight to Liverpool.
					To Great Britain.	To France.	North of Europe.	Other Fo'n Ports	Total Exports.	
June 17..	16@20	15½@19								
" 21..	16@20	15@18½	2,200	4,947	5,583	924	689	393	7,589	
" 24..	16@20	15@18½								
" 28..	16@20	15@18½	3,000	1,031	2,816	1,876	103	447	5,242	
July 1..	16@20	15@18½								5-16@3/8d
" 5..	15½@20	15@19	1,750	3,657	3,220	614	841	110	4,785	
" 8..	15½@20	15@19								
" 12..	15½@21	15@20	3,400	1,772	....	....	....	....	....	
" 15..	15½@21	15@20								
" 19..	15½@21	15@20	2,000	2,076	....	....	....	....	....	
" 22..	15½@21	15@20								
" 26..	15½@21	15@20	2,100	3,040	*14,817	2,640	1,097	134	18,688	
" 29..	15½@21	15@20								
August 2..	15½@21	15@20	1,800	2,239	77	390	495	....	962	¼@5-16d
" 5..	15½@21	15@20								
" 9..	15½@21	15@20	1,250	2,047	888	....	....	26	914	
" 12..	15½@21	15@20								
" 16..	15½@21	15@20	1,050	680	976	101	....	....	1,077	
" 19..	15@21	14@20								
" 23..	15@21	13@20	1,500	1,072	719	....	791	....	1,510	
" 26..	14@21	12½@20								
" 30..	14@21	12½@20	2,600	788	336	348	....	52	736	
Septem. 2..	14@21	12½@20								3-16@¼d
" 6..	14@21	12½@20	1,450	2,432	1,637	....	102	....	1,739	
" 9..	14@21	12½@20								
" 13..	14@21	12½@20	950	56	1,696	....	....	....	1,696	
" 16..	14@21	12½@20								
" 20..	14@21	12½@20	2,850	2,112	1,476	....	....	363	1,839	
" 23..	14@21	12½@20								
" 27..	14@21	12½@20	2,300	787	862	938	40	....	1,840	
" 30..	14@21	12½@20								3/8@7-16d
Average prices and total sales, receipts and exports.	17.72	16.50	123,650	215,546	89,270	32,236	9,929	3,225	134,660	

General Remarks.

Exchange.

Foreign bills were steady during the greater part of the year.

The opening quotation October 1, 1835, was 9@9¼ per cent. premium; toward the latter part of the month it was advanced to 9½@10; in November the price was 9¼@9¾; in December the range was from 8½@9½; in January the quotation went up from 8¼@8½ to 9½; in February it fell back to 9@9½; in March it further declined to 8@8¼; in April prices steadily receded from 7¾@8, down to 5½@6; recovering toward the close to 6½@7; May opened at 7¾@8, went down to 6½@7; and then advanced again to 7@7¼; in June 6½@8 was the range; July and August 7¼@7¾; September 7½@8; closing October 1, at 8@8½ per cent. premium.

*These figures include the shipments of the previous two weeks, no statement being made up for those two weeks.

LIVERPOOL STATEMENT FOR 1836.

UNITED STATES—1835–1836.

		Export	1,117,000
Stock, 1st October, 1835	42,000	Consumption	243,000
Crop, 1835–36	1,361,000	Stock, 1st October, 1836	43,000
Bags	1,403,000	Bags	1,403,000

1836.	Liverpool.	Gt. Britain.	France.	Continent.	Tot. Europe.
United States Bags	96,340	109,270	19,909	14,081	143,260
Brazil	32,970	33,810	8,261	741	42,812
West Indies	4,840	6,160	2,226	3,176	11,562
East Indies	32,270	60,540	....	6,733	67,273
Egypt and Levant	18,280	20,230	9,700	9,203	39,133
Bags	184,700	230,010	40,096	33,934	304,040

CONSUMPTION. / IMPORT.

CONSUMPTION. Tot. Europe.	Continent.	France.	Gt. Britain.	Liverpool.	IMPORT.	Liverpool.	Gt. Britain.	France.	Continent.	Tot. Europe.
1,106 381	90,797	247,798	767,786	717,563	United States	712,343	765,236	269,695	93,539	1,107,670
170,409	12,861	22,100	135,448	135,408	Brazil	147,268	148,098	22,684	16,172	185,154
72,655	19,813	25,989	26,853	21,958	West Indies	29,528	35,753	30,690	22,907	87,450
152,933	72,356	....	80,577	59,091	East Indies	102,801	219,157	....	79,010	220,067
150,858	59,404	57,118	34,336	32,812	Egypt and Levant	31,682	32,946	74,385	87,160	180,391
1,653,236	255,231	353,005	1,045,000	966,832	Bags	1,023,622	1,201,190	397,454	298,788	1,780,732
....	15,000	4,500	97,200	36,900	Export.					
431,536	62,491	80,045	289,000	204,590	Stock Dec. / Stock above,	184,700	230,010	40,096	33,934	304,040
2,084,772	332,722	437,550	1,431,200	1,208,322	Total supply, bags	1,208,322	1,431,200	437,550	332,722	2,084,772

COTTON AT LIVERPOOL. YEAR 1836.

Week Ending.	Receipts.						Sales.				Stocks.			Prices.			Actual Export	Con-sumption.
	Americ'n.	E. I.	Egypt.	Brazil.	Other.	Total.	Con-sumption.	Specu-lation.	Export	Total.	Amer'n.	Other.	Total.	Mid. Up.	Mid. Orl.	Dhol		
Jan. 8...	10,994	......	1,700	1,467	1,842	16,003	24,790	5,200	350	30,340	89,000	86,000	175,000	7¾	7⅞	5¾	350	24,790
" 15...	17,432	273	......	1,088	......	18,793	23,757	5,700	700	30,157	87,500	82,500	169,000	7⅞	8	5¾	700	48,547
" 22...	7,709	......	1,543	6,620	139	16,011	25,590	13,570	900	40,060	76,000	82,000	158.000	8⅛	8½	5¾	900	74,137
" 29...	25,705	621	......	4,118	1,334	31,778	20,930	3,700	690	25,320	84,500	81,000	165,500	8½	8⅛	5½	690	95,067
Feb. 5...	8,930	......	892	......	......	9,822	10,607	860	670	12,137	85,500	77,000	162,500	8	8	5½	670	105,674
" 12...	8,719	926	......	2,441	......	12,086	27,277	5,150	1,300	33,727	70,500	75,500	146,000	8¼	8¼	5½	1,300	132,951
" 19...	13,442	......	......	1,551	705	15,698	23,653	11,700	2,700	38,053	67,500	68,000	135,500	8⅞	8⅞	5½	2,700	156,604
" 25...	1,207	2,476	562	1,200	586	6,031	22,240	4,550	1,290	28,080	54,500	63,500	118,000	8⅞	8⅞	5½	1,290	178,844
Mch. 4...	44,835	820	170	......	......	45,825	9,727	1,750	630	12,107	91,500	57,000	148,500	8⅞	8⅞	5½	630	188,571
" 11...	14,983	......	336	......	86	15,405	13,880	1,000	750	15,630	92,000	54,500	146,500	8⅞	8⅞	5½	750	202,451
" 18...	21,875	7,839	......	4,210	1,838	35,762	27,450	8,800	1,850	38,100	89,500	60,500	150,000	9¼	9⅛	5½	1,850	229,901
" 25...	15,179	4,552	......	2,876	275	22,882	15,382	6,300	670	22,352	93,000	62,000	158,000	9½	9½	5¾	670	245,283
" 31...	8,576	1,411	......	......	25	10,012	18,755	16,350	1,700	36,805	90,000	54,500	144,500	10	10	6	1,700	264,038
Apr. 8...	18,457	4,332	......	4,288	1,547	28,624	15,330	3,950	1,150	20,430	98,000	58,500	156,500	9⅞	9⅞	5¾	1,150	276,368
" 15...	4,363	1,005	......	1,003	1,380	7,751	18,145	2,800	600	21,545	88,000	58,000	146,000	9⅞	9⅞	5¾	600	297,513
" 22...	18,098	5,159	......	4,201	177	27,635	14,700	1,500	......	16,200	94,000	63,000	157,000	9⅝	9¾	5¾	......	312,213
" 29...	16.392	......	......	......	927	17,319	8,120	500	450	9,070	102,000	62,000	164,000	9½	9½	5¾	450	320,333
May 6...	4,862	......	......	......	......	4,862	16,237	1,500	450	18,187	96,000	55,500	151,500	9½	9½	5¾	450	336,570
" 13...	10,438	......	......	......	586	11,024	9,080	......	550	9,630	99,000	54 500	153,500	9⅜	9½	5¾	550	345,650
" 20...	24,841	7,602	1,000	10,561	249	44,253	14,260	200	150	14,610	111,000	69,000	180,000	9⅜	9½	5¾	150	359,910
" 27...	511	650	......	......	......	1,161	23,480	1,500	360	25,340	92,500	65,000	157,500	9⅜	9½	5¾	360	383,390
June 3...	9,721	......	......	......	80	9,801	12,660	500	700	13,860	91,500	61,000	152,500	9⅜	9½	5¾	700	396,050
" 10...	45,795	9,162	......	5,996	1,708	62,661	10,930	......	400	11,330	126,000	74,500	200,500	9¼	9⅜	5¼	400	406,980
" 17...	33,963	1,819	800	4,550	423	41,555	14,300	......	500	14,800	144,000	79,000	223,000	9¼	9¼	5¼	500	421,280
" 24...	36,929	......	......	......	111	37,040	13,160	1,500	1,200	14,860	171,500	74,500	246,000	9⅛	9⅛	5¼	1,200	434,440
July 1...	16,737	......	2,505	......	2	19,244	16,280	350	250	16,880	174,500	72,500	247,000	9	9	5¼	250	450,720
" 8...	34,333	1,876	990	11,693	297	49,189	13,280	300	850	14,430	197,500	83,500	281,000	8¾	8¾	5¼	850	464,000
" 15...	26,396	......	......	......	343	26,739	15,920	2,600	1,950	20,470	210,500	97,500	290,000	8¾	8¾	5¼	1,950	479,920
" 22...	5,406	......	......	......	976	6,382	24,740	4,700	580	30,020	196 000	75.000	271,000	8⅞	9	5¼	580	504,660
" 29...	26,136	......	......	7,107	1,623	34,866	13,110	1,200	680	15,990	212,500	79.000	291,500	8⅞	9	5¼	680	517,770
Aug. 5...	27,939	925	......	2,717	1,168	32,749	17,365	......	2,100	19,465	283,500	78,500	301,500	8¾	8¾	5¼	2,100	535,135
" 12...	8,547	......	......	......	1,677	10,224	30,930	2,100	3,770	36,800	204,000	70,500	274,500	8¾	8¾	4¾	3,770	566,065
" 19...	15,693	......	......	3,743	23	19,459	29,227	5,150	3,190	37,567	193,500	67,000	260,500	8⅞	8⅞	5	3,190	595.292
" 26...	19,553	......	......	3,566	870	23,989	27,930	6,500	2,450	36,880	191,500	62,500	254,000	8⅞	8⅞	5	2,450	623,222

Sept. 2...	15,106	......	......	7,887	4,460	27,453	12,390	1,000	600	13,990	197,500	71,000	268,500	$8\frac{7}{8}$	$8\frac{7}{8}$	5	600	635,612
" 9...	4,511	7,402	......	5,498	1,916	19,327	13,520	1,500	740	15,760	192,000	81,500	273,500	$8\frac{7}{8}$	$8\frac{7}{8}$	5	740	649,132
" 16...	1,545	......	......	......	......	1,545	18,600	3,000	2,100	23,700	177,000	77,000	254,000	$8\frac{7}{8}$	$8\frac{7}{8}$	5	2,100	667,732
" 23...	4,922	1,071	612	......	1,309	7,914	22,320	5,950	700	28,970	166,500	72,500	239,000	$8\frac{7}{8}$	$8\frac{7}{8}$	$4\frac{1}{2}$	700	690,052
" 30...	17,590	12,514	......	3,755	4,516	38,375	11,040	7,700	500	19,240	168,000	88,000	256,000	$8\frac{7}{8}$	$8\frac{7}{8}$	$4\frac{1}{2}$	500	701,092
Oct. 7...	5,315	4,522	750	2,529	10	13,126	12,485	1,350	360	14,195	166,000	91,000	257,000	$8\frac{7}{8}$	$8\frac{7}{8}$	$4\frac{1}{2}$	360	713,577
" 14...	4,895	4,165	2,732	2,948	1,547	16,287	14,453	750	110	15,313	159,500	99,000	258,500	$8\frac{3}{4}$	$8\frac{3}{4}$	$4\frac{1}{2}$	110	728,030
" 21...	1,173	189	......	......	76	1,438	7,550	......	690	8,240	155,500	96,000	251,500	$8\frac{5}{8}$	$8\frac{5}{8}$	$4\frac{1}{2}$	690	735,580
" 28...	207	......	1,511	2,929	130	4,777	8,430	500	400	9,330	149,500	98,000	247,500	$8\frac{1}{2}$	$8\frac{1}{2}$	$4\frac{1}{2}$	400	744,010
Nov. 4...	2,460	......	......	971	4,257	7,688	11,595	200	420	12,215	143,000	99,500	242,500	$8\frac{1}{2}$	$8\frac{1}{2}$	$4\frac{1}{2}$	420	755,605
" 11...	3,503	2,167	2,029	7,762	1,080	16,541	11,470	500	300	12,270	138,000	109,500	247,500	$8\frac{1}{4}$	$8\frac{1}{4}$	$4\frac{1}{2}$	300	767,075
" 18...	6,431	10,688	......	4,642	......	21,761	13,660	......	600	14,260	134,500	120,000	254,500	$8\frac{1}{8}$	$8\frac{1}{8}$	$4\frac{1}{4}$	600	780,735
" 25...	......	......	800	4,843	777	6,420	17,496	1,600	420	19,516	122,000	121,000	243,000	8	$8\frac{1}{8}$	4	420	788,231
Dec. 2...	4,955	5,821	......	6,675	172	17,623	18,061	2,500	......	20,561	114,500	128,000	242,500	8	$8\frac{1}{8}$	4	......	806,292
" 9...	8,852	1,300	......	8,127	1,904	20,183	18,120	......	......	18,120	112,500	132,500	245,000	$7\frac{7}{8}$	$8\frac{1}{8}$	$3\frac{3}{4}$	......	824,412
" 16...	13,345	......	2,075	......	......	15,420	24,210	1,500	680	26,390	110,000	125,000	235,000	$7\frac{7}{8}$	$8\frac{1}{8}$	$3\frac{3}{4}$	680	848,622
" 23...	8,086	2,961	......	......	3,182	14,229	19,250	1,600	460	21,310	107,000	123,000	230,000	$7\frac{7}{8}$	$8\frac{1}{8}$	$3\frac{7}{8}$	460	867,872
" 30...	1,405	......	390	......	......	1,795	19,420	1,200	......	20,620	......	......	......	$7\frac{7}{8}$	$8\frac{1}{8}$	$3\frac{7}{8}$	......	887,292
Average prices & total sales, receipts & stocks.	708,994	103248	21,397	143,761	46,233	1,023,633	887,292	152330	45,610	1,085,232				8.79	8.59	5.10	45,610	170,633

1837.

Number of cotton mills in Massachusetts this year, 282, with 565,031 spindles.

A third mill built on Factory Island, Saco, Maine. (*See* years 1830, 1833 and 1835.)

The following table sets forth the state of the cotton manufacture in Massachusetts at this period. The figures are official:

Number of mills	282
" Spindles	565,031
Pounds of cotton consumed yearly	37,275,917
Yards of cotton manufactured yearly	126,319,221
Value of cotton goods manufactured yearly	$13,056,659
Males employed	4,997
Females "	14,757
Capital invested	$14,369,719

COTTON CROP OF THE UNITED STATES.

Statement and Total Amount of the Cotton Crop of the United States, for the Year ending September 30, 1837.

	Bales.	Bales.	Total.	Same period 1836.
NEW ORLEANS.				
Export—				
To Foreign Ports	509,393			
Coastwise	85,145			
Burnt, &c., estimated	1,837			
Estimated quantity exported coastwise not cleared at Custom House	1,500			
Stock on hand 1st October, 1837	15,302			
Deduct—		613,177		
Stock on hand 1st October, 1836	8,702			
Received from Mobile	7,655			
" Florida	1,053			
" Texas	2,645			
		20,055		
NATCHEZ.			593,122	474,647
Export—				
To Liverpool	....	6,995		
New York	....	760		
			7,755	6,889
FLORIDA.				
Export—				
To Foreign Ports	19,387			
Coastwise	60,851			
Burnt on board steamer Reindeer, and on wharf	514			
Stock on hand 1st October, 1837, estimated	4,000			
Deduct—		84,752		
Stock on hand 1st October, 1836	....	1,049		
			83,703	79,762
ALABAMA.				
Export from Mobile—				
To Foreign Ports	172,124			
Coastwise	58,648			
Burnt	410			
Stock in Mobile 1st October, 1837	1,977			
		233,159		
Deduct—				
Stock on hand 1st October, 1836	64			
Received from Florida	852			
		916		
GEORGIA.			232,243	236,715
Export from Savannah—				
To Foreign Ports—Uplands	153,304			
" Sea Island	3,906			
	157,210			
Coastwise	72,150			
		229,360		
From Darien—				
To Foreign Ports	4,578			
New York	10,458			
		15,036		
Stock in Savannah 1st October, 1837	4,533			
" Augusta and Hambro', 1st October, 1837	19,528			
		24,061		
Deduct—				
Stock on hand 1st October, 1836, in Sav. & Augusta	....	5,486		
			262,971	270,121

Statement and Total Amount of the Cotton Crop of the United States, for the Year ending September 30, 1837.—*Concluded.*

	Bales.	Bales.	Total.	Same period 1836.
SOUTH CAROLINA.				
Export from Charleston—				
To foreign ports—Uplands	153,987			
" Sea Island	12,154			
	166,141			
Coastwise	33,230			
		199,371		
From Georgetown—				
To New York	....	8,670		
Stock in Charleston 1st October, 1837	....	10,743		
		218,784		
Deduct—				
Stock in Charleston 1st October, 1836	2,997			
Received from Savannah	16,957			
" Florida	2,453			
		22,407		
			196,377	231,237
NORTH CAROLINA.				
Export—				
To Foreign Ports	1,403			
Philadelphia, estimated at	1,200			
New York	13,550			
Boston	111			
Stock on hand 1st October, 1837	2,140			
		18,404		
Deduct—				
Stock on hand 1st October, 1836	....	400		
			18,004	[illegible]
VIRGINIA.				
Export—				
To Foreign Ports	14,547			
Coastwise	5,000			
Taken by Manufacturers	8,000			
Stock on hand 1st October, 1837	3,547			
		31,094		
Deduct—				
Quantity received coastwise	547			
Stock on hand 1st October, 1836	1,929			
		2,476		
			28,618	29,197
Received at New York from Grand Gulf, Miss	....	....	137	100
Total crop of the United States	....	....	1,422,930	1,360,725
Crop of last year	....	....	1,360,725	
Increase	....	....	62,205	

Export to Foreign Ports, from October 1, 1836, *to September* 30, 1837.

	To Great Britain.	To France.	North of Europe.	Other Foreign Ports.	Total.
New Orleans (bales)	355,096	133,641	6,198	14,458	509,393
Natchez	6,995	....	....	....	6,995
Alabama	139,756	29,406	....	2,962	172,124
Florida	17,650	1,737	....	....	19,387
Georgia (Savannah and Darien)	139,503	20,450	1,099	696	161,748
South Carolina	107,369	45,467	9,967	3,338	166,141
North Carolina	768	635	....	....	1,403
Virginia	7,077	2,760	325	4,385	14,547
Baltimore	50	....	127	....	177
Philadelphia	5,287	176	....	1,117	6,580
New York	71,235	26,450	7,942	3,423	109,050
Boston	....	....	779	101	880
Grand total	850.786	260,722	26,437	30,480	1,168,425
Total last year	771,148	266,188	54,889	24,378	1,116,603
Increase	79,638	....	....	6,102	51,822
Decrease	....	5,466	28,452	....	....

Growth.

Total crop of 1824–5,	560,000	bales.
" 1825–6,	710,000	"
" 1826–7,	937,000	"
" 1827–8,	712,000	"
" 1828–9,	857,744	"
" 1829–30,	976,845	"
" 1830–1,	1,038,848	"
" 1831–2,	987,477	"
" 1832–3,	1,070,438	"
" 1833–4,	1,205,394	"
" 1834–5,	1,254,328	"
" 1835–6,	1,360,725	"
" 1836–7,	1,422,930	"

Consumption.

To estimate the quantity manufactured in the United States, we take the growth of the year		1,422,930 bales.	
Add—Stocks on hand at the commencement of the year October 1, 1836, in the Southern ports	20,627		
" Northern ports	22,714		
		43,341	
			1,466,271
Deduct therefrom the export to Foreign ports		1,168,425	
Stocks on hand at the close of the year October 1, 1837. In the Southern ports	61,770		
In the Northern ports	14,050		
		75,820	
Burnt and lost at New Orleans	1,837		
" St. Joseph's	514		
" New York, April 16th	400		
		2,751	
		1,246,996	
Less Foreign cotton included in the export		3,265	
			1,243,731

Quantity consumed and in the hands of manufacturers,	1836–7	222,540
" " "	1835–6	236,733
" " "	1834–5	216,888
" " "	1833–4	196,413
" " "	1832–3	194,412
" " "	1831–2	173,800
" " "	1830–1	182,142
" " "	1829–30	126,512
" " "	1828–9	118,853
" " "	1827–8	120,593
" " "	1826–7	103,483

Note.—It is a fact well understood that the quantity of cotton remaining in the interior is much larger than usual—the low price of the article, compared with former years, having prevented it from coming forward. Of the quantity thus held we have no means sufficiently accurate to form an estimate.

It will be seen also that we have deducted from the New Orleans statement, the quantity received at that port from Texas—that being a foreign country.

Our estimate of the quantity taken for consumption, does not include any cotton manufactured in the States south and West of Virginia, nor any in that State, except in the vicinity of Petersburg and Richmond.

LIVERPOOL STATEMENT FOR 1837.

UNITED STATES, 1836–1837.

Stock 1st October, 1836	43,000	Export	1,168,000
Crop	1,423,000	Consumption	222,000
		Stock 1st Oct., 1837	76,000
Bags	1,466,000	Bags	1,466,000

Stock 1st Jan., 1837, in......	Liverpool.	Gt. Britain.	France	Continent.	Tot. Europe.
United States..........Bags.	78,920	90,420	37,306	16,823	144,549
Brazil	43,230	44,660	8,845	4,052	57,557
West Indies	11,910	14,160	6,927	5,270	26,357
East Indies	53,480	121,020	....	13,387	134,407
Egypt and Levant	17,050	18,740	26,967	22,959	68,666
Bags	204,590	289,000	80,045	62,491	431,536

CONSUMPTION. / IMPORT.

Tot. Europe.	Continent.	France.	Gt. Britain.	Liverpool.	IMPORT.	Liverpool.	Gt. Britain.	France	Continent.	Tot. Europe.
1,160,524	99,071	255,805	805,648	752,577	United States	786,907	845,188	250,457	100,743	1,150,588
160,998	8,755	22,638	129,605	126,861	Brazil	113,121	116,605	16,010	9,282	138,697
69,478	17,813	22,437	29,228	24,575	West Indies	25,925	30,088	27,373	24,913	79,874
152,193	66,270	....	85,923	61,880	East Indies	74,650	145,063	....	68,213	142,326
172,534	76,646	56,809	39,079	39,130	Egypt and Levant	35,700	39,329	50,346	71,325	158,400
1,715,727	268,555	357,689	1,089,483	1,005,023	Bags	1,036,303	1,176,273	344,186	274,476	1,669,885
....	5,600	3,000	116,450	65,050	Export.					
385,694	62,812	63,542	259,340	170,820	Stock Dec. Stock above,	204,590	289,000	80,045	62,491	431,536
2,101,421	336,967	424,231	1,465,273	1,240,893	Total supply, bags	1,240,893	1,465,273	424,231	336,967	2,101,421

COTTON AT LIVERPOOL. YEAR 1837.

Week Ending.		Receipts.						Sales.				Stocks.			Prices.			Actual Export	Con-sumption.
		Americ'n.	E. I.	Egypt.	Brazil.	Other.	Total.	Con-sumption.	Specu-lation.	Export.	Total.	Amer'n.	Other.	Total.	Mid. Up.	Mid. Orl.	Dhol		
Jan.	6..	5,854	......	......	....	191	6,045	22,725	4,700	330	27,755	68,500	119,000	187,500	8	8 1/8	4	330	22,725
"	13..	9,954	3,276	......	1,158	358	14,746	22,610	4,530	670	27,810	64,500	114,500	178,000	8	8 1/8	4	670	45,335
"	20..	16,626	142	......	....	419	17,187	13,550	250	100	13,900	70,000	110,500	180,500	8	8 1/8	4	100	58,885
"	27..	43,009	4,056	......	6,548	258	53,871	11,556	......	.....	11,556	104,000	119,500	223,500	7 7/8	8 1/8	4	......	70,435
Feb.	3..	15,604	......	......	3,984	......	19,588	13,950	......	200	14,150	106,000	119,000	225,000	7 3/4	8	4	200	84,385
"	10..	21,095	......	......	13,428	691	35,214	15,430	250	......	15,680	111,500	130,500	242,000	7 1/2	7 1/2	4	......	99,815
"	17..	33,619	1,880	......	2,022	188	37,709	14,310	......	.....	14,310	130,000	131,000	261,000	7 3/8	7 3/8	4	......	114,125
"	24..	22,546	......	......	971	770	24,287	20,620	500	400	21,520	130,000	128,500	258,500	7 1/2	7 1/2	3 3/4	400	134,745
Mch.	3..	15,416	......	......	2,849	......	18,265	16,343	420	480	17,243	133,000	127,000	260,000	7 3/8	7 3/8	3 3/4	480	151,088
"	10..	8,033	3,176	1,112	....	405	12,726	17,470	......	1,150	18,620	126,500	127,000	253,500	7 1/8	7 1/8	3 3/4	1,150	168,558
"	17..	15,102	......	......	....	3,711	18,813	14,750	350	600	15,700	129,000	126,500	255,500	6 5/8	6 5/8	3 1/2	600	183,308
"	23..	2,433	......	......	....	111	2,544	13,590	3,000	450	17,040	120,000	124,000	244,000	6 1/8	6 1/8	3 1/2	450	198,898
"	31..	36,261	4,197	......	....	301	40,759	15,020	1,200	1,000	17,220	143,000	123,000	266,000	5 3/4	5 7/8	3 1/4	1,000	211,918
April	7..	6,608	......	......	....	......	6,608	10,070	1,000	850	11,920	140,500	121,000	261,500	5 5/8	5 5/8	3 1/4	850	221,988
"	14..	9,371	......	......	1,319	2,156	12,846	14,355	2,000	1,800	18,155	137,000	121,000	258,000	5 5/8	5 5/8	3 1/4	1,800	236,343
"	21..	44,887	6,300	470	5,184	908	57,749	12,370	750	2,000	15,120	169,000	130,000	299,000	5 1/4	5 3/8	3	2,000	248,713
"	28..	52,565	3,995	......	2,345	380	59,285	21,170	6,500	1,300	28,970	198,500	130,000	329,000	5 1/2	5 1/2	3	1,300	269,883
May	5..	29,262	......	1,694	4,709	1,192	36,857	22,313	2,800	900	26,013	212,000	133,500	345,500	5 1/2	5 1/2	3	900	292,196
"	12..	11,740	2,743	......	5,644	429	20,556	11,770	500	1,500	13,770	212,000	138,000	350,000	5	5	3	1,500	303,966
"	19..	28,891	7,691	4,308	4,793	186	45,869	12,545	2,500	2,915	17,960	223,500	152,000	375,500	5	5	2 3/4	2,915	316,511
"	26..	11,576	......	......	594	......	12,170	12,913	4,500	2,550	29,963	214,000	146,000	360,000	5	5	2 3/4	2,550	329,424
June	2..	12,694	......	......	1,461	40	14,195	28,090	9,400	2,800	40,290	203,000	138,500	341,500	5 1/4	5 1/4	2 3/4	2,800	357,514
"	9..	3,412	......	......	....	2,758	6,070	17,706	6,500	1,650	25,856	195,500	136,000	331,500	5 1/4	5 1/4	2 3/4	1,650	375,220
"	16..	58,063	......	......	857	3,270	62,190	18,297	4,000	830	23,180	237,500	135,000	372,500	5 1/8	5 1/8	2 5/8	830	393,517
"	23..	41,985	......	......	1,936	209	44,130	23,510	9,500	3,390	36,400	258,500	129,500	388,000	5 1/4	5 1/4	2 5/8	3,390	417,027
"	30..	12,612	......	747	......	173	13,532	22,000	3,600	4,100	29,700	249,500	125,500	375,000	5 1/4	5 1/4	2 5/8	4,100	439,027
July	7..	2,665	......	......	......	453	3,118	15,330	3,000	3,000	21,330	239,000	120,500	359,500	5 1/8	5 1/8	2 5/8	3,000	454,357
"	14..	8,758	......	5,385	....	......	14,143	18,450	1,000	4,850	24,300	230,000	120,500	350,500	5 1/8	5 1/8	2 5/8	4,850	472,807
"	21..	40,473	4,158	......	6,086	2,542	53,259	14,700	1,000	1,300	17,000	258,500	129,500	388,000	5	5	2 5/8	1,300	487,507
"	28..	9,155	1,539	......	810	104	11,608	13,140	2,500	2,730	18,370	255,000	129,000	384,000	5	5	2 5/8	2,730	500,647
Aug.	4..	9,472	......	1,675	....	744	11,891	28,110	4,800	2,730	35,670	240,500	124,000	364,500	5	5	2 5/8	2,730	528,787
"	11..	3,551	......	......	....	504	4,055	32,535	4,720	3,025	40,280	219,000	114,000	333,000	5 1/8	5 1/4	2 3/4	3,025	561,322
"	18..	4,551	2,249	......	1,310	433	8,543	33,640	11,500	1,950	47,090	198,500	107,500	306,000	5 3/8	5 1/2	3	1,950	594,962
"	25..	16,483	5,407	......	....	2,271	24,161	21,430	4,700	2,900	29,030	197,500	108,500	306,000	5 1/2	5 1/2	3	2,900	616,392

Sept. 1..	407	......	......	....	42	449	24,360	3,500	950	28,810	190,500	106,000	296,500	$5\frac{3}{4}$	$5\frac{3}{4}$	3	950	640,752
" 8..	7,439	1,298	2,414	5,726	......	16,877	27,720	7,100	1,100	35,920	174,500	110,000	284,500	6	6	3	1,100	668,472
" 15..	18,885	4,726	......	3,482	742	27,835	12,503	1,000	630	14,133	185,500	113,500	299,000	$5\frac{7}{8}$	$5\frac{7}{8}$	3	630	680,975
" 22..	2,257	......	3,883	....	912	7,052	21,450	2,400	2,750	26,600	170,000	112,000	282,000	$5\frac{7}{8}$	$5\frac{7}{8}$	3	2,750	702,425
" 29..	....	......	......	....	1,626	1,626	10,470	1,000	1,590	13.060	161,000	110,000	271,000	$5\frac{7}{8}$	$5\frac{7}{8}$	3	1,590	712,895
Oct. 6..	9,452	1,633	1,312	....	655	13,052	11,810	750	520	13,080	160,500	110,500	271,000	$5\frac{5}{8}$	$5\frac{5}{8}$	3	520	724,705
" 13..	6,788	6,032	......	3,879	1,075	17,774	11,930	.....	1,300	13,230	157,500	118,000	275,500	$5\frac{5}{8}$	$5\frac{5}{8}$	3	1,300	736,635
" 20..	338	......	......	....	570	908	19,570	2,500	500	22,570	142,500	113,500	256,000	$5\frac{5}{8}$	$5\frac{5}{8}$	3	500	756,205
" 27..	1,861	.....	......	....	1,485	3,346	21,050	500	1,200	22,750	127,500	110,000	237,500	$5\frac{5}{8}$	$5\frac{5}{8}$	3	1,200	777,255
Nov. 3..	3,803	......	.....	1,472	255	5,530	31,050	14,400	250	45,700	110,000	101,000	211,000	$5\frac{3}{4}$	$5\frac{7}{8}$	3	250	808,305
" 10..	6,365	......	......	2,071	3,665	12,101	24.540	7,550	750	32,840	99.000	98,500	197,500	6	6	$3\frac{1}{2}$	750	832,845
" 17..	4,402	1,298	......	3,118	7	8,825	25,730	13,300	250	39,280	86,000	96,000	182,000	$6\frac{3}{8}$	$6\frac{3}{8}$	$3\frac{3}{4}$	250	858 575
" 24..	7,985	3,888	1,504	3,043	64	16,484	23,000	14,550	......	37,550	79,000	95,500	174,500	$6\frac{7}{8}$	7	4	......	881,575
Dec. 1..	5,541	......	......	......	2,028	7,569	10,190	1,850	......	12,040	76,500	94,500	171,000	$6\frac{3}{4}$	$6\frac{7}{8}$	4	......	891,765
" 8..	1,676	.....	......	....	2,880	4,556	15,680	13,500	......	29,180	67,000	93,000	160,000	7	$7\frac{1}{8}$	4	......	907,445
" 15..	958	......	1,313	4,049	1,069	7,389	20,880	2,200	......	23,080	58,000	92,500	150,500	7	$7\frac{1}{8}$	4	......	928 325
" 22..	7,143	......	......	1,873	1,855	10,871	13,100	5,150	......	18,250	55,500	93,000	148,500	$7\frac{1}{4}$	$7\frac{3}{8}$	$4\frac{1}{4}$	......	941,425
" 29..	23,762	......	......	....	3,638	27,400	12,400	1,150	......	13,550	70,500	92,500	163,000	$7\frac{1}{4}$	$7\frac{3}{8}$	$4\frac{1}{4}$	......	953,825
Average price & total sales, receipts and stocks.	769,408	69,684	25,817	97,701	48,321	1,010,931	953,825	191170	66,240	1,211,235				6.09	6.14	3.28	66,240	1,834,278

The semi-weekly Price and Weekly Sales and Receipts at New York, Weekly Exports from New York and Rates of Freight to Liverpool 1st of each month, for the Crop Year ending October 1, 1837.

1836.	Price of New Orleans	Price of Upland.	Sales for week.	Receipts for week.	EXPORTS FOR THE WEEK.					Rates of Freight to Liverpool.	GENERAL REMARKS.
					To Great Britain.	To France.	North of Europe.	Other Fo'n Ports	Total Exports.		
October 4..	14@21	12½@20	1,900	922	....	....	....	....	....	⅜@7-16d	
" 7..	14@21	12½@20									This crop year proved to be an eventful one. The first five months prices were subject only to about the usual changes, but, early in March, '37, indications became apparent of a great disturbance in financial circles; business was very unsettled. A severe stringency set in, which continued, with more or less intensity, through March, April and May, culminating finally on the 10th May in the suspension of specie payments. In the meantime, prices of Cotton rapidly receded, until the shrinkage was fully 8½@9 cents per lb.; many staunch houses were ruined, and embarrassment was general throughout Cotton circles.
" 11..	14@21	12½@20	1,850	736	....	....	....	....	....		
" 14..	14@21	12½@20									
" 18..	14@21	12½@20	900	1,437	....	....	....	....	....		
" 21..	14@21	12½@20									
" 25..	14@21	12½@20	1,910	2,549	....	....	....	....	....		
" 28..	14@21	12½@20									
Novem. 1..	14@21	12½@20	1,550	1,386	6,520	1,721	707	....	8,948	⅜@7-16d	
" 4..	14@21	12½@20									
" 8..	14@21½	12½@20	2,250	243	2,008	796	....	....	2,804		
" 11..	14@21½	12½@20									
" 15..	14@21½	12½@20	1,300	6,795	1,300	598	....	....	1,898		
" 18..	14@21½	14@19½									
" 22..	14@21½	14@19½	1,727	3,968	1,407	530	....	34	1,971		
" 25..	14@21	14@19¼									
" 29..	14@21	14@19¼	2,610	2,442	1,767	514	....	....	2,281		
Decem. 2..	14@21	14@19¼								⅜@7-16d	
" 6..	14@20½	14@19	1,550	4,119	1,302	714	25	132	2,173		
" 9..	16@19½	16@19									Foreign exchange was difficult to negotiate; bills sold at very high rates; and the premium on gold advanced to 14@16 per cent., with large shipments. For a time there was a general demoralization in business affairs, which continued, with but little abatement, until the close of the year.
" 13..	16@19½	16@19	1,300	2,598	558	144	....	....	702		
" 16..	16@19½	15¾@18									
" 20..	16@19½	15¾@18	1,250	6,303	463	980	....	....	1,443		
" 23..	16@19	15¾@18									
" 27..	16@19	15¾@18	1,270	4,517	....	....	150	....	150		
" 30..	16@19	15½@17½									
1837.											
January 3..	16@19	15½@17½	1,150	2,162	587	855	....	....	1,442	¾@⅞d.	
" 6..	16@19	15½@17½									
" 10..	16@19	15½@17½	1,400	1,814	532	869	139	....	1,540		
" 13..	15½@18¾	15@17¼									
" 17.	15½@18¾	15@17¼	2,050	4,656	3,187	630	....	154	3,971		These figures embrace the shipments of the previous three weeks.
" 20..	15½@18¾	15@17¼									

January 24..	15½@18¾	15@17½	2,150	2,092	1,064	1,022	83	....	2,169	
" 27..	15½@18¾	15@17½								
" 31..	15½@18¾	15@17½	1,300	6,310	714	920	....	....	1,634	
February 3..	15½@18¾	14½@17½								½@⅝d.
" 7..	15½@18¾	14½@17½	2,050	3,321	1,347	....	139	....	1,486	
" 10..	15½@18¾	14½@17½								
" 14..	15½@18¾	14½@17½	2,480	3,178	216	246	185	....	647	
" 17..	15½@18¾	14½@17½								
" 21..	15½@18¾	14½@17½	1,250	4,099	624	412	....	....	1,036	
" 24..	15½@19	14½@17½								
" 28..	15@19	14½@17½	1,650	4,737	595	....	51	....	646	
March 3..	14¼@19	14½@17½								⅜@½d.
" 7..	14¼@19	14½@17½	1,450	2,668	1,756	....	220	....	1,976	
" 10..	14¼@19	14@17½								
" 14..	14¼@19	14@17½	1,200	6,193	861	....	....	....	861	
" 17..	14¼@19	14@17½								
" 21..	12½@18	12½@16½	550	7,555	694	524	6	6	1,230	
" 24..	12½@18	12½@16½								
" 28..	12½@18	12½@16½	910	1,965	1,110	....	283	56	1,449	
" 31..	11½@16	11@15								
April 4..	11½@16	11@15	4,250	7,280	2,144	311	....	....	2,455	½@11-16
" 7..	11½@15½	10½@14½								
" 11..	11@15½	10½@14½	1,150	5,548	668	587	198	....	1,453	
" 14..	10@14½	10@13½								
" 18..	10@14½	10@13½	900	3,805	5,507	1,067	....	....	6,574	
" 21..	9½@14	9½@13								
" 25..	9@13½	9@13	650	6,123	3,630	861	485	....	4,976	
" 28..	8@13	8@12½								
May 2..	8@13	8@12½	800	5,144	773	663	....	....	1,436	½@⅝d.
" 5..	7@12	7@11½								
" 9..	7@12	7@11½	2,250	3,135	3,147	705	45	99	3,996	
" 12..	7½@12	7½@11½								
" 16	8@12	8@11½	1,700	1,775	2,081	1,034	475	....	3,590	
" 19..	8½@12½	8½@12								
" 23..	8½@12½	8½@12	1,750	2,463	....	....	....	....	....	
" 26..	8½@12½	8½@12								
" 30..	8½@12½	8½@12	1,800	3,392	....	....	....	....	....	
June 2..	8½@12½	8½@12								½@⅝d.
" 6..	8½@12½	8½@12	1,250	1,005	†7,374	305	804	784	9,267	
" 9..	8½@12½	8½@12								
" 13..	8½@12½	8½@12	1,400	1,921	555	48	194	43	840	
" 16..	8½@12	8½@11½								

† Embracing the shipments of the previous fortnight.

New York Statement for 1837.—*Concluded.*

1837.	Price of New Orleans	Price of Upland.	Sales for week.	Receipts for week.	EXPORTS FOR THE WEEK.					Rates of Freight to Liverpool.
					To Great Britain.	To France.	North of Europe.	Other Fo'n Ports	Total Exports.	
June 20..	$8\frac{1}{2}$@12	$8\frac{1}{2}$@$11\frac{1}{2}$	1,900	4,206	157	22	47	115	341	
" 23..	$8\frac{1}{2}$@12	$8\frac{1}{2}$@$11\frac{1}{2}$								
" 27..	$8\frac{1}{2}$@12	$8\frac{1}{2}$@$11\frac{1}{2}$	2,200	4,272	1,176	650	202	129	2,157	
" 30..	$8\frac{1}{2}$@12	$8\frac{1}{2}$@$11\frac{1}{2}$								
July 4.	$8\frac{1}{2}$@12	$8\frac{1}{2}$@$11\frac{1}{2}$	2,900	4,370	1,742	1,360	587	55	3,744	$\frac{1}{4}$@$\frac{3}{8}$d.
" 7..	$8\frac{1}{2}$@12	$8\frac{1}{2}$@$11\frac{1}{2}$								
" 11..	$8\frac{1}{2}$@12	$8\frac{1}{2}$@12	1,900	1,6-7	....	....	216	73	289	
" 14..	$8\frac{1}{2}$@12	$8\frac{1}{2}$@12								
" 18..	$8\frac{1}{2}$@12	$8\frac{1}{2}$@12	1,200	1,428	534	29	16	45	624	
" 21..	$8\frac{1}{2}$@12	$8\frac{1}{2}$@12								
" 25..	$8\frac{1}{2}$@$12\frac{1}{2}$	$8\frac{1}{2}$@12	2,250	4,834	1,023	417	54	30	1,524	
" 28..	$8\frac{1}{2}$@13	$8\frac{1}{2}$@13								
August 1..	$8\frac{1}{2}$@13	$8\frac{1}{2}$@13	3,000	4,708	1,618	7	100	559	2,284	$\frac{1}{4}$@$\frac{3}{8}$d.
" 4..	$8\frac{1}{2}$@13	$8\frac{1}{2}$@13								
" 8..	$8\frac{1}{2}$@13	$8\frac{1}{2}$@13	4,300	1,887	1,686	513	22	....	2,221	
" 11..	$8\frac{1}{2}$@13	$8\frac{1}{2}$@13								
" 15..	$8\frac{1}{2}$@13	$8\frac{1}{2}$@13	2,600	2,545	635	879	218	....	1,732	
" 18..	$8\frac{1}{2}$@13	$8\frac{1}{2}$@13								
" 22..	$7\frac{1}{2}$@13	$7\frac{1}{2}$@$12\frac{1}{4}$	1,500	883	1,810	589	143	504	3,046	
" 25..	$7\frac{1}{2}$@13	$7\frac{1}{2}$@$12\frac{1}{4}$								
" 29..	$7\frac{1}{2}$@13	$7\frac{1}{2}$@$12\frac{1}{4}$	2,500	2,551	590	1,163	160	....	1,913	
Septem. 1..	$7\frac{1}{2}$@13	$7\frac{1}{2}$@$12\frac{1}{4}$								$\frac{1}{4}$@$\frac{3}{8}$d.
" 5..	$7\frac{1}{2}$@13	$7\frac{1}{2}$@$12\frac{1}{4}$	1,850	3,061	501	814	701	....	2,016	
" 8..	$7\frac{1}{2}$@13	$7\frac{1}{2}$@$12\frac{1}{4}$								
" 12..	$7\frac{1}{2}$@13	$7\frac{1}{2}$@$12\frac{1}{4}$	2,100	1,926	687	531	649	....	1,867	
" 15..	8@13	8@$12\frac{1}{4}$								
" 19..	8@13	8@$12\frac{1}{4}$	2,250	1,008	165	....	....	....	165	
" 22..	8@13	8@$12\frac{1}{4}$								
" 26..	8@13	8@$12\frac{1}{4}$	2,100	2,641	1,714	941	197	381	3.233	
" 29..	8@13	8@$12\frac{1}{4}$								
October 3..	8@13	8@$12\frac{1}{4}$	1,350	3,152	2,705	1,479	447	31	4,662	$\frac{3}{8}$@$\frac{1}{2}$d.
Average price and total sales, receipts and exports.	13.87	13.25	94,757	175,515	71,234	26,450	7,948	3,230	108,862	

GENERAL REMARKS.

Exchange.

From causes noted above, the market for Foreign Bills was in a semi-chaotic state for a considerable portion of the year.

The opening quotation, October 1, '36, for 60 days' Bills on London was 8@$8\frac{1}{2}$ per cent. premium, afterwards declining to $7\frac{1}{2}$@8; the range in November was from $8\frac{1}{4}$ up to $9\frac{3}{4}$; in December, from $9\frac{1}{2}$ down to $6\frac{1}{2}$@$7\frac{1}{4}$; in January, from $7\frac{1}{2}$ up to $10\frac{1}{4}$; in February, from $10\frac{1}{4}$ down to $8\frac{1}{2}$@9. At this time the market began to be unsettled. The opening price in March was $8\frac{1}{2}$@$9\frac{1}{4}$, but soon ran up to 10@$11\frac{1}{2}$; in April the advance progressed until $11\frac{1}{2}$@$12\frac{1}{2}$ was touched; in May, 12@16, falling off, late in the month, to 10@$12\frac{1}{2}$; in June the quotation went up again from 12@16 to 20@22; in July it fell back to 18@21, and ranged the same in August; in September prices began to decline, and went down from 20@21 to 14@15, the year closing, October 1, '37, with 14@15 per cent. premium the quotation.

1838.

Total number of cotton factories in Great Britain and Ireland, 1,815; England and Wales, 1,599; Scotland, 192; and Ireland, 24. Total number of hands employed, 206,061—114,129 males and 145,934 females.

COTTON CROP OF THE UNITED STATES.

Statement and Total Amount of the Cotton Crop of the United States, for the Year ending September 30, 1838.

	Bales.	Bales.	Total.	1837.
NEW-ORLEANS.				
Export—				
To Foreign Ports	631,437			
Coastwise	105,749			
Burnt, lost in repacking, &c	12,491			
Stock on hand, 1st October, 1838	8,843			
		758,520		
Deduct—				
Stock on hand, 1st October, 1837	15,302			
Received from Mobile	22,900			
" " Florida	5,437			
" " Texas	3,300			
		46,939		
			711,581	593,122
MISSISSIPPI.				
Export from Natchez—				
To Foreign Ports	....	15,246		
Coastwise	....	2,359		
Vicksburg to New York	....	900		
Burnt on board steamer Vicksburg	....	1,170		
			19,675	7,892
ALABAMA.				
Export from Mobile—				
To Foreign Ports	225,060			
Coastwise	86,035			
Burnt at Selma	1,600			
Stock in Mobile, 1st October, 1838	59			
		312,754		
Deduct—				
Stock in Mobile, 1st October, 1837	1,977			
Received from Florida	937			
" " Texas	33			
		2,947		
			309,807	232,243
FLORIDA.				
Export—				
To Foreign Ports	34,154			
Coastwise	75,017			
Stock on hand, 1st October, 1838	1,000			
		110,171		
Deduct—				
Stock on hand, 1st October, 1837	....	4,000		
			106,171	83,703
GEORGIA.				
Export from Savannah—				
To Foreign Ports—Uplands	220,327			
Sea Island	5,680			

Statement and Total Amount of the Cotton Crop of the United States, for the Year ending September 30, 1838.—*Concluded.*

	Bales.	Bales.	Total.	1837.
Coastwise	74,084			
	300,091			
From Darien—				
To Foreign Ports	3,791			
New York and Providence	14,651			
Stock in Savannah, 1st October, 1838	3,199			
" Augusta and Hambro', 1st October, 1838	6,738			
		328,470		
Deduct—				
Stock in Savannah and Augusta, 1st October, 1837	24,061			
Received from Florida	199			
		24,260		
			304,210	262,971
SOUTH CAROLINA.				
Export from Charleston—				
To Foreign Ports—Uplands	229,755			
Sea Island	16,712			
Coastwise	57,270			
	303,737			
From Georgetown—				
To New York	16,276			
Burnt at Columbia	340			
Stock in Charleston, 1st October, 1838	3,169			
		323,522		
Deduct—				
Stock in Charleston, 1st October, 1837	10,743			
Received from Savannah	9,899			
" " Florida	8,375			
Wrecked Cotton	171			
		29,188		
			294,334	196,377
NORTH CAROLINA.				
Export—				
To Foreign Ports	4,279			
Coastwise	18,500			
Stock on hand, 1st October, 1838	800			
		23,579		
Deduct—				
Stock on hand, 1st October, 1837	....	2,140		
			21,439	18,004
VIRGINIA.				
Export—				
To Foreign Ports	19,438			
Coastwise	6,800			
Manufactured	8,000			
Stock on hand, 1st October, 1838	762			
		35,000		
Deduct—				
Stock on hand, 1st October, 1837	....	3,000		
			32,000	28,618
Received at Philadelphia and Baltimore overland from Pittsburg	....	2,175		
Received at New York from Franklin, Lou.	....	105		
			2,280	
Total Crop of the United States	....	....	1,801,497	1,422,930
Crop of last year	....	....	....	1,422,930
Increase	....	....	....	378,567

Export to Foreign Ports, from October 1, 1837, *to September* 30, 1838.

FROM	To Great Britain.	To France.	To North of Europe.	Other F'n Ports.	Total.
New Orleans	481,501	127,828	7,580	14,528	631,437
Mississippi (Natchez)	15,246	....	....	....	15,246
Alabama	158,029	61,123	3,998	1,910	225,060
Florida	31,902	2,240	....	12	34,154
Georgia (Savannah and Darien)	201,582	27,024	560	632	229,798
South Carolina	158,212	55,685	28,853	3,717	246,467
North Carolina	4,279	....	....	....	4,279
Virginia	12,205	4,136	2,446	651	19,438
Baltimore	2,240	....	78	....	2,318
Philadelphia	2,954	465	905	282	4,606
New York	97,005	42,929	18,196	3,820	161,950
Boston	....	50	483	343	876
Grand total	1,165,155	321,480	63,099	25,895	1,575,629
Total last year	850,786	260,722	26,437	30,480	1,168,425
Increase	314,369	60,758	36,662	....	407,204
Decrease	....	....	....	4,585	....

Growth.

Total crop of 1824–5,	560,000	bales.
" 1825–6,	710,000	"
" 1826–7,	937,000	"
" 1827–8,	712,000	"
" 1828–9,	857,744	"
" 1829–30,	976,845	"
" 1830–31,	1,038,848	"
" 1831–32,	987,477	"
" 1832–33,	1,070,438	"
" 1833–34,	1,205,394	"
" 1834–35,	1,254,328	"
" 1835–36,	1,360,725	"
" 1836–37,	1,422,930	"
" 1837–38,	1,801,497	"

Consumption.

Total crop of the United States, as above stated			1,801,497 bales.
Add—Stocks on hand at the commencement of the year, 1st October, 1837.—In the Southern ports	61,770		
" Northern "	14,050		
		75,820	
Makes a supply of			1,877,317
Deduct therefrom—The export to Foreign ports	1,575,629		
Off foreign cotton included	281		
		1,575,348	
Stocks on hand at the close of the year, 1st Oct., 1838.—			
In the Southern ports	24,570		
" Northern "	15,735		
		40,305	
Burnt and lost at New Orleans	12,491		
" Vicksburg	1,170		
" Selma	1,600		
" Columbia	340		
		15,601	
			1,631,254

Quantity consumed by and in the hands of manufacturers,	1837–8	246,063
" " "	1836–7	222,540
" " "	1835–6	236,733
" " "	1834–5	216,888
" " "	1833–4	196,413
" " "	1832–3	194,412
" " "	1831–2	173,800
" " "	1830–1	182,142
" " "	1829–30	126,512
" " "	1828–9	118,853
" " "	1827–8	120,593
" " "	1826–7	103,483

Note.—Our present statement of the cotton crop shows a large increase over that of last year; but it will be recollected, that the crisis of that year prevented a large quantity from coming to market, equal probably to 150,000 bales, which remained in the interior, and actually belonged to the crop of 1836–7. Under this view of the subject, the crop of 1836–7 would have been about 1,570,000, and that of 1837–8 about 1,650,000 bales.

It will be seen, also, that we have deducted from the New Orleans statement the quantity received at that port from Texas—that being a foreign country.

Our estimate of the quantity taken for consumption, does not include any cotton manufactured in the States south and west of Virginia, nor any in that State, except in the vicinity of Petersburg and Richmond.

ANNUAL REVIEW.

From the New Orleans Price Current, 1837–38.

The occasion is one which naturally leads us to take a retrospective view of the commercial operations of the past year, and we find that while there has been much well grounded cause of complaint on account of a general want of animation in business, and the deranged state of money matters, and of exchanges, there is still some room left for congratulation, in the steady progress which has been made towards a return of confidence, which was so completely prostrated by the disasters of the previous season, and in the cheering hopes which present circumstances would seem to warrant the indulgence of, in regard to the approaching business season. In an other column we insert a resolution adopted at a meeting of the presidents of the banks of this city, designating the first Monday in January next, as a period for a general and unconditional resumption of specie payments by the banks, of all their obligations. Of the ability of these institutions fully to resume at the time appointed, there can be no question, and the promulgation of their resolution to that effect, cannot fail of having a very beneficial influence, as it will serve to pave the way for a complete restoration of confidence at home, and also exalt our credit abroad.

LIVERPOOL STATEMENT FOR 1838.

UNITED STATES, 1837–1838.

Stock, Oct., '37..	76,000	Export	1,575,000
Crop, " " ..	1,801,000	Consumption....	262,000
		Stock, Oct., '38..	40,000
Bags............	1,877,000	Bags	1,877,000

Stock 1st Jan., 1838, in......	Liverpool.	Gt. Britain.	France.	Continent.	Tot. Europe.
United States..........Bags.	73,230	88,160	28,958	17,495	134,613
Brazil	26,290	28,460	2,217	4,579	35,256
West Indies.................	13,110	14,520	11,863	10,370	36,753
East Indies	44,550	109,210	....	15,330	124,540
Egypt and Levant...........	13,620	18,990	20,504	15,038	54,532
Bags	170,820	259,340	63,542	62,812	385,694

CONSUMPTION, 1838. — IMPORT, 1838.

Tot. Europe.	Continent.	France.	Gt. Britain.	Liverpool.	IMPORT, 1838.	Liverpool.	Gt. Britain.	France.	Continent.	Tot. Europe.
1,391,900	117,552	309,926	964,422	922,157	....United States,.........Bags.	1,073,007	1,124.192	316,181	116,134	1,517,607
156,544	11,757	11.748	133,039	132,303	....Brazil	137,313	137,499	11,594	9,271	155,664
80.608	23,581	20,209	36,818	33.727	....West Indies................	28,317	32,198	14,290	18,239	63,927
159,523	53,673	811	105,039	80,457	....East Indies	64,427	108,879	2,276	46.559	111,314
147,668	55,457	50,850	41,361	36,333	....Egypt and Levant..........	28,033	28,461	46,661	87,120	162,242
1,936,243	262,020	393,544	1,280,679	1,204,977	........................Bags.	1,331,097	1,431,229	391,002	277,323	2,010,754
....	....	....	88,800	48,600	Export.					
460,205	78,115	61,000	321,090	248,340	Stock, Dec. Stock above,	170,820	259,340	63,542	62,812	385,694
2,396,448	340,135	454,544	1,690,569	1,501,917	Total supply, bags.............	1,501,917	1,690,569	454,544	340,135	2,396,448

COTTON AT LIVERPOOL. YEAR 1838.

Week Ending.	Receipts.						Sales.				Stocks.			Prices.			Actual Export	Consumption.
	Americ'n.	E. I.	Egypt.	Brazil.	Other.	Total.	Consumption.	Speculation.	Export	Total.	Amer'n.	Other.	Total.	Mid. Up.	Mid. Orl.	Dhol		
Jan. 5..	4,941	.....	553	3,758	99	9,351	7,620	....	......	7,620	72,500	100,500	173,000	7¼	7¼	4¼	......	7,620
" 12..	10,446	......	3,240	1,818	1,800	17,304	17,160	1,100	......	18,260	68,500	104,000	172,500	7¼	7¼	4	......	24,780
" 19..	12,674	......	......	....	14	12,688	12,400	....	......	12,400	70,500	101,500	172,000	7	7½	4	......	37,180
" 28..	2,467	......	950	379	......	3,796	20,240	1,500	......	21,740	55,500	100,500	155,000	7¼	7¼	4	......	57,420
Feb. 2..	9,210	......	925	548	2,130	12,813	17,440	600	......	18,040	51,500	99,000	150,500	7¼	7¼	4	......	74,860
" 9..	31,650	1,964	......	1,364	73	35,051	20,280	500	......	20,780	67,000	99,000	166,000	7¼	7¼	4	......	95,140
" 16..	20,279	941	......	4,728	......	25,948	18,550	....	......	18,550	73,500	99 500	173,000	6⅞	7	4	......	113,690
" 23..	47,013	......	......	....	6,397	53,410	30,840	1,100	......	31,940	96,500	100,500	196,000	6⅞	7	4	......	144,530
Mch 2..	3,480	......	2,263	....	......	5,743	20,590	500	250	21,340	83,000	98,500	181,500	6½	6¾	4	250	165,120
" 9..	119,544	......	760	6,517	2,281	129,102	21,460	500	450	22,410	184,500	104,500	288,000	6¼	6⅜	4	450	186,580
" 16..	21,714	......	......	3,054	2,643	27,411	27,620	2,500	550	30,670	182,000	103,500	285,500	6⅜	6⅜	4	550	214,200
" 23..	28,554	1,691	......	....	943	31,188	35,680	2,800	1,600	40,080	181,500	98,000	279,500	6¼	6¼	4	1,600	249,880
" 30..	33,152	......	......	....	165	33,317	18,700	4,300	1,700	24,700	198,000	103,000	291,000	6⅛	6⅛	3¾	1,700	268,580
April 6..	9,229	......	......	3,709	1,286	14,224	23,910	3,000	1,250	28,160	188,000	102,000	280,000	6⅛	6⅛	3¾	1,250	292,490
" 13..	52,987	......	......	4,553	808	58,348	17,450	1,500	1,550	20,500	225,000	92,000	317,000	6	6	3¾	1,550	309,940
" 20..	32,663	2,439	......	....	747	35,849	13,560	1,500	1,000	16,060	244,500	95,500	339,000	6	6	3¾	1,000	323,500
" 27..	13,474	......	......	....	987	14,461	34,040	4,000	1,200	39,240	228,000	87,500	315,500	6	6⅛	3¾	1,200	357,540
May 4..	35,230	8,668	......	3,463	192	47,553	29,070	7,500	1,600	38,170	238,000	90,500	328,500	6	6⅛	3¾	1,600	386,610
" 11..	1,820	......	......	....	906	2,726	19,080	3,600	1,000	23,680	224,000	77 500	311,500	6	6⅛	3¾	1,000	405,690
" 18..	11,544	4,316	2,490	2,107	92	20,549	31,910	14.000	1,690	47,600	229.500	88,500	318,000	6⅛	6⅛	3¾	1,690	437,600
" 24..	68,845	......	......	7,265	1,206	77,316	26,580	13,860	1,100	41,540	275,500	90,500	365,000	6⅛	6⅛	3¾	1,100	464,180
June 1..	40,801	2,248	1,080	....	112	44,241	27,500	7,180	1.500	36,180	294,000	84,000	378,000	6⅛	6⅛	3¾	1,500	491,680
" 8..	16,231	......	......	....	3,382	19,613	30,780	5,200	3,350	39,330	285,000	78.000	363,000	6⅛	6⅛	3¾	3,350	522,460
" 15..	26,677	......	......	....	1,685	28,362	17,730	2,300	1,150	21,180	297,500	76,500	373,000	6	6	3¾	1,150	540,190
" 22..	74,852	......	......	3,936	2,259	81,047	17,210	2,000	1,500	20,710	355,000	77,500	432,500	6	6	3¾	1,500	557,400
" 29..	37,746	1,276	......	3,700	1,997	44,719	13,370	300	450	14,120	381,500	83,000	464,500	6	6	3¾	450	570,770
July 6..	15,676	......	......	....	7,952	23,628	19,860	1,000	2,300	23,160	378,500	87,500	465,000	5⅞	5⅞	3¾	2,300	590,630
" 13..	47,565	......	......	....	3,683	51,248	18,640	250	1,150	20,040	410,500	86,500	497,000	5¾	5¾	3¾	1,150	609,2[illegible]0
" 20..	41,979	2,503	2,318	1,041	76	47,917	18,260	300	1,200	19,760	437.000	98,000	525,000	5¾	5¾	3¾	1,200	627,530
" 27..	24,267	1,781	2,029	2,909	116	31,092	21,260	2,200	1,500	24,960	443,500	90,000	533,500	5¾	5¾	3¾	1,500	648,790
Aug. 3..	22,272	......	......	....	1,610	23,882	30,360	800	1,450	32.610	439,500	84,000	323.500	5⅞	5⅞	3¾	1,450	679,150
" 10..	15,562	......	......	....	3,718	19,280	44,670	300	2,570	47,540	422,500	76,000	498,500	6	6	3¾	2,570	723,820
" 17..	6,949	2,588	......	2,926	419	12,882	13,820	1,000	1,000	15,820	417,000	78,000	495,000	6	6	3¾	1,000	737,640
" 24..	22,774	3,550	......	4,910	161	31,395	18,000	1,500	1,400	20,900	424,000	83,000	507,000	6	6	3¾	1,400	755,640

Aug. 31..	6,603	8,736	......	....	491	15,830	17,950	1,500	1,650	21,100	416.000	88,000	504,000	6	6	3¾	1,650	773,590
Sep. 7..	6,900	......	......	....	......	6,900	19,470	....	1,250	20,720	407,000	83,500	490,500	6	6	3¾	1,250	793,060
" 14..	13,492	7,459	......	414	......	21,365	18,460	....	1,240	19,700	404,500	88,500	492,000	6	6	3¾	1,240	811,520
" 21..	13,807	......	......	2.227	......	16,034	20,080	400	2,200	22.680	401,000	84,500	485,500	6	6	3¾	2,200	831,600
" 28..	9,420	......	......	6,037	3,211	18.668	19,650	....	3,100	22,750	393,000	88,500	481,500	6	6	3¾	3,100	851,250
Oct. 5..	456	2,505	......	....	853	3,814	15,680	....	1,750	17,430	380,000	87,000	467,000	5⅞	5⅞	3¾	1,750	866,930
" 12..	....	......	498	2,490	94	3.082	18,450	....	1,180	19,630	363,000	81,500	444,500	5¾	5¾	3¾	1,180	885,380
" 19..	9,404	3,621	......	10,598	122	24,045	24,710	7,500	3,750	35,960	354,500	89,500	443,000	5⅞	6	3¾	3,750	910.090
" 26..	5,543	......	579	....	3,481	9,603	28,290	14,550	270	43,110	339.000	85,000	424,000	6	6	3¾	270	938,380
Nov. 2..	500	......	......	....	8,269	8,769	33,380	15,000	260	48,640	325,000	86,500	411,500	6	6⅛	4	260	971,760
" 9..	2,190	......	......	....	......	2,190	28,760	2,800	200	31,760	305,500	80,500	385,000	6	6⅛	4	200	1,000,520
" 16..	1,413	3,010	......	2,566	......	6,989	38,070	6,500	1,500	46,070	280,000	72,000	352,000	6¼	6¼	4	1,500	1,038,590
" 23..	....	......	1,975	....	......	1,975	36,490	13,000	......	49,490	252,500	64,500	317,000	6⅜	6⅜	4	......	1,075,080
" 30..	643	......	......	660	511	1,814	17,630	30,500	......	48,130	241,000	60,000	301,000	6½	6½	4	......	1,092,710
Dec. 7..	2,854	1,296	......	2,038	1,073	7,261	20,750	40,000	......	60,750	228,500	59,000	287,500	6¾	6¾	4½	......	1,113,460
" 14..	1,369	......	2,350	4,753	161	8,633	17,760	6,500	200	24,460	216,500	63,500	279,000	6¾	6¾	4½	200	1,131,220
" 21..	1,206	......	......	....	1,236	2,442	21,220	19,500	......	40,720	201,500	58,000	259,500	6⅞	6⅞	4½	......	1,152,440
" 28..	6,962	......	820	....	4,846	12,628	31,920	38,450	......	70,370	186,000	33,500	240,500	7½	7½	5	......	1,184,360
Average prices & total sales, receipts & stocks.	1,066,790	60, 592	22, 820	94,743	74, 287	1,319,232	1,184,360	284,890	54, 010	1,523,260				6.28	6.31	3.9	54, 010	2,277,615

The semi-weekly Price and Weekly Sales and Receipts, at New York, Weekly Exports from New York and Rates of Freight to Liverpool 1st of each Month, for the Crop Year ending October 1, 1838.

1837.	Price of New Orleans	Price of Upland.	Sales for week.	Receipts for week.	EXPORTS FOR WEEK.					Rates of Freight to Liverpool.
					To Great Britain	To France.	North of Europe.	Other Fo'n Ports	Total Exports.	
October 5..	8@13	8@$12\frac{1}{4}$								$\frac{3}{8}$@$\frac{1}{2}$d.
" 10..	8@13	8@$12\frac{1}{4}$	2,400	1,764						
" 13..	8@13	8@$12\frac{1}{4}$								
" 17..	8@13	8@$12\frac{1}{4}$	1,450	2,116						
" 20..	8@13	8@$12\frac{1}{4}$								
" 24..	8@13	8@$12\frac{1}{4}$	2,100	1,410						
" 27..	8@13	8@$12\frac{1}{4}$								
" 31..	8@13	8@$12\frac{1}{4}$	2,100	2,224	*4,243	3,196	1,421	332	9,192	
Novem. 3..	8@13	8@$12\frac{1}{4}$								$\frac{3}{8}$@$\frac{1}{2}$d.
" 7..	8@13	8@$12\frac{1}{4}$	1,300	5,419						
" 10..	8@13	8@$12\frac{1}{4}$								
" 14..	8@13	8@$12\frac{1}{4}$	2,550	5,626	911	182	101		1,194	
" 17..	8@14	8@13								
" 21..	8@14	8@13	2,000	5,264	483	540	245		1,268	
" 24..	8@$13\frac{1}{2}$	8@$12\frac{1}{2}$								
" 28..	8@$13\frac{1}{2}$	8@$12\frac{1}{2}$	4,350	3,225	633	464	19		1,116	
Decem. 1..	8@$13\frac{1}{2}$	8@$12\frac{1}{2}$								$\frac{3}{8}$@$\frac{1}{2}$d.
" 5..	8@$13\frac{1}{2}$	8@$12\frac{1}{2}$	2,150	6,329	1,448	843			2,291	
" 8..	8@13	8@12								
" 12..	8@13	8@12	2,500	6,790	686	440			1,126	
" 15..	8@13	8@12								
" 19..	8@13	8@12	2,900	3,398		592		62	654	
" 22..	$8\frac{1}{2}$@13	$8\frac{1}{2}$@12								
" 26..	$8\frac{1}{2}$@13	$8\frac{1}{2}$@12	2,250	11,116	992		150		1,142	
" 29..	$8\frac{1}{2}$@13	$8\frac{1}{2}$@12								
1838.										
January 2..	$8\frac{1}{2}$@13	$8\frac{1}{2}$@12	1,900	5,428	2,188	757	632		3,577	$\frac{1}{2}$@$\frac{5}{8}$d.
" 6..	$8\frac{1}{2}$@13	$8\frac{1}{2}$@12								
" 9..	$8\frac{1}{2}$@13	$8\frac{1}{2}$@12	9,050	4,847	2,678	673			3,351	
" 12..	9@13	9@12								
" 16..	9@13	9@12	5,950	7,009	1,046	1,373		59	2,478	
" 19..	9@13	9@12								
" 23..	9@13	9@12	2,750	6,742	6,896	832	210		7.938	

GENERAL REMARKS.

*These figures embrace the shipments of the previous three weeks.

This year will be ever memorable in the annals of the commercial world as it witnessed the establishment of regular lines of communication between the old world and the new, by means of vessels propelled by steam. It was the dawn of a new epoch, and the "Sirius" and the "Great Western," were the heralds that ushered the vast steam fleet that now covers every ocean, and pass from hemisphere to hemisphere, almost with the frequency and regularity of our river ferries. The "Sirius," 700 tons, arrived here from London via Cork, April 22, 1838, in 18 days passage from the latter port, followed immediately by the "Great Western" on the 23d, in 15 days and 5 hours from Bristol. In relation to this event, the chronicle of the times says, "that the interest excited by the presence of the European steamers in our waters has rendered them objects of universal curiosity, and tended to withdraw attention very generally from the markets." Contrary, however, to popular belief, these vessels were not the first to solve the problem of

January 26..	9@13	9@12								
" 30..	9@13	9@12	5,800	5,210	2,157	890	1,022	....	4,069	
February 2..	9@13	9@12								9-16@$\frac{11}{16}$
" 6..	9@13	9@12	3,650	2,137	5,103	756	597	....	6,456	
" 9..	9@13	9@12								
" 13..	9@13	9@12	1,100	16,048	4,495	1,110	123	290	6,018	
" 16..	8½@12½	8½@11½								
" 20..	8½@12½	8½@11½	2,750	4,545	2,314	941	....	249	3,504	
" 23..	8½@12½	8¼@11¼								
" 27..	8½@12½	8¼@11¼	3,350	7,180	684	....	190	....	874	
March 2..	8@12½	8@11								¾@⅞d.
" 6..	8@12½	8@11	2,800	4,489	3,388	1,202	705	....	5,295	
" 9..	8@12½	8@11								
" 13..	8@12½	8@11	7,500	14,927	1,129	1,015	679	....	2,823	
" 16..	8@12½	8@11								
" 20..	8@12½	8@11	4,850	2,060	....	1,172	....	285	1,457	
" 23..	8@12½	8@11								
" 27..	7¾@12	7¾@10½	3,000	19,715	4,176	1,583	2,289	66	8,114	
" 30..	7½@12	7½@10½								
April 3..	7½@12	7½@10½	4,650	8,016	3,771	2,124	540	....	6,435	⅞@1d.
" 6..	7½@12	7½@10½								
" 10..	7½@12	7½@10½	5,100	7,278	5,442	1,617	293	....	7,352	
" 13..	7½@12	7½@10½								
" 17..	7½@12	7½@10½	3,900	12,126	3,537	3,379	1,361	5	8,282	
" 20..	7½@12	7½@10½								
" 24..	7½@12	7¼@10½	4,300	4,016	1,457	1,671	237	308	3,673	
" 27..	7½@12	7¼@10½								
May 1..	7½@12	7¼@10½	4,200	5,891	1,393	1,731	60	....	3,184	⅝@⅞d.
" 4..	7½@12½	7¼@10¾								
" 8..	7½@12½	7¼@10¾	3,850	4,360	2,594	1,421	500	....	4,515	
" 11..	7½@12½	7¼@10¾								
" 15..	7½@12½	7¼@10¾	4,050	1,771	5,081	712	405	153	6,351	
" 18..	7½@12½	7½@11								
" 22..	7¾@12¾	7¾@11	6,600	5,491	2,976	1,010	1,063	....	5,049	
" 25..	8@13	7¾@11½								
" 29..	8@13	8¼@12	4,500	4,056	2,600	1,791	1,426	312	6,129	
June 1..	8@13	8¼@12								¼@½d.
" 5..	8@13	8¼@12	3,550	2,492	....	187	951	705	1,843	
" 8..	8@13	8¼@12								
" 12..	8@13	8¼@12	2,400	5,163	3,249	876	752	....	4,877	
" 15..	8@13	8¼@12								
" 19..	8@13	8¼@12	5,050	3,066	1,947	....	24	436	2,407	
" 22..	8¼@13	8½@12¼								

the feasibility of navigating the ocean by the means of steam; an American vessel, the "Savannah," has indisputable claims to the honor and eclat that arises from that, then novel achievement. The "Savannah" was built in New York, measured 350 tons, and was launched August 22, 1818, her first voyage was from here to Savannah, under the command of Captain Moses Rogers, at which port she arrived April 19, 1819, in seven days' passage; on the 25th of May, she left there for St. Petersburgh, via Liverpool, arriving at the latter port in 25 days' passage. The voyage was continued to St. Petersburgh, and in the fall of that year, she returned to Savannah, after a run of 50 days. It is probable that the venture was not remunerative, owing to the small size of the vessel, or that she was sent abroad in the expectation of finding a profitable sale; at all events no further attempts in this direction were made until the successful voyages of the "Sirius" and "Great Western" noted above

In regard to the cotton market, there was no feature of especial interest; prices were quite steady throughout the greater part of the crop year, the difference between the highest and lowest points being only 1½@3 cents per pound.

The unsettled state of financial affairs noted in the previous year, still prevailed, though in a modified form, and business matters wore an unsatisfactory aspect. At one period, in early March, the latest

The semi-weekly Price and Weekly Sales and Receipts at New York, Weekly Exports from New York and Rates of Freight to Liverpool 1st of each Month, for the Crop Year ending October 1, 1838.—Concluded.

1838.		Price of New Orleans	Price of Upland.	Sales for week.	Receipts for week.	EXPORTS FOR WEEK.					Rates of Freight to Liverpool.	GENERAL REMARKS.
						To Great Britain.	To France.	North of Europe.	Other Fo'n Ports	Total Exports.		
June	26..	8½@13½	8½@13	5,950	2,680	1,534	843	....	....	2,377		advices from Europe were over 60 days old, owing to long passages and the cotton market, as regards purchases for export, was at a stand for several weeks for the lack of intelligence, vessels and freight room. *Exchange.* The market for 60 days' bills on London was quite fluctuating, owing to the disturbed condition of monetary affairs; the opening price October 1, 1837, was 14@15 per cent. premium. Gold commanding 11@12 per cent. premium. The range for bills throughout October, was 14@16 per cent. premium; in November. 14¾@15½; in December, the quotatiou fell from 14@14½ to 9¼@10; the range in January, was from 9¼@10½; in February the quotation declined until 5½@6 was touched; in March it advanced again to 7¼@7¾ and then fell back at the close, to 5@6; the first fortnight in April, the price was 4½@5½, subsequently advancing to 6½@7; throughout May, the quotation varied but little, say from 6½@7½; in June, the range was from 8@9½; in July, 7½@8½; August, 8@9½; September, 9¼@9¾; closing Otober 1, at 10@10¼ premium.
"	29..	8¾@13½	8½@13									
July	3..	8¾@13½	8½@13	3,950	4,992	5,886	929	236	....	7,051	¼@⅜d.	
"	6..	8¾@13½	8½@13									
"	10..	8¾@13½	8½@13	2,550	3,606	....	273	61	....	334		
"	13..	8¾@13½	8½@13									
"	17..	8¾@13½	8½@13	3,200	3,065	4,074	904	543	....	5,521		
"	20..	8¾@13½	8½@13									
"	24..	8¾@13½	8½@13	1,500	1,501	860	298	71	169	1,398		
"	27..	8¾@13½	8½@13									
"	31..	8¾@13½	8½@13	1,300	2,701	810	94	175	....	1,079		
August	3..	8¾@13½	8½@13								5-16@7/16	
"	7..	8¾@13½	8½@13	1,750	3,860	423	523	100	....	1,046		
"	10..	8½@13½	8½@13									
"	14..	8½@13½	8½@13	2,200	3,124	336	54	26	30	446		
"	17..	8½@13½	8½@13									
"	21..	8½@13½	8½@13	2,450	4,398	766	1,009	38	....	1,813		
"	24..	8½@13½	8½@13									
"	28..	8½@13½	8½@13	3,050	1,040	503	....	63	....	566		
"	31..	8½@13½	8½@13									
Septem.	4..	8½@13½	8½@13	3,750	3,106	276	914	....	35	1,225	¼@⅜d.	
"	7..	8½@13½	8½@13									
"	11..	8½@13½	8½@13	3,700	433	....	....	426	....	426		
"	14..	8¾@13½	8½@13¼									
"	18..	9@13¾	8¾@13½	5,550	37	....	1.153	....	324	1,477		
"	21..	9@14	9@14									
"	25 .	9@14	9@14	3,450	2,003	1,049	1,081	....	....	2,130		
"	28..	9@14	9@14									
October	2..	9@14	9@14	2,200	1,327	791	210	22	....	1,023	¼@⅜d.	
Average price and total sales, receipts and exports.		10.54	10.14	181,200	256,617	97,005	43,365	17,756	3,820	161,946		

1839.

A machine, called a Dynamometer, affording better means of ascertaining the power for driving cotton and other machinery, either by water or steam, than any other instrument, was designed and built, during this year, at Saco, Me., by Samuel Batchelder. (*See* years 1832 and 1835.)

COTTON CROP OF THE UNITED STATES.

Statement and Total Amount of the Cotton Crop of the United States, for the Year ending September 30, 1839.

	Bales.	Bales.	Total.	Same period 1838.
NEW ORLEANS.				
Export—				
To Foreign Ports	442,706			
Coastwise	138,111			
Burnt, lost in repacking, &c	1,000			
Stock on hand 1st October, 1839	16,307			
Deduct—		598,124		
Stock on hand 1st October, 1838	8,843			
Received from Mobile	16,768			
" Florida	1,080			
" Texas	2,871			
		29,562		
			568,562	711,581
MISSISSIPPI.				
Export from Natchez, &c.—				
To Foreign Ports		2,009		
Coastwise		14,423		
			16,432	19,675
ALABAMA.				
Export from Mobile—				
To Foreign Ports	149,945			
Coastwise	99,784			
Burnt	1,195			
Stock in Mobile, 1st October, 1839	1,380			
Deduct—		252,304		
Stock on hand 1st October, 1838	59			
Received from Florida	285			
" Texas	218			
		562		
			251,742	309,807
FLORIDA.				
Export—				
To Foreign Ports	14,767			
Coastwise	60,760			
Stock on hand 1st October, 1839	650			
Deduct—		76,177		
Stock on hand 1st October, 1838		1,000		
			75,177	106,171
GEORGIA.				
Export from Savannah—				
To Foreign Ports—Uplands	106,342			
" Sea Island	4,225			
Coastwise	88,609			
	199,176			

Statement and Total Amount of the Cotton Crop of the United States, for the Year ending September 30, 1839—Concluded.

	Bales.	Bales.	Total.	Same period 1838.
From Darien—				
To New York and Boston	8,373			
Lost in steamer Clarendon	316			
Stock in Savannah, 1st October, 1839	1,641			
" Augusta and Hambro', 1st Oct., 1839	6,000			
		215,506		
Deduct—				
Stock in Savannah and Augusta 1st October, 1838	9,937			
Received from Florida	457			
		10,394		
			205,112	304,210
SOUTH CAROLINA.				
Export from Charleston—				
To Foreign Ports—Uplands	148,285			
" Sea Island	9,975			
Coastwise	54,454			
	212,714			
From Georgetown—				
To New York	11,861			
Stock in Charleston, 1st October, 1839	4,706			
		229,281		
Deduct—				
Stock in Charleston, 1st October, 1838	3,169			
Received from Savannah	14,104			
Received from Florida and Key West	1,837			
		19,110		
			210,171	294,334
NORTH CAROLINA.				
Export—				
To Foreign Ports	21			
Coastwise	11,315			
Stock on hand, 1st October, 1839	600			
		11,936		
Deduct—				
Stock on hand 1st October, 1838	....	800		
			11,136	21,439
VIRGINIA.				
Export—				
To Foreign Ports	7,800			
Coastwise	5,700			
Manufactured	9,000			
Stock on hand, 1st October, 1839	500			
		23,000		
Deduct—				
Stock on hand 1st October, 1838	....	800		
			22,200	32,000
Received at other ports	....	....		2,280
Total crop of the United States	....	....	1,360,532	
Crop of last year	....	....	....	1,801,497
Decrease	....	....		440,965

Export to Foreign Ports, from October 1, 1838, *to September* 39, 1839.

FROM	To Great Britain.	To France.	To North of Europe.	Other F'n Ports.	Total.
New Orleans........................Bales	309,768	122,452	1,446	9,040	442,703
Mississippi (Natchez)	2,009	....	....	....	2,009
Alabama	125,633	22,304	1,728	280	149,945
Florida	11,928	2,562	277	....	14,767
Georgia (Savannah and Darien)	97,853	10,480	2,234	....	110 567
South Carolina	119,486	30,665	7,733	376	158,260
North Carolina	....	21	....	....	21
Virginia	6,648	98	1,054	....	7,800
Baltimore	186	400	....	....	586
Philadelphia	1,767	138	37	55	1,997
New York	122,674	53,123	6,453	2,760	185,010
Boston	466	....	555	....	1,021
Grand total	798,418	242,243	21,517	12,511	1,074,689
Total last year	1,165,155	321,480	63,099	25,895	1,575,629
Decrease	366,737	79,237	41,582	13,384	500,940

Growth.

Total crop of 1824–5,	560,000	bales.
" 1825–6,	710,000	"
" 1826–7,	937,000	"
" 1827–8,	712,000	"
" 1828–9,	857,744	"
" 1829–30,	976,845	"
" 1830–1,	1,038,848	"
" 1831–2,	987,477	"
" 1832–3,	1,070,438	"
" 1833–4,	1,205,394	"
" 1834–5,	1,254,328	"
" 1835–6,	1,360,725	"
" 1836–7,	1,422,930	"
" 1837–8,	1,801,497	"
" 1838–9,	1,360,532	"

Consumption.

To estimate the quantity manufactured in the United States, we take the growth of the year	1,360,532 bales.		
Add—Stocks on hand at the commencement of the year (October 1, 1838)—In the Southern ports	24,570		
" Northern ports	15,735		
		40,305	
			1,400,837
Deduct therefrom—The Export to Foreign ports	1,074,689		
Less Texas and other foreign included.	4,625		
		1,070,064	
Stocks on hand at the close of the year (October 1, 1839)—			
In the Southern ports	31,784		
" Northern ports	20,460		
		52,244	
Burnt and Lost at New Orleans	1,000		
" Mobile	1,195		
" Darien	316		
		2,511	
			1,124,819

Quantity consumed by and in the hands of manufacturers.	1838-9	276,018
" " "	1837-8	246,063
" " "	1836-7	222,540
" " "	1835-6	236,733
" " "	1834-5	216,888
" " "	1833-4	196,413
" " "	1832-3	194,412
" " "	1831-2	173,800
" " "	1830-1	182,142
" " "	1829-30	126,512
" " "	1828-9	118,853
" " "	1827-8	120,593
" " "	1826-7	103,483

Note.—It will be observed by the above statement that there is a decrease in the crop, compared with last year, of 440,965 bales—but if we deduct 150,000 bales, included in that year, which was believed to have belonged to the previous one—the actual difference of this year, compared with last, will be 290,965 bales.

It will be seen, also, that we have deducted from the New Orleans statement the quantity received at that port from Texas—Texas being a foreign country.

Our estimate of the quantity taken for consumption does not include any Cotton manufactured in the States south and west of Virginia, nor any in that State, except in the vicinity of Petersburg and Richmond.

ANNUAL REVIEW.

From the New Orleans Price Current—1838-39.

The year just closed has been a most disastrous one to those engaged in the shipment of Cotton. Early in the season it was well known that the crop would fall short of that of the previous years, and various were the estimates formed on the subject, the lowest being about 1,370,000 bales, which is now almost ascertained to have been nearest the truth. In consequence of this, the market gradually advanced until the extreme rates for Louisiana and Mississippi Cottons, ordinary to choice, were 12 to 18 cents per lb., it being with reason anticipated that a corresponding rise in foreign markets would be the consequence of the very great diminution in the production. This expectation, however, has been entirely frustrated by the depressed state of the money markets and successful combination among the spinners to withstand a rise in prices in Liverpool; and the effect must be that enormous losses will be sustained by shippers from this country. It is yet too early to form any correct estimate of the crop of this season, everything depending on whether there is any early frost or not. It appears to be generally allowed that in the lower Cotton-growing districts the crops are very promising, but heavy complaints are daily received from Mississippi, Tennessee, and North and South Alabama, of the great injury done through long drought and the ravages committed by the worm.

LIVERPOOL STATEMENT FOR 1839.

UNITED STATES, 1838–1839.

Stock, October 1, 1838	40,000	Export	1,075,000
Crop	1,360,000	Consumption	273,000
		Stock, Oct. 1, 1839	52,000
Bags	1,400,000	Bags	1,400,000

CONSUMPTION. Tot. Europe.	Continent.	France.	Gt. Britain.	Liverpool.	Stock 1st Jan., 1839, in	Liverpool.	Gt. Britain.	France.	Continent.	Tot. Europe.
					United States .... Bags	189,100	209,030	35,213	16,077	260,320
					Brazil	29,100	30,220	2,063	2,093	34,376
					West Indies	7,300	9,100	5,944	5,028	20,072
					East Indies	17,520	66,650	1,465	8,216	76,331
					Egypt and Levant	5,320	6,090	16,315	46,701	69,106
					Bags	248,340	321,090	61,000	78,115	460,205
					IMPORT.					
1,105,215	90,277	235,013	779,925	751,961	United States .... Bags	782,861	813,125	251,600	92,100	1,088,825
140,712	13,543	12,463	114,706	113,233	Brazil	97,033	97,656	17,200	14,400	126,056
94,659	22,528	34,244	37,887	33,073	West Indies	30,123	38,077	33,800	24,100	90,577
149,072	40,466	3,565	105,041	64,510	East Indies	74,640	131,731	3,700	35,450	137,581
118,542	54,501	39,015	25,026	24,240	Egypt and Levant	30,520	31,576	34,600	63,800	117,176
1,608,200	221,315	324,300	1,062,585	987,017	Bags	1,015,177	1,112,165	340,900	229,850	1,560,215
....	15,500	2,000	105,200	70,500	Export.					
412,220	71,150	75,600	265,470	206,000	Stock, Dec. .... Stock above,	248,340	321,090	61,000	78,115	460,205
2,020,420	307,965	401,900	1,433,255	1,263,517	Total supply, bags,	1,263,517	1,433,255	401,900	307,965	2,020,420

COTTON AT LIVERPOOL. YEAR 1839.

WEEK ENDING.	RECEIPTS.						SALES.				STOCKS.			PRICES.			ACTUAL EXPORT	CON-SUMPTION.
	Americ'n.	E. I.	Egypt.	Brazil.	Other.	Total.	Con-sumption.	Specu-lation.	Export.	Total.	Amer'n.	Other.	Total.	Mid. Up.	Mid. Orl.	Dhol		
Jan. 4..	1,715	......	1,751	2,919	130	6,515	10,990	9,600	......	20,590	182,000	61,000	243,000	7½	7½	5	......	10,990
" 11..	11,630	276	......	1,781	1,679	15.366	12,150	23,500	......	35,650	184,000	63,000	247,000	7½	7½	5	......	23,140
" 18..	6,951	5,588	......	3,585	138	16,262	12,530	5,700	......	18 230	182,500	68,000	250.500	7½	7½	5	......	35,670
" 25..	937	......	1,128	2,012	246	4,323	20,470	16,800	......	37,270	170,000	63,500	233,500	7½	7½	5	......	56.140
Feb. 1..	3,086	......	......	......	1,391	4,477	13,540	5,700	......	19,240	162,500	63,000	225,500	7½	7½	5	......	69,680
" 8..	17,557	......	292	2,930	126	20,905	13,810	6,950	......	20,760	172,000	60,500	232,500	7½	7½	5	......	83,490
" 15..	31,430	......	810	7,926	1,922	42.088	11,430	8,000	......	19,430	195,000	68,500	263,500	7½	7½	5	......	94,920
" 22..	15,844	......	......	......	......	15,844	13,060	1,000	......	14.060	202,500	64,000	266,500	7½	7½	5	......	107,980
Mch 1..	23,836	......	1,100	2,270	1,143	28,349	19,920	4,200	120	24,240	208,500	66,000	274,500	7½	7⅝	5	120	127,900
" 8..	600	......	......	......	7	607	46,900	40,500	600	88,000	178,500	50,500	229,000	8	8	5½	600	174,800
" 15..	8,859	275	......	2,209	908	12,251	17,510	43,000	220	60,730	175.500	48,500	224,000	8⅜	8½	5¾	220	192,310
" 22..	33,836	6,871	......	620	675	42,002	28,820	19,000	......	47,820	185,000	50,000	235,000	8¾	8¾	6	......	221,130
" 28..	16,406	......	......	......	903	17,309	13,930	4,500	......	18,430	193,000	45,500	238,500	8¾	8¾	6	......	235,060
April 5..	5,129	......	784	1,416	......	7,329	7,650	......	......	7.650	193,000	45,500	238,500	8⅜	8⅜	6	......	242,710
" 12..	28,188	......	......	2,598	1,255	32,041	13,340	6,000	......	19,340	210,500	46,500	257,000	8½	8½	6	......	256,050
" 19..	36,558	......	......	1,788	1,050	39,396	9,620	750	400	10,770	240,500	45,500	286,000	8½	8½	6	400	265,670
" 26..	5,795	1,992	1,200	3,332	......	12,319	7,430	......	370	7,800	241,000	50,000	291,000	8¼	8¼	6	370	273,100
May 3..	12,145	......	......	......	1,684	13,829	7,410	620	850	8,880	247,000	49,500	296,500	8⅛	8⅛	5½	850	280,510
" 10..	17,225	......	......	......	......	17,225	8.590	100	......	8,690	253,500	51,000	304,500	7⅞	7⅞	5¼	......	289,100
" 17..	14,087	......	......	......	175	14,262	7,650	......	300	7,950	261,500	49,500	311,000	7⅞	7⅞	5¼	300	296,750
" 24..	28,382	......	......	2,593	3,754	34,729	8,990	......	450	9,440	282,500	53,500	336,000	7½	7½	5	450	305.740
" 31..	22,868	1,731	382	......	313	25,294	15,220	3,700	930	19,850	293,500	51.500	345,000	7½	7½	5	930	320,960
June 7..	25,672	......	1,071	......	842	27,585	20,880	500	1,650	23,030	303,000	47,500	350,500	7¾	7¾	4¾	1,650	341,840
" 14..	92,015	3,368	......	4,549	641	100,573	20,800	300	400	21.500	379,000	50,000	429,000	7¾	7¾	4¾	400	362,640
" 21..	12,379	......	......	......	1,411	13,790	11,030	......	120	11,150	383,500	48,000	431,500	7½	7½	4¾	120	373,670
" 28..	87,952	1,194	2,103	......	803	92,052	10,620	......	150	10,770	459,500	53,500	513,000	7½	7½	4½	150	384,290
July 5..	25,002	2,133	......	......	2,634	29,769	11,700	......	250	11,950	475,500	55,500	531,000	7⅛	7⅛	4¼	250	395,990
" 12..	31,204	......	......	......	3,615	34,819	17,310	1,500	2,670	21,480	492,000	53,500	545,500	6¼	6⅜	4	2,670	413,300
" 19..	7,643	......	......	......	2,446	10,089	12.610	3,000	5,900	21,510	481,500	52,000	533,500	6¼	6⅜	4	5,900	425,910
" 26..	37,811	12,696	......	3,325	2,102	55,934	35,940	5,500	......	41,440	494,000	59,500	535,500	6⅝	7	4¼	......	461,850
Aug. 2..	3.069	3,310	......	......	524	6,903	21,940	2,750	1,900	26.590	476,500	58,000	534.500	6¾	6⅞	4½	1,900	483,790
" 9..	36.288	4,536	......	5,852	1,604	48,280	28,330	3,200	2,250	34,080	494,500	58,500	553,000	7	7	4½	2,550	512,120
" 16..	9,129	......	......	......	3,368	12,497	16,770	1,500	4.800	23,090	484,000	56,500	540,500	6¾	6⅞	4½	4,800	528,890
" 23..	6,336	......	......	......	130	6,466	25,450	4,200	10,100	39,750	463,000	57,000	512,000	6¾	6⅞	4½	10,100	554,340

Aug. 30..	13,768	4,333	......	......	1,479	19,580	28,330	1,500	4,010	33,840	454,000	46,000	500,000	$6\frac{3}{8}$	$6\frac{3}{4}$	$4\frac{1}{2}$	4,010	582,670
Sep. 6..	7,090	......	......	3,737	......	10,827	16,670	500	3,400	20,570	446,000	44,500	490,500	$6\frac{1}{2}$	$6\frac{5}{8}$	$4\frac{1}{2}$	3,400	599,340
" 13..	6,083	......	......	......	1,158	7,241	34,130	2,250	5,000	41,380	420,500	38,000	458,500	$6\frac{1}{2}$	$6\frac{5}{8}$	$4\frac{1}{2}$	5,000	633,470
" 20..	7,673	2,520	957	......	1,091	12,241	24,890	......	2,160	27,050	408,500	35,500	444,000	$6\frac{1}{2}$	$6\frac{5}{8}$	$4\frac{1}{2}$	2,160	658,360
" 27..	928	2,912	......	3,674	267	7,781	23,480	850	2,650	26,980	388,500	38,000	426,500	$6\frac{1}{2}$	$6\frac{5}{8}$	$4\frac{3}{4}$	2,650	681,840
Oct. 4..	5,223	4,356	......	......	......	9,579	22,000	1,500	3,000	26,500	374,500	36,500	411,000	$6\frac{1}{2}$	$6\frac{1}{2}$	5	3,000	703,840
" 11..	3,934	......	......	......	2,025	5,959	20,800	1,000	7,820	29,620	353,000	35,500	388,500	$6\frac{1}{2}$	$6\frac{1}{2}$	5	7,820	724,640
" 18..	3,654	2,145	1,851	......	......	7,650	18,720	2,000	2,870	23,590	338,500	36,000	374,500	$6\frac{3}{8}$	$6\frac{5}{8}$	5	2,870	743,360
" 25..	1,563	5,699	1,653	......	153	9,068	20,570	500	3,450	24,520	320,000	39,500	359,500	$6\frac{1}{4}$	$6\frac{1}{4}$	$4\frac{1}{2}$	3,450	763,930
Nov. 1..	......	......	......	......	......	......	20,660	4,000	3,200	27,860	300,000	36,000	336,000	$6\frac{1}{4}$	$6\frac{1}{4}$	$4\frac{1}{2}$	3,200	784,590
" 8..	298	......	......	......	......	298	18,770	2,500	1,300	22,570	283,500	33,600	316,500	$6\frac{3}{8}$	$6\frac{3}{8}$	$4\frac{1}{2}$	1,300	803,360
" 15..	2,011	......	......	......	4,688	6,699	26,060	7,100	150	33,310	263,000	33,500	296,500	$6\frac{1}{2}$	$6\frac{5}{8}$	$4\frac{1}{2}$	150	829,420
" 22..	4,993	5,264	......	3,783	1,171	15,211	20,840	10,500	250	31,590	252,000	38,500	290,500	$6\frac{3}{4}$	$6\frac{3}{4}$	$4\frac{1}{2}$	250	850,260
" 29..	2,265	2,269	......	......	......	4,534	17,895	1,650	250	19,795	240,000	37,000	277,000	$6\frac{5}{8}$	$6\frac{5}{8}$	$4\frac{1}{2}$	250	868,155
Dec. 6..	3,806	......	1,947	......	......	5,753	21,750	400	250	22,400	226,500	34,500	261,000	$6\frac{1}{2}$	$6\frac{1}{2}$	$4\frac{1}{2}$	250	889,905
" 13..	2,789	......	......	......	4,726	7,515	17,560	1,500	700	19,760	214,000	36,000	250,000	$6\frac{3}{8}$	$6\frac{1}{2}$	$4\frac{1}{2}$	700	907,465
" 20..	2,261	8,923	......	3,192	723	15,099	20,200	1,700	2,100	24,000	197,500	45,500	243,000	$6\frac{1}{4}$	$6\frac{3}{8}$	$4\frac{1}{2}$	2,100	927,665
" 27..	2,011	409	......	558	2,200	5,178	17,770	500	1,000	19,270	184,500	45,000	229,500	$6\frac{1}{4}$	$6\frac{3}{8}$	$4\frac{1}{4}$	1,000	945,435
Average prices & total sales, receipts & stocks.	787,900	82,800	17,029	66,749	57,300	1,011,778	945,435	262,020	78,290	1,285,745				7.19	7.25	4.89	78,290	1,818,144

The semi-weekly Price and Weekly Sales and Receipts at New York, Weekly Exports from New York and Rates of Freight to Liverpool 1st of each month, for the Crop Year ending October 1, 1839.

1838.	Price of New Orleans	Price of Upland.	Sales for week.	Receipts for week.	EXPORTS FOR THE WEEK. To Great Britain.	To France.	North of Europe.	Other Fo'n Ports	Total Exports.	Rates of Freight to Liverpool.	General Remarks.
October 5..	9½@14½	9½@14½								¼@⅜d.	
" 9..	9½@14½	9½@14½	3,650	637	....	428	....	....	428		The cotton market, for a large part of this year, was very much unsettled and depressed by the unfavorable accounts constantly received from Europe, though specie payments having been resumed, prices rallied considerably from the low points of the previous year. The difference between the highest and the lowest prices of the year was 4¼@4½ cents per lb. It was during this period that the Bank of the United States, through its agents, endeavored, with such disastrous results, to control the article.
" 12..	9½@14½	9½@14½									
" 16..	9½@14½	9½@14½	2,350	3,253	411	898	....	....	1,309		
" 19..	9½@14½	9½@14½									
" 23..	9½@14½	9½@14½	2,650	1,483	423	605	....	....	1,028		
" 26..	10@15½	10@14½									
" 30..	10@15½	10@14½	2,800	2,828	860	467	....	....	1,327		
Novem. 2..	10@16	10@14¾								¼@⅜d.	
" 6..	10@16	10@14¾	1,700	1,586	65	653	....	....	718		
" 9..	10@16	10@14¾									
" 13..	10@16	10@14¾	3,400	8,747	14	....	....	....	14		
" 16..	10@16	10@14¾									
" 20..	10@15½	10@14½	2,050	618	315	167	....	....	482		
" 23..	10@15½	10@14½									
" 27..	10@15½	10@14½	3,950	5,997	209	349	....	....	558		
" 30..	10@15½	10@14									On the 6th of June, 1839, a remarkable circular was issued by Messrs. Humphreys & Biddle, which excited universal attention. The circular stated "that an arrangement had been made by which an advance of three-fourths on what may be estimated as the present market value of cotton, say 14 cents per lb., will be made on every bale of cotton in the country, at all the shipping ports, to all holders, the consignments to go forward to Humphreys & Biddle, who, sustained by adequate means on both sides the water, will be able to hold on until prices vigorously rally."
Decem. 4..	10@15	10@14	3,200	3,545	1,206	879	....	....	2,085	¼@⅜d.	
" 7..	11½@15½	11@14½									
" 11..	11½@15½	11@14½	3,900	7,534	334	....	....	....	334		
" 14..	11½@16	11@14½									
" 18..	11½@16½	11@15	3,600	2,841	894	2,034	....	328	3,256		
" 21..	12@16½	11@15									
" 25..	12@16½	11@15	1,600	1,993	566	639	....	....	1,205		
" 28..	12@16½	11½@15¼									
1839.											
January 1..	12@16½	11½@15¼	4,500	11,862	1,218	962	216	....	2,396	5-16@7/16.	
" 4..	12@16½	11½@15¼									
" 8..	12@16½	11½@15¼	6,950	7,913	....	....	....	....	....		
" 11..	12@16½	11½@15¼									
" 15..	12@16½	12@15½	5,900	12,881	81	468	....	....	549		
" 18..	12@16½	12@15½									
" 22..	12½@16¾	12½@15½	4,800	5,046	744	710	....	....	1,454		

January 25..	12½@16¾	12½@15½									This extraordinary effort, however, was fruitless, for values continued to decline, and, before the close of the crop year, fell off further 3@3¼ cents per lb.
" 29..	12½@16¾	12½@15½	4,850	9,153	311	1,229	....	....	1,540		
February 1..	13½@17	12½@15½								9-16@11/16.	
" 5..	13½@17½	13@16	6,350	16,726	3,839	628	52	....	4,519		
" 8..	13½@17½	13@16									
" 12..	13½@17½	13@16	4,400	8,279	233	75	....	....	308		
" 15..	13½@17½	13½@16½									
" 19..	13½@17½	13½@16½	8,000	5,848	785	1,004	....	....	1,789		
" 22..	13½@17½	13½@16½									
" 26..	13½@17½	13½@16½	6,500	15,394	2,629	378	308	....	3,315		
March 1..	13½@17½	13½@16½								⅝@¾d.	
" 5..	13½@17½	13½@16½	1,850	14,093	4,214	4,910	105	....	9,229		
" 8..	14@17½	13½@16½									
" 12..	14@17½	13½@16½	6,500	8,842	2,173	....	280	....	2,453		
" 15..	14@17½	13¾@16½									
" 19..	14@17½	13¾@16½	7,000	11,030	3,266	4,129	32	....	7,427		
" 22..	14@17½	13¾@16½									
" 26..	14@17½	13¾@16½	2,900	12,203	3,356	3,295	....	....	6,651		
" 29..	14@17½	13¼@16									
April 2..	14@17	13@15½	2,450	18,845	3,639	1,979	107	77	5,802	⅝@¾d.	
" 5..	13½@17	13@15½									
" 9..	13½@17	13@15½	900	1,967	2,432	2,473	105	....	5,010		
" 12..	13½@17	13@15½									
" 16..	13½@17	13@15½	5,050	1,284	2,200	1,241	....	....	3,441		
" 19..	14@17½	13¾@16½									
" 23..	14@18	14@17	6,750	14,646	1,777	2,193	256	....	4,226		
" 26..	14@18	14@17									
" 30..	14@18	14@17	9,200	8,310	15,961	1,732	....	....	17,693		
May 3..	14@18	14@17								½@⅝d.	
" 7..	14@18	14@17	2,400	6,579	2,721	....	....	....	2,721		
" 10..	13¾@17½	13¾@16½									
" 14..	14@17¾	13¾@16½	5,300	4,051	7,702	2,277	349	....	10,328		
" 17..	14@17¾	13¾@16½									
" 21..	14@17¾	13¾@16½	7,000	6,807	5,024	....	50	....	5,074		
" 24..	14@17¾	13¾@16½									
" 28..	14@17¾	13¾@16½	2,500	3 590	5,640	135	....	16	5,791		
" 31..	13½@17¾	13¼@16									
June 4..	13@17	12¾@15½	2,500	4,988	7,322	393	....	1	7,716	¼@⅜d.	
" 7..	12¼@16½	12@15½									
" 11..	12@16	12@15	2,800	8,059	5,212	205	30	....	5,447		
" 14..	12@16	12@15									
" 18..	12@16	12@15	3,500	5,831	3,089	1,795	222	....	5,106		

New York Statement for Year 1839—*Concluded*

1839.		Price of New Orleans	Price of Upland.	Sales for week.	Receipts for week.	EXPORTS FOR THE WEEK.					Rates of Freight to Liverpool.
						To Great Britain.	To France.	North of Europe.	Other Fo'nPorts	Total Exports.	
June	21..	12@16	11½@14¾								
"	25..	11½@15½	11½@14¾	2,450	2,375	1,617	306	73	1,129	3,125	
"	28..	11½@15½	11½@14¾								
July	2..	11½@15½	11½@14¾	2,050	3,616	5,873	739	596	763	7,971	¼@⅜d.
"	5..	11½@15½	11½@14¾								
"	9..	11½@15½	11½@14¾	3,050	3,467	....	....	....	....	....	
"	12..	11½@15½	11½@14¾								
"	16..	11½@15½	11½@14¾	4,050	1,540	1,001	....	464	....	1,465	
"	19..	11½@15½	11½@14¾								
"	23..	11½@15½	11½@14¾	2,400	1,896	5,764	1,513	1,069	....	8,346	
"	26..	10¼@14½	10¼@14								
"	30..	10¼@14½	10¼@14	2,650	2,021	3,359	346	156	....	3,861	
August	2..	10@14½	10@14								¼@5-16d
"	6..	10@14½	10@14	2,600	885	2,980	1,001	37	....	4,018	
"	9..	10@14½	10@14								
"	13..	10@14½	10@14	3,500	1,596	1,710	1,156	233	....	3,099	
"	16..	10@14½	10@14								
"	20..	10@14½	10@14	2,450	235	3,062	1,291	281	....	4,634	
"	23..	10@14½	10@14								
"	27..	9½@14	9½@13½	2,950	1,387	823	1,023	106	....	1,952	
"	30..	9½@14	9½@13½								
Septem.	3..	9½@14	9½@13½	3,050	62	1,481	969	615	447	3,512	¼@5-16d
"	6..	9½@14	9½@13½								
"	10..	9½@14	9½@13½	2,450	1,503	....	1,062	....	....	1,062	
"	13..	9½@14	9½@13½								
"	17..	9½@14	9½@13½	5,100	622	1,209	1,703	250	....	3,162	
"	20..	9½@14	9½@13½								
"	24..	9½@14	9½@13½	3,950	865	1,219	1,600	120	....	2,939	
"	27..	9½@14	10@13½								
October	1 .	9½@14	10@13½	2,475	195	1,718	1,355	300	2	3,375	¼@5-16d
Average prices and total sales, receipts and exports.		13.92	13.36	200,875	287,554	122,681	53,394	6,412	2,763	185,250	

GENERAL REMARKS.

Exchange.

Foreign bills were unusually steady; the quotation October 1, '38, was 10@10¼ per cent. premium., and declined later in the month to 9¼@9½; in November the range was from 9¼@10¼; in December, 9¼@10; through January the price was \$4.82@\$4.87 per £; in February, \$4.82@\$4.84; in March, \$4.83@\$4.87; in April the range was from \$4.81 to \$4.85½; in May, \$4.83@\$4.87; in June, from \$4.84 to \$4.90; July, \$4.85@\$4.88; August, \$4.80@\$4.86; September, \$4.82@\$4.90, closing, October 1, at \$4.87@\$4.90.

1840.

No yarn finer than No. 350—the finest spinning seldom exceeds 300 hanks to the pound—was made in England previous to this year. (*See* years 1841 and 1851.)

The following is the number of spindles in use for cotton manufactures alone in the New England States at this time. (*See* year 1854.)

Maine	29,736
New Hampshire	195,173
Massachusetts	665,095
Vermont	7,254
Rhode Island	518,817
Connecticut	181,319
Total	1,597,394

Computed number of lbs. of cotton yarn and twine spun in Great Britain this year, 407,000,000. (*See* year 1835.) Exports of cotton piece goods, 791,000,000 yards.

Cotton imported into Great Britain, 592,000,000 lbs. (*See* year 1700.) No less than 76 per cent. of this quantity was from the United States; 14 per cent. being from the East Indies; 6 per cent. from Brazil; 2½ per cent. from Egypt, and 1½ per cent. from the West Indies and miscellaneous countries.

The census report of this year fixes the number of spindles in operation in the United States, at 2,285,337, of which 1,598,198 were in New England. (*See* years 1850 and 1860.)

North Providence, R. I., now (*see* year 1767) has ten cotton mills in operation.

COTTON CROP OF THE UNITED STATES.

Statement and Total Amount of the Cotton Crop of the United States, for the Year ending September 30, 1840.

	Bales.	Bales.	Total.	Same period 1839.
NEW ORLEANS.				
Export—				
To Foreign Ports	832,625			
Coastwise	124,061			
Stock on hand 1st October, 1840	27,911			
		984,597		
Deduct—				
Stock on hand 1st October, 1839	15,824			
Received from Mobile	15,386			
" Florida	2,568			
" Texas	3,914			
		37,692		
			946,905	568,562
MISSISSIPPI.				
Export from Natchez, &c.—				
To Foreign Ports	....	2,208		
Coastwise	....	4,559		
			6,767	16,432
ALABAMA.				
Export from Mobile—				
To Foreign Ports	354,708			
Coastwise	85,394			
Burnt and lost	6,400			
Stock in Mobile 1st October, 1840	1,737			
		448,239		
Deduct—				
Stock in Mobile 1st October, 1839	1,464			
Received from Florida	1,050			
		2,514		
			445,725	251,742
FLORIDA.				
Export—				
To Foreign Ports	61,049			
Coastwise	75,558			
Stock on hand 1st October, 1840	300			
		136,907		
Deduct—				
Stock on hand 1st October, 1839	....	650		
			136,257	75,177
GEORGIA.				
Export from Savannah—				
To Foreign Ports—Uplands	199,842			
" Sea Island	8,108			
Coastwise	76,299			
	284,249			
From Darien—				
To New York	10,537			
Stock in Savannah 1st October, 1840	2,011			
" Augusta and Hambro', 1st October, 1840	3,730			
		300,527		
Deduct—				
Stock in Savannah and Augusta, 1st October, 1839	....	7,834		
			292,693	205,112

Statement and Total Amount of the Cotton Crop of the United States, for the Year ending September 30, 1840.—*Concluded.*

	Bales.	Bales.	Total.	Same period 1839.
SOUTH CAROLINA.				
Export from Charleston—				
To foreign ports—Uplands	228,191			
" Sea Island	19,310			
Coastwise	60,178			
	307,679			
From Georgetown—				
To New York	13,200			
Stock in Charleston 1st October, 1840	4,153			
		325,032		
Deduct—				
Stock in Charleston 1st October, 1839	4,706			
Received from Savannah	4,663			
" Florida and Key West	2,469			
		11,838		
			313,194	210,171
NORTH CAROLINA.				
Export—				
To Foreign Ports	65			
Coastwise	9,729			
Stock on hand 1st October, 1840	200			
		9,994		
Deduct—				
Stock on hand 1st October, 1839		600		
			9,394	11,136
VIRGINIA.				
Export—				
To Foreign Ports	7,987			
Coastwise	6,263			
Manufactured	9,000			
Stock on hand 1st October, 1840	900			
		24,150		
Deduct—				
Stock on hand 1st October, 1839		500		
			23,650	22,200
Received at Philadelphia and Baltimore, overland			3,250	
Total crop of the United States			2,177,835	1,360,532
Crop of last year			1,360,532	
Increase			817,303	

Export to Foreign Ports, from October 1, 1839, *to September* 30, 1840.

FROM	To Great Britain.	To France.	To North* of Europe.	Other Foreign Ports.	Total.
New Orleans (bales)	510,690	239,774	23,204	58,957	832,625
Mississippi (Natchez)	*2,208	....	....	....	2,208
Alabama	257,985	80,528	11,824	4,371	354,708
Florida	49,952	11,097	....	....	61,049
Georgia (Savannah and Darien)	189.372	17,942	....	636	207,950
South Carolina	153,042	62,917	29,453	2,089	247,501
North Carolina	65	....	....	....	65
Virginia	4,455	2,676	830	26	7,987
Baltimore	1,707	41	753	....	2,501
Philadelphia	3,076	30	175	404	3,685
New York	73,611	32,092	34,590	11,923	152,216
Boston	628	368	2,403	109	3,508
Grand total	1,246,791	447,465	103,232	78,515	1,876,003
Total last year	798,418	242,243	21,517	12,511	1,074,689
Increase	448,373	205.222	81,715	66,004	801,314

* The remainder of the shipments from Mississippi are included in the export from New Orleans.

Growth.

Total crop of 1824–5,	560,000	bales.
" 1825–6,	710,000	"
" 1826–7,	937,000	"
" 1827–8,	712,000	"
" 1828–9,	857,744	"
" 1829–30,	976,845	"
" 1830–1,	1,038,848	"
" 1831–2,	987,477	"
" 1832–3,	1,070,438	"
" 1833–4,	1,205,394	"
" 1834–5,	1,254,328	"
" 1835–6,	1,360,725	"
" 1836–7,	1,422,930	"
" 1837–8,	1,801,497	"
" 1838–9,	1,360,532	"
" 1839–40,	2,177,835	"

Consumption.

Total crop of the United States, as above stated			2,177,835 bales.
Add—Stocks on hand at the commencement of the year, October 1, 1839 in the Southern ports	31,784		
" Northern ports	20,460		
		52,244	
Makes a supply of			2,230,079
Deduct therefrom the export to Foreign ports	1,876,003		
Less Texas and other foreign, included	6,509		
		1,869,494	
Stocks on hand at the close of the year, October 1, 1840.			
In the Southern ports	40,942		
In the Northern ports	17,500		
		58,442	
Burnt and lost at Mobile	6,400		
" New York	550		
		6,950	
			1,934,886

Quantity consumed by and in the hands of manufacturers,			1839–40..........bales.	295,193
"	"	"	1838–9................	276,018
"	"	"	1837–8................	246,063
"	"	"	1836–7................	222,540
"	"	"	1835–6................	236,733
"	"	"	1834–5................	216,888
"	"	"	1833–4................	196,413
"	"	"	1832–3................	194,412
"	"	"	1831–2................	173,800
"	"	"	1830–1................	182,142
"	"	"	1829–30...............	126,512
"	"	"	1828–9................	118,853
"	"	"	1827–8................	120,593
"	"	"	1826–7................	103,483

Note.—It will be observed by the above statement, that there is a very large increase in the crop compared with last year; the quantity also exceeds that of any previous year by 376,338 bales. Of the new crop, now gathering, about 30,000 bales were received previous to 1st inst. principally at New Orleans.

It will be seen also that we have deducted from the New Orleans statement, the quantity received at that port from Texas—Texas being a foreign country.

Our estimate of the quantity taken for consumption, does not include any cotton manufactured in the States south and west of Virginia, nor any in that State, except in the vicinity of Petersburg and Richmond.

ANNUAL REVIEW.

From the New Orleans Price Current, 1839—40.

The production of cotton last season has been unprecedented, far exceeding the most sanguine estimates of the period last year. A combination of favorable circumstances during its growth, and almost entire absence of frost in the succeeding Fall and Winter months, enabling planters to continue the picking for a long time beyond the usual period of its termination, are the causes of the immense yield, and it may be many years ere a season combining so many advantages may again occur. With such a vast supply, low prices have, of course, prevailed throughout the season, the periods at which they were lowest being the months of December, March and April, when "fair" cottons could have been purchased at 7½ to 7¾ cents per pound, but more favorable accounts from Europe, and the very heavy amounts taken out of the market by speculators, and stored for low rates of freight to Europe, produced gradual improvement, until the prices of fair cotton reached 10 cents per pound. With respect to the extent of the coming crop, it would be difficult, at this early period, to form an opinion, so much depending on the weather for some time to come. As much ground has been used in the cultivation of the plant as last year, but various causes have combined to produce very serious injury. In the low lands, a large extent of country was overflowed, and much cotton destroyed.

Throughout the whole of Louisiana, great injury has been done by the army worm. From the Atlantic States we hear complaints of heavy rains, by which the crops will be much retarded. What the extent of damage done from the various causes will be, we must leave time to determine.

LIVERPOOL STATEMENT FOR 1840.

UNITED STATES—1839–1840.

Stock, 1st October, 1839	52,000	Export	1,881,000
Receipts	2,178,000	Consumption	291,000
		Stock, 1st October, 1840	58,000
Bales	2,230,000	Bales	2,230,000

CONSUMPTION. Tot. Europe.	Continent.	France.	Gt. Britain.	Liverpool.		Liverpool.	Gt. Britain.	France.	Continent.	Tot. Europe.
					Stock Jan. 1, 1840, in					
					United States, Bales	158,000	176,000	50,000	18,000	244,000
					Brazil	10,000	10,000	7,000	3,000	20,000
					West Indies	4,000	7,000	5,000	4,000	16,000
					East Indies	22,000	60,000	2,000	3,000	65,000
					Egypt	12,000	12,000	12,000	43,000	67,000
					Bales	206,000	265,000	76,000	71,000	412,000
					IMPORT.					
1,599 000	151,000	377.000	1,071,000	1,011,000	United States	1,166,000	1,245,000	412,000	202,000	1,803,000
84,000	5,000	9,000	70,000	70,000	Brazil	84,000	84,000	3,000	4,000	90,000
50,000	20,000	15,000	15,000	10,000	West Indies	20,000	25,000	15,000	22,000	61,000
184,000	58,000	5,000	121,000	86,000	East Indies	109,000	216,000	9,000	58,000	226,000
98,000	42,000	28,000	28,000	28,000	Egypt	37,000	37,000	27,000	32,000	96,000
2,015,000	276,000	434,000	1,305,000	1,205,000	Bales	1,416,000	1,607,000	466,000	318,000	2,276,000
....	4,000	8,000	103,000	51,000	Export.					
673,000	109,000	100,000	464,000	366,000	Stock Dec. 31. Stock above,	206,000	265,000	76,000	71,000	412,000
2,688,000	389,000	542,000	1,872,000	1,622,000	Total supply, bales	1,622,000	1,872,000	542,000	389,000	2,688,000

COTTON AT LIVERPOOL. YEAR 1840.

Week Ending.	Receipts.						Sales.				Stocks.			Prices.			Actual Export	Con-sumption.
	Americ'n.	E. I.	Egypt.	Brazil.	Other.	Total.	Con-sumption.	Specu-lation.	Export	Total.	Amer'n.	Other.	Total.	Mid. Up.	Mid. Orl.	Dhol		
Jan. 3..	11,974	......	......	....	2,693	14,667	10,900	2,000	1,000	13,900	154,500	46,000	200.500	6½	6½	4¼	1,000	10,900
" 10..	6,545	......	1,708	....	638	8,891	24,050	1,500	250	25,800	140,500	45,000	185,500	6⅜	6⅜	4¼	250	34,950
" 17..	35,116	......	......	4,412	2,226	41,754	16,320	....	2,600	18,920	170,000	47,500	217.500	5⅞	5⅞	3¾	2,600	51,270
" 24..	12,932	......	1,964	2,847	3,675	21,418	24,130	6,000	2,150	32,280	158,500	54,000	212,500	5¾	5¾	3¾	2,150	75,400
" 31..	18.342	......	1,115	2,413	1,034	22,904	26,150	3,500	800	30.450	151,500	54,500	206,000	5¾	5⅞	3¾	800	101,550
Feb. 7..	19,502	......	......	....	77	19.579	25,060	3,500	2,000	30,560	148,000	51,000	199.000	5¾	5⅞	3¾	2,000	126,610
" 14..	37,036	......	......	....	3,595	40,631	22,880	3,300	1,500	27,680	161,500	52,000	213,500	5¾	5⅞	3¾	1,500	149,490
" 21..	5,139	3,103	1,240	....	......	9,482	25,920	4,700	400	31,020	144,000	52,000	196,000	5¾	5⅞	3¾	400	175,410
" 28..	2,865	1,854	1,887	....	......	6,606	21,330	1,450	1,900	24,680	128,500	50.500	179,000	5⅝	5¾	3¾	1,900	196,740
Mch 6..	3,686	......	2,532	....	......	6,218	17,900	1,000	1,350	20,250	117,000	49,000	166,000	5½	5⅝	3¾	1,350	214,640
" 13..	14 916	2,245	564	....	444	18,169	13,020	500	530	14,050	121,500	49,000	170,500	5½	5⅝	3¾	530	227.660
" 20..	64,414	......	......	....	6,171	70,585	18,090	1,200	880	20,170	171,000	52.000	223,000	5½	5⅝	4	880	245,750
" 27..	11,062	......	......	....	2,366	13,428	32,800	1,500	1,000	35,300	153.000	50,000	203,000	5½	5½	4	1,000	278,550
April 3..	35,092	4,635	1,701	337	712	42,477	22,590	5,150	4,500	32,240	162,000	51,000	313,000	5½	5½	4	4,500	301,140
" 10..	19,732	6,644	......	1,104	51	27,531	24,170	2,000	1,500	27,670	159,500	65,500	215,000	5⅜	5⅜	4	1,500	325,310
" 16..	49,522	......	......	....	184	49,706	22,800	2,000	2,500	27,300	188,000	51,500	239,500	5⅜	5⅜	4	2,500	348,110
" 24..	58,136	2,572	......	....	171	60,879	25,430	8,350	1,000	34,780	221,500	53,000	274,500	5½	5½	4	1,000	373,540
May 1..	30,446	......	......	....	4,418	34,864	28,280	4,500	1,000	33,780	220,500	53,500	274,000	5½	5½	4	1,000	401,820
" 8..	14,314	......	......	2,350	60	16.724	31,090	8,500	1,200	40,790	204,000	54,500	258,500	5½	5½	4	1,200	432,910
" 15..	15,085	......	......	....	1,569	16,654	23,230	4,800	1,000	29,030	198,500	52,500	251,000	5½	5½	4	1,000	456,140
" 22..	60,451	......	......	....	592	61,043	18,930	2,000	2,500	23,430	241,000	48,500	290,500	5½	5½	4	2,500	475,070
" 29..	84,724	1,860	......	2,158	326	89,068	19,960	500	600	21,060	304,500	51,500	356,000	5¼	5¼	4	600	495,030
June 5..	12,586	8,058	......	1,595	40	22,279	22,075	2,100	1,250	25,425	295,000	57.500	352 500	5¼	5⅛	4	1,250	517,105
" 12..	8,386	......	......	1,117	2,059	11,562	25,320	4,000	1,700	31.020	279,000	58,500	337,500	5⅛	5⅛	3¼	1,700	542,425
" 19..	70,511	......	1,089	....	......	71,600	26,090	10,600	1,570	38,260	326,000	56.500	382,500	5⅛	5⅛	3¼	1.570	568,515
" 26..	67.109	......	......	....	2,275	69,384	24.790	8.500	1,100	34,390	274,000	54,000	428,000	5⅛	5⅛	3¼	1,100	593,305
July 3..	49,639	......	......	....	30	49,669	26,460	16,500	1,530	44,490	394 000	50,000	444,000	5¼	5⅜	3¼	1,530	619,765
" 10..	36,606	......	......	2,413	......	39,019	21,000	5,500	1,550	28,050	411,000	50,000	461,000	5¼	5⅜	3¼	1,550	640,765
" 17..	5,446	......	......	2,007	......	7,453	22,660	7,550	940	31,150	395,000	49,500	444.500	5¼	5⅜	3¼	940	663,425
" 24..	90,783	......	......	3,946	957	95,686	22,920	3,300	1,100	27,320	464,500	51,500	516,000	5⅛	5¼	3¼	1,100	686,345
" 31..	17,379	2,479	......	2,707	174	22,739	22,780	6,500	1,500	30,780	456,500	54,000	510,500	5⅛	5¼	3¼	1,500	709,125
Aug. 7..	4,905	2,320	......	....	1,464	8,689	27,200	11,500	1,250	39,950	435,500	55,500	+90,000	5⅛	5¼	3½	1,250	736,325
" 14..	28,338	......	......	....	3,079	31,417	24,450	18,500	3,000	45,950	440,000	53,500	493.500	5⅜	5½	3¾	3,000	760,775
" 21..	20,388	......	......	4,193	1,765	26,346	19,900	2,200	250	22,350	443,500	56,000	499,500	5¼	5⅜	3⅞	250	780,675

Aug. 28..	27,761	5,630	704	643	586	35,324	15,600	1,500	600	17,700	459,000	60,000	519,000	5¼	5⅜	3⅞	600	796,275
Sep. 4..	6,408	7,190	......	......	80	13,678	19,250	900	750	20,900	446,000	63,500	509,500	5¼	5⅜	3⅞	750	815,525
" 11..	5,314	......	......	3,139	......	8,453	13,030	1,100	700	14,830	440,500	63,000	503,500	5¼	5⅜	3¾	700	828,555
" 18..	12,416	3,424	......	3,186	1,644	20,670	19,480	3,000	480	22,960	437,000	68,000	505,000	5¼	5⅜	3¾	480	848,035
" 25..	4,585	......	......	1,047	......	5,632	25,320	11,050	250	36,620	420,500	64,500	485.000	5⅜	5½	3¾	250	873,355
Oct. 2..	27,799	7,566	......	1,620	517	37,502	18,660	8,300	100	27,060	432,500	70,000	502,500	5⅜	5½	3¾	100	892,015
" 9..	795	2,622	......	1,934	430	5,781	22,930	1,200	......	24,130	415,500	70,500	486,000	5⅜	5½	3¾	......	914,945
" 16..	......	3,626	......	......	115	3,741	16,640	350	140	17,130	402,000	70,500	472,500	5⅜	5⅜	3¾	140	931,585
" 23..	4,982	6,255	......	......	359	11,596	14,920	1,500	......	16,420	395,500	74,000	469,500	5¼	5⅜	3¾	......	945,505
" 30..	4,919	4,663	11,486	1,263	......	22,331	25,280	3,000	200	28,480	375,500	88,000	463,500	5¼	5⅜	3¾	200	971,785
Nov. 6..	1,928	1,911	......	660	2,707	7.206	16,430	1,350	......	17,780	364,500	89.000	453,500	5¼	5⅜	3¼	......	988,215
" 13..	16,159	4,153	......	4,032	3,061	27,405	21,090	750	......	21,840	364,000	96,000	460,000	5¼	5⅜	3⅝	......	1,009,305
" 20..	2,095	......	1,891	3,103	......	7,089	17,260	500	......	17,760	352.000	97,500	449.500	5¼	5⅜	3⅝	......	1,026,565
" 27..	210	4,554	......	2,569	......	7,333	18,500	500	350	19,350	338,500	99,500	438,000	5¼	5⅜	3⅝	350	1,055,065
Dec. 3..	9,983	......	......	4,056	1,150	15,189	22,090	3,500	......	25,590	328,000	101,000	429,000	5¼	5⅜	3⅝	......	1,077,155
" 11..	14.289	2,106	3,424	3,184	490	23,493	27,560	3,000	......	30,560	320,500	104,500	425,000	5⅜	5½	3⅝	......	1,104,715
" 18..	2,520	3,173	......	......	......	5,693	17,800	4,500	......	22,300	302 500	100,000	402,500	5½	5⅝	3⅝	......	1,121,515
" 24..	......	......	3,289	......	......	3,289	46,760	12,000	350	49,110	280,000	98,500	378,500	5¾	5⅞	3¾	350	1,168,275
" 31..	......	......	......	......	......	......	......	6,200	......	25,350	......	......	......	......	......	......	......	......
Average prices & total sales, receipts & stocks.	1,155,270	92,643	34,594	64,035	53,984	1,400,528	1,168,290	229,850	52,820	1,450,940				5.42	5.5	3.72	52,820	2,246,673

The semi-weekly Price and Weekly Sales and Receipts at New York, Weekly Exports from New York and Rates of Freight to Liverpool 1st of each month, for the Crop Year ending October 1, 1840.

1839.	Price of New Orleans	Price of Upland.	Sales for week.	Receipts for week.	EXPORTS FOR THE WEEK.					Rates of Freight to Liverpool.
					To Great Britain.	To France.	North of Europe.	Other Fo'n Ports	Total Exports.	
October 4..	9½@14	10@13½								¼@5-16d
" 8..	9½@14	10@13½	2,400	1,213	....	....	....	....	....	
" 11..	9½@14	10@13½								
" 15..	9½@14	10@13½	1,900	429	404	2,406	1,175	....	3,985	
" 18..	9½@14	10@13½								
" 22..	9½@14	10@13½	1,750	2,268	39	946	218	....	1,203	
" 25..	9½@14	10@13½								
" 29..	9½@14	10@13	950	1,317	161	1,266	....	41	1,468	
Novem. 1..	9½@13	9½@12¼								¼@5-16d
" 5..	9½@13	9½@12¼	2,100	1,984	632	548	612	....	1,792	
" 8..	9@12½	9@12								
" 12..	9@12½	9@11½	3,200	6,969	655	104	16	....	775	
" 15..	9@12½	9@11¼								
" 19..	9@12	9@10¾	2,850	4,207	258	852	197	....	1,307	
" 22..	9@12	9@10¾								
" 26..	9@12	9@10¾	3,000	1,367	703	1,517	....	....	2,220	
" 29..	9@12	9@10¾								
Decem. 3..	9@12	9@10¾	2,350	2,342	2,421	2,982	144	562	6,109	¼@5-16d
" 6..	9@12	9@10¾								
" 10..	9@12	9@10¾	1,400	3,250	....	....	65	....	65	
" 13..	9@12	9@10¾								
" 17..	9@12	9@10¾	3,050	6,189	470	1,208	205	....	1,883	
" 20..	9@12	9@10¾								
" 24..	9@12	9@10¾	3,900	396	244	440	....	....	684	
" 27..	9@12	9@10¾								
1840.										
January 1..	9@12	9@10¾	1,950	2,920	1,003	878	1,072	594	3,547	½@⅝d.
" 3..	9@12	9@10¾								
" 7..	9@12	9@10¾	2,900	7,824	650	195	425	....	1,270	
" 10..	9@11½	8½@10½								
" 14..	9@11½	8½@10½	3,950	4,345	716	....	....	540	1,256	
" 17.	8¼@11	8¼@10								
" 21..	8@11	8@9¾	4,500	7,286	880	....	....	....	880	

GENERAL REMARKS.

This was a year of great depression and low prices. Early in October, the entanglements with which the Bank of the United States was hampered, already referred to, compelled it to suspend specie payments, and ultimately to close its doors. Subsequently, the suspension became quite general through the country, though the banks of New York and some of the stronger institutions elsewhere were enabled to weather the financial storm. Exchange was disturbed, and altogether these perturbations operated to the prejudice of trade. Manufacturers generally suspended operations, wholly or in part, and shippers purchased to a comparatively limited extent. At one period, in March, sixty days intervened between the dates of European advices; vessels became scarce, and freights ranged unusually high.

January 24..	8@11	8@10								
" 28..	8@11	8@10	4,400	2,692	464	1,508	....	....	1,972	
" 31..	8@11	8@10								
February 4..	7½@11	7½@10	4,250	3,673	1,776	989	858	....	3,623	⅝@¾d.
" 7..	7½@11	7½@10								
" 11..	7½@11	7½@10	4,150	3,026	2,733	....	557	825	4,115	
" 14..	7½@11	7@10								
" 18..	7½@11	7@10	4,350	10,153	1,112	672	1,152	545	3,481	
" 21..	7½@11	7@10								
" 25..	7½@11	7@10	3,300	3,389	2,784	379	2,327	50	5,540	
" 28..	7½@11	7@9¾								
March 3..	7½@11	7@9¾	6,950	2,656	1,621	....	809	102	2,532	¾@⅞d.
" 6..	7½@11	7@9¾								
" 10..	7½@11	7@9¾	2,350	8,341	436	159	439	....	1,034	
" 13..	6¾@10½	6¾@9½								
" 17..	6¾@10½	6¾@9¼	3,400	11,784	2,164	....	1,572	1,190	4,926	
" 20..	6¾@10½	6¾@9¼								
" 24..	6¾@10½	6¾@9¼	4,200	15,262	2,634	374	2,332	....	5,340	
" 27..	6½@10¼	6¾@9¼								
" 31..	6¼@10¼	6¼@9¼	7,500	9,776	1,975	....	264	....	2,239	
April 3..	6¼@10¼	6¼@9¼								¾@⅞d.
" 7..	6¼@10¼	6¼@9¼	5,600	11,618	602	....	1,308	....	1,910	
" 10..	6@10	6@9								
" 14..	6@10	6@9	5,800	3,842	866	1,976	562	440	3,844	
" 17..	6@10	6@9								
" 21..	6@10	6@9	6,900	7,393	3,785	1,026	1,850	....	6,661	
" 24..	6¼@10¼	6¼@9								
" 28..	6¼@10¼	6¼@9	7,050	8,587	3,335	....	1,927	....	5,262	
May 1..	6¼@10¼	6@9								9-16@11/16.
" 5..	6@10¼	6@9	6,750	8,949	6,343	1,231	736	1,157	9,467	
" 8..	6@10¼	6@9								
" 12..	6@10¼	6@9	6,050	1,811	1,270	1,755	721	1,734	5,480	
" 15	6@10¼	6@9								
" 19..	6@10¼	6@9	6,900	5,699	4,142	285	1,416	....	5,843	
" 22..	6@10½	6@9½								
" 26..	6@10½	6@9½	5,100	8,131	1,188	....	2,438	....	3,626	
" 29..	6@11	6@10								
June 2..	6@11	6@10	6,450	1,562	5,214	324	796	29	6,363	½@⅝d.
" 5..	6@11	6@10								
" 9..	6@11	6@10	5,600	6,668	415	142	1,079	754	2,390	
" 12..	6@11	6@10								
" 16..	6@11	6@10	5,750	5,708	1,344	1,253	779	....	3,376	

New York Statement for 1840.—*Concluded.*

1840.		Price of New Orleans	Price of Upland.	Sales for week.	Receipts for week.	EXPORTS FOR THE WEEK.					Rates of Freight to Liverpool.
						To Great Britain.	To France.	North of Europe.	Other Fo'n Ports	Total Exports.	
June	19..	6@11	6@10								
"	23..	6@11	6@10	5,150	6,639	3,196	277	224	164	3,861	
"	26..	6@11	6@10								
"	30..	6½@11	6½@10	5,800	5,801	4,019	601	531	....	5,151	
July	3..	6½@11	6½@10								⅜@½d.
"	7..	6½@11	6½@10	3,020	3,025	3,305	749	327	....	4,381	
"	10..	6½@11	6½@10								
"	14..	6½@11	6½@10	4,400	4,115	1,426	51	54	....	1,531	
"	17..	6½@11	6½@10								
"	21..	7@11	7@10	3,000	2,554	283	1,179	873	870	3,205	
"	24..	7@11	7@10								
"	28..	7@11	7@10	5,200	1,050	628	....	365	770	1,763	
"	31..	7@11	7@10								
August	4..	7@11	7@10	3,800	5,578	777	1,418	1,646	....	3,841	⅜@½d.
"	7..	7@11	7@10								
"	11..	7@11	7@10	2,250	1,680	649	....	....	....	649	
"	14..	7@11	7@10								
"	18..	7@11	7@10	3,000	887	631	820	525	947	2,923	
"	21..	7@11	7@10								
"	25..	7@11	7@10	3,700	887	1,520	390	488	....	2,398	
"	28..	7½@11½	7½@10½								
Septem	1..	7½@11½	7½@10½	4,250	1,393	448	....	100	116	664	¼@⅜d.
"	4..	7½@11½	7½@10½								
"	8..	7½@11½	7½@10½	2,100	2,500	606	....	470	....	1,076	
"	11..	7½@11½	7½@10½								
"	15..	7½@11½	7½@10½	1,100	3,429	317	615	....	5	937	
"	18..	7½@11½	7½@10½								
"	22..	7½@11½	7½@10½	1,750	2,066	....	160	358	....	518	
"	25..	7½@11½	7½@10½								
"	29..	7½@11½	7½@10½	1,950	2,230	387	417	1,066	....	1,870	5-16@7/16.
October	2..	7½@11½	7½@10½								
Average price and total sales, receipts and exports.		9.43	8.92	205,420	238,860	73,631	32,092	35,078	11,435	152,236	

GENERAL REMARKS.

Exchange.

Bills on London opened, October 1, '39, at \$4.87@4.90 per £, declining gradually to \$4.77@\$4.82; in November the range was from \$4.58@\$4.68 up to \$4.80@\$4.84; in December, \$4.83@\$4.87; January, \$4.75@\$4.82; February, \$4.80@\$4.84; March, \$4.75@\$4.80; April, \$4.78@\$4.80; May, \$4.77@\$4.81; June, \$4.74@\$4.78; July, \$4.73@\$4.76; September, \$4.76@\$4.81, closing, October 1, at \$4.80@\$4.81 per £.

1841.

The Messrs. Houldsworth of Manchester spun yarn as fine as No. 450. (*See* years 1840 and 1851.)

A steam factory was incorporated at Beverly, Mass., this year, with large capital. (*See* year 1808.)

Cotton exported from the United States into Great Britain, 530,204,100 lbs., of which 6,237,424 lbs. were Sea Island. (*See* years 1790 and 1791.)

COTTON CROP OF THE UNITED STATES.

Statement and Total Amount of the Cotton Crop of the United States, for the Year ending September 30, 1841.

	Bales.	Bales.	Total.	1840.
NEW-ORLEANS.				
Export—				
To Foreign Ports	656,816			
Coastwise	161,448			
Burnt and damaged	2,000			
Stock on hand, 1st October, 1841	31,576			
		851,840		
Deduct—				
Stock on hand, 1st October, 1840	27,911			
Received from Mobile	5,418			
" " Florida	508			
" " Texas	4,408			
		38,245		
			813,595	946,905
MISSISSIPPI.				
Export from Natchez, etc—				
Coastwise (Remainder included in New Orleans)	....	....	1,085	6,767
ALABAMA.				
Export from Mobile—				
To Foreign Ports	216,239			
Coastwise	103,837			
Burnt and lost	1,170			
Stock in Mobile, 1st October, 1841	1,831			
		323,077		
Deduct—				
Stock in Mobile, 1st October, 1840	1,737			
Received from Florida	486			
" " Texas	153			
		2,376		
			320,701	445,725
FLORIDA.				
Export—				
To Foreign Ports	32,297			
Coastwise	59,555			
Burnt and lost	1,400			
Stock on hand, 1st October, 1841	600			
Deduct—		93,852		
Stock on hand, 1st October, 1840	....	300		
			93,552	136,257

Statement and Total Amount of the Cotton Crop of the United States, for the Year ending September 30, 1841.—*Concluded.*

	Bales.	Bales.	Total.	1840.
GEORGIA.				
Export from Savannah—				
To Foreign Ports—Uplands	80,496			
Sea Island	5,100			
Coastwise—Uplands	56,412			
Sea Island	867			
	142,875			
From Darien—				
To New York	5,630			
Burnt	600			
Stock in Savannah, 1st October, 1841	1,456			
" Augusta and Hambro', 1st October, 1841	4,127			
		154,688		
Deduct—				
Stock in Savannah and Augusta, 1st October, 1840		5,741		
			148,947	292,693
SOUTH CAROLINA.				
Export from Charleston—				
To Foreign Ports—Uplands	149,272			
Sea Island	12,991			
Coastwise—Uplands	62,989			
Sea Island	970			
	226,222			
From Georgetown—				
To New York	12,043			
Burnt and lost	750			
Stock in Charleston, 1st October, 1841	3,708			
		242,723		
Deduct—				
Stock in Charleston, 1st October, 1840	4,153			
Received from Savannah	9,562			
" " Florida and Key West	1,608			
		15,323		
			227,400	313,194
NORTH CAROLINA.				
Export—				
All Coastwise	7,765			
Stock on hand, 1st October, 1841	300			
		8,065		
Deduct—				
Stock on hand, 1st October, 1840		200		
			7,865	9,394
VIRGINIA.				
Export—				
To Foreign Ports	4,732			
Coastwise	4,500			
Manufactured	15,000			
Stock on hand, 1st October, 1841	420			
Deduct—		24,652		
Stock on hand, 1st October, 1840	900			
Received from Southern ports	2,952			
		3,852		
			20,800	23,650
Received at Philadelphia and Baltimore, overland			1,000	3,250
Total Crop of the United States			1,634,945	2,177,835
Crop of last year				2,177,835
Decrease				542,890

Export to Foreign Ports, from October 1, 1840, *to September* 30, 1841.

FROM	To Great Britain.	To France.	To North of Europe.	Other F'n Ports.	Total.
New Orleans	427,472	182,310	10,091	36,943	656,816
* Mississippi (Natchez)	......	......	......	......	......
Alabama	149,854	57,204	4,357	4,824	216,239
Florida	20,113	11,349	740	95	32,297
Georgia (Savannah and Darien)	82,842	2,283	......	471	85,596
South Carolina	101,564	35,886	22,305	2,520	162,275
North Carolina	......	......	......	......	......
Virginia	2,800	1,724	150	58	4,732
Baltimore	177	......	40	......	217
Philadelphia	1,556	11	138	229	1,934
New York	71,696	57,847	16,315	3,711	149,569
Boston	668	162	2,143	629	3,602
Grand total	858,742	348,776	56,279	49,480	1,313,277
Total last year	1,246,791	447,465	103,232	78,515	1,876,003
Decrease	388,049	98,689	46,953	29,035	562,726

* The shipments from Mississippi are included in the export from New Orleans.

Growth

Total crop of 1824–5, 560,000 bales.
" 1825–6, 710,000 "
" 1826–7, 937,000 "
" 1827–8, 712,000 "
" 1828–9, 857,744 "
" 1829–30, 976,845 "
" 1830–31, 1,038,848 "
" 1831–32, 987,477 "
" 1832–33, 1,070,438 "
" 1833–34, 1,205,394 "
" 1834–35, 1,254,328 "
" 1835–36, 1,360,725 "
" 1836–37, 1,422,930 "
" 1837–38, 1,801,497 "
" 1838–39, 1,360,532 "
" 1839–40, 2,177,835 "
" 1840–41, 1,634,945 "

Consumption.

Total crop of the United States, as above stated .... 1,634,945 bales

Add—Stocks on hand at the commencement of the year, 1st October, 1840.—In the Southern ports 40,942
" Northern " 17,500
58,442

Makes a supply of 1,693,387

Deduct therefrom—The export to Foreign ports 1,313,277
Less Texas and other foreign 5,900
1,307,377

Stocks on hand at the close of the year 1st October, 1841.—
In the Southern ports 44,018
" Northern " 38,050
82,068

Brought forward		1.389,445	
Burnt and lost at New Orleans	2,000		
" Mobile	1,170		
" Apalachicola	1,400		
" Savannah	600		
" Charleston	750		
" New-York	734		
		6,654	
			1,396,099

Quantity consumed by and in the hands of manufacturers,	1840–41	297,288
" " "	1839–40	295,193
" " "	1838–9	276,018
" " "	1837–8	246,063
" " "	1836–7	222,540
" " "	1835–6	236,733
" " "	1834–5	216,888
" " "	1833–4	196,413
" " "	1832–3	194,412
" " "	1831–2	173,800
" " "	1830–1	182,142
" " "	1829–30	126,512
" " "	1828–9	118,853
" " "	1827–8	120,593
" " "	1826–7	103,483

It will be seen that we have deducted from the New Orleans statement the quantity received at that port from Texas—Texas being a foreign country.

Our estimate of the quantity taken for consumption, does not include any cotton manufactured in the States south and west of Virginia, nor any in that State, except in the vicinity of Petersburg and Richmond.

Of the new crop, now gathering, about 32,000 bales were received previous to 1st inst.; of which 28.175 were received at New Orleans.

It is our intention hereafter to make up our statement of the crop to the 1st September, in conformity with the plan adopted in the Southern ports.

ANNUAL REVIEW.

From the New Orleans Price Current, 1840–41.

The history of the Cotton trade during the year just closed again affords a most conclusive proof of the futility of calculations, based solely with reference to a short production; and in this respect it assimilates closely to the disastrous season of 1839, the crop of the year previous to which had been the most productive hitherto known. The last four years have been remarkable for great fluctuation in the yield of the plant, as will be perceived by the following statement: The crop of the year 1837–38 produced 1,800,000 bales—increase, 26½ per cent.; 1838–39, 1,360,000 bales—decrease, 24 per cent.; 1839–40, 2,182,000 bales—increase, 60¼ per cent.; 1840–41, 1,600,000—decrease, 27 per cent., giving an average for the last four years of 1,735,500 bales.

At the commencement of this season the conviction had already become general that the growth would fall considerably short of the extraordinary production of the year previous, and about 17 to 1,800,000 bales was the general estimate of its extent. Under this influence the market for new Cotton opened on the 9th of September at 10 cents per lb. for "fully fair," but, in the face of a succession of discouraging advices from England, it was found impossible to support the opening rates, and prices gradually declined until they reached the extreme point of depression on the 31st of October, when "fair" Cotton ruled at 9 cents per lb.

Business continued in this state until the 24th of November, when the increasing certainty of a greatly diminished production, combined with more favorable intelligence from Europe, caused an advance in our own market, which continued gradually improving, under the influences above alluded to, until the 25th of May, when ordinary Cotton ranged from 9½ to 9¾, and fair was quoted at 12 cents per lb., being the highest which the market touched. A combination of events, among which the hostile position of the United States and England, and the presumed adjustment of the China question, are most prominent, tended to inflate prices in Liverpool until fair Uplands reached 7⅛ to 7¼ per lb.; but on the removal of the causes of excitement by the pacific turn which the affairs of the two countries took, and the improbability of any immediate termination of the war in China, all speculative demand was checked; the large stocks in the hands of speculators were again thrown on the market, which rapidly assumed a declining tendency; trade in Manchester and other manufacturing districts, already bad, became alarmingly so; and in the beginning of May the expedient of working short time was partially resorted to by many of the leading spinners. Under these circumstances the demand for the raw material was confined to the immediate wants of consumptions; trade progressed from bad to worse; working short time was adopted as a general measure, and failure amongst spinners became a daily occurrence. Stocks of the raw material in Liverpool accumulated in the meantime in a manner almost beyond precedent, and prices, of course, gave way. As a result, consequent on this state of trade in England, our own market gradually drooped, until ordinary Cottons fell to 8 a 8¼ cents on the 29th of June, being a decline of 1½ cents on previous quotations; and it is only attributable to the very small stock remaining on hand that prices have not been more seriously affected. With regard to the growing crops, the accounts from various sections of the country are contradictory, though, from a general review of them, we are led to the belief that the production, in all probability, will at least equal that of the season now closed. In the rich bottom lands of Louisiana and lower sections of Mississippi the crops generally are in a most promising condition, and planters are very forward with the picking, but in the high lands the plant has sustained very serious injury from the long drought which has prevailed during the Summer. This, however, may be remedied in some degree by the late rains, and, with a favorable Autumn, the second growth of bolls may compensate for the previous loss. From Tennessee and North Alabama there are some complaints, but nothing of a serious nature; any injury which may have been sustained, however, more particularly in the former State, will be counterbalanced by the increase of cultivation. From South Alabama the advices speak confidently of a full crop, and in the Atlantic States the prospects are also good. In South Carolina, it is true, the late heavy rains have inflicted great injury, but this extends chiefly to the crop of Sea Island Cotton.

LIVERPOOL STATEMENT FOR 1841.

UNITED STATES, 1840–1841.

Stock 1st October, 1840	58,000	Export	1,314,000
Receipts	1,635,000	Consumption	297,000
		Stock 1st Oct., 1841	82,000
Bales	1,693,000	Bales	1,693,000

CONSUMPTION.

Tot. Europe.	Continent.	France.	Gt. Britain.	Liverpool.	Stock Jan. 1, 1841, in	Liverpool.	Gt. Britain.	France.	Continent.	Tot. Europe.
					United States	271,000	305,000	78,000	64,000	477,000
					Brazil	22,000	23,000	1,000	2,000	26,000
					West Indies	13,000	16,000	5,000	6,000	27,000
					East Indies	39,000	98,000	5,000	2,000	105,000
					Egypt	21,000	22,000	11,000	35,000	68,000
					Bales	366,000	464,000	100,000	109,000	673,000
					IMPORT.					
1,384,000	133,000	364,000	887,000	818,000	United States	844,000	902,000	380,000	125,000	1,367,000
75,000	5,000	4,000	66,000	66,000	Brazil	90,000	91,000	6,000	5,000	100,000
59,000	18,000	19,000	22,000	17,000	West Indies	28,000	34,000	21,000	22,000	75,000
218,000	58,000	6,000	154,000	108,000	East Indies	162,000	275,000	1,000	58,000	271,000
112,000	48,000	33,000	31,000	31,000	Egypt	39,000	40,000	50,000	34,000	123,000
1,848,000	262,000	426,000	1,160,000	1,040,000	Bales	1,163,000	1,342,000	458,000	244,000	1,936,000
....	....	....	108,000	59,000	Export.					
761,000	91,000	132,000	538,000	430,000	Stock Dec. 31. Stock above,	366,000	464,000	100,000	109,000	673,000
2,609,000	353,000	558,000	1,806,000	1,529,000	Total supply, bales	1,529,000	1,806,000	558,000	353,000	2,609,000

COTTON AT LIVERPOOL. YEAR 1841.

Week Ending.	Receipts.						Sales.				Stocks.			Prices.			Actual Export	Con-sumption.
	Americ'n.	E. I.	Egypt.	Brazil.	Other.	Total.	Con-sumption.	Specu-lation.	Export.	Total.	Amer'n.	Other.	Total.	Mid. Up.	Mid. Orl.	Dhol		
Jan. 8..	25.586	2,396	......	1,025	1,561	31,568	15,400	15,300	100	30,800	281,500	96,000	377,500	6	6	4	100	15,400
" 15..	29,189	3,054	2,342	6,756	601	41,942	48,380	2,600	400	51,380	287,500	100,500	388,000	$6\frac{1}{4}$	$6\frac{3}{8}$	$4\frac{1}{4}$	400	63,780
" 22..	2.390	2,107	......	......	176	4,673	23,350	12,700	......	36,050	273,000	96,500	369,500	$6\frac{1}{4}$	$6\frac{3}{8}$	$4\frac{1}{4}$	......	87,130
" 29..	30.416	3,443	1,676	3,432	961	39,928	21,320	3,300	......	24,620	286,500	102.500	388,000	$6\frac{1}{4}$	$6\frac{3}{8}$	$4\frac{1}{4}$	......	108,450
Feb. 5..	5,137	......	......	1,389	810	7,336	21,830	6,000	200	28,030	272,500	99,000	371,500	$6\frac{1}{4}$	$6\frac{1}{4}$	$4\frac{1}{4}$	200	130,280
" 12..	135	......	2,211	......	......	2,346	14,120	......	......	14,120	260,500	98,500	359,000	$6\frac{1}{4}$	$6\frac{1}{4}$	$4\frac{1}{4}$	......	144,400
" 19..	34,798	1,247	2,075	4,827	1,376	44,323	45,500	1,800	450	47,750	268,000	104,000	372,000	$6\frac{3}{8}$	$6\frac{3}{8}$	$4\frac{1}{4}$	450	189,900
" 26..	14,297	1,810	1,575	......	1,465	19,147	21,090	10,600	100	31,790	265,000	104,000	369,000	$6\frac{3}{8}$	$6\frac{3}{8}$	$4\frac{1}{4}$	100	210,990
Mch 5..	8,636	2,648	......	......	216	11,500	23,000	7,500	......	30,500	256,000	101,500	357,500	$6\frac{3}{8}$	$6\frac{3}{8}$	$4\frac{1}{4}$	......	233,990
" 12..	10,233	2,881	1,507	2,832	......	17,453	25,920	10,500	150	36,570	246,000	103,500	349,500	$6\frac{5}{8}$	$6\frac{5}{8}$	$4\frac{1}{4}$	150	259,910
" 19..	10,694	2 301	......	2,172	1,028	16,195	12,160	9,200	......	21,360	247,000	106,500	353,500	$6\frac{1}{2}$	$6\frac{1}{2}$	$4\frac{3}{8}$	......	272,070
" 26..	24,317	......	......	722	1,092	26,131	12,070	3,400	180	15,650	261,500	104,500	366,000	$6\frac{1}{2}$	$6\frac{1}{2}$	$4\frac{1}{4}$	180	284,140
April 2..	19,713	......	1,748	......	365	21,826	15,600	1,700	1,300	18,600	267,500	103,500	371,000	$6\frac{3}{4}$	$6\frac{3}{4}$	$4\frac{1}{4}$	1,300	299,740
" 8..	29,150	4,114	......	2,122	420	35,806	10,420	......	1,600	12,020	289,500	107,000	396,500	$6\frac{1}{4}$	$6\frac{1}{4}$	4	1,600	310,160
" 16..	15,770	2,008	......	......	733	18,511	21,360	1,500	3,400	26,260	285,000	106,000	391,000	$6\frac{1}{4}$	$6\frac{1}{4}$	4	3,400	331,520
" 23..	19,152	......	1,168	1,731	937	22,988	14,620	......	......	14,620	291,000	107,500	398,500	$6\frac{1}{4}$	$6\frac{1}{4}$	4	......	346,140
" 30..	29,495	1,874	1,875	3,210	355	36,809	9.040	1,300	1,000	11,340	312,000	112,000	424,000	$6\frac{1}{8}$	$6\frac{1}{4}$	4	1,000	355,180
May 7..	18,177	3,404	......	637	......	22,218	15,100	1,000	1,200	17,300	317,000	113.000	430,000	6	$6\frac{1}{8}$	4	1,200	370,280
" 14..	17,152	3,761	......	1,000	......	21,913	12,860	500	500	13,860	322,000	116,000	438,000	6	6	4	500	393,140
" 21..	74,272	......	......	......	30	74,302	15,420	300	1,300	17,020	383,000	191,500	494,500	6	6	4	1,300	408,560
" 28..	21,427	......	......	......	248	21,675	11,430	......	800	12,230	394,500	108,000	502,500	$5\frac{7}{8}$	$5\frac{7}{8}$	$3\frac{7}{8}$	800	419,990
June 4..	70,061	3,292	1,370	7,070	1,495	83,288	11,760	......	250	12,010	455,000	118,500	573,500	$5\frac{3}{4}$	$5\frac{7}{8}$	$3\frac{7}{8}$	250	431,750
" 11..	9,394	1,416	386	......	173	11,569	18,180	1,200	500	19,880	440,500	117,500	567,000	$5\frac{3}{8}$	$5\frac{3}{8}$	$3\frac{3}{4}$	500	449,930
" 18..	10,621	......	......	......	148	10,769	23,170	2,130	800	26,100	438,500	114,500	553,000	$5\frac{1}{2}$	$5\frac{1}{2}$	$3\frac{3}{4}$	800	473,100
" 25.	66,790	1,728	......	5,966	......	74,484	26,390	2,000	400	28,790	482,500	117,500	600,000	$5\frac{1}{2}$	$5\frac{1}{2}$	$3\frac{3}{4}$	400	499,490
July 2..	17,192	......	......	4,690	......	21.882	27,230	3,000	200	30,430	475,500	116,500	592,000	$5\frac{1}{2}$	$5\frac{1}{2}$	$3\frac{3}{4}$	200	526,720
" 9..	30,774	4,669	3,476	878	1,413	41,210	24,370	3,200	700	28,270	487,500	101,000	608,500	$5\frac{1}{2}$	$5\frac{1}{2}$	$3\frac{3}{4}$	700	551,090
" 16..	26,093	3,298	......	2,666	41	32,098	16,240	700	1,300	18,240	498,000	133,000	621,000	$5\frac{1}{2}$	$5\frac{1}{2}$	$3\frac{3}{4}$	1,300	567,330
" 23..	23,087	......	......	1,253	552	24,892	16,490	350	2,400	19,240	504,000	121,000	504,000	$5\frac{1}{2}$	$5\frac{1}{2}$	$3\frac{3}{4}$	2,400	583,820
" 30..	15.968	5,422	917	4,167	......	26,474	16,920	850	2,450	20,220	504,500	127,000	631,500	$5\frac{1}{2}$	$5\frac{1}{2}$	$3\frac{3}{4}$	2,450	600,740
Aug. 6..	10,604	......	532	......	76	11,212	20,510	1,000	2,000	22,510	497,000	123,500	620,500	$5\frac{1}{2}$	$5\frac{1}{2}$	$3\frac{3}{4}$	2,000	621,250
" 13..	13,967	3,022	......	5,363	595	22,947	16,550	400	1,750	18,700	497,000	128,000	625,000	$5\frac{1}{4}$	$5\frac{1}{4}$	$3\frac{3}{4}$	1,750	637,800
" 20..	8,798	3,341	1,019	......	927	14,085	22,810	7,000	2,660	32,470	485,000	128,500	613,500	$5\frac{3}{8}$	$5\frac{3}{8}$	$3\frac{3}{4}$	2,660	660.610
" 27..	15,057	6,516	1,622	2,352	320	25,867	21,440	500	1,700	23,640	481,000	134,000	615,000	$5\frac{1}{4}$	$5\frac{3}{8}$	$3\frac{3}{4}$	1,700	682,050

Sep. 3..	6,264	5,439	462	1,145	......	13,310	2,200	1,200	800	4,200	468,500	137,000	605,500	5¼	5⅝	3½	800	684,250
" 10..	6,289	2,463	1,388	1,410	......	11,550	20,680	500	1,400	22,580	456,000	139,000	595,000	5⅛	5¼	3½	1,400	704,930
" 17..	2,052	6,600	......	2,466	573	11,691	22,900	4,750	2,450	30,100	438,000	143,500	581,500	5¼	5⅜	3½	2,450	727,830
" 24..	2,209	3,789	......	2,564	204	8,766	25,350	3,600	3,500	32,450	414,500	146,500	561,000	5¼	5⅜	3½	3,500	753,180
Oct. 1..	1,329	......	......	1,862	313	3,504	23,020	3,100	1,000	27,120	397,000	142,000	539,000	5¼	5⅜	3½	1,000	776,200
" 8..	....	......	813	1,000	......	1,813	25,280	4,000	900	30,180	377,000	137,500	514,500	5⅜	5⅜	3½	900	801,480
" 15..	10,029	13,219	......	2,170	373	25,791	19,500	3,200	400	23,190	371,000	149,000	520,000	5½	5¾	3½	400	820,980
" 22..	4,144	1,849	600	4,103	168	10,864	20,540	5,700	......	26,240	358,500	152,000	510,500	5⅝	5⅞	3½	......	841,520
" 29..	2,726	1,043	......	......	369	4,138	20,560	1,750	600	22,910	346,500	147,000	493,500	5½	5¾	3½	600	862,080
Nov. 5..	1,909	3,507	1,142	1,985	1,040	9,583	17,300	1,200	1,000	19,500	335,500	150,500	486,000	5⅜	5⅝	3⅜	1,000	879,380
" 12..	863	5,158	115	5,364	334	11,834	23,300	500	600	24,400	318,500	155,000	473,500	5⅜	5⅝	3¼	600	902,680
" 19..	1,335	1,628	......	......	820	3,783	19,590	200	50	19,840	304,500	154,500	458,000	5¼	5⅝	3¼	50	922,270
" 26..	2,541	10,095	......	......	1,567	14,203	20,910	2,900	1,000	24,810	291,000	159,000	450,000	5¼	5⅝	3	1,000	943,180
Dec. 3..	5,225	17,443	3,627	1,946	121	28,362	24,880	2,500	1,070	28,450	276,500	155,500	452,000	5¼	5½	3	1,070	968,060
" 10..	14,793	5,059	......	702	1,310	21,864	20,500	1,200	200	21,900	275,500	177,500	453,000	5¼	5½	3	200	988,560
" 17..	4,005	......	1,317	6,949	1,210	13,481	27,950	2,500	400	30,850	265,500	182,000	447,500	5¼	5⅜	3	400	1,016,510
" 24..	6,998	4,942	369	......	1,019	13,328	30,680	2,700	500	33,880	249,000	181,000	430,000	5¼	5⅜	3	500	1,047,190
" 31..	12,502	1,560	......	1,256	1,384	16,702	21,680	6,000	500	28,180	243,000	180,000	423,000	5⅜	5½	3¼	500	1,068,870
Average prices & total sales, receipts & stocks.	843,755	153,396	35,332	101,182	28,919	1,162,584	1,068,870	159,030	46,160	1,274,060				5.73	5.81	3.76	46,160	2,055,518

The semi-weekly Price and Weekly Sales and Receipts, at New York, Weekly Exports from New York and Rates of Freight to Liverpool 1st of each Month, for the Crop Year ending October 1, 1841.

1840.	Price of New Orleans	Price of Upland.	Sales for week.	Receipts for week.	EXPORTS FOR WEEK.					Rates of Freight to Liverpool.	GENERAL REMARKS.
					To Great Britain	To France.	North of Europe.	Other Fo'n Ports	Total Exports.		
October 6..	$7\frac{1}{2}@11\frac{1}{2}$	$7\frac{1}{2}@10\frac{1}{2}$	2,300	4,157	....	....	....	....	....	$\frac{5}{16}@\frac{7}{16}$d.	There were but few features of interest this crop year. On February 1, accounts were received from England to January 7, advising a settlement of the difficulty between Great Britain and China, known as the "Opium War," and for a time this gave an impetus to the demand here, and the business was large for home use, export and on speculation, at an advance of $\frac{3}{8}@\frac{1}{2}$ cent. on the prices previously current; but this improved feeling was subsequently checked by the stoppage, later in the month, of the United States Bank and the other Philadelphia banks, with most of the southern banks. Foreign and domestic exchanges became unsettled, and business in consequence was much deranged; the demand for cotton fell off, and prices receded, notwithstanding the receipts at the ports showed a falling off, up to that time, of about 200,000 bales, as compared with the previous year; toward the last of March, however, a very active demand set in for export and speculation, based on the small receipts at the ports, then 278,234 bales less than at the same time the previous season, afterward increased to over half a million
" 9..	$7\frac{1}{2}@11\frac{1}{2}$	$7\frac{1}{2}@10\frac{1}{2}$									
" 13..	$7\frac{1}{2}@11\frac{1}{2}$	$7\frac{1}{2}@10\frac{1}{2}$	2,800	408	....	....	....	....	....		
" 16..	$7\frac{1}{2}@11\frac{1}{2}$	$7\frac{1}{2}@10\frac{1}{2}$									
" 20..	$7\frac{1}{2}@11\frac{1}{2}$	$7\frac{1}{2}@10$	1,600	4,439	146	178	48	....	372		
" 23..	$7\frac{1}{2}@11$	$7\frac{1}{2}@10$									
" 27..	$7\frac{1}{2}@11$	$7\frac{1}{2}@10$	3,050	3,749	1,243	466	....	....	1,709		
" 30..	$7\frac{1}{2}@11$	$7@9\frac{3}{4}$									
Novem. 3..	$7\frac{1}{2}@11$	$7@9\frac{3}{4}$	3,000	3,349	2,709	888	200	1,072	4,869	$\frac{3}{8}@\frac{1}{2}$d.	
" 6..	$7\frac{1}{2}@11$	$7@9\frac{3}{4}$									
" 10..	$7\frac{1}{2}@11$	$7@9\frac{3}{4}$	3,800	408	....	78	....	....	78		
" 13..	$7\frac{1}{2}@11$	$7\frac{1}{2}@10$									
" 17..	$7\frac{1}{2}@11$	$7\frac{1}{2}@10$	2,550	2,899	674	509	....	59	1,242		
" 20..	$7\frac{1}{2}@11$	$7\frac{1}{2}@10$									
" 24..	$7\frac{1}{2}@11$	$7\frac{1}{2}@10$	3,650	7,564	883	375	....	....	1,258		
" 27..	$7\frac{1}{2}@11$	$7\frac{1}{2}@10$									
Decem. 1..	$7\frac{1}{2}@11$	$7\frac{1}{2}@10$	3,800	6,205	629	753	43	....	1,425	$\frac{1}{4}@\frac{3}{8}$d.	
" 4..	$7\frac{1}{2}@11$	$7\frac{1}{2}@10$									
" 8..	$8@11$	$8@10\frac{1}{4}$	3,700	173	....	....	....	....	....		
" 11..	$8@11$	$8@10\frac{1}{4}$									
" 15..	$8@11$	$8@10\frac{1}{4}$	4,000	1,676	152	1,307	140	....	1,599		
" 18..	$8@11$	$8@10\frac{1}{4}$									
" 22..	$8@11$	$8@10\frac{1}{4}$	1,100	4,341	200	620	128	40	988		
" 25..	$8@11$	$8@10\frac{1}{4}$									
" 29..	$8@11$	$8@10\frac{1}{4}$	2,150	1,780	1,086	....	....	....	1,086		
1841.											
January 5..	$8@11$	$8@10\frac{1}{4}$	1,450	3,319	183	231	....	413	827	$\frac{3}{16}@\frac{5}{16}$d.	
" 8..	$8@11$	$8@10\frac{1}{4}$									
" 12..	$8\frac{3}{4}@11$	$8\frac{3}{4}@10\frac{3}{8}$	2,150	11,608	135	592	....	....	727		
" 15..	$8\frac{3}{4}@11$	$8\frac{3}{4}@10\frac{3}{8}$									
" 19..	$9@12$	$9@11$	7,350	2,079	....	....	....	....	....		
" 22..	$9\frac{1}{4}@12$	$9\frac{1}{4}@11\frac{1}{2}$									
" 26..	$9\frac{1}{2}@12$	$9\frac{1}{2}@11\frac{1}{2}$	7,950	6,471	926	583	....	....	1,509		

January 29..	9½@12	9½@11½								
February 2..	10@12½	10@11½	11,900	5,644	1,121	2,504	50	....	3,675	⅜@½d.
" 5..	10@12½	10@11½								
" 9..	10@12½	10@11½	5,200	3,881	....	81	....	....	81	
" 12..	9¾@12½	9¾@11½								
" 16..	9¾@12½	9¾@11½	1,250	698	....	1,480	29	....	1,509	
" 19..	9¾@12½	9¾@11½								
" 23..	9¾@12½	9¾@11½	9,000	10,235	1,296	....	....	....	1,296	
" 26..	9¾@12½	9¾@11½								
Marcl 2..	9¾@12¼	9½@11¼	6,850	17,622	998	1,197	66	....	2,261	⅜@½d.
" 5..	9¼@11¾	9¼@10¾								
" 9..	9¼@11¾	9¼@10¾	3,850	4,635	488	....	....	....	488	
" 12..	9¼@11¾	9¼@10¾								
" 16..	9@11¾	9@10¾	6,000	10,955	2,095	4,188	....	....	6,283	
" 19..	9@11¾	9@10¾								
" 23..	9@11¾	9@10¾	8,200	10,887	574	1,726	154	....	2,454	
" 26..	9¼@12	9¼@11								
" 30..	9½@12½	9½@11	9,000	16,518	462	2,256	79	....	2,797	
April 2..	9½@12½	9½@11								$\frac{7}{16}$@$\frac{9}{16}$d.
" 6..	9½@12½	9½@11	5,000	10,978	2,460	3,112	1,751	....	7,323	
" 9..	9½@12½	9½@11								
" 13..	9½@12½	9½@11	5,600	4,608	4,652	4,568	675	185	0,080	
" 16..	9½@12½	9½@11								
" 20..	9½@12½	9½@11	8,200	7,231	3,317	2,610	2,063	....	7,990	
" 23..	9½@12½	9½@11								
" 27..	9½@12½	9½@11	4,700	963	5,402	2,185	617	....	8,204	
" 30..	9½@12½	9½@11								
May 4..	9½@13	9½@11½	5,950	10,071	1,462	958	....	118	2,538	¼@⅜d.
" 7..	9½@13	9½@11½								
" 11..	9½@13	9½@11½	5,100	7,350	4,070	....	....	....	4,070	
" 14..	9½@13	9½@11½								
" 18..	9½@13	9½@11½	3,400	3,002	3,064	670	1,326	....	5,060	
" 21..	9@13	9@11½								
" 25..	9@13	9@11½	3,750	9,528	1,591	1,192	593	1,287	4,663	
" 28..	9@13	9@11½								
June 1..	9@13	9@11½	5,100	1,831	3,110	1,266	628	....	5,004	¼@⅜d.
" 4..	9@13	9@11½								
" 8..	9@13	9@11½	4,600	4,565	....	....	....	....	....	
" 11..	9@13	9@11½								
" 15..	9@13	9@11½	3,350	7,725	2,636	572	1,001	555	4,764	
" 18..	9@13	9@11½								
" 22..	9@13	9@11½	2,300	7,573	7,750	1,824	78	....	9,652	
" 25..	8½@12½	8½@11								

bales, but the foreign accounts were unfavorable, and this market did not respond with vigor to the statistical position of the article, the year closing with receipts at the ports 542,890 bales short of the previous year, and exports 562,726 bales less.

The semi-weekly Price and Weekly Sales and Receipts at New York, Weekly Exports from New York and Rates of Freight to Liverpool 1st of each Month, for the Crop Year ending October 1, 1841—*Concluded.*

1841.		Price of New Orleans	Price of Upland.	Sales for week.	Receipts for week.	EXPORTS FOR WEEK.					Rates of Freight to Liverpool.
						To Great Britain.	To France.	North of Europe.	Other Fo'n Ports	Total Exports.	
June	29..	$8\frac{1}{2}$@$12\frac{1}{2}$	$8\frac{1}{2}$@11	3,900	3,949	3,210	657	598	....	4,465	
July	2..	$8\frac{1}{2}$@$12\frac{1}{2}$	$8\frac{1}{2}$@11								$\frac{1}{8}$@$\frac{1}{4}$d.
"	6..	$8\frac{1}{2}$@$12\frac{1}{2}$	$8\frac{1}{2}$@11	3,500	3,282	1,903	....	....	....	1,903	
"	9..	$8\frac{1}{2}$@$12\frac{1}{2}$	$8\frac{1}{4}$@$10\frac{3}{4}$								
"	13..	$8\frac{1}{2}$@$12\frac{1}{2}$	$8\frac{1}{4}$@$10\frac{3}{4}$	3,900	3,688	279	2,686	218	....	3,183	
"	16..	$8\frac{1}{2}$@$12\frac{1}{2}$	$8\frac{1}{4}$@$10\frac{3}{4}$								
"	20..	$8\frac{1}{2}$@$12\frac{1}{2}$	$8\frac{1}{4}$@$10\frac{3}{4}$	3,350	2,092	424	1,915	889	....	3,228	
"	23..	$8\frac{1}{2}$@$12\frac{1}{2}$	$8\frac{1}{4}$@$10\frac{3}{4}$								
"	27..	$8\frac{1}{2}$@$12\frac{1}{2}$	$8\frac{1}{4}$@$10\frac{3}{4}$	4,550	3,813	546	843	477	27	1,893	
"	30..	$8\frac{1}{2}$@$12\frac{1}{2}$	$8\frac{1}{4}$@$10\frac{3}{4}$								
August	3..	$8\frac{1}{2}$@$12\frac{1}{2}$	$8\frac{1}{4}$@$10\frac{3}{4}$	2,600	1,822	1,682	2,743	311	....	4,736	$\frac{1}{4}$@$\frac{3}{8}$d.
"	6..	$8\frac{1}{2}$@$12\frac{1}{2}$	$8\frac{1}{4}$@$10\frac{3}{4}$								
"	10..	$8\frac{1}{2}$@$12\frac{1}{2}$	$8\frac{1}{4}$@$10\frac{3}{4}$	4,150	2,703	....	210	66	....	276	
"	13..	$8\frac{1}{2}$@$12\frac{1}{2}$	$8\frac{1}{4}$@$11\frac{1}{4}$								
"	17..	$8\frac{1}{2}$@$12\frac{1}{2}$	$8\frac{1}{4}$@$11\frac{1}{4}$	2,700	2,019	2,171	856	391	....	3,418	
"	20..	$8\frac{1}{2}$@$12\frac{1}{2}$	$8\frac{1}{4}$@$11\frac{1}{4}$								
"	24..	$8\frac{1}{4}$@$12\frac{1}{2}$	8@11	2,450	1,075	1,739	805	625	....	3,169	
"	27..	8@$12\frac{1}{2}$	8@11								
"	31..	8@$12\frac{1}{2}$	$7\frac{3}{4}$@$10\frac{1}{2}$	3,250	757	162	1,710	888	102	2,862	
Septem.	3..	8@$12\frac{1}{2}$	$7\frac{1}{2}$@$10\frac{1}{2}$								$\frac{1}{8}$@$\frac{3}{16}$d.
"	7..	7@$11\frac{1}{2}$	7@10	2,700	816	187	1,165	404	....	1,756	
"	10..	7@$11\frac{1}{2}$	7@10								
"	14..	$7\frac{1}{4}$@$12\frac{1}{2}$	7@$11\frac{1}{4}$	3,600	405	205	711	503	....	1,419	
"	17..	$7\frac{1}{4}$@$12\frac{1}{2}$	7@$11\frac{1}{4}$								
"	21..	$7\frac{1}{4}$@$12\frac{1}{2}$	7@$11\frac{1}{4}$	3,850	802	1,064	2,491	190	....	3,745	
"	24.	$7\frac{1}{2}$@$11\frac{1}{2}$	7@$10\frac{1}{2}$								
"	28..	$7\frac{1}{2}$@$11\frac{1}{2}$	7@$10\frac{1}{2}$	3,100	2,480	2,554	2,086	930	9	5,579	
October	1..	$7\frac{1}{2}$@$11\frac{1}{2}$	7@$10\frac{1}{2}$								$\frac{1}{4}$@$\frac{5}{16}$d.
Average price and total sales, receipts and exports.		10.31	9.50	222,300	250,828	71,640	57,847	16,159	3,867	149,513	

General Remarks

Exchange.

Bills on London were quoted in October, $8\frac{1}{2}$@9 per cent. premium; in November, $8\frac{3}{4}$@9 down to $8\frac{1}{4}$@$8\frac{1}{2}$; in December, $8\frac{1}{2}$@9; in January, between 8 and 9; in February, from $8\frac{3}{4}$@9 down to 8; in March, $7\frac{1}{2}$@8; in April, $6\frac{1}{2}$@8; in May, 8@$8\frac{1}{4}$; in June, $8\frac{1}{4}$@$8\frac{3}{4}$; in July, $7\frac{3}{4}$@$8\frac{1}{4}$; in August, 8@9; in September, 9@$9\frac{3}{4}$, closing October 1 at $9\frac{3}{4}$@10 per cent. premium.

On the 12th September 1840, I first visited a cotton plantation. and learned something of the vicissitudes to which the cotton plant is subject. I then heard from the planter that the crop was exceedingly promising until the last week in August. On the 20th August, he considered it past all danger; and, to use his own words, he would not have given ten dollars to have his crop guaranteed. Within three days one-fourth of it was entirely destroyed. The result proved that this was general throughout the whole cotton growing country. The crop showed a falling off from the previous one of about one-fourth. The speculation based upon this was disastrous, because of the great depression resulting from the revulsion of 1837. In fact, that revulsion did not terminate, that is, we did not touch bottom, until 1842. It is a remarkable fact that periods of expansion and periods of contraction—I mean such as are general throughout the world—seem to last about five years. The next break-down was in 1847—to be succeeded by the revulsion in 1857. The depression, commencing in 1847, was checked and modified by the discovery of gold in California. The result of this was a partial reaction in 1854.

Never in this country was there so little enterprise or activity as during the five years from 1837 to 1843; never were prices so low and property so difficult to sell. The degree of stagnation is indicated by the fact that in 1842 the circulation and deposits of our banks had declined to $10\frac{40}{100}$ per capita of the population; whereas in 1837, just before the panic, they reached $17\frac{40}{100}$ per capita of the population.

I believe they are now, in 1872, more than $30 per capita. Just before the outbreak of our civil war in 1860 they were a little over $15 per capita.

These facts are a key to the times.

1842.

This year (1841–42) the crop was again very short. I remember it well. The Fall and Winter were remarkably wet; but there were other causes at work, quite as powerful as the weather, to keep down production. At least half the planters were deeply involved in debt and struggling hard to extricate themselves by one expedient or another. It was extremely difficult to borrow money at any price. Of course, there was no spirit of enterprise; at least half the people regarded any such spirit with extreme suspicion. The prevailing opinion favored what people called "hard money," low prices, and strict economy. People were constantly boasting of their cautiousness and deprecating speculation. In this tone of the public mind was probably the main secret of General Jackson's wonderful popularity with the masses of the people, after he had, by his crude and violent financial measures, plunged his country into bankruptcy. Jackson's ideas of trade and finance were little more than barter. He was a backwoods' Statesman, literally and truly. In this he fairly represented the thought of the common people, who attributed all their misfortunes to speculators and paper money, and mainly to the old United States Bank. The effect of such ideas was to reduce trade to the smallest dimensions.

There was also at that time much talk of India as a cotton-producing country, and great fears expressed that the United States would not be able to compete with her cheap and abundant labor in raising cotton.

How changed is the public sentiment now! The South fears no competition in cotton-raising, and the whole people have learned to love paper money and national banking.

COTTON CROP OF THE UNITED STATES.

Statement and Total Amount of the Cotton Crop of the United States, for the Year ending August 31, 1842.

	Bales.	Bales.	Total.	Same period 1841.
NEW ORLEANS.				
Export—				
To Foreign Ports	649,435			
Coastwise	99,832			
Burnt and damaged	950			
Stock on hand 1st September, 1842	4,428			
		754,645		
Deduct—				
Stock on hand 1st September, 1841	14,490			
Received from Mobile	4,565			
" Florida	2,831			
" Texas	5,101			
		26,987	727,658	813,595
MISSISSIPPI.				
Export from Natchez, &c.—				
Included in New Orleans	....	....	....	1,085
ALABAMA.				
Export from Mobile—				
To Foreign Ports	241,877			
Coastwise	77,161			
Stock in Mobile, 1st September, 1842	422			
		319,460		
Deduct—				
Stock in Mobile, 1st September, 1841	360			
Received from Florida	632			
" Texas	153			
		1,145	318,315	320,701
FLORIDA.				
Export—				
To Foreign Ports	46,518			
Coastwise	68,048			
Stock on hand 1st September, 1842	250			
		114,816		
Deduct—				
Stock on hand 1st September, 1841	....	400	114,416	93,552
GEORGIA.				
Export from Savannah—				
To Foreign Ports—Uplands	135,410			
" Sea Island	6,976			
Coastwise—Uplands	79,194			
Sea Island	674			
	222,254			
From Darien—				
To New York	8,724			
Burnt	450			
Stock in Savannah, 1st September, 1842	2,651			
" Augusta and Hambro', 1st September, 1842	2,459			
		236,538		
Deduct—				
Stock in Savannah and Augusta 1st September, 1841	....	4,267	232,271	148,947

Statement and Total Amount of the Cotton Crop of the United States, for the Year ending August 31, 1842—*Concluded.*

	Bales.	Bales.	Total.	Same period 1841.
SOUTH CAROLINA.				
Export from Charleston—				
To Foreign Ports—Uplands	184,705			
" Sea Island	14,119			
Coastwise—Uplands	70,442			
Sea Island	341			
	269,607			
Export from Georgetown—				
To New York	12,617			
Burnt and lost	140			
Stock in Charleston, 1st September, 1842	2,747			
		285,111		
Deduct—				
Stock in Charleston, 1st September, 1841	4,552			
Received from Savannah	16,258			
Received from Florida and Key West	4,137			
		24,947		
			260,164	227,400
NORTH CAROLINA.				
Export—				
All Coastwise	9,787			
Stock on hand, 1st September, 1842	250			
		10,037		
Deduct—				
Stock on hand 1st September, 1841	....	300		
			9,737	7,865
VIRGINIA.				
Export—				
To Foreign Ports	6,341			
Coastwise	4,500			
Manufactured	9,000			
Stock on hand, 1st September, 1842	100			
		19,941		
Deduct—				
Stock on hand 1st September, 1841	....	928		
			19,013	20,800
Received at Philadelphia and Baltimore, overland	....	....	2,000	1,000
Total crop of the United States	....	....	1,683,574	
Crop of last year	....	....	1,634,945	
Increase	....	....	48,629	

Export to Foreign Ports, from September 1, 1841, *to August* 31, 1842.

FROM		To Great Britain.	To France.	To North of Europe.	Other F'n Ports.	Total.
New Orleans	Bales	421,450	183,272	21,207	23,506	649,435
*Mississippi (Natchez)		....	....	....	....	....
Alabama		185,414	49,544	1,351	5,568	241,877
Florida		29,412	14,097	....	3,009	46,518
Georgia (Savannah and Darien)		124,296	15,590	1,192	1,308	142,386
South Carolina		98,305	75,504	21,417	3,598	198,824
North Carolina		....	....	....	....	....
Virginia		5,031	650	183	477	6,341
Baltimore		724	....	594	....	1,318
Philadelphia		1,217	79	329	50	1,675
New York		69,548	59,393	30,578	13,519	173,038
Boston		234	....	3,105	498	3,837
Grand total		935,631	398,129	79,956	51,533	1,465,249
Total last year		858,742	348,776	56,279	49,480	1,313,277
Increase		76,889	49,353	23,677	2,053	151,972

* The shipments from Mississippi are included in the export from New Orleans.

Growth.

Total crop of	1824–5,	560,000	bales.
"	1825–6,	710,000	"
"	1826–7,	937,000	"
"	1827–8,	712,000	"
"	1828–9,	857,744	"
"	1829–30,	976,845	"
"	1830–1,	1,038,848	"
"	1831–2,	987,477	"
"	1832–3,	1,070,438	"
"	1833–4,	1,205,394	"
"	1834–5,	1,254,328	"
"	1835–6,	1,360,725	"
"	1836–7,	1,422,930	"
"	1837–8,	1,801,497	"
"	1838–9,	1,360,532	"
"	1839–40,	2,177,835	"
"	1840–1,	1,634,945	"
"	1841–2,	1,683,574	"

Consumption.

To estimate the quantity manufactured in the United States, we take the growth of the year			bales..1,683,574
Add—Stocks on hand at the commencement of the year (September 1, 1841)—In the Southern ports	27,479		
" Northern ports	45,000		
		72,479	
			1,756,053
Deduct therefrom—The Export to Foreign ports	1,465,249		
Less Texas and other foreign included.	10,393		
		1,454,856	

Brought forward....................................		1,454,856	
Stocks on hand at the close of the year (September 1, 1842)—			
In the Southern ports..............	13,307		
" Northern ports...............	18,500		
		31,807	
Burnt and Lost at New Orleans.......	950		
" Savannah..........	450		
" Charleston.........	140		
		1,540	
			1,488,203
		bales..	267,850

Quantity consumed by and in the hands of manufacturers,			1841–2 ....bales..	267,850
"	"	"	1840–1	297,288
"	"	"	1839–40..............	295,193
"	"	"	1838–9	276,018
"	"	"	1837–8	246,063
"	"	"	1836–7	222,540
"	"	"	1835–6	236,733
"	"	"	1834–5	216,888
"	"	"	1833–4	196,413
"	"	"	1832–3	194,412
"	"	"	1831–2	173,800
"	"	"	1830–1	182,142
"	"	"	1829–30..............	126,512
"	"	"	1828–9	118,853
"	"	"	1827–8	120,593
"	"	"	1826–7	103,483

NOTE.—It will be seen that we have deducted from the New Orleans statement the quantity received at that port from Texas—that being a foreign country.

Our estimate of the quantity taken for consumption, does not include any cotton manufactured in the States south and West of Virginia, nor any in that State, except in the vicinity of Petersburg and Richmond.

Of the new crop, now gathering, about 3,000 bales were received previous to 1st inst.; of which 1,734 were received at New Orleans.

The general tenor of the accounts from the cotton-growing States leads to the conclusion that the crop now coming in will exceed that of last year by several hundred thousand bales ; but the article is subject to so many vicissitudes that no certain calculation can be made as to the quantity that may reach the market.

ANNUAL REVIEW.

From the New Orleans Price Current—1841-42.

The market opened in the beginning of the season under rather unfavorable circumstances, owing to the heavy stocks of the preceding crop which remained on hand, not only in Europe, but in Northern ports, and on account of the general stagnation in trade on both sides of the Atlantic. In addition to these causes of depression, sufficiently important in themselves, there was another, which likewise exercised a very unfavorable influence on the market, viz., an over-estimate of the crop of the United States, the opinion being pretty generally entertained, until the season was considerably advanced, that it would reach above 1,800,000 bales, while, by reference to our table, it will be seen that the actual receipts at the various receiving ports will fall short of 1,700,000 bales. From the scarcity of fair Cotton, and qualities above, they have experienced but little fluctuation in value during the season, our quotations having seldom varied more than a quarter of a cent; but for the lower grades prices have continued gradually to recede, and ordinary Cotton, which was worth from 8 to 8¼ cents in the latter part of October, would not, since the 1st of May, command above 4½ to 6 cents. Several causes combined to produce this great reduction in the lower qualities. Among the most prominent were the unusual proportion of these descriptions in the crop of the United States, the unprecedented depression in the manufacturing trade of Great Britain, where the poorer sorts of Cotton find their principal market, and the competition of unusually large supplies from India, growing out of increased production and the war with China. Until recently, the accounts from every section of country tributary to this market were of the most flattering character in regard to the appearance of the crops, and a most abundant yield was very generally anticipated; but we regret to say that the tone of the letters from our correspondents in the interior has of late materially changed. It is now said that the late heavy and protracted rains, which have extended a large distance above this, have introduced the boll-worm and rust into many districts, and have also caused many of the forms to shed, in consequence of which the crop will be considerably diminished. The labor of picking has likewise been much retarded, and the early Cotton greatly deteriorated in quality from the same cause. A return of fair weather soon, however, would do much to lessen the effects of these unseasonable rains.

The semi-weekly Price and Weekly Sales and Receipts at New York, Weekly Exports from New York and Rates of Freight to Liverpool 1st of each month, for the Crop Year ending September 1, 1842.

1841.	Price of New Orleans	Price of Upland.	Sales for week.	Receipts for week.	EXPORTS FOR THE WEEK.					Rates of Freight to Liverpool.	GENERAL REMARKS.
					To Great Britain.	To France.	North of Europe.	Other Fo'n Ports	Total Exports.		
October 5..	7½@11½	7@10½	1,950	240	....	....	....	....	....	¼@5-16d	This crop year was devoid of anything of interest. After March, most of the suspended banks resumed specie payments, which imparted a healthier tone to business generally, though the tendency of prices throughout the year was steadily downward. The crop year was changed at this time from October 1 to September 1.
" 8..	7½@11½	7@10½									
" 12..	7½@11½	7@10½	2,400	909	4,269	5,288	1,724	9	11,290		
" 15..	7¾@12	7½@10¼									
" 19..	7¾@12	7½@10¼	4,300	980	1,079	1,247	....	....	2,326		
" 22..	7¾@12	7½@10¼									
" 26..	7¾@12	7½@10¼	2,300	3,494	2,391	846	1,106	....	4,343		
" 29..	7¾@12	7½@10¼									
Novem. 2..	7¾@12	7½@10¼	3,600	3,263	1,997	2,003	691	82	4,773	¼@5-16d	
" 5..	7¾@12	7½@10¼									
" 9..	7¾@12	7½@10¼	3,750	2,638	....	1,525	118	....	1,643		
" 12..	7¾@12	7½@10¼									
" 16..	8@12	7½@10¼	3,700	3,840	354	....	555	566	1,475		
" 19..	8@12	7½@10¼									
" 23..	8@12	7½@10¼	2,700	8,142	522	413	479	45	1,459		
" 26..	8@12	7½@10									
" 30..	8@12	7½@10	2,250	1,438	1,127	2,833	....	....	3,960		
Decem. 3..	8@12	7½@10								5-16@⅜d	
" 7..	8@11½	7¼@9¾	4,850	7,592	980	....	63	....	1,043		
" 10..	8@11½	7¼@9½									
" 14..	8@11½	7¼@9½	4,450	6,698	813	2,389	471	....	3,673		
" 17..	7½@11	7@9½									
" 21..	7½@11	7@9½	5,350	1,154	809	888	....	....	1,697		
" 24..	7½@11	7@9½									
" 28..	7½@11	7@9½	3,400	4,341	564	1,704	....	....	2,268		
" 31..	7½@11	7@9½									
1842.											
January 4..	7½@11	7@9½	2,700	8,730	2,690	1,728	163	781	5,362	5-16@$\frac{7}{16}$.	
" 7..	7¼@11	7@9½									
" 11..	7¼@11	7@9½	4,100	4,774	929	1,284	....	....	2,213		
" 14..	7¼@11	7@9½									
" 18..	7¼@11	7@9½	7,400	6,625	374	2,706	188	962	4,230		
" 21..	7¼@11	7@9½									

January 25..	7¼@11	7@9½	3,800	6,241	2,417	1,206	464	....	4,087	
,, 28..	7¼@11	7@9½								
February 1..	7¼@11	7@9½	4,350	6,312	733	3,701	132	158	4,724	5-16@$\frac{1}{16}$
" 4..	6¾@11	6½@9½								
" 8..	6¾@11	6½@9½	4,850	4,925	620	1,532	446	....	2,598	
" 11..	6¾@11	6½@9½								
" 15..	6¾@11	6@9½	4,600	4,458	2,305	1,723	822	....	4,850	
" 18..	6¾@11	6@9½								
" 22..	6¾@11	6@9½	3,650	6,620	2,442	700	679	88	3,909	
" 25..	6¾@11	6@9½								
March 1..	6¼@11	5¾@9½	4,600	9,669	3,197	1,851	....	....	5,048	⅜@½d.
" 4..	6¼@11	5½@9½								
" 8..	6¼@11	5½@9½	4,200	6,663	1,269	....	217	....	1,486	
" 11..	6@11	5¼@9½								
" 15..	5½@11	5@9¾	6,300	2,618	2,132	3,449	611	89	6,281	
" 18..	5½@11	5@9¾								
" 22..	5½@11	5@9¾	5,800	6,803	1,378	1,000	126	....	2,504	
" 25..	5½@11	5@9¾								
" 29..	5½@11	5@9¾	4,300	6,941	3,564	1,446	426	....	5,436	
April 1..	5½@11	5@9¾								
" 5..	5½@11	5@9¾	4,350	10,876	2,387	110	705	915	4,117	5-16@⅜d
" 8..	5½@11	5@9¾								
" 12..	5½@11	5¼@9¾	7,350	3,999	1,504	1,264	2,014	....	4,782	
" 15..	5½@11	5¼@9¾								
" 19..	5½@10½	5¼@9¼	4,050	10,157	2,010	....	741	137	2,888	
" 22..	5½@10½	5¼@9¼								
" 26..	5½@10½	5¼@9¼	5,950	12,150	1,254	1,889	2,024	....	5,167	
" 29..	5½@10½	5¼@9¼								
May 3..	5½@10½	5¼@9¼	6,500	6,106	4,282	1,414	322	73	6,091	¼@5-16d
" 6..	5½@10½	5¼@9¼								
" 10..	5½@10½	5¼@9¼	5,400	2,393	1,688	694	1,344	....	3,726	
" 13..	5½@10½	5¼@9¼								
" 17..	5½@10½	5¼@9¼	6,750	7,743	10	734	1,741	2,261	4,746	
" 20..	5½@10½	5¼@9¼								
" 24..	5½@10½	5¼@9¼	3,400	1,953	716	656	563	1,000	2,935	
" 27..	5½@10½	5¼@9¼								
" 31..	5½@10½	5¼@9¼	6,030	8,494	2,862	542	1,951	1,054	6,409	
June 3..	5½@10½	5¼@9¼								—@3-16d
" 7..	5½@10½	5¼@9¼	5,022	3,557	921	....	726	....	1,647	
" 10..	5½@10½	5¼@9¼								
" 14..	5½@10½	5¼@9¼	4,600	6,264	607	1,069	993	880	3,549	
" 17..	5½@10½	5¼@9¼								

New York Statement for Year 1842—*Concluded.*

1842.		Price of New Orleans	Price of Upland.	Sales for week.	Receipts for week.	EXPORTS FOR THE WEEK					Rates of Freight to Liverpool.
						To Great Britain.	To France.	North of Europe.	Other Fo'nPorts	Total Exports.	
June	21..	5½@10½	5¼@9¼	5,500	9,298	1,545	87	1,601	132	3,365	
"	24..	5½@10½	5½@9¼								
"	28..	6@11	5¾@9½	6,700	4,882	1,479	26	179	655	2,339	
July	1..	6@11	5¾@9½								⅛@3-16d
"	5..	6@11	5¾@9½	2,600	2,445	2,130	3,113	300	605	6,148	
"	8..	6@11	5¾@9½								
"	12..	6@11	5¾@9½	4,700	2,562	....	432	165	563	1,160	
"	15..	6@11	5¾@9½								
"	19..	6@11	5¾@9½	4,000	2,268	1,568	1,662	325	1,036	4,591	
"	22..	6@11	5¾@9½								
"	26..	6@11	5¾@9½	2,650	1,820	1,024	1,254	646	....	2,924	
"	29..	6@11	5¾@9½								
August	2..	6@11	5¾@9½	3,550	2,737	921	787	1,385	....	3,093	⅛@–d.
"	5..	6@10½	5¾@9								
"	9..	6@10½	5¾@9	4,800	2,794	238	678	....	....	916	
"	12..	6@10½	5¾@9								
"	16..	6@10½	5¾@9	2,600	1,267	504	471	....	681	1,656	
"	19..	6@10½	5¾@9								
"	23..	6@10½	5¾@9	2,800	2,221	1,046	386	1,179	1,428	4,039	
"	26..	6@10½	5¾@9								
"	30..	6@10½	5¾@9	2,425	1,976	1,118	400	1,150	....	2,668	
Septem.	2..	6@10½	5¾@9								3-16@¼d
"	6..	6@10½	5¾@9	3,850	783	779	263	362	....	1,404	
Average prices and total sales, receipts and exports.		8.75	7.85	211,227	234,853	69,548	59,393	29,897	14,200	173,038	

GENERAL REMARKS.

Exchange.

Sixty days' Bills on London opened in September at 9@9¼ per cent. premium, and subsequently advanced to 9¼@9¾; in October the range was from 9¾ to 10½; in November, 9½@10; in December, 8½@9½; in January, 8¾@7¾; in February, 7½@8½; in March, from 7½@8 down to 6½@7; in April, 6½@7¼; in May, 7½@8¾; in June, 8@6½; in July the quotation declined from 6¾@7¼ to 5½@6¼; in August, steady at 6½@7; and in September, 7@7½.

LIVERPOOL STATEMENT FOR 1842.

UNITED STATES, 1841–1842.			
Stock, October 1, 1841	82,000	Export	1,466,000
Receipts	1,684,000	Consumption	268,000
		Stock, Oct. 1, 1842	32,000
Bales	1,766,000	Bales	1,766,000

Tot. Europe.	Continent.	France.	Gt. Britain.	Liverpool.	Stock 1st Jan., 1842, in	Liverpool.	Gt. Britain.	France.	Continent.	Tot. Europe.
					United States .... Bales	254,000	279,000	94,000	57,000	430,000
					Brazil	44,000	44,000	3,000	3,000	50,000
					West Indies	23,000	27,000	6,000	9,000	42,000
					East Indies	80,000	157,000	1,000	3,000	161,000
					Egypt	29,000	31,000	28,000	19,000	78,000
					Bales	430,000	538,000	132,000	91,000	761,000
CONSUMPTION.					IMPORT.					
1,488,000	160,000	366,000	962,000	898,000	United States .... Bales	964,000	1,019,000	381,000	159,000	1,506,000
90,000	7,000	13,000	70,000	70,000	Brazil	86,000	86,000	12,000	8,000	103,000
69,000	18,000	28,000	23,000	19,000	West Indies	16,000	20,000	34,000	18,000	70,000
240,000	68,000	5,000	167,000	135,000	East Indies	170,000	255,000	6,000	71,000	264,000
118,000	62,000	29,000	27,000	27,000	Egypt	18,000	18,000	14,000	76,000	108,000
2,005,000	315,000	441,000	1,249,000	1,149,000	Bales	1,254,000	1,398,000	447,000	332,000	2,051,000
......	......	......	126,000	79,000	Export.					
807,000	108,000	138,000	561,000	456,000	Stock, Dec. 31 .... Stock above,	430,000	538,000	132,000	91,000	761,000
2,812,000	423,000	579,000	1,936,000	1,684,000	Total supply, bales	1,684,000	1,936,000	579,000	423,000	2,812,000

COTTON AT LIVERPOOL. YEAR 1842.

Week Ending.	Receipts.						Sales.				Stocks.			Prices.			Actual Export	Con-sumption.
	Americ'n.	E. I.	Egypt.	Brazil.	Other.	Total.	Con-sumption.	Specu-lation.	Export	Total.	Amer'n.	Other.	Total.	Mid. Up.	Mid. Orl.	Dhol		
Jan. 7..	14,883	......	2,022	1,028	643	18,576	22,880	750	300	23,930	251,500	173,910	425,500	5 3/8	5 1/2	3 1/4	300	22,880
" 14..	13,068	9,705	......	4,217	......	26,990	17,840	3,100	200	21,140	249,500	182,500	432,000	5 1/8	5 3/8	3	200	40,720
" 21..	27,489	12,074	1,468	639	678	42,348	19,340	500	1,000	20,840	259,000	193,500	452,500	5	5 1/8	3	1,000	60,060
" 28..	20,044	6,063	......	4,466	23	30,596	18,510	450	1,190	20,150	262,500	199,000	461,500	5	5 1/8	3	1,190	78,570
Feb. 4..	23,389	......	......	674	......	24,063	24,380	2,500	2,830	29,710	264,500	193,000	457,500	4 7/8	5	3	2,830	102,950
" 11..	9,418	......	......	1,916	......	11,334	23,970	2,900	850	27,720	253,500	191,500	445,000	4 7/8	5	3	850	126,920
" 18..	29,174	5,655	1,312	796	467	37,404	18,870	500	260	19,630	269,000	199,500	463,500	4 7/8	5	3	260	145,790
" 25..	38,336	......	350	1,816	330	40,832	17,590	2,000	3,600	23,190	291,000	192,500	483,500	4 3/4	4 7/8	3	3,600	163,380
Mch 4..	17,703	......	......	......	......	17,703	23,900	3,600	1,100	28,600	287,500	187,500	475,000	4 3/4	4 7/8	3	1,100	187,280
" 11..	43,159	4,368	911	4,812	772	54,022	20,150	4,200	2,450	26,800	314,000	192,500	506,500	4 3/4	4 7/8	3	2,450	207,430
" 18..	13,719	6,753	......	997	97	21,566	19,590	2,000	1,800	23,390	308,000	196,500	504,500	4 3/4	4 7/8	3	1,800	227,020
" 24..	16,404	4,339	......	......	8	20,751	15,370	1,800	800	17,970	311,500	197,500	509,000	4 3/4	4 7/8	3	800	242,390
April 1..	43,890	8,235	489	......	455	53,069	16,690	1,200	1,050	18,940	342,000	200,000	542,000	4 5/8	4 3/4	3	1,050	259,080
" 8..	5,142	......	......	3,059	......	8,201	3,650	3,000	19,110	25,760	328,500	145,000	527,500	4 5/8	4 3/4	3	19,110	262,730
" 15..	6,595	......	......	1,345	......	7,940	24,930	12,000	2,100	39,030	314,000	184,000	508,000	4 3/4	4 7/8	2 7/8	2,100	287,660
" 22..	3,592	3,302	......	2,388	696	9,978	23,570	14,050	2,300	39,920	300,500	192,000	492,500	4 3/4	4 7/8	2 7/8	2,300	311,230
" 29..	40,431	1,666	......	......	221	42,318	16,210	2,000	1,700	19,910	327,000	188,500	515,500	4 3/4	4 7/8	2 7/8	1,700	327,440
May 6..	88,642	......	......	10,759	......	99,401	16,330	1,200	1,000	18,530	402,500	195,000	597.500	4 5/8	4 3/4	2 7/8	1,000	343,770
" 13..	56,285	3,200	......	......	......	59,485	23,390	1,050	1,800	26,240	436,000	194,000	630,000	4 5/8	4 3/4	2 3/4	1,800	367,160
" 20..	23,045	......	......	......	875	23,920	24,150	5,000	2,100	31,250	434,500	180,500	625,000	4 5/8	4 3/4	2 3/4	2,100	391,310
" 27..	44,237	1,479	......	90	.114	45,920	32,030	9,200	1,100	42,330	452,000	186,000	638,000	4 3/4	4 7/8	2 3/4	1,100	423,340
June 3..	29,954	......	......	1,335	......	31,289	26,810	6,500	1,140	34,450	457,500	182,000	639,500	4 3/4	4 7/8	2 3/4	1,140	450,150
" 10..	6,470	......	......	......	214	6,684	22,400	2,500	1,710	26,610	443,000	178,500	621,500	4 3/4	4 7/8	2 7/8	1,710	472,550
" 17..	3,181	......	......	......	......	3,181	11,850	5,500	3,560	20,910	434,000	176,000	610,000	4 3/4	4 7/8	2 7/8	3,560	484,400
" 24..	36,535	12,104	......	......	......	48,639	22,990	1,400	2,150	26,540	450,500	182,500	633,000	4 5/8	4 3/4	2 7/8	2,150	507,390
July 1..	....	4,773	......	......	415	5,188	24,410	1,500	1,540	27,450	456,500	188,000	644,500	4 3/8	4 3/4	2 7/8	1,540	531,800
" 8..	36,070	5,379	......	2,507	......	43,956	21,340	1,500	1,400	24,240	473,500	192,500	666,000	4 5/8	4 3/4	2 7/8	1,400	553,140
" 15..	33,006	......	......	......	663	33,669	24,250	4,300	2,000	30,550	486,500	187,500	647,000	4 5/8	4 3/4	2 7/8	2,000	577,390
" 22..	10,255	......	1,539	1,202	127	13,123	23,190	11,800	3,000	37,990	476,000	184,500	660,500	4 5/8	4 3/4	2 7/8	3,000	600,580
" 29..	9,637	......	......	2,041	2,053	13,731	30,950	3,700	820	35,470	458,500	183,000	641,500	4 5/8	4 3/4	2 7/8	820	631,530
Aug. 5..	1,030	1,420	......	......	10	2,460	40,360	13,000	700	54,060	428,000	175,000	603,000	4 7/8	4 7/8	3	700	671,890
" 12..	28,122	2,554	1,046	1,232	......	32,954	23,010	11,250	2,200	36,460	436,500	174,000	610,500	5	5	3	2,200	694,900
" 19..	5,754	......	......	......	408	6,162	10,740	14,400	1,030	26,170	433,000	171,500	604,500	5 1/8	5 1/8	3 1/8	1,030	705,640
" 26..	11,110	4,505	......	......	......	15,615	22,680	20.400	690	43,770	427,500	170,000	597,500	5 1/4	5 1/4	3 1/4	670	728,320

Sep. 2..	1,480	2,536	......	781	292	5,089	15,470	4,950	680	21,100	418,000	168,500	586,500	$5\frac{1}{8}$	$5\frac{1}{4}$	$3\frac{1}{4}$	680	743,790
" 9..	225	4,990	......	2,762	92	8,069	11,670	1,200	1,200	14,070	408,000	172,500	580,500	$5\frac{1}{8}$	$5\frac{1}{4}$	$3\frac{1}{4}$	1,200	755,460
" 16..	3,852	16,765	......	......	1,966	22,583	11,410	500	500	12,410	402,500	188,000	590,500	$5\frac{1}{8}$	$5\frac{1}{4}$	$3\frac{1}{4}$	500	766,870
" 23..	3,606	5,573	......	3,697	2,124	15,000	12,010	......	700	12,710	396,000	197,000	593.000	$5\frac{1}{8}$	$5\frac{1}{4}$	$3\frac{1}{4}$	700	778,880
" 30..	1,575	......	2,294	1,089	452	5,410	15,070	1,300	450	16,820	350,500	190,000	540,500	5	$5\frac{1}{8}$	$3\frac{1}{4}$	450	793,950
Oct. 7..	......	......	1,180	......	1,116	1,296	13,280	2,500	700	16,480	340,000	189.500	529.500	5	$5\frac{1}{8}$	3	700	807,230
" 14..	2,431	6,678	2,852	......	153	12,114	15,290	1,600	350	17,240	331,000	194,500	525,500	$4\frac{7}{8}$	5	3	350	822,520
" 21..	3,661	3,295	......	......	3,354	10,310	13,930	1,600	500	16,030	323,500	198,000	521,500	$4\frac{7}{8}$	5	3	500	836,450
" 28..	4,770	3,070	......	3,676	52	11,568	17,560	10,000	300	27,860	316,000	196,000	512,000	5	$5\frac{1}{8}$	3	300	854,010
Nov. 4..	532	......	......	......	15	547	19,590	3,700	470	23,760	300,000	192,500	492,500	5	$5\frac{1}{8}$	3	470	873,600
" 11..	1,263	1,968	......	......	......	3,231	29,120	3,500	200	32,820	278,500	189.500	467,000	$4\frac{7}{8}$	5	3	200	902,720
" 18..	5,077	6,203	......	6,653	1,154	19,087	24,360	2,000	400	26,760	265,500	195,000	460,500	$4\frac{7}{8}$	5	3	400	927,080
" 25..	6,698	6,081	1,070	3,688	......	17,537	33,690	15,000	......	48,690	247,000	198,000	445,000	$5\frac{1}{8}$	$5\frac{1}{4}$	3	......	960,770
Dec. 2..	11,917	1,593	......	2,095	54	15,659	27,850	19,100	100	47,050	240,000	190,500	430,500	$5\frac{1}{8}$	$5\frac{1}{4}$	3	100	988,620
" 9..	27,838	5,241	1,807	1,620	374	35,880	16,150	6,500	500	23,150	254,000	195,500	449,500	5	$5\frac{1}{8}$	3	500	1,004,770
" 16..	21,463	3,459	......	320	884	26,126	20,320	1,750	200	22,270	260,500	184,500	445,000	$4\frac{7}{8}$	5	3	200	1,025,090
" 23..	14,481	......	......	6,962	929	22,372	26,800	6,500	150	33,450	254.000	197,500	451,500	$4\frac{7}{8}$	5	3	150	1,051,890
" 30..	30,874	......	......	......	330	31,204	29,300	6,500	300	36,100	255,000	194,500	449,500	$4\frac{7}{8}$	5	3	300	1,081,190
Average prices & total sales, receipts & stocks.	931,612	165026	17,340	80,662	22,620	1,217,260	1,081,190	258,950	78,280	1,418,420				4.86	4.98	2.98	78,280	2,079,300

1843.

COTTON CROP OF THE UNITED STATES.

Statement and Total Amount of the Cotton Crop of the United States, for the Year ending August 31, 1843.

	Bales.	Bales.	Total.	Same period 1842.
NEW ORLEANS.				
Export—				
To Foreign Ports	954,738			
Coastwise	134,132			
Burnt and damaged	500			
Stock on hand 1st September, 1843	4,700			
		1,094,070		
Deduct—				
Stock on hand 1st September, 1842	4,428			
Received from Mobile	10,687			
" Florida	3,381			
" Texas	15,328			
		33,824		
			1,060,246	727,658
ALABAMA.				
Export from Mobile—				
To Foreign Ports	366,012			
Coastwise	115,882			
Stock in Mobile 1st September, 1843	1,128			
		483,022		
Deduct—				
Stock in Mobile 1st September, 1842	422			
Received from Florida	886			
		1,308		
			481,714	318,315
FLORIDA.				
Export—				
To Foreign Ports	58,901			
Coastwise	102,237			
Stock on hand 1st September, 1843	200			
		161,338		
Deduct—				
Stock on hand 1st September, 1842	....	250		
			161,088	114,416
GEORGIA.				
Export from Savannah—				
To Foreign Ports—Upland	186,655			
" Sea Island	6,444			
Coastwise—Uplands	86,681			
Sea Islands	1,046			
	280,826			
Export from Darien—				
To New York and Providence	13,656			
Stock in Savannah 1st September, 1843	3,347			
" Augusta and Hambro', 1st September, 1843	7,401			
		305,230		
Deduct—				
Stock in Savannah and Augusta, 1st September, 1842.	5,110			
Received from Florida	629			
		5,739		
			299,491	232,271

Statement and Total Amount of the Cotton Crop of the United States, for the Year ending August 31, 1843.—*Concluded.*

	Bales.	Bales.	Total.	Same period 1842.
SOUTH CAROLINA.				
Export from Charleston—				
To foreign ports—Uplands	257,035			
" Sea Island	16,351			
Coastwise—Uplands	78,523			
Sea Island	681			
	352,590			
Export from Georgetown—				
To New York and Providence	13,042			
Stock in Charleston 1st September, 1843	8,274			
		373,906		
Deduct—				
Stock in Charleston 1st September, 1842	2,747			
Received from Savannah	14,916			
" Florida and Key West	4,585			
		22,248		
			351,658	260,164
NORTH CAROLINA.				
Export—				
To Foreign Ports	512			
Coastwise	8,577			
Stock on hand 1st September, 1843	200			
		9,289		
Deduct—				
Stock on hand 1st September, 1842	....	250		
			9,039	9,737
VIRGINIA.				
Export—				
To Foreign Ports	1,917			
Manufactured	9,347			
Stock on hand 1st September, 1843	975			
		12,239		
Deduct—				
Stock on hand 1st September, 1842	....	100		
			12,139	19,013
Received at Philadelphia and Baltimore, overland	....	....	3,500	2,000
Total crop of the United States	....	....	2,378,875	1,683,574
Crop of last year	....	....	1,683,574	
Increase	....	....	695,301	

Export to Foreign Ports, from September 1, 1842, *to August* 31, 1843.

FROM	To Great Britain.	To France.	To North of Europe.	Other Foreign Ports.	Total.
New Orleans (bales)	679,438	180,875	50,882	43,543	954,738
Alabama	283,382	55,421	8,032	19,177	366,012
Florida	53,005	4,196	....	1,700	58,901
Georgia (Savannah and Darien)	169,676	15,126	6,621	1,676	193,099
South Carolina	201,645	53,725	15,646	2,370	273,386
North Carolina	512	....	....	....	512
Virginia	1,735	....	182	....	1,917
Baltimore	....	....	246	....	246
Philadelphia	1,059	....	....	....	1,059
New York	79,259	36,796	35,340	6,311	157,706
Boston	....	....	845	1,716	2,561
Grand total	1,469,711	346,139	117,794	76,493	2,010,137
Total last year	935,631	398,129	79,956	51,531	1,465,249
Increase	534,080	....	37,838	24,962	544,888
Decrease	....	51,990	....	....	....

☞ The shipments from Mississippi are included in the export from New Orleans.

Growth.

Total crop of 1824–5...... bales.	560,000	
" 1825–6	710,000	
" 1826–7	937,000	
" 1827–8	712,000	
" 1828–9	857,744	
" 1829–30	976,845	
" 1830–1	1,038,848	
" 1831–2	987,477	
" 1832–3	1,070,438	
" 1833–4	1,205,394	
Total crop of 1834–5...... bales.	1,254,328	
" 1835–6	1,360,725	
" 1836–7	1,422,930	
" 1837–8	1,801,497	
" 1838–9	1,360,532	
" 1839–40	2,177,835	
" 1840–1	1,634,945	
" 1841–2	1,683,574	
" 1842–3	2,378,875	

Consumption.

Total crop of the United States, as above stated................bales.			2,378,875
Add—Stocks on hand at the commencement of the year, September 1, 1842, in the Southern ports		13,307	
" Northern ports		18,500	
			31,807
Makes a supply of			2,410,682
Deduct therefrom—The export to Foreign ports	2,010,137		
Less—Texas and other foreign	20,070		
		1,990,067	
Stocks on hand at the close of the year, September 1, 1843.			
In the Southern ports	26,225		
In the Northern ports	68,261		
		94,486	
Burnt and lost at New Orleans	500		
" New York	500		
		1,000	
			2,085,553
			325,129

Quantity consumed by and in the hands of Manufacturers.

1842–3 ....bales.	325,129	1833–4 ....bales.	196,413
1841–2..........................	267,850	1832–3..........................	194.412
1840–1..........................	297,288	1831–2..........................	173,800
1839–40........................	295,193	1830–1..........................	182,142
1838–9..........................	276,018	1829–30........................	126,512
1837–8..........................	246,063	1828–9..........................	118,853
1836–7..........................	222,540	1827–8..........................	120,593
1835–6..........................	236,733	1826–7..........................	103,483
1834–5..........................	216,888		

Note.—It will be seen, that we have deducted from the New Orleans statement, the quantity received at that port from Texas—Texas being a foreign country.

Our estimate of the quantity taken for consumption, does not include any cotton manufactured in the States, south and west of Virginia, nor any in that State, except in the vicinity of Petersburg and Richmond.

Of the new crop, now gathering, but little over 300 bales had been received previous to 1st inst.

The general tenor of the accounts from the cotton growing States leads to the conclusion, that the crop now coming in will not reach that of last year by several hundred thousand bales. The article is subject to so many vicissitudes, that no certain calculation can be made as yet as to the quantity that may reach the market.

ANNUAL REVIEW.

From the New Orleans Price Current, 1842—43.

Among the peculiar features developed during the past season, we may notice the increased activity and extension of our home manufactures, consequent upon a more favorable adjustment of the tariff and the opening of a trade in goods and cotton with China. This last new resource has already attained some considerable importance, in view of the brief period that has elapsed since its commencement; and a spirited rivalry appears to be maintained between our own country and Great Britain for ascendency in the markets of the Celestial Empire. Already, as we see stated in the Northern papers, have cotton goods to the extent of 15,000,000 yards been shipped from this country to China, while only 12,000,000 are known to have gone from England; and if the success of the American manufacturer in this enterprise be equal to that which has attended him in other parts of the world, in his competition with the British for the supply of heavy fabrics, the rivalry will not be of long duration, unless some modification of the duties on the raw material should give a more favorable position to the manufacturers of Great Britain. From our own port, two cargoes of raw cotton, amounting to 4,303 bales, have been shipped to Canton, and other shipments, though we know not to what extent, have been made from the ports of the North. With regard to the coming crop, we shall venture a few brief remarks, confining ourselves to those sections of the country which find an outlet in this market. We believe it to be generally conceded that the late spring which prevented planting at the proper period, and subsequent heavy and long continued rains during the growing season, have retarded the maturing of the plant some three to four weeks, a fact which may prove of essential detriment, as we believe it to be established that, however favorable other circumstances may be, the extent and quality of the crop mainly depend upon the duration and character of the picking season. From some sections also, complaints are made of attacks from the caterpillar, etc.; but these are only partial, though there seems to be a very general impression that, under all the circumstances which are known to have transpired, and to which we have above referred, no future union of incidents even the most favorable is likely to swell the production to an amount equal to the extraordinary yield of the past season.

LIVERPOOL STATEMENT FOR 1843.

UNITED STATES, 1842–1843.

Stock, Oct., '42..	32,000	Export.........	1,992,000
Receipts....	2,379,000	Consumption....	325,000
		Stock, Oct. 1, '43.	94,000
Bales............	2,411,000	Bales..........	2,411,000

Stock 1st Jan., 1843, in......	Liverpool.	Gt. Britain.	France.	Continent.	Tot. Europe.
United States.........Bales.	260,000	282,000	109,000	57,000	448,000
Brazil....................	57,000	57,000	2,000	4,000	63,000
West Indies................	18,000	22,000	12,000	9,000	43,000
East Indies................	101,000	178,000	2,000	6,000	186,000
Egypt....	20,000	22,000	13,000	32,000	67,000
Bales....................	456,000	561,000	138,000	108,000	807,000

CONSUMPTION, 1843.					IMPORT, 1843.					
Tot. Europe.	Continent.	France.	Gt. Britain.	Liverpool.		Liverpool.	Gt. Britain.	France.	Continent.	Tot. Europe.
1,689,000	193,000	351,000	1,145,000	1,045,000	....United States.........Bales.	1,287,000	1,397,000	334,000	200,000	1,880,000
99,000	8,000	8,000	83,000	83,000	....Brazil....................	99,000	99,000	12,000	9,000	115,000
61,000	13,000	20,000	28,000	23,000	....West Indies................	15,000	20,000	17,000	12,000	49,000
171,000	52,000	4,000	115,000	89,000	....East Indies................	111,000	182,000	5,000	51,000	186,000
135,000	71,000	23,000	41,000	39,000	....Egypt......................	46,000	46,000	21,000	116,000	173,000
2,155,000	337,000	406,000	1,412,000	1,279,000	.......................Bales,	1,558,000	1,744,000	389,000	388,000	2,403,000
....	10,000	....	108,000	81,000	Export.					
1,055,000	149,000	121,000	785,000	654,000	Stock, Dec. 31. Stock above,	456,000	561,000	138,000	108,000	807,000
3,210,000	496,000	527,000	2,305,000	2,014,000	Total supply, bales..............	2,014,000	2,305,000	527,000	496,000	3,210,000

COTTON AT LIVERPOOL. YEAR 1843.

Week Ending.	Receipts.						Sales.				Stocks.			Prices.			Actual Export	Con-sumption.
	Americ'n.	E. I.	Egypt.	Brazil.	Other.	Total.	Con-sumption.	Specu-lation.	Export.	Total.	Amer'n.	Other.	Total.	Mid. Up.	Mid. Orl.	Dhol		
Jan. 6..	18,201			459	959	19,619	19,470	4,000	500	23,970	263,500	193,000	456,500	$4\frac{7}{8}$	5	3	500	19,470
" 13..	16,116	1,714	2,458	7,438		27,726	22,200	4,600	450	27,250	255,000	200,000	455,000	$4\frac{7}{8}$	5	3	450	41,670
" 20..	24,221					24,221	17,140	3,500	550	21,190	261,500	196,500	458,000	$4\frac{3}{4}$	$4\frac{7}{8}$	3	550	58,810
" 27..	47,568			897	79	48,544	23,280	2,000	1,400	26,680	286,500	194,500	481,000	$4\frac{5}{8}$	$4\frac{3}{4}$	3	1,400	82,090
Feb. 3..	63,451			5,448	515	69,414	23,600	4,000	1,200	28,800	326,500	197,000	523,500	$4\frac{3}{8}$	$4\frac{5}{8}$	3	1,200	105,690
" 10..	9,910		1,258			11,168	28,130	9,500	700	38,330	309,500	194,500	504,000	$4\frac{3}{8}$	$4\frac{5}{8}$	3	700	133,820
" 17..	907		987	1,178		3,072	22,910	4,000	1,000	27,910	289,500	193,500	483,000	$4\frac{1}{4}$	$4\frac{3}{8}$	3	1,000	156,730
" 24..			1,297	1,220		2,517	31,480	14,000	400	45,880	264,000	190,000	454,000	$4\frac{1}{4}$	$4\frac{3}{8}$	3	400	188,210
Mch 3..	5,923		1,297	1,220		8,440	12,390	3,000	1,300	16,690	258,500	189,500	448,000	$4\frac{1}{4}$	$4\frac{3}{8}$	3	1,300	200,600
" 10..	60,237			3,437	164	63,838	18,130	4,000	500	22,630	301,500	190,500	492,000	$4\frac{1}{4}$	$4\frac{3}{8}$	3	500	218,730
" 17..	177,256	2,172		1,625	2,362	183,415	23,160	3,300	400	26,860	456,000	192,500	648.500	$4\frac{1}{4}$	$4\frac{3}{8}$	3	400	241,890
" 24..	28,581	3,143		1,480	15	33,219	37,850	14,000	300	52,150	446,000	193,500	639,500	$4\frac{1}{4}$	$4\frac{3}{8}$	3	300	279,740
" 31..	4,333	3,607				7,940	29,630	23,000	3,400	56,030	420,000	189,000	609,000	$4\frac{1}{4}$	$4\frac{3}{8}$	3	3,400	309,370
April 7..	102,680			2,625	1,108	106,413	17,690	5,300	1,920	24,910	504,500	189,000	693,500	$4\frac{1}{8}$	$4\frac{1}{4}$	$2\frac{3}{4}$	1,920	327,060
" 13..	35,611		3,493	2,348	590	42,042	19,480	9,000	1,200	29,680	523,000	189,500	712,500	$4\frac{1}{8}$	$4\frac{1}{4}$	$2\frac{3}{4}$	1,200	346,540
" 21..	56,726	2,216	2,042	2,658		63,642	26,230	5,000	2,800	34,030	553,500	191,000	744,500	$4\frac{1}{8}$	$4\frac{1}{4}$	$2\frac{3}{4}$	2,800	372,770
" 28..	76,673			3,137	1,189	80,999	19,230	5,000	1.900	26,130	607,500	181,500	799,000	4	$4\frac{1}{8}$	$2\frac{3}{4}$	1,900	392,000
May 5..	11,859				877	12,736	25,850	12,000	3,370	41,220	591,500	188,000	779,500	4	$4\frac{1}{8}$	$2\frac{3}{4}$	3,370	417,850
" 12..	13,588	1,639		2,767	87	18,081	16,600	31,700	700	49,000	592,000	187,500	779,500	$4\frac{1}{8}$	$4\frac{1}{4}$	$2\frac{5}{8}$	700	434,450
" 19..	18,165	5,770	2,661		554	27,150	22,930	11,500	750	35,180	587,500	191,000	778,500	$4\frac{1}{8}$	$4\frac{1}{4}$	$2\frac{5}{8}$	750	457,380
" 26..	68,520	2,489	1,400	1,258	106	73,773	17,860	4,700	1,250	23,810	638,500	193,000	831,500	$4\frac{1}{8}$	$4\frac{1}{4}$	$2\frac{5}{8}$	1,250	475,240
June 2..	38,908	8,729		3,971	316	51,924	25,620	2,000	1,000	28,620	650,000	198,000	848,000	$4\frac{1}{8}$	$4\frac{1}{4}$	$2\frac{5}{8}$	1,000	500,860
" 9..	77,093	2,681	3,290	1,997	760	85,821	22,700	1,000	1,000	24,700	706,000	204,000	910,000	$4\frac{1}{8}$	$4\frac{1}{4}$	$2\frac{5}{8}$	1,000	523,560
" 16..	43,488				1,400	44,888	21,800	500	900	23,200	730,000	201,500	931,500	4	$4\frac{1}{8}$	$2\frac{5}{8}$	900	545,360
" 23..	23,295				1,836	25,131	20,410	1,700	300	22,410	736,000	199,500	935,500	4	$4\frac{1}{8}$	$2\frac{5}{8}$	300	565,570
" 30..	22,090		1,752		20	23,862	22,870	1,000	2,500	26,370	714,000	198,000	912,000	$3\frac{7}{8}$	4	$2\frac{1}{2}$	2,500	588,640
July 7..	74,066	1,909		2,815	227	79,017	25,930	750	1,350	27,030	764,500	199,000	963,500	$3\frac{7}{8}$	4	$2\frac{1}{2}$	1,350	614,570
" 14..	8,728	14,350		4,060		27,138	20,200	2,600	3,000	25,800	754,065	213,349	967,414	$3\frac{7}{8}$	4	$2\frac{1}{2}$	3,000	634,770
" 21..	17,961		1,760	3,280		23,001	26,700	2,500	7,350	36,550	744,500	211,500	956,000	4	$4\frac{1}{8}$	$2\frac{1}{2}$	7,350	661,470
" 28..	22,713		3,103	3,759		29,575	22,870	3,600	1,900	28,370	748,500	213,000	961,500	4	$4\frac{1}{8}$	$2\frac{1}{2}$	1,900	684,340
Aug. 4..	12,554		1,110	1,128	121	14,913	27,110	3,500	850	31,460	732,500	208,500	941,000	4	$4\frac{1}{8}$	$2\frac{1}{2}$	850	711,450
" 11..	19,683		1,499	900		22,082	20,340	6,600	1,650	28,590	735,500	205,000	940,500	4	$4\frac{1}{8}$	$2\frac{1}{2}$	1,650	731,790
" 18..	5,277	5,454	1,616		622	12,969	36,940	9,500	750	47,190	710,000	204,000	914,000	$4\frac{1}{8}$	$4\frac{1}{4}$	$2\frac{7}{8}$	750	768,730
" 25..	5,401	4,095	2,256	1,830	105	13,687	30.460	17.000	1,150	48,610	690,500	197,500	898,000	$4\frac{1}{4}$	$4\frac{3}{8}$	$2\frac{5}{8}$	1.150	799,190

Sep. 1..	6,168	......		......	3,020	9,188	31,680	29,300	2,150	63,130	670,500	201,000	871,500	$4\frac{3}{8}$	$4\frac{1}{2}$	$2\frac{3}{4}$	2,150	830,870
" 8..	6,698	2,457	......	600	......	9,755	25,430	12,200	1,830	39,460	654,500	200,000	854,500	$4\frac{3}{8}$	$4\frac{1}{2}$	$2\frac{3}{4}$	1,830	856,300
" 15..	2,637	4,152	......	1,042	1,353	9,184	27,990	20,200	1,500	49,690	638,000	196,000	834,000	$4\frac{3}{8}$	$4\frac{1}{2}$	$2\frac{3}{4}$	1,500	884,290
" 22..	1,762	12,644	......	2,958	450	17,814	21,910	18,000	570	40,480	622,500	207,000	829,500	$4\frac{3}{8}$	$4\frac{1}{2}$	$2\frac{3}{4}$	570	906,200
" 29..	......	1,698	......	3,315	......	5,013	21,090	2,500	700	24,290	604,500	208,000	812,500	$4\frac{3}{8}$	$4\frac{1}{2}$	$2\frac{3}{4}$	700	927,290
Oct. 6..	750	......	5,376	952	880	7,958	41,580	31,000	1,500	74,080	571,500	206,000	777,500	$4\frac{5}{8}$	$4\frac{3}{4}$	3	1,500	968,870
" 13..	721	4,337	......	1,666	......	6,724	39,180	52,400	......	91,580	545,500	200,000	745,500	5	$5\frac{1}{8}$	$3\frac{1}{8}$	......	1,008,050
" 20..	1,817	......	......	1,803	......	3,620	17,770	11,000	300	29,070	533,000	198,000	731,000	$4\frac{7}{8}$	5	$3\frac{1}{8}$	300	1,025,820
" 27..	632	2,763	1,377	3,342	......	8,114	15,080	3,000	200	18,280	521,000	202,000	723,000	$4\frac{3}{4}$	$4\frac{7}{8}$	3	200	1,040,900
Nov. 3..	716	2,622	......	4,532	......	7,870	11,050	5,000	......	16,050	511,000	167,500	718,500	$4\frac{3}{4}$	$4\frac{7}{8}$	3	......	1,051,950
" 10..	1,284	2,699	......	1,831	268	6,082	24,240	3,450	300	27,990	490,500	209,500	700,000	$4\frac{3}{4}$	$4\frac{7}{8}$	3	300	1,076,190
" 17..	......	......	......	2,906	62	2,968	19,580	11,000	......	30,580	474,000	209,500	683,500	$4\frac{3}{4}$	$4\frac{7}{8}$	3	......	1,095,770
" 24..	5,411	3,149	3,672	5,414	......	17,646	20,630	11,000	......	31,630	463,000	227,000	680,000	$4\frac{3}{4}$	$4\frac{7}{8}$	3	......	1,116,400
Dec. 1..	6,090	......	385	614	643	7,732	25,360	17,000	......	42,360	450,000	211,500	661,500	$4\frac{7}{8}$	5	3	......	1,141,760
" 8..	9,868	4,212	......	1,228	......	15,308	19,710	5,000	......	24,710	443,500	203,500	757,000	$4\frac{7}{8}$	5	3	......	1,161,470
" 15..	7,611	6,478	......	......	......	14,089	26,140	8,100	......	34,240	429,000	216,500	645,500	$4\frac{7}{8}$	5	3	......	1,187,610
" 22..	9,528	1,550	1,560	......	......	12,638	19,730	4,000	550	24,280	423,500	204,000	637,500	$4\frac{7}{8}$	5	3	550	1,207,340
" 29..	18,231	......	......	1,876	1,988	22,095	30,660	12,700	......	43,360	416,000	211,000	627,000	5	$5\frac{1}{8}$	3	......	1,238,000
Average prices & total sales, receipts & stocks.	1,291,807	108729	45,649	97,004	19,679	1,562,868	1,238,000	486,200	59,290	1,775,490				4.37	4.5	2.83	59,290	2,380,769

The semi-weekly Price and Weekly Sales and Receipts at New York, Weekly Exports from New York and Rates of Freight to Liverpool 1st of each month, for the Crop Year ending September 1, 1843.

1842.		Price of Fair New Orleans	Price of Fair Upland.	Sales for week.	Receipts for week.	EXPORTS FOR THE WEEK.					Rates of Freight to Liverpool.
						To Great Britain.	To France.	North of Europe.	Other Fo'n Ports	Total Exports.	
Septem.	6..	0@9¼	0@8½	3,850	783	....	....	....	....	....	3-16@¼d
"	9..	0@9¼	0@8½								
"	13..	0@9¼	0@8½	3,250	357	1,543	....	440	....	1,983	
"	16..	0@9¼	0@8½								
"	20..	0@9¼	0@8½	3,000	4,215	673	....	1,079	....	1,752	
"	23.	0@9¼	0@8½								
"	27..	0@9¼	0@8½	2,950	414	1,124	....	....	237	1,361	
"	30..	0@9¼	0@8½								
October	4..	0@9¼	0@8½	2,300	1,226	1,156	133	614	134	2,037	3-16@¼d
"	7..	0@9¼	0@8½								
"	11..	0@9¼	0@8½	2,800	2,595	1,062	754	125	....	1,941	
"	14..	0@9¼	0@8½								
"	18..	0@9¼	0@8½	1,500	5,250	878	406	103	....	1,387	
"	21..	9@9¼	8@8½								
"	25..	9@9¼	8@8½	2,400	2,823	1,121	99	207	175	1,602	
"	28..	9@9¼	8@8½								
Novem.	1..	9@9¼	8@8¼	3,220	2,940	1,204	491	209	....	1,904	¼@5-16d
"	4..	8¾@9	7¾@8								
"	8..	8½@8¾	7¾@8	3,100	1,694	....	....	....	....	....	
"	11..	8¼@8½	7¾@8								
"	15..	8¼@8½	7⅞@8	5,550	13,772	927	779	533	358	2,597	
"	18..	8⅛@8⅜	7⅝@7⅞								
"	22..	8⅛@8⅜	7⅝@7⅞	4,700	7,326	1,090	492	142	....	1,724	
"	25..	8@8¼	7¼@7½								
"	29..	8@8¼	7¼@7½	4,850	5,754	1,513	918	279	....	2,710	
Decem.	2..	7½@7¾	7@7¼								⅜@½d.
"	6..	7½@7¾	7@7¼	5,050	8,095	1,385	1,062	48	....	2,495	
"	9..	7½@7¾	7@7¼								
"	13..	7½@0	7@0	2,940	8,247	526	608	50	....	1,184	
"	16..	7½@0	7@0								
"	20..	7½@0	7@0	6,550	6,230	1,614	1,260	272	....	3,146	
"	23..	7⅝@7⅞	7@7¼								
"	27..	7⅝@7⅞	7@7¼	8,950	7,238	3,118	602	....	482	4,202	

GENERAL REMARKS.

In December, 1842, advices were received from England of a conclusion of a treaty of peace between that nation and China, and the consequent opening of five of the northern ports of China, hitherto closed to foreigners; this enlargement of markets to manufactured goods gave an impetus to the demand for cotton, and prices at once advanced ⅜ cents per pound.

The early part of the crop year was very unfavorable for the staple, the season being wet and very backward, the bloom not appearing in Georgia until June 12, and in Louisiana June 9; speculation in consequence was rife respecting a short crop, as the crop of 1838, which did not bloom through the cotton belt until June 14, was 446,000 bales less than the crop of the previous year; later in the season frosts occurred early, October 14 and 18, which assisted in curtailing the crop, and through a considerable portion of the year there was a strong speculative feeling, though prices did not vary very greatly.

Decem. 30 .	7⅝@7⅞	7@7¼								
1843.										
January 3..	7⅝@7⅞	7@7¼	4,350	13,066	4,827	2,710	989	927	9,453	½@⅝d.
" 6..	7⅝@7⅞	7@7¼								
" 10..	7⅜@7⅝	6¾@7	7,650	12,742	....	....	....	....	....	
" 13..	7⅜@7⅝	6¾@7								
" 17..	7⅜@7⅝	6¾@7	6,200	13,874	4,831	1,208	635	....	6,674	
" 20..	7@7¼	6⅝@6⅞								
" 24..	7@7¼	6⅝@6⅞	7,650	8,977	3,929	1,255	1,005	....	6,189	
" 27..	7@7¼	6⅝@6⅞								
" 31..	7@7¼	6⅝@6⅞	9,300	4,966	3,368	1,369	842	....	5,579	
February 3..	7@7¼	6⅝@6⅞								⅝@¾d.
" 7..	7@7¼	6⅝@6⅞	7,250	9,666	3,038	320	3,755	....	7,113	
" 10..	7@7¼	6⅝@6⅞								
" 14..	7@7¼	6⅝@6⅞	5,950	10,539	3,040	2,444	987	62	6,533	
" 17..	7@7¼	6⅝@6⅞								
" 21..	7@7¼	6⅝@6⅞	5,000	5,277	4,891	891	2,791	876	9,449	
" 24..	6¾@7	6¼@6½								
" 28..	6¾@7	6⅜@6¾	6,500	4,931	997	673	1,590	....	3,260	
March 3..	6¾@7	6⅜@6¾								⅝@¾d.
" 7..	6¾@7	6⅜@6¾	5,150	1,145	2,584	....	377	....	2,961	
" 10..	6¾@7	6⅜@6¾								
" 14..	6¾@7	6⅛@6½	2,700	8,770	1,627	1,753	699	121	4,200	
" 17..	6¾@7	6⅛@6½								
" 21..	6¾@7	0@6½	5,800	7,745	1,007	395	2,205	....	3,607	
" 24..	6¾@7	0@6½								
" 28..	6¾@7	6½@6¾	7,150	3,236	4,041	....	805	....	4,846	
" 31..	6¾@7	6½@6¾								⅜@½d.
April 4..	6¾@7	6½@6¾	4,800	7,594	2,668	1,169	1,729	80	5,646	
" 7..	6¾@7	6½@6¾								
" 11..	6¾@7	6½@6¾	7,050	4,427	774	960	1,737	....	3,471	
" 14..	6¾@7	6½@6¾								
" 18..	7@7¼	6½@6¾	5,950	2,143	887	791	602	95	2,375	
" 21..	7½@0	6¾@7								
" 25..	7¼@7½	6¾@7	8,050	5,175	3,878	1,298	544	....	5,720	
" 28..	7¼@7½	6¾@7								
May 2..	7½@7¾	7@7¼	6,550	3,942	1,816	1,407	910	1,419	5,552	¼@5-16d
" 5..	7½@7¾	7@7¼								
" 9..	7½@7¾	7@7¼	6,550	2,898	....	817	466	....	1,283	
" 12..	7½@7¾	7@7¼								
" 16	7½@7¾	7@7¼	2,950	9,184	1,088	62	450	370	1,970	
" 19..	7½@7¾	7@7¼								

New York Statement for 1843.—Concluded.

1843.		Price of Fair New Orleans	Price of Fair Upland.	Sales for week.	Receipts for week.	EXPORTS FOR THE WEEK.					Rates of Freight to Liverpool.
						To Great Britain.	To France.	North of Europe.	Other Fo'n Ports	Total Exports.	
May	23..	7½@7¾	7@7¼	3,450	7,091	595	94	1,960	....	2,649	
"	26..	7½@7¾	7@7¼								
"	30..	7¾@8	7¼@7½	8,750	6,138	437	980	526	....	1,943	
June	2..	7¾@8	7¼@7½								3-16@¼d
"	6..	7¾@8	7¼@7½	7,150	8,138	907	879	60	....	1,846	
"	9..	7¾@8	7¼@7½								
"	13..	7½@7¾	7@7¼	4,250	7,850	31	577	413	322	1,343	
"	16..	7½@7¾	7@7¼								
"	20..	7½@7¾	7@7¼	2,200	7,736	416	....	135	618	1,169	
"	23..	7½@7¾	7@7¼								
"	27..	7½@7¾	7@7¼	4,450	6,253	1,195	10	982	....	2,187	
"	30..	7½@7¾	7@7¼								
July	4..	7½@7¾	7@7¼	3,600	6,750	575	412	860	45	1,892	3-16@¼d
"	7..	7¼@7½	6¾@7								
"	11..	7¼@7½	6¾@7	3,350	4,763	494	1,561	270	....	2,325	
"	14..	7¼@7½	6¾@7								
"	18..	7¼@7½	6¾@7	3,500	5,541	615	61	459	....	1,135	
"	21..	7¼@7½	6¾@7								
"	25..	7¼@7½	6¾@7	4,750	7,804	1,312	846	99	....	2,257	
"	28..	7¼@7½	6¾@7								
August	1..	7¼@7½	6¾@7	1,950	4,495	1,382	2,622	256	....	4,260	3-16@¼d
"	4..	7¼@7½	6¾@7								
"	8..	7½@7¾	7@7¼	4,650	6,378	....	....	....	....	....	
"	11..	7½@7¾	7@7¼								
"	15..	7⅝@7⅞	7¼@7½	6,800	2,891	358	587	1,294	....	2,239	
"	18..	7¾@8	7¼@7½								
"	22	8@8¼	7½@7¾	6,300	533	1,205	523	537	....	2,265	
"	25..	8⅛@8⅜	7⅝@7⅞								
"	29..	8⅛@8⅜	7⅝@7⅞	12,700	2,903	1,502	518	270	....	2,290	
Septem.	1..	8¼@8½	7¾@8								3-16@¼d
Average price and total sales, receipts and exports.		7.74	7.25	267,360	304,320	79,249	36,796	35,340	6,321	157,706	

GENERAL REMARKS.

Exchange

Bills on London ruled, through September, between 7 and 8¼ per cent. premium; in October the quotation fell from 7½@8 to 5½@6; in November, the range was from 5¾ to 6¾; in December, 6½ to 5½; in January, 5¼@5¾; in February, 5¼ to 6¼; in March, 5¼@6; in April, rates went up from 5½@6 to 6¾@7; in May, the range was from 7½ up to 9; in June, the quotation was steady at 8½@8¾; in July, 8¾@9; in August, 8¾@9¼; and in September, 9@9¼ per cent premium.

The expansion to which I have already alluded, as having commenced in 1842, showed itself at once by building cotton factories. In New England this became the rage.

Of course the politicians who passed the highly protective tariff of 1842, like the fly on the chariot wheel, thought they had done it all. It was even thought that under the fostering care of a high protective tariff, we could beat England in the markets of the world, especially in China, the trade of which was then attracting much attention. We might have done so if England had maintained still higher import duties than we ; but by abolishing all protective duties, she has not only driven our manufactures out of nearly all foreign markets, but has replaced our ships with her own in nearly every sea.

The politician is fortunate who passes his measures and incorporates his policy into the laws at the commencement of a period of expansion. Reputations are sometimes gained in that way, but they do not endure. Sir Robert Peel effected the repeal of the corn laws in England a short time previous to the revulsion and famine of 1847. It was a severe ordeal, but both the statesman and his policy stood it triumphantly.

The largest increase in the cotton crop in any one year, in proportion to its total, was in 1839, when it reached nearly 64 per cent. The percentage of increase in 1842, was about the same as in 1870.

1844.

COTTON CROP OF THE UNITED STATES.

Statement and Total Amount of the Cotton Crop of the United States, for the Year ending August 31, 1844.

	Bales.	Bales.	Total.	Same period 1843.
NEW-ORLEANS.				
Export—				
To Foreign Ports	718,417			
Coastwise	176,958			
Burnt and lost	7,245			
Stock on hand, 1st September, 1844	12,934			
		915,554		
Deduct—				
Stock on hand, 1st September, 1843	4,700			
Received from Mobile	47,596			
" " Florida	12,916			
" " Texas	18,170			
		83,382		
			832,172	1,060,246
ALABAMA.				
Export from Mobile—				
To Foreign Ports	269,526			
Coastwise	195,679			
Stock in Mobile, 1st September, 1844	4,175			
Burnt	473			
		469,853		
Deduct—				
Stock in Mobile, 1st September, 1843	1,128			
Received from Florida	735			
		1,863		
			467,990	481,714
FLORIDA.				
Export—				
To Foreign Ports	29,393			
Coastwise	116,069			
Stock on hand, 1st September, 1844	300			
		145,762		
Deduct—				
Stock on hand, 1st September, 1843	....	200		
			145,562	161,088
GEORGIA.				
Export from Savannah—				
To Foreign Ports—Uplands	127,410			
Sea Island	3,554			
Coastwise—Uplands	111,460			
Sea Island	2,151			
	244,575			
Export from Darien—				
To New York	1,411			
Burnt	700			
Stock in Savannah, 1st September, 1844	2,161			
" Augusta and Hambro', 1st September, 1844	17,498			
		266,345		
Deduct—				
Stock in Savannah and Augusta, 1st September, 1843	....	10,748		
			255,597	299,491

Statement and Total Amount of the Cotton Crop of the United States, for the Year ending August 31, 1844.—*Concluded.*

	Bales.	Bales.	Total.	Same period 1843.
SOUTH CAROLINA.				
Export from Charleston—				
To Foreign Ports—Uplands	166,290			
Sea Island	15,043			
Coastwise—Uplands	123,023			
Sea Island	1,148			
	305,504			
Export from Georgetown—				
To New York	15,391			
Burnt	1,066			
Stock in Charleston, 1st September, 1844	13,536			
		335,497		
Deduct—				
Stock in Charleston, 1st September, 1843	8,284			
Received from Savannah	20,699			
" " Florida, Key West, &c	1,644			
		30,627		
			304,870	351,658
NORTH CAROLINA.				
Export—				
Coastwise	8,618			
Stock on hand, 1st September, 1844	200			
		8,818		
Deduct—				
Stock on hand, 1st September, 1843	....	200		
			8,618	9,039
VIRGINIA.				
Export—				
To Foreign Ports	2,638			
Manufactured	10,687			
Stock on hand, 1st September, 1844	2,150			
		15,475		
Deduct—				
Stock on hand, 1st September, 1843	....	975		
			14,500	12,139
Received at Philadelphia and Baltimore, overland	....	....	1,100	3,500
Total Crop of the United States	....	....	2,030,409	2,378,875
Crop of last year	....	....	....	2,030,409
Decrease	....	....	....	348,466

Export to Foreign Ports, from September 1, 1843, *to August* 31, 1844.

FROM	To Great Britain.	To France.	To North of Europe.	Other F'n Ports.	Total.
New Orleans	527,675	119,980	17,907	52,855	718,417
Alabama	204,140	53,005	6,578	5,803	269,526
Florida	26,537	1,507	....	1,349	29,393
Georgia (Savannah and Darien)	119,873	8,134	1,617	1,340	130,964
South Carolina	137,389	36,620	7,324	....	181,333
North Carolina	....	....	....	....	....
Virginia	2,330	....	308	....	2,638
Baltimore	....	....	147	....	147
Philadelphia	5,022	503	595	454	6,574
New York	176,778	62,936	34,179	13,408	287,301
Boston	2,754	....	398	45	3,197
Grand total	1,202,498	282,685	69,053	75,254	1,629,490
Total last year	1,469,711	346,139	117,794	76,493	2,010,137
Decrease	267,213	63,454	48,741	1,239	380,647

The shipments from Mississippi are included in the export from New Orleans.

Growth.

Total crop of 1824–5	bales.	560,000	Total crop of 1834–5	bales.	1,254,328
" 1825–6		710,000	" 1835–6		1,360,725
" 1826–7		937,000	" 1836–7		1,422,930
" 1827–8		712,000	" 1837–8		1,801,497
" 1828–9		857,744	" 1838–9		1,360,532
" 1829–30		976,845	" 1839–40		2,177,835
" 1830–1		1,038,848	" 1840–1		1,634 945
" 1831–2		987,477	" 1841–2		1,683,574
" 1832–3		1,070,438	" 1842–3		2,378,875
" 1833–4		1,205,394	" 1843–4		2,030,409

Consumption.

Total crop of the United States, as above stated		2,030,409 bales.		
Add—Stocks on hand at the commencement of the year, 1st September, 1843.—In the Southern ports			26,225	
" Northern "			68,261	
				94,486
Makes a supply of				2,124,895
Deduct therefrom—The export to Foreign ports	1,629,490			
Less Texas and other foreign	21,695			
		1,607,795		
Stocks on hand at the close of the year, 1st September, 1844—				
In the Southern ports	52,954			
" Northern "	106,818			
		159,772		
Burnt and lost at New Orleans	7,245			
" Mobile	473			
" Savannah	700			
" Charleston	1,066			
" New-York	1,100			
		10,584		
				1,778,151
Bales				346,744

Quantity consumed by and in the hands of Manufacturers :—

1843–4 .... bales.	346,744	1834–5 .... bales.	216,888
1842–3	325,129	1833–4	196,413
1841–2	267,850	1832–3	194,412
1840–1	297,288	1831–2	173,800
1839–40	295,193	1830–1	182,142
1838–9	276,018	1829–30	126,512
1837–8	246,063	1828–9	118,853
1836–7	222,540	1827–8	120,593
1835–6	236,733	1826–7	103,483

It will be seen that we have deducted from the New Orleans statement the quantity received at that port from Texas—Texas being a foreign country.

Our estimate of the quantity taken for consumption, does not include any cotton manufactured in the States south and west of Virginia, nor any in that State, except in the vicinity of Petersburg and Richmond.

Of the crop now coming in, it may be remarked that the picking commenced three weeks earlier than last year. An increased quantity of land was put under cultivation—and unless some extraordinary fatality should occur during the remainder of the season, the quantity will probably exceed that of any previous year.

The quantity of new cotton received at the shipping ports up to the first inst. amounted to about 7,500 bales, against only 300 at the same time last year.

ANNUAL REVIEW.

From the New Orleans Price Current, 1843—44.

In regard to prices, it may be said that the market opened with favorable prospects; and had business, particularly in our great export staple, been suffered to move forward in its legitimate course, there is reason to believe that a steady, though moderate improvement upon the rates of the previous year would have been maintained. But at an early period of the season's progress low estimates of the amount of production, founded upon accounts from the interior, of extensive damage from various causes, induced a series of improvident speculations, which were soon found to have been started upon a mistaken basis; and the natural consequence has already been seen in a great and disastrous reaction. About the period in November, just referred to above, a speculative demand, which had its origin in New York, and which was based upon estimates of a very short crop, sprung up and continued to prevail to a greater or less extent until about the 1st of March. Under the influence of the excitement created by this demand, and aided by some operations on European account, prices rapidly advanced, attaining their highest elevation on the 20th December, middling being then quoted at 8¾ and 9, and fair at 9¾ and 10¼ cents, and they did not recede more than half a cent before the last of February, although the demand at times was very limited, as factors, for the reasons previously stated above, held on to their stocks with great tenacity, notwithstanding that the market was burthened with a supply, large, we believe, beyond precedent; extraordinary facilities being afforded for sending the crop forward, owing to the fine stage of water in all the streams tributary to the Mississippi. Judging from the information now in our possession, there is every prospect of good crops being produced in all the sections of country tributary to this port, except on the rich bottom lands on and adjacent to the Mississippi, where irreparable damage, and to an enormous extent, has been done to the crops, in consequence of the almost unprecedented overflow of the rivers.

LIVERPOOL STATEMENT FOR 1844.

UNITED STATES—1843–1844.			
Stock, 1st October, 1843	94,000	Export	1,617,000
Receipts	2,030,000	Consumption	347,000
		Stock, 1st October, 1844	160,000
Bales	2,124,000	Bales	2,124,000

20

Tot. Europe.	Continent.	France.	Gt. Britain.	Liverpool.		Liverpool.	Gt. Britain.	France.	Continent.	Tot. Europe.
					Stock Jan. 1, 1844, in					
					United States.........Bales.	441,000	483,000	92,000	67,000	642,000
					Brazil	68,000	68,000	6,000	5,000	79,000
					West Indies	10,000	14,000	9,000	8,000	31,000
					East Indies	108,000	193,000	3,000	5,000	201,000
					Egypt	27,000	27,000	11,000	64,000	102,000
					Bales	654,000	785,000	121,000	149,000	1,055,000
CONSUMPTION.					IMPORT.					
1,643 000	179,000	335,000	1,129,000	1,038,000	United States	1,159,000	1,248,000	295,000	179,000	1,662,000
133,000	8,000	10,000	115,000	115,000	Brazil	112,000	112,000	8,000	5,000	123,000
44,000	13,000	13,000	18,000	13,000	West Indies	15,000	18,000	17,000	12,000	47,000
198,000	65,000	6,000	127,000	94,000	East Indies	143,000	239,000	5,000	66,000	242,000
109,000	33,000	24,000	52,000	52,000	Egypt	63,000	66,000	20,000	40,000	99,000
2,127,000	298,000	388,000	1,441,000	1,312,000	Bales	1,492,000	1,683,000	345,000	302,000	2,173,000
....	27,000	....	130,000	85,000	Export.					
1,101,000	126,000	78,000	897,000	749,000	Stock Dec. 31. Stock above,	654,000	785,000	121,000	149,000	1,055,000
3,228,000	451,000	466,000	2,468,000	2,146,000	Total supply, bales	2,146,000	2,468,000	466,000	451,000	3,228,000

COTTON AT LIVERPOOL. YEAR 1844.

Week Ending.	Receipts.						Sales.				Stocks.			Prices.			Actual Export	Con-sumption.
	Americ'n.	E. I.	Egypt.	Brazil.	Other.	Total.	Con-sumption.	Specu-lation.	Export.	Total.	Amer'n.	Other.	Total.	Mid. Up.	Mid. Orl.	Dhol		
Jan. 5..	6,359	......	......	2,219	2,779	11,357	40,920	23,300	100	64,320	417,500	207,000	624,500	5 1/8	5 1/4	3 1/4	100	40,920
" 12..	22,446	......	......	2,210	1,197	25,853	13,100	32,800	450	46,350	431,000	205,500	636,500	5 1/4	5 3/8	3 3/8	450	54,029
" 19..	25,025	......	......	......	2,781	27,806	36,580	56,000	......	92.580	434,500	196,000	630,500	5 3/8	5 1/2	3 1/2	......	90,600
" 26..	6,416	......	......	......	1,912	8,328	25,880	16,900	......	42,780	420,500	192,000	612,500	5 3/8	5 1/2	3 1/2	......	116.480
Feb. 2..	28,896	4,458	......	5.262	522	39,138	25,370	84,200	......	109,570	430,000	194,000	624,000	5 3/4	5 7/8	3 7/8	......	141,850
" 9..	25,600	3,394	......	3,780	1,108	33,882	30,680	24.600	200	55,480	431.500	195,500	627,000	5 3/4	5 7/8	3 7/8	200	172,530
" 16..	9,899	......	......	......	1,338	11,237	23,410	17,100	......	40,510	422,500	192,000	422,500	5 3/4	5 7/8	3 7/8	......	195,940
" 23..	18.702	......	......	1,000	......	19,702	12,930	4 500	100	17,530	430,000	190,500	620,500	5 3/8	5 3/4	3 3/4	100	208,870
Mch 1..	16,635	2,112	......	4,559	338	23,644	21,890	9,500	......	31,390	427,500	193,000	620,500	5 5/8	5 3/4	3 3/4	......	230,760
" 8..	6,576	......	......	6.623	......	13,199	11,870	4,000	160	16,030	425,000	197,000	622,000	5 3/8	5 3/4	3 3/4	160	242,630
" 15..	5,014	......	......	2,741	......	7,755	18,800	7,500	500	26,800	413,000	196,000	609,000	5 1/2	5 5/8	3 5/8	500	261,430
" 22..	11,601	......	......	1,674	......	13,275	12.800	4,100	700	17,600	414,000	195,000	609,000	5 1/2	5 5/8	3 1/2	700	274.230
" 29..	21,351	3,498	......	2,900	480	28,229	2,410	11,200	500	14,110	425,000	200,000	625,000	5 1/4	5 3/8	3 3/8	500	276,640
April 4..	12,705	......	......	810	2,181	15,696	11,470	3,200	......	14,670	428,000	201,000	629,000	5 1/8	5 1/4	3 3/8	......	288,110
" 12..	45,390	1,223	......	3,253	1,293	51,159	26,580	7,000	500	34,080	448,000	203,500	651,500	5 1/8	5 1/4	3 3/8	500	314,690
" 19..	46,679	......	......	......	......	46,679	17,060	1,200	800	19,060	478,000	200,500	678,500	5	5 1/8	3 1/4	800	331,750
" 26..	9,547	188	......	980	16	10,731	18,290	4,500	1,300	24,090	470,000	199,500	669,500	5	5 1/8	3 1/4	1,300	350,040
May 3..	1,767	2,711	......	......	1	4,479	24,470	4,000	1,200	29,670	450,500	197,500	648,000	5	5 1/8	3 1/4	1,200	374,510
" 10..	5,803	4,521	......	962	......	11,286	32,640	20,200	1,400	54,240	428,000	197,000	625,000	5	5 1/8	3 1/4	1,400	407,150
" 17..	26,637	6,534	......	9,529	443	43,143	15,250	3,000	350	18,600	442,500	210,500	653,000	4 3/4	5	3 1/8	350	422,400
" 24..	21,367	3,021	......	4,201	......	28,589	18,690	1,000	300	19,990	448,500	214,000	662,500	4 5/8	4 7/8	2 7/8	300	441,090
" 31..	28,503	50	389	4,591	......	33,523	15,860	1,000	1,800	18,660	472,500	215,500	688,000	4 5/8	4 3/4	2 7/8	1,800	456,950
June 7..	111,640	7,192	3,426	3,985	1,059	127,302	33,370	500	800	34,670	560,500	225,000	785.500	4 5/8	4 3/4	2 3/4	800	490,320
" 14..	139,869	7,827	......	3,395	868	151,959	24,810	4,400	1,500	30,710	679,500	231,000	910,500	4 5/8	4 3/4	2 3/4	1,500	515,130
" 21..	39,495	......	1,500	......	......	40,995	27,690	10,600	2,760	51,050	683,500	224,000	907,500	4 5/8	4 7/8	2 3/4	2,760	552,720
" 28..	51,700	2,747	......	685	182	55,314	28,480	11,500	1,520	41,500	704,500	217,000	921,500	4 5/8	5	2 3/4	1,520	581,300
July 5..	8,753	......	......	1,222	......	9,975	33,060	8,800	2,170	44,030	688,000	209,500	897,500	4 5/8	5	2 3/4	2,170	614,360
" 12..	117,872	2.809	......	9,408	1,809	131,898	28,030	1,500	3,500	33,030	778,000	220,500	998,500	4 5/8	4 7/8	2 3/4	3,500	642,390
" 19..	14,746	......	......	......	136	14.882	27,970	500	1,150	29,620	766,500	216,500	983,000	4 1/2	4 3/4	2 3/4	1,150	670,360
" 26..	27,124	11,604	4,508	2,559	254	46,049	43,080	7,100	2,550	52,730	756,000	227,500	983,500	4 5/8	4 7/8	2 3/4	2,550	713.440
Aug. 2..	35,870	8,358	4,752	878	217	50,075	32,610	3,000	1,120	36,730	756,500	236,500	993,000	4 5/8	4 7/8	2 3/4	1,120	746,050
" 9..	17,801	3,496	757	1,630	......	23,684	35,610	350	6,700	42,660	739,000	235,000	974,000	4 5/8	4 7/8	2 3/4	6,700	781,660
" 16..	21,935	3.635	......	2.840	731	29,141	25,710	......	500	26,210	741,000	236,000	977,000	4 5/8	4 7/8	2 3/4	500	807,370
" 23..	16,299	......	......	......	......	16,299	28,340	......	750	29,090	736,500	228,000	964,500	4 1/2	4 3/4	2 5/8	750	835,710

Aug. 30..	6,542	1,998	......	1,253	......	9,793	25,550	2,000	6,300	33,850	712,000	225,500	937,500	4½	4¾	2⅝	6,300	861,260
Sep. 6..	1,780	4,353	2,766	......	200	9,099	17,990	......	2,000	19,990	697,500	229,000	926,500	4⅜	4⅝	2½	2,000	879,250
" 13..	6,415	10,687	2,088	......	90	19,280	20,640	3,500	1,880	26,020	686,500	237,000	923,500	4⅜	4⅝	2½	1,880	899,890
" 20..	13,013	......	831	......	1,919	15,763	16,980	1,500	2,650	21,130	684,000	235,500	919,500	4¼	4½	2½	2,650	916,870
" 27..	2,481	......	1,276	......	......	3,757	21,700	2,600	1,800	26,100	668,500	231,000	899,500	4¼	4½	2½	1,800	938,570
Oct. 4..	9,472	4,579	926	5,116	416	20,509	22,200	1,400	1,370	24,970	658,500	238,000	896,500	4¼	4½	2½	1,370	960,770
" 11..	6,838	13,864	119	5,689	627	27,157	21,990	3,200	1,350	26,540	646,000	254,000	900,500	4¼	4½	2½	1,350	982,760
" 18..	5,425	8,952	4,115	1,565	270	20,327	31,910	1,400	1,300	34.610	626,500	262,000	887,500	4⅛	4⅜	2½	1,300	1,014,670
" 25..	6,883	4,104	1,284	3,354	......	15,625	20,200	4,900	1,120	26,220	615,500	266,000	881,500	4⅛	4⅜	2¼	1,120	1,034,870
Nov. 1..	714	......	2,453	2,121	......	5,288	34,000	2,700	200	36,900	590,000	262,500	852,500	4⅛	4⅜	2¼	200	1,068,870
" 8..	....		1,615	......	......	1,615	30,630	5,000	500	36,130	564,500	258,000	822,500	4⅛	4¼	2¼	500	1,099,500
" 15..	18,426	9,260	......	10,837	......	38,523	33,080	8,500	260	41,840	556,500	271,500	828,000	4¼	4½	2¼	260	1,132,580
" 22..	9,961	2,108	......	2,592	664	15,325	36,390	3,000	300	29,690	545,500	272,000	817,500	4	4⅛	2¼	300	1,168,970
" 29..	16,346	3,484	......	......	372	20,202	27,160	1,200	800	29,160	540,500	257,500	808,000	3⅞	4	2¼	800	1,196,130
Dec. 6..	3,041	1,788	2,450	......	......	7,279	31,270	12,000	......	43,270	514,500	265,500	780,000	3⅞	4	2¼	......	1,227,400
" 13..	7,380	......	......	......	865	8,245	27,120	10,000	......	37,120	499,500	262,000	761,500	3⅞	4	2¼	......	1,254,520
" 20..	1,766	......	553	......	510	2,829	24,600	10,500	950	36,050	483,000	256,000	739,000	3⅞	4	2¼	950	1,279,120
" 27..	15,636	100	1,733	......	240	17,709	31,400	16,000	100	47,500	476,000	249,000	725,000	4	4⅛	2¼	100	1,310,520
Average prices & total sales, receipts & stocks.	1,028,811	145165	37,551	116,333	28,018	1,355,878	1,310,520	475,750	58,260	1,844,539				4.71	4.91	2.9	58,260	252,023

The semi-weekly Price and Weekly Sales and Receipts, at New York, Weekly Exports from New York and Rates of Freight to Liverpool 1st of each Month, for the Crop Year ending September 1, 1844.

1843.	Price of Fair New Orleans Liverpool Classificati'n	Price of Fair Upland. Liverpool Classificati'n	Sales for week.	Receipts for week.	EXPORTS FOR WEEK.					Rates of Freight to Liverpool.
					To Great Britain.	To France.	North of Europe.	Other Fo'n Ports	Total Exports.	
Septem. 5..	8¼@8½	8@8¼	11,078	1,004	....	....	....	....	....	$\frac{3}{16}$@¼d.
" 8..	8½@8¾	8¼@8½								
" 12..	8½@8¾	8¼@8½	18,500	2,718	....	810	....	....	810	
" 15..	8¾@9	8¼@8½								
" 19..	8¾@9	8¼@8½	7,650	2,780	....	442	1,230	112	1,784	
" 22..	9¼@9½	8¾@9								
" 26..	9¼@9½	8¾@9	18,560	1,769	117	....	295	....	412	
" 29..	9¼@9½	8¾@9								
October 3..	9¼@9½	8¾@9	6,560	4,791	852	206	67	....	1,125	$\frac{3}{16}$@¼d.
" 6..	7¾@8	7½@7¾								
" 10..	7½@7¾	7¼@7½	5,550	2,871	....	....	....	....	....	
" 13..	7¼@7½	7@7¼								
" 17..	7¼@7½	7@7¼	2,850	4,956	323	1,570	....	....	1,893	
" 20..	7¼@7½	7@7¼								
" 24..	7¼@7½	7@7¼	4,600	5,473	991	698	....	....	1,689	
" 27..	7¼@7½	7¼@7½								
" 31..	7¼@7½	7¼@7½	7,800	2,651	908	1,523	....	....	2,431	
Novem. 3..	7¼@7½	7@7¼								$\frac{5}{16}$@⅜d.
" 7..	7¼@7½	7@7¼	4,350	9,100	2,278	2,799	....	....	5,077	
" 10..	7½@7¾	7⅜@7½								
" 14..	7½@7¾	7⅜@7½	10,200	3,881	2,731	1,263	....	....	3,994	
" 17..	7½@7¾	7⅜@7½								
" 21..	7½@7¾	7⅜@7½	3,000	9,787	1,301	814	609	....	2,724	
" 24..	7½@8	7½@7¾								
" 28..	7½@8	7½@7¾	10,300	8,019	1,843	1,262	618	....	3,723	
Decem. 1..	7¾@8¼	7½@8								$\frac{5}{16}$@$\frac{7}{16}$d.
" 5..	8@8⅜	7¾@8	12,650	3,359	2,299	2,197	....	799	5,295	
" 8..	8@8⅜	7¾@8								
" 12..	8@8⅜	7¾@8	10,350	9,565	4,154	464	....	....	4,618	
" 15..	8@8½	8@8¼								
" 19..	8¼@8¾	8¼@8¾	15,050	7,036	4,418	1,028	....	....	5,446	
" 22..	8½@8¾	8¼@8¾								
" 26..	8¾@9	8½@8¾	16,700	6,707	2,736	826	400	....	3,962	

General Remarks.

The speculative feeling noted in the previous crop year, based upon the short yield of this year, ran through a considerable portion of the crop year under review, most marked in November, December, January and February; prices were more or less unsettled, and the advanced rates interfered with the foreign demand, so that stocks accumulated throughout the country more than ordinarily. It will be seen that the result justified general expectation, as when the crop was made up, September 1, '44, it proved to be 2,030,409 bales, against a crop of 2,378,875 bales the pervious year, being a decrease of 348,466 bales.

* From October 6, the quotation changed to "Middling" on both descriptions.

Decem. 29..	8¾@9	8½@8¾									
1844.											
January 2..	8¾@9	8½@8¾	13,300	3,720	2,930	186	27	....	3,143	⅜@½d.
" 5..	9¼@9½	9@9¼									
" 9..	9½@9¾	9¼@9½	36,400	20,522	6,255	862	....	....	7,117	
" 12..	9½@9¾	9¼@9½									
" 16..	9½@9¾	9¼@9½	32,950	16,796	3,507	187	77	....	3,771	
" 19..	9¼@9½	9@9¼									
" 23..	9½@9¾	9@9¼	7,500	15,248	2,448	771	....	....	3,219	
" 26..	9½@9¾	9¼@9½									
" 30..	9½@9¾	9¼@9½	15,950	7,498	2,061	1,450	....	....	3,511	
February 2..	9½@9¾	9¼@9½								⅜@½d.
" 6..	9¼@9½	9¼@—	11,000	6.238	3,594	1,464	....	....	5,058	
" 9..	9½@9¾	9½@—									
" 13..	9½@9¾	9½@—	10,350	21,280	832	1,186	....	....	2,018	
" 16..	9¼@9½	9¼@9½									
" 20..	9¼@9½	9¼@9½	14,327	32,360	2,060	879	164	....	3,103	
" 23..	9¼@9½	9¼@9½									
" 27..	9@9⅛	8⅞@9	21,700	28,212	1,533	1,398	531	....	3,462	
March 1..	8¾@9	8½@8¾								½@⅝d.
" 5..	8¼@8½	8¼@8⅜	11,400	12,608	4,579	....	371	250	5,200	
" 8..	8¼@8½	8¼@8⅜									
" 12..	8¼@8½	8¼@8⅜	14,650	21,489	13,176	2,549	196	956	16,877	
" 15..	8⅛@8¼	8⅛@8¼									
" 19..	8⅛@8¼	8⅛@8¼	11,450	22,672	6,214	3,069	1,062	....	10,345	
" 22..	8⅛@8⅜	8⅛@8¼									
" 26..	7⅝@7⅞	7½@7¾	8,800	8,002	12,549	3,795	1.400	847	18,591	
" 29..	7½@7¾	7½@—									
April 2..	7¾@—	7¾@—	15,450	15,726	4,189	3,056	2,175	....	9,420	$\frac{7}{16}$@$\frac{9}{16}$d.
" 5..	8@8¼	7¾@8									
" 9..	7¾@8	7⅝@7¾	14,919	13,667	5,892	2,309	1,233	....	9,434	
" 12..	7⅝@7¾	7½@7⅝									
" 16..	7⅝@7¾	7½@7⅝	10,850	11,184	8,024	2,033	2,289	....	12,346	
" 19..	7⅝@7¾	7½@7⅝									
" 23..	7⅜@7½	7¼@7⅜	11,000	1,692	10,595	5,855	748	242	17,440	
" 26..	7⅜@7½	7¼@7⅜									
" 30..	7⅜@7½	7¼@—	13,800	10,116	5,647	2,364	646	1,744	10,401	⅜@$\frac{7}{16}$d.
May 3..	7¼@7⅜	7¼@—									
" 7..	7¼@7⅜	7@7⅛	9,100	10,400	1,713	1,790	543	1,038	5,084	
" 10..	7⅛@7¼	7@7⅛									
" 14..	7⅛@7¼	7@7⅛	10,600	5,829	4,920	433	804	....	6,157	
" 17..	7⅛@7¼	7@7⅛									
" 21..	6⅞@7⅛	6¾@7	11,900	9,374	10,259	579	819	4,381	16,038	

The semi-weekly Price and Weekly Sales and Receipts at New York, Weekly Exports from New York and Rates of Freight to Liverpool 1st of each Month, for the Crop Year ending September 1, 1844—Concluded.

1844.		Price of Fair New Orleans Liverpool Classificati'n	Price of Fair Upland. Liverpool Classificati'n	Sales for week.	Receipts for week.	EXPORTS FOR WEEK.					Rates of Freight to Liverpool.
						To Great Britain	To France.	North of Europe.	Other Fo'n Ports	Total Exports.	
May	24..	$6\frac{7}{8}@7\frac{1}{8}$	$6\frac{3}{4}@7$								
"	28..	$6\frac{7}{8}@7\frac{1}{8}$	$6\frac{3}{4}@7$	10,150	7,669	5,772	1,190	1,193	902	9,057	
"	30..	$6\frac{7}{8}@7\frac{1}{8}$	$6\frac{3}{4}@7$								
June	4..	$6\frac{7}{8}@7\frac{1}{8}$	$6\frac{3}{4}@7$	10,450	3,622	5,336	1,537	651	....	7,524	$\frac{9}{32}@\frac{3}{8}$d.
"	7..	$6\frac{7}{8}$@—	$6\frac{3}{4}$@—								
"	11..	$6\frac{7}{8}$@—	$6\frac{3}{4}$@—	7,900	5,486	5,031	834	1,281	202	7,348	
"	14..	$7@7\frac{1}{4}$	$6\frac{3}{4}@7$								
"	18..	$7@7\frac{1}{4}$	$6\frac{3}{4}@7$	9,300	3,741	3,479	1,275	1,788	....	6,542	
"	21..	$7@7\frac{1}{4}$	$6\frac{3}{4}@7$								
"	25..	$6\frac{7}{8}@7$	$6\frac{3}{4}@7$	3,850	6,485	2,571	898	1,275	....	4,744	
"	28..	$6\frac{7}{8}@7$	$6\frac{3}{4}@7$								
July	2..	$6\frac{7}{8}@7$	$6\frac{3}{4}@7$	3,550	3,750	4,425	....	1,305	....	5,730	$\frac{1}{4}$@–d.
"	5..	$6\frac{7}{8}@7$	$6\frac{3}{4}@7$								
"	9..	$6\frac{7}{8}@7$	$6\frac{3}{4}@7$	1,600	1,467	536	204	259	277	1,276	
"	12..	$6\frac{7}{8}@7$	$6\frac{3}{4}@7$								
"	16..	$6\frac{7}{8}@7$	$6\frac{3}{4}@7$	8,500	2,584	674	234	393	....	1,301	
"	19..	$6\frac{7}{8}@7$	$6\frac{3}{4}@7$								
"	23..	$6\frac{7}{8}@7$	$6\frac{3}{4}@7$	5,350	2,086	3,552	90	1,329	....	4,971	
"	26..	$6\frac{7}{8}@7$	$6\frac{3}{4}@7$								
"	30..	$6\frac{7}{8}@7$	$6\frac{3}{4}@7$	4,400	2,077	1,755	....	2,362	980	5,097	$\frac{1}{4}$@–d.
August	2..	$6\frac{7}{8}@7$	$6\frac{3}{4}@7$								
"	6..	$6\frac{3}{4}$@—	$6\frac{3}{4}$@—	3,600	935	1,320	810	736	....	2,866	
"	9..	$6\frac{5}{8}@6\frac{7}{8}$	$6\frac{1}{2}@6\frac{3}{4}$								
"	13..	$6\frac{5}{8}@6\frac{7}{8}$	$6\frac{1}{2}@6\frac{3}{4}$	5,200	5,521	1,002	1,181	1,168	....	3,351	
"	16..	$6\frac{5}{8}@6\frac{7}{8}$	$6\frac{1}{2}@6\frac{3}{4}$								
"	20..	$6\frac{5}{8}@6\frac{7}{8}$	$6\frac{1}{2}@6\frac{3}{4}$	3,000	2,406	1,128	605	1,521	280	3,534	
"	23..	$6\frac{5}{8}@7$	$6\frac{1}{2}$@—								
"	27..	$6\frac{1}{4}@6\frac{1}{2}$	$6@\frac{1}{4}$	6,200	3,641	1,335	1,314	1,825	.....	4,474	
"	30..	$6\frac{1}{4}@6\frac{1}{2}$	$6@\frac{1}{4}$								
Septem.	3..	$5\frac{3}{4}@6$	$5\frac{3}{4}@6$	5,250	2,540	935	647	789	398	2,769	$\frac{5}{16}@\frac{3}{8}$d.
Average price and total sales, receipts and exports.		7.90	7.73	561,444	435,110	176,779	62,936	34,179	13,408	287,302	

General Remarks.

Exchange.

The quotation for 60 days' Bills on London through September was 9@9½ per cent. premium ; in October it fell from 9@9¼ to 8½@8¾; in November it further declined to 7½@8, at one time afterwards rising to 8¼@8½ ; in December the range was from 8 to 9¼ ; in January, 8¾@9¼ ; February, 9½ to 8¾ ; March, 7¾ to 8½ ; April, 8¼@8¾ ; in May it was steady at 8¾@9 ; June, 8¾@9½ ; July, 9@9½ ; August, 9¼@10, closing September 1 at 9¾@10 per cent. premium.

1845.

This was a harvest year for cotton manufacturers. They paid £10,000,000 for raw cotton, receiving £45,000,000 for yarn and manufactured cotton goods, leaving a margin of £35,000,000 for machinery, fuel, dyeing, bleaching, printing, wages, interest of capital and profit. (*See* year 1857.)

Importation of cotton into Great Britain, 722,000,000 lbs., of which 43,000,000 lbs. were re-exported; cotton yarn and twist made 495,000,000 lbs., of which 131,000,000 lbs. were exported; piece goods exported, 1,092,000,000 yards; value of exports, £26,000,000.

The price of cotton fell in Great Britain this year to a level never before known, which was one cause of the large importation (722,000,000 lbs.): Average United States cotton was $3\frac{92}{100}$d.; Brazillian $6\frac{1}{2}$d, and East India $2\frac{11}{100}$d. per lb. A year of large profits among the Lancashire manufacturers; they had an abundance of cotton at low prices and sent out their manufactured goods to all accessible quarters.

Number of cotton mills in Massachusetts, this year, 302, with 817,483 spindles.

The convention of American Geologists and Naturalists, who met at New Haven, Conn., in May of this year, were invited, together with their ladies, by Mrs. Whitney, the widow of the inventor of the cotton gin (*see* years 1793 and 1825), to attend an evening party at her house, which invitation was accepted.

Cotton crop of the United States, 2,100,537 bales.

The supply of American cotton imported into Great Britain this year was 1,499,600 bales. (*See* year 1846.)

COTTON CROP OF THE UNITED STATES.

Statement and Total Amount of the Cotton Crop of the United States, for the Year ending August 31, 1845.

	Bales.	Bales.	Total.	Same period 1844.
NEW ORLEANS.				
Export—				
To Foreign Ports	836,401			
Coastwise	148,215			
Stock on hand 1st September, 1845	7,556			
		992,172		
Deduct—				
Stock on hand 1st September, 1844	12,934			
Received from Mobile	12,123			
" Florida	12,830			
" Texas	25,159			
		63,046		
			929,126	832,172
MOBILE.				
Export—				
To Foreign Ports	390,714			
Coastwise	131,282			
Stock, 1st September, 1845	609			
		522,605		
Deduct—				
Stock, 1st September, 1844	4,175			
Received from Florida	485			
" Texas	718			
" New Orleans	31			
		5,409		
			517,196	467,990
FLORIDA.				
Export—				
To Foreign Ports	64,853			
Coastwise	124,040			
Stock on hand 1st September, 1845	100			
		188,993		
Deduct—				
Stock on hand 1st September, 1844		300		
			188,693	145,562
GEORGIA.				
Export from Savannah—				
To Foreign Ports—Uplands	175,965			
" Sea Island	6,108			
Coastwise—Uplands	120,570			
Sea Island	1,901			
	304,544			
Burnt in Savannah	1,900			
Stock in Savannah, 1st September, 1845	2,736			
" Augusta and Hambro', 1st September, 1845	5,919			
		315,099		
Deduct—				
Stock in Savannah and Augusta 1st September, 1844		19,659		
			295,440	255,597
SOUTH CAROLINA.				
Export from Charleston—				
To Foreign Ports—Uplands	288,870			
" Sea Island	20,905			
Coastwise—Uplands	111,698			
Sea Island	423			
	421,896			

Statement and Total Amount of the Cotton Crop of the United States, for the Year ending August 31, 1845—*Concluded.*

	Bales.	Bales.	Total.	Same period 1844.
Export from Georgetown—				
To New York	15,395			
Burnt in Charleston	3,481			
Stock in Charleston, 1st September, 1845	10,879			
		451,651		
Deduct—				
Stock in Charleston, 1st September, 1844	13,536			
Received from Savannah	10,911			
Received from Florida, Key West, &c	843			
		25,290		
			426,361	304,870
NORTH CAROLINA.				
Export—				
Coastwise	12,587			
Stock on hand, 1st September, 1845	100			
		12,687		
Deduct—				
Stock on hand 1st September, 1844	....	200		
			12,487	8,618
VIRGINIA.				
Export—				
To Foreign Ports	3,823			
Coastwise	6,609			
Manufactured	14,500			
Stock on hand, 1st September, 1845	2,418			
		27,350		
Deduct—				
Stock on hand 1st September, 1844	....	2,150		
			25,200	14,500
Received at Philadelphia and Baltimore, overland	....	....	....	1,100
Total crop of the United States	....	....	2,394,503	2,030,409
Crop of last year	....	....	2,030,409	
Increase	....	....	364,094	

Export to Foreign Ports, from September 1, 1844, *to August* 31, 1845.

FROM	To Great Britain.	To France.	To North of Europe.	Other F'n Ports.	Total.
New Orleans ....Bales	585,888	125,020	33,035	92,458	836,401
Mobile	268,849	68,929	24,843	28,093	390,714
Florida	49,460	7,660	....	7,733	64,853
Georgia (Savannah and Darien)	164,085	14,071	1,214	2,703	182,073
South Carolina	218,618	72,221	15,877	3,059	309,775
North Carolina	....	....	....	....	....
Virginia	1,158	423	2,242	....	3,823
Baltimore	246	....	375	....	621
Philadelphia	2,237	183	....	641	3,061
New York	145,614	69,962	49,795	14,173	279,544
Boston	3,151	888	7,120	1,732	12,891
Grand total	1,439,306	359,357	134,501	150,592	2,083,756
Total last year	1,202,498	282,685	69,053	75,254	1,629,490
Increase	236,808	76,672	65,448	75,338	454,266

☞ The shipments from Mississippi are included in the export from New Orleans.

Growth.

Total crop of 1825–6 ......bales.	710,000	Total crop of 1835–6 ......bales.	1,360,725
" 1826–7............	937,000	" 1836-7............	1,422,930
" 1827–8............	712,000	" 1837–8............	1,801,497
" 1828–9............	857,744	" 1838–9............	1,360,532
" 1829–30...........	976,845	" 1839–40...........	2,177,835
" 1830–1............	1,038,848	" 1840–1............	1,634,945
" 1831–2............	987,477	" 1841–2............	1,683,574
" 1832–3............	1,070,438	" 1842–3............	2,378,875
" 1833–4............	1,205,394	" 1843–4............	2,030,409
" 1834–5............	1,254,328	" 1844–5............	2,394,503

Consumption.

Total crop of the United States, as above stated ....bales.			2,394,503
Add—Stocks on hand at the commencement of the year, September 1, 1844—			
In the Southern ports..		52,954	
In the Northern ports..		106,818	
			159,772
Makes a supply of..			2,554,275
Deduct therefrom—The Export to Foreign ports............	2,083,756		
Less Texas and other foreign............	29,194		
		2,054,562	
Stocks on hand at the close of the year, Sept. 1, 1845—			
In the Southern ports	30,317		
In the Northern ports	63,809		
		94,126	
Burnt at Savannah..................................	1,900		
" Charleston..................................	3,481		
" New York..................................	11,200		
		16,581	
			2,165,269
Taken for home use..bales.			389,006

Quantity consumed by and in the hands of Manufacturers.

1844–5 ....bales.	389,006	1835–6 ....bales.	236,733
1843–4........................	346,744	1834–5	216,888
1842–3	325,129	1833–4	196,413
1841–2	267,850	1832–3	194,412
1840–1	297,288	1831–2	173,800
1839–40........................	295,193	1830–1	182,142
1838–9	276,018	1829–30	126,512
1837–8	246,063	1828–9	118,853
1836–7	222,540	1827–8	120,593

NOTE.—It will be seen that we have deducted from the New Orleans and Mobile statements, the quantity received at those ports from Texas—Texas being a foreign country. Our next annual statement will probably include Texas in the crop of the United States.

Our estimate of the quantity taken for consumption, does not include any cotton manufactured in the States south and west of Virginia, nor any in that State, except in the vicinity of Petersburg and Richmond.

The quantity of new cotton received at the shipping ports up to the first inst., amounted to about 7,500 bales, same as last year.

In regard to the crop now gathering, we have loud complaints of injury from drought in certain sections, while in others the yield is represented as good. It is too early yet to form any reliable conclusion as to the quantity that may reach the market.

In the New Orleans statement, we notice an allowance of 6,000 bales for cotton sent up the river to the Western States. As it is probable some of this cotton reaches Philadelphia and Baltimore "overland," we omit the overland item in our statement of the crop for this year.

ANNUAL REVIEW.

From the New Orleans Price Current—1844–45.

The season commencing on the 1st of September last, may be said to have opened under circumstances not at all encouraging to the planting interest, and the progress of the market for a lengthened period, but too clearly verified the most gloomy anticipations. A period of inordinate speculation had been succeeded by disastrous reaction, and heavy stocks with large (and, in some instances, extravagant) estimates of the coming crop, enabled consumers to dictate their own terms—all speculative spirit being prostrated, and the markets of this country and of Europe at the mercy, so to speak, of the manufacturers. As regards the extent of production, it was early conceded, notwithstanding the immense damage known to have accrued from overflows, in some sections, that the crop would in all probability prove a large one; and although at an early period some estimates carried the amount much beyond that of any former year, yet as the prospects became more fully developed the general impression seemed to be that the production would not greatly exceed that of 1842–43, or at all events not be more than 2,400,000 bales in the United States. About the 1st of January, the greatest depression took place, the immediate cause being further discouraging accounts from Europe, and the prices of middling to middling fair, fell to 4⅜ to 5½ cents, the extreme being 3 to 7½ cents, for inferior to good and fine. These remarkably low rates, however, soon induced a disposition to operate more freely, and under the influence of an active demand, assisted by more favorable accounts from Liverpool, the market began gradually to recover from its deep depression, and by the latter part of March, prices had improved to 4 to 8 cents extremes, the quotations for middling to middling fair being 5⅜ to 6½. At this point the important intelligence was received from England, that the British Ministry had recommended a total repeal of the duty on raw cotton, and as the mere recommendation was looked upon as tantamount to its passage into the law, the immediate effects here were increased activity in the demand, and an advance of ⅜ to ½ a cent per lb. From this point the market continued steady, with well sustained prices, until the early part of May, when increasing supplies of Tennesssee and North Alabama cottons, and advices of a decline in the Liverpool market, caused prices to recede again to 5½ to 6½ cents for middling to middling fair Louisianas and Mississippis, the extremes for Tennessees and North Alabamas being 4½ to 6⅝ cents. Thus the crop of 1844, has been disposed of; and it affords us unfeigned pleasure to state that the market opens upon the new crop with more encouraging prospects for ready sales, at fair prices in the regular and legitimate course of this important trade, than have marked the advent of any similar period, at least for many years gone by. All the mills of England were working full time, with orders in advance of production, and new machinery was being put in rapid operation, the manufacturing trade appearing to be in a more healthy and prosperous state than at any former period known in its history. A corresponding degree of prosperity is also observable in the cotton manufactures of the continent, and there seems a strong probability that, notwithstanding the material increase in the supply during the past season, the leading markets of Europe will enter upon 1846 with smaller stocks than they possessed at the commencement of the year. To this gratifying picture of the state of the cotton trade in Europe, we take pleasure in adding that the manufactures of our own country present a condition of activity and prosperity, which is without a parallel in the annals of any former period. Every loom is in active and profitable employment, and new mills are springing up in nearly every section of our wide extended land. Nevertheless it is our province to speak of present prospects, and we have to remark that though excessive drought throughout the season in Georgia and South Carolina, and latterly in Alabama, will, doubtless, considerably diminish the crops in those States, particularly the two former, yet from present indications, it seems likely that, with a favorable picking season (a contingency upon which the extent of the crop now mainly depends), the production of the rich region of the Mississippi Valley will fully compensate for any falling off in the States on or near the Atlantic.

LIVERPOOL STATEMENT FOR 1845.

UNITED STATES, 1844–1845.			
Stock 1st October, 1844	160,000	Export	2,071,000
Receipts	2,394,000	Consumption	389,000
		Stock 1st Oct., 1845	94,000
Bales	2,554,000	Bales	2,554,000

CONSUMPTION.										
Tot. Europe.	Continent.	France.	Gt. Britain.	Liverpool.	Stock Jan. 1, 1845, in	Liverpool.	Gt. Britain.	France	Continent.	Tot. Europe.
					United States	493,000	542,000	52,000	67,000	661,000
					Brazil	62,000	63,000	4,000	2,000	69,000
					West Indies	12,000	14,000	13,000	7,000	34,000
					East Indies	144,000	237,000	2,000	6,000	245,000
					Egypt	38,000	41,000	7,000	44,000	92,000
					Bales	749,000	897,000	78,000	126,000	1,101,000
					IMPORT.					
1,870,000	228,000	351,000	1,291,000	1,179,000	United States	1,376,000	1,502,000	353,000	220,000	2,013,000
131,000	6,000	7,000	118,000	116,000	Brazil	111,000	111,000	3,000	5,000	115,000
51,000	13,000	21,000	17,000	13,000	West Indies	7,000	9,000	12,000	12,000	33,000
166,000	62,000	3,000	101,000	74,000	East Indies	79,000	155,000	3,000	61,000	166,000
138,000	48,000	36,000	54,000	55,000	Egypt	79,000	81,000	38,000	42,000	147,000
2,356,000	357,000	418,000	1,581,000	1,437,000	Bales	1,652,000	1,858,000	409,000	340,000	2,474,000
....	14,000	....	119,000	79,000	Export.					
1,219,000	95,000	69,000	1,055,000	885,000	Stock Dec. 31. Stock above,	749,000	897,000	78,000	126,000	1,101,000
3,575,000	466,000	487,000	2,755,000	2,401,000	Total supply, bales	2,401,000	2,755,000	487,000	466,000	3,575,000

COTTON AT LIVERPOOL. YEAR 1845.

Week Ending.	Receipts.						Sales.				Stocks.			Prices.			Actual Export	Con-sumption.
	Americ'n.	E. I.	Egypt.	Brazil.	Other.	Total.	Con-sumption.	Specu-lation.	Export	Total.	Amer'n.	Other.	Total.	Mid. Up.	Mid. Orl.	Dhol		
Jan. 3..	1,187					1,187	5,640	3,500	150	9,290				3 7/8	4 1/8	2 1/4	150	5,640
" 10..	64,368		1,085	8,057	16	73,526	27,000	1,000	300	28,300	530,500	257,500	788,000	3 7/8	4 1/8	2 1/4	300	32,640
" 17..	24,087	2,367	2,178	5,854	238	34,724	37,610	6,000	300	43,910	523,000	260,000	783,000	3 7/8	4 1/8	2 1/4	300	70,250
" 24..	37,391	2,285		3,627		43,303	41,300	3,000	300	44,600	527,000	257,000	784,000	3 7/8	4 1/8	2 1/4	300	111,550
" 31..	21,849			1,018		22,867	27,120	4,800	1,600	33,520	521,500	250,500	772,000	3 7/8	4 1/8	2 1/4	1,600	138,670
Feb. 7..	7,355	2,223	432	774		10,784	30,150	8,900	800	39,850	501,500	249,500	751,000	3 7/8	4 1/8	2 1/4	800	168,820
" 14..	36,466	1,674	2,035	2,649	770	43,594	30,500	25,300	150	55,950	513,500	249,000	762,500	4	4 3/8	2 1/4	150	199,320
" 21..	35,956			3,476		39,432	20,270	15,000	450	35,720	533,000	249,000	782,000	4	4 3/8	2 1/4	450	219,590
" 28..	28,213	6,045				34,258	23,880	17,750	450	42,080	538,500	249,500	788,000	4	4 3/8	2 1/4	450	243,470
Mch 7..	28,274	4,401			60	32,735	37,020	33,000	1,500	71,520	540,500	241,500	782,000	4 1/8	4 3/8	2 1/4	1,500	280,490
" 14..	5,065					5,065	21,400	12,000	300	33,700	529,500	236,000	765,500	4 3/8	4 1/2	2 1/4	300	301,890
" 20..	5,142		1,257			6,399	12,750	5,500	700	18,950	524,000	234,000	758,000	4 1/4	4 3/8	2 1/4	700	314.640
" 28..	74,437	3,812		1,107	114	79,470	26,070	2,500	350	28,920	574,000	234,500	808,500	4	4 1/8	2 1/4	350	340,710
April 4..	39,359			4,876	57	44,292	38,500	9,500	2,450	50,450	576,500	234,500	811,000	4	4 1/8	2 1/4	12,450	379,210
" 11..	7,005	3,106	152	19	181	10,463	29,700	8,700	3,130	41,530	549,500	232,500	782,000	4	4 1/8	2 1/4	3,130	408,910
" 18..	68,655	2,921		475		72,051	31,910	7,500	2,500	41,910	585,000	230,000	815,000	3 7/8	4 1/8	2 1/4	2,500	440,820
" 25..	24,694	611		1,408		26,713	38,040	39,100	1,060	78,200	584,000	224,500	808,500	4 1/8	4 3/8	2 3/8	1,060	478,860
May 2..	71,622	816	1,662	4,678	642	79,420	20,670	29,700	850	51,220	638,000	227,500	865,500	4	4 3/8	2 1/4	850	499,530
" 9..	17,675	7,741	1,677	2,648	219	29,960	32,130	27,000	1,980	61,110	628,000	234,000	862,000	4	4 3/8	2 1/4	1,980	531,660
" 16..	48,390			1,337		49,727	18,800	5,200	1,400	25,400	658,000	232,000	890,000	3 3/4	4 1/8	2 1/4	1,400	550,460
" 23..	21,340	1,155	1,953	2,805		27,253	31,420	5,000	1,000	37,420	649,000	234,000	883,000	3 7/8	4 1/8	2 1/4	1,000	581,880
" 30..	20,829					20,829	33,990	5,000	1,200	40,190	639,000	229,500	868,500	3 7/8	4 1/8	2 1/4	1,200	615,870
June 6..	105,723	5,236		5,731	57	116,747	25,540	1,750	1,710	29,000	724,500	233,500	958,000	3 7/8	4 1/8	2 1/4	1,710	641,410
" 13..	102,581		915		38	103,534	32,670	8,100	3,100	43,870	799,000	227,500	1,026,500	3 7/8	4 1/8	2 1/4	3,100	674,080
" 20..	44,284	663	1,800	3,002	150	49,899	35,370	11,000	1,500	47,870	812,000	227,000	1,039,000	3 7/8	4 1/8	2	1,500	709,450
" 27..	8,638		4,082	1,568		14,288	27,070	11,900	2,700	41,670	798,500	225,000	1,023,500	3 7/8	4 1/8	2	2,700	736,520
July 4..	60,554		1,033	1,412	928	63,927	39,910	19,000	2,080	60,990	820,000	218,000	1,038,000	4	4 1/4	2	2,080	776,430
" 11..	35,531	2,213	3,746	3,236	1,129	45,855	35,320	26,900	2,800	65,020	823,500	220,500	1,044,000	4	4 3/8	2	2,800	811,750
" 18..	39,415	124	4,061	1,011	221	44,832	27,360	37,100	8,230	72,690	835,500	219,000	1,054,500	4 1/4	4 3/8	2	8,230	839,110
" 25..	19,723		1,800		350	21,873	29,300	12,300	5,180	46,780	826,000	214,000	1,040,000	4 1/8	4 3/8	2	5,180	868,410
Aug. 1..	37,631	3,410		9,215	202	50,458	22,600	7,600	2,770	32,970	844,000	223,000	1,067,000	4 1/8	4 3/8	2	2,770	891,010
" 8..	17,560		100		799	18,459	24,860	8,400	3,700	36,960	837,000	220,000	1,057,000	4 1/8	4 3/8	2	3,700	915,870
" 15..	18,610		1,878	1,677		22,165	31,290	8,400	2,300	41,990	826,000	219,000	1,045,000	4 1/8	4 3/8	2	2,300	947,160
" 22..	7,527		950			8,477	23,540	6,350	3,150	33,040	812,000	216,000	1,028,000	4 1/8	4 3/8	2	3,150	970,700

Aug. 29..	22,354	2,982	3,542	742	534	30,154	33,330	13,700	1,900	48,930	804,500	218,500	1,023,000	$4\frac{1}{8}$	$4\frac{3}{8}$	2	1,900	1,004,030
Sep. 5..	....	......	1,871	1,169	364	3,404	31,750	31,300	1,200	64,250	777,000	214,000	991,000	$4\frac{1}{4}$	$4\frac{1}{2}$	2	1,200	1,035,780
" 12..	589	......	2,289	895	......	3,773	18,530	14,000	450	32,980	765,000	211,000	976,000	$4\frac{1}{4}$	$4\frac{1}{2}$	2	450	1,054,310
" 19..	9,938	2,655	2,847	2,706	2,672	20,818	26,300	11,700	500	38,500	753,000	216,500	969,500	$4\frac{1}{4}$	$4\frac{1}{2}$	2	500	1,080,610
" 26..	....	3,050	......	....	63	3,113	19,100	6,500	800	26,400	743,000	207,500	950,500	$4\frac{1}{4}$	$4\frac{1}{2}$	2	800	1,099,710
Oct. 3..	1,424	3,889	1,866	2,616	275	10,070	22,230	1,500	200	23,930	726,500	220,500	947,000	$4\frac{1}{8}$	$4\frac{3}{8}$	2	200	1,121,940
" 10..	2,016	......	......	385	......	2,401	13,550	2,500	700	16,750	717,500	218,000	935,500	$4\frac{1}{8}$	$4\frac{3}{8}$	2	700	1,135,490
" 17..	1,618	1,693	843	4,019	228	8,401	14,800	5,000	200	20,000	706,000	222.500	928,500	$4\frac{1}{8}$	$4\frac{3}{8}$	2	200	1,150,290
" 24..	3,233	3,807	......	2,938	......	9,978	12,840	4,500	......	17,340	700,000	226,000	926,000	$4\frac{1}{8}$	$4\frac{1}{4}$	2	......	1,163,130
" 31..	3,121	7,898	2,360	1,789	......	15,168	12,420	5,500	150	18,070	693,000	235,500	928,500	$4\frac{1}{8}$	$4\frac{1}{4}$	2	150	1,175,530
Nov. 7..	5,203	......	1,918	....	......	7,121	14,870	1,000	100	15,970	686,500	235,000	921,500	4	$4\frac{1}{4}$	2	100	1,190,400
" 14..	6,932	2,972	3,430	....	80	13,414	23,560	12,000	......	35,560	674,000	237,500	911,500	4	$4\frac{1}{4}$	2	......	1,213,960
" 21..	5,164	1,030	......	4,874	30	11,098	12,540	3,000	......	15,540	671,000	239,500	910,500	4	$4\frac{1}{4}$	2	......	1,226,500
" 28..	30.240	1,660	......	6,982	507	39,389	25,370	12,500	......	37,870	678,500	245,500	924,000	$3\frac{7}{8}$	$4\frac{1}{8}$	2	......	1,251,870
Dec. 5..	19,543	......	2,232	3,528	148	25,451	24,960	500	......	25,460	677,000	247,500	924,500	$3\frac{3}{4}$	$4\frac{1}{8}$	2	......	1,276,830
" 12..	11,374	......	1,344	....	487	13,205	25,840	500	100	26,440	667,000	245,500	912,500	$3\frac{3}{8}$	$4\frac{1}{8}$	2	100	1,302,670
" 19..	7,736	2,848	3,982	1,193	......	15,759	19,690	740	......	20,430	658,500	249,500	908,000	$3\frac{3}{4}$	$4\frac{1}{8}$	2	......	1,322,360
" 26..	19,825	......	1,457	867	......	22,149	21,920	1,500	100	23,520	659,000	249,000	908,000	$3\frac{7}{8}$	$4\frac{1}{8}$	2	100	1,344,280
" 31..	33,009	1,601	1,250	688	......	36,548	16,970	2,500	......	19,470	....	....	....	4	$4\frac{1}{4}$	2	......	1,361,250
Average prices & total sales, receipts & stocks.	1,370,455	86,888	64,127	107,051	11,549	1,640,070	1,361,250	563,990	68,540	1,993,780				3.92	4.31	2.11	68,540	2,568,396

The semi-weekly Price and Weekly Sales and Receipts at New York, Weekly Exports from New York and Rates of Freight to Liverpool 1st of each month, for the Crop Year ending September 1, 1845.

1844.	Price of Middling New Orleans Liverpool Classification.	Price of Middling Upland, Liverpool Classification.	Sales for week.	Receipts for week.	EXPORTS FOR THE WEEK.					Rates of Freight to Liverpool.
					To Great Britain.	To France.	North of Europe.	Other Fo'n Ports	Total Exports.	
Septem. 6..	5¾@6	5¾@6								5-16@⅜d
" 10..	5¾@6	5¾@6	15,200	2,343	....	....	....	....	....	
" 13..	5¾@6	5⅝@5¾								
" 17..	5¾@6⅛	5⅝@5¾	9,250	....	5,470	3,375	5,995	1,010	15,850	
" 20..	5¾@6⅛	5⅝@5¾								
" 24..	5¾@6⅛	5⅝@5¾	12,900	3,336	4,046	1,275	2,104	20	7,445	
" 27..	5¾@6	5¾@5⅞								
October 1..	5¾@6	5¾@5⅞	4,300	764	6,177	3,434	1,078	....	10,689	5-16@⅜d
" 4..	5¾@6	5¾@5⅞								
" 8..	5¾@6⅛	5¾@6	3,800	2,111	3,000	4,353	2,684	424	10,461	
" 11..	5¾@6⅛	5¾@6								
" 15..	5¾@6⅛	5¾@6	4,750	589	1,667	1,279	638	619	4,203	
" 18..	6@6¼	5¾@6								
" 22..	6@6¼	5¾@6	6,500	1,411	703	....	1,212	....	1,915	
" 25..	6@6¼	5¾@6								
" 29..	6@6¼	5¾@6	3,200	210	944	1,007	905	439	3,295	
Novem. 1..	6@6¼	5¾@6								¼@5-16d
" 5..	6@6¼	5¾@6	3,800	6,839	1,325	1.043	529	26	2,923	
" 8..	6@6¼	5¾@6								
" 12..	6@6¼	5¾@6	5,350	10,863	2,824	88	381	889	4,182	
" 15..	5¾@6	5⅝@5¾								
" 19..	5½@5¾	5½@5⅝	4,500	2,145	457	518	234	....	1,209	
" 22..	5¼@5½	5¼@5½								
" 26..	5¼@5½	5¼@5½	10,250	6,875	1,417	891	191	25	2,524	
" 29..	5¼@5½	5¼@5½								
Decem. 3..	5¼@5½	5¼@5½	5,350	7,176	1,926	1,663	287	230	4,106	5-16@⅜d
" 6..	5¼@5½	5¼@5½								
" 10..	5¼@5½	5¼@5½	7,650	7,089	1,728	1,304	363	436	3,831	
" 13..	5¼@5½	5¼@5½								
" 17..	5¼@5½	5¼@5½	5,550	5,592	1,880	1,291	....	754	3,925	
" 20..	5¼@5½	5¼@5½								
" 24..	5@5¼	5@5¼	4,650	6,691	2,802	1,829	204	....	4,835	
" 27..	5@5⅛	5@5¼								

GENERAL REMARKS.

This crop year was one of unmitigated dullness and depression, prices touching lower points than ever before, or since, known in the annals of the trade. The stagnation that ruled for much of the time was only relieved by two events, one of which was the reception of advices, on the 20th March, from England, relative to a proposed abolition of duty there on the article, which caused large speculative purchases in Liverpool, and put prices up here ¼@½ cent, and the other was the excitement that prevailed in England consequent upon President Polk's course on the Oregon question, fears being entertained that hostilities would ensue; subsequently, however, the public mind became more calm, and the market returned to its former lethargic condition.

Decem. 31..	5¼@5⅜	5@5¼	4,450	12,758	1,501	....	....	151	1,652	
1845.										
January 3..	5¼@5⅜	5@5¼								⅜@7-16d
" 7..	5¼@5⅜	5@5¼	5,650	5,476	964	762	40	....	1,766	
" 10..	5¼@5⅜	5@5¼								
" 14..	5¼@5⅜	5@5¼	7,450	8,675	2,307	778	722	797	4,604	
" 17..	5½@5¾	5¼@5½								
" 22..	5⅝@5¾	5¼@5½	9,500	3,044	1,295	759	1,299	....	3,353	
" 24..	5⅝@5¾	5¼@5½								
" 28..	5¾@6⅛	5½@5¾	6,550	10,093	1,589	1,620	333	656	4,198	
" 31..	5⅞@6¼	5⅝@6								
February 4..	5⅞@6¼	5⅝@6	7,800	3,080	1,793	1,901	897	963	5,554	⅜@7-16d
" 7..	5⅞@6¼	5⅝@6								
" 11..	5⅝@6	5⅜@5¾	4,150	6,586	1,285	1,180	570	214	3,249	
" 14..	5⅝@6	5⅜@5¾								
" 18..	5⅝@6	5⅜@5¾	5,500	24,227	1,310	1,605	1,255	....	4,170	
" 21..	5⅝@6	5⅜@5¾								
" 25..	5⅝@6	5⅜@5¾	12,650	11,103	2,042	1,668	254	844	4,808	
" 28..	5⅝@6⅛	5½@5⅞								
March 4..	5⅝@6⅛	5½@5⅞	10,550	10,974	3,803	1,662	444	....	5,909	⅜@7-16d
" 7..	5⅝@6⅛	5½@5⅞								
" 11..	5½@5⅞	5⅜@5⅝	10,150	18,774	2,818	869	222	66	3,975	
" 14..	5½@5⅞	5⅜@5⅝								
" 18..	5⅝@6⅛	5½@5¾	9,150	8,521	4,118	....	926	....	5,044	
" 21..	6¼@6⅝	5¾@6								
" 25..	6¼@6⅝	5¾@6	18,700	7,908	7,065	1,526	2,401	49	11,041	
" 28..	6¼@6⅝	5¾@6								
April 1..	6¼@6⅝	5¾@6	12,500	25,001	9,837	1,007	420	900	12,164	5-16@⅜d
" 4..	6¼@6⅝	5¾@6⅛								
" 8..	6⅜@6¾	5¾@6¼	15,600	4,591	3,409	3,338	2,185	56	8,988	
" 11..	6⅜@6¾	5¾@6¼								
" 15..	6⅜@6¾	5¾@6½	5,000	9,658	6,618	1,618	623	....	8,859	
" 18..	6⅛@6½	5½@6								
" 22..	6@6½	5½@6	7,550	11,687	4,620	....	1,622	97	6,339	
" 25..	6@6½	5½@6								
" 29..	6@6½	5½@6	13,500	13,400	3,244	845	1,286	758	6,133	
May 2..	6@6½	5½@6								¼@5-16d
" 6..	6@6½	5½@6	6,750	13,058	6,485	2,033	1,769	75	10,362	
" 9..	6@6½	5⅝@6⅛								
" 13..	6@6½	5⅝@6⅛	10,775	4,936	2,821	1,374	394	927	5,516	
" 16..	6@6½	5⅝@6⅛								
" 20..	6@6½	5⅝@6⅛	9,800	7,835	4,732	518	126	500	5,876	

New York Statement for Year 1845—*Concluded.*

1845.		Price of Middling New Orleans Liverpool Classification.	Price of Middling Upland, Liverpool Classification.	Sales for week.	Receipts for week.	EXPORTS FOR THE WEEK.					Rates of Freight to Liverpool.
						To Great Britain.	To France.	North of Europe.	Other Fo'n Ports	Total Exports.	
May	23..	6@6½	5⅝@6⅛								
"	27..	6@6⅝	5⅝@6⅛	13,800	5,618	2,462	1,141	2,468	287	6,358	
"	30..	6@6⅝	5⅝@6⅛								
June	3..	6@6⅝	5⅝@6⅛	8,050	4,915	6,334	1,477	2,660	845	11,316	¼@5-16d
"	6..	6¼@6⅞	5⅞@6⅜								
"	10..	6¼@6⅞	5⅞@6⅜	13,750	8,803	3,642	3,508	1,031	....	8,181	
"	13..	6⅜@7	6@6½								
"	17..	6⅜@7	6@6½	5,850	6,740	4,096	1,173	1,580	365	7,214	
"	20..	6⅜@7	6@6½								
"	24..	6⅜@7	6@6½	4,650	719	3,525	3,077	842	257	7,701	
"	27..	6⅜@7	6@6½								
July	1..	6⅜@7	6¼@7	5,250	4,092	2,463	1,238	831	....	4,532	¼@5-16d
"	4..	6⅜@7	6¼@7								
"	8..	7@7⅜	6½@7	6,850	4,357	2,125	481	757	....	3,363	
"	11..	7@7⅜	6½@7								
"	15..	7@7⅜	6½@7	4,550	6,415	2,407	1,163	1,346	51	4,967	
"	18..	7@7⅜	6½@7								
"	22..	7¼@7¾	7@7¼	5,850	5,061	1,542	....	860	443	2,845	
"	25..	7¼@7¾	7¼@7½								
"	29..	7¼@7¾	7¼@7½	4,600	2,568	302	1,605	469	1	2,377	
August	1..	7¼@7¾	7¼@7½								⅛@3-16d
"	5..	7¼@7¾	7¼@7½	4,500	6,446	589	1,714	325	....	2,628	
"	8..	7¼@7¾	7¼@7½								
"	12..	7¼@7¾	7¼@7½	4,750	4,843	1,795	1,561	471	....	3,827	
"	15..	7¼@7¾	7¼@7½								
"	19..	7¼@7¾	7¼@7½	3,300	2,739	924	771	545	....	2,240	
"	22..	7¼@7¾	7¼@7½								
"	26..	7¼@7¾	7¼@7½	2,600	3,914	1,630	959	616	....	3,205	
"	29..	7¼@7½	7@7¼								
Septem.	2..	7¼@7½	7@7¼	5,450	3,472	1,755	1,681	411	....	3,847	¼@5-16d
Average prices and total sales, receipts and exports.		6.22	5.63	394,525	356,171	145,613	69,962	49,805	14,174	279,554	

General Remarks.

Exchange.

Bills on London ruled steady through September at 9¾@10 per cent. premium; in October, the range was from 9¾@10½; in November, 9¾@10½; December, 10@10¼; January, 9¾@10¼; February, 9¾@10¼; March, 9½@9¾; April, 9¼@9⅝; May, 9½@10; June, 9½@10; July, 9½@10; August, 9¾@10¼, closing September 1 at that quotation.

1846.

Remarkable ravages by the cotton worm this year. (*See* years 1800, 1804 and 1825.)

Messrs. Du Fay & Co., of Manchester, England, at considerable personal expense, obtained some valuable statistics referring to the cotton manufacture, from which it appears there were in Great Britain, this year, 17,500,000 spindles at work. (*See* year 1856.)

Estimated number of operators employed in cotton mills in Great Britain, this year, 316,327.

The supply of East India cotton received in Great Britain this year reached but 49,000 bales, while the imports from this country was but 932,000. (*See* year 1845.)

COTTON CROP OF THE UNITED STATES.

Statement and Total Amount of the Cotton Crop of the United States, for the Year ending August 31, 1846.

	Bales.	Bales.	Total.	Same period 1845.
NEW ORLEANS.				
Export—				
To Foreign Ports	834,775			
Coastwise	220,082			
Stock on hand 1st September, 1846	6,332			
		1,061,189		
Deduct—				
Stock on hand 1st September, 1845	7,556			
Received from Mobile	6,356			
" Florida	5,884			
" Texas	4,249			
		24,045		
MOBILE.			1,037,144	929,126
Export—				
To Foreign Ports	301,735			
Coastwise	115,898			
Stock on hand 1st September, 1846	7,476			
		425,109		
Deduct—				
Stock on hand 1st September, 1845	609			
Received from wrecked ships	1,275			
" Texas	666			
" New Orleans	593			
		3,143		
FLORIDA.			421,966	517,196
Export—				
To Foreign Ports	49,981			
Coastwise	90,215			
Stock on hand 1st September, 1846	1,088			
		141,284		
Deduct—				
Stock on hand 1st September, 1845	....	100		
			141,184	188,693
TEXAS.				
Export—				
To Foreign Ports	11,324			
Coastwise	14,184			
Stock on hand 1st September, 1846	1,500			
			27,008	

Statement and Total Amount of the Cotton Crop of the United States, for the Year ending August 31, 1846.—*Concluded.*

	Bales.	Bales.	Total.	Same period 1845.
GEORGIA.				
Export from Savannah—				
To Foreign Ports—Uplands	69,380			
" Sea Island	8,472			
Coastwise—Uplands	106,229			
Sea Island	2,225			
	186,306			
Burnt in Savannah	1,848			
Stock in Savannah 1st September, 1846	5,922			
" Augusta and Hambro', 1st September, 1846	9,906			
		203,982		
Deduct—				
Stock in Savannah and Augusta, 1st September, 1845	....	9,071		
			194,911	295,440
SOUTH CAROLINA.				
Export from Charleston—				
To foreign ports—Uplands	160,233			
" Sea Island	19,527			
Coastwise—Uplands	87,841			
Sea Island	476			
	268,077			
Export from Georgetown—				
To New York	3,852			
Stock in Charleston 1st September, 1846	8,709			
		280,638		
Deduct—				
Stock in Charleston 1st September, 1845	10,879			
Received from Savannah	16,397			
" Florida, Key West, etc	1,957			
		29,233		
			251,405	426,361
NORTH CAROLINA.				
Export—				
Coastwise	....	....	10,637	12,487
VIRGINIA.				
Export—				
To Foreign Ports	1,308			
Coastwise	3,505			
Manufactured	10,787			
Stock on hand 1st September, 1846	100			
		15,700		
Deduct—				
Stock on hand 1st September, 1845	....	2,418		
			13,282	25,200
Received at Philadelphia and Baltimore, overland	....	....	3,000	
Total crop of the United States	....	....	2,100,537	2,394,503
Crop of last year	....	....	2,394,503	
Decrease	....	....	293,966	

Export to Foreign Ports, from September 1, 1845, *to August* 31, 1846.

FROM	To Great Britain.	To France.	To North of Europe.	Other Foreign Ports.	Total.
New Orleans (bales)	562,320	159,528	28,841	84,086	834,775
Mobile	208,082	66,821	15,974	10,858	301,735
Florida	42,844	7,137	....	....	49,981
Georgia (Savannah and Darien)	67,117	8,813	....	1,922	77,852
South Carolina	117,070	50,980	5,118	6,592	179,760
North Carolina	....	....	....	....	....
Virginia	630	250	378	50	1,308
Baltimore	1,494	....	319	....	1,813
Philadelphia	1,723	....	....	....	1,723
New York	94,292	65,438	26,556	11,464	197,750
Boston	3,973	736	3,506	556	8,771
Texas	2,824	....	6,000	2,500	11,324
Grand total	1,102,369	359,703	86,692	118,028	1,666,792
Total last year	1,439,306	359,357	134,501	150,592	2,083,756
Increase	....	346	....	....	....
Decrease	336,937	....	47,809	32,564	416,964

☞ The shipments from Mississippi are included in the export from New Orleans.

Growth.

Total crop of 1826-7...... bales.	937,000	Total crop of 1836-7...... bales.	1,422,930
" 1827-8	712,000	" 1837-8	1,801,497
" 1828-9	857,744	" 1838-9	1,360,532
" 1829-30	976,845	" 1839-40	2,177,835
" 1830-1	1,038,848	" 1840-1	1,634,945
" 1831-2	987,477	" 1841-2	1,683,574
" 1832-3	1,070,438	" 1842-3	2,378,875
" 1833-4	1,205,394	" 1843-4	2,030,409
" 1834-5	1,254,328	" 1844-5	2,394,503
" 1835-6	1,360,725	" 1845-6	2,100,537

Consumption.

Total crop of the United States, as above stated................bales.			2,100,537
Add—Stocks on hand at the commencement of the year, September 1, 1845, in the Southern ports		30,733	
" Northern ports		67,687	
			98,420
Makes a supply of			2,198,957
Deduct therefrom—The export to Foreign ports	1,666,792		
Less—Foreign included	349		
		1,666,443	
Stocks on hand at the close of the year, September 1, 1846.			
In the Southern ports	41,033		
In the Northern ports	66,089		
		107,122	
Burnt at Savannah	1,848		
" Philadelphia	347		
" New York	600		
		2,795	
			1,776,360
Taken for home use			422,597

Quantity consumed by and in the hands of Manufacturers.

1845–6 ....bales.	422,597	1836–7 ....bales.	222,540
1844–5	389,006	1835–6........................	236,733
1843–4	346,744	1834–5........................	216,888
1842–3	325,129	1833–4	196,413
1841–2........................	267,850	1832–3........................	194,412
1840–1........................	297,288	1831–2........................	173,800
1839–40........................	295,193	1830–1........................	182,142
1838–9........................	276,018	1829–30........................	126,512
1837–8........................	246,063	1828–9........................	118,853

*** By the foregoing statement it will be seen that the crop falls short of last year's, by 293,966 bales—add to this the quantity put down for Texas, and the actual difference is 320,974.

Our estimate of the quantity taken for consumption, does not include any cotton manufactured in the States, south and west of Virginia, nor any in that State, except in the vicinity of Petersburg and Richmond.

The quantity of new cotton received at the shipping ports up to the 1st inst. amounted only to about 200 bales, against 7,500 bales last year.

In regard to the crop now commenced picking, it may be remarked that it is about three weeks later than last year, and consequently more exposed to injury—while the fact is very generally conceded that in parts of Alabama, and in Mississippi, Louisiana and Texas, much injury has already been done by wet and by worms. In the Atlantic States, however, appearances indicate a full average crop.

ANNUAL REVIEW.

From the New Orleans Price Current, 1845–46.

As the season advanced the cotton demonstrated a very material falling off in the amount of production, as compared with the previous year, a circumstance which would unquestionably have led to an important enhancement of prices, under a continuance of the favorable features which marked the opening of the commercial year. But before this conviction became strong enough to act upon to any important extent, a variety of adverse elements were brought into operation, which tended in a material degree to disappoint the flattering anticipations that were indulged in in the early part of the season. The most prominent of these obstacles, to a prosperous progress in the market, were the partial failure of the grain and potato crops of Europe; the inordinate speculations in railway shares, both in England and on the continent, and the consequent tightness in the money market; the agitation of the Oregon question; the glutted condition of the Asiatic markets, and a variety of other impediments which were not calculated on, and which need not here be enumerated. The excessive depression of the previous year, however, has been avoided, and the closing rates are a fraction higher than those which prevailed at the opening of the season. As regards the crop of 1845, it was early manifest that excessive drought, at an early period, had materially curtailed the extent of production in South Carolina, Georgia, Alabama, etc.; but at the same time an impression prevailed that the increased yield in the sections of country bordering on the Mississippi River would go far to counterbalance the deficiency in the States on the Atlantic. The result, however, proves that this calculation was widely erroneous, and is another among the many instances on record which evidence the futility of early estimates. Apart from this backwardness of the plant—a point which is fully established—there appeared, up to some four or five weeks past, a tolerable prospect for something like an average yield, provided a favorable picking season should ensue; but since then the caterpillar, or army worm, and boll worm have made their appearance, to a greater or less extent, in most sections of the cotton-growing regions, including Texas, and are said to be making great ravages, on some plantations nearly destroying the whole crops. The appearance of these destructive agents seems to have created more alarm among the planters this season than usual, from the fact that their advent has been about a month earlier than in previous years, while the crop generally is said to be full three weeks later, thus making a difference of seven weeks in the position of the plant when first attacked, and rendering it much more susceptible of injury. Heavy rains have somewhat destroyed the caterpillars, by beating them off the plant on to the ground, and the prospects thus far present a marked contrast to last year.

LIVERPOOL STATEMENT FOR 1846.

UNITED STATES, 1845–1846.

Stock, October 1, 1845	94,000	Export	1,665,000
Receipts	2,100,000	Consumption	422,000
		Stock, Oct. 1, 1846	107,000
Bales	2,194,000	Bales	2,194,000

Stock 1st Jan., 1846, in	Liverpool.	Gt. Britain.	France.	Continent.	Tot. Europe.
United States ... Bales	624,000	691,000	54,000	59,000	804,000
Brazil	52,000	52,000	....	1,000	53,000
West Indies	6,000	6,000	4,000	6,000	16,000
East Indies	141,000	238,000	2,000	5,000	245,000
Egypt	62,000	68,000	9,000	24,000	101,000
Bales	885,000	1,055,000	69,000	95,000	1,219,000

CONSUMPTION. Tot. Europe.	Continent.	France.	Gt. Britain.	Liverpool.	IMPORT.	Liverpool.	Gt. Britain.	France.	Continent.	Tot. Europe.
1,859,000	220.000	369,000	1,270,000	1,170,000	United States ... Bales	932,000	991,000	340,000	188,000	1,410,000
114.000	8,000	1,000	105,000	105,000	Brazil	84,000	84.000	1,000	8,000	86.000
41,000	12,000	15,000	14,000	12,000	West Indies	9,000	13,000	11,000	8,000	31,000
181,000	65,000	4,000	112.000	87,000	East Indies	50,000	95.000	4,000	63,000	98,000
128,000	40,000	16,000	72,000	70,000	Egypt	59,000	61,000	9,000	31,000	101,000
2,323,000	345,000	405,000	1,573,000	1,444,000	Bales	1,134,000	1,244,000	365,000	298,000	1,726,000
....	....	....	181,000	136,000	Export.					
622,000	48,000	29,000	545,000	439,000	Stock, Dec. 31 .... Stock above,	885,000	1,055,000	69,000	95,000	1,219,000
2,945,000	393,000	434,000	2,299,000	2,019,000	Total supply, bales	2,019,000	2,299,000	434,000	393,000	2,945,000

COTTON AT LIVERPOOL. YEAR 1846.

Week Ending.	Receipts.						Sales.				Stocks.			Prices.			Actual Export	Consumption.
	American	E. I.	Egypt.	Brazil.	Other.	Total.	Consumption.	Speculation.	Export.	Total.	Amer'n	Other.	Total.	Mid. Up.	Mid. Orl.	Dhol		
Jan. 9..	41,321	9,080	......	1,072	1,045	52,518	34,590	14,400	......	48,990	636,500	266,000	902,500	4⅛	4⅜	2	......	34,590
" 16..	13,981	2,191	999	1,962	800	19,933	28,500	8,000	100	36,600	626,500	265,500	892,000	4⅛	4¾	2	100	63,090
" 23..	11,404	......	1,524	....	648	13,576	29,520	8,000	......	37,520	611,000	262,500	873,500	4⅛	4½	2	......	92,610
" 30..	29,313	......	2,253	4,853	......	36,419	32,760	6,000	300	39,060	614,500	262,000	876,500	4⅛	4⅜	2	300	125,370
Feb. 6..	22,757	2,488	2,092	2,270	......	29,607	38,580	7,000	700	46,280	602,500	263,500	866,000	4⅛	4⅜	2	700	163,950
" 13..	15,833	......	1,788	....	......	17,621	26,450	2,400	1,210	30,060	595,000	260,000	855,000	4⅛	4⅜	2	1,210	190,400
" 20..	1,990	......	1,631	....	41	3,662	25,340	......	1,700	27,040	573,000	259,000	832,000	4	4¼	2	1,700	215,740
" 27..	5,107	......	1,434	....	......	6,541	26,410	3,700	2,700	32,810	551,500	256,500	808,000	4	4¼	2	2,700	242,150
Mch 6..	26,185	......	1,860	5,736	307	34,088	18,720	6,000	2,800	27,520	502,000	259,000	821,000	4⅛	4⅜	2	2,800	260,870
" 13..	37,542	......	3,532	....	......	41,074	19,020	3,900	4,400	27,320	578,000	258,500	836,500	4⅛	4⅜	2	4,400	279,800
" 20..	9,792	6,955	1,268	2,915	......	20,930	24,230	2,250	2,200	28,680	565,000	266,000	831,000	4⅛	4⅜	2	2,200	304,120
" 27..	4,737	2,468	2,348	650	120	10,323	22,930	1,500	3,200	27,630	547,000	268,000	815,000	4⅛	4⅜	2	3,200	327,050
April 3..	6,674	1,672	1,229	....	425	10,000	30,830	3,000	5,200	39,030	525,500	263,000	788,500	4¼	4½	2	5,200	357,880
" 9..	4,851	......	22	629	......	3,502	22,930	6,000	1,750	30,680	510,500	259,000	769,500	4¼	4½	2	1,750	380,810
" 17..	34,787	......	3,574	....	......	38,361	32,860	10,600	9,900	53,360	510,000	255,000	765,000	4⅜	4½	2¼	9,900	413,670
" 24..	100,580	3,860	5,868	....	......	110,308	25,350	5,800	4,100	35,250	584,500	259,500	844,000	4⅜	4½	2¼	4,100	439,020
May 1..	6,861	......	......	2,610	......	9,471	6,250	33,000	6,410	45,660	551,000	254,000	805,000	4⅜	4½	2¼	6,410	445,270
" 8..	25,649	......	......	....	230	25,879	30,790	4,900	4,110	39,800	515,000	279,500	794,500	4½	4⅜	2¼	4,110	476,060
" 15..	18,288	......	1,415	1,274	60	21,037	28,470	22,000	5,400	55,870	539,000	243,000	782,000	4⅝	4¾	2¼	5,400	504,530
" 22..	28,468	......	6,220	....	......	34,688	23,690	4,330	1,320	29,340	545,500	245,500	791,000	4½	4¾	2¼	1,320	528,220
" 29..	11,542	......	......	1,448	......	12,990	20,850	3,400	2,400	26,650	539,000	241,500	780,500	4⅝	4¾	2¼	2,400	549,070
June 5..	1,325	......	......	....	......	1,325	26,020	10,000	2,000	38,020	515,000	239,000	754,000	4⅝	4¾	2¼	2,000	575,090
" 12..	36,814	195	......	8,609	484	46,102	24,705	4,000	1,370	30,075	528,500	245,500	774,000	4⅝	4¾	2¼	1,370	599,795
" 19..	....	......	1,303	....	......	1,303	29,500	1,500	1,900	32,900	503,500	239,000	742,500	4⅝	4¾	2¼	1,900	629,295
" 26..	44,787	......	2,718	....	77	47,582	20,980	500	3,190	24,670	528,000	238,500	766,500	4½	4⅝	2¼	3,190	650,275
July 3..	28,529	......	......	5,953	......	34,482	34,590	3,250	2,500	40,340	523,500	237,500	761,000	4½	4⅝	2¼	2,500	684,865
" 10..	51,300	......	......	1,242	......	52,542	37,470	1,000	4,650	43,120	537,500	234,000	771,500	4½	4⅝	2¼	4,650	722,335
" 17..	19,109	1,210	2,601	35	149	23,104	30,230	3,700	2,850	36,780	530,500	234,000	764,500	4½	4⅝	2¼	2,850	752,565
" 24..	60,226	1,319	......	2,509	......	64,054	29,810	3,700	3,200	36,710	559,000	234,500	793,500	4½	4⅝	2¼	3,200	782,375
" 29..	17,965	......	866	....	2,970	21,801	18,270	4,200	4,030	26,500	560,000	233,000	793,000	4½	4⅝	2¼	4,030	800,645
Aug. 7..	11,357	......	1,924	1,100	129	14,510	31,100	8,000	4,510	43,610	536,500	229,500	766,000	4½	4¾	2¼	4,510	831,745
" 14..	33,326	2,855	1,745	4,343	......	42,269	23,719	2,500	2,620	28,839	546,000	233,000	779,000	4½	4¾	2¼	2,620	855,464
" 21..	31,008	......	101	1,276	75	32,460	24,080	6,000	3,170	33,250	557,500	228,500	786,000	4⅝	4¾	2¼	3,170	879,544
" 28..	15,677	976	883	....	376	17,912	27,450	3,800	4,020	35,270	546,500	225,500	771,000	4⅝	4¾	2¼	4,020	906,994

Sep. 4..	16,208	......	2,344	863	360	19,775	46,460	16,400	3,870	66,730	521,000	220,500	741,500	4¾	5	2¼	3,870	953,454
" 11..	21,496	3,220	......	3,657	326	28,699	32,070	8,500	3,740	44,310	516,000	219,000	735,000	4¾	5	2¼	3,740	985,524
" 18..	3,237	......	2,936	......	......	6,173	38,540	41,800	1,810	82,150	491,000	209,000	700,000	5⅛	5¼	2½	1,810	1,024,064
" 25..	6,690	2,163	......	662	90	9,605	35,280	34,810	2,720	72,810	467,000	205,000	672,000	5⅛	5⅜	2½	2,720	1,059,344
Oct. 2..	9,796	......	......	4,532	......	14,628	27,910	21,700	2,180	51,790	452,000	205,000	657,000	5⅛	5⅜	2¾	2,180	1,087,254
" 9..	16,677	......	......	522	661	17,860	27,570	22,500	2,730	52,800	445,500	199,500	645,000	5¼	5½	2⅞	2,730	1,114,824
" 16..	4,005	580	......	1,146	96	5,827	30,240	38,700	1,960	70,900	426,000	192,000	618,000	5½	5¾	3	1,960	1,145,064
" 23..	1,019	753	......	2,868	128	4,768	27,170	32,300	390	59,860	406,000	192,000	598,000	5½	5¾	3	390	1,172,234
" 30..	5,219	......	......	1,030	......	6,249	23,550	23,600	1,550	48,700	393,000	186,500	579,500	5⅝	5⅞	3	1,550	1,195,784
Nov. 6..	5,121	32	......	2,689	239	8,081	12,460	6,300	1,400	20,160	387,500	187,500	575,000	5½	5¾	3	1,400	1,208,244
" 13..	357	2,164	......	......	287	2,808	22,460	8,000	1,750	32,210	369,500	184.500	554,000	5½	5¾	3	1,750	1.230,704
" 20..	1,857	1,174	249	2,587	......	5,867	21,100	11,800	950	33,850	354,500	182,000	536,500	5½	5¾	3	950	1,251,804
" 27..	992	......	277	710	149	2,128	25,390	5,300	730	31,420	335,500	177,500	513,000	5½	5¾	3	730	1,277,194
Dec. 4..	......	......	1,630	......	......	1,630	30,890	46.600	530	78,020	311,500	172,000	483,500	5⅞	6⅛	3¼	530	1,308,084
" 11..	......	......	702	173	......	875	54,320	67,450	200	121,970	272,000	158,000	430,000	6⅜	6⅝	3½	200	1,362,404
" 18..	3,169	......	......	1,723	......	4,892	36,520	93,700	100	130,320	266,000	152,000	418,000	6⅞	7	3½	100	1,398,924
" 24..	19,850	......	......	2,169	303	22,322	8,120	13,170	50	21,340	281,500	151,000	432,500	6¾	7	3½	50	1,407,044
" 31..	8,957	4,166	1,431	2,461	......	17,015	11,660	19,900	250	31,810	......	......	......	7	7¼	3¾	250	1,418,704
Average prices & total sales, receipts & stocks.	933,833	49,521	60,767	77,998	10,578	1,132,694	1,418,704	720,860	126820	2,266,384				4.8	5.00	2.44	126820	2,728,276

The semi-weekly Price and Weekly Sales and Receipts at New York, Weekly Exports from New York and Rates of Freight to Liverpool 1st of each month, for the Crop Year ending September 1, 1846.

1845.	Price of Mid. Fair to Fair New Orleans Liverpool Classificat'n.	Price of Mid. Fair to Fair Upland, Liverpool Classificat'n.	Sales for week.	Receipts for week.	EXPORTS FOR THE WEEK. To Great Britain.	To France.	North of Europe.	Other Fo'n Ports	Total Exports.	Rates of Freight to Liverpool.	GENERAL REMARKS.
Septem. 5..	7¼@7½	7@7¼								¼@5-16d	
" 9..	7¼@7½	7@7¼	5,450	4,561	1,846	877	400	....	3,123		
" 12..	7¼@7½	7@7¼									
" 16..	7¼@7½	7@7¼	3,400	2,059	2,807	645	1,363	420	5,235		An event that had an important bearing upon the future cotton supply, occurred during this crop year; on December 22, 1845, the Republic of Texas was formally admitted into, and became one of, the United States; this measure was followed by a declaration of war between this country and Mexico on the 13th May, 1846, and this had an unfavorable effect upon business generally; for a time, cotton being much depressed; the unsettled feeling, however, was much mitigated by the fact, that early in the season, it became apparent, that the crop coming in would not come up to previous expectations; the receipts at the ports steadily fell off when compared with the previous year, and prices, with one or two brief exceptions, steadily appreciated, closing at an advance of 1¾@2 cents per pound above those current at the opening of the crop year.
" 19..	8@8¾	7½@7⅞									
" 23..	8¼@9	7¾@8¼	10,875	4,308	3,105	1,119	374	822	5,420		
" 25..	8¼@9	7¾@8¼									
" 30..	8¼@9	7¾@8¼	8,400	3,464	1,834	2,981	614	....	5,429		
October 3..	8¼@9¼	7¾@8¼								5-16@⅜d	
" 7..	8¼@9⅛	7¾@8¼	9,200	3,728	3,123	....	1,685	....	4,808		
" 10..	8¼@9¼	7¾@8¼									
" 14..	8¼@9¼	7¾@8⅜	6,000	3,757	747	826	447	....	2,020		
" 17..	8¼@9¼	7¾@8¼									
" 21..	8¼@9	7⅝@8	4,350	959	1,551	1,673	264	....	3,488		
" 24..	8½@9¼	7¾@8¼									
" 28..	8½@9⅛	7¾@8¼	5,750	771	5,216	3,115	1,772	110	10,213		
" 31..	8½@9	7⅝@8⅛									
Novem. 4..	8@8¾	7½@8	2,850	6,248	1,403	1,024	591	819	3,837	5-16@⅜d	
" 7..	8@8¾	7½@7⅞									
" 11..	8@8¾	7⅜@7¾	5,450	8,324	840	506	580	....	1,926		
" 14..	8@8⅝	7¼@7⅝									
" 18..	7¾@8½	7@7½	4,450	6,292	2,211	1,702	111	536	4,560		
" 21..	7¾@8½	7@7½									
" 25..	7¾@8½	7@7⅜	5,800	5,056	559	854	....	....	1,413		
" 28..	8@8⅝	7¼@7½									
Decem. 2..	8@8⅝	7¼@7½	6,000	2,931	2,378	2,008	174	771	5,331	5-16@⅜d	
" 5..	8@8⅝	7¼@7½									
" 9..	8@8¾	7⅜@7⅝	3,400	4,826	1,503	1,345	....	644	3,492		
" 12..	8¼@8¾	7⅜@7¾									
" 16..	8¼@8¾	7⅝@8	3,900	8,383	1,677	428	93	1,134	3,332		
" 19..	8¼@8¾	7½@8									
" 23..	8@8½	7⅜@7¾	3,750	4,453	326	....	....	....	326		
" 26..	8@8½	7⅜@7¾									

Decem. 30..	8@8½	7⅜@7¾	1,800	6,052	1,181	768	....	....	1,949	
1846.										
January 2..	7⅞@8⅜	7⅜@7¾								3-16@¼d
" 6..	7⅞@8⅜	7¼@7¾	2,450	3,892	59	1,260	....	....	1,319	
" 9..	7⅞@8⅜	7⅜@7¾								
" 13..	7¾@8⅜	7¼@7⅝	2,400	5,840	471	304	....	28	803	
" 16..	7⅞@8⅜	7¼@7⅝								
" 20..	7⅞@8⅜	7¼@7⅝	4,309	1,400	394	978	....	....	1,372	
" 23..	8@8½	7¼@7⅝								
" 27..	7⅞@8½	7¼@7⅝	3,370	7,403	801	....	....	....	801	
" 30..	8@8½	7¼@7⅝								
February 3..	8@8½	7¼@7⅝	5,100	6,130	1,195	1,479	....	....	2,674	3-16@¼d
" 6..	8@8½	7¼@7⅝								
" 10..	8@8½	7¼@7⅝	5,750	2,037	1,849	1,665	135	....	3,649	
" 13..	8@8½	7¼@7⅝								
" 17..	8@8⅝	7¼@7⅝	5,400	7,316	1,589	934	26	....	2,549	
" 20..	8@8⅝	7¼@7⅝								
" 24..	8@8¾	7⅜@7¾	7,700	5,683	264	1,615	81	....	1,960	
" 27..	8¼@9	7⅝@8								
March 3..	8¼@9	7⅝@8	7,000	4,799	1,253	1,184	....	....	2,437	¼@5-16d
" 6..	8½@9¼	7⅞@8¼								
" 10..	8½@9¼	7⅞@8⅜	6,300	6,994	947	1,299	....	524	2,770	
" 13..	8¾@9⅜	8@9½								
" 17..	8¾@9½	8⅛@8⅝	5,100	11,890	106	1,350	132	532	2,120	
" 20..	8¾@9½	8⅛@8⅝								
" 24..	8⅝@9⅜	8@8⅜	3,500	4,172	....	....	147	....	147	
" 27..	8⅝@9⅜	8@8⅜								
" 31..	8½@9¼	8@8⅜	4,150	17,364	....	....	233	69	302	
April 3..	8½@9¼	8@8⅜								3-16@¼d
" 7..	8½@9⅜	8@8⅜	4,450	1,732	433	934	696	....	2,063	
" 10..	8½@9¼	8@8⅜								
" 14..	8½@9¼	8@8¼	5,000	22,169	565	2,585	88	....	3,238	
" 17..	8½@9¼	8@8¼								
" 21..	8½@9¼	7⅞@8¼	2,950	10,575	804	1,222	....	....	2,026	
" 24..	8⅜@9¼	7¾@8⅛								
" 28..	8½@9⅛	7¾@8⅛	6,850	12,684	554	698	94	....	1,346	
May 1..	8½@9¼	7⅞@8¼								¼@5-16d
" 5..	8½@9¼	8@8¼	9,500	10,925	1,471	2,031	736	....	4,238	
" 8..	8½@9⅛	7⅞@8¼								
" 12..	8¼@9⅛	7¾@8½	9,300	11,280	2,231	848	143	....	3,222	
" 15	8¼@9	7¾@8½								
" 19..	8⅜@9	7⅝@8⅛	7,700	9,806	1,434	3,581	769	....	5,784	

The semi-weekly Price and Weekly Sales and Receipts at New York, Weekly Exports from New York and Rates of Freight to Liverpool 1st of each Month, for the Crop Year ending September 1, 1846—Concluded.

1846.	Price of Mid. Fair to Fair New Orleans Liverpool Classificat'n.	Price of Mid Fair to Fair Upland, Liverpool Classificat'n.	Sales for week.	Receipts for week.	EXPORTS FOR WEEK. To Great Britain	To France.	North of Europe.	Other Fo'n Ports	Total Exports.	Rates of Freight to Liverpool.
May 22..	8⅜@9	7⅝@8⅛								
" 26..	8¼@8⅞	7⅝@8	6,400	8,461	3,328	1,538	891	713	6,470	
" 29..	8¼@8⅞	7⅝@8								
June 2..	8¾@9	7¾@8⅛	10,500	3,219	4,201	1,083	1,082	1,228	7,594	7-16@$\frac{9}{16}$
" 5..	8¼@9	7¾@8⅛								
" 9..	8¼@9	7¾@8⅛	13,400	11,381	7,640	2,181	596	....	10,417	
" 12..	8½@9	7¾@8⅛								
" 16..	8½@9	7¾@8¼	9,550	1,964	4,066	2,460	750	776	8,052	
" 19..	8⅜@9	7¾@8¼								
" 23..	8½@9	7¾@8¼	5,600	5,115	4,196	1,446	1,260	102	7,004	
" 26..	8½@9	7⅞@8¼								
" 30..	8½@9	7⅞@8¼	6,200	2,914	6,015	1,353	2,536	462	10,366	
July 3..	8½@9	7⅞@8⅜								¼@5-16d
" 7..	8½@9	7⅞@8⅜	1,500	6,791	1,277	1,013	298	....	2,588	
" 10..	8½@9	7¾@8⅜								
" 14..	8½@9	7¾@8⅜	5,700	3,877	3,219	1,129	834	....	5,182	
" 17..	8½@9	7¾@8⅜								
" 21..	8½@9	7⅞@8⅜	6,100	1,652	5,098	1,220	1,551	....	7,869	
" 24..	8⅝@9⅛	8@8½								
" 28..	8⅝@9¼	8@8½	8,250	4,289	780	....	643	....	1,423	
" 31..	8⅝@9¼	8@8½								
August 4..	8⅝@9¼	8@8½	6,700	2,619	1,627	1,965	521	1,265	5,378	7-32@¼
" 7..	8¾@9¼	8⅛@8½								
" 11..	8⅝@9¼	8⅛@8½	4,700	3,908	2,141	1,882	669	509	5,201	
" 14..	8¾@9¼	8⅛@8½								
" 18..	8¾@9¼	8¼@8½	4,700	9,989	546	1,206	517	....	2,269	
" 21..	8⅞@9⅜	8⅜@8⅝								
" 25..	9@9½	8⅜@8¾	8,300	2,286	858	1,738	1,676	....	4,272	
" 28..	9@9⅝	8½@9								
Septem. 1..	9@9¾	8⅝@9⅛	10,800	4,335	570	1,386	984	....	2,940	¼@5-16d
Average price and total sales, receipts and exports.	8.59	7.87	307,504	313,063	94,289	65,438	26,556	11,464	197,747	

GENERAL REMARKS.

Exchange.

The course of exchange varied but little this year; the range for bills on London in September, was 9¾@10¼ per cent. premium; October, 9¼@10; November, 9½@8½; December, 8@9; January, 8¼@8¾; February, 8@8½; March, 8¼ up to 9¾; April, 9¼@10; May, 8@9¾; June, 8¾@7½; July, between 7 and 8; and in August, the quotation advanced from 7¼@7¾, up to 8½@9, which latter was the closing rate, September 1, 1846.

1847.

COTTON CROP OF THE UNITED STATES.

Statement and Total Amount of the Cotton Crop of the United States, for the Year ending August 31, 1847.

	Bales.	Bales.	Total.	Same period 1846.
NEW-ORLEANS.				
Export—				
To Foreign Ports	565,007			
Coastwise	159,501			
Stock on hand, 1st September, 1847	23,493			
		748,001		
Deduct—				
Stock on hand, 1st September, 1846	6,332			
Received from Mobile	16,379			
" " Florida	16,966			
" " Texas	2,345			
		42,022		
			705,979	1,037,144
MOBILE.				
Export—				
To Foreign Ports	190,221			
Coastwise	116,801			
Stock on hand, 1st September, 1847	24,172			
		331,194		
Deduct—				
Stock on hand, 1st September, 1846	7,476			
Received from New Orleans	256			
		7,732		
			323,462	421,966
FLORIDA.				
Export—				
To Foreign Ports	36,726			
Coastwise	90,006			
Burnt at Apalachicola	100			
Stock on hand, 1st September, 1847	2,108			
		128,940		
Deduct—				
Stock on hand, 1st September, 1846	....	1,088		
			127,852	141,184
TEXAS.				
Export—				
To Foreign Ports	543			
Coastwise	9,242			
Stock on hand, 1st September, 1847	32			
		9,817		
Deduct—				
Stock on hand, 1st September, 1846	....	1,500		
			8,317	27,008
GEORGIA.				
Export from Savannah—				
To Foreign Ports—Uplands	113,656			
Sea Island	5,665			
Coastwise—Uplands	113,300			
Sea Island	1,530			
	234,151			
Export from Darien—				
To New York	5			
Stock in Savannah, 1st September, 1847	7,787			
" Augusta and Hambro', 1st September, 1847	17,233			
		259,176		

Statement and Total Amount of the Cotton Crop of the United States, for the Year ending August 31, 1847.—*Concluded.*

	Bales.	Bales.	Total.	Same period 1846.
Deduct—				
Stock in Savannah and Augusta, 1st September, 1846.	15,828			
Received from Florida......	559			
		16,387		
			242,789	194,911
SOUTH CAROLINA.				
Export from Charleston—				
To Foreign Ports—Uplands......	179,467			
Sea Island......	10,869			
Coastwise—Uplands......	156,064			
Sea Island......	698			
	347,098			
Export from Georgetown—				
To New York, Boston, &c......	2,000			
Stock in Charleston, 1st September, 1847......	29,655			
		378,753		
Deduct—				
Stock in Charleston, 1st September, 1846......	8,709			
Received from Savannah......	18,408			
" " Florida......	1,436			
		28,553		
			350,200	251,405
NORTH CAROLINA.				
Export—				
Coastwise......	......	......	6,061	10,637
VIRGINIA.				
Export—				
To Foreign Ports......	152			
Coastwise......	3,000			
Manufactured—Taken from the Ports......	10,491			
Stock on hand, 1st September, 1847......	448			
		14,091		
Deduct—				
Stock on hand, 1st September, 1846......	......	100		
			13,991	13,282
Received overland last year......	......	......		3,000
Total Crop of the United States......	......	......	1,778,651	2,100,537
Received at Philadelphia and Baltimore, overland......	......	......	1,828	
Crop of last year......	......	......	......	2,100,537
Decrease......	......	......	......	321,886

Export to Foreign Ports, from September 1, 1846, *to August* 31, 1847.

FROM	To Great Britain.	To France.	To North of Europe.	Other F'n Ports.	Total.
New Orleans........................Bales.	385,368	95,719	26,297	57,623	565,007
Mobile....................................	131,154	39,293	5,293	14,481	190,221
Florida..................................	30,896	2,592	....	3,238	36,726
Texas....................................	....	....	543	....	543
Georgia..................................	107,227	11,150	....	944	119,321
South Carolina.........................	121,662	51,452	8,794	8,428	190,336
North Carolina.........................	....	....	....	....	....
Virginia..................................	152	....	....	....	152
Baltimore................................	30	425	....	....	455
Philadelphia............................	433	....	....	....	433
New York................................	53,638	40,798	32,074	7,998	134,508
Boston....................................	349	57	2,688	426	3,520
Grand total..............................	830,909	241,486	75,689	93,138	1,241,222
Total last year.........................	1,102,369	359,703	86,692	118,028	1,666,792
Decrease.................................	271,460	118,217	11,003	24,890	425,570

☞ The shipments, if any, from Mississippi are included in the export from New Orleans.

Growth.

Total crop of 1827–8...... bales.	712,000	
" 1828–9............	857,744	
" 1829–30............	976,845	
" 1830–1............	1,038,848	
" 1831–2............	987,477	
" 1832–3............	1,070,438	
" 1833–4............	1,205,394	
" 1834–5............	1,254,328	
" 1835–6............	1,360,725	
" 1836–7............	1,422,930	
Total crop of 1837–8 ....bales.	1,801,497	
" 1838–9..............	1,360,532	
" 1839–40..............	2,177,835	
" 1840–1..............	1,634 945	
" 1841–2..............	1,683,574	
" 1842–3..............	2,378,875	
" 1843–4..............	2,030,409	
" 1844–5..............	2,394,503	
" 1845–6..............	2,100,537	
" 1846–7..............	1,778,651	

Consumption.

Total crop of the United States, as above stated......................			1,778,651 bales.
Add—Stocks on hand at the commencement of the year, 1st September, 1846.—In the Southern ports..................		41,033	
" Northern "		66,089	
			107,122
Makes a supply of....................................			1,885,773
Deduct therefrom—The export to Foreign ports.............	1,241,222		
Less, foreign included..................	353		
		1,240,869	
Stocks on hand at the close of the year, 1st September, 1847—			
In the Southern ports..................	104,928		
" Northern "	109,909		
		214,837	
Burnt at Apalachicola..................................	100		
" New-York..................................	2,000		
		2,100	
			1,457,806
Taken for home use.................Bales.			427,967

Quantity consumed by and in the hands of Manufacturers.

1846–7 .... bales.	427,967	1836–7 .... bales.	222,540
1845–6	422,597	1835–6	236,733
1844–5	389,006	1834–5	216,888
1843–4	346,744	1833–4	196,413
1842–3	325,129	1832–3	194,412
1841–2	267,850	1831–2	173,800
1840–1	297,288	1830–1	182,142
1839–40	295,193	1829–30	126,512
1838–9	276,018	1828–9	118,853
1837–8	246,063	1827–8	120,593

NOTE.—By the foregoing statement, it will be seen that the crop is 321,886 bales less than last year, and 615,852 less than the year before.

Our estimate of the quantity taken for consumption in the cotton growing States, does not include any cotton manufactured in the States south and west of Virginia.

The quantity of new cotton received at the shipping ports up to the 1st inst. amounted to 1,121 bales, against about 200 bales last year.

We have this year made up our Statement of the Crop of the United States, without including the quantity received overland at Baltimore and Philadelphia, it being almost universally conceded that it has already been included in the shipments up the Mississippi from New Orleans; to show a fair comparison, however, with last year, we have appended the amounts so received at those ports.

The shipments given in the above statement from Texas, are those by sea only; a considerable portion of the crop of that State finds its way to market via Red River, and is included in the receipts at New Orleans—upwards of 4,000 bales, it is supposed, have thus been received during the season.

ANNUAL REVIEW.

From the New Orleans Price Current, 1846—47.

It is, of course, familiarly known to all that a large deficiency in the crops of Great Britain and Ireland, and of many portions of the Continent of Europe, produced an extraordinary demand upon this country for various articles of food; among the most prominent of which are flour, wheat, Indian corn, and corn meal; and the remarkable increase in the receipts and exports of these articles —as shown by our tables, and particularly referred to in another place, under the head of Western Produce—forms a feature of commanding interest in the trade of the last season. This famine, however, in foreign lands, while it has added largely to the wealth of the producers of grain in the Northern, Western, and Middle States, has had a contrary influence upon the prospects of the cotton planters of the South, particularly those of Louisiana and Mississippi, who being deprived of full crops by a dispensation of Providence, had looked to a large advance in prices to remunerate them for deficiency in quantity. In this, however, their too sanguine hopes have been disappointed, and must ever be, under similar circumstances; for dear food and dear clothing cannot be maintained at the same time among such a population as constitute the consuming masses of Europe. As a general remark, their means are too limited to purchase both; and, as food is the first necessity of life, it absorbs all the earnings of the laborer.

The backwardness of the plant generally, and the destructive ravages of the caterpillar, which had but just commenced, led to serious apprehensions of an important curtailment of the product, particularly in the Southern Valley of the Mississippi, and also in South Alabama and Texas. These apprehensions rapidly gained strength as the season advanced; and it soon became evident that the hopes of the planter had been blasted to an extent which had never before occurred by a similar agency. For a considerable period many parties continued sceptical in regard to the representations of damage to the crop; but their correctness has long since been conceded, and the demonstration is established by a reference to the receipts at this port, which show a falling off of 334,069 bales, as compared with the previous year: The conviction of a large deficiency in the expected yield soon led to speculative movements, and induced many planters to place high limits upon their crops, as they forwarded them to market, under the reasonable expectation that when this fact should be known in Europe—where the stocks were already comparatively limited—a material enhancement of prices would be the natural result. These expectations have been partially attained, and under ordinary circumstances might, perhaps, have been realized to their fullest extent. But it has happened, in the course of events, that the same cause which was most prominent among those which produced disappointment and disaster in 1845–6, has exercised a most powerful and extended influence during the past season, viz.: famine in Europe. It is beyond question that this has been the main obstacle to a much more considerable advance than has actually occurred in the market abroad, for, as we have already remarked in another place, dear food and dear clothing cannot be maintained at the same time among the consuming masses of Europe—those who usually take the great bulk of the supply; and consequently we see that the quantity taken for consumption in Great Britain, including all descriptions, from January 1st to July 1st of the present year, has been only 590,657 bales, against 775,509 bales during the same period in 1846, showing a decrease for the six months of 184,852 bales, or an average weekly decrease of 7,100 bales. During a considerable portion of the time, however, the ratio of the decrease has been much greater, and for some weeks—say in May and June—the sales for consumption scarcely amounted to one-half the average of the previous year. A similar state of depression, and from similar causes, existed on the Continent; and our home consumption, which it was expected would be similarly increased, will be found to have scarcely reached the extent of last year, when it was put down at 422,597 bales.

LIVERPOOL STATEMENT FOR 1847.

UNITED STATES—1846–1847.

Stock, 1st September, 1846	98,000	Export	1,241,000
Crop	1,779,000	Consumption	421,000
		Stock, 1st September, 1847	215,000
Bales	1,877,000	Bales	1,877,000

CONSUMPTION. / IMPORT.

Tot. Europe.	Continent.	France.	Gt. Britain.	Liverpool.	Stock Jan. 1, 1847, in	Liverpool.	Gt. Britain.	France.	Continent.	Tot. Europe.
					United States........ Bales.	270,000	303,000	25,000	24,000	352,000
					Brazil	24,000	24,000	....	1,000	25,000
					West Indies	2,000	4,000	....	2,000	6,000
					East Indies	92,000	157,000	2,000	2,000	161,000
					Egypt	51,000	57,000	3,000	14,000	74,000
					Bales	439,000	545,000	30,000	43,000	618,000
					IMPORT.					
1,265 000	190,000	249,000	826,000	792,000	United States	831,000	874,000	266,000	209,000	1,237,000
74,000	8,000	2,000	64,000	62,000	Brazil	110,000	111,000	4,000	8,000	111,000
21,000	8,000	4,000	9,000	7,000	West Indies	6,000	7,000	7,000	9,000	23,000
258,000	87,000	5,000	166,000	133,000	East Indies	121,000	222,000	4,000	90,000	228,000
127,000	45,000	33,000	49,000	47,000	Egypt	20,000	20,000	44,000	57,000	119,000
1,745,000	338,000	293,000	1,114,000	1,041,000	Bales	1,088,000	1,234,000	325,000	373,000	1,718,000
....	....	....	214,000	122,000	Export.					
591,000	78,000	62,000	451,000	364,000	Stock Dec. 31. Stock above,	439,000	545,000	30,000	43,000	618,000
2,336,000	416,000	355,000	1,779,000	1,527,000	Total supply, bales	1,527,000	1,779,000	355,000	416,000	2,336,000

COTTON AT LIVER

Week Ending.	Receipts.						Sales.			
	Americ'n.	E. I.	Egypt.	Brazil.	Other.	Total.	Consumption.	Speculation.	Export	Total.
Jan. 8..	5,126	......	......	......	......	5,126	22,880	1,750	550	23,430
" 15..	10,623	......	......	3,883	......	14,506	25,800	1,600	1,850	29,250
" 22..	20,089	......	1,532	1,182	......	22,803	14,850	3,600	850	19,300
" 29..	39,122	......	......	4,440	202	43,764	22,320	1,100	1,170	24,590
Feb. 5..	14,977	......	......	......	......	14,977	17,270	8,200	1,570	27,040
" 12..	10,300	1,313	......	677	......	12,290	10,660	8,000	1,200	19,860
" 19..	39,848	5,744	......	2,240	259	48,091	11,270	1,200	1,050	13,520
" 26..	14,823	......	......	......	4,004	18,827	18,290	6,500	1,400	26,190
Mch 6..	3,702	2,414	317	800	......	7,233	18,000	7,000	1,200	26,200
" 12..	2,354	......	......	......	......	2,354	14,950	2,000	1,150	18,100
" 19..	11,168	1,674	......	4,129	......	16,971	12,850	1,000	2,600	16,450
" 26..	32,374	3,452	1,810	5,292	1,138	44,066	19,910	9,700	3,200	32,810
April 1..	33,710	1,784	......	1,797	......	37,391	18,100	9,600	3,100	30,800
" 9..	19,790	2,125	......	1,408	......	23,323	19,310	5,000	1,200	25,510
" 16..	36,704	2,169	......	1,351	......	40,224	33,860	30,200	2,100	66,160
" 23..	36,579	......	......	3,215	30	39,824	18,150	4,800	700	23,650
" 30..	46,318	3,172	......	130	100	49,720	21,550	1,900	350	23,800
May 7..	9,710	2,315	1,232	1,605	......	14,862	23,270	3,300	500	27,070
" 14..	10,306	4,744	......	......	16	15,066	25,690	800	1,800	28,290
" 21..	29,563	8,279	1,155	1,712	104	40,813	20,120	1,850	7,520	29,490
" 28..	1,916	......	......	......	150	2,066	23,620	11,000	4,380	39,000
June 4..	18,818	2,891	......	2,207	645	24,561	23,660	5,400	4,280	33,340
" 11..	15,493	1,814	......	716	......	18,023	16,015	5,850	5,265	27,130
" 18..	30,378	3,576	......	2,235	224	36,413	26,400	5,600	4,400	36,400
" 25..	15,112	1,628	......	59	185	16,984	39,070	9,800	5,530	54,400
July 2..	7,864	......	3	......	25	7,892	43,280	28,700	4,720	76,700
" 9..	8,725	4,122	......	2,652	......	15,499	26,560	10,660	2,050	39,270
" 16..	11,946	3,325	......	2,751	66	18,088	21,170	11,300	7,700	40,170
" 23..	10,434	......	......	123	3	10,560	21,420	1,800	2,370	25,590
" 30..	21,360	......	......	1,005	48	22,413	14,470	2,700	960	18,130
Aug. 6..	7,286	1,753	......	20	......	9,059	24,200	3,900	1,870	29,970
" 13..	17,197	2,287	......	2,695	......	22,179	17,840	1,000	3,690	22,530
" 20..	5,552	......	......	......	......	5,552	18,730	2,900	2,500	24,130
" 27..	21,174	1,818	......	3,261	16	26,269	17,130	1,900	1,680	20,710
Sept. 3..	21,137	5,525	......	1,784	1	28,447	18,120	650	1,830	20,600
" 10..	5,978	......	......	672	......	6,650	14,000	700	3,810	18,510
" 17..	30,454	7,867	321	2,066	......	40,708	11,710	120	3,050	14,880
" 24..	2,956	5,890	......	1,000	......	9,846	19,500	2,000	3,200	24,700
Oct. 1..	3,738	5,583	......	3,373	11	12,705	19,210	800	2,200	22,210
" 8..	2,964	......	2,185	87	23	5,259	17,720	3,500	1,150	22,370
" 15..	31,103	10,864	......	5,619	......	47,586	16,520	3,900	1,280	21,700
" 22..	17,442	......	......	19	2	17,463	10,840	1,900	2,420	15,160
" 29..	12,237	......	......	5,326	85	17,648	15,070	3,000	4,130	22,200
Nov. 5..	7,035	......	......	3,273	83	10,391	16,510	1,100	3,980	21,590
" 12..	3,210	11,343	4,521	7,670	......	27,244	19,920	1,850	4,630	26,400
" 19..	2,273	......	......	2,100	540	4,913	17,920	1,000	3,510	22,430
" 26..	2,281	1,942	......	5,080	100	9,403	2,560	700	2,500	5,760
Dec. 3..	19,500	3,376	3,005	1,464	......	27,345	21,800	......	800	22,600
" 10..	7,286	1,204	1,482	......	21	9,993	21,250	......	650	21,900
" 17..	13,504	4,030	......	7,903	......	25,457	20,850	......	750	21,600
" 24..	5,158	......	......	7,638	......	12,796	19,870	......	650	20,520
" 31..	11,212	1,529	......	2,915	20	15,676	17,450	......	......	17,450
Average prices & total sales, receipts & stocks.	809,809	114730	21,712	113,747	8,099	1,068,147	1,023,485	232,180	126990	1,382,655

POOL. YEAR 1847.

STOCKS.			PRICES.			ACTUAL EXPORT.	CONSUMPTION.	REMARKS.
American.	Other.	Total.	Mid. Up.	Mid. Orl.	Surats.			
258,186	162,480	420,666	6 @7⅞	6 @9	4¼@6½	550	22,880	
257,666	162,673	420,339	6¾@7	7⅛@7⅜	5	1,850	48,680	
265,715	161,727	427,442	6⅝	6⅞	4¾	850	63,530	
293,527	162,289	455,816	6½	6¾	4¾	1,170	85,850	
293,714	155,709	449,423	6½	6¾	4¾	1,570	103,120	
153,069	296,784	449,853	6⅜	6⅝	4⅞	1,200	113,780	
327,492	157,132	484,624	6¼	6½	4¾	1,050	125,050	
326,345	157,426	483,771	6¼	6½	4¾	1,400	143.340	
308,137	152,667	460,804	6¼	6½	4⅝	1,200	161,340	
295,281	151,777	447,058	6⅛	6⅜	4⅝	1,150	176,290	
295,499	153,672	449,171	5⅞	6	4½	2,600	189,140	
310.283	159,844	470,127	6¼	6⅜	4¾	3,200	209.050	
325,673	157,775	483,448	6⅜	6½	4¾	3,100	227,150	
329,893	156.368	486,261	6½	6⅝	4¾	1,200	246,460	
338,077	151,548	489,625	6⅞	7⅛	4⅞	2,100	280,320	
367,529	149,325	516,854	6½	6⅝	4⅞	700	298,470	
390,947	148,757	539,704	6¼	6⅜	4¾	350	320,020	
382.207	148,589	530,796	5⅞	6	4⅝	500	343,290	Financial difficulties, &c.
369,683	148,689	518,372	5¾	5⅞	4⅜	1,800	368,980	
378,306	153,239	531,545	5⅞	6⅛	4⅜	7,520	389,100	
359,222	146,389	505,611	6½	6⅝	4⅝	4,380	412,720	Favorable weather, &c.
358,210	141,472	499,682	6½	6⅝	4⅝	4,280	436,380	
359.103	137,322	496,425	6⅝	6¾	4⅝	5,265	452,395	
369 541	132,497	502,038	6¾	6⅞	4¾	4,400	478,795	
355.293	119,039	474,332	6⅞	7	4⅞	5,530	517,865	
331,767	103,367	437,104	7	7⅛	5	4,720	561,145	
325,652	105,341	430,993	6⅞	7	5	2,050	587,705	
317,456	102,733	420,189	7	7⅛	5⅛	7,700	608,875	
310,730	96,611	407,341	6⅞	7	5⅛	2,370	630,295	
318,320	94,545	412,865	6⅞	7	5⅛	960	644,765	
306,356	89,498	395,854	6⅞	7	5⅛	1,870	668,965	
307,683	88.820	396,503	6⅞	7	5⅛	3,690	686,805	
296,775	84,050	380.825	6⅞	7	5⅛	2,500	705,535	
303,679	84,605	388,284	6¾	6⅞	5	1,680	722,665	
318,986	87,435	406,421	6¾	6⅞	5	1,830	740,785	
312,074	83,187	395,261	6⅝	6¾	5	3,810	754,785	
330,618	90,591	421,209	6⅜	6½	5	3,050	766,495	
314,914	93.441	408,355	6	6⅛	4¾	3,200	785,995	
300,732	98,368	399,100	5¾	5⅞	4⅝	2,200	805,205	
287,356	98,133	385,489	5⅜	5⅝	4¼	1,150	822,925	
304,949	110,326	415,275	5¼	5½	4¼	1,280	839,445	Many failures in London and elsewhere.
314,021	105,457	419,478	4¾	5	4	2,420	850,285	
307,228	110,651	417,879	4⅞	5	4	4,130	865,355	
294 693	109,507	404,200	4⅝	4¾	4	3,980	881,865	
278,603	128,291	406,894	4⅝	4¾	4	4,630	901,785	
263.747	126,631	390,378	4½	4⅝	3¾	3,510	919,705	
245,698	126,523	372,221	4⅜	4⅞	3¾	2,500	922,265	
246,988	129,978	376,966	4½	4⅝	3¾	800	944,065	
236,804	128,255	365,059	4⅜	4½	3¾	650	965,315	
233,828	134,068	368,896	4⅜	4½	3¾	750	986,165	
223,346	137,826	361,172	4⅜	4½	3⅝	650	1.006,035	
....	....	363,530	4⅜	4½	3⅝	....	1,023,485	
			6.03	6.18	4.57	126,990	196,824	

The semi-weekly Price and Weekly Sales and Receipts, at New York, Weekly Exports from New York and Rates of Freight to Liverpool 1st of each Month, for the Crop Year ending September 1, 1847.

1846.	Price of Mid. Fair to Fair New Orleans Liverpool Classificat'n.	Price of Mid. Fair to Fair Upland. Liverpool Classificati'n	Sales for week.	Receipts for week.	EXPORTS FOR WEEK.					Rates of Freight to Liverpool.	General Remarks.
					To Great Britain.	To France.	North of Europe.	Other Fo'n Ports	Total Exports.		
Septem. 4..	9@9¾	8⅝@9⅛								¼@5/16d.	
" 8..	9@9½	8½@9	3,150	4,533	178	2,123	1,211	546	4,058		This was the period of the great famine in Ireland and suffering in England occasioned by the failure of the crops. Shipments of breadstuffs hence to the United Kingdom were of unprecedented magnitude. Exchange rapidly fell, and the ordinary ebb tide of specie from this country was arrested, the current now flowing in full volume this way. These causes brought on a stagnation of trade in England; money affairs became much deranged, and the rates of interest were higher there than before, since 1837 and 1839. The manufacturers curtailed largely their working hours, and the consumption of cotton fell off 10,000 bales per week, though early in the season advices from the Southern States represented the crop as at least three weeks later than usual; that the army worm had generally appeared, which, together with heavy rains and floods, augured ill for the incoming crop. The English markets, however, under the difficulties mentioned, responded very feebly to the speculative feeling inaugurated here, until toward the latter part of the season, when it became apparent
" 11..	9@9½	8½@9									
" 15..	9@9½	8½@9	3,950	974	1,663	1,146	276	....	3,085		
" 18..	9½@10	9@9¼									
" 22..	9½@10	9@9½	11,100	3,030	680	101	735	653	2,169		
" 25 .	9½@10¼	9¼@9¾									
" 29..	9½@10¼	9¼@9¾	11,450	3,026	1,189	1,268	587	....	3,044		
October 2..	9¾@10½	9½@10								¼@5/16d.	
" 6..	10½@11	10@10½	14,900	5,409	549	1,452	530	524	3,055		
" 9..	10¼@10¾	9¾@10⅜									
" 13..	10@10½	9½@10⅛	3,800	5,340	796	215	218	81	1,310		
" 16..	10@10½	9½@10⅛									
" 20..	10@10½	9½@10	3,550	3,966	191	653	967	....	1,811		
" 23..	10@10¾	9⅝@10⅛									
" 27..	10¼@10¾	9¾@10¼	13,100	4,174	137	1,335	156	1	1,629		
" 30..	10¼@10¾	9¾@10¼									
Novem. 3..	10¼@10¾	9¾@10¼	5,600	1,420	90	1,173	312	....	1,575	¼@⅜d.	
" 6..	10¼@10¾	9¾@10¼									
" 10..	10¼@10¾	9¾@10⅜	5,850	5,503	1,068	1,526	362	....	2,956		
" 13..	10@10½	9⅝@10⅛									
" 17..	10@10¾	9⅝@10⅛	4,750	6,069	2,364	1,035	747	....	4,146		
" 20..	9¾@10½	9½@10									
" 24..	9¾@10½	9⅜@9¾	9,900	9,218	1,802	....	272	57	2,131		
" 27..	9¾@10¼	9⅜@9¾									
Decem. 1..	10@10½	9½@10	6,700	3,788	1,843	1,772	573	....	4,188	—@7-16d	
" 4..	10@10½	9½@10									
" 8..	10@10⅝	9⅝@10⅛	9,400	10,550	1,772	1,097	234	....	3,103		
" 11..	10¼@10¾	10@10⅜									
" 15..	10½@11	10@10½	13,000	2,649	3,227	681	1,029	....	4,937		
" 18..	10½@11	10@10½									
" 22..	11@11½	10½@11	11,800	4,896	1,467	....	1,504	1,804	4,775		
" 25..	11@11½	10½@11									

Decem. 29..	11@11½	10⅝@11	4,300	9,594	3,717	2,437	163	....	6,317	
1847.										
January 1..	11@11½	10⅝@11								⅜@7/16d.
" 5..	11@11¾	10⅞@11¼	6,200	4,441	1,691	784	732	....	3,207	
" 8..	11½@12	11@11½								
" 12..	11½@12	11⅛@11½	8,300	3,368	1,582	147	166	....	1,895	
" 15..	11⅝@12¼	11¼@11⅝								
" 19..	11¾@12¼	11⅜@11¾	11,800	7,177	916	1,050	50	2,471	4,487	
" 22..	12¼@13	11¾@12¼								
" 26..	12½@13½	12¼@12¾	18,200	7,450	562	100	....	....	662	
" 29..	13¼@13¾	12¾@13¼								
February 2..	13¼@13¾	12¾@13¼	23,500	6,190	713	1,245	128	453	2,539	7/16@½d.
" 5..	13@13½	12½@13								
" 9..	13@13½	12½@13	8,100	7,354	500	856	661	....	2,017	
" 12..	12½@12¾	11¾@12¼								
" 16..	12@12½	11½@12	5,500	9,890	1,128	925	597	....	2,650	
" 19..	11¼@12	11@11½								
" 23..	11¼@12	11@11½	5,000	5,089	1,420	....	66	1,131	2,617	
" 26..	11@11½	10¼@10¾								
March 2..	11¼@12	10⅝@11⅛	9,700	19,063	687	187	622	....	1,496	¾@⅞d.
" 5..	11½@12¼	10⅞@11⅜								
" 9..	11¾@12½	11@11½	9,400	11,335	1,656	943	163	....	2,762	
" 12..	11¾@12¾	11¼@11¾								
" 16..	11¾@12¾	11¼@11¾	5,400	3,567	2,811	818	1,148	....	4,777	
" 19..	12¼@13	11⅝@12⅛								
" 23..	12¼@13	11⅝@12⅛	5,300	10,029	....	1,593	829	....	2,422	
" 26..	12@12¾	11⅜@12								
" 30..	12¼@12¾	11½@12	7,800	13,741	334	....	765	....	1,099	
April 2..	12¼@12¾	11½@12⅛								½@⅝d.
" 6..	12¼@12¾	11⅝@12¼	6,600	3,561	293	104	842	....	1,239	
" 9..	12¼@12¾	11⅝@12¼								
" 13..	12@12¾	11½@12⅛	2,950	5,812	589	408	107	....	1,104	
" 16..	12@12¾	11½@12⅛								
" 20..	12¼@13	11½@12⅛	4,700	2,575	673	432	807	....	1,912	
" 23..	12¾@13½	12¼@12¾								
" 27..	12¾@13½	12½@13	12,450	6,874	136	....	335	....	471	
" 30..	12¾@13¾	12¼@13¼								
May 4..	12¾@13¾	12½@13¾	6,550	5,657	762	410	225	....	1,397	¼@5/16d.
" 7..	12¾@13¾	12½@13¾								
" 11..	12½@13½	12½@13	5,250	4,498	....	....	2,303	....	2,303	
" 14..	12½@13¼	12@12¾								
" 18..	12⅛@13	11¾@12½	2,700	168	10	....	....	....	10	

that the crop would not come up to early expectations. There was then a more lively movement in the English markets, being assisted by some little alleviation in financial affairs there, though the improvement was very brief, as will be seen by the events of the succeeding crop year.

New York Statement for 1847.—Concluded.

1847.		Price of Mid. Fair to Fair New Orleans Liverpool Classificat'n.	Price of Mid. Fair to Fair Upland. Liverpool Classificat'n.	Sales for week.	Receipts for week.	EXPORTS FOR THE WEEK. To Great Britain.	To France.	North of Europe.	Other Fo'n Ports	Total Exports.	Rates of Freight to Liverpool.
May	21..	12⅛@13	11¾@12½								
"	25..	12@13	11¾@12½	2,800	5,582	163	....	635	....	798	
"	28..	12@13	11¾@12½								
June	1..	12@13	11¾@12½	3,700	14,182	331	....	381	....	712	1/16@¼d.
"	4..	11¾@12½	11⅜@12								
"	8..	11½@12¼	11@11⅝	3,650	4,869	855	....	123	....	978	
"	11..	11½@12½	11@11¾								
"	15..	11½@12½	11@11¾	3,450	6,420	276	....	1	....	277	
"	18..	12@12¾	11⅝@12¼								
"	22..	12@12¾	11⅝@12¼	4,650	5,408	61	....	277	....	338	
"	25..	11¾@12½	11¾@12½								
"	29..	11¾@12½	11⅜@12	4,400	2,396	329	447	....	....	776	
July	2.	11¾@12½	11⅜@12								5/16@⅜d.
"	6..	11¾@12½	11½@12	3,950	3,166	183	....	644	....	827	
"	9..	11½@12¼	11¼@11⅞								
"	13..	11½@12¼	11¼@11¾	4,000	15,373	151	479	455	....	1,085	
"	16..	11½@12¼	11¼@11⅞								
"	20..	12@12¾	11¾@12¼	10,900	6,650	1,305	793	207	....	2,305	
"	23..	12@12¾	11⅞@12⅜								
"	27..	12@12¾	11¾@12⅜	9,500	5,658	1,466	1,450	28	277	3,221	
"	30..	12@12¾	11¾@12⅜								
August	3..	12½@13	12¼@12⅝	9,400	3,644	2,403	1,152	2,377	....	5,932	⅜@½d.
"	6..	12¾@13½	12½@13								
"	10..	12¾@13½	12½@13	10,800	9,518	784	1,561	663	....	3,008	
"	13..	12½@13¼	12½@12⅞								
"	17..	12¾@13¼	12½@12⅞	9,300	1,892	1,937	2,213	687	....	4,837	
"	20..	12¾@13¼	12½@13								
"	24..	12½@13¼	12⅜@12⅞	7,100	2,753	1,231	1,306	3,015	....	5,552	
"	27..	12½@13	12¼@12¾								
"	31..	12⅜@12⅞	12¼@12⅝	7,700	2,959	2,967	3,381	2,159	....	8,507	1/16@¼d.
Average price and total sales, receipts and exports.		11.71	11.21	397,000	312,448	53,638	40,798	32,074	7,998	134,508	

General Remarks.

Exchange.

Bills on London were quoted, through September, at from 8¾ to 9½ per cent. premium; in October the quotation fell from 8½@9 down to 6½@7½; in November the range was between 7⅝ and 6 per cent.; in December the quotation dropped to 5@5½; in January it ranged from 5 to 6½; in February, between 6 and 4¾; in March the quotation fell from 4½@5¼ down to 3½@4; in April it began to rise, and went up from 4½@5 to 6½@7; in May the range was 6 to 7¼; in June, 5½@7¼; in July, 5¼ to 6½; in August, rising from 5¾@6¼ up to 7½@8 per cent. premium, which latter was the rate current on September 1, 1847.

1848.

COTTON CROP OF THE UNITED STATES.

Statement and Total Amount of the Cotton Crop of the United States, for the Year ending August 31, 1848.

	Bales.	Bales.	Total.	Same period 1847.
NEW ORLEANS.				
Export—				
To Foreign Ports	949,858			
Coastwise	252,039			
Stock on hand 1st September, 1848	37,401			
		1,239,298		
Deduct—				
Stock on hand 1st September, 1847	23,493			
Received from Mobile	10,857			
" Florida	4,208			
" Texas	10,007			
		48,565		
			1,190,733	705,979
MOBILE.				
Export—				
To Foreign Ports	319,081			
Coastwise	118,168			
Stock on hand 1st September, 1848	23,584			
		460,833		
Deduct—				
Stock on hand 1st September, 1847	24,172			
Received from New Orleans	275			
" Key West	50			
		24,497		
			436,336	323,462
FLORIDA.				
Export—				
To Foreign Ports	50,050			
Coastwise	105,327			
Stock on hand 1st September, 1848	507			
		155,884		
Deduct—				
Stock on hand 1st September, 1847	....	2,108		
			153,776	127,852
TEXAS.				
*Export to Foreign Ports	772			
Coastwise	38,255			
Stock on hand 1st September, 1848	747			
		39,774		
Deduct—				
Stock on hand 1st September, 1847	....	32		
			39,742	8,317
GEORGIA.				
Export from Savannah—				
To Foreign Ports—Uplands	120,502			
" Sea Island	7,258			
Coastwise—Uplands	114,220			
Sea Island	1,253			
	243,233			
Export from Darien—				
To New York ... 9				
Stock in Savannah, 1st September, 1848 ... 10,050				
" Augusta and Hambro', 1st Sept., 1848. 26,553				
	36,612			
		279,845		
Deduct—				
Stock in Savannah and Augusta, 1st September, 1847	....	25,020		
			254,825	242,789

Statement and Total Amount of the Cotton Crop of the United States, for the Year ending August 31, 1848—Concluded.

		Bales.	Bales.	Total.	Same period 1847.
SOUTH CAROLINA.					
Export from Charleston—					
To Foreign Ports—Uplands		183,501			
" Sea Island		15,345			
Coastwise—Uplands		98,061			
Sea Island		685			
		297,592			
Burnt at Charleston		1,392			
Export from Georgetown—					
To New York and Boston	228				
Stock in Charleston, 1st September, 1848	14,085				
		14,313			
			313,297		
Deduct—					
Stock in Charleston, 1st September, 1847		29,655			
Received from Savannah		20,851			
" Florida		1,039			
			51,545		
NORTH CAROLINA.				261,752	350,200
Export—					
Coastwise		....	....	1,518	6,061
VIRGINIA.					
Export—					
To Foreign Ports		556			
Coastwise		520			
Manufactured (Taken from the Ports)		7,880			
Stock on hand, 1st September, 1848		444			
			9,400		
Deduct—					
Stock on hand 1st September, 1847		....	448		
				8,952	13,991
Total crop of the United States		....	....	2,347,634	1,778,651

Total crop of 1848, as above	bales.	2,347,634
Crop of last year		1,778,651
Crop of year before		2,100,537
Increase over last year		568,983
Increase over year before		247,097

Export to Foreign Ports, from September 1, 1847, *to August* 31, 1848.

FROM	To Great Britain.	To France.	To North of Europe.	Other F'n Ports.	Total.
New Orleans.......Bales	654,083	140,968	50,056	104,751	949,858
Mobile	228,179	61,832	16,153	12,917	319,081
Florida	42,376	2,212	1,732	3,730	50,050
Texas	....	....	772	....	772
Georgia	121,172	5,177	424	987	127,760
South Carolina	153,090	29,579	11,390	4,787	198,846
North Carolina	....	....	....	....	....
Virginia	268	....	254	34	556
Baltimore	60	....	....	....	60
Philadelphia	3,375	....	....	80	3,455
New York	116,061	37,992	37,541	6,650	198,244
Boston	5,601	1,412	2,026	540	9,579
Grand total	1,324,265	279,172	120,348	134,476	1,858,261
Total last year	830,909	241,486	75,689	93,138	1,241,222
Increase	493,356	37,686	44,659	41,338	617,039

Growth.

Total crop of 1828–9......bales.	857,744	Total crop of 1838–9......bales.	1,360,532
" 1829–30	976,845	" 1839–40	2,177,835
" 1830–1	1,038,848	" 1840–1	1,634,945
" 1831–2	987,477	" 1841–2	1,683,574
" 1832–3	1,070,438	" 1842–3	2,378,875
" 1833–4	1,205,394	" 1843–4	2,030,409
" 1834–5	1,254,328	" 1844–5	2,394,503
" 1835–6	1,360,725	" 1845–6	2,100,537
" 1836–7	1,422,930	" 1846–7	1,778,651
" 1837–8	1,801,497	" 1847–8	2,347,634

Consumption.

Total crop of the United States, as above stated.......bales.			2,347,634
Add—Stocks on hand at the commencement of the year, September 1, 1847—			
In the Southern ports		104,928	
In the Northern ports		109,909	
			214,837
Makes a supply of			2,562,471
Deduct therefrom—The Export to Foreign ports	1,858,261		
Less Foreign included	372		
		1,857,889	
Stocks on hand, September 1, 1848—			
In the Southern ports	113,471		
In the Northern ports	57,997		
		171,468	
Burnt at Charleston		1,392	
			2,030,749
Taken for home use.......bales.			**531,772**

Quantity consumed by and in the hands of Manufacturers.

1847-8 ....bales.	531,772	1837-8 ....bales.	246,063
1846-7	427,967	1836-7	222,540
1845-6	422,597	1835-6	236,733
1844-5	389,006	1834-5	216,888
1843-4	346,744	1833-4	196,413
1842-3	325,129	1832-3	194,412
1841-2	267,850	1831-2	173,800
1840-1	297,288	1830-1	182,142
1839-40	295,193	1829-30	126,512
1838-9	276,018	1828-9	118,853

NOTE.—Our estimate in this statement of the quantity taken for consumption, in the cotton-growing States, does not include any cotton manufactured in the States south and west of Virginia, but it cannot have escaped observation that the consumption at the South and West is gradually increasing, and it seems proper in making up an account of the production of the country, that some notice should be taken of it. The following estimate, from a judicious and careful observer at the South, of the quantity so consumed (and not included in the receipts at all), may not be devoid of interest. Thus, in—

North Carolina ....bales.	15,500	
South Carolina	6,000	
Georgia	6,000	
Alabama	5,000	
		32,500
Sent up the western rivers and consumed, say—		
Received at Cincinnati	12,500	
" Pittsburg and Wheeling	12,500	
" Kentucky	5,000	
		30,000
" Missouri, Tennessee, Indiana, Illinois, &c.		12,500
Total ....bales.		75,000

To which may be added the quantity burnt in the interior, and that lost on its way to market; these, added to the crop as given above, received at the shipping ports, will show very nearly the amount raised in the United States the past season.

The quantity of new cotton received at the shipping ports up to the 1st inst., amounted to about 3,000 bales, against 1,121 bales last year.

The shipments given in the above statement from Texas, are those by sea only; a considerable portion of the crop of that State finds its way to market via Red River, and is included in the receipts at New Orleans.

The receipts at Philadelphia and Baltimore, overland from the West this season, were 1,479 bales, against 1,828 bales last year.

ANNUAL REVIEW.

From the New Orleans Price Current, 1847—48.

The early prices obtained in this market were highly satisfactory, and the trade gave high promise, until the latter part of October, when the commercial revolution which prostrated credit in Great Britain, and which subsequently spread to nearly all parts of the Continent of Europe, and to the Indies, put a sudden check to our prosperous course, and produced a more rapid depreciation of prices than we remember ever to have witnessed, in an experience of many years, as an observer of the varying phases of this most important and most sensitive of our commercial interests. After recovering materially from the shock produced by the state of things just enumerated, a still more severe blow was given by the startling intelligence of a revolution in France, and the overthrow of the monarchy. This movement of the people in favor of popular rights rapidly spread to other countries of Europe, and in the tumultuous state of political affairs, commercial credit was completely overthrown, and trade in a measure annihilated. In this general prostration of credit and commerce, probably no interest, connecting our own country with Europe, was more severely affected than the cotton trade, and prices here were at times depressed to within a fraction of the lowest point reached in 1843, while at Liverpool sales were made at lower rates than were ever before known for American cotton.

This depression, too, both in its causes and effects, presents a marked contrast to that of 1843; for in the latter instance it was attributable solely to an excessive accumulation of the raw material, which gave the manufacturers, particularly those of Great Britain, wholly the advantage, and enabled them to prosecute an active trade, at immense profits; while during the past year the stocks in Great Britain, at least for the greater portion of the period, have been unusually low; but so great have been the derangements of trade, that the manufacturers could not work with profit, even upon a lower cost of the raw material than was ever before known; and many mills were stopped, while many others were compelled to resort to short-time working. Speculation has been comparatively unknown, as will readily be seen by the fact that, during the first six months of the present year, the quantity taken by speculators at Liverpool was only 27,800 bales, against 228,400 bales for the same period in 1847. Having thus given a rapid summary of what we conceive to have been the prominent causes of the extraordinary depression of the past season, we proceed to a brief review of the course of our own market. We may venture a few remarks touching the prospects of supply in this region, as they appear at present, avoiding—as it has ever been our custom to do—anything like a definite estimate in regard to a matter that is involved in so much uncertainty, at this early period of the season. We may then state that, up to within a few weeks, the crops gave highly favorable promise generally. The plant was well advanced and healthy and little or no complaint was heard from any section. True, the rains commenced early in June, but they did not appear to be of that general and severe character to cause injury, but on the contrary, while the plant was in progress, their influence was favorable, particularly in the uplands. When, however, the plant was well matured, and the season for the commencement of picking arrived—say in the latter part of July—the rains lost their beneficial character ; and as they have since continued, and become more general, attended in one or to instances by severe storms of wind, they have for some weeks been productive of injury, by retarding the ripening of the bolls, beating the cotton from those fully opened, and promoting the ravages of the boll worms, which are said to be quite destructive in several districts.

LIVERPOOL STATEMENT FOR 1848.

UNITED STATES, 1847–1848.

Stock 1st September, 1847	215,000	Export	1,858,000
Crop	2,348,000	Consumption	534,000
		Stock 1st Sept., '48.	171,000
Bales	2,563,000	Bales	2,563,000

CONSUMPTION. Tot. Europe.	Continent.	France.	Gt. Britain.	Liverpool.		Liverpool.	Gt. Britain.	France	Continent.	Tot. Europe.
					Stock Jan. 1, 1848, in					
					United States	215,000	239,000	42,000	43,000	324,000
					Brazil	59,000	59,000	2,000	1,000	62,000
					West Indies	1,000	2,000	3,000	3,000	8,000
					East Indies	66,000	125,000	1,000	5,000	131,000
					Egypt	23,000	26,000	14,000	26,000	66,000
					Bales	364,000	451,000	62,000	78,000	591,000
					IMPORT.					
1,745,000	250,000	277,000	1,218,000	1,157,000	United States	1,297,000	1,374,000	255,000	246,000	1,752,000
96,000	12,000	4,000	80,000	79,000	Brazil	100,000	100,000	2,000	12,000	103,000
26,000	9,000	10,000	7,000	6,000	West Indies	6,000	8,000	8,000	8,000	24,000
221,000	58,000	1,000	162,000	119,000	East Indies	136,000	228,000	1,000	58,000	232,000
71,000	22,000	11,000	38,000	38,000	Egypt	28,000	29,000	4,000	9,000	42,000
2,159,000	351,000	303,000	1,505,000	1,399,000	Bales	1,567,000	1,739,000	270,000	333,000	2,153,000
....	....	....	189,000	139,000	Export.					
585,000	60,000	29,000	496,000	393,000	Stock Dec. 31. Stock above,	364,000	451,000	62,000	78,000	591,000
2,744,000	411,000	332,000	2,190,000	1,931,000	Total supply, bales	1,931,000	2,190,000	332,000	411,000	2,744,000

COTTON AT LIVER

Week Ending.	Receipts.						Sales.			
	American	E. I.	Egypt.	Brazil.	Other.	Total.	Con-sumption.	Specu-lation.	Export.	Total.
Jan. 7..	17,593	2,193	262	4,035	32	24,115	24,580	300	......	24,880
" 14..	3,476	3,170	.....	784	151	7,581	24,975	1,250	75	26,300
" 21..	6,047	......	......	1,500	707	8,254	28,510	100	100	28,710
" 28..	13,236	2,767	......	5,203	......	21,106	25,010	....	570	25,580
Feb. 4..	7,277	296	......	967	100	8,640	31,140	3,300	190	34,630
" 11..	8,309	930	945	2,238	3,458	15,880	26,940	3,100	1,630	31,670
" 18..	14,500	2,151	......	2,708	306	19,665	17,890	2,400	1,550	21,840
" 25..	32,132	1,410	......	....	......	33,542	18,380	500	350	19,230
Mch 3..	12,213	8,361	10	900	926	22,410	16,410	1,000	400	17,810
" 10..	15,802	254	......	1,453	......	17,509	23,080	400	560	24,040
" 17..	14,462	5,249	......	1,000	......	20,711	22,010	300	700	23,010
" 24..	17,331	......	1,225	1,012	......	19,568	22,000	600	2,180	24,780
" 31..	12,464	9,889	1,774	....	12	24,139	20,570	600	3,500	24,670
April 7..	54,902	......	146	1,599	......	56,647	19,660	100	970	20,730
" 14..	40,964	4,218	......	74	......	45,256	21,680	250	3,420	25,350
" 21..	26,914	9,833	778	9,273	......	46,798	18,880	500	5,070	24,450
" 28..	40,522	......	......	....	100	40,622	26,100	....	4,840	30,940
May 5..	13,316	......	......	....	......	13,316	33,160	1,000	5.050	39,210
" 12..	73,200	5,595	......	2,716	987	82,498	22,440	....	1,680	24,120
" 19..	40,513	......	......	....	21	40,534	26,020	1,200	2,500	29,720
" 26..	79,257	......	1,284	2,942	299	83.782	25,750	1,200	1,240	28,190
June 2..	68.699	......	3,751	3,736	12	76,198	25,830	2,700	1,550	30,080
" 9..	20,625	......	......	....	747	21,372	28,260	1,700	1.400	31,360
" 16..	12,571	......	1,285	20	......	13,876	21,350	1,300	3,150	25,800
" 23..	43,491	......	1,172	....	......	44,663	24,315	2,800	5,855	32,970
" 30..	149,630	2,076	......	....	371	152,077	22,210	1,200	3,420	26,830
July 7..	19,340	6,285	......	2,855	371	28,851	28,810	3,300	7,100	39,210
" 14..	22,046	4,523	1,507	769	......	28,845	37,590	8.500	5,240	51,330
" 21..	14,911	3,382	1,753	1,290	......	21,336	40,990	5,150	4,490	50,630
" 28..	39,088	2,897	1,898	900	241	45,024	25,380	2,700	6,100	34,180
Aug. 4..	37,191	......	......	2,107	......	39,298	25,300	2,800	3,700	31,800
" 11..	30,123	......	1,557	607	51	32,338	21,200	1,800	6,290	29,290
" 18..	7,310	......	......	....	......	7,310	24,895	500	3,045	28,440
" 25..	35,686	......	1,723	1,667	......	39,076	27,500	1,100	4.850	33,450
Sep. 1..	11,977	1,428	......	....	......	13,405	32,240	1,000	6,170	39,410
" 8..	18,526	......	500	4,333	241	23,600	21,930	1,100	3,700	26,730
" 15..	14,602	4,081	1,682	2,657	48	23,070	21.120	1,300	6,400	28,820
" 22..	1,553	......	199	2,394	22	4,168	20,700	500	3,830	25,030
" 29..	24,994	......	......	1,479	6	26.479	21,020	800	4,450	26,270
Oct. 6..	20,908	7,786	......	....	......	28,694	19,860	600	3,330	23,790
" 13..	29,586	18,327	2,772	2,928	13	53,626	21,080	500	6,310	27,890
" 20..	3,076	1,247	......	....	......	4,323	19,820	100	4,940	24,860
" 27..	9,548	......	......	2,237	50	11,835	26.650	500	1,450	28,600
Nov. 3..	9,776	5,555	......	5,490	......	20.821	28,230	5,500	1,450	35,180
" 10..	3,126	......	......	2,196	......	5,322	24,900	2,700	520	28,120
" 17..	462	......	......	....	......	462	29,210	1,450	1,700	32,360
" 24..	18,656	......	......	6,812	581	26,049	37,870	8,150	520	46,940
Dec. 1..	29,157	18,265	365	8,682	......	56,469	26,800	7,000	100	33,900
" 8..	13,894	......	530	1,958	......	16,382	19,700	7,500	1,450	28,650
" 15..	11.000	......	......	3,815	104	14,919	50,980	2,600	2,050	55,630
" 22..	19,077	......	722	2,131	......	21,930	29,760	15,500	1,490	46,750
" 29..	....	......	......	....	......	....	....	....	......	....
Average prices & total sales, receipts & stocks.	1,284,689	133168	27,840	99,467	9,948	1,555,112	1,271,965	110,450	144625	1,527,040

POOL. YEAR 1848.

STOCKS.			PRICES.			Actual Export.	Con-sumption.	Remarks.
Amer'n.	Other.	Total.	Mid. Up.	Mid. Orl.	Dhol.			
213,133	149,932	363,065	4 1/2	4 5/8	3 5/8	....	24,580	
198,459	147,137	345,596	4 1/2	4 5/8	3 3/4	75	49,555	
182,196	143,044	325,240	4 3/8	4 1/2	3 3/4	100	78,065	
175,622	145,144	320,766	4 1/2	4 7/8	3 3/4	570	103,075	
156,529	138,777	295,306	4 3/4	4 3/4	3 7/8	190	134,215	
143,328	139,278	282,606	4 7/8	5	3 7/8	1,630	161,155	
142,988	139,843	282,831	4 7/8	5	3 7/8	1,550	179,045	
160,830	136,813	297,643	4 3/4	4 7/8	3 7/8	350	197,425	
157,933	144,110	302,043	4 1/2	4 5/8	3 3/4	400	213,835	
154,525	141,387	295,912	4 1/2	4 5/8	3 3/4	560	236,915	
150,697	143,216	293,913	4 1/4	4 3/8	3 3/4	700	258,925	
148,118	141,183	289,301	4 1/8	4 1/4	3 5/8	2,180	280,925	
140,302	147,828	288,130	4 1/4	4 3/8	3 5/8	3,500	301,495	
177,874	146,273	324,147	4 1/8	4 1/8	3 5/8	970	320,155	Manchester districts, dull accounts.
197,968	146,335	344,303	4	4 1/8	3 5/8	3,420	341,835	
205,032	162,119	367,151	3 7/8	3 7/8	3 1/2	5,070	350,715	
216,944	157,609	374,553	3 3/4	3 7/8	3 3/8	4,840	376,815	
197,030	152,629	349,659	4	4 1/8	3 3/8	5,050	409,975	
249,450	158,587	408,037	4	4 1/8	3 3/8	1,680	432,415	
275,003	145,048	420,051	3 7/8	4	3 3/8	2,500	458,435	
321.580	155,263	476,843	3 3/8	3 7/8	3 3/8	1,240	484,185	Heavy receipts, large shipments to Britain.
362,529	158,502	521,031	3 3/4	3 7/8	3 3/8	1,550	510,015	
355,425	157,007	512,432	3 5/8	3 7/8	3 3/8	1,400	538,275	
346,816	154,992	501,808	3 5/8	3 3/4	3 3/8	3,150	559,625	
364,307	151,994	516,301	3 3/4	3 7/8	3 3/8	5,855	583,940	
487,177	148,821	635,998	3 5/8	3 3/4	3 3/8	3,420	606,150	Heavy receipts.
475,937	152,831	628,768	3 3/4	3 7/8	3 3/8	7,100	634,060	
460,553	154,230	614,783	3 7/8	4	3 3/8	5,240	672,550	Cheerful accounts from Continent.
436,304	154,335	590,639	3 7/8	4	3 3/8	4,490	713,550	
433,712	152,771	586,483	3 7/8	4	3 3/8	6,100	738,930	
446,693	150,898	597,591	3 7/8	4	3 3/8	3,700	764,230	
453,136	149,293	602,429	3 3/4	3 7/8	3 3/8	6,290	785,430	Bad harvest accounts, bad weather, etc.
436,116	145,693	581,809	3 3/4	3 7/8	3 3/8	3,045	810,325	
444,172	144,363	588,535	3 3/4	3 3/4	3 3/8	4,850	837,825	
419,899	140,461	560,360	3 3/4	3 3/4	3 3/8	6,170	870,065	
417,315	141,015	558,330	3 3/4	3 3/4	3 3/8	3,700	891,995	
409,367	144,513	553,880	3 5/8	3 3/4	3 3/8	6,400	913,115	
390,550	142,968	533,518	3 5/8	3 3/4	3 3/8	3,830	933,815	
387,534	140,303	527,837	3 5/8	3 3/4	3 3/8	4,450	954,835	
388,208	145,096	533,304	3 5/8	3 5/8	3 3/8	3,330	974,695	
398,684	160,856	559,540	3 5/8	3 5/8	3 1/4	6,310	995,775	
376,990	162,113	539,103	3 3/8	3 1/2	3 1/4	4,940	1,015,595	
361,358	161,480	522,838	3 3/8	3 1/2	3 3/8	1,450	1,042,245	Better accounts from Manchester.
344,244	168,045	512,289	3 1/2	3 5/8	3 3/8	1,450	1,070,475	
325,836	164,896	490,732	3 1/2	3 3/8	3 3/8	520	1,095,375	
299,828	160,456	460,284	3 5/8	3 5/8	3 3/8	1,700	1,124,585	
286,894	160,049	446,943	3 5/8	3 3/4	3 3/8	520	1,162,455	
294,731	179,961	474,692	3 3/4	3 3/4	3 3/8	100	1,189,255	
290,075	179,849	469,924	3 3/4	3 7/8	3 3/8	1,400	1,191,225	Better feeling in manufacturing districts.
268,065	176,348	444,413	3 7/8	4	3 3/8	2,050	1,242,205	
267,692	167,401	435,093	4	4 1/8	3 3/8	1,490	1,271,965	
....	....	....	......	......	......	....	....	
			3.93	4.1	3.48	144,625	2,494,048	

The semi-weekly Price and Weekly Sales and Receipts at New York, Weekly Exports from New York and Rates of Freight to Liverpool 1st of each month, for the Crop Year ending September 1, 1848.

1847.		Price of Mid. Fair to Fair New Orleans Liverpool Classification.	Price of Mid. Fair to Fair Upland, Liverpool Classification.	Sales for week.	Receipts for week.	EXPORTS FOR THE WEEK.					Rates of Freight to Liverpool.	General Remarks.
						To Great Britain.	To France.	North of Europe.	Other Fo'n Ports	Total Exports.		
Septem.	3..	12@12⅝	11¾@12⅜								3/16@7/32d.	This crop year was crowded with the most stirring events that influenced, more or less directly, the value of the staple. The dullness that was a leading feature during the previous crop year, caused by the excessive money pressure in England, was intensified early in the crop year under review, by a renewal of the stagnation and embarrassments before prevailing in Great Britain. Business there became paralyzed, and the pressure and derangement was unexampled, so that failures throughout the Kingdom were numerous, even the oldest houses, and those of most wealthy repute, succumbing, one after another, to the severity of the times, the Royal Bank of Liverpool, among others, suspending payment. At this time ordinary to middling upland cotton was dull of sale in Liverpool at 3d @4d per lb., the Bank of England having advanced its discounts to 8 @9 per cent.
"	7..	12@12½	11¾@12¼	9,300	4,810	....	....	....	....	....		
"	10..	12¼@12¾	12@12½									
"	14..	12⅝@13¼	12½@13	12,400	401	2,808	2,532	848	....	6,188		
"	17..	12¾@13¼	12½@13									
"	21..	12⅜@12⅞	12@12⅝	4,050	1,936	5,289	2,066	499	....	7,854		
"	24..	12¼@12¾	12@12½									
"	28..	12@12½	11¾@12⅜	10,150	934	313	1,368	2,866	....	4,547		
October	1..	12@12⅝	11⅞@12⅜								⅛@3-16d	
"	5..	11½@12⅛	11⅜@11⅞	5,800	4,924	911	2,988	405	....	4,304		
"	8..	11⅛@11⅝	11@11½									
"	12..	11@11⅝	10⅞@11⅜	6,050	3,476	1,080	5,203	42	....	6,325		
"	15..	11@11½	10¾@11¼									
"	19..	9½@10	9¼@9½	4,150	1,447	2,622	1,623	3,025	....	7,270		
"	22..	9½@10	9¼@9½									
"	26..	9¼@9¾	9@9½	12,200	3,417	373	1,683	1,433	....	3,489		
"	29..	9¼@9¾	9@9½									
Novem.	2..	9@9½	8¾@9¼	6,700	2,365	1,180	1,195	3,119	1,060	6,554	—@⅛d.	
"	5..	9@9½	8⅝@9⅛									
"	9..	8¾@9¼	8⅜@8¾	5,600	3,036	388	1,467	....	573	2,428		
"	12..	8⅜@8⅞	8¼@8½									
"	16..	8¼@8¾	7⅞@8¼	6,500	8,585	1,733	201	1,859	1,188	4,981		
"	19..	8@8½	7½@8									
"	23..	7¾@8¼	7½@7⅞	5,500	2,059	1,618	366	1,636	....	3,620		
"	26..	7½@8	7½@7¾									
"	30..	8@8½	7¾@8¼	5,600	2,930	715	462	2,383	484	4,044		
Decem.	3..	8@8½	7¾@8¼								⅛@3-16d	During September, October, November, and December, these disasters continued causing the return here of large lines of protested paper; a feeling of depression and gloom in this market was now universal, leading to the failure of sev-
"	7..	8@8⅜	7⅝@8⅛	2,950	6,737	480	926	....	....	1,406		
"	10..	7⅞@8⅜	7⅝@8⅛									
"	14..	8@8⅜	7¾@8⅛	6,900	8,327	1,201	1,000	125	715	3,041		
"	17..	8¼@8⅝	7⅞@8¼									
"	22..	8@8⅝	7⅞@8¼	5,700	1,853	782	540	15	1,210	2,547		
"	24..	8¼@8⅝	7⅞@8¼									

Decem. 28..	8¼@8¾	8@8⅜	4,600	213	412	188	....	....	600	
1848.										
January 1..	8½@9	8¼@8½								⅛@—d.
" 4..	8½@9	8¼@8½	3,000	1,185	680	....	168	....	848	
" 7..	8⅜@8⅞	8@8⅜								
" 11..	8⅜@8⅞	8@8⅜	4,650	1,696	1,211	1,482	383	....	3,076	
" 14..	8¾@8⅞	8⅛@8⅞								
" 18..	8¾@8⅞	8⅛@8⅞	5,000	8,234	611	....	30	257	898	
" 21..	8½@9	8¼@8½								
" 25..	8½@8⅞	8⅛@8⅜	6,950	5,558	1,848	1,976	784	....	4,608	
" 28..	8½@8⅞	8⅛@8⅜								
February 1..	8½@8⅞	8⅛@8⅜	4,100	6,225	1,198	336	....	91	1,625	3-16@—d
" 4..	8½@8¾	8@8¼								
" 8..	8⅜@8¾	8@8⅛	7,300	7,779	673	1,840	330	....	2,843	
" 11..	8¼@8¾	7⅞@8⅛								
" 15..	8¼@8½	7¾@8	3,600	7,351	1,044	....	....	15	1,059	
" 18..	8⅛@8½	7¾@8								
" 22..	8@8⅜	7⅝@7⅞	6,550	15,275	1,812	1,761	209	....	3,782	
" 25..	8@8⅜	7½@7¾								
" 28..	8@8½	7½@7⅞	8,600	7,082	3,485	1,081	713	50	5,329	
March 3..	8¼@8½	7⅝@8								⅛@5-32d
" 7..	8½@8⅞	8¼@8½	12,600	10,149	2,261	1,175	526	....	3,962	
" 10..	8¼@8⅝	8@8¼								
" 14..	8¼@8⅝	8@8⅜	6,700	15,985	5,633	....	312	....	5,945	
" 17..	8⅜@8¾	8⅛@8⅜								
" 21..	7¾@8¼	7½@7¾	5,800	8,491	1,332	2,540	1,041	....	4,913	
" 24..	7¾@8¼	7½@7¾								
" 28..	7½@7¾	7⅛@7⅜	6,150	14,196	3,155	1,308	623	....	5,086	
" 31..	7¼@7½	6⅞@7⅛								
April 4..	7⅛@7½	6⅞@7⅛	10,100	11,536	3,680	....	1,123	....	4,803	3-16@¼d
" 7..	7⅛@7½	6⅞@7⅛								
" 11..	6⅝@7	6⅝@6¾	5,200	4,469	4,569	438	573	653	6,233	
" 14..	6½@7	6½@6¾								
" 18..	6½@7	6½@6¾	7,200	8,195	4,201	....	1,092	....	5,293	
" 21..	6½@7	6½@6¾								
" 25..	6¼@6¾	6¼@6½	5,900	16,198	3,913	....	1,729	....	5,642	
" 28..	6⅛@6⅝	6@6¼								
May 2..	6¼@6¾	6@6⅜	5,950	8,631	5,390	....	691	118	6,199	$\frac{5}{32}$@$\frac{3}{16}$d.
" 5..	6¼@6¾	6@6⅜								
" 9..	6½@7	6¼@6⅝	10,500	12,213	5,156	....	368	....	5,524	
" 12..	6½@7	6⅜@6¾								
" 16..	6⅝@7	6⅜@6¾	8,300	4,424	4,038	....	104	....	4,142	

eral banks, banking houses, and mercantile firms, though to a considerable extent confined to parties engaged in the grain trade. After December and January, the worst appeared to be over, and business began to improve in Great Britain, the Bank of England reducing its discount to 4@5 per cent.

As regards cotton, the article so far had stood up pretty well amid the general prostration, the market receiving some strength by continued reports of injury to the growing crop, so that the foreign accounts were in a measure counteracted. As the season advanced, however, the receipts became larger, the Fall being one of the most favorable on record for field work, and prices receded very rapidly, say 4@4½ cents per lb., from those current at the opening of the crop year. The improvement in business affairs, on the other side, noted in January, was of short duration. On the 18th March, accounts came to hand relative to a revolution in France and abdication of Louis Philippe; affairs in France became very much unsettled, and a money and commercial panic set in throughout Europe; the Bank of France and other banking houses suspended payments, and the financial state of that country had not been in such an alarming condition since 1789, so that there was scarcely any business going forward in France or the Continent generally This alarming state of affairs was reflected throughout Great Britain, aggra-

New York Statement for Year 1848—Concluded.

1848.		Price of Mid. Fair to Fair New Orleans Liverpool Classification.	Price of Mid. Fair to Fair Upland, Liverpool Classification.	Sales for week.	Receipts for week.	EXPORTS FOR THE WEEK To Great Britain.	To France.	North of Europe.	Other Fo'n Ports	Total Exports.	Rates of Freight to Liverpool.
May	19..	$6\frac{5}{8}@7$	$6\frac{3}{8}@6\frac{3}{4}$								
"	23..	$6\frac{3}{4}@7\frac{1}{4}$	$6\frac{1}{2}@6\frac{7}{8}$	10,800	7,646	2,718	....	232	....	2,950	
"	26..	$6\frac{3}{4}@7\frac{1}{4}$	$6\frac{1}{2}@7$								
"	30..	$6\frac{3}{4}@7\frac{1}{8}$	$6\frac{5}{8}@7$	8,300	6,159	2,008	....	120	....	2,128	
June	2..	$6\frac{7}{8}@7\frac{1}{2}$	$6\frac{3}{4}@7$								–@5-32d
"	6..	$6\frac{7}{8}@7\frac{1}{2}$	$6\frac{3}{4}@7$	4,650	5,779	4,609	31	71	87	4,798	
"	9..	$6\frac{7}{8}@7\frac{1}{2}$	$6\frac{3}{4}@7$								
"	13..	$6\frac{3}{4}@7\frac{1}{2}$	$6\frac{1}{2}@6\frac{7}{8}$	6,400	1,893	2,767	187	696	....	3,650	
"	16..	$6\frac{3}{4}@7\frac{1}{2}$	$6\frac{5}{8}@6\frac{7}{8}$								
"	20..	$6\frac{3}{4}@7\frac{1}{2}$	$6\frac{5}{8}@6\frac{7}{8}$	6,300	6,041	2,084	....	....	....	2,084	
"	23..	$6\frac{7}{8}@7\frac{1}{2}$	$6\frac{5}{8}@7$								
"	27..	$6\frac{7}{8}@7\frac{1}{2}$	$6\frac{5}{8}@7$	5,000	8,103	2,288	....	359	....	2,647	
"	30..	$6\frac{7}{8}@7\frac{1}{2}$	$6\frac{5}{8}@7$								
July	4..	$6\frac{7}{8}@7\frac{1}{2}$	$6\frac{5}{8}@7$	3,350	3,202	1,167	....	930	....	2,097	–@5-32d
"	7..	$6\frac{3}{4}@7\frac{1}{4}$	$6\frac{1}{2}@6\frac{7}{8}$								
"	11..	$6\frac{3}{4}@7\frac{1}{2}$	$6\frac{1}{2}@6\frac{7}{8}$	5,700	4,216	850	....	295	39	1,184	
"	14..	$6\frac{3}{4}@7\frac{1}{2}$	$6\frac{1}{2}@6\frac{7}{8}$								
"	18..	$6\frac{7}{8}@7\frac{1}{2}$	$6\frac{1}{2}@6\frac{7}{8}$	14,300	6,005	1,780	....	....	....	1,780	
"	21..	$6\frac{7}{8}@7\frac{1}{2}$	$6\frac{1}{2}@6\frac{7}{8}$								
"	25..	$7@7\frac{5}{8}$	$6\frac{3}{4}@7$	5,900	3,853	2,865	....	1,297	....	4,162	
"	28..	$7\frac{1}{4}@7\frac{3}{4}$	$6\frac{7}{8}@7\frac{1}{4}$								
August	1..	$7\frac{1}{4}@7\frac{3}{4}$	$6\frac{7}{8}@7\frac{1}{4}$	7,100	4,625	3,801	....	669	....	4,470	–@5-32d
"	4..	$7\frac{1}{4}@7\frac{3}{4}$	$6\frac{7}{8}@7\frac{1}{4}$								
"	8..	$7\frac{1}{4}@7\frac{3}{4}$	$6\frac{7}{8}@7\frac{1}{4}$	3,700	4,197	2,638	29	1,056	110	3,833	
"	11..	$7\frac{1}{4}@7\frac{3}{4}$	$6\frac{7}{8}@7\frac{1}{8}$								
"	15..	$7\frac{1}{4}@7\frac{3}{4}$	$6\frac{7}{8}@7\frac{1}{8}$	3,000	5,604	679	....	301	....	980	
"	18..	$7\frac{1}{4}@7\frac{3}{4}$	$6\frac{7}{8}@7\frac{1}{8}$								
"	22..	$7\frac{1}{4}@7\frac{3}{4}$	$6\frac{7}{8}@7\frac{1}{8}$	6,700	5,303	2,267	....	835	....	3,102	
"	25..	$7@7\frac{1}{2}$	$6\frac{3}{4}@7$								
"	29..	$7@7\frac{1}{2}$	$6\frac{5}{8}@6\frac{7}{8}$	4,250	5,187	4,667	....	1,292	....	5,959	
Septem.	1..	$7@7\frac{1}{2}$	$6\frac{5}{8}@6\frac{7}{8}$								–@3-16d
Average prices and total sales, receipts and exports.		8.39	8.03	343,750	310,135	114,983	37,992	37,207	6,650	196,832	

General Remarks.

vated by gatherings of Chartists and the unruly spirit manifested by the population of Ireland; all these causing the greatest uneasiness and general stagnation of trade.

These foreign difficulties continued with more or less intensity, until about the close of the crop year, when, quiet being restored to France, and the war between Prussia and Denmark apparently drawing to a close, a more cheerful feeling obtained in manufacturing circles in Great Britain, and was responded to here. Our own war with Mexico was also now brought to a successful close by the cession of California, &c., and the markets here assumed a much more cheerful tone.

Exchange.

The quotation for bills on London through September ranged between $7\frac{1}{2}$ and 9 per cent. premium; in October, $8\frac{1}{4}@9\frac{1}{2}$; in December, $9@10\frac{1}{2}$; in January, $9\frac{1}{2}@11$; in February, $9@10\frac{1}{4}$; in March, $10\frac{1}{2}$ down to $8\frac{3}{4}$; in April, up from $8\frac{1}{2}$ to $10@11$; in May, $10\frac{1}{2}@11\frac{1}{4}$; in June, $10\frac{1}{2}@11\frac{1}{4}$ to $9\frac{1}{2}@10$; in July, $9\frac{1}{2}@10\frac{1}{4}$; in August, $8\frac{3}{4}@9\frac{1}{2}$, closing September 1 at $9@9\frac{3}{8}$ per cent. premium.

1849.

COTTON CROP OF THE UNITED STATES.

Statement and Total Amount of the Cotton Crop of the United States, for the Year ending August 31, 1849.

	Bales.	Bales.	Total.	Same period 1848.
NEW ORLEANS.				
Export—				
To Foreign Ports	961,492			
Coastwise	205,811			
Stock on hand 1st September, 1849	15,480			
		1,182,783		
Deduct—				
Stock on hand 1st September, 1848	37,401			
Received from Mobile	35,164			
" Florida	5,065			
" Texas	11,356			
		88,986		
MOBILE.				
Export—			1,093,797	1,190,733
To Foreign Ports	396,341			
Coastwise	141,090			
Burnt at Mobile	400			
Stock on hand 1st September, 1849	5,046			
		542,877		
Deduct—				
Stock on hand 1st September, 1848	23,584			
Received from New Orleans	587			
		24,171		
FLORIDA.				
Export—			518,706	436,336
To Foreign Ports	79,739			
Coastwise	120,339			
Stock on hand 1st September, 1849	615			
		200,693		
Deduct—				
Stock on hand 1st September, 1848	....	507		
TEXAS.			200,186	153,776
Export—				
To Foreign Ports	2,495			
Coastwise	36,627			
Stock on hand 1st September, 1849	452			
Deduct—		39,574		
Stock on hand 1st September, 1848	....	747		
GEORGIA.			38,827	39,742
Export from Savannah—				
To Foreign Ports—Uplands	207,043			
" Sea Island	10,622			
Coastwise—Uplands	186,853			
Sea Island	938			
Export from Darien—	405,456			
To New York .... none.				
Stock in Savannah 1st September, 1849 11,500				
" Augusta and Hambro', 1st Sept., 1849. 13,819				
	25,319	430,775		
Deduct—				
Stock in Savannah and Augusta, 1st September, 1848.	36,603			
Received from Florida	2,800			
		39,403		
			391,372	254,825

Statement and Total Amount of the Cotton Crop of the United States, for the Year ending August 31, 1849.—*Concluded.*

	Bales.	Bales.	Total.	Same period 1848.
SOUTH CAROLINA.				
Export from Charleston—				
To foreign ports—Uplands	280,671			
" Sea Island	18,111			
Coastwise—Uplands	163,356			
Sea Island	813			
	462,951			
Burnt at Charleston	150			
Export from Georgetown—				
To New York and Boston 3,285				
Stock in Charleston 1st September, 1849 23,806				
	27,091			
		490,192		
Deduct—				
Stock in Charleston 1st September, 1848	14,085			
Received from Savannah	17,990			
		32,075		
NORTH CAROLINA.			458,117	261,752
Export—				
Coastwise	....	....	10,041	1,518
VIRGINIA.				
Export—				
To Foreign Ports	1,406			
Coastwise and Manufactured—taken from the ports	14,838			
Stock on hand 1st September, 1849	1,750			
		17,994		
Deduct—				
Stock on hand 1st September, 1848	....	444		
			17,550	8,952
Total crop of the United States	....	....	2,728,596	2,347,634

Total crop of 1849, as above .... bales	2,728,596
Crop of last year	2,347,634
Crop of year before	1,778,651
Increase over last year	380,962
Increase over year before	949,945

Export to Foreign Ports, from September 1, 1848, *to August* 31, 1849.

FROM	To Great Britain.	To France.	To North of Europe.	Other Foreign Ports.	Total.
New Orleans (bales)	645,018	154,647	61,062	100,765	961,492
Mobile	290,[illegible]83	61,597	16,822	27,539	396,341
Florida	62,734	5,721	6,836	4,448	79,739
Texas	......	750	1,745	....	2,495
Georgia	195,443	18,458	3,764	....	217,665
South Carolina	206,109	48,768	26,242	17,663	298,782
North Carolina	....	....	....	....	....
Virginia	242	108	1,056	....	1,406
Baltimore	106	....	....	....	106
Philadelphia	2,819	....	....	484	3,303
New York	132,612	78,037	44,893	5,101	260,643
Boston	2,435	173	3,038	226	5,872
Grand total	1,537,901	368,259	165,458	156,226	2,227,844
Total last year	1,324,265	279,172	120,348	134,476	1,858,261
Increase	213,636	89,087	45,110	21,750	369.583

Growth.

Total crop of 1823–4	bales.	509,158	Total crop of 1836–7	bales.	1,422,930
" 1824–5		569,249	" 1837–8		1,801,497
" 1825–6		720,027	" 1838–9		1,360,532
" 1826–7		957,281	" 1839–40		2,177,835
" 1827–8		720,593	" 1840–1		1,634,945
" 1828–9		857,744	" 1841–2		1,683,574
" 1829–30		976,845	" 1842–3		2,378,875
" 1830–1		1,038,848	" 1843–4		2,030,409
" 1831–2		987,477	" 1844–5		2,394,503
" 1832–3		1,070,438	" 1845–6		2,100,537
" 1833–4		1,205,394	" 1846–7		1,778,651
" 1834–5		1,254,328	" 1847–8		2,347,634
" 1835–6		1,360,725	" 1848–9		2,728,596

Consumption.

Total crop of the United States, as above stated ......bales.			2,728,596
Add—Stocks on hand at the commencement of the year, September 1, 1848, in the Southern ports		113,471	
" Northern ports		57,997	
			171,468
Makes a supply of			2,900,064
Deduct therefrom—The export to Foreign ports	2,227,844		
Less—Foreign included	1,122		
		2,226,722	
Stocks on hand September 1, 1849.			
In the Southern ports	72,468		
In the Northern ports	82,285		
		154,753	
Burnt at Charleston and Mobile	550		
		2,382,025	
Taken for home use			518,039

Quantity consumed by and in the hands of Manufacturers.

1848–9 ....bales.	518.039	1836–7 ....bales.	222,540
1847–8	531,772	1835–6	236,733
1846–7	427,967	1834–5	216,888
1845–6	422,597	1833–4	196,413
1844–5	389,006	1832–3	194,412
1843–4	346,744	1831–2	173,800
1842–3	325,129	1830–1	182,142
1841–2	267,850	1829–30	126,512
1840–1	297,288	1828–9	118,853
1839–40	295,193	1827–8	120.593
1838–9	276,018	1826–7	149,516
1837–8	246,063		

NOTE.—In our last annual statement, the estimate of cotton taken for consumption for the year ending Sept. 1st, 1848, in the States south and west of Virginia, was probably below the mark. The following for the past year is believed to be very nearly correct. The number of mills has increased since that time, and is still increasing. The following estimate is from a judicious and careful observer at the South, of the quantity so consumed, and not included in the receipts: thus in

North Carolina..........................bales..	20,000
South Carolina	15,000
Georgia	20,500
Alabama	7,000
Tennessee	12,000
Kentucky	5,000
Ohio	9,000
Pittsburg, Wheeling, &c.	12,500
Missouri, Indiana, Illinois, &c..	9,000
Total to Sept. 1st, 1849	110,000
Same time, 1848	75,000

Virginia manufactures more than 20,000 bales, and obtains a portion of it by importations from the southern and northern ports.

To which should be added the quantity burnt in the interior, and that lost on its way to market; these added to the crop as given above, received at the shipping ports, will show very nearly the amount raised in the United States the past season, say, in round numbers, 2,840,000 bales.

The quantity of new cotton received at the shipping ports up to the 1st inst. amounted to about 575 bales, against about 3,000 bales last year.

The shipments given in the above statement from Texas, are those by sea only, a considerable portion of the crop of that State finds its way to market via Red River, and is included in the receipts at New Orleans.

ANNUAL REVIEW.

From the New Orleans Price Current—1848-49.

As regards demand and the course of prices, they are, as must be apparent to every one, involved in a variety of contingencies which preclude the possibility of definite calculations. The unsettled state of the continent, which seems tending more and more to a general war among the great nations of Europe, and which by possibility may yet involve our own country in the spreading vortex, is a formidable obstacle in the path of commercial enterprise, and greatly increases the maze of uncertainty which ordinarily envelops the chances of the future. Apart from this difficulty, however, there would seem to be fair encouragement for the anticipation of a rather prosperous trade, if prudently conducted. The consumption of our great staple, cotton, even under what appears not to have been the most favorable circumstances, has within the past six or eight months somewhat exceeded the maximum of any previous period; and, with a reduced supply of the coming year—of which there seems to be little or no doubt—and the promise of abundant food crops in the principal countries of Europe, there would seem to be a fair prospect for the realization of more remunerating prices for this grand product of Southern industry. We think it may be safely asserted, however, that, what with the frosts of April, which, by rendering replanting necessary to an important extent, retarded the progress of cultivation several weeks in some sections, and the excessive rains of July, which washed the hills, flooded the lowlands, and kept the grass and weeds rank in the fields—and still further, the recent destructive overflow on Red River—there can hardly be a doubt that the total product of the present season will fall short of the last by several hundred thousand bales.

The semi-weekly Price and Weekly Sales and Receipts at New York, Weekly Exports from New York and Rates of Freight to Liverpool 1st of each month, for the Crop Year ending September 1, 1849.

1848.		Price of Mid. Fair to Fair New Orleans Liverpool Classificat'n.	Price of Mid. Fair to Fair Upland, Liverpool Classificat'n.	Sales for week.	Receipts for week.	EXPORTS FOR THE WEEK.					Rates of Freight to Liverpool.
						To Great Britain.	To France.	North of Europe.	Other Fo'n Ports	Total Exports.	
Septem.	5..	7@7⅝	6¾@7	7,900	4,197	....	....	....	....	....	3-16d.
"	8..	6⅞@7½	6⅝@6⅞								
"	12..	6¾@7½	6½@6¾	5,600	5,571	3,289	2,441	1,169	....	6,899	
"	15..	6¾@7½	6⅝@6⅞								
"	19..	7@7¾	6¾@7	8,400	7,875	1,242	....	906	4	2,152	
"	22.	6⅞@7⅜	6⅝@6⅞								
"	26..	6¾@7⅜	6⅝@6⅞	9,200	6,997	1,149	1,383	73	....	2,605	
"	29..	6¾@7⅜	6⅝@6⅞								
October	3..	6¾@7⅜	6½@6⅞	6,000	1,953	3,430	2,804	1,184	195	7,613	$\frac{5}{16}$@7 32
"	6..	6¾@7¼	6½@6¾								
"	10..	6⅝@7¼	6½@6¾	6,100	6,197	1,963	....	1,599	99	3,661	
"	13..	6⅝@7	6½@6¾								
"	17..	6½@7	6½@6¾	4,400	2,412	1,544	2,239	2,362	....	6,145	
"	20..	6⅜@6⅞	6¼@6½								
"	24..	6⅜@6¾	6¼@6½	7,100	13,755	1,200	661	3,024	558	5,443	
"	27..	6¼@6⅝	6⅛@6⅜								
"	31..	6¼@6¾	6⅛@6¼	8,600	6,142	795	....	483	50	1,328	
Novem.	3..	6¼@6⅝	6@6¼								¼d.
"	7..	6@6½	5⅞@6⅛	6,000	8,099	1,614	1,226	546	348	3,734	
"	10..	6@6½	5¾@6								
"	14..	6¼@6⅝	5⅞@6⅛	11,000	6,858	2,011	1,676	1,682	....	5,369	
"	17..	6⅜@6¾	6⅛@6¼								
"	21..	6⅝@7	6⅜@6⅝	6,750	1,929	1,622	998	2,321	....	4,941	
"	24..	6¾@7¼	6½@6⅝								
"	28..	6½@7	6¼@6½	4,800	16,344	1,382	....	1,749	....	3,131	
Decem.	1..	6⅜@6⅞	6⅛@6⅜								5-32@$\frac{3}{16}$
"	5..	6½@6⅞	6⅛@6⅜	6,800	12,540	2,976	1,085	493	....	4,554	
"	8..	6½@6⅞	6¼@6⅜								
"	12..	6½@7	6¼@6⅜	8,300	9,429	1,154	2,516	163	975	4,808	
"	15..	6⅝@7	6⅜@6½								
"	19..	6⅝@7	6⅜@6½	10,300	5,688	1,460	1,802	....	....	3,262	
"	22..	6⅝@7	6⅜@6½								
"	26..	6⅝@7	6⅜@6½	5,100	4,319	2,712	1,574	268	....	4,554	

GENERAL REMARKS.

Though this crop year opened more auspiciously, still the markets of Great Britain had not yet recovered from the shock of the previous year, the unsettled state of political affairs on the continent operated to the prejudice of the staple, the stock of which had accumulated to large figures owing to the stagnation existing in manufactured goods from the disorder reigning in much of Europe, and the markets there for them being partially or wholly closed at this period, Oct., 1848. Good ordinary and middling Orleans was quoted in Liverpool at 3¼@3½d., while the inferior descriptions of North Alabama and Tennessee, were unsaleable at 2¾d. In November and December, the Liverpool market with more settled state of continental affairs, recovered from the previous great depression, and an improved feeling both there and here was visible; our market at once beginning to rise and the period of the great paralysis passed The gold discoveries of California at this time, also imparted much activity and buoyancy to trade generally

Decem. 29..	6⅝@7	6⅜@6½								
1849.										
January 2..	7@7½	6¾@7	7,800	11,238	700	....	....	....	700	¼d.
" 5..	7¼@7⅝	6⅞@7¼								
" 9..	7⅜@7¾	7@7⅜	14,100	4,658	1,898	2,486	....	111	4,495	
" 12..	7½@8	7¼@7½								
" 16..	7⅜@7¾	7@7⅜	9,500	20,552	1,608	2,206	905	312	5,031	
" 19..	7⅜@7¾	7@7⅜								
" 23..	7½@7¾	7¼@7⅜	12,000	21,640	3,926	2,754	1,092	....	7,772	
" 26..	7½@8	7¼@7½								
" 30..	7½@8	7¼@7½	7,950	8,046	2,097	2,010	203	....	4,310	
February 2..	7⅜@7¾	7⅛@7⅜								5-16d.
" 6..	7⅜@8	7⅛@7⅜	8,300	9,399	1,088	....	510	126	1,724	
" 9..	7¼@8	7@7⅜								
" 13..	7⅝@8¼	7¼@7⅝	10,500	10,441	1,100	1,819	1,078	....	3,997	
" 16..	7½@8⅛	7¼@7½								
" 20..	7¼@8	7@7⅜	9,400	4,643	9,440	1,380	3,004	....	13,824	
" 23..	7½@8	7¼@7½								
" 27..	7½@8¼	7¼@7½	12,000	7,374	4,178	1,787	1,045	246	7,256	
March 2..	7½@8	7¼@7⅜								5-16d.
" 6..	7½@8	7¼@7⅜	9,800	1,086	3,989	1,261	390	....	5,640	
" 9..	7⅝@8¼	7½@7⅝								
" 13..	7½@8	7¼@7½	10,500	17,503	5,397	....	355	....	5,752	
" 16..	7½@8	7¼@7½								
" 20..	7½@8	7⅜@7½	13,700	33,070	7,660	1.664	852	....	10,176	
" 23..	7½@8¼	7⅜@7⅝								
" 27..	7½@8	7¼@7½	10,200	17,772	6,070	2,022	1,630	455	10,177	
" 30..	7⅝@8	7⅜@7⅝								
April 3..	7½@8	7¼@7⅝	11,450	2,566	8,926	179	852	....	9,957	7-32@¼d
" 6..	7½@8	7⅛@7⅜								
" 10..	7¼@7¾	7⅛@7⅜	6,000	23,543	4,770	6,312	....	437	11,519	
" 13..	7¼@7¾	7@7¼								
" 17..	7¼@7¾	7@7¼	8,900	8,137	3,016	....	102	....	3,118	
" 20..	7⅜@7¾	7⅛@7⅜								
" 24..	7¾@8¼	7½@7¾	18,100	8,353	2,945	4,374	3,021	....	10,340	
" 27..	7½@8	7¼@7½								
" 30..	7½@8	7¼@7½	5,200	11,802	3,327	2,049	1,013	20	6,409	
May 4..	7½@8	7⅜@7⅝								5-16d.
" 8..	7½@8	7⅜@7⅝	6,600	11,100	3,558	....	1,212	38	4,808	
" 11..	7½@8	7⅜@7¾								
" 15..	7⅝@8¼	7⅜@7¾	8,900	9,257	3,978	4.452	163	47	8,640	
" 18	7¾@8¼	7½@7¾								

On April 20, accounts were received from Georgia and Alabama, of a frost so severe as to kill the plant, and the market became excited, prices at once advancing ½ cent per pound. Subsequently, however, the rising tendency was checked by advices from Liverpool, unfavorable, owing to a renewal of hostilities between Denmark and Prussia, troubles in South Italy and grave differences between Austria and Sardinia, supplemented by the blockade of Venice.

These discouraging advices were, however, partially relieved by the short crop accounts that were current throughout the year, affecting this and the Liverpool market to a greater or less extent, and causing an advance in the prices of the staple of 4¼@4¾ cents, from the lowest points touched during the crop year.

The semi-weekly Price and Weekly Sales and Receipts at New York, Weekly Exports from New York and Rates of Freight to Liverpool 1st of each Month, for the Crop Year ending September 1, 1849—Concluded.

1849.		Price of Mid. Fair to Fair New Orleans Liverpool Classificat'n.	Price of Mid. Fair to Fair Upland, Liverpool Classificat'n.	Sales for week.	Receipts for week.	EXPORTS FOR WEEK.					Rates of Freight to Liverpool.
						To Great Britain.	To France.	North of Europe.	Other Fo'n Ports	Total Exports.	
May	22..	$7\frac{7}{8}@8\frac{1}{2}$	$7\frac{5}{8}@7\frac{7}{8}$	11,500	7,256	4,805	1,869	1,265	....	7,939	
"	25..	$8@8\frac{1}{2}$	$7\frac{3}{4}@8$								
"	29..	$8@8\frac{1}{2}$	$7\frac{3}{4}@8$	6,200	3,170	3,182	242	809	....	4,233	
June	1..	$8@8\frac{1}{2}$	$7\frac{3}{4}@8$								$\frac{1}{8}$@5-32d
"	5..	$8@8\frac{1}{2}$	$7\frac{3}{4}@8$	5,800	8,849	779	1,567	45	....	2,391	
"	8..	$8@8\frac{1}{2}$	$7\frac{3}{4}@8$								
"	12..	$8@8\frac{1}{2}$	$7\frac{3}{4}@8$	4,500	6,107	612	2,981	741	99	4,433	
"	15..	$8@8\frac{1}{2}$	$7\frac{3}{4}@8$								
"	19..	$8@8\frac{1}{2}$	$7\frac{3}{4}@8$	3.800	6,943	1,734	....	417	54	2,205	
"	22..	$8\frac{1}{4}@8\frac{3}{4}$	$8@8\frac{3}{8}$								
"	26..	$8\frac{1}{4}@9$	$8\frac{1}{8}@8\frac{3}{8}$	11,000	4,637	1,638	552	985	20	3,195	
"	29..	$8\frac{3}{8}@9$	$8@8\frac{3}{8}$								
July	3..	$8\frac{1}{2}@9\frac{1}{4}$	$8\frac{1}{4}@8\frac{5}{8}$	10,700	9,990	3,195	1,125	824	406	5,550	$\frac{1}{8}$d.
"	6..	$8\frac{3}{4}@9\frac{1}{2}$	$8\frac{1}{2}@9$								
"	10..	$8\frac{3}{4}@9\frac{1}{2}$	$8\frac{5}{8}@9$	16,500	2,035	557	100	283	....	940	
"	13..	$9\frac{1}{4}@9\frac{3}{4}$	$9@9\frac{1}{4}$								
"	17..	$9\frac{1}{2}@10$	$9@9\frac{3}{8}$	13,200	7,714	1,344	2,675	1,188	....	5,207	
"	20..	$9\frac{3}{8}@9\frac{3}{4}$	$9@9\frac{3}{8}$								
"	24..	$9\frac{1}{2}@10\frac{1}{4}$	$9\frac{1}{4}@9\frac{1}{2}$	11,300	4,798	1,275	1,629	492	....	3,396	
"	27..	$9\frac{3}{4}@10\frac{1}{2}$	$9\frac{1}{2}@9\frac{5}{8}$								
"	30..	$10@10\frac{1}{2}$	$9\frac{3}{4}@10$	14,600	3,517	795	1,522	....	263	2,580	
August	3..	$10@10\frac{1}{2}$	$9\frac{3}{4}@10$								$\frac{1}{8}$d.
"	7..	$10\frac{1}{2}@11$	$10@10\frac{3}{8}$	7.400	3,918	540	2,535	....	....	3,075	
"	10..	$10\frac{1}{2}@11$	$10@10\frac{3}{8}$								
"	14..	$10\frac{1}{2}@11$	$10\frac{1}{8}@10\frac{1}{2}$	14,400	5,978	178	1,536	942	....	2,656	
"	17..	$10\frac{1}{2}@11$	$10\frac{1}{8}@10\frac{1}{2}$								
"	21..	$10\frac{1}{8}@10\frac{3}{4}$	$10@10\frac{1}{4}$	4,550	1,932	583	1,464	269	....	2,316	
"	24..	$10\frac{1}{2}@11$	$10\frac{1}{4}@10\frac{3}{8}$								
"	28..	$10\frac{1}{2}@11\frac{1}{4}$	$10\frac{1}{4}@10\frac{5}{8}$	8,600	3,058	818	....	1,084	....	1,902	
"	31..	$10\frac{3}{4}@11\frac{1}{4}$	$10\frac{1}{4}@10\frac{3}{4}$								
Septem.	4..	$10\frac{3}{4}@11\frac{1}{4}$	$10\frac{1}{4}@10\frac{5}{8}$	10,500	4,798	1,363	1,080	100	238	2,781	$\frac{1}{8}$d.
Average price and total sales, receipts and exports.		7.98	7.55	477,800	447,185	132,612	78,037	44,893	5,101	260,643	

GENERAL REMARKS.

Exchange.

Bills on London ruled steady through September, at 9@$9\frac{1}{2}$ per cent. premium; in October, the quotation fell from $9\frac{1}{4}$@$9\frac{3}{4}$ to $8\frac{1}{4}$@$8\frac{1}{2}$; in November, the range was $8\frac{1}{4}$@9; in December, the same; in January, the quotation dropped to $7\frac{1}{2}$@$8\frac{1}{4}$; in February, 7@8; in March, it further declined to $6\frac{1}{2}$@7; April opened at 5@6, but closed at 7@$7\frac{3}{4}$; in May, the quotation advanced from $7\frac{1}{2}$@8 to $8\frac{1}{4}$@$8\frac{3}{4}$; in June, the range was from $8\frac{1}{4}$@9; July, $8\frac{1}{4}$@$8\frac{3}{4}$; advancing in August, from $8\frac{1}{4}$@$8\frac{3}{4}$ to $9\frac{1}{2}$@$9\frac{3}{4}$ premium, which was the quotation September 1, 1849.

LIVERPOOL STATEMENT FOR 1849.

UNITED STATES, 1848–1849.			
Stock, September 1, 1848	171,000	Export	2,227,000
Crop	2,729,000	Consumption	518,000
		Stock, Sept. 1, 1849	155,000
Bales	2,900,000	Bales	2,900,000

Stock 1st Jan., 1849, in	Liverpool.	Gt. Britain.	France.	Continent.	Tot. Europe.
United States......Bales	235,000	272,000	20,000	39,000	331,000
Brazil	68,000	68,000	....	1,000	69,000
West Indies	2,000	3,000	1,000	2,000	6,000
East Indies	74,000	136,000	1,000	5,000	142,000
Egypt	14,000	17,000	7,000	13,000	37,000
Bales	393,000	496,000	29,000	60,000	585,000

CONSUMPTION.						IMPORT.				
Tot. Europe.	Continent.	France.	Gt. Britain.	Liverpool.		Liverpool.	Gt. Britain.	France.	Continent.	Tot. Europe.
1,958,000	315,000	364,000	1,279,000	1,199,000	United States......Bales	1,383,000	1,478,000	379,000	297,000	2,000,000
144,000	20,000	4,000	120,000	120,000	Brazil	163,000	163,000	8,000	20,000	175,000
25,000	6,000	9,000	10,000	8,000	West Indies	8,000	9,000	10,000	5,000	24,000
228,000	99,000	3,000	126,000	97,000	East Indies	107,000	182,000	2,000	96,000	194,000
122,000	52,000	19,000	51,000	51,000	Egypt	71,000	73,000	17,000	56,000	145,000
2,477,000	492,000	399,000	1,586,000	1,475,000	Bales	1,732,000	1,905,000	416,000	474,000	2,538,000
....	....	....	257,000	182,000	Export.					
646,000	42,000	46,000	558,000	468,000	Stock, Dec. 31......Stock above,	393,000	496,000	29,000	60,000	585,000
3,123,000	534,000	445,000	2,401,000	2,125,000	Total supply, bales	2,125,000	2,401,000	445,000	534,000	3,123,000

COTTON AT LIVER

Week Ending.	Receipts.						Sales.			
	American	E. I.	Egypt.	Brazil.	Other.	Total.	Consumption.	Speculation.	Export	Total.
Jan. 5..	27,682	7	2,403	1,755	......	31,847	27,860	13,900	2,300	44,060
" 12..	38,574	......	......	1,053	......	39,627	22,210	5,300	1,100	28,610
" 19..	62,952	2,316	1,068	11,613	60	78,009	25,640	9,800	1,880	37,320
" 26..	11,527	4,134	100	3,701	58	19,520	40,630	14,200	3,460	58,290
Feb. 2..	26,561	400	......	3,530	......	30,491	39,740	26,500	1,610	67,850
" 9..	24,832	1.595	2,836	3,172	......	32,435	35,520	19,500	6,100	61,120
" 16..	12,978	932	......	2,884	......	16,794	43,260	41;100	4,860	89,220
" 23..	31,746	......	3,469	....	......	35,215	25,950	6,140	1,810	33,900
March 2..	....	......	......	....	......	....	....	......	......	....
" 9..	44,825	......	1,295	1,518	615	48,253	17,650	3,560	1,000	22,210
" 16..	16,806	......	......	1,060	......	17,866	20,330	8,090	3,270	31,690
" 23..	18,517	......	......	4,780	......	23,297	22,890	2,570	3,830	29,290
" 30..	8,198	......	......	2,030	......	10,228	18,650	2,050	5,310	26,010
April 5..	74,947	......	......	7,236	......	82,183	29,670	1,930	2,980	34,580
" 13..	35,064	......	......	2,090	......	37,154	17,220	580	4,300	22,100
" 20..	27,311	......	......	1,100	......	28,411	25,570	2,380	4,850	32,800
" 27..	98,407	4,023	......	3,664	......	106,094	28,650	2,500	3,580	34,730
May 4..	104,248	3,907	......	4,419	703	113,277	41,100	2,260	2,830	46,190
" 11..	22,429	2,180	......	....	......	24,609	34,040	3,670	4,970	42,680
" 18..	52,401	......	14,053	6,225	......	72,679	28,940	6,340	4,320	39,600
" 25..	113,463	1,542	3,820	6,674	230	125,729	30,490	4,470	5,360	40,320
June 1..	15,961	......	70	180	54	16,265	30,550	760	4,480	35,790
" 8..	21,896	194	60	2,406	845	25,401	48,730	24,300	4,790	77,820
" 15..	7,828	......	......	6,914	16	14.758	24,330	13,800	2,560	40,690
" 22..	27,690	......	12,560	2,912	......	43,162	23,110	14,450	4,530	42,090
" 29..	101,690	3,445	2,846	4,057	342	112,380	50,110	20,020	4,060	74,190
July 6..	52,740	1,350	193	8,121	93	62,497	36,930	16,530	5,390	58,850
" 13..	23,206	2,660	......	2,360	60	28,286	50,450	27,360	4,720	82,530
" 20..	29,188	......	2,676	....	948	32,812	45,430	23,560	7,500	76,490
" 27..	32,493	3,730	941	2,602	90	39,856	31,080	16,740	7,520	55,340
Aug. 3..	29,006	2,381	......	2,393	......	33,780	27,080	9,360	5,760	42,200
" 10..	5,217	2,661	......	2,559	......	10,437	48,220	24,600	5,990	78,810
" 17..	16,619	11,878	2,430	3,919	172	35,018	37,770	38,330	7,090	83,190
" 24..	5,096	......	......	6	......	5,102	28,410	32,090	6,480	66,980
" 31..	10,800	4,263	......	747	301	16,111	20,360	10,370	5,100	35,830
Sep. 7..	6,044	......	......	....	49	6,093	18,080	10,230	7,970	36,280
" 14..	5,188	2,753	1,490	2,589	89	12,109	22,520	5,240	3,940	31,700
" 21..	2,492	12,272	1,435	1,363	......	17,562	18,150	7,600	2,450	28,200
" 28..	322	......	......	....	......	322	17,150	13,200	2,670	33,020
Oct. 5..	5,639	......	......	7,246	76	12,961	21,490	5,700	1,800	28,990
" 12..	6,321	1,298	1,935	10,027	460	20,041	66,750	48,300	1,720	116,770
" 19..	2,742	......	47	1,216	......	4,005	64,520	65,900	1,170	131,590
" 26..	13,001	2,858	......	1,443	......	17,302	27,810	33,700	1,200	62,710
Nov. 2..	1,711	12,292	1,113	2,952	374	18,442	19,320	19,810	2,790	41,920
" 9..	2,520	......	3,571	5,538	521	12,150	25,690	31,180	2,960	59,830
" 16..	3,768	......	834	1,086	......	5,688	19,474	30,766	730	50,970
" 23..	5,270	2,981	863	1,402	......	10,516	11,930	9,340	160	21,430
" 30..	533	......	4,082	....	46	4,661	13,880	4,540	580	19,000
Dec. 7..	5,221	2,041	3,927	5,782	307	17,278	12,780	5,830	300	18,910
" 14..	6,388	6,699	......	4,962	......	18,049	25,150	7,390	90	32,630
" 21..	40,141	10,175	......	710	......	51,026	19,860	8,230	250	28,340
" 28..	2,402	......	......	1,000	561	3,962	31,470	17,130	420	49,020
Average prices & total sales, receipts & stocks.	1,342,771	106177	70,117	168,046	7,070	1,694,181	1,516,594	786,196	176890	2,479,680

’OOL. YEAR 1849.

STOCKS.			PRICES.			Actual Export.	Con-sumption.	REMARKS.
Amer’n	Other.	Total.	Mid. Up.	Mid. Orl.	Dhol.			
237,442	157,585	395,027	4⅛	4¼	3½	2,300	27,860	
258,076	153,268	411,344	4	4⅛	3½	1,100	50,070	Large receipts at United States Ports.
297,928	163,905	461,833	4	4⅛	3½	1,880	75,710	
271,675	165,588	437,263	4⅛	4¼	3½	3,460	116,340	
256,526	159,078	415,604	4¼	4¼	3⅝	1,610	156,080	
246,098	162,087	408,185	4⅜	4½	3¾	6,100	191,600	
221,156	155,803	376,959	4⅝	4¾	3¾	4,860	234,860	
231,352	153,062	384,414	4½	4⅝	3¾	1,810	260,810	
......	......	......	......	......	......	......	......	
265,893	151,717	417,610	4⅜	4½	3⅝	1,000	278,460	Better accounts from Manchester, &c.
263,629	148,247	411,876	4⅜	4⅜	3⅝	3,270	298,790	
259,086	149,367	408,453	4¼	4⅜	3⅝	3,830	366,970	
245,614	149,207	394,821	4⅛	4¼	3½	5,310	317,440	
292,571	151,783	444,354	4¼	4⅜	3½	2,980	347,110	Hostilities between Denmark and Germany.
308,055	151,933	459,988	4	4¼	3⅝	4,300	364,330	
309,416	148,473	457,889	4	4⅛	3⅝	4,850	389,900	
379,053	152,690	531,743	4	4⅛	3⅝	3,580	418,550	
443,691	157,399	601,090	4⅛	4⅛	3⅝	2,830	459,650	Frost accounts damaging crop.
433,560	153,129	586,689	4⅛	4¾	3⅝	4,970	495,690	
456,381	169,727	626,108	4⅛	4¼	3⅝	4,320	524,630	
542,114	173,873	715,987	4⅛	4⅛	3⅝	5,360	555,120	
526,245	170,977	697,222	4	4⅛	3⅝	4,480	585,670	
501,291	167,812	669,103	4¼	4⅜	3⅝	4,790	634,400	
488,449	168,522	656,971	4¼	4½	3⅝	2,560	658,730	Crop accounts more favorable.
492,839	179,654	672,493	4¼	4½	3⅝	4,530	681,840	
549,699	181,004	730,703	4½	4⅜	3⅞	4,060	731,950	
568,029	182,851	750,880	4⅝	4¾	3⅞	5,390	768,880	Unfavorable weather for crop.
550,145	173,871	724,016	4¾	4⅞	3⅞	4,720	819,330	
537,463	166,435	703,898	4⅞	5	3⅞	7,500	864,760	
538,923	167,613	706,536	4⅞	4⅞	4	7,520	895,840	
541,556	165,947	707,503	4⅞	4⅞	4	5,760	922,920	Better accounts from manufacturing districts.
508,463	155,267	663,730	5	5⅛	4	5,990	971,140	
492,752	161,136	653,888	5¼	5⅜	4¼	7,090	1,008,910	
471,988	152,112	624,100	5⅜	5½	4⅜	6,480	1,037,320	
462,818	152,333	615,151	5⅜	5½	4⅜	5,100	1,057,680	Small estimates of crop.
448,552	146,642	595,194	5⅜	5½	4⅜	7,970	1,075,760	
434,050	146,793	580,843	5¼	5⅜	4⅜	3,940	1,098,280	
419,012	158,928	577,935	5¼	5⅜	4⅜	2,450	1,116,430	
404,694	153,743	558,437	5¼	5⅜	4⅜	2,670	1,133,580	
391,253	156,855	548,108	5¼	5⅜	4⅜	1,800	1,155,070	More encouraging accounts from Manchester.
348,074	151,605	499,679	5¾	5⅞	4⅝	1,720	1,221,820	
298,186	139,808	437,994	6	6⅛	4¾	1,170	1,286,340	
288,247	138,039	426,286	6⅛	6¼	4¾	1,200	1,314,150	Better feeling.
273,638	148,980	422,618	6⅛	6¼	4¾	2,790	1,333,470	
254,488	151,234	405,722	6⅜	6½	4⅞	2,960	1,359,160	
246,176	145,024	391,200	6⅜	6½	4⅞	730	1,378,634	
242,176	147,650	389,826	6¼	6¼	4¾	160	1,390,564	Large demand.
231,399	148,628	380,027	6⅛	6⅛	4¾	580	1,404,444	
226,840	157,405	384,245	5⅞	6	4⅝	300	1,417,224	
214,348	162,706	377,054	5⅞	6	4⅝	90	1,442,374	
240,159	168,411	408,570	6¼	6½	4⅝	250	1,462,234	
219,970	161,672	381,642	6⅛	6¼	4⅝	420	1,493,704	
			4.09	4.95	4.02	176,890	1,516,594	

1850.

Census returns for this year place average crop of cotton and seed per acre in United States as follows: South Carolina, 320 lbs.; Georgia, 500; Florida, 250; Tennessee, 300; Alabama, 525; Louisiana, 550; Mississippi, 650; Arkansas, 700; Texas, 750. Seed constituting 50 to 60 per cent. of above averages.

Cotton consumed in Great Britain, 588,200,000 lbs.

Importation of cotton into Great Britain, 664,000,000 lbs., of which 102,000,000 lbs. were re-exported; cotton yarn and twist manufactured, 529,000,000 lbs., of which 124,000,000 lbs. were exported; piece goods exported, 1,358,000,000 yards; value of exports, £28,000,000.

Cotton mills in Great Britain, 1,932; moving power therein, 83,000 horse power; spindles, 21,000,000; power looms, 247,000; children employed, 14,993; total hands employed, 330,924, of whom 189,423 were females. (*See* year 1856.)

Number of cotton factories in 25 of the United States this year....	1,074
" spindles running " " "	3,633,693
" hands employed " " "	94,956
(32,295 males and 62,661 females.)	
Value of raw material consumed....	$37,778,014
" manufactured cotton products....	65,501,687
" " mixed "	3,693,731
Number yards cotton cloth made....	763,678,407
" pounds " yarn batting....	30,000,000
" factories making mixed goods....	103
Value of mixed goods made....	$3,693,731
Aggregate capital invested....	76,032,578

It was in this year that the French Government first began to direct attention to the cultivation of cotton in Algeria, a portion of the coast having been found to possess a soil and climate fitted for the plant. (*See* years 1855 and 1856.)

The number of spindles in New England this year has been estimated upon reliable authority to have been 2,751,078. (*See* years 1840, 1860 and 1861.) Population 2,728,106, an average of 1,008 spindles to 1,000 inhabitants. In England, Scotland, and Wales the population was 20,793,552, and the number of spindles 20,857,062, an average of 1,003 spindles to 1,000 inhabitants.

The following table sets forth the actual crop per acre, on the average, in the States named, this year:

South Carolina	320 lbs.
Georgia	500 "
Tennessee	300 "
Florida	250 "
Alabama	525 "
Louisiana	550 "
Mississippi	650 "
Arkansas	700 "
Texas	750 "

Frost in Spring, heavy rains, overflow of the Mississippi, Red River, etc., during the Summer, in the United States.

Home consumption of cotton by the United States this year, in bales: States north of Virginia, 487,769; States south and west of Virginia, 107,500; total, 595,269; weekly average, 11,447.

"Self-acting temples" unknown in England at this period only as a novelty. (*See* years 1805, 1816, 1825, and 1855.)

Total number of slaves in the United States this year, 3,204,000, 1,943,000 of whom were in the cotton States alone. (*See* years 1790 and 1830.)

COTTON CROP OF THE UNITED STATES.

Statement and Total Amount of the Cotton Crop of the United States, for the Year ending August 31, 1850.

	Bales.	Bales.	Total.	Same period 1849.
NEW-ORLEANS.				
Export—				
To Foreign Ports	624,748			
Coastwise	213,843			
Stock on hand, 1st September, 1850	16,612			
		855,203		
Deduct—				
Stock on hand, 1st September, 1849	15,480			
Received from Mobile and Montgomery, Ala	41,148			
" " Florida	10,601			
" " Texas	6,088			
		73,317		
			781,886	1,093,797
ALABAMA.				
Export—				
To Foreign Ports	214,164			
Coastwise	128,872			
Stock on hand, 1st September, 1850	12,962			
		355,998		
Deduct—				
Stock on hand, 1st September, 1849	....	5,046		
			350,952	518,706
FLORIDA.				
Export—				
To Foreign Ports	48,934			
Coastwise	131,877			
Stock on hand, 1st September, 1850	1,148			
		181,959		
Deduct—				
Stock on hand, 1st September, 1849	....	615		
			181,344	200,186
TEXAS.				
Export—				
To Foreign Ports	513			
Coastwise	30,937			
Stock on hand, 1st September, 1850	265			
		31,715		
Deduct—				
Stock on hand, 1st September, 1849	....	452		
			31,263	38,827
GEORGIA.				
Export from Savannah—				
To Foreign Ports—Uplands	144,540			
Sea Island	8,603			
Coastwise—Uplands	186,721			
Sea Island	1,839			
	341,703			
Export from Darien—				
To New York	22			
Stock in Savannah, 1st September, 1850	9,599			
" Augusta and Hambro', 1st September, 1850	19,470			
	29,091			
		370,794		

Statement and Total Amount of the Cotton Crop of the United States, for the Year ending August 31, 1850.—*Concluded.*

	Bales.	Bales.	Total.	Same period 1849.
Deduct—				
Stock in Savannah and Augusta, 1st September, 1849.	25,319			
Received from Florida	1,840			
		27,159		
			343,635	391,372
SOUTH CAROLINA.				
Export from Charleston—				
To Foreign Ports—Uplands	213,205			
Sea Island	14,366			
Coastwise—Uplands	152,122			
Sea Island	2,071			
	381,764			
Burnt at Charleston	6,146			
Export from Georgetown—				
To New York and Boston	1,449			
Stock in Charleston, 1st September, 1850	30,698			
	32,147			
		420,057		
Deduct—				
Stock in Charleston, 1st September, 1849	23,806			
Received from Savannah	11,647			
" " Florida	339			
		35,792		
			384,265	458,117
NORTH CAROLINA.				
Export—				
Coastwise	....	....	11,861	10,041
VIRGINIA.				
Export—				
To Foreign Ports	183			
Coastwise, and Manufactured—Taken from the Ports	12,067			
Stock on hand, 1st September, 1850	1,000			
		13,250		
Deduct—				
Stock on hand, 1st September, 1849	....	1,750		
			11,500	17,550
Total Crop of the United States	....	....	2,096,706	2,728,596

Total crop of 1850, as above	bales.	2,096,706
Crop of 1849		2,728,596
Crop of 1848		2,347,634
Crop of 1847		1,778,651
Decrease from last year	bales.	631,890
Decrease from year before		250,928

Export to Foreign Ports, from September 1, 1849, *to August* 31, 1850.

FROM	To Great Britain.	To France.	To North of Europe.	Other F'n Ports.	Total.
New Orleans....................... Bales.	397,189	117,413	25,196	84,950	624,748
Mobile..............................	162,219	39,968	....	11,977	214,164
Florida..............................	39,594	....	7,165	2,175	48,934
Texas................................	513	....	....	....	513
Georgia..............................	137,185	14,110	1,848	....	153,143
South Carolina......................	165,623	33,082	8,944	19,922	227,571
North Carolina......................	....	....	....	....	....
Virginia..............................	....	....	133	50	183
Baltimore............................	202	....	230	....	432
Philadelphia........................	3,454	....	....	599	4,053
New York............................	200,113	85,054	27,726	1,907	314,800
Boston................................	679	....	914	21	1,614
Grand total..........................	1,106,771	289,627	72,156	121,601	1,590,155
Total last year......................	1,537,901	368,259	165,458	156,226	2,227,844
Decrease..............................	431,130	78,632	93,302	34,625	637,689

Growth.

Total crop of 1823–4...... bales.	509,158	Total crop of 1837–8 ....bales.	1,801,497
" 1824–5............	569,249	" 1838–9.............	1,360,532
" 1825–6............	720,027	" 1839–40............	2,177,835
" 1826–7............	957,281	" 1840–1.............	1,634,945
" 1827–8............	720,593	" 1841–2.............	1,683,574
" 1828–9............	857,744	" 1842–3.............	2,378,875
" 1829–30...........	976,845	" 1843–4.............	2,030,409
" 1830–1............	1,038,848	" 1844–5.............	2,394,503
" 1831–2............	987,477	" 1845–6.............	2,100,537
" 1832–3............	1,070,438	" 1846–7.............	1,778,651
" 1833–4............	1,205,394	" 1847–8.............	2,347,634
" 1834–5............	1,254,328	" 1848–9.............	2,728,596
" 1835–6............	1,360,725	" 1849–50............	2,096,706
" 1836–7............	1,422,930		

Consumption.

Total crop of the United States, as above stated......................				2,096,706 bales.
Add—Stocks on hand at the commencement of the year, 1st September, 1849.—In the Southern ports..............................			72,468	
" Northern "			82,285	
				154,753
Makes a supply of..				2,251,459
Deduct therefrom—The export to Foreign ports............	1,590,155			
Less, foreign included....................	1,341			
		1,588,814		
Stocks on hand at the close of the year, 1st September, 1850—				
In the Southern ports..................	91,754			
" Northern "	76,176			
		167,930		
Burnt at New York and Charleston........................		6,946		
				1,763,690
Taken for home use................Bales.				487,769

Quantity consumed by and in the hands of Manufacturers.

1849–50 bales.	487,769	1837–8 bales.	246,063
1848–9	518,039	1836–7	222,540
1847–8	531,772	1835–6	236,733
1846–7	427,967	1834–5	216,888
1845–6	422,597	1833–4	196,413
1844–5	389,006	1832–3	194,412
1843–4	346,744	1831–2	173,800
1842–3	325,129	1830–1	182,142
1841–2	267,850	1829–30	126,512
1840–1	297,288	1828–9	118,853
1839–40	295,193	1827–8	120,593
1838–9	276,018	1826–7	149,516

NOTE.—In our last Annual Statement, the Estimate of Cotton taken for Consumption for the year ending Sept. 1, 1849, in the States south and west of Virginia, was probably over-estimated—the following for the past year is believed to be very nearly correct. The number of mills has increased since that time, and is still increasing, but the quantity consumed, as far as we can learn, is, owing to high prices, &c., less than the year previous. The following estimate is from a judicious and careful observer at the South, of the quantity so consumed, and not included in the receipts: Thus, in—

	Mills.	Spindles.	Quantity consumed.
North Carolina	30	......	20,000 bales.
South Carolina	16	36,500	15,000 "
Georgia	36	51,150	27,000 "
Alabama	11	16,960	6,000 "
Tennessee	30	36,000	12,000 "
On the Ohio, &c	30	102,220	27,500 "
Total to Sept. 1, 1850			107,500 bales.
Total to Sept. 1, 1849			110,000 "
Total to Sept. 1, 1848			75,000 "

To which should be added the stocks in the interior towns, the quantity burnt in the interior, and that lost on its way to market; these, added to the crop as given above, received at the shipping ports, will show very nearly the amount raised in the United States the past season—say, in round numbers, 2,212,000 bales.

The quantity of new cotton received at the shipping ports up to the 1st inst. amounted to about 255 bales, against about 575 bales last year.

The shipments given in this Statement from Texas, are those by sea only; a considerable portion of the crop of that State finds its way to market via Red River, and is included in the Receipts at New Orleans.

ANNUAL REVIEW.

From the New Orleans Price Current, 1849–50.

Cotton—This great staple of the South, and leading article of our own varied commerce, has been an object of unusual interest during the past season. We closed the year upon a crop of nearly 2,730,000 bales, or an excess of 330,000 bales over any previous crop, and yet the aggregate of the stock at Liverpool and at the shipping ports of the country did not vary materially from the quantity on hand at the same time the year previous; thus demonstrating the important fact that consumption had more than kept pace with production; for it is well understood that fully 200,000 bales of the crop above referred to was of the product of 1847. The demonstration of this interesting fact respecting consumption, coupled with the certainty that the then growing crop would fall short of that which immediately preceded it, caused planters generally to raise their expectations of prices to a high, and, in some instances, to an extravagant standard; and thus it is probable that some parties have met with disappointment, though the average return to the producer has been greater than any year except 1839. The same considerations that operated with the planter early gave rise to a speculative feeling, but the opening prices were so much above those of the previous year, and so much uncertainty existed in regard to the real extent of the crop, though known to be materially short, that it was not until near January that speculators began to operate freely, at which time many parties had made a considerable reduction in their estimates of the supply. In the latter part of December, the estimates of the crop being quite generally reduced, speculators entered the market freely, and prices continued to advance until in the latter part of January the quotations were 11½ and 12 cents for low middling to good middling. At this critical period of the market, when parties were looking with confidence for a material advance in Liverpool, with a stock in that port reduced below the figures of many previous years, the annual cotton circulars, which are there made up to the close of the calendar year, came to hand. By these it appeared that upon investigation it had been discovered that the actual amount of cotton on hand at Liverpool on the 31st of December, 1849, exceeded by nearly 100,000 bales the quantity stated from week to week in the General Brokers' and other circulars, upon the faith of which, speculations and operations to an important extent were based. Instead, therefore, of a stock reduced to a lower point than for many previous years, this extraordinary discrepancy showed the current year to have commenced with 489,000 bales at Liverpool, or an excess over the previous year of nearly 80,000 bales. This gave an important advantage to consumers, and was productive of disappointment on this side, where confidence had been placed in the accuracy of the published statistics, as they had reached us by each successive steamer. And thus it was, notwithstanding the same circulars that noted this great discrepancy in the stock, also advised of an advance of a ¼ and ⅜d. per lb. at Liverpool; and, notwithstanding further, that there was unshaken confidence here in the reduced estimates of the crop, which had formed one of the bases of the then recent operations, our market exhibited a slight reaction, as speculators mostly retired for the time being, and nearly all European orders were at limits too low to admit of their being executed.

The semi-weekly Price and Weekly Sales and Receipts, at New York, Weekly Exports from New York and Rates of Freight to Liverpool 1st of each Month, for the Crop Year ending September 1, 1850.

1849.	Price of Mid. Fair to Fair New Orleans Liverpool Classificat'n.	Price of Mid. Fair to Fair Upland. Liverpool Classificati'n	Sales for week.	Receipts for week.	EXPORTS FOR WEEK.					Rates of Freight to Liverpool.
					To Great Britain.	To France.	North of Europe.	Other Fo'n Ports	Total Exports.	
Septem. 7..	10¾@11¼	10¼@10⅝								⅛d.
" 11..	10½@11	10¼@10½	9,000	3,637	....	....	....	....	....	
" 14..	10¼@10¾	10@10⅜								
" 18..	10½@11	10¼@10⅝	13,900	5,443	2,454	1,415	2,348	357	6,574	
" 21..	11@11½	10½@11								
" 25 .	11¼@11¾	11@11¼	16,370	3,190	4,721	1,534	999	....	7,254	
" 28..	11¼@11¾	11@11¼								
October 2..	11¼@11¾	11@11¼	7,900	10,074	1,597	3,858	1,482	....	6,937	$\frac{5}{32}$d.
" 5..	11¼@11¾	11@11¼								
" 9..	11½@12	11@11⅜	8,500	5,793	1,243	....	174	....	1,417	
" 12..	11½@12	11⅛@11½								
" 16..	11½@12	11⅛@11½	10,900	2,919	1,384	2,404	94	....	3,882	
" 19..	11¼@11¾	10⅞@11¼								
" 23..	11@11¾	10¾@11	7,500	6,883	1,628	2,116	145	....	3,889	
" 26..	11½@12	11@11⅜								
" 30..	11½@12	11@11⅜	16,900	4,779	1,711	....	693	....	2,404	
Novem. 2..	11½@12	11⅛@11⅜								¼d.
" 6..	11½@12	11¼@11½	9,600	8,805	1,213	1,121	810	....	3,144	
" 9..	11½@11¾	11⅛@11⅜								
" 13..	11½@12	11⅛@11⅜	12,800	12,597	533	5,770	281	....	6,584	
" 16..	11¼@11¾	11@11⅛								
" 20..	11@11½	10¾@10⅞	4,700	2,682	8,991	1 786	794	....	11,571	
" 23..	10¾@11¼	10½@10¾								
" 27..	11¼@11½	10¾@11⅛	16,150	11,710	2,859	....	1,282	....	4,141	
" 30..	11⅜@11¾	11@11¼								$\frac{7}{32}$@¼d.
Decem. 4..	11⅜@11¾	11@11⅜	7,000	11,167	5,438	3,990	....	....	9,428	
" 7..	11¼@11¾	11@11⅜								
" 11..	11@11½	10¾@11	5,500	9,885	2,459	2,626	....	....	5,085	
" 14..	11@11½	10¾@11								
" 18..	11@11¼	10¾@10⅞	3,300	4,684	6,209	115	....	....	6,324	
" 21..	11½@11¾	11⅛@11⅜								
" 25..	11½@12	11¼@11½	10,700	9,290	2,704	3,543	319	....	6,566	
" 28..	11⅞@12⅛	11⅝@11⅞								

GENERAL REMARKS.

Though the foreign markets remained very dull, this crop year opened with a very firm feeling, based upon the unfavorable reports from the growing districts, noted in the previous year; early in the season, the receipts at our Ports showed a large falling off as compared with those of the season before, and at one time, (January,) prices here were 2 cents per pound above those current in Liverpool, the advanced rates having largely checked the consumption in England. This market, for much of this year, acted independently of foreign accounts, and at times, notably in January, was much excited

The falling off in receipts continued to grow larger as the year wore on, but the Liverpool market for the most part responded very feebly, owing to the low prices ruling for manufactured goods, the markets of India being stagnant, so that spinners bought only for immediate wants, and consuming much Surat and Brazil descriptions, at the same time very generally curtailing work.

In June, however, when the falling off in the receipts had reached over half-a-million bales, there set in an

1850.											
January 1..	11¾@12¼	11½@11¾	6,000	14,304	3,105	388	....	78	3,571	5/32d.	active speculative demand in Liverpool, with considerable excitement, and a decided advance in prices both there and here was realized. The spring months of 1850 were also characterized by late ungenial weather, with rains and floods, which, in connection with the short crop now being marketed, proved an important element in strengthening prices.
" 4..	12¼@12¾	11¾@12									
" 8..	12½@13	12¼@12½	12,500	4,766	1,828	1,867	....	....	3,695		
" 11..	13@13¼	12⅝@13									
" 15..	13½@14	13¼@13½	25,100	29,567	2,328	1,669	840	....	4,837		
" 19..	13¾@14¼	13½@13¾									
" 22..	13½@14	13¼@13½	21,100	3,370	....	1,831	....	....	1,831		
" 25..	13½@13¾	13@13⅜									
" 29..	13½@13¾	13⅛@13⅜	9,400	18,981	1,867	....	337	113	2,317		
February 1..	13⅜@13¾	13⅛@13⅜								5/32@3/16d.	
" 5..	13⅜@13¾	13⅛@13⅜	11,400	12,615	1,833	1,254	259	29	3,375		
" 8..	13¼@13¾	13@13¼									
" 12..	13@13½	12¾@13	8,600	17,472	2,908	1,135	74	....	4,117		
" 15..	13@13¼	12½@12¾									
" 19..	13¼@13¾	13@13¼	15,300	17,632	3,626	2,469	....	....	6,095		
" 22..	13½@14	13¼@13½									
" 26..	13@13¾	12⅞@13⅛	12,100	14,699	6,229	2,926	529	....	9,684		
March 1..	13@13½	12⅝@13								5/32@3/16d.	
" 5..	12¾@13¼	12½@12⅞	8,100	16,803	628	1,103	....	....	1,731		
" 8..	12½@12¾	12¼@12½									
" 12..	12¼@12¾	12@12⅜	7,800	7,112	4,225	1,820	897	....	6,942		
" 15..	12¼@12¾	12@12⅜									
" 19..	12¾@13½	12½@12¾	12,700	12,608	247	....	51	....	298		
" 22..	12½@13	12¼@12½									
" 26..	12¼@12¾	12@12¼	5,500	8,221	6,277	4,217	213	....	10,707		
" 29..	12½@13¼	12¼@12⅝									
April 2..	12½@13½	12¼@12¾	10,100	13,719	1,808	....	535	....	2,343	1/8@5/32d.	
" 5..	12½@13¼	12@12½									
" 9..	12½@13¼	12@12½	5,400	19,524	3,722	2,034	....	....	5,756		
" 12..	12¼@13	12@12½									
" 16..	12½@13¼	12¼@12⅝	9,200	11,220	2,561	1,365	1,358	....	5,284		
" 19..	12½@13¼	12¼@12½									
" 23..	12¾@13¼	12½@13	9,500	7,320	2,940	1,122	252	....	4,314		
" 26..	12¾@13½	12½@13									
" 30..	12¾@13½	12½@12¾	8,300	10,726	2,497	1,025	112	....	3,634		
May 3..	12¾@13½	12⅝@13								1/8d.	
" 7..	12¾@13½	12½@13	10,900	7,493	3,890	1,050	428	....	5,368		
" 10..	13@13½	12⅝@13									
" 14..	13@13¾	12¾@13⅛	16,400	6,316	3,234	439	17	....	3,690		
" 17..	13¼@13¾	12¾@13⅛									
" 21..	13¼@13¾	12¾@13⅛	13,800	8,164	5,329	1,281	910	....	7,223		

New York Statement for 1850.—*Concluded.*

1850.		Price of Mid. Fair to Fair New Orleans Liverpool Classificat'n.	Price of Mid. Fair to Fair Upland. Liverpool Classificat'n.	Sales for week.	Receipts for week.	EXPORTS FOR THE WEEK.					Rates of Freight to Liverpool.
						To Great Britain.	To France.	North of Europe.	Other Fo'n Ports	Total Exports.	
May	24..	13¼@13¾	12¾@13⅛								
"	28..	13⅛@13¾	12¾@13⅛	17,400	2,869	5,523	1,412	288	....	7,223	
"	31..	12½@13½	12½@13								
June	4..	12¾@13½	12½@13	8,200	7,403	5,126	1,188	455	101	6,870	⅛d.
"	7..	12¾@13½	12½@13								
"	11..	13@13½	12⅝@13	21,600	9,763	5,567	1,795	741	....	8,103	
"	14..	13¼@13¾	12¾@13⅛								
"	18..	13¼@14	13@13¼	21,000	6,302	9,547	....	593	....	10,140	
"	21..	13½@14	13@13⅜								
"	25..	13½@14	13⅛@13⅜	18,900	9,763	5,700	1,409	409	219	7,737	
"	28..	13½@14	13⅛@13⅜								
July	2..	13¾@14¼	13¼@13⅝	13,100	10,288	13,656	2,858	535	....	17,049	5/32d.
"	5..	13¾@14¼	13¼@13⅝								
"	9..	13¾@14¼	13¼@13⅝	5,100	8,816	6,366	....	....	....	6,366	
"	12..	13⅝@14	13¼@13⅝								
"	16..	13½@14	13¼@13⅝	13,300	7,344	1,552	1,969	1,013	....	4,534	
"	19..	13½@14	13¼@13⅝								
"	23..	13½@14	13¼@13⅝	14,100	5,776	1,455	1,626	330	....	3,411	
"	26..	13½@14	13¼@13⅝								
"	30..	13½@14	13⅛@13⅝	16,000	14,092	3,787	2,135	59	....	5,981	
August	2..	13½@14¼	13¼@13⅝								
"	6..	13½@14	13¼@13⅝	17,800	5,976	10,412	2,211	2,421	100	15,144	3/16@¼d.
"	9..	13¼@14	13@13½								
"	13..	13½@14	13@13½	16,600	2,646	6,915	1,395	959	....	9,269	
"	16..	13½@14¼	13¼@13⅝								
"	20..	13½@14¼	13¼@13⅝	19,000	4,192	8,508	1,667	2,612	910	13,697	
"	23..	13⅝@14¼	13¼@13¾								
"	27..	13¾@14¼	13½@13⅞	19,300	7,841	3,812	2,573	311	....	6,696	
"	30..	13¾@14¼	13½@13⅞								
Septem.	3..	13¾@14¼	13½@13⅞	6,900	2,697	5,961	3,543	767	....	10,271	¼d.
Average price and total sales, receipts and exports.		12.76	12.34	628,020	481,938	200,113	85,054	27,726	1,907	314,800	

GENERAL REMARKS.

Exchange.

The range for Bills on London in September was from 9¼@10 per cent. premium; in October, 9½@10½; November, 9½@10⅛ down to 8½@9; in December the quotation fell from 8½@9 to 7½@8¼, which latter was the price through January; in February the range was 8¼@9¾; March, 8¼@8¾; April, 8½@9¾; May, 9¼@10; June, 9¾@10¼; July, 9¾@10½; August, the same; closing at 10@10½ per cent. premium September 1, 1850.

LIVERPOOL STATEMENT FOR 1850.

UNITED STATES—1849–1850.

		Export	1,590,000
Stock, 1st September, 1849	155,000	Consumption	494,000
Crop	2,097,000	Stock, 1st September, 1850	168,000
Bales	2,252,000	Bales	2,252,000

Stock Jan. 1, 1850, in	Liverpool.	Gt. Britain.	France.	Continent.	Tot. Europe.
United States .... Bales.	278,000	317,000	35,000	21,000	373,000
Brazil	95,000	95,000	4,000	1,000	100,000
West Indies	2,000	2,000	2,000	1,000	5,000
East Indies	58,000	106,000	....	2,000	108,000
Egypt	35,000	38,000	5,000	17,000	60,000
Bales	468,000	558,000	46,000	42,000	646,000

CONSUMPTION.										
Tot. Europe.	Continent.	France.	Gt. Britain.	Liverpool.	IMPORT.	Liverpool.	Gt. Britain.	France.	Continent.	Tot. Europe.
1,714,000	332,000	305,000	1,077,000	1,004,000	United States	1,126,000	1,183,000	313,000	336,000	1,682,000
229,000	45,000	10,000	174,000	168,000	Brazil	171,000	172,000	7,000	48,000	203,000
36,000	8,000	23,000	5,000	5,000	West Indies	4,000	5,000	23,000	7,000	34,000
276,000	100,000	1,000	175,000	129,000	East Indies	193,000	309,000	1,000	101,000	314,000
196,000	85,000	29,000	82,000	80,000	Egypt	79,000	79,000	31,000	86,000	196,000
2,451,000	570,000	368,000	1,513,000	1,386,000	Bales	1,573,000	1,748,000	375,000	578,000	2,429,000
....	....	....	272,000	200,000	Export.					
624,000	50,000	53,000	521,000	455,000	Stock Dec. 31 Stock above,	468,000	558,000	46,000	42,000	646,000
3,075,000	620,000	421,000	2,306,000	2,041,000	Total supply, bales	2,041,000	2,306,000	421,000	620,000	3,075,000

COTTON AT LIVER

Week Ending.	Receipts.						Sales.			
	Americ'n.	E. I.	Egypt.	Brazil.	Other.	Total.	Consumption.	Speculation.	Export	Total.
Jan. 4..	20,959	......	......	955	......	21,914	13,640	11,360	60	25,060
" 11..	27,834	3,221	1,437	9,710	......	42,206	28,570	39,920	770	69,260
" 18..	21,887	......	......	......	......	21,887	28,200	12,090	770	41,060
" 25..	29,113	2,176	2,849	5,738	24	39,900	40,820	40,860	1,710	83,390
Feb. 1..	25,267	2,881	......	5,569	......	33,717	17,600	13,520	700	31,820
" 8..	29,955	4,386	1	6,084	......	40,436	27,190	17,400	1,730	46,320
" 15..	21,729	......	2,726	15	52	24,522	15,930	5,140	950	22,020
" 22..	14,424	......	1,482	7,295	......	23,201	13,020	6,110	950	20,080
Mch 1..	19,270	......	......	398	172	19,840	20,985	4,885	2,070	27,940
" 8..	13,288	......	......	606	......	13,894	18,710	3,210	3,700	25,620
" 15..	1,222	......	671	......	......	1,893	19,060	3,980	2,730	25,770
" 22..	9,747	......	......	3,435	......	13,182	19,880	3,590	7,020	30,490
" 29..	16,036	......	2,549	1,049	......	19,634	17,920	2,610	1,890	22,420
April 5..	24,651	5,254	11,629	5,508	12	47,054	25,600	14,600	3,680	43,880
" 12..	59,051	17,561	6,339	10,146	64	93,161	28,900	10,210	3,990	43,100
" 19..	34,563	8,735	236	4,746	......	48,280	40,480	12,380	8,330	61,190
" 26..	22,577	......	......	600	4	23,181	45,660	7,020	4,150	56,830
May 3..	22,017	10,596	......	2,660	......	35,273	48,460	37,570	4,610	90,640
" 10..	5,737	7,220	......	......	......	12,957	34,630	12,190	3,350	50,170
" 17..	27,773	9,055	2,919	2,196	......	41,943	30,110	15,730	2,800	48,640
" 24..	691	......	3,287	......	......	3,978	23,660	5,800	2,500	31,960
" 31..	63,892	2,055	4,481	10,947	168	81,543	40,080	21,750	3,510	65,340
June 7..	48,319	......	......	7	......	48,326	41,950	13,060	5,290	60,300
" 14..	22,711	......	......	4,313	50	27,074	16,640	3,920	4,790	25,350
" 21..	12,601	6,350	943	2,571	......	22,465	18,630	5,070	4,840	28,540
" 28..	21,610	......	1,938	......	......	23,548	32,620	22,000	10,750	65,370
July 5..	17,862	7,340	4,483	7,515	1	37,201	36,710	17,620	6,120	60,450
" 12..	29,451	6,567	5,321	......	34	41,373	47,710	28,290	9,610	85,610
" 19..	6,092	......	......	......	23	6,115	42,430	52,010	16,600	111,040
" 26..	40,056	6,715	1,302	8,994	169	57,236	17,640	13,740	8,330	39,710
Aug. 2..	34,102	4,468	......	3,571	90	42,231	33,990	70,910	11,810	116,710
" 9..	19,346	2,849	1,393	600	......	24,188	23,630	27,230	8,160	59,020
" 16..	21,996	6,351	......	9,213	1	37,561	18,280	6,800	14,210	39,290
" 23..	40,602	375	......	2,240	......	43,217	19,080	15,150	2,820	37,050
" 30..	21,294	......	1,242	5,861	103	28,500	16,260	5,270	1,680	23,210
Sept. 6..	9,054	......	4,373	3,792	......	17,219	20,170	3,290	7,380	30,840
" 13..	10,199	......	......	1,580	......	11,779	17,560	3,550	3,530	24,640
" 20..	6,352	......	1,629	1,860	172	10,013	34,280	25,980	6,660	66,920
" 27..	44,280	4,166	6,276	2,547	12	57,281	22,330	15,610	2,420	40,360
Oct. 4..	20,768	10,045	757	1,386	38	32,994	31,250	10,700	2,650	44,600
" 11..	14,762	11,060	......	2,297	72	28,191	29,760	11,150	2,550	43,460
" 18..	3,653	2,996	......	2,607	199	9,455	21,930	3,950	4,400	30,280
" 25..	12,711	518	1,860	......	......	15,089	26,740	8,290	3,360	38,390
Nov. 1..	28,100	5,084	3,813	......	106	37,103	28,410	7,190	4,180	39,780
" 8..	23,389	5,703	711	......	259	30,062	21,860	1,350	2,590	25,800
" 15..	3,853	2,506	722	1,300	......	8,381	23,300	7,960	1,480	32,740
" 22..	3,286	3,645	548	2,641	......	10,120	29,180	5,500	1,950	36,630
" 29..	7,434	10,187	1,232	6,870	11	25,734	19,190	1,720	1,400	22,310
Dec. 6..	5,361	2,350	......	1,122	967	9,800	28,630	11,530	1,190	41,350
" 13..	16,406	10,861	497	4,668	170	32,602	32,730	6,500	1,950	41,180
" 20..	31,319	14,862	1,427	2,819	501	50,928	50,700	13,680	1,280	65,660
" 27..	19,358	......	......	1,067	346	20,771	34,190	8,590	1,210	43,990
Average prices & total sales, receipts & stocks.	1,084,644	198138	83,052	152,498	3,820	1,522,152	1,436,885	722,365	217160	2,376,410

POOL. YEAR 1850.

STOCKS.			PRICES.			ACTUAL EXPORT.	CONSUMPTION.	REMARKS.
American.	Other.	Total.	Mid. Up.	Mid. Orl.	Surats.			
289,371	187,018	476,389	6¼	6⅜	4⅝	60	13,640	
296,419	192,836	489,255	6½	6⅝	4¾	770	42,210	
298,056	184,116	482,172	6½	6⅝	4¾	770	70,410	
298,809	180,733	479,542	6⅝	6⅞	5	1,710	111,230	
308,916	184,043	492,959	6⅝	6⅞	5	700	128,830	
318,451	186,027	504,478	6⅝	6¾	5	1,730	156,020	
327,980	184,140	512,120	6½	6⅝	5	950	171,950	
331,244	190,146	521,390	6⅜	6½	4⅞	950	184,970	
332,894	185,281	518,175	6⅜	6½	4⅞	2,070	205,955	
343,320	181,110	524,430	6¼	6⅜	4⅞	3,700	224,265	Bad accounts from Manchester.
329,690	174,840	504,530	6⅛	6¼	4⅞	2,730	243,725	
319,930	170,880	490,810	6	6⅛	4⅞	7,020	263,605	
320,470	170,160	490,630	5⅞	6⅛	4⅞	1,890	281,525	
327,420	183,610	511,030	6¼	6⅜	4⅞	3,680	307,125	
362,050	209,250	571,300	6⅜	6½	4⅞	3,990	336,025	
365,290	205,480	570,770	6⅝	6¾	4⅞	8,330	376,505	Deficiency in receipts, &c.
359,930	189,710	549,640	6⅝	6¾	4⅞	4,150	422,165	
350,240	181,600	531,840	6⅞	7	5⅛	4,610	470,625	
330,450	176,090	506,540	6⅞	7	5¼	3,350	505,255	
333,030	182,560	515,590	6⅞	7	5¼	2,800	535,365	
313,690	179,720	493,410	6⅞	7	5¼	2,500	559,025	
345,960	185,410	531,370	7	7⅛	5½	3,510	599,105	
362,280	170,180	532,460	7	7⅛	5½	5,290	641,055	Better feeling in Manchester.
370,920	169,480	540,400	6⅞	7	5½	4,790	657,695	
364,350	175,070	539,420	6⅞	7	5½	4,840	676,325	
356,530	163,090	519,620	7	7⅛	5½	10,750	708,945	
343,090	170,980	514,070	7⅛	7¼	5¾	6,120	745,655	
332,570	167,020	499,590	7⅜	7½	5⅞	9,610	793,365	
299,140	147,550	446,690	7⅝	7¾	6	16,600	835,795	Excitement in Manchester.
320,870	156,720	477,590	7½	7⅝	6	8,330	853,435	
324,110	150,270	474,380	7⅞	7⅞	6⅛	11,810	887,425	
323,020	143,760	466,780	7¾	7⅞	6⅛	8,160	911,055	
321,240	150,980	472,220	7⅝	7¾	6	14,210	929,335	
344,000	149,590	493,590	7½	7⅝	5⅞	2,820	948,415	
352,010	152,140	504,150	7¼	7⅜	5¾	1,680	964,675	
339,740	153,980	493,720	7¼	7⅜	5⅝	7,380	984,845	
333,940	150,470	484,410	7⅛	7¼	5⅝	3,530	1,002,405	
311,540	141,960	453,500	7½	7⅝	5¾	6,660	1,036,685	
337,340	148,691	486,031	7½	7⅝	5¾	2,420	1,059,015	
334,468	150,657	485,125	7⅝	7¾	5¾	2,650	1,090,265	
326,350	154,656	481,006	7⅝	7¾	5¾	2,550	1,120,025	
311,433	152,698	464,131	7½	7 11/16	5¾	4,400	1,141,955	
304,604	144,516	449,120	7⅝	7 13/16	5¾	3,360	1,168,695	Account of frost in Tennessee.
310,504	143,129	453,633	7½	7⅝	5¾	4,180	1,197,105	
315,463	144,382	459,845	7¼	7½	5¾	2,590	1,218,965	
301,976	141,470	443,446	7⅜	7 9/16	5¾	1,480	1,242,265	
283,702	138,734	422,436	7⅜	7 9/16	5¾	1,950	1,271,445	
276,699	151,601	428,300	7¼	7 7/16	5⅞	1,400	1,290,635	Unsettled state of things in Germany.
259,367	148,913	408,280	7½	7⅝	5⅞	1,190	1,319,265	
250,043	156,119	406,162	7 9/16	7 11/16	5⅞	1,950	1,351,995	
242,642	162,468	405,110	7 11/16	7 13/16	6	1,280	1,402,695	Better accounts from Continent.
237,850	152,631	390,481	7 11/16	7⅞	6	1,210	1,436,885	
			7.1	7.16	5.22	217,160	276,324	

1851.

Yarn as fine as No. 600, was exhibited at the Great Exhibition by the Messrs. Houldsworth of Manchester, England. (*See* years 1840 and 1841.)

Cotton mills in Great Britain employed 470,317 operatives.

Spring backward, partial overflow of the Mississippi, drought in summer, picking late.

Number of hands engaged in the cotton manufacture of Great Britain, 470,317—149,214 men, 143,268 women, and 104,437 girls; beside, 13,263 men as cotton and calico printers and 3,024 men as dyers.

18,811 power-looms in Lanarkshire, Scotland, 85 factories; 864,088 spindles; hands employed, 22,463—5,013 males, 17,450 females.

COTTON CROP OF THE UNITED STATES.

Statement and Total Amount of the Cotton Crop of the United States, for the Year ending August 31, 1851.

	Bales.	Bales.	Total.	Same period 1850.
NEW ORLEANS.				
Export—				
To Foreign Ports	844,641			
Coastwise	152,817			
Stock on hand 1st September, 1851	15,390			
		1,012,848		
Deduct—				
Stock on hand 1st September, 1850	16,612			
Received from Mobile and Montgomery, Ala	42,524			
" Florida	11,091			
" Texas	9,252			
		79,479		
			933,369	781,886
ALABAMA.				
Export—				
To Foreign Ports	321,777			
Coastwise	114,451			
Consumed in Mobile	685			
Stock on hand 1st September, 1851	27,797			
		464,710		
Deduct—				
Stock on hand 1st September, 1850	....	12,962		
			451,748	350,952
FLORIDA.				
Export—				
To Foreign Ports	70,547			
Coastwise	111,532			
Stock on hand 1st September, 1851	273			
		182,352		
Deduct—				
Stock on hand 1st September, 1850	....	1,148		
			181,204	181,344

Statement and Total Amount of the Cotton Crop of the United States, for the Year ending August 31, 1851—Concluded.

		Bales.	Bales.	Total.	Same period 1850.
TEXAS.					
Export—					
To Foreign Ports		2,261			
Coastwise		43,014			
Stock on hand 1st September, 1851		596			
			45,871		
Deduct—					
Stock on hand 1st September, 1850		....	51		
				45,820	31,263
GEORGIA.					
Export from Savannah—					
To Foreign Ports—Uplands		145,150			
" Sea Island		8,497			
Coastwise—Uplands		160,642			
Sea Island		3,145			
Stock in Savannah, 1st September, 1851		4,500			
" Augusta, 1st September, 1851		29,511			
			351,445		
Deduct—					
Stock in Savannah and Augusta, 1st September, 1850		....	29,069		
				322,376	343,635
SOUTH CAROLINA.					
Export from Charleston—					
To Foreign Ports—Uplands		254,442			
" Sea Island		13,576			
Coastwise—Uplands		138,429			
Sea Island		2,210			
		408,657			
Export from Georgetown—					
To New York	1,812				
Stock in Charleston, 1st September, 1851	10,953				
		12,765			
			421,422		
Deduct—					
Stock in Charleston, 1st September, 1850		30,698			
Received from Savannah		3,649			
			34,347		
				387,075	384,265
NORTH CAROLINA.					
Export—					
Coastwise		....	....	12,928	11,861
VIRGINIA.					
Export—					
Coastwise, and Manufactured (taken from the ports)		20,320			
Stock on hand, 1st September, 1851		620			
			20,940		
Deduct—					
Stock on hand 1st September, 1850		....	1,000		
				19,940	11,500
Received here by New York and Erie Canal		....	....	797	
Total crop of the United States		....	....	2,355,257	2,096,706

Increase from last year .... bales.	258,551
Decrease from year before	373,339

Export to Foreign Ports, from September 1, 1850, *to August* 31, 1851.

FROM	To Great Britain.	To France.	To North of Europe.	Other F'n Ports.	Total.
New Orleans........Bales	582,373	130,362	47,786	84,120	844,641
Mobile	249,897	45,460	6,084	20,336	321,777
Florida	56,167	7,805	6,575	....	70,547
Texas	....	....	2,261	....	2,261
Georgia	137,143	11,826	2,993	1,685	153,647
South Carolina	203,970	25,608	13,159	25,281	268,018
North Carolina	....	....	....	....	....
Virginia	....	....	....	....	....
Baltimore	206	....	200	75	481
Philadelphia	2,691	....	....	....	2,691
New York	184,815	80,297	48,713	7,970	321,795
Boston	1,003	....	1,721	128	2,852
Grand total	1,418,265	301,358	129,492	139,595	1,988,710
Total last year	1,106,771	289,627	72,156	121,601	1,590,155
Increase	311,494	11,731	57,336	17,994	398,555

Growth.

Total crop of 1850–1 ......bales.	2,355,257	Total crop of 1836-7 ......bales.	1,422,930
" 1849–50	2,096,706	" 1835–6	1,360,725
" 1848–9	2,728,596	" 1834–5	1,254,328
" 1847–8	2,347,634	" 1833–4	1,205,394
" 1846–7	1,778,651	" 1832–3	1,070,438
" 1845–6	2,100,537	" 1831–2	987,477
" 1844–5	2,394,503	" 1830–1	1,038,848
" 1843–4	2,030,409	" 1829–30	976,845
" 1842–3	2,378,875	" 1828–9	857,744
" 1841–2	1,683,574	" 1827–8	720,593
" 1840–1	1,634,945	" 1826–7	957,281
" 1839–40	2,177,835	" 1825–6	720,027
" 1838–9	1,360,532	" 1824–5	569,249
" 1837–8	1,801,497	" 1823–4	509,158

Consumption.

Total crop of the United States, as before stated ....bales.			2,355,257
Add—Stocks on hand at the commencement of the year, September 1, 1850—			
In the Southern ports		91,754	
In the Northern ports		76,176	
			167,930
Makes a supply of			2,523,187
Deduct therefrom—The Export to Foreign ports	1,988,710		
Less Foreign included	1,077		
		1,987,633	
Stocks on hand, September 1, 1851—			
In the Southern ports	89,044		
In the Northern ports	39,260		
		128,304	
Burnt at New York, Boston and Baltimore		3,142	
			2,119,079
Taken for home use ....bales.			404,108

Quantity consumed by and in the hands of Manufacturers, North of Virginia.

1850-1 ....bales.	404,108	1837-8 ....bales.	246,063
1849-50	487,769	1836-7	222,540
1848-9	518,039	1835-6	236,733
1847-8	531,772	1834-5	216,888
1846-7	427,967	1833-4	196,413
1845-6	422,597	1832-3	194,412
1844-5	389,006	1831-2	173,800
1843-4	346,744	1830-1	182,142
1842-3	325,129	1829-30	126,512
1841-2	267,850	1828-9	118,853
1840-1	297,288	1827-8	120,593
1839-40	295,193	1826-7	149,516
1838-9	276,018		

NOTE.—It will be seen that we have materially reduced our estimate of the amount of cotton consumed the past year in the States south and west of Virginia—the capacity of the mills has been very nearly the same as before, but the high prices of the raw material for the greater part of the season, and the low rates obtained for the manufactured article, have rendered the business unprofitable. The following estimate is from a judicious and careful observer at the South, of the quantity so consumed, and not included in the receipts: Thus, in—

	Mills.	Spindles.	Quantity consumed.	
North Carolina..............	30	......	13,000 bales,	of 400 lb.
South Carolina..............	16	36,500	10,000 "	"
Georgia	36	51,400	13,000 "	"
Alabama	10	12,580	4,000 "	of 500 lb.
Tennessee	30	36,000	8,000 "	"
On the Ohio, &c.............	30	100,000	12,000 "	"

Total to September 1, 1851	60,000 bales.
" " 1850	107,500 "
" " 1849	110,000 "
" " 1848	75,000 "

To which should be added the stocks in the interior towns, etc., the quantity burnt in the interior, and that lost on its way to market these, added to the crop as given above, received at the shipping ports, will show very nearly the amount raised in the United States the past season—say, in round numbers, 2,450,000 bales.

During the year just closed, there have been received here, chiefly, it is believed, from Tennessee, 797 bales by way of the New York and Erie Canal, which we have added in another place to the crop of the country. This route, however, is not a favorite one, and no further supplies of moment are expected.

It may be remarked in this connection, that some of the cotton received overland at Philadelphia and Baltimore is doubtless unaccounted for elsewhere, not being counted in the receipts at New Orleans, but as we have of late years omitted this item from the crop, in deference to the views of judicious friends, it is not now added, though it may be advisable to introduce it hereafter.

The quantity of new cotton received at the shipping ports up to the 1st inst., amounted to about 3,200 bales, against 255 bales last year.

The shipments given in this statement from Texas, are those by sea only; a considerable portion of the crop of that State finds its way to market via Red River, and is included in the receipts at New Orleans.

ANNUAL REVIEW.

From the New Orleans Price Current, 1850—51.

It is well known that in this leading branch of our commerce, the season opened with high hopes on the part of both producers and dealers. The previous year had closed upon greatly enhanced prices which had given large profits to shippers, and this success, together with calculations of another short crop, stimulated speculations to an imprudent degree, and the result has been a reaction more disastrous than any that has occurred in the cotton trade since 1825. For a month succeeding, the rates fluctuated between $10\frac{5}{8}$ to 11 cents, when in early May the market was again affected by the character of the foreign advices, and also by the large increase in the receipts at the ports, as compared with the previous year; and as nearly every circumstance that has arisen since, has been of a nature to increase the depression, there has been a constant yielding of prices, until they reached $6\frac{3}{4}$ cents for middling Louisianas and Mississippis, or a decline on this description of nearly 7 cents per pound from the highest point; being more than fifty per cent. We have thus rapidly sketched the course of the market during a season of extraordinary vicissitudes, and such a one we hope never to witness again. In glancing at the peculiarities of the season, it may be safely remarked that its prominent feature has been an under-estimate of the production. This, as we have already intimated, led to the opening of the market at unfortunately high prices, which, under speculative action, were subsequently carried to a higher point than they have reached since 1839. These under-estimates were to a greater or less extent general, and we think it may be safely asserted that a large majority placed the crop at or under 2,200,000 bales, while the bulk of the business during the first six or seven months of the season was done upon a basis of 2,100,000 to 2,150,000 bales. The estimates of very few parties were beyond what the actual crop is likely to be, and these were looked upon as so extravagant that their opinion provoked discussion and animadversion to a degree that has given them wide spread notoriety. In viewing the causes of this astonishing reaction, the leading ones, of course, are the under-estimates of the crop, and the consequent elevation of prices to what has proved to be an extravagant point. But as a collateral one, growing out of these, we may mention that the entire or partial stoppage of many of our home mills, owing to the high prices of the raw material, and excessive stocks of manufactured articles on hand, threw an undue proportion of the supply on the European markets. Thus Great Britain alone has not only taken the whole excess of our receipts over those of last year, but nearly 100,000 bales more, that with moderate prices, would have been consumed in the United States.

To Great Britain, therefore, the crop has been equal to one of about 2,450,000 bales, while at the same time there has been a material increase in her imports from Brazil, Egypt, and the East Indies. It is understood that there was considerable increase in the breadth of land planted, but an unusually cold and backward spring retarded the growth of the plant; and it had made comparatively little progress up to the early part of May, when a favorable change in the character of the weather gave an impulse to vegetation. From this time up to the 1st of July the accounts from the country, with some exceptions, were favorable, though from the uplands there were some complaints of a lack of sufficient rain. The plant generally, though small, was said to look healthy, and to give good promise; besides which, the crops were usually "clean" the very lack of rain complained of, favored cultivation by preventing any excessive growth of grass and weeds.

The semi-weekly Price and Weekly Sales and Receipts at New York, Weekly Exports from New York and Rates of Freight to Liverpool 1st of each month, for the Crop Year ending September 1, 1851.

1850.	Price of Mid. Fair to Fair New Orleans Liverpool Classification.	Price of Mid. Fair to Fair Upland, Liverpool Classification.	Sales for week.	Receipts for week.	EXPORTS FOR THE WEEK. To Great Britain.	To France.	North of Europe.	Other Fo'n Ports	Total Exports.	Rates of Freight to Liverpool.
Septem. 6..	14@14½	13⅝@14								$\frac{3}{16}$@$\frac{7}{32}$d.
" 10..	14¼@14¾	14@14¼	17,500	287	....	....	....	....	....	
" 13..	14⅜@14¾	14@14⅜								
" 17..	14½@15	14⅛@14½	6,000	7,959	11,199	3,293	7,040	987	22,519	
" 20..	14½@15	14¼@14½								
" 24..	14½@15	14¼@14½	9,200	5,367	5,654	2,075	1,493	....	9,222	
" 27..	14½@15	14¼@14½								
October 1..	14½@15	14¼@14½	6,300	3,690	4,259	1,130	196	....	5,585	3-16d.
" 4..	14½@15	14¼@14½								
" 8..	15@15½	14½@14⅞	13,000	5,552	799	4,676	....	119	5,594	
" 11..	14¾@15¼	14½@14⅞								
" 15..	14¾@15¼	14½@14⅞	8,500	4,350	5,249	....	584	....	5,833	
" 18..	14¾@15½	14½@15								
" 22..	14¾@15½	14½@14¾	6,800	4,751	2,248	3,288	1,361	628	7,525	
" 25..	14¾@15½	14¼@14½								
" 29..	15@15½	14½@14¾	12,200	9,634	1,147	2,275	381	....	3,803	
Novem. 1..	15@15½	14½@14¾								⅛d.
" 5..	14¾@15¼	14½@14¾	8,100	2,893	1,757	973	....	....	2,730	
" 8..	14¾@15¼	14¼@14½								
" 12..	14¾@15¼	14⅜@14⅝	6,250	2,210	948	1,871	354	134	3,307	
" 15..	14¾@15¼	14⅜@14⅝								
" 19..	14¾@15¼	14⅜@14⅝	5,800	11,365	2,936	1,611	855	....	5,402	
" 22..	14¾@15¼	14¼@14½								
" 26..	14¾@15¼	14¼@14½	4,100	13,355	1,659	3,878	50	200	5,787	
" 29..	14¼@15	14@14¼								
Decem. 3..	14¼@15	14@14¼	5,200	10,700	142	2,659	....	....	2,801	⅛d.
" 6..	14¼@14¾	14@14¼								
" 10..	14¼@14¾	13¾@14	8,100	5,738	1,054	494	687	....	2,235	
" 13..	14@14¼	12¾@13								
" 17..	14@14¼	12¾@13	5,500	8,290	828	4,622	....	....	5,450	
" 20..	14@14½	13@13¼								
" 24..	14¼@14½	13½@13¾	8,600	10,244	562	....	....	....	562	
" 27..	14½@14⅝	13⅝@13⅞								

GENERAL REMARKS.

Early in September, 1850, heavy rain storms prevailed at the South, and considerable injury to the crop was reported, which had some effect upon prices, although the foreign accounts were unfavorable. Subsequently, however, upon the receipt of these advices at Liverpool, that market became excited, and higher rates ruled there as well as here.

On October 26th and 27th, frosts occurred in Georgia, Alabama, &c.; which gave a further stimulus to the market. In the mean time political affairs on the Continent had become unsettled, and the European markets, for a time, did not respond to the advanced rates ruling here, and prices yielded in this market. In December, German affairs wore a more pacific aspect and a buoyant feeling prevailed with higher values ruling toward the last of the month.

On the 1st of January the account of stock was taken, as usual, in Liverpool, and was found to be about 70,000 bales larger than the previous running account; this had a depressing influence, the effect of which was felt there and here for two or three months.

In February, the receipts at our ports

Decem. 31..	14½@14¾	13⅞@14⅛	9,100	8,977	2,117	4,162	519	117	6,915	
1851.										
January 3..	14½@14¾	13⅞@14⅛								⅛@5-32d
" 7..	14¾@15¼	14⅛@14⅜	10,000	12,142	637	1,734	53	198	2,622	
" 10..	14¾@15¼	14⅛@14⅜								
" 14..	14½@15⅛	14@14¼	9,000	15,090	3,486	1,352	129	....	4,967	
" 17..	14¾@15⅛	14@14¼								
" 21..	14¾@15⅛	14@14¼	11,000	6,138	437	1,953	1,034	....	3,424	
" 24 .	14½@15	13⅞@14⅛								
" 28..	14½@15	13⅝@13⅞	3,000	10,151	3,231	1,686	667	....	5,584	
" 31..	14¼@14¾	13½@13⅞								
February 4..	13¾@14¼	13@13⅜	5,200	7,435	1,533	3,713	208	....	5,454	⅛d.
" 7..	14@14½	13¼@13½								
" 11..	13¾@14½	13⅛@13½	5,200	20,246	3,210	835	871	242	5,158	
" 14..	13¾@14¼	13@13⅜								
" 18..	13¼@14	12½@12⅞	10,200	14,835	1,755	....	1,495	....	3,250	
" 21..	13@13½	12@12½								
" 25..	12½@13¼	11¾@12¼	7,800	8,677	5,881	762	....	....	6,643	
" 28..	11¾@12¾	11@11½								
March 4..	11½@12½	10¾@11⅜	10,500	19,339	1,181	2,950	2,043	544	6,718	¼@5-16d
" 7..	12½@13½	11⅝@12¼								
" 11..	12¾@13½	11⅞@12⅜	15,100	7,209	1,944	....	842	419	3,205	
" 14..	12¾@13¾	11⅞@12⅜								
" 18..	12¾@13¾	11⅞@12⅜	15,000	14,029	8,744	2,299	2,774	304	14,121	
" 21..	12¾@13¾	11⅞@12⅜								
" 25..	12¾@13¾	11⅞@12⅜	17,000	11,572	7,512	779	561	....	8,852	
" 28..	12½@13¾	11⅞@12⅜								
April 1..	12½@13½	11¾@12¼	8,500	14,442	9,842	....	1,184	....	11,026	3-16@¼d
" 4..	12¾@13¾	11⅞@12½								
" 8..	13@14	12@12½	21,300	8,031	6,824	1,191	918	122	9,055	
" 11..	12½@13¾	11⅞@12⅜								
" 15..	12½@13¾	11⅞@12⅜	11,454	3,638	10,620	100	1,253	....	11,973	
" 18..	12½@13¾	11¾@12⅜								
" 22..	12¼@13½	11½@12¼	12,300	7,668	2,142	432	1,970	1,255	5,799	
" 25..	12@13¼	11@11¾								
" 29..	11½@13	10⅞@11⅜	12,100	15,739	9,400	3,034	1,063	154	13,651	
May 2..	11¾@13	11@11⅜								⅛@5-32d
" 6..	12@13½	11⅜@11⅝	18,100	6,118	7,870	1,152	2,352	568	11,942	
" 9..	11@12¾	10¾@11⅛								
" 13..	11@12¾	10¾@11¼	6,000	11,791	5,341	1,243	291	....	6,875	
" 16..	11@12	10¼@10¾								
" 20..	11@12	10½@11	12,300	8,313	2,784	1,458	907	....	5,149	

were large, the crop estimates advanced, and with dull foreign markets prices here rapidly receded.

In March, the receipts fell off and the market rallied somewhat. The weather in early May was very unfavorable for the plant, with frosts in some districts, and prices went up ⅜@½ cent; toward the latter part of this month, however, the immense receipts at our ports, and the probabilities that hitherto the crop estimates had been much too low, caused an uneasy feeling, a panic prevailed in Liverpool, and prices gave way there and here.

Prices in June having touched low points in Liverpool, that market and this became more steady for a time, but in July there was again much depression abroad, owing to the continued free receipts at the shipping ports, and accumulation of stock in Liverpool, now larger than before in two years; but in August there was an improved demand, and with considerable activity, both on the other side and here, there was at the close some advance established on the previous quotations.

New York Statement for Year 1851—*Concluded.*

1851.		Price of Mid. Fair to Fair New Orleans Liverpool Classification.	Price of Mid. Fair to Fair Upland, Liverpool Classification.	Sales for week.	Receipts for week.	EXPORTS FOR THE WEEK.					Rates of Freight to Liverpool.
						To Great Britain.	To France.	North of Europe.	Other Fo'n Ports	Total Exports.	
May	23..	10¾@11¾	10¼@10¾								
"	27..	10¾@11½	10@10⅝	5,800	6,909	11,095	1,832	1,849	....	14,776	
"	30..	10½@11¼	10@10½								
June	3..	10½@11¼	9¾@10¼	9,200	9,044	2,486	1,015	976	....	4,477	7-32d.
"	6..	10@11	9@9¾								
"	10..	10½@11¼	9½@10¼	7.800	9,889	4,103	1,371	984	211	6,669	
"	13..	10½@11¼	9½@10¼								
"	17..	10½@11¼	9½@10¼	5,500	5,293	8,222	548	1,085	934	10,789	
"	20..	10½@11¼	9½@10¼								
"	24..	10¾@12	9¾@10¼	9,700	5,229	3,633	68	291	....	3,992	
"	27..	10¾@12	9¾@10¼								
July	1..	10¾@12	9½@10¼	8,600	7,737	1,855	630	1,126	....	3,611	⅛@3-16d
"	4..	10¾@12	9½@10¼								
"	8..	10¾@12	9½@10	6,100	3,383	1,673	....	373	216	2,262	
"	11..	10½@11¾	9¼@9¾								
"	15..	10½@11¾	9@9¾	9,300	3,302	765	562	1,668	269	3,264	
"	18..	10¼@11½	9@9½								
"	22..	10@11	8¾@9½	7,100	12,275	1,885	86	629	....	2,600	
"	25..	9¾@10¾	8½@9⅜								
"	29..	9⅞@10⅞	8¾@9½	10,400	10,489	1,971	1,944	815	....	4,730	
August	1..	9@9⅞	8¾@9¼								⅛d.
"	5..	9¼@10¼	8½@9	5,700	3,830	2,097	1,503	1,832	....	5,432	
"	8.	9¼@10½	8¾@9								
"	12..	9@10½	8¾@9¼	10,175	3,023	2,453	863	1,939	....	5,255	
"	15..	9½@10¾	8⅞@9⅜								
"	19..	9½@10¾	8⅞@9⅜	7,700	2,883	3,306	897	376	....	4,579	
"	22..	9¾@10¾	9@9½								
"	26..	9½@10¾	9@9½	11,500	6,131	559	3,813	518	349	5,239	
"	29..	10@11	9¼@9½								
Septem.	2..	10½@11	9½@10	6,700	2,360	1,482	1,485	117	....	3,084	⅛d.
Average prices and total sales, receipts and exports.		12.99	12.14	482,579	429,742	184,517	80,297	48,713	7,970	321,497	

GENERAL REMARKS.

Exchange.

Sixty days' bills on London were steady through September at 10@10½ per cent. premium; in October, 10¼@10¾; in November the quotation fell from 10¼@10¾ down to 9@9½; in December it advanced again from 9½@9⅞ to 10@10⅝; in January rates were steady at 10@10½, declining in February to 9⅝@10; the range in March was from 9⅝ to 10⅜; in April, 9¾@10½; in May, 10¼@10⅝; in June, 10¼@10¾; in July, 9¾@10½; and in August, 9½@10¼.

LIVERPOOL STATEMENT FOR 1851.

UNITED STATES, 1850–1851.

Stock 1st September, 1850	168,000	Export	1,989,000
Crop	2,355,000	Consumption	406,000
		Stock 1st Sept., '51.	128,000
Bales	2,523,000	Bales	2,523,000

CONSUMPTION.										
Tot. Europe.	Continent.	France.	Gt. Britain.	Liverpool.		Liverpool.	Gt. Britain.	France	Continent.	Tot. Europe.
					Stock Jan. 1, 1851, in					
					United States	261,000	273,000	43,000	25,000	341,000
					Brazil	69,000	69,000	1,000	4,000	74,000
					West Indies	1,000	1,000	2,000	......	3,000
					East Indies	92,000	143,000	....	3,000	146,000
					Egypt	32,000	35,000	7,000	18,000	60,000
					Bales	455,000	521,000	53,000	50,000	624,000
					IMPORT.					
1,959,000	363,000	324,000	1,272,000	1,220,000	United States	1,339,000	1,397,000	297,000	361,000	1,903,000
154,000	30,000	8,000	116,000	115,000	Brazil	109,000	109,000	9,000	32,000	137,000
18,000	5,000	8,000	5,000	5,000	West Indies	7,000	8,000	10,000	7,000	25,000
301,000	104,000	3,000	194,000	152,000	East Indies	233,000	326,000	9,000	103,000	335,000
186,000	81,000	30,000	75,000	75,000	Egypt	62,000	64,000	32,000	88,000	183,000
2,618,000	583,000	373,000	1,662,000	1,567,000	Bales	1,750,000	1,904,000	357,000	591,000	2,583,000
....	....	....	269,000	214,000	Export.					
589,000	58,000	37,000	494,000	424,000	Stock Dec. 31. Stock above,	455,000	521,000	53,000	50,000	624,000
3,207,000	641,000	410,000	2,425,000	2,205,000	Total supply, bales	2,205,000	2,425,000	410.000	641,000	3,207,000

COTTON AT LIVER

Week Ending.	Receipts.						Sales.			
	American	E. I.	Egypt.	Brazil.	Other.	Total.	Consumption.	Speculation.	Export.	Total.
Jan. 10..	13,137	8,662	773	1,707	74	24,353	25,890	1,590	1,280	28,760
" 17..	17,653	2,988	......	1,453	1	22,095	18,700	1,920	480	21,100
" 24..	30,386	......	180	2,622	218	33,406	22,150	1,250	230	23,630
" 31..	8,444	......	2	1,037	89	9,572	22,960	2,290	1,290	26,540
Feb. 7..	....	......	......	....	......	20,947	20,620	1,720	1,010	23,350
" 14..	10,963	3,144	1,324	....	27	15,458	22,410	1,390	900	24,700
" 21..	29,934	9,160	983	2,841	54	42,972	27,390	5,430	790	33,610
" 28..	9,159	1,958	100	2,112	75	13,404	27,100	9,810	2,550	39,460
Mch. 7..	47,990	......	4,325	2,884	72	55,271	27,240	6,610	1,790	35,640
" 14..	54,530	6,484	750	2,419	128	64,311	26,940	2,990	1,250	31,180
" 21..	13,326	7,051	166	....	......	20,543	30,350	10,100	2,180	42,630
" 28..	48,476	10,897	837	43	32	60,285	31,390	14,360	1,960	47,710
April 4..	49,028	3,100	......	6,420	38	58,586	29,670	4,420	3,290	37,380
" 11..	17,885	......	83	520	108	18,596	26,110	1,600	890	28,600
" 17..	22,888	......	......	1,250	......	24,138	21,650	2,500	1,250	25,400
" 25..	39,074	13,734	7,108	5,055	29	65,000	23,010	2,700	3,060	28,770
May 2..	6,184	......	......	....	57	6,241	23,020	3,510	2,520	29,050
" 9..	92,648	5,430	2,658	2,256	467	103,459	26,860	3,010	2,830	32,700
" 16..	68,531	200	......	827	76	69,434	33,200	8,120	4,430	45,750
" 23..	78,872	11,929	3,649	7,045	16	101,511	31,730	4,860	4,680	41,270
" 30..	15,079	......	......	....	13	15,092	37,130	4,950	9,000	51,080
June 6..	18,541	......	......	100	......	18,641	34,870	4,630	5,730	45,230
" 13..	70,898	901	8,382	1,002	332	81,515	37,620	3,200	5,990	46,810
" 20..	62,093	4,277	6,185	5,982	5	78,542	37,230	8,010	8,360	53,600
" 27..	18,023	......	253	4,243	20	22,539	25,930	1,450	6,970	34,350
July 4..	17,614	......	......	426	25	18,065	24,190	1,130	5,350	30,670
" 11..	25,279	1,752	2,821	1,262	407	31,521	26,470	1,300	6,460	34,230
" 18..	73,047	......	3,934	1,477	22	78,480	31,490	470	7,250	39,210
" 25..	33,319	......	876	1,225	136	35,556	35,810	1,480	8,330	45,620
Aug. 1..	35,352	8,545	......	3,392	25	47,314	49,290	9,730	10,750	69,770
" 8..	26,642	......	1,431	953	64	29,090	41,430	8,580	9,180	59,190
" 15..	13,521	......	2,936	3,198	132	19,787	28,730	7,830	10,530	47,090
" 22..	26,388	20,562	884	3,711	36	51,581	41,210	10,270	16,500	67,980
" 29..	24,128	8,801	955	....	6	33,890	34,850	14,290	8,130	57,270
Sep. 5..	2,481	5,222	......	....	76	7,779	24,585	3,585	8,400	36,570
" 12..	2,605	......	......	....	15	2,620	30,845	11,985	8,810	51,640
" 19..	9,067	2,703	150	1,679	85	13,684	27,170	2,320	7,370	36,860
" 26..	17,491	13,181	2,021	6,276	42	39,011	25,990	3,580	9,790	39,360
Oct. 3..	6,497	9,234	1,496	2,022	68	19,317	22,930	2,900	7,850	33,680
" 8..	10,825	7,205	1,240	4,258	....	23,528	24,170	1,820	4,000	29,990
" 17..	13,941	18,428	801	2,788	45	36,003	23,970	680	4,960	29,610
" 24..	4,926	2,783	281	4,590	36	12,616	33,010	1,960	5,970	40,940
" 31..	....	1,953	1,355	1,100	34	4,442	38,550	1,850	5,700	46,100
Nov. 7..	3,424	4,607	1,156	....	......	9,187	48,000	5,880	4,940	58,820
" 14..	3,565	......	1,517	....	......	5,082	49,830	15,550	5,260	70,640
" 21..	10,943	2,754	......	4,248	......	17,945	28,920	6,960	4,770	40,650
" 28..	28,801	21,611	1,222	2,891	732	55,257	39,540	17,390	2,310	59,240
Dec. 5..	4,576	4,589	756	....	......	9,921	29,580	8,710	1,080	39,370
" 12..	25,543	4,264	682	2,584	......	33,073	24,120	6,380	1,280	31,780
" 19..	22,828	......	909	1,183	......	24,920	19,510	6,120	2,400	28,030
" 24..	19,960	......	593	3,214	......	23,767	26,110	5,920	2,330	34,360
Average prices & total sales, receipts & stocks.	1,346,505	227604	65,674	108,295	3,917	2,772,942	1,521,470	259,120	244410	2,025,000

* Circular missing for this week.

POOL. YEAR 1851.

STOCKS.			PRICES.			ACTUAL EXPORT.	CON-SUMPTION.	REMARKS.
Amer'n.	Other.	Total.	Mid. Up.	Mid. Orl.	Dhol.			
255,601	196,461	452,062	$7\frac{5}{8}$	$7\frac{3}{4}$	6	1,280	25,890	
258,644	196,333	454,977	$7\frac{3}{8}$	$7\frac{9}{16}$	$5\frac{7}{8}$	480	44,590	Under estimated Stock in Liverpool.
273,880	192,123	466,008	$7\frac{3}{8}$	$7\frac{9}{16}$	$5\frac{7}{8}$	230	66,740	
263,744	187,579	451,323	$7\frac{1}{4}$	$7\frac{3}{8}$	$5\frac{3}{4}$	1,290	89,700	
....	....	450,640	7	$7\frac{1}{4}$	$5\frac{3}{4}$	1,010	110,320	
257,540	185,238	442,778	$6\frac{3}{4}$	7	$5\frac{3}{8}$	900	132,730	
266,914	190,666	457,580	$6\frac{7}{8}$	$7\frac{1}{8}$	$5\frac{3}{8}$	790	160,120	
254,103	187,231	441,334	$7\frac{1}{4}$	$7\frac{5}{16}$	$5\frac{1}{2}$	2,550	187,220	Better demand.
282,913	184,712	467,625	7	$7\frac{5}{16}$	$5\frac{1}{2}$	1,790	214,460	
317,413	186,333	503,746	$6\frac{13}{16}$	$7\frac{1}{16}$	$5\frac{1}{2}$	1,250	241,400	
306,619	185,140	491,759	7	$7\frac{1}{4}$	$5\frac{1}{2}$	2,180	271,750	
332,385	186,309	518,694	$7\frac{1}{8}$	$7\frac{3}{8}$	$5\frac{1}{2}$	1,960	303,140	
358,203	186,117	544,320	7	$7\frac{1}{4}$	$5\frac{3}{8}$	3,290	332,810	
356,908	178,952	535,860	$6\frac{3}{4}$	7	$5\frac{3}{8}$	890	353,920	Smaller demand.
361,966	175,192	537,158	$6\frac{5}{8}$	$6\frac{7}{8}$	$5\frac{3}{8}$	1,250	380,570	
381,380	194,708	576,088	$6\frac{1}{4}$	$6\frac{1}{2}$	$5\frac{1}{4}$	3,060	403,580	
367,434	189,355	556,789	$5\frac{7}{8}$	$6\frac{1}{8}$	$5\frac{1}{4}$	2,520	426,600	
411,588	190,141	601,729	$5\frac{5}{8}$	$5\frac{7}{8}$	$5\frac{1}{8}$	2,830	453,460	
450,239	185,874	636,113	$5\frac{5}{16}$	$5\frac{9}{16}$	5	4,430	486,660	
502,381	198,833	701,214	5	$5\frac{1}{8}$	$4\frac{3}{4}$	4,680	518,390	Heavy receipts U.S. Ports.
479,120	191,146	670,266	5	$5\frac{1}{4}$	$4\frac{5}{8}$	9,000	555,520	
465,691	182,616	648,307	5	$5\frac{1}{4}$	$4\frac{3}{8}$	5,730	590,390	
501,009	185,203	686,212	5	$5\frac{1}{4}$	$4\frac{3}{8}$	5,990	628,010	
529,582	189,582	719,164	$5\frac{1}{8}$	$5\frac{3}{8}$	$4\frac{3}{8}$	8,360	665,240	
522,825	185,978	708,803	5	$5\frac{1}{4}$	$4\frac{1}{2}$	6,970	691,190	
518,289	179,109	697,398	$4\frac{3}{4}$	5	$4\frac{1}{4}$	5,350	715,360	
516,178	179,802	695,980	$4\frac{5}{8}$	$4\frac{13}{16}$	$4\frac{1}{4}$	6,460	741,830	
557,065	178,664	735,729	$4\frac{1}{2}$	$4\frac{3}{4}$	$4\frac{1}{4}$	7,250	773,320	
554,474	172,671	727,145	$4\frac{1}{4}$	$4\frac{3}{8}$	4	8,330	809,130	Small manufacturing demand.
538,886	175,533	714,419	$4\frac{3}{8}$	$4\frac{9}{16}$	$3\frac{3}{4}$	10,750	858,420	
524,578	168,321	692,899	$4\frac{7}{16}$	$4\frac{9}{16}$	$3\frac{3}{4}$	9,180	899,850	
505,419	168,007	673,426	$4\frac{7}{16}$	$4\frac{3}{8}$	$3\frac{3}{4}$	10,530	928,580	
483,747	183,550	667,297	$4\frac{3}{4}$	5	$3\frac{3}{4}$	16,500	969,790	
474,571	183,632	658,203	5	$5\frac{1}{8}$	$3\frac{3}{4}$	8,130	1,004,640	
449,351	172,698	622,049	$4\frac{7}{8}$	$5\frac{1}{8}$	$3\frac{7}{8}$	8,400	1,029,225	
420,616	162,893	583,509	5	$5\frac{3}{16}$	$3\frac{7}{8}$	8,810	1,060,070	
404,143	158,510	562,653	$4\frac{15}{16}$	$5\frac{1}{8}$	4	7,370	1,087,240	
394,594	171,290	565,884	$4\frac{15}{16}$	$5\frac{1}{8}$	4	9,790	1,113,230	
376,481	177,940	554,421	$4\frac{13}{16}$	5	4	7,850	1,136,160	
365,176	184,603	549,779	$4\frac{13}{16}$	5	4	4,000	1,160,330	Royal visit to Liverpool.
357,647	199,205	556,852	$4\frac{15}{16}$	$4\frac{7}{8}$	$3\frac{7}{8}$	4,960	1,184,300	(holiday market.)
334,353	196,135	530,488	$4\frac{1}{2}$	$4\frac{15}{16}$	$3\frac{3}{4}$	5,970	1,217,310	
298,173	192,507	490,680	$4\frac{1}{2}$	$4\frac{11}{16}$	$3\frac{3}{4}$	5,700	1,255,860	
260,447	186,480	446,927	$4\frac{5}{8}$	$4\frac{3}{4}$	$3\frac{5}{8}$	4,940	1,303,860	
224,652	172,267	396,919	$4\frac{3}{4}$	$4\frac{7}{8}$	$3\frac{5}{8}$	5,260	1,353,690	
210,505	170,669	381,174	$4\frac{3}{4}$	$4\frac{7}{8}$	$3\frac{5}{8}$	4,770	1,382,610	
206,826	187,755	394,581	$4\frac{15}{16}$	$5\frac{1}{16}$	$3\frac{3}{4}$	2,310	1,422,150	
190,162	183,680	373,842	$4\frac{13}{16}$	$4\frac{15}{16}$	$3\frac{3}{4}$	1,080	1,451,730	
196,495	185,027	381,522	$4\frac{11}{16}$	$4\frac{13}{16}$	$3\frac{3}{4}$	1,280	1,475,350	French Revolution.
201,403	183,129	384,532	$4\frac{9}{16}$	$4\frac{3}{4}$	$3\frac{5}{8}$	2,400	1,495,300	
199,273	180,586	379,859	$4\frac{5}{8}$	$4\frac{13}{16}$	$3\frac{5}{8}$	2,330	1,521,470	
			5.51	5.67	4.44	235,230	29,832.74	

NEW ORLEANS—SEASON 1850–51.

1850.	Receipts.	EXPORTS.				Stock.	Middling.	London Exchange.	Liverpool Freights.
		Great Britain.	Continent.	North.	Total Exports.				
Sep. 7..	1,943	3,950	....	14	3,964	14,591	12½@12¾	8½@9½ pr.	13/32@7/16d.
" 14..	2,698	......	803	1	804	16,485	12½@13		
" 21..	4,914	4,919	....	82	5,001	16,898	12½@13		
" 28..	8,992	....	......	406	406	24,984	12¼@12¾		
Oct. 5..	13,478	5,706	2,757	....	8,463	29,999	12¾@13¼	8½@9½ pr.	⅜@13-32d
" 12..	17,546	2,396	1,245	1,483	5,124	42,421	13⅜@13⅝		
" 19..	22,445	1,779	2,247	1,817	5,843	59,023	13⅛@13⅜		
" 26..	32,218	9,163	7,237	2,721	19,121	71,120	13⅜@13½		
Nov. 2..	23,729	1,503	2,779	3,343	7,625	87,224	13¼@13⅜	7@8 prem.	11-32@⅜d
" 9..	38,579	8,606	8,855	2,981	20,442	105,697	13¼@13⅜		
" 16..	21,654	12,225	5,232	5,217	22,674	104,677	13@13¼		
" 23..	31,354	13,617	8,093	2,112	23,822	111,309	13⅛@13⅜		
" 30..	33,853	8,180	4,666	2,585	15,431	129,931	13⅛@13⅜		
Dec. 7..	28,482	1,063	11,623	9,430	22,116	136,297	13@13¼	7¾@8½ pr.	13/32@7/16d.
" 14..	21,751	2,210	9,437	1,524	13,171	144,877	12⅞@13⅛		
" 21..	14,969	10,048	9,654	3,786	23,488	161,609	12⅜@12⅝		
" 28..	35,584	6,449	10,367	3,598	20,404	175,653	12½@12¾		
1851.									
Jan. 4..	25,738	8,468	4,481	2,416	15,365	185,570	12¾@13	7¼@8 pr.	⅜@13-32d
" 11..	38,522	11,642	5,043	5,770	22,455	201,637	12⅞@13⅛		
" 18..	47,455	14,203	7,539	2,673	24,415	227,783	12½@12¾		
" 25..	40,767	18,365	7,041	5,242	30,648	237,932	12⅜@12⅝		
Feb. 1..	44,512	22,254	11,384	3,255	36,893	245,451	12@12⅜	7@7¾ pr.	½@9-16d.
" 8..	41,227	3,160	6,426	6,896	16,482	270,196	11⅜@11¾		
" 15..	33,536	13,949	7,693	5,597	27,239	276,593	11@11⅝		
" 22..	37,956	22,267	12,118	1,113	35,498	279,051	10½@11		
March 1..	30,374	14,168	7,735	9,559	31,462	277,963	9½@10⅛	7½@8¼ pr.	¾@13-16d
" 8..	26,166	20,138	9,045	3,266	32,449	265,740	9½@10⅛		
" 15..	25,254	10,207	10,814	3,752	23,773	263,221	9½@10¼		
" 22..	22,732	22,704	5,036	1,687	29,427	256,526	10½@11¼		
" 29..	24,638	14,746	11,782	5,490	32,018	249,146	10@10⅝		
April 5..	25,814	20,205	913	1,898	22,116	252,874	10⅜@11	9½@10½ pr	⅜d.
" 12..	15,977	24,062	2,800	2,659	29,521	239,390	10⅜@11		
" 19..	23,559	11,118	9,736	9,455	30,309	232,641	10@10⅝		
" 26..	19,350	16,421	5,857	1,395	23,673	228,317	9½@10⅛		
May 3..	13,918	25,164	5,663	3,763	34,590	207,729	9⅛@9⅝	9½@10½ pr	½d.
" 10..	18,104	18,106	4,882	922	23,910	201,923	8¾@9¼		
" 17..	10,139	20,422	4,317	3,869	28,708	183,454	8¼@8⅞		
" 24..	7,412	13,412	4,716	1,073	19,261	171,605	8¼@8⅞		
" 31..	7,901	26,084	2,687	2,102	30,873	148,633	8¼@9		
June 7..	3,192	18,387	5,189	3,344	26,920	125,505	8@8½	9½@11 pr.	⅜@7-16d.
" 14..	3,134	13,776	3,548	1,366	18,690	109,949	8@8½		
" 21..	4,510	9,575	5,879	4,481	19,935	94,524	8@8½		
" 28..	2,808	8,889	1,811	4,250	14,950	82,382	7¾@8¼		
July 5..	4,139	9,382	1,421	3,577	13,880	72,641	7¾@8¼	8½@10¼ pr	5-16@⅜d.
" 12..	1,894	7,330	3,981	1,234	12,545	61,990	7½@8		
" 19..	1,112	8,696	2,632	1,913	13,241	49,831	7¼@8		
" 26..	1,367	9,556	2,138	1,391	13,085	38,113	7@7¾		
Aug. 2..	1,147	7,414	963	1,415	9,792	29,468	6½@7¼	8½@10 pr.	7-16d.
" 9..	642	4,972	851	2,020	7,843	22,267	6¼@7		
" 16..	711	6,100	872	13	6,985	18,885	6¼@7		
" 23..	1,013	5,374	21	335	5,730	14,168	6½@7¼		
Totals..	995,036	582,373	262,268	152,817	997,458			Av. pr. 11c.	

I give the New Orleans movement and quotations this season, because that market was the centre of a very remarkable and very disastrous short crop speculation, in which the whole trade of that city, led by the factors, engaged with perfect unanimity of conviction and purpose. Early in the season they had formed their estimate of the crop, based upon the correspondence of planters with their factors, in which liberal allowance was made for the bad weather of August and September and the early killing frost.

At no time during the season did the receipts seem to justify the small crop estimate, but there were always plausible reasons at hand to explain the discrepancy. Long after, Liverpool, New York and nearly all other centres of the cotton trade began to doubt, New Orleans laughed their doubts to scorn. New Orleans had had two or three years of great prosperity and was therefore puffed up—in fact infallible in her own opinion. The leading house in the trade in New Orleans at that time, was Hill, McLean & Co. The head of the firm, Harry Hill, as he was generally called, was quite a noted personage. He was at least as eccentric as he was supposed to be rich. He was the leading spirit in the short crop speculation of that season, and he had a large following, a great many of whom were influenced mainly by their confidence in his judgment. Instead of yielding to the current, accepting a moderate loss, factors and speculators generally refused to sell; prefering to consign to northern cities and to Europe. The losses in consequence were very heavy. Mr. Hill about the 1st of July sold a lot of about 1,000 bales, deliverable in Philadelphia to R. R. Graves of New York at 6½ cents for middling. I find that no such price as this is shown in the recorded quotations, but I know it to be a fact, that cotton cost 13¼ cents in New Orleans. The losses were of course very heavy, especially in New Orleans. I know of cotton shipped from New Orleans to Liverpool losing over 70 dollars per bale. How it was is more than I have ever been able to understand clearly, though I saw the account sales and paid the loss. One cause of the total failure of the speculation this season was, no doubt, the fact that it was in the midst of a period of contraction, which in the natural course of events ended in 1852, having commenced in 1847. There was considerable perturbation in financial circles in 1854, the immediate cause of which was the commencement of the Crimean war, but the recovery was immediate and rapid, the period of expansion having commenced in 1852 to 1853.

The mining speculations of 1850 and 1851, caused by the gold discoveries in California were short lived and generally abortive.

If they had commenced with the general expansion in 1852, they would probably have been more successful. The main element in all such enterprises is the condition and tendency of the public mind, which seems to have successive moods of elevation and depression, of sanguine hope and unreasoning distrust, in successive periods of about five years. It will be seen hereafter, that cotton is not always under the dominion of these general influences. The period from 1857 to 1860 is a notable exception, which will be considered when we come to it.

1852.

Mr. Thomas Clegg began encouraging the natives of Sierra Leone to raise cotton, providing them with gins and other apparatus. Result of the first six years' work:—Shipped to Manchester as follows—1852, 1,810 lbs.; 1853, 4,617; 1854, 1,588; 1855, 1,651; 1856, 11,492; 1857, 35,419; 1858, to April 15th, 96,470.

Lowell, Mass., had 12 manufacturing companies in operation (see year 1822), 51 mills, and gave employment to 12,633 operatives.

The following Table sets forth the number of hands and acres employed in the United States this year in the cultivation of cotton:

STATES.	Acres in Cotton.	Hands Employed.	Acres susceptible of cultivation in Cotton.
Florida	160,000	20,000	6,000,000
Texas	200,000	25,000	10,000,000
Arkansas	200,000	25,000	3,000,000
Louisiana	400,000	50,000	3,000,000
Tennessee	440,000	55,000	2,000,000
South Carolina	620,000	77,500	200,000
Mississippi	1,300,000	162,500	6,000,000
Georgia	1,480,000	185,000	3,000,000
Alabama	1,500,000	187,500	6,000,000
Total	6,300,000	787,500	39,200,000

COTTON CROP OF THE UNITED STATES.

Statement and Total Amount of the Cotton Crop of the United States, for the Year ending August 31, 1852.

	Bales.	Bales.	Total.	Same period 1851.
NEW ORLEANS.				
Export—				
To Foreign Ports	1,179,103			
Coastwise	256,712			
Stock on hand 1st September, 1852	9,758			
		1,445,573		
Deduct—				
Stock on hand 1st September, 1851	15,390			
Received from Mobile and Montgomery, Ala.	37,366			
" Florida	4,807			
" Texas	14,546			
		72,109	1,373,464	933,369
ALABAMA.				
Export—				
To Foreign Ports	430,846			
Coastwise	143,804			
Consumed in Mobile	842			
Stock on hand 1st September, 1852	2,319			
		577,811		
Deduct—				
Wrecked Cotton returned	344			
Received from Texas and New Orleans	221			
Stock on hand 1st September, 1851	27,797			
		28,362	549,449	451,748
TEXAS.				
Export—				
To Foreign Ports	7,235			
Coastwise	57,096			
Stock on hand 1st September, 1852	317			
		64,648		
Deduct—				
Stock on hand 1st September, 1851	....	596		
			64,052	45,820
FLORIDA.				
Export—				
To Foreign Ports	64,492			
Coastwise	123,829			
Stock in Apalachicola, 1st September, 1852	451			
		188,772		
Deduct—				
Stock in Apalachicola, 1st September, 1851	....	273		
			188,499	181,204
GEORGIA.				
Export—				
To Foreign Ports—Uplands	111,249			
" Sea Island	7,605			
Coastwise—Uplands	224,958			
Sea Island	3,656			
Burnt at Savannah	5,600			
Stock in Savannah, 1st September, 1852	2,950			
" Augusta, 1st September, 1852	3,707			
		359,725		
Deduct—				
Stock in Savannah and Augusta, 1st September, 1851.	....	34,011		
			325,714	322,376

Statement and Total Amount of the Cotton Crop of the United States, for the Year ending August 31, 1852.—*Concluded.*

		Bales.	Bales.	Total.	Same period 1851.
SOUTH CAROLINA.					
Export from Charleston—					
To foreign ports—Uplands		270,427			
" Sea Island		19,008			
Coastwise—Uplands		199,605			
Sea Island		3,305			
Burnt at Charleston		300			
		492,645			
Export from Georgetown—					
To New York and Boston	2,535				
Stock in Charleston, 1st September, 1852	11,146				
		13,681	506,326		
Deduct—					
Stock in Charleston, 1st September, 1851		10,953			
Received from Savannah		18,759			
			29,712		
				476,614	387,075
NORTH CAROLINA.					
Export—					
To foreign ports		424			
Coastwise		15,818			
				16,242	12,928
VIRGINIA.					
Export—					
To Foreign Ports		35			
Coastwise and Manufactured—taken from the ports		20,955			
Stock on hand 1st September, 1852		450			
			21,440		
Deduct—					
Stock on hand 1st September, 1851		......	620		
				20,820	19,940
Received by New York & Erie Canal				175	797
Total crop of the United States		......	......	3,015,029	2,355,257

Increase from last year .... bales .. 659,722
Increase from year before 918,323

Growth.

Total crop of 1823–4...... bales.	509,158	Total crop of 1838–9 ...... bales.	1,360,532
" 1824–5	569,249	" 1839–40	2,177,835
" 1825–6	720,027	" 1840–1	1,634,945
" 1826–7	957,281	" 1841–2	1,683,574
" 1827–8	720,593	" 1842–3	2,378,875
" 1828–9	857,744	" 1843–4	2,030,409
" 1829–30	976,845	" 1844–5	2,394,503
" 1830–1	1,038,848	" 1845–6	2,100,537
" 1831–2	987,477	" 1846–7	1,778,651
" 1832–3	1,070,438	" 1847–8	2,347,634
" 1833–4	1,205,394	" 1848–9	2,728,596
" 1834–5	1,254,328	" 1849–50	2,096,706
" 1835–6	1,360,725	" 1850–1	2,355,257
" 1836–7	1,422,930	" 1851–2	3,015,029
" 1837–8	1,801,497		

Export to Foreign Ports, from September 1, 1851, *to August* 31, 1852.

FROM	To Great Britain.	To France.	To North of Europe.	Other Foreign Ports.	Total.
New Orleans (bales)	772,242	196,254	75,950	134,657	1,179,103
Mobile	306,002	97,753	8,826	18,265	430,846
Texas	1,338	3,202	2,695	....	7,235
Florida	48,638	1,560	9,840	4,454	64,492
Georgia	109,378	12,593	2,483	....	124,454
South Carolina	207,220	43,950	16,240	22,025	289,435
North Carolina	419	....	5	....	424
Virginia	....	35	....	....	35
Baltimore	71	....	100	....	171
Philadelphia	4,619	55	....	422	5,096
New York	218,772	65,973	50,536	4,491	339,772
Boston	50	....	2,200	333	2,583
Grand total	1,668,749	421,375	168,875	184,647	2,443,646
Total last year	1,418,265	301,358	129,492	139,595	1,988,710
Increase	250,484	120,017	39,383	45,052	454.936

Consumption.

Total crop of the United States, as above stated............bales.			3,015,029
Add—Stocks on hand at the commencement of the year, September 1, 1851, in the Southern ports		89,044	
" Northern ports		39,260	
			128,304
Makes a supply of			3,143,333
Deduct therefrom—The export to Foreign ports	2,443,646		
Less—Foreign included	543		
		2,443,103	
Stocks on hand September 1, 1852.			
In the Southern ports	31,098		
In the Northern ports	60,078		
		91,176	
Burnt at Savannah, Charleston and Providence	6,025		
		2,540,304	
Taken for home use			603,029

Quantity consumed by and in the hands of Manufacturers, North of Virginia.

Year	Bales	Year	Bales
1851-2........bales.	603,029	1838-9........bales.	276,018
1850-1	404,108	1837-8	246,063
1849-50	487,769	1836-7	222,540
1848-9	518.039	1835-6	236,733
1847-8	531,772	1834-5	216,888
1846-7	427,967	1833-4	196,413
1845-6	422,597	1832-3	194,412
1844-5	389,006	1831-2	173,800
1843-4	346,744	1830-1	182,142
1842-3	325,129	1829-30	126,512
1841-2	267,850	1828-9	118,853
1840-1	297,288	1827-8	120,593
1839-40	295,193	1826-7	149,516

NOTE.—We give below our usual Table of the amount of Cotton consumed the past year in the States south and west of Virginia, and not included in the receipts at the ports.

We have increased the estimate somewhat from the year previous, though the number and capacity of the mills have been about the same, but give it only for what it purports to be, an *estimate*, which we believe approximates correctness: Thus, in—

North Carolina	15,000	bales, of 400 lbs.
South Carolina	10,000	" "
Georgia	22,000	" "
Alabama	5,000	" of 500 lbs.
Tennessee	7,000	" "
On the Ohio, &c.	16,000	" "
Total to Sept. 1st, 1852	75,000	bales.
" " 1851	60,000	"
" " 1850	107,500	"
" " 1849	110,000	"
" " 1848	75,000	"

To which, if we add the stocks in the interior towns, &c., the quantity burnt in the interior, and that lost on its way to market, to the crop as given above, received at the shipping ports, the aggregate will show very nearly the amount raised in the United States the past season—say, in round numbers, 3,100,000 bales, against 2,450,000 bales the year previous.

During the year just closed, there was received at an Eastern port, 175 bales by way of the New York & Erie Canal, which we have added in another place to the crop of the country.

It may be remarked in this connection, that some of the cotton received overland at Philadelphia and Baltimore is doubtless unaccounted for elsewhere, not being counted in the receipts at New Orleans, but as we have of late years omitted this item from the crop, it is not now added.

The quantity of new cotton received at the shipping ports up to the 1st inst. amounted to about 5,125 bales, against about 3,200 bales last year.

The shipments given in this statement from Texas, are those by sea only, a considerable portion of the crop of that State finds its way to market via Red River, and is included in the receipts at New Orleans.

ANNUAL REVIEW.

From the New Orleans Price Current, 1851—52.

Cotton has long been, and is likely long to be, the leading staple of our commerce; and that its importance is not waning is evident from the fact that the receipts of the past year, at our own port alone, reach nearly a million and a half of bales; or an excess over any previous year of nearly two hundred and fifty thousand bales. Yet, with this large increase, we have the pleasure of saying that there probably has never been, in the whole history of the cotton trade, a season more satisfactory in its general course and results than the one just closed. Thus the largest crop ever produced in the United States has been disposed of, and with results more satisfactory than we remember to have witnessed any previous year. The circumstances which have tended to these results present some remarkable peculiarities, and we propose to touch briefly upon a few of the most prominent, among which we may mention the policy of the factors, generally, of meeting the market freely, and thus guarding against any unwieldy accummulation of stock, which would tend to break down the market. In this course they have been aided by circumstances, which to many were a momentary evil of magnitude, though they contributed favorably to the general result. We allude to the remarkable drought, which, while constituting a season of the most favorable character for picking, at the same time kept all the tributary streams too low for the purposes of navigation; and thus the great bulk of the supplies which come from the banks of the main river had been received and disposed of before the tributaries were in a condition to contribute to the stock. We would also refer to the great abundance and cheapness of money in Europe, which brought speculators into competition with spinners, and to the remarkable increase in the consumption. This is most prominently shown by the half-yearly returns from Great Britain, by which it appears that the quantity taken for consumption, for the six months ending on the 1st of July, was 1,031,763 bales, against 776,120 bales for the corresponding six months of the previous year. This made a weekly average of 39,683 bales, or an increase of about 5,000 bales per week over any previous period.

The semi-weekly Price and Weekly Sales and Receipts at New York, Weekly Exports from New York and Rates of Freight to Liverpool 1st of each month, for the Crop Year ending September 1, 1852.

1851.		Price of Mid. Fair to Fair New Orleans Liverpool Classificat'n.	Price of Mid. Fair to Fair Upland, Liverpool Classificat'n.	Sales for week.	Receipts for week.	EXPORTS FOR THE WEEK.					Rates of Freight to Liverpool.	General Remarks.
						To Great Britain.	To France.	North of Europe.	Other Fo'n Ports	Total Exports.		
Septem.	5..	10½@11¼	9½@10								⅛d.	This crop year was without points of interest; the market here was remarkably void of sensational crop, &c., reports, the weather, both in the spring and autumn, being favorable for the plant.
"	9..	10¾@11¼	10@10½	8,500	2,983	2,719	1,615	10	44	4,388		
"	12..	10¾@11¼	10¼@10½									
"	16..	10¾@11	10@10½	6,600	1,075	1,895	1,773	....	436	4,104		
"	19..	10½@11	10@10½									
"	23.	10½@11	10@10½	4,500	1,898	886	435	....	....	1,321		
"	26..	10½@11	9¾@10¼									
"	30..	10½@11	9⅝@10	4,000	1,864	265	803	....	....	1,068		
October	3..	10½@11	9⅝@10								⅛d.	
"	7..	10@10½	9½@9¾	3,600	4,403	675	689	509	....	1,873		
"	10..	9¾@10¼	9¼@9½									
"	14..	9½@10	9@9¼	3,200	3,877	297	830	....	20	1,147		
"	17..	9½@10¼	8¾@9¼									
"	21..	9¾@10¼	9@9⅜	7,900	10,112	2,355	1,465	266	....	4,086		
"	24..	9¾@10½	9@9½									
"	28..	9¾@10½	9@9½	9,600	3,878	853	1,044	51	....	1,948		
"	31..	9½@10¼	8¾@9¼									
Novem.	4..	9¼@10	8¾@9	8,000	9,739	1,680	....	777	....	2,457	⅛d.	
"	7..	9½@10	8⅞@9⅛									
"	11..	9½@10	8⅞@9⅛	9,500	9,502	554	2,018	28	150	2,750		
"	14..	9½@10	8⅞@9⅛									
"	18..	9½@10	8¾@9	9,400	8,865	1,325	5,276	....	1,129	7,730		
"	21..	9½@10	8¾@9									
"	25..	9½@10	8⅞@9⅛	9,300	11,694	1,078	....	373	752	2,203		
"	28..	9½@10	8⅞@9⅛									
Decem.	2..	9½@10	8¾@9	5,000	6,902	4,043	2,185	38	212	6,478	3-16d.	
"	5..	9½@10	8¾@9									
"	9..	9½@10	8⅞@9⅛	11,800	15,685	1,310	2,096	....	....	3,406		
"	12..	9½@10	8⅞@9⅛									
"	16..	9½@10	8⅞@9⅛	14,000	19,679	2,723	582	473	50	3,828		
"	19..	9¼@9¾	8¾@9									
"	23..	9¼@9½	8⅜@8¾	6,100	13,987	5,994	2,273	90	....	8,357		
"	26..	9¼@9½	8½@8¾									

Decem. 30 .	$8\frac{7}{8}$@$9\frac{1}{2}$	$8\frac{1}{2}$@$8\frac{3}{4}$	10,200	17,809	1,443	....	786	....	2,229	
1852.										
January 2..	$8\frac{7}{8}$@$9\frac{1}{2}$	$8\frac{1}{2}$@$8\frac{5}{8}$								$\frac{1}{8}$@3-16d
" 6..	9@$9\frac{1}{2}$	$8\frac{1}{2}$@$8\frac{3}{4}$	9,000	13,469	4,193	1,510	....	....	5,703	
" 9..	$8\frac{3}{4}$@$9\frac{1}{4}$	$8\frac{1}{2}$@$8\frac{5}{8}$								
" 13..	9@$9\frac{1}{2}$	$8\frac{5}{8}$@$8\frac{3}{4}$	19,600	16,259	2,936	1,803	....	....	4,739	
" 16..	9@$9\frac{1}{2}$	$8\frac{5}{8}$@$8\frac{7}{8}$								
" 20..	9@$9\frac{1}{2}$	$8\frac{5}{8}$@$8\frac{3}{4}$	6,600	18,227	11,200	3,106	1,212	....	15,518	
" 23..	9@$9\frac{1}{2}$	$8\frac{1}{2}$@$8\frac{3}{4}$								
" 27..	9@$9\frac{1}{2}$	$8\frac{1}{2}$@$8\frac{3}{4}$	7,000	13,144	1,384	2,276	104	654	4,418	
" 30..	9@$9\frac{1}{2}$	$8\frac{1}{2}$@$8\frac{3}{4}$								
February 3..	9@$9\frac{1}{2}$	$8\frac{1}{2}$@$8\frac{3}{4}$	12,100	13,409	2,645	4,954	1,174	24	8,797	5-32@$\frac{3}{16}$
" 6..	9@$9\frac{1}{2}$	$8\frac{1}{2}$@$8\frac{3}{4}$								
" 10..	$8\frac{3}{4}$@$9\frac{1}{2}$	$8\frac{3}{8}$@$8\frac{5}{8}$	13,400	18,748	6,935	1,772	269	....	8,976	
" 13..	9@$9\frac{1}{2}$	$8\frac{1}{2}$@$8\frac{3}{4}$								
" 17..	9@$9\frac{1}{2}$	$8\frac{1}{2}$@$8\frac{3}{4}$	20,000	17,359	6,204	1,734	987	....	8,925	
" 20..	$9\frac{1}{4}$@$9\frac{1}{2}$	$8\frac{1}{2}$@$8\frac{3}{4}$								
" 24..	$9\frac{1}{4}$@$9\frac{1}{2}$	$8\frac{1}{2}$@$8\frac{3}{4}$	16,000	13,121	3,571	1,259	439	....	5,269	
" 27..	$9\frac{1}{4}$@$9\frac{1}{2}$	$8\frac{1}{2}$@$8\frac{3}{4}$								
March 2..	$9\frac{1}{4}$@$9\frac{1}{2}$	$8\frac{1}{2}$@$8\frac{3}{4}$	15,500	34,403	4,105	....	3,101	....	7,206	$\frac{1}{4}$@5-16d
" 5..	$9\frac{1}{4}$@$9\frac{1}{2}$	$8\frac{1}{2}$@$8\frac{3}{4}$								
" 9..	$9\frac{1}{4}$@$9\frac{3}{4}$	$8\frac{5}{8}$@$8\frac{7}{8}$	21,500	11,738	11,747	2,361	27	100	14,235	
" 12..	$9\frac{1}{4}$@$9\frac{3}{4}$	$8\frac{5}{8}$@$8\frac{7}{8}$								
" 16..	$9\frac{1}{4}$@$9\frac{3}{4}$	$8\frac{1}{2}$@$8\frac{7}{8}$	19,200	19,742	7,949	2,040	6,418	....	16,407	
" 19..	$9\frac{1}{4}$@$9\frac{3}{4}$	$8\frac{1}{2}$@$8\frac{7}{8}$								
" 23..	$9\frac{1}{4}$@$9\frac{3}{4}$	$8\frac{1}{4}$@$8\frac{3}{4}$	16,500	10,193	7,633	2,037	856	443	10,969	
" 26..	9@$9\frac{1}{2}$	$8\frac{3}{8}$@$8\frac{5}{8}$								
" 30..	9@$9\frac{1}{2}$	$8\frac{3}{8}$@$8\frac{5}{8}$	17,500	18,657	7,683	....	3,787	370	11,840	
April 2..	9@$9\frac{1}{2}$	$8\frac{3}{8}$@$8\frac{3}{4}$								$\frac{1}{4}$@9-32d
" 6..	9@$9\frac{1}{2}$	$8\frac{1}{2}$@$8\frac{3}{4}$	21,000	20,506	9,735	2,538	659	10	12,942	
" 9..	9@$9\frac{3}{4}$	$8\frac{1}{2}$@$8\frac{3}{4}$								
" 13..	9@$9\frac{3}{4}$	$8\frac{1}{2}$@$8\frac{3}{4}$	12,400	7,950	11,978	2,188	5,710	....	19,876	
" 16..	$9\frac{1}{4}$@$9\frac{3}{4}$	$8\frac{5}{8}$@$8\frac{7}{8}$								
" 20..	$9\frac{1}{4}$@$9\frac{3}{4}$	$8\frac{5}{8}$@$8\frac{7}{8}$	9,600	16,254	6,504	1,758	3,581	....	11,843	
" 23..	$9\frac{1}{2}$@10	$8\frac{3}{4}$@$9\frac{1}{8}$								
" 27..	$9\frac{1}{2}$@10	9@$9\frac{3}{8}$	24,000	9,988	10,715	....	3,643	....	14,358	
" 30..	$9\frac{3}{4}$@$10\frac{1}{4}$	$9\frac{1}{4}$@$9\frac{1}{2}$								
May 4..	$9\frac{3}{4}$@$10\frac{1}{2}$	$9\frac{1}{4}$@$9\frac{5}{8}$	17,700	5,867	13,760	2,041	729	....	16,530	$\frac{3}{16}$@7-32
" 7..	$9\frac{1}{2}$@$10\frac{1}{2}$	$9\frac{1}{4}$@$9\frac{1}{2}$								
" 11..	$9\frac{3}{4}$@$10\frac{3}{4}$	$9\frac{1}{2}$@10	13,300	7,224	5,980	1,680	3,357	....	11,017	
" 14..	$10\frac{1}{4}$@11	$9\frac{3}{4}$@10								
" 18.	$10\frac{1}{2}$@11	$9\frac{3}{4}$@$10\frac{1}{8}$	21,000	10,787	4,714	1,716	2,147	....	8,577	

In September, the domestic and foreign markets were very dull, and with a stringency in financial affairs here, aggravated by the failure of several country banks, and very fine weather for gathering and marketing the crop, prices generally were weak until November, when there was a better feeling, though no advance in prices. About the latter part of December, accounts came to hand from Europe, advising an unsettled state of political affairs in France, with an attempted revolution, and prices of cotton at once fell off $\frac{1}{4}$@$\frac{1}{2}$ cent per pound; but the difficulties proved to be of short duration, and in January, the markets partially recovered their tone and remained pretty steady until the latter part of March, when the crop estimates began to enlarge their proportions, and the markets on both sides the water became for a time quite dull. During the last three or four months of the crop year, there was considerable activity, easy money in England, low prices for the staple and cheap food stimulated the consumption, the Manchester mills were full of orders and prices closed higher here than at any time during the year.

The semi-weekly Price and Weekly Sales and Receipts at New York, Weekly Exports from New York and Rates of Freight to Liverpool 1st of each Month, for the Crop Year ending September 1, 1852—Concluded.

1852.		Price of Mid. Fair to Fair New Orleans Liverpool Classificat'n.	Price of Mid. Fair to Fair Upland, Liverpool Classificat'n.	Sales for week.	Receipts for week.	EXPORTS FOR WEEK.					Rates of Freight to Liverpool.
						To Great Britain.	To France.	North of Europe.	Other Fo'n Ports	Total Exports.	
May	21..	10¾@11¼	10@10½								
"	25..	10¾@11¼	10¼@10⅝	18,500	13,111	7,727	100	1,485	....	9,312	
"	28..	11@11½	10¼@10⅝								
June	1..	11@11½	10@10⅝	10,400	12,266	8,836	858	2,204	....	11,898	⅛@5-32d
"	4..	10¾@11¼	10@10½								
"	8..	10¾@11½	10⅛@10½	4,500	6,986	3,009	1,662	216	97	4,984	
"	11..	10¾@11½	10@10½								
"	15..	10½@11¼	9¾@10¼	6,900	7,378	1,365	1,128	392	....	2,885	
"	18..	10½@..	9¾@10⅜								
"	22..	11@11¾	10¼@10⅝	6,600	6,778	5,227	835	1,560	....	7,622	
"	24..	11@11¾	10¼@10⅝								
"	29..	11@11¾	10¼@10⅝	5,900	16,613	1,748	772	678	....	3,198	
July	2..	11@11½	10⅛@10½								⅛@5-32d
"	6..	11@11½	10¼@10⅝	4,800	7,378	2,557	202	1,430	....	4,189	
"	9..	11@11½	10¼@10⅝								
"	13..	11@11½	10¼@10⅝	9,000	6,795	2,337	46	448	....	2,831	
"	16..	11@11½	10¼@10⅝								
"	20..	11@11½	10¼@10⅝	4,600	1,832	2,534	48	....	....	2,582	
"	23..	11@11½	10¼@10⅝								
"	27..	11@11½	10¼@10¾	7,800	4,009	2,776	....	60	....	2,836	
"	30..	11@11½	10½@10¾								
August	3..	11¼@11¾	10⅝@10⅞	9,000	5,879	1,465	445	221	....	2,131	⅛d.
"	6..	11¼@11¾	10⅝@11								
"	10..	11½@..	10¾@11¼	16,000	1,595	1,293	....	40	....	1,333	
"	13..	11⅝@..	11@11⅜								
"	17..	11⅝@..	11@11⅜	11,000	2,310	2,029	....	59	....	2,088	
"	20..	11⅝@..	11@11⅜								
"	24..	11½@12	11@11¼	4,700	2,583	2,741	15	51	....	2,807	
"	27..	11¼@12	10⅞@11¼								
"	31..	11¼@11¾	11@11¼	5,000	605	1,461	5	91	....	1,557	⅛d.
Average price and total sales, receipts and exports.		10.20	9.50	568,800	537,115	218,771	65,973	50,536	4,491	339,771	

GENERAL REMARKS.

Exchange.

The quotation for bills on London in September, was 10¼@10½ per cent. premium; in October, the extremes were 9¾@10⅜; steady in November, at 10@10½; December, the same; January, 9¾@10¼; March, the range was from 9¾@10¼; April, between 8⅞ and 9¾; May, 9¾@10½; June, 10@10½; July, 10¼@10½; and in August, 10¼@10⅝.

LIVERPOOL STATEMENT FOR 1852.

UNITED STATES, 1851–1852.

Stock, September 1, 1851	128,000	Export	2,444,000
Crop	3,015,000	Consumption	608,000
		Stock, Sept. 1, 1852	91,000
Bales	3,143,000	Bales	3,143,000

Stock 1st Jan., 1852, in	Liverpool.	Gt. Britain.	France.	Continent.	Tot. Europe.
United States ... Bales	236,000	246,000	16,000	23,000	285,000
Brazil	50,000	49,000	2,000	6,000	57,000
West Indies	4,000	4,000	4,000	2,000	10,000
East Indies	114,000	172,000	6,000	2,000	180,000
Egypt	20,000	23,000	9,000	25,000	57,000
Bales	424,000	494,000	37,000	58,000	589,000

CONSUMPTION. — IMPORT.

Tot. Europe.	Continent.	France.	Gt. Britain.	Liverpool.	IMPORT.	Liverpool.	Gt. Britain.	France.	Continent.	Tot. Europe.
2,391,000	495,000	388,000	1,508,000	1,455,000	United States ... Bales	1,715,000	1,789,000	393,000	500,000	2,516,000
162,000	31,000	7,000	124,000	124,000	Brazil	144,000	144,000	6,000	28,000	163,000
41,000	16,000	15,000	10,000	9,000	West Indies	10,000	12,000	14,000	16,000	42,000
272,000	104,000	8,000	160,000	135,000	East Indies	151,000	222,000	2,000	103,000	226,000
246,000	86,000	51,000	109,000	102,000	Egypt	186,000	190,000	46,000	73,000	308,000
3,112,000	732,000	469,000	1,911,000	1,825,000	Bales	2,206,000	2,357,000	461,000	720,000	3,255,000
....	....	....	283,000	227,000	Export.					
732,000	46,000	29,000	657,000	578,000	Stock, Dec. 31 ... Stock above,	424,000	494,000	37,000	58,000	589,000
3,844,000	778,000	498,000	2,851,000	2,630,000	Total supply, bales	2,630,000	2,851,000	498,000	778,000	3,844,000

COTTON AT LIVER

WEEK ENDING.	RECEIPTS.						SALES.			
	American	E. I.	Egypt.	Brazil.	Other.	Total.	Consumption.	Speculation.	Export	Total.
Jan. 2..	....	......	......	....	......	....	....	......	......	....
" 9..	47,389	3,116	788	1,156	......	52,449	41,210	3,960	4,770	49,940
" 16..	6,053	150	......	5,934	......	12,137	33,380	4,370	2,840	40,590
" 23..	33,458	......	3,940	3,804	......	41,202	34,550	4,960	4,220	43,730
" 30..	48,411	1,815	158	1,610	374	52,368	42,830	16,340	9,510	68,680
Feb. 6..	43,443	2,711	878	....	......	47,032	30,040	6,130	6,380	42,550
" 13..	23,441	2,611	......	2,017	......	28,069	43,280	7,320	6,720	57,320
" 20..	24,623	.....	852	....	......	25,475	46,800	22,810	4,690	74,300
" 27..	13,288	.....	......	2,015	25	15,328	40,930	10,170	2,940	54,040
March 5..	43,093	100	1,062	600	.....	44,855	53,810	19,500	4,230	77,540
" 12..	30,267	......	405	1,775	......	32,447	34,683	17,030	4,190	55,903
" 19..	10,706	2,766	220	....	38	13,730	30,200	8,600	6,000	44,800
" 26..	12,506	......	8,013	....	17	20,536	27,521	6,300	3,029	36,850
April 2..	32,493	2,715	5,988	1,272	......	42,468	24,850	2,170	3,730	30,750
" 8..	32,246	......	7,247	3,360	......	42,853	20,960	760	2,660	24,380
" 16..	34,380	.....	......	1,240	258	35,878	38,010	4,450	6,590	49,050
" 23..	75,701	2,189	3,539	4,608	292	86,329	48,280	13,130	6,490	67,900
" 30..	67,490	......	6,175	2,767	......	76,432	41,700	18,100	10,110	69,910
May 7..	136,549	......	2,128	6,073	105	144,855	53,340	23,650	11,190	88,180
" 14..	131,218	4,529	3,517	5,497	353	145,114	51,510	41,710	18,680	111,900
" 21..	62,080	343	2,156	2,632	25	67,236	46,990	28,780	13,660	89,430
" 28..	8,409	......	......	....	......	8,409	56,524	31,790	16,406	104,720
June 4..	20,575	......	1,154	6,564	179	28,472	34,250	20,580	8,010	62,840
" 11..	25,736	......	10,958	....	......	36,694	33,280	23,670	12,060	69,010
" 18..	107,588	......	722	3,941	319	112,670	28,400	15,540	6,600	50,540
" 25..	77,351	......	792	2,111	1,222	81,476	29,970	10,100	4,340	44,410
July 2..	41,434	2,700	1,254	1,902	6	47,296	35,770	9,700	7,280	52,750
" 9..	27,160	2,124	56	2,172	133	31,645	24,620	6,140	5,960	36,720
" 16..	35,596	2,124	929	1,514	39	40,202	30,780	6,310	4,960	42,050
" 23..	48,284	......	16,605	10.357	666	75,912	39,330	17,250	7,610	64,190
" 30..	8,149	......	9,150	3,788	192	21,279	93,124	8,336	13.290	114,750
Aug. 6..	44,515	......	7,117	3,654	34	55,320	37,950	9,840	9,270	57,060
" 13..	27,150	5,337	9,830	1,770	40	44,127	37,300	16,340	5,110	58,750
" 20..	16,002	......	2,214	4,449	60	22,725	44,980	36,760	10,080	91,820
" 27..	16,426	2,843	......	4,943	......	24,212	30,920	11,140	5,470	47,530
Sep. 3..	10,106	5,636	......	1,252	458	17,452	32,440	12,970	5,460	50,870
" 10..	5,791	2,599	3,273	....	134	11,797	31,020	6,400	3,590	41,010
" 17..	4,211	......	6,426	922	71	11,630	26,330	17.470	3,050	46,850
" 24..	6,166	5,372	12,740	2,282	350	26,910	28,910	8,340	3,180	40,430
Oct. 1..	10,947	7,928	4,739	......	76	23,690	48,550	43,550	4,130	96,230
" 8..	169	17,541	2,890	4,074	115	24,789	46,000	30,690	5,060	81,750
" 15..	4,159	10,024	2,163	2,070	54	18,470	43,410	26,310	4,410	74,130
" 22..	1,378	......	10,082	3,789	5	15,254	51,060	40,720	3,090	94,870
" 29..	11,658	11,330	11,074	9,831	73	43,966	56,180	57.220	4,160	117,560
Nov. 5..	9,198	10,028	9,626	1,540	......	30,392	30,780	19,800	3,820	54,400
" 12..	10,429	17,874	649	2,060	962	31,974	17,230	12,040	2,810	32,080
" 19..	1,068	......	......	....	......	1,068	12,400	5,590	2,810	20,800
" 26..	5,797	3,843	1,696	47	......	11,383	20,080	2,490	870	23,440
Dec. 3..	22,567	.....	1,453	8,677	......	32,697	29,130	16,810	1,920	47,860
" 10..	29,693	10,192	187	5,255	.....	45,327	21,110	3,180	910	25,200
" 17..	36,396	10,158	......	....	1,303	47,857	28,950	6,620	1,300	36,870
" 24..	63,863	286	2,243	4,234	......	70,626	29,320	2,210	1,300	32,830
Average prices & total sales, receipts & stocks.	1,646,804	150064	176088	139,558	7,978	2,121,402	1,894,969	796,146	300945	2,992,060

POOL. YEAR 1852.

STOCKS.			PRICES.			ACTUAL EXPORT.	CONSUMPTION.	REMARKS.
Amer'n	Other.	Total.	Mid. Up.	Mid. Orl.	Dhol.			
248,229	183,970	432,199	4 5/8	4 3/4	3 5/8	4,770	39,210	Favorable crop accounts from United States.
225,022	183,094	408,116	4 5/8	4 3/4	3 5/8	2,840	74,590	
226,540	184,008	410,548	4 5/8	4 3/4	3 5/8	4,220	109,140	
233,911	176,665	410,576	4 3/4	4 7/8	3 3/4	9,510	151,970	
249,144	172,044	421,188	4 3/4	4 13/16	3 3/4	6,380	182,010	
250,601	165,958	416,559	4 3/4	4 3/16	3 3/4	6,720	225,290	
234,803	157,452	392,255	4 7/8	5	4	4,690	272,090	
214,591	152,122	366,713	4 15/16	5 1/16	4 1/8	2,940	313,020	
209,664	139,864	349,528	5 1/16	5 3/16	4 1/4	4,230	366,830	
205,927	131,840	337,767	4 13/16	5 1/16	4 1/4	4,190	401,510	
191,959	125,894	317,853	4 13/16	5 1/16	4 1/8	6,000	431,710	
179,916	128,757	308,673	4 3/4	4 13/16	4	3,029	459,231	
190,849	132,712	323,561	4 3/4	4 13/16	4 1/8	3,730	484,081	
204,325	139,255	343,580	4 3/4	4 15/16	4 1/8	2,660	505,041	
203,393	134,009	337,302	4 3/4	5	4 1/8	6,590	543,051	
235,685	137,256	372,941	4 15/16	5 1/8	4	6,490	591,331	
269,510	144,760	414,270	4 15/16	5 1/8	4	10,110	633,031	
359,569	141,003	500,572	5 1/16	5 1/4	4	11,190	686,371	
448,165	140,747	588,912	5 1/8	5 5/16	4 1/8	18,680	737,881	
469,517	132,743	602,260	5 1/16	5 1/4	4 1/4	13,660	784,871	Encouraging reports of growing crop from United States
429,496	107,971	547,467	5 3/16	5 3/4	4 1/4	16,406	841,395	
413,601	115,938	529,539	5 3/16	5 3/8	4 3/8	8,010	875,645	
399,997	120,606	520,603	5 5/16	5 1/2	4 3/8	12,060	908,925	
478,135	126,618	604,753	5 3/16	5 7/16	4 3/8	6,600	937,325	
524,297	123,836	648,133	5 1/4	5 3/8	4 3/8	4,340	967,295	
538,331	118,010	656,341	5 1/4	5 3/8	4 3/8	7,280	1,003,065	
541,691	116,165	657,856	5 1/4	5 3/8	4 3/8	5,960	1,027.655	
550,940	110,648	661,588	5 3/16	5 3/16	4 3/8	4,960	1,058,465	
564,290	130,610	694,900	5 1/4	5 7/16	4 1/2	7,610	1,097,795	
521,073	130,897	651,970	5 7/16	5 5/8	4 1/2	13,290	1,190,919	Less encouraging reports of crop.
534,005	132,238	666,243	5 3/8	5 1/2	4 1/2	9,270	1,228,869	
526,955	139,365	666,320	5 3/8	5 9/16	4 1/2	5,110	1,266,169	
504,732	135,908	640,640	5 9/16	5 11/16	4 3/4	10,080	1,311,149	
486,708	131,754	618,462	5 9/16	5 5/8	4 3/4	5,470	1,342,069	
470,900	146,490	617,390	5 9/16	5 13/16	4 3/4	5,460	1,374,509	
442,920	144,451	587,371	6 1/4	5 5/8	4 3/4	3,590	1,405,529	
422,716	144,370	567,086	5 9/16	5 3/4	4 3/4	3,050	1,431,859	
405,352	156,724	562,076	5 9/16	5 3/4	4 3/4	3,180	1,460.769	Damage done to crop through terrific storm.
378,979	154,107	533,086	5 3/4	5 7/8	4 3/4	4,130	1,509,319	
337,039	169,655	506,694	5 3/4	5 7/8	4 3/4	5.060	1,555,319	
306,424	170,456	476,884	5 13/16	6	4 3/4	4,410	1,598,729	
268,486	169,504	437,990	5 7/8	6 1/16	4 3/4	3,090	1,649,789	
236,964	186,252	423,216	6 1/8	6 1/4	5	4,160	1,705,969	
223,543	199,474	423,017	6	6 1/8	5	3,820	1,736,749	Heavy receipts at United States Ports.
219,742	215,506	435,248	5 3/4	5 7/8	4 7/8	2,810	1,753,979	
211,235	209,860	421,095	5 5/8	5 3/4	4 3/4	2,810	1,766,379	
200,642	210,386	411,028	5 3/8	5 9/16	4 5/8	870	1,786,459	
200,769	211,856	412,625	5 5/8	5 3/4	4 3/4	1,920	1,815,589	Report of frost among crops.
214,462	121,470	435,932	5 3/16	5 7/16	4 3/8	910	1,836,699	
226,248	227,291	453,539	5 3/8	5 9/16	4 3/8	1,300	1,865,649	
265,531	228,014	493,545	5 1/2	5 5/8	4 3/8	1,300	1,894,969	
			5.05	5.39	4.37	300,945	36,441.7	

1853.
COTTON CROP OF THE UNITED STATES.

Statement and Total Amount of the Cotton Crop of the United States, for the Year ending August 31, 1853.

	Bales.	Bales.	Total.	Same period 1852.
NEW-ORLEANS.				
Export—				
To Foreign Ports	1,378,285			
Coastwise	266,696			
Burnt at New Orleans	20,000			
Stock on hand, 1st September, 1853	10,522			
		1,675,503		
Deduct—				
Stock on hand, 1st September, 1852	9,758			
Received from Mobile and Montgomery, Ala	62,319			
" " Florida	7,866			
" " Texas	14,685			
		94,628		
			1,580,875	1,373,464
ALABAMA.				
Export—				
To Foreign Ports	345,930			
Coastwise	195,271			
Consumed in Mobile	1,239			
Stock on hand, 1st September, 1853	7,516			
		549,956		
Deduct				
Wrecked Cotton returned	2,530			
Received from Texas	78			
Stock on hand, 1st September, 1852	2,319			
		4,927		
			545,029	549,449
TEXAS.				
Export—				
To Foreign Ports	16,346			
Coastwise	69,333			
Stock on hand, 1st September, 1853	428			
		86,107		
Deduct—				
Stock on hand, 1st September, 1852	....	317		
			85,790	64,052
FLORIDA.				
Export—				
To Foreign Ports	54,397			
Coastwise	125,007			
Stock in Apalachicola, 1st September, 1853	523			
		179,927		
Deduct—				
Stock in Apalachicola, 1st September, 1852	....	451		
			179,476	188,499
GEORGIA.				
Export—				
To Foreign Ports—Uplands	135,565			
Sea Island	6,731			
Coastwise—Uplands	194,727			
Sea Island	6,140			
Stock in Savannah, 1st September, 1853	5,150			
Stock in Augusta, 1st September, 1853	7,834			
		356,147		

Statement and Total Amount of the Cotton Crop of the United States, for the Year ending August 31, 1853.—Concluded.

	Bales.	Bales.	Total.	Same period 1852.
Deduct—				
Stock in Savannah and Augusta, 1st September, 1852	....	6,657		
			349,490	325,714
SOUTH CAROLINA.				
Export from Charleston—				
To Foreign Ports—Uplands	279,961			
Sea Island	17,848			
Coastwise—Uplands	166,649			
Sea Island	2,128			
Burnt at Charleston	325			
Stock in Charleston, 1st September, 1853	15,126			
	482,037			
Export from Georgetown—				
To Northern Ports, &c	5,000			
		487,037		
Deduct—				
Stock in Charleston, 1st September, 1852	11,146			
Received from Savannah	12,688			
		23,834		
			463,203	476,614
NORTH CAROLINA.				
Export—				
To Foreign Ports	....			
Coastwise	23,496			
			23,496	16,242
VIRGINIA.				
Export—				
To Foreign Ports	....			
Coastwise, and Manufactured—taken from the ports	25,833			
Stock on hand, 1st September, 1853	400			
		26,233		
Deduct—				
Stock on hand, 1st September, 1852	....	450		
			25,783	20,820
Received at Boston by New York and Erie Canal	....	....	....	175
" New York by New York and Erie R. R.	....	....	640	....
" Baltimore and Philadelphia, overland	....	....	9,100	....
Total Crop of the United States	....	....	3,262,882	3,015,029
Increase from last year......bales	....	....	....	247,853
Increase from year before	....	....	....	907,625

Export to Foreign Ports, from September 1, 1852, *to August* 31, 1853.

FROM	To Great Britain.	To France.	To North of Europe.	Other F'n Ports.	Total.
New Orleans........................bales.	922,086	211,526	95,635	149,038	1,378,285
Mobile....................................	237,292	87,824	8,447	12,367	345,930
Texas......................................	5,617	6,342	4,387	....	16,346
Florida....................................	43,708	5,565	5,124	....	54,397
Georgia	122,492	15,059	3,481	1,264	142,296
South Carolina...........................	191,306	59,502	19,319	27,682	297,809
North Carolina...........................	....	....	....	....	....
Virginia..................................	....	....	....	....	....
Baltimore................................	175	....	471	....	646
Philadelphia.............................	3,514	....	....	137	3,651
New York.................................	207,647	40,910	32,720	2,763	284,040
Boston.....................................	3,023	....	1,592	385	5,000
Grand total..............................	1,736,860	426,728	171,176	193,636	2,528,400
Total last year..........................	1,668,749	421,375	168,875	184,647	2,443,646
Increase..................................	68,111	5,353	2,301	8,989	84,754

Growth.

Total crop of 1823–4...... bales.	509,158	Total crop of 1838–9 ....bales.	1,360,532
" 1824–5............	569,249	" 1839–40............	2,177,835
" 1825–6............	720,027	" 1840–1.............	1,634,945
" 1826–7............	957,281	" 1841–2.............	1,683,574
" 1827–8............	720,593	" 1842–3.............	2,378,875
" 1828–9............	857,744	" 1843–4.............	2,030,409
" 1829–30..........	976,845	" 1844–5.............	2,394,503
" 1830–1............	1,038,848	" 1845–6.............	2,100,537
" 1831–2............	987,477	" 1846–7.............	1,778,651
" 1832–3............	1,070,438	" 1847–8.............	2,347,634
" 1833–4............	1,205,394	" 1848–9.............	2,728,596
" 1834–5............	1,254,328	" 1849–50............	2,096,706
" 1835–6............	1,360,725	" 1850–1.............	2,355,257
" 1836–7............	1,422,930	" 1851–2.............	3,015,029
" 1837–8............	1,801,497	" 1852–3.............	3,262,882

Consumption.

Total crop of the United States, as above stated......................		3,262,882	bales.
Add—Stocks on hand at the commencement of the year, 1st September, 1852.—In the Southern ports		31,098	
" Northern "		60,078	
			91,176
Makes a supply of..			3,354,058
Deduct therefrom—The export to Foreign ports..............	2,528,400		
Less, foreign included..................	1,855		
		2,526,545	
Stocks on hand at the close of the year, 1st September, 1853—			
In the Southern ports..................	47,499		
" Northern "	88,144		
		135,643	
Burnt at New York, Charleston and New Orleans...........		20,861	
			2,683,049
Taken for home use...............Bales.			**671,009**

Quantity consumed by and in the hands of Manufacturers, North of Virginia.

Year	Bales	Year	Bales
1852-3 .... bales.	671,009	1838-9 .... bales.	276,018
1851-2	603,029	1837-8	246,063
1850-1	404,108	1836-7	222,540
1849-50	487,769	1835-6	236,733
1848-9	518,039	1834-5	216,888
1847-8	531,772	1833-4	196,413
1846-7	427,967	1832-3	194,412
1845-6	422,597	1831-2	173,800
1844-5	389,006	1830-1	182,142
1843-4	346,744	1829-30	126,512
1842-3	325,129	1828-9	118,853
1841-2	267,850	1827-8	120,593
1840-1	297,288	1826-7	149,516
1839-40	295,193		

NOTE.—We give below our usual table of the amount of cotton consumed the past year in the States south and west of Virginia, and not included in the receipts at the ports. We have increased the estimate, as a whole, from the year previous, being satisfied that our figures for the consumption "On the Ohio, &c"., have heretofore been considerably too low, while on the other hand, for some other parts of the country they are slightly reduced, as it is well known that less mills have been in operation the past, than the previous year; but give it only for what it purports to be, an estimate, which we believe approximates correctness: Thus—

	Quantity Consumed. 1853.	1852.		
North Carolina	20,000	15,000	bales,	of 400 lb.
South Carolina	10,000	10,000	"	"
Georgia	20,000	22,000	"	"
Alabama	5,000	5,000	"	of 500 lb.
Tennessee	5,000	7,000	"	"
On the Ohio, &c	30,000	16,000.	"	"
Total to Sept. 1	90,000	75,000		
Total to Sept. 1, 1851		60,000	bales.	
Total to Sept. 1, 1850		107,500	"	
Total to Sept. 1, 1849		110,000	"	
Total to Sept. 1, 1848		75,000	"	

To which, if we add the stocks in the interior towns, &c., the quantity burnt in the interior, and that lost on its way to market, to the crop as given above, received at the shipping ports, the aggregate will show very nearly the amount raised in the United States the past season—say, in round numbers, 3,360,000 bales, against 3,100,000 for the year 1851-2, and 2,450,000 for the year before.

During the year just closed, there was received here 640 bales from Louisville by way of the New York and Erie Railroad, and 7,000 at Baltimore, and 2,100 at Philadelphia, overland, from the West, nearly if not quite all of which, it is believed, came from Tennessee. This last item we have of late years omitted from our Annual Statement of the Cotton Crop, owing to its insignificance; but the increased facilities of transportation afforded by the Baltimore and Ohio Railroad, having rendered this a favorite route for shipments from Tennessee, &c., the amount coming from that quarter, the past year, has largely increased, and we have now added it in another place to the crop of the country.

The quantity of new cotton received at the shipping ports up to the 1st inst. amounted to 716 bales, against 5,125 bales last year.

The shipments given in this Statement from Texas, are those by sea only; a considerable portion of the crop of that State finds its way to market via Red River, and is included in the Receipts at New Orleans.

ANNUAL REVIEW.

From the New Orleans Price Current, 1852—53.

Thus the largest crop ever produced in the United States has been disposed of, and at a very favorable average of prices, though besides the material increase in our crop, the lower grades of American cotton have had to contend with unusual imports into Great Britain from India, the quantity received from that source during the first six months of the present year being 266,603 bales, against 44,019 bales in same period last year. According to the semi-annual circular of Messrs. Hollinhead, Tetly & Co., Liverpool, which we have been accustomed to take as authority, it would appear that the total supply of cotton in Great Britain, for the six months ended on the 30th of June, 1853, was 2,182,250 bales, against 1,895,963 bales for the same period last year, and that of this quantity 1,496,595 bales were American, against 1,470,662 bales last year. The quantity taken for consumption in the same time was 1,131,763 bales, against 1,040,150 bales last year, which shows slight increase, though in the quantity of American amounts being 825,412 bales in 1852, and 806,295 bales, in 1853. As to the quality of the last crop, the great bulk of it was of a low average, and we had occasion frequently through the season to remark upon the unusually wide difference in price between the lower and better grades, owing to the abundance of the former and comparative scarcity of the latter.

LIVERPOOL STATEMENT FOR 1853.

UNITED STATES—1852-1853.

Stock, 1st September, 1852	91,000	Export	2,528,000
Crop	3,263,000	Consumption	691,000
		Stock, 1st September, 1853	135,000
Bales	3,354,000	Bales	3,354,000

CONSUMPTION.

Tot. Europe.	Continent.	France.	Gt. Britain.	Liverpool.		Liverpool.	Gt. Britain.	France.	Continent.	Tot. Europe.
					Stock Jan. 1, 1853, in					
					United States. Bales	339,000	361,000	21,000	28,000	410,000
					Brazil	54,000	54,000	1,000	3,000	58,000
					West Indies	6,000	6,000	3,000	2,000	11,000
					East Indies	80,000	133,000	....	1,000	134,000
					Egypt	99,000	103,000	4,000	12,000	119,000
					Bales	578,000	657,000	29,000	46,000	732,000
					IMPORT.					
2,260,000	463,000	389,000	1,408,000	1,341,000	United States	1,460,000	1,532,000	394,000	473,000	2,223,000
153,000	30,000	4,000	119,000	119,000	Brazil	132,000	132,000	3,000	28,000	145,000
34,000	11,000	12,000	11,000	7,000	West Indies	8,000	9,000	11,000	11,000	31,000
344,000	147,000	1,000	196,000	162,000	East Indies	324,000	486,000	1,000	151,000	486,000
222,000	61,000	41,000	120,000	120,000	Egypt	104,000	105,000	59,000	59,000	220,000
3,013,000	712,000	447,000	1,854,000	1,749,000	Bales	2,028,000	2,264,000	468,000	722,000	3,105,000
....	....	....	349,000	260,000	Export.					
824,000	56,000	50,000	718,000	597,000	Stock Dec. 31. Stock above,	578,000	657,000	29,000	46,000	732,000
3,837,000	768,000	497,000	2,921,000	2,606,000	Total supply, bales	2,606,000	2,921,000	497,000	768,000	3,837,000

COTTON AT LIVER

Week Ending.	Receipts.						Sales.			
	Americ'n.	E. I.	Egypt.	Brazil.	Other.	Total.	Consumption.	Speculation.	Export	Total.
Jan. 7..	26,233	3,997	2,705	1,600	......	44,535	31,860	2,190	850	34,900
" 14..	43,287	......	2,266	2,858	......	48,411	34,480	6,860	2,360	43,700
" 21..	20,978	2,712	5,851	12,707	......	42,248	40,614	11,090	6,686	58,390
" 28..	21,162	4,918	1,506	44	50	27,680	36,880	14,200	4,570	55,650
Feb. 4..	29,776	2,864	759	1,909	......	35,308	39,410	18,820	7,090	65,320
" 11..	68,936	......	......	......	......	68,936	27,120	19,540	3,490	50,150
" 18..	7,077	1,095	......	......	......	8,172	28,520	7,190	4,170	39,880
" 25..	28,391	......	......	......	......	28,391	26,690	4,880	2,860	34,430
Mch. 4..	77,511	280	269	8,794	......	86,854	27,960	3,220	4,390	35,570
" 11..	107,491	24,366	2,506	6,207	414	140,984	34,890	1,410	4,580	40,880
" 18..	63,491	2,838	5,977	1,500	......	73,806	44,323	10,230	3,787	58,340
" 24..	9,386	......	......	......	......	9,386	32,200	2,920	8,420	43,540
April 1..	18,327	......	......	4,226	......	22,553	39,350	28,560	14,710	82,620
" 8..	65,279	33,607	2,679	5,835	151	107,551	30,340	9,390	9,690	49,420
" 15..	36,502	3,521	1,220	......	45	41,288	37,710	10,800	11,460	59,970
" 22..	14,054	......	3,133	398	......	17,585	33,130	5,250	3,110	41,490
" 29..	69,761	9,104	2,078	4,038	......	84,981	34,240	2,940	5,160	42,340
May 6..	22,231	10,382	1,163	......	381	34,157	45,600	4,950	4,120	54,670
" 13..	18,574	3,350	3,604	......	67	25,595	34,880	6,360	5,620	46,860
" 20..	20,927	......	5,439	......	7	26.373	31,970	13,490	5,870	51,330
" 27..	64,228	14,237	240	2,256	2,696	83,657	42,410	9,380	2,750	54,540
June 3..	84,851	12,935	1,242	474	103	99,605	37,480	9,820	3,470	50,770
" 10..	42,553	......	1,863	3,373	284	48,073	29,350	11,190	2,970	43,510
" 17..	27,029	18,361	100	17,057	......	62,547	39,870	20,390	6,240	66,500
" 24..	48,358	......	2,499	6,215	......	57,072	41,530	8,010	7,440	56,980
July 1..	40,417	15,404	3,507	3,794	35	63,217	45,510	8,710	6,100	60,320
" 8..	62,492	203	8,026	2,740	15	73,476	40,050	3,630	12,050	55,730
" 15..	46,919	8,647	5,906	1,599	......	63,071	33,180	3,560	10,030	46,770
" 22..	22,411	11,741	2,758	2,863	442	40,215	57,310	16,630	12,560	86,500
" 29..	55,670	......	4,614	560	150	60,994	49,070	16,870	11,950	77,890
Aug. 5..	13,451	2,368	2,277	2,964	7	21,067	33,390	3,960	6,860	44,210
" 12..	5,575	......	1,749	1,786	7	9,117	28,020	6,060	7,410	41,490
" 19..	14,846	2,868	1,360	3,887	......	22,961	25,400	3,680	5,820	34,900
" 26..	33,259	5,616	5,767	2,081	15	46,738	22,590	4,240	4,880	31,710
Sept. 2..	16,404	......	503	......	......	16,907	27,760	3,480	5,730	36,970
" 9..	16,404	......	503	615	......	17,522	24,470	2,490	5,630	32,590
" 16..	19,031	33,379	1,910	10,795	39	65,154	20,190	3,610	4,400	28,200
" 23..	16,073	5,524	1,740	230	16	23,583	20,740	1,960	4,330	27,030
" 30..	5,854	7,181	2,570	2,527	183	18,315	26,810	2,820	4,420	34,050
Oct. 7..	1,004	7,604	481	3,310	10	12,409	31,120	3,450	1,600	36,170
" 14..	5,384	5,640	......	98	......	11,122	30,470	3,660	3,170	37,300
" 21..	13,368	4,473	2,399	1,800	62	22,102	32,910	8,120	4,690	45,720
" 28..	5,406	7,881	3,184	5,385	86	21,942	27,520	9,030	3,540	40,090
Nov. 4..	2,940	281	535	1,541	......	5,297	24,610	4,250	4,020	32,880
" 11..	6,159	3,011	......	......	......	9,170	12,160	9,100	3,060	24,320
" 18..	3,296	3,689	......	......	577	7,562	31,350	4,090	1,910	37,350
" 25..	10,823	17,379	1,428	4,301	37	33,968	32,690	5,740	2,750	41,180
Dec. 2..	8,190	12,886	4,202	6,676	......	31,954	37,310	6,310	2,760	46,380
" 9..	3,219	11,450	179	3,809	403	19,060	32,610	4,250	2,220	39,080
" 16..	4,744	......	......	672	......	5,416	34,910	15,860	3,930	54,700
" 23..	....	......	......	......	......	....	32,930	9,740	2,790	45,460
Average prices & total sales, receipts & stocks.	1,479,731	315652	102697	143,524	6,282	2,047,886	1,697,887	408,380	274473	2,380,740

POOL. YEAR 1853.

STOCKS.			PRICES.			ACTUAL EXPORT.	CONSUMPTION.	REMARKS.
American.	Other.	Total.	Mid. Up.	Mid. Orl.	Dhol.			
349,083	240,882	589,965	$5\frac{3}{8}$	$5\frac{9}{16}$	$4\frac{3}{8}$	850	31,860	
363,500	239,536	603,036	$5\frac{3}{8}$	$5\frac{9}{16}$	$4\frac{3}{8}$	2,360	66,340	
354,708	248,859	603,567	$5\frac{9}{16}$	$5\frac{3}{4}$	$4\frac{3}{8}$	6,686	106,954	
345,840	247,237	593 077	$5\frac{13}{16}$	$5\frac{13}{16}$	$4\frac{3}{8}$	4,570	143,834	Favorable acc'ts from manufacturing districts.
346,866	240,819	587,685	$5\frac{13}{16}$	$5\frac{7}{8}$	$4\frac{1}{2}$	7,090	183,244	
392,902	234,109	627,011	$5\frac{3}{4}$	$5\frac{7}{8}$	$4\frac{1}{2}$	3,490	210,364	
377,209	227,244	604,453	$5\frac{11}{16}$	$5\frac{13}{16}$	$4\frac{1}{2}$	4,170	238,884	Heavy receipts at United States ports
385,020	216,274	601,294	$5\frac{9}{16}$	$5\frac{11}{16}$	$4\frac{3}{8}$	2,860	265,574	
436,551	220,247	656,798	$5\frac{1}{2}$	$5\frac{5}{8}$	$4\frac{5}{8}$	4,390	293,534	
514,682	247,130	761,812	$5\frac{7}{16}$	$5\frac{1}{2}$	$4\frac{1}{2}$	4,580	328,424	
532,293	247,814	780,107	$5\frac{1}{2}$	$5\frac{5}{8}$	$4\frac{1}{2}$	3,787	372,747	Good crop accounts from United States.
527,248	241,237	768,485	$5\frac{1}{2}$	$5\frac{5}{8}$	$4\frac{1}{2}$	8,420	404,947	
514,124	234,450	748,574	$5\frac{3}{4}$	$5\frac{7}{8}$	$4\frac{1}{2}$	14,710	444,297	
553,593	266,302	819,895	$5\frac{3}{4}$	$5\frac{7}{8}$	$4\frac{1}{2}$	9,690	474,637	
558,634	259,757	818,391	$5\frac{7}{8}$	$5\frac{15}{16}$	$4\frac{1}{2}$	11,460	512,347	
542,419	251,717	794,136	$5\frac{3}{4}$	$5\frac{15}{16}$	$4\frac{3}{8}$	3,110	545,477	Smaller estimates of crop.
516,034	253,537	769,571	$5\frac{3}{4}$	$5\frac{15}{16}$	$4\frac{3}{8}$	5,160	579,717	
498,015	250,364	748,379	$5\frac{13}{16}$	$5\frac{7}{8}$	$4\frac{3}{8}$	4,120	625,317	
480,934	248,061	728,995	$5\frac{7}{8}$	$5\frac{15}{16}$	$4\frac{3}{8}$	5,620	660,197	
467,471	243,027	710,498	$5\frac{13}{16}$	6	$4\frac{3}{8}$	5,870	692,167	
497,819	247,416	745,235	$5\frac{15}{16}$	6	$4\frac{3}{8}$	2,750	734,572	
549,520	251,370	800,890	$5\frac{15}{16}$	$5\frac{15}{16}$	$4\frac{3}{8}$	3,470	772,057	
569,213	246,230	815,443	$5\frac{1}{4}$	$5\frac{15}{16}$	$4\frac{1}{2}$	2,970	801,407	
561,792	256,331	818,123	$5\frac{15}{16}$	$6\frac{1}{16}$	$4\frac{1}{2}$	6,240	841.277	Probabilty of war in Eastern Europe.
557,130	255,235	812,365	$5\frac{13}{16}$	$6\frac{1}{16}$	$4\frac{5}{8}$	7,440	882,807	
558,667	267,905	826,572	6	$6\frac{1}{8}$	$4\frac{1}{2}$	6,100	928,317	
583,740	269,708	853,448	6	$6\frac{1}{8}$	$4\frac{1}{2}$	12,050	968,367	
600,908	277,921	878,829	6	$6\frac{1}{8}$	$4\frac{1}{2}$	10,030	1,001,547	Unfavorable reports of growing crop.
576,391	280,444	856,834	$6\frac{1}{8}$	$6\frac{13}{16}$	$4\frac{1}{2}$	12,560	1,058,857	
588,021	274,588	862,609	$6\frac{1}{2}$	$6\frac{3}{16}$	$4\frac{1}{2}$	11,950	1,107,927	
576,072	271,457	847,529	$6\frac{1}{16}$	$6\frac{1}{8}$	$4\frac{1}{2}$	6,860	1,141,317	
541,589	258,032	799,621	$6\frac{1}{16}$	$6\frac{1}{8}$	$4\frac{1}{2}$	7,410	1,169,337	
529,115	255,517	784,632	6	$6\frac{1}{8}$	$4\frac{1}{2}$	5,820	1,194,737	
540,049	262,779	802,828	6	$6\frac{1}{8}$	$4\frac{1}{2}$	4,880	1,217,327	
524,400	255,590	779,990	6	$6\frac{1}{8}$	$4\frac{1}{2}$	5,730	1,245,087	Unfavorable political news.
507,510	251,972	759,482	$5\frac{15}{16}$	$6\frac{1}{8}$	$4\frac{1}{2}$	5,630	1,269,557	
509,561	291,970	801,531	$5\frac{3}{16}$	6	$4\frac{3}{8}$	4,400	1,289,747	
508,554	294,200	802,754	$5\frac{11}{16}$	$5\frac{7}{8}$	$4\frac{1}{4}$	4,330	1,310,487	
493,218	299,381	792,599	$5\frac{5}{8}$	$5\frac{3}{4}$	$4\frac{1}{8}$	4,420	1,337,297	
468,632	299,136	767,768	$5\frac{1}{2}$	$5\frac{5}{8}$	$4\frac{1}{8}$	1,600	1,368,417	Bad harvest accounts, &c.
452,084	301,618	753,702	$5\frac{7}{8}$	$5\frac{7}{8}$	$4\frac{1}{8}$	3,170	1,398,887	
436,082	302,922	739,004	$5\frac{13}{16}$	6	$4\frac{1}{4}$	4,690	1,431,797	
419,618	309,568	729,186	$5\frac{13}{16}$	6	$4\frac{1}{4}$	3,540	1,459,317	Wages agitation in manufacturing districts.
400,218	307,035	707,253	$5\frac{3}{4}$	$5\frac{7}{8}$	$4\frac{1}{8}$	4,020	1,483,927	
380,727	300,236	680 965	$5\frac{3}{4}$	6	$4\frac{1}{8}$	3,060	1,496,087	
358,943	296,322	655,265	$5\frac{3}{4}$	6	$4\frac{1}{8}$	1,910	1,527,437	Frost in cotton districts.
345,856	308,337	654,193	$5\frac{3}{4}$	6	$4\frac{1}{8}$	2,750	1,560,127	
324,156	322,921	647,077	$5\frac{7}{8}$	6	$4\frac{1}{8}$	2,760	1,597,437	
302,125	329,412	631,537	$5\frac{7}{8}$	6	$4\frac{1}{8}$	2,220	1,630,047	Outbreak of war in Eastern Europe.
280,939	316,964	597,903	$5\frac{7}{8}$	6	$4\frac{1}{8}$	3,930	1,664,957	
354,280	303,660	557,940	$5\frac{7}{8}$	6	$4\frac{1}{8}$	2,790	1,697,887	
			5.54	5.94	4.38	274,473	332,919	

The semi-weekly Price and Weekly Sales and Receipts, at New York, Weekly Exports from New York and Rates of Freight to Liverpool 1*st of each Month, for the Crop Year ending September* 1, 1853.

1852.	Price of Mid. Fair to Fair New Orleans Liverpool Classificat'n.	Price of Mid. Fair to Fair Upland, Liverpool Classificati'n	Sales for week.	Receipts for week.	EXPORTS FOR WEEK.					Rates of Freight to Liverpool.
					To Great Britain.	To France.	North of Europe.	Other Fo'n Ports	Total Exports.	
Septem. 3..	11¼@11¾	11@11¼								⅛d.
" 7..	11@11¾	10¾@11	3,000	3,800	....	....	....	....	....	
" 10..	11@11¾	10¾@11								
" 14..	11@11¾	10⅝@11⅛	6,800	1,415	3,782	....	124	....	3,906	
" 17..	11½@11½	11@11¼								
" 21..	11¼@11¾	11@11¼	9,300	2,041	2,461	709	....	435	3,605	
" 24.	11¼@11¾	11@11¼								
" 28..	11¼@11¾	10⅞@11⅛	3,350	2,581	1,738	332	293	....	2,363	⅛d.
October 1..	10¾@11¾	10¾@11⅛								
" 5..	11@11¾	10¾@11	5,100	1,209	4,147	930	310	....	5,387	
" 8..	11@11½	10⅝@10⅞								
" 12..	11@11½	10⅝@10⅞	4,600	7,226	1,676	971	....	551	3,198	
" 15..	11@11¾	10¾@11								
" 19..	11@11¾	10¾@11	11,000	2,951	726	205	60	....	991	
" 22..	11@11¾	10⅝@10⅞								
" 26..	11@11½	10½@10¾	6,100	2,979	3,097	961	167	....	4,225	
" 29..	11@11½	10½@10¾								
Novem. 2..	11@11½	10½@10¾	8,200	8,920	4,437	....	703	....	5,140	¼d.
" 5..	11@11½	10⅝@10¾								
" 9..	11¼@11¾	10⅝@10¾	9,300	4,593	2,860	....	319	....	3,179	
" 12..	11½@12	10⅞@11⅛								
" 16..	11¼@11¾	10¾@11	8,900	8,772	6,151	772	....	683	7,606	
" 19..	11@11½	10½@10¾								
" 23..	11@11½	10½@10⅝	7,900	9,366	1,968	1,363	....	....	3,331	
" 26..	10½@11	10⅛@10⅜								
" 30..	10½@11	10⅛@10⅜	4,600	15,452	2,012	....	565	....	2,577	
Decem. 3..	10½@11	10⅛@10⅜								⅜d.
" 7..	10½@11	10@10¼	7,600	10,029	4,255	....	....	25	4,280	
" 10..	10½@11	9⅞@10⅛								
" 14..	10¼@11	9⅝@9⅞	8,600	15,192	2,745	1,665	379	....	4,789	
" 17..	10¼@11	9¾@10								
" 21..	10½@11¼	9¾@10⅛	11,500	7,192	4,226	....	....	80	4,306	
" 24..	10¾@11¼	9⅞@10¼								

GENERAL REMARKS.

The fluctuations in prices were not many or wide this crop year. Early in September there were reports current of injury to the crop, and again in October, but the amount of damage proved to have been exaggerated. In January, again, there was found to be another error in the Liverpool stock of 46,000 bales, which had an unfavorable effect The receipts at the shipping ports fell off considerably in April, and prices advanced both here and abroad. Through the next four months the market was very steady, though a feeling of caution began to obtain in the latter part of June, and continued during the residue of the crop year, consequent upon the menacing aspect of affairs between Russia and the Porte.

Decem. 28..	$10\frac{1}{2}$@$11\frac{1}{4}$	$9\frac{5}{8}$@$9\frac{7}{8}$	6,800	9,891	3,946	....	26	....		
" 31..	$10\frac{3}{4}$@$11\frac{1}{4}$	$9\frac{3}{4}$@10							3,972	
1853.										
January 4..	11@$11\frac{1}{2}$	10@$10\frac{1}{4}$	8,400	7,204	3,550	....	....	36	3,586	$\frac{5}{8}$d.
" 7..	11@$11\frac{3}{4}$	10@$10\frac{3}{8}$								
" 11..	11@12	$10\frac{1}{4}$@$10\frac{1}{2}$	6,500	10,761	3,092	2,037	....	....	5,129	
" 14..	11@12	$10\frac{1}{4}$@$10\frac{5}{8}$								
" 18..	$11\frac{1}{2}$@$12\frac{1}{4}$	$10\frac{1}{2}$@11	8,800	1,742	2,196	165	....	....	2,361	
" 21..	$11\frac{1}{2}$@$12\frac{1}{4}$	11@$11\frac{1}{4}$								
" 25..	$11\frac{1}{2}$@$12\frac{1}{2}$	$11\frac{1}{8}$@$11\frac{3}{8}$	7,800	16,217	1,330	289	691	....	2,310	
" 28..	$11\frac{1}{2}$@$12\frac{1}{2}$	11@$11\frac{3}{8}$								
February 1..	$11\frac{1}{4}$@$12\frac{1}{2}$	$10\frac{7}{8}$@$11\frac{1}{4}$	4,850	13,063	2,663	....	1,325	594	4,582	$\frac{3}{16}$@$\frac{1}{4}$d.
" 4..	$11\frac{1}{4}$@$12\frac{1}{2}$	$10\frac{3}{4}$@$11\frac{1}{8}$								
" 8..	$11\frac{1}{4}$@$12\frac{1}{2}$	$10\frac{7}{8}$@$11\frac{1}{4}$	9,100	10,006	1,578	1,380	569	....	3,527	
" 11..	$11\frac{1}{4}$@$12\frac{1}{2}$	$10\frac{5}{8}$@11								
" 15..	11@$12\frac{1}{4}$	$10\frac{1}{2}$@11	5,700	13,502	....	....	....	....	....	
" 18..	11@$12\frac{1}{4}$	$10\frac{1}{2}$@$10\frac{3}{4}$								
" 22..	$10\frac{3}{4}$@12	$10\frac{1}{4}$@$10\frac{3}{4}$	6,400	6,311	3,219	1,676	729	....	5,624	
" 25..	$10\frac{3}{4}$@$11\frac{3}{4}$	$10\frac{1}{4}$@$10\frac{5}{8}$								
March 1..	$10\frac{3}{4}$@$11\frac{3}{4}$	$10\frac{1}{2}$@$10\frac{3}{4}$	8,300	22,380	776	2,120	......	54	2,950	$\frac{7}{32}$@$\frac{1}{4}$d.
" 4..	11@12	$10\frac{5}{8}$@11								
" 8..	11@12	$10\frac{1}{2}$@11	11,800	19,608	3,920	....	1,124	....	5,044	
" 11..	11@12	$10\frac{1}{2}$@11								
" 15..	11@12	$10\frac{1}{2}$@11	8,400	14,909	49	694	1,296	50	2,089	
" 18..	$11\frac{1}{2}$@$12\frac{1}{2}$	11@$11\frac{1}{2}$								
" 22..	$11\frac{1}{2}$@$12\frac{1}{2}$	11@$11\frac{1}{2}$	17,300	13,474	5,425	6,089	253	....	11,767	
" 25..	$11\frac{1}{2}$@$12\frac{1}{2}$	11@$11\frac{1}{2}$								
" 29..	$11\frac{1}{2}$@$12\frac{1}{2}$	11@$11\frac{1}{2}$	8,200	18,692	3,577	2,767	1,788	67	8,199	
April 1..	$11\frac{1}{2}$@$12\frac{1}{2}$	$11\frac{1}{4}$@$11\frac{1}{2}$								$\frac{9}{32}$@$\frac{5}{16}$d.
" 5..	$11\frac{3}{4}$@$12\frac{3}{4}$	$11\frac{1}{4}$@$11\frac{3}{4}$	13,300	5,740	5,686	2,888	1,351	....	9,925	
" 8..	$11\frac{3}{4}$@$12\frac{3}{4}$	$11\frac{1}{4}$@$11\frac{3}{4}$								
" 12..	$11\frac{3}{4}$@$12\frac{3}{4}$	$11\frac{1}{4}$@$11\frac{3}{4}$	7,000	15,416	9,634	635	658	....	10,927	
" 15..	$11\frac{3}{4}$@$12\frac{3}{4}$	$11\frac{3}{8}$@$11\frac{3}{4}$								
" 19..	12@13	$11\frac{3}{8}$@$11\frac{3}{4}$	16,200	12,377	5,681	215	894	....	6,790	
" 22..	$11\frac{3}{4}$@13	$11\frac{1}{4}$@$11\frac{3}{4}$								
" 26..	$11\frac{3}{4}$@13	$11\frac{1}{4}$@$11\frac{3}{4}$	10,000	14,417	5,895	141	152	45	6,233	
" 29..	$11\frac{3}{4}$@13	$11\frac{1}{4}$@$11\frac{3}{4}$								
May 3..	$11\frac{1}{2}$@$12\frac{3}{4}$	11@$11\frac{1}{2}$	17,900	10,986	5,314	....	1,776	....	7,090	$\frac{1}{4}$@$\frac{9}{32}$d.
" 6..	$11\frac{3}{4}$@$12\frac{3}{4}$	11@$11\frac{1}{2}$								
" 10..	$11\frac{1}{2}$@$12\frac{1}{2}$	11@$11\frac{1}{4}$	13,000	17,775	4,151	1,797	2,100	....	8,048	
" 13..	$11\frac{1}{2}$@$12\frac{1}{2}$	11@$11\frac{3}{8}$								
" 17..	$11\frac{1}{2}$@$12\frac{1}{2}$	11@$11\frac{3}{8}$	10,300	12,760	4,791	2,721	2,024	43	9,579	

New York Statement for 1853.—*Concluded.*

1853.		Price of Mid. Fair to Fair New Orleans Liverpool Classificat'n.	Price of Mid. Fair to Fair Upland, Liverpool Classificat'n.	Sales for week.	Receipts for week.	EXPORTS FOR THE WEEK.					Rates of Freight to Liverpool.	General Remarks.
						To Great Britain.	To France.	North of Europe.	Other Fo'n Ports	Total Exports.		
May	20..	11½@12½	11@11⅜									
"	24..	11½@12½	11@11⅜	11,100	12,614	7,425	....	2,827	....	10,252		*Exchange.*
"	27..	11¾@12¾	11¼@11⅝									Bills on London were steady throughout September at 10⅛@10⅝ per cent. premium; October, 10@10½; November, 9¾@10¼; December, 9½@10¼; January the range was from 9 to 9⅞; February, 9⅜@10¼; in March, the quotation fell from 9¾@10 to 8¾@9¼; in April it was steady at 9@9¾; May, 9⅜@10; June, the same; July, 9½@9⅞; and in August the range was between 8 and 9¾ per cent. premium.
"	31..	12@13	11½@11¾	30,100	16,171	4,950	626	1,531	....	7,107		
June	3..	12@13	11½@11¾								¼@$\frac{9}{32}$d.	
"	7..	12@13	11½@11¾	14,500	10,659	8,827	1,760	1,465	....	12,052		
"	10..	12@13	11½@11¾									
"	14..	12@13	11½@11¾	7,600	7,483	7,024	65	1,351	....	8,440		
"	17..	12@13	11½@11¾									
"	21..	12@13	11⅜@11¾	9,200	13,437	10,933	....	788	50	11,771		
"	24..	12@13	11¼@11¾									
"	28..	12@13	11¼@11¾	6,300	9,125	3,799	318	75	....	4,192		
July	1.	12@13	11¼@11¾								⅛@$\frac{5}{32}$d.	
"	5..	12@13	11¼@11¾	10,500	8,184	4,596	1,606	728	....	6,930		
"	8..	12@13	11⅜@11¾									
"	12..	12@13	11⅜@11¾	9,000	8,391	7,847	....	1,078	....	8,925		
"	15..	12@13	11⅜@11¾									
"	19..	12@13	11⅜@11¾	3,400	5,062	5,908	....	594	....	6,502		
"	22..	12@13	11⅜@11¾									
"	26..	12@13	11¼@11¾	10,400	6,995	6,283	66	219	....	6,568		
"	29..	12@13	11¼@11¾									
August	2..	12@13	11¼@11¾	10,550	7,673	1,350	278	100	....	1,728	¼d.	
"	5..	12@13	11½@11¾									
"	9..	12@13	11½@11¾	8,400	3,332	3,807	298	514	50	4,669		
"	12..	12@13	11½@11¾									
"	16..	12@13	11½@11¾	5,600	2,591	4,182	....	....	....	4,182		
"	19..	12@13	11⅜@11⅝									
"	23..	12@13	11⅜@11⅝	3,500	3,734	3,579	....	484	....	4,063		
"	26..	12@13	11⅜@11⅝									
"	30..	12@13	11⅜@11¾	8,100	682	4,322	801	1,124	....	6,247	$\frac{3}{16}$@¼d.	
Septem.	2..	12@13	11¼@11¾									
Average price and total sales, receipts and exports.		11.81	11.02	470,150	487,082	207,586	39,340	32,554	2,763	282,243		

1854.

The following table shows the number of mills, spindles and looms, using cotton wholly, in operation in New England at this time (*see* year 1840):

STATES.	Mills.	Looms.	Spindles.
Maine	15	3,439	113,900
New Hampshire	40	12,462	440,401
Massachusetts	165	32,655	1,288,091
Vermont	12	345	31,736
Rhode Island	166	28,233	624,138
Connecticut	109	6,506	252,812
Total	507	82,640	2,754,078

The cotton manufactures of Ireland—centred almost entirely in Belfast and its neighborhood—at this time employed 5,000 persons. (*See* years 1770 and 1790.)

During the twenty years preceding this date, the American settlers of Liberia established communication across the country to Timbuctoo, and have found there a considerable market for cotton goods.

The value of the annual production of the cotton manufacture of Great Britain this year, was estimated at £54,000,000 sterling, of which nearly £33,000,000 was the value of goods and yarn made for exportation.

Backward spring, unseasonable rains, yellow fever very prevalent, average picking season in the United States.

COTTON CROP OF THE UNITED STATES.

Statement and Total Amount of the Cotton Crop of the United States, for the Year ending August 31, 1854.

	Bales.	Bales.	Total.	Same period 1853.
NEW ORLEANS.				
Export—				
To Foreign Ports	1,236,653			
Coastwise	192,527			
Stock on hand 1st September, 1854	24,121			
		1,453,301		
Deduct—				
Stock on hand 1st September, 1853	10,522			
Received from Mobile, Montgomery, &c	64,806			
" Florida	9,368			
" Texas	21,680			
		106,376		
			1,346,925	1,580,875
MOBILE.				
Export—				
To Foreign Ports	336,963			
Coastwise	178,668			
Consumed in Mobile	1,465			
Stock on hand 1st September, 1854	29,278			
		546,374		
Deduct—				
Received from New Orleans	63			
" Texas	111			
Stock on hand 1st September, 1853	7,516			
		7,690		
			538,684	545,029
TEXAS.				
Export—				
To Foreign Ports	18,467			
Coastwise	90,081			
Stock on hand 1st September, 1854	2,205			
		110,753		
Deduct—				
Stock on hand 1st September, 1853	....	428		
			110,325	85,790
FLORIDA.				
Export—				
To Foreign Ports	49,190			
Coastwise	106,194			
Stock in Apalachicola, 1st September, 1854	583			
		155,967		
Deduct—				
Stock in Apalachicola, 1st September, 1853	....	523		
			155,444	179,476
GEORGIA.				
Export—				
To Foreign Ports—Uplands	98,580			
" Sea Island	3,861			
Coastwise—Uplands	203,363			
Sea Island	11,667			
Stock in Savannah, 1st September, 1854	3,200			
" Augusta, 1st September, 1854	8,318			
		328,989		
Deduct—				
Stock in Savannah and Augusta, 1st September, 1853	....	12,984		
			316,005	349,490

Statement and Total Amount of the Cotton Crop of the United States, for the Year ending August 31, 1854—*Concluded.*

	Bales.	Bales.	Total.	Same period 1853.
SOUTH CAROLINA.				
Export from Charleston—				
To Foreign Ports—Uplands	217,603			
" Sea Island	18,154			
Coastwise—Uplands	190,675			
Sea Island	6,612			
Stock in Charleston, 1st September, 1854	17,031			
Export from Georgetown—	450,075			
To Northern Ports	3,209			
		453,284		
Deduct—				
Stock in Charleston, 1st September, 1853	15,126			
Received from Florida	4,133			
" Savannah	17,271			
		36,530		
			416,754	463,203
NORTH CAROLINA.				
Export—				
To Foreign Ports	....			
Coastwise	11,524			
		....	11,524	23,496
VIRGINIA.				
Export—				
To Foreign Ports	500			
Coastwise, and Manufactured (taken from the ports)	21,086			
Stock on hand, 1st September, 1854	750			
		22,336		
Deduct—				
Stock on hand 1st September, 1853	....	400		
			21,936	25,783
Received at New York by New York and Erie Canal	....	....	1,182	....
" New York by New York & Erie R.R.	....	....	2,258	640
" Baltimore & Phila. from Tennessee, &c	....	....	8,990	9,100
Total crop of the United States	....	....	2,930,027	3,262,882

Decrease from last year .... bales. 332,855
Decrease from year before 85,002

Export to Foreign Ports, from September 1, 1853, *to August* 31, 1854.

FROM	To Great Britain.	To France.	To North of Europe.	Other F'n Ports.	Total.
New Orleans .... Bales	813,736	193,571	93,375	135,971	1,236,653
Mobile	231,230	76,752	14,466	14,515	336,963
Texas	6,191	4,275	8,001	....	18,467
Florida	43,086	1,965	2,429	1,710	49,190
Savannah	92,363	6,487	2,921	670	102,441
Charleston	162,970	41,245	12,641	18,901	235,757
North Carolina	....	....	....	....	....
Virginia	500	....	....	....	500
Baltimore	2,159	....	200	52	2,411
Philadelphia	3,490	....	....	1,472	4,962
New York	245,746	49,763	29,845	2,742	328,096
Boston	2,279	....	1,294	135	3,708
Grand total	1,603,750	374,058	165,172	176,168	2,319,148
Total last year	1,736,860	426,728	171,176	193,636	2,528,400
Decrease	133,110	52,670	6,004	17,468	209,252

Consumption.

Total crop of the United States, as before stated .... bales.			2,930,027
Add—Stocks on hand at the commencement of the year, September 1, 1853—			
In the Southern ports		47,499	
In the Northern ports		88,144	
			135,643
Makes a supply of			3,065,670
Deduct therefrom—The Export to Foreign ports	2,319,148		
Less Foreign included	1,565		
		2,317,583	
Stocks on hand, September 1, 1854—			
In the Southern ports	85,486		
In the Northern ports	50,117		
		135,603	
Burnt at New York and Philadelphia		1,913	
			2,455,099
Taken for home use .... bales.			610,571

ANNUAL REVIEW.

From the New Orleans Price Current—1853–54.

Towards the close of November, however, the market rallied again, under the influence of an improved demand, which was instigated by more favorable advices from Europe, and by accounts of frosts through a large portion of the cotton region, which would materially reduce the crop in quantity, besides injuring it in quality. At this juncture the prices again gave way, under pressure of the unfavorable aspect of European affairs and unusually high rates of freight, and with various fluctuations, taking an extreme range of $1\frac{3}{4}$ cents per pound; the lowest point of the market, was reached in the latter part of May, when the quotations were for low middling $6\frac{3}{4}$ to 7, middling $7\frac{1}{4}$ to $7\frac{1}{2}$, good middling $8\frac{1}{4}$ to $8\frac{1}{2}$ cents per pound. At this period the quotation for freight of cotton to Liverpool was $\frac{15}{16}$ to 1d. per pound, with little or no room immediately available, even at these high rates, and the operations of purchasers were checked by the impossibility to effect prompt shipments; while, at the same time, a large stock had accumulated in the hands of exporters, who had bought from time to time and held their purchases in store, in the hope of shipping on more favorable terms. Prices rallied again in the latter part of May, and during June; there were some sales in July, which showed a recovery of $\frac{1}{2}$ cent from the lowest point, the stock on sale being much reduced; the advices from abroad, rather more favorable, and freight to Liverpool down to $\frac{11}{16}$d. In August the transactions were comparatively unimportant; and thus closed a season which, we suppose, has proved little satisfactory to any of the parties interested, the perplexities and uncertainties growing out of the European war question having led to fluctuations that baffled all commercial calculations. The crop in quality, as we have already intimated, was of a very low average; resulting partly from unseasonable rains and partly from frost damage, but mainly from careless and hasty picking, which looked more to quantity than quality, and thus the proportion of the finer grades has been unusually small, while the lower qualities have been abundant. The proportion of frost-stained cotton has been greater than ever before, and factors have found much difficulty in disposing of it, as most orders prohibited its purchase. The import of Surats into Great Britain in 1853 exceeded the import in 1852 by 264,114 bales.

The semi-weekly Price and Weekly Sales and Receipts at New York, Weekly Exports from New York and Rates of Freight to Liverpool 1st of each month, for the Crop Year ending September 1, 1854.

1853.	Price of Mid. Fair to Fair New Orleans * Liverpool Classification.	Price of Mid. Fair to Fair Upland, * Liverpool Classification.	Sales for week.	Receipts for week.	EXPORTS FOR THE WEEK.					Rates of Freight to Liverpool.
					To Great Britain.	To France.	North of Europe.	Other Fo'n Ports	Total Exports.	
Septem. 6..	12½@13	11¼@11¾	10,600	3,420	2,588	236	348	....	3,172	¼@9-32d
" 9..	12½@13	11¼@11¾								
" 13..	12@13	11¼@11¾	4,300	2,550	3,072	505	291	25	3,893	
" 16..	12@13	11¼@11¾								
" 20..	12@13	11¼@11¾	2,400	6,097	7,473	674	2,703	2	10,852	
" 23..	12@13	11¼@11¾								
" 27..	12@13	11⅜@11¾	6,950	2,732	7,546	28	192	....	7,766	
" 30..	12@13	11⅜@11¾								
October 4..	12@13	11¼@11¾	3,700	3,498	2,751	12	1,279	276	4,318	¼d.
" 7..	12@13	11¼@11¾								
" 11..	12@13	11¼@11¾	2,900	4,726	2,962	....	405	....	3,367	
" 14..	—@—	10¾@11¼								
" 17..	—@—	—@—	9,204	6,029	5,005	51	1,083	....	6,139	
" 21..	--@—	10¼@10½								
" 25..	—@—	10¼@10¾	10,204	5,798	3,935	606	....	....	4,541	
" 28..	11½@12	10⅝@11¼								
Novem. 1..	11¾@12½	11@11½	11,841	4,870	4,203	....	289	....	4,492	¼@9-32d
" 4..	11¾@12½	11@11½								
" 8..	11¾@12½	11@11½	7,528	3,797	2,610	231	573	....	3,414	
" 11..	11¾@12½	11⅛@11½								
" 15..	11½@12¼	11@11⅜	9,114	10,097	3,010	44	....	....	3,054	
" 18..	11½@12¼	11@11⅜								
" 22..	11¾@12½	11@11½	11,568	7,688	5,614	37	64	....	5,715	
" 25..	11¾@12½	11@11½								
" 29..	11¾@12½	11¼@11⅝	9,693	4,698	2,170	....	278	....	2,448	
Decem. 2..	11¾@12½	11¼@11⅝								⅛@5-32d
" 6..	11⅝@12¼	11@11⅜	9,976	10,039	4,960	787	141	....	5,888	
" 9..	11⅝@12¼	11@11⅜								
" 13..	11¾@12½	11@11⅜	15,566	8,965	2,933	450	....	....	3,383	
" 16..	11¾@12½	11@11⅜								
" 20..	12@12½	11@11⅜	16,059	7,771	1,516	256	148	....	1,920	
" 23..	12@12½	11@11⅜								
" 27..	12@12⅝	11@11⅜	10,023	17,727	5,981	136	....	....	6,117	

GENERAL REMARKS.

After November 1, 1853, New York Classification.

The heavy rains in August and September, 1853, with picking two weeks later than usual, and then early frosts—a killing frost in Georgia, October 25—for a time, in a measure, sustained the market against other adverse influences.

Early in October, the Russian-Turkish imbroglio assumed a grave aspect, besides which the poor crops, both of England and France, proved to be below an average, all tending to repress buoyancy in the cotton markets; the money market here became more stringent, and a want of confidence was observable; on the 18th of this month (October), accounts reached us of a declaration of war between Turkey and Russia, and business at once came to a stand, prices falling off ½@1 cent per lb.

It was during October a Cotton Brokers' Association was first formed here, and market quotations thenceforward were based upon a New York classification, which classed middling half a grade higher than before.

Decem.	30..	12@12⅝	11@11⅜								In November, many mills in the manufacturing districts of England closed on account of strikes, and all through the Fall, yellow fever prevailed with more or less virulence in New Orleans and Mobile.
1854.											
January	3..	11¾@12½	10⅞@11¼	7,007	12,006	1,443	48	728	....	2,219	5-16d.
"	6..	11¾@12½	10⅞@11¼								
"	10..	11⅝@12¼	10¾@11⅛	9,933	18,782	1,355	2,056	732	....	4,143	
"	13..	11⅝@12¼	10¾@11⅛								
"	17..	11½@12¼	10⅝@11	13,142	8,344	5,140	823	259	231	6,453		In January again, another error was ascertained to have been made in the Liverpool stock, of 50,000 bales, the supply being that much in excess of the previous estimates.
"	20..	11½@12¼	10⅝@11								
"	24..	11½@12¼	10¾@11⅛	13,133	13,838	5,249	149	534	202	6,134	
"	27..	11½@12¼	10¾@11⅛								
"	31..	11½@12¼	10¾@11⅛	9,715	5,131	2,952	1,456	....	19	4,427	
February	3..	11½@12¼	10¾@11⅛							$\frac{5}{16}$@$\frac{11}{32}$d.	In February, dear food, and the complications of the Eastern question, caused great dullness in the European markets, though the effects were partially counteracted here by the large falling off in the receipts at the ports.
"	7..	11⅜@12¼	10¾@11⅛	13,085	12,384	4,470	901	694	....	6,065	
"	10..	11⅝@12¼	10¾@11⅛								
"	14..	11½@12¼	10½@11	4,111	17,210	9,541	32	1,230	....	10,803	
"	17..	11½@12¼	10½@11								
"	21..	11½@12¼	10⅜@10⅞	10,624	12,783	5,914	....	140	....	6,054	
"	24..	11½@12¼	10⅜@10⅞								
"	28..	11¾@12½	10¾@11⅛	18,839	10,391	5,457	70	886	....	6,413		On the 4th of April accounts were received of the rejection, by the Czar, of the ultimatum of England and France, and on the 14th, of the declaration of war between those powers and Russia; which had a depressing effect upon the market, followed by a considerable decline in prices.
March	3..	11¾@12½	10¾@11⅛							$\frac{7}{16}$@$\frac{15}{32}$d.	
"	7..	11¾@12½	11@11¼	19,244	8,418	7,130	683	654	....	8,467	
"	10..	11¾@12½	11@11¼								
"	14..	11½@12¼	10¾@11¼	10,407	13,176	5,887	388	677	....	6,952	
"	17..	11½@12¼	10¾@11¼								
"	21..	11¾@12½	11@11½	15,445	13,722	4,433	....	79	....	4,512	
"	24..	11¾@12½	11@11½								
"	28..	11½@12½	10¾@11¼	7,787	10,167	6,609	....	579	....	7,188	
"	31..	11½@12½	10¾@11¼								Frosts occurred in Georgia, and as far south as Louisiana, in the latter part of April, so severe as to make replanting a large portion of the plantations a necessity; but the very heavy stock in Liverpool, and dull markets in Europe generally, militated against any substantial rise; these unfavorable crop prospects continued through May and June, the plant suffering up to that time from excessive rains, and the course of prices was gradually upward.
April	4..	11½@12½	10¾@11⅛	8,396	19,529	7,290	1,020	....	17	8,327	⅜d.	
"	7..	11½@12½	10¾@11⅛								
"	11..	11@12¼	10⅜@10⅞	5,610	18,807	5,675	575	131	....	6,381	
"	14..	11@12¼	10⅜@10⅞								
"	18..	10½@11½	10@10½	10,467	3,793	8,039	....	656	....	8,695	
"	21..	10½@11½	10@10½								
"	25..	11@12¼	10½@11	10,060	18,117	2,192	....	....	....	2,192	
"	28..	11@12¼	10½@11								
May	2..	11@12	10¼@10¾	7,092	10,714	3,024	1,078	99	179	4,380	5-16@⅜d
"	5..	11@12	10¼@10¾								
"	9..	11@12¼	10¼@11	8,420	4,456	4,325	625	1,444	....	6,394	
"	12..	11@12¼	10¼@11								
"	16..	11⅜@12¼	10½@11⅛	11,896	15,498	8,723	1,212	1,607	....	11,542	
"	19..	11⅜@12¼	10½@11⅛								

New York Statement for Year 1854—*Concluded.*

1854.	Price of Mid. Fair to Fair New Orleans Liverpool Classification.	Price of Mid. Fair to Fair Upland, Liverpool Classification.	Sales for week.	Receipts for week.	EXPORTS FOR THE WEEK.					Rates of Freight to Liverpool.
					To Great Britain.	To France.	North of Europe.	Other Fo'n Ports	Total Exports.	
May 23..	11⅛@12¼	10¼@11	9,450	8,257	4,464	....	1,132	....	5,596	
" 26..	11⅛@12¼	10¼@11								
" 30..	11¼@12¼	10¼@11	7,166	5,261	5,642	2,496	511	....	8,649	
June 2..	11¼@12¼	10¼@11								
" 6..	10¾@12	10¼@10¾	10,000	3,398	9,443	2,446	330	....	12,219	¼d.
" 9..	10¾@12	10¼@10¾								
" 13..	10½@11¾	10@10½	9,500	12,431	2,372	3,961	1,319	....	7,652	
" 16..	11@12	10¼@10¾								
" 20..	11@12	10¼@10¾	15,000	12,395	5,426	160	437	....	6,023	
" 23..	11@12	10¼@10¾								
" 27..	11¼@12¼	10½@11	7,000	1,683	11,085	3,925	315	697	16,022	
July 1..	11¼@12¼	10½@11								3-16d.
" 4..	11¼@12¼	10⅝@11	9,400	3,647	4,884	2,781	503	42	8,210	
" 7..	11¼@12¼	10⅝@11								
" 11..	11½@12½	10¾@11¼	12,500	4,915	3,628	4,289	127	869	8,913	
" 14..	11½@12½	10¾@11¼								
" 18..	11½@12½	10¾@11¼	14,000	3,493	3,234	68	513	....	3,815	
" 21..	11½@12½	10¾@11¼								
" 25..	11½@12½	10⅞@11⅜	9,000	7,903	424	....	1,907	72	2,403	
" 28..	11¼@12½	10¾@11⅛								
August 1..	11½@12½	10⅝@11¼	6,100	4,915	7,352	4,158	458	....	11,968	⅛d.
" 4..	11½@12½	10⅝@11¼								
" 8..	11½@12½	10¾@11¼	6.500	5,947	6,390	2,987	968	....	10,345	
" 11..	11½@12½	10¾@11¼								
" 15..	11½@12½	10¾@11¼	9,000	4,114	2,722	....	927	....	3,649	
" 18..	11½@12½	10¾@11¼								
" 22..	11½@12½	10⅞@11⅜	4,000	9,209	773	2,863	663	....	4,299	
" 25..	11½@12½	10⅞@11⅜								
" 29..	11¼@12¼	10¾@11¼	4,300	5,137	4,605	4.460	812	111	9,988	
Septem. 1..	11@12	10½@11								⅛d.
Average prices and total sales, receipts and exports.	11.85	10.97	498,955	450,473	245,621	49,763	29,845	2,742	327,971	

GENERAL REMARKS.

About this time, say July 8, a money panic was precipitated, owing to the discovery of the large issue of stock of the New Haven Railroad; confidence, for a time, was unsettled, and business restricted.

August opened with a subsidence of the stringency in financial affairs, and the market became steady, though at the close, the crop prospects being good and the foreign markets dull, prices fell off a little.

Exchange.

Bills on London were steady through September at 8½@9¾ per cent. premium; October, 9@10; in November the range was from 8¾@9¼ to 9¾@10; in December the extremes were 9 to 10; January, 8½@9½; February, 8½@9⅛; March, 8¼@9; April, 8¼@9½; May, 9@9⅝; June, 9@9¾; July, 9@9⅝; and August, 9¼@9⅞.

LIVERPOOL STATEMENT FOR 1854.

UNITED STATES, 1853–1854.

Stock Sept. 1, 1853	136,000	Export	2,319,000
Crop	2,928,000	Consumption	611,000
		Stock Sept. 1, 1854	134,000
Bales	3,064,000	Bales	3,064,000

				Stock Jan. 1, 1854, in	Gt. Britain.	France.	Continent.	Tot. Europe.
				United States	309,000	26,000	38,000	373,000
				Brazil	49,000	....	1,000	50,000
				West Indies	4,000	2,000	2,000	8,000
				East Indies	271,000	....	5,000	276,000
				Egypt	85,000	22,000	10,000	117,000
				Bales	718,000	50,000	56,000	824,000
CONSUMPTION.				IMPORT.				
Tot. Europe.	Continent.	France.	Gt. Britain.					
2,381,000	469,000	385,000	1,527,000	United States	1,667,000	432,000	468,000	2,430,000
121,000	18,000	2,000	101,000	Brazil	107,000	2,000	25,000	126,000
30,000	10,000	11,000	9,000	West Indies	9,000	10,000	12,000	31,000
371,000	163,000	1,000	207,000	East Indies	308,000	1,000	169,000	308,000
213,000	65,000	43,000	105,000	Egypt	81,000	27,000	59,000	165,000
3,116,000	725,000	442,000	1,949,000	Bales	2,172,000	472,000	733,000	3,060,000
....	....	....	317,000	Export.				
768,000	64,000	80,000	624,000	Stock Dec. 31. Stock above,	718,000	50,000	56,000	824,000
3,884,000	789,000	522,000	2,890,000	Total supply, bales	2,890,000	522,000	789,000	3,884,000

COTTON AT LIVER

Week Ending.	Receipts.						Sales.			
	American	E. I.	Egypt.	Brazil.	Other.	Total.	Consumption.	Speculation.	Export.	Total.
Jan. 6..	6,637	6,722	......	2,883	......	16,242	23,450	4,230	2,210	29,890
" 13..	30,448	5,212	1,127	7,591	......	44,378	33,330	3,920	1,790	39,040
" 20..	85,348	17,035	......	6,040	......	108,423	35,890	5,590	3,620	45,100
" 27..	12,018	667	838	4,415	225	18,163	29,340	3,350	3,930	36,620
Feb. 3..	21,619	......	......	1,171	429	23,219	34,850	3,370	4,350	42,570
" 10..	27,830	3,028	......	......	308	31,166	41,530	12,340	5,630	59,500
" 17..	16,454	2,895	......	201	......	19,550	45,360	20,800	3,440	69,600
" 24..	38,677	5,689	1,778	1,918	......	48,062	31,340	5,580	3,120	40,040
Mch. 3..	36,131	5,924	2,282	23	......	44,360	26,210	1,680	3,830	31,720
" 10..	31,798	17,566	696	......	......	50,060	27,970	6,140	7,210	41,320
" 17..	55,632	6,615	763	......	......	63,010	27,660	2,620	2,170	32,450
" 24..	6,625	5,022	1,336	......	......	12,983	28,940	1,760	1,490	32,190
" 31..	44,533	5,111	512	4,162	......	54,318	25,910	1,120	3,440	30,470
April 7..	68,814	11,428	2,821	102	......	83,165	29,610	2,810	2,060	34,480
" 13..	12,235	......	319	3,173	......	15,727	30,490	7,450	2,450	40,390
" 21..	43,683	5,872	......	......	......	49,555	33,080	5,570	1,910	40,560
" 28..	21,641	1,879	991	356	......	24,867	28,380	7,060	2,030	37,470
May 5..	171,480	8,627	2,696	6,299	......	189,102	30,120	2,240	1,990	34,350
" 12..	52,898	4,416	1,327	347	9	58,997	42,430	4,560	7,570	54,560
" 19..	11,874	2,019	3,048	......	......	16,941	37,940	2,900	4,290	45,130
" 26..	61,774	......	1,860	751	......	64,385	33,730	2,370	4,950	41,050
June 2..	14,116	9,600	2,282	3,908	......	29,906	43,730	6,330	6,410	56,470
" 9..	1,436	......	394	......	......	1,830	46,290	19,190	7,450	72,930
" 16..	107,119	2,155	5,944	1,850	......	117,068	31,350	4,570	6,920	42,840
" 23..	41,021	4,440	788	1,290	......	47,539	45,400	6,530	8,300	60,230
" 30..	71,405	3,965	1,514	3,763	......	80,647	54,970	16,500	5,780	77,250
July 7..	25,514	4,725	......	4,260	......	34,499	38,880	5,540	5,240	49,660
" 14..	32,795	......	1,188	3,511	......	37,494	29,800	3,890	6,830	40,520
" 21..	71,481	......	7,057	3,832	58	82,428	37,710	5,680	7,850	51,240
" 28..	10,531	......	524	2,899	......	13,954	48,520	5,140	8,300	61,960
Aug. 4..	14,234	6,313	1,801	......	......	22,348	42,950	3,420	6,870	53,240
" 11..	12,234	5,496	1,174	1,100	......	20,004	37,780	4,860	4,000	46,640
" 18..	70,533	3,989	5,553	3,500	89	83,664	30,400	2,500	4,910	37,810
" 25..	19,828	3,095	1,500	3,506	......	27,929	27,210	2,050	4,450	33,710
Sep. 1..	9,448	......	......	......	34	9,482	41,830	8,290	7,120	57,240
" 8..	19,790	......	1,336	1,910	1	23,037	34,040	4,350	4,930	43,320
" 15..	21,711	20,997	3,547	2,803	1,247	50,305	31,520	1,160	8,460	41,140
" 22..	24,535	3,448	5,197	3,526	205	36,911	30,090	670	4,750	35,510
" 29..	4,318	2,570	813	575	......	8,276	44,000	2,190	7,540	53,730
Oct. 6..	4,856	......	......	......	......	4,856	45,050	7,600	8,310	60,960
" 13..	10,615	4,496	7,853	4,063	......	27,027	31,770	3,130	4,520	39,420
" 20..	13,548	9,871	733	1,920	43	26,115	36,930	4,110	5,360	46,400
" 27..	12,041	2,765	1,515	4,271	616	21,208	33,160	2,670	4,200	40,030
Nov. 3..	8,054	7,165	1,324	1,680	......	18,223	40,190	2,550	4,180	46,920
" 10..	4,406	4,421	110	1,100	33	10,070	41,990	2,580	4,500	49,070
" 17..	....	......	2,272	......	......	2,272	32,930	1,950	4,250	39,130
" 24..	10,444	5,397	1,279	2,175	540	19,835	30,090	1,280	4,380	35,750
Dec. 1..	11,026	......	746	69	......	11,841	32,900	920	3,940	37,760
" 8..	38,154	9,173	374	1,766	587	50,054	38,020	1,480	3,110	42,610
" 15..	28,815	......	......	......	403	29,218	33,190	1,410	3,690	38,290
" 22..	12,345	2,556	569	6,912	......	22,382	37,430	1,020	3,460	41,910
Average prices & total sales, receipts & stocks.	1,584,502	232364	79,781	105,620	4,824	2,007,091	1,811,680	241,020	243490	2,296,190

POOL. YEAR 1854.

STOCKS.			PRICES.			ACTUAL EXPORT.	CON-SUMPTION.	REMARKS.
Amer'n.	Other.	Total.	Mid. Up.	Mid. Orl.	Dhol.			
275,887	312,295	588,182	5 7/8	6 1/8	4 1/8	2,210	23,450	
281,135	316,303	597,440	6	6 1/8	4 1/8	1,790	56,780	
338,303	330,180	668,483	5 7/8	6 1/16	4 1/8	3,620	92,670	Heavy arrivals of cotton.
327,401	328,675	656,076	5 3/4	5 7/8	4	3,930	122,010	
322,650	319,445	642,095	5 5/8	5 7/8	4	4,350	156,860	
314,100	315,226	629,326	5 3/4	6	4	5,630	198,390	
294,954	304,941	599,895	5 7/8	6	4 1/8	3,440	243,750	
308,391	306,906	615,297	5 3/4	5 7/8	4 1/8	3,120	275,090	
320,982	310,135	631,117	5 11/16	5 13/16	4 1/8	3,830	301,300	
328,580	317,417	645,997	5 5/8	5 3/4	4 1/8	7,210	329,270	
357,802	717,628	675,430	5 5/8	5 3/4	4 1/8	2,170	356,930	
339,835	316,558	656,393	5 1/2	5 5/8	4 1/8	1,490	385,870	Unfavorable reports, from United States, of the cotton crop.
363,578	318,274	681,852	5 5/16	5 1/2	4	3,440	411,780	
406,602	326,574	733,176	5 5/16	5 1/2	4	2,060	441,390	
392,377	323,586	715,963	5 3/8	5 9/16	4	2,450	471,880	
409,330	321,198	730,528	5 7/16	5 5/8	4	1,910	504,960	
404,861	317,924	722,785	5 3/8	5 9/16	4	2,030	533,340	Better accounts from United States.
551,321	327,456	878,777	5 3/16	5 3/8	3 7/8	1,990	563,460	
569,059	324,715	893,774	5 3/16	5 3/8	3 7/8	7,570	605,890	
548,543	322,442	870,985	3 1/8	5 1/4	3 3/4	4,290	643,830	
581,647	317,633	899,280	5	5 1/8	3 3/4	4,950	677,560	
555,743	323,253	878,996	5 1/8	5 1/4	3 3/4	6,410	721,290	
515,910	311,186	827,096	5 9/16	5 3/16	3 3/4	7,450	770,580	
610,718	312,946	923,664	5 3/16	5 5/16	3 3/4	6,920	801,930	
610,199	307,944	918,143	5 1/4	5 7/16	3 3/4	8,300	847,330	
634,154	305,866	940,020	5 1/4	5 7/16	3 3/4	5,780	902,300	
625,068	305,491	930,559	5 1/4	5 3/8	3 3/4	5,240	941,180	
630,873	303,350	934,223	5 3/16	5 3/16	3 7/8	6,830	970,980	
668,734	301,567	970,301	5 3/16	5 3/16	3 7/8	7,850	1,008,690	
637.115	294,320	931,434	5 1/4	5 3/8	3 7/8	8,300	1,057,210	
613,639	290,124	903,763	5 1/4	5 3/8	3 7/8	6,870	1,100,160	
594,363	286,144	880,507	5 3/16	5 3/16	3 7/8	4,000	1,137,940	Favorable accounts of crop from United States.
636,736	289,755	926,491	5 1/8	5 1/4	3 7/8	4,910	1,168,340	
632,204	290,396	922,600	5	5 1/8	3 7/8	4,450	1,195,550	
604,562	281,560	886,122	5 1/8	5 1/4	3 7/8	7,120	1,237,380	
594,122	277,407	871,529	5 1/8	5 1/4	3 7/8	4,930	1,271,420	
587,403	296,141	883,544	5 1/8	5 1/4	3 7/8	8,460	1,302,940	Contradictory accounts of crop, &c.
584,878	300,137	885,015	5 1/8	5 1/4	3 7/8	4,750	1,333,030	
550,886	293,065	843,951	5 1/4	5 1/4	3 3/4	7,540	1,377,030	
520,252	280,195	800.447	5 3/16	5 3/16	3 3/4	8,310	1,422,080	
504,977	286,307	791,284	5 3/16	5 3/16	3 3/4	4,520	1,454,850	News of victory of Allies in the Crimea.
480,835	281,974	762,809	5 3/16	5 3/16	3 3/4	5,360	1,491,780	
457,496	279,261	736,757	5 3/16	5 3/16	3 3/4	4,200	1,524,940	
431,560	279,060	710,620	5 1/4	5 3/8	3 3/4	4,180	1,565,130	Rumors of heavy failures about to take place, &c.
398,016	273,184	671,200	5 1/4	5 3/8	3 3/4	4,500	1,607,120	
371,566	264,726	636,292	5 3/16	5 5/16	3 3/4	4,250	1,640,050	
354,824	264,801	619,625	5 1/16	5 3/16	3 3/4	4,380	1,670,140	
337,130	256,996	594,126	5	5 1/8	3 5/8	3,940	1,703,040	Increased estimates of crop in New York circulars, &c.
342,974	260,076	603,050	4 15/16	5 1/8	3 5/8	3,110	1,741,060	
342,059	253,329	595,388	4 13/16	5	3 5/8	3,690	1,774,250	
321,704	254,486	576,190	4 13/16	5	3 5/8	3,460	1,811,680	
			5.31	5.43	3.88	243,490	35,523.14	

1855.

Mr. John D. Prince, who superintended the introduction of calico printing into the United States (*see* year 1826), was retired with an annuity of $2,000, settled upon him by the company who had employed him.

Cotton consumed in Great Britain, 839,100,000 lbs.

Number of cotton mills in Massachusetts, this year, 294, with 1,519,527 spindles.

There were 150 exhibitors of Algerian cotton at the Paris Industrial Exhibition, and about 9,000 acres of land in Algeria were under cultivation. (*See* years 1850 and 1856.)

COTTON CROP OF THE UNITED STATES.

Statement and Total Amount of the Cotton Crop of the United States, for the Year ending August 31, 1855.

	Bales.	Bales.	Total.	Same period 1854.
NEW ORLEANS.				
Export—				
To Foreign Ports	1,067,947			
Coastwise	202,317			
Stock on hand 1st September, 1855	39,425			
		1,309,689		
Deduct—				
Received from Mobile Montgomery, &c	32,087			
" Florida	4,147			
" Texas	16,690			
Stock on hand 1st September, 1854	24,121			
		77,045		
			1,232,644	1,346,925
MOBILE.				
Export—				
To Foreign Ports	340,311			
Coastwise	112,792			
Consumed in Mobile	1,683			
Burnt at Mobile	603			
Stock on hand 1st September, 1855	28,519			
		483,908		
Deduct—				
Received from Texas and New Orleans	35			
Stock on hand 1st September, 1854	29,278			
		29,313		
			454,595	538,684
TEXAS.				
Export—				
To Foreign Ports	16,160			
Coastwise	64,720			
Stock on hand 1st September, 1855	2,062			
		82,942		
Deduct—				
Stock on hand 1st September, 1854		2,205		
			80,737	110,325

Statement and Total Amount of the Cotton Crop of the United States, for the Year ending August 31, 1855.—*Concluded.*

	Bales.	Bales.	Total.	Same period 1854.
FLORIDA.				
Export—				
To Foreign Ports	35,018			
Coastwise	101,996			
Stock on hand 1st September, 1855	166			
		137,180		
Deduct—				
Stock in Apalachicola, 1st September, 1854	....	583		
			136,597	155,444
GEORGIA.				
Export—				
To Foreign Ports—Uplands	176,194			
" Sea Island	6,993			
Coastwise—Uplands	195,714			
Sea Island	7,474			
Stock in Savannah, 1st September, 1855	2,130			
" Augusta, 1st September, 1855	1,707			
		390,212		
Deduct—				
Stock in Savannah and Augusta, 1st September, 1854.	....	11,518		
			378,694	316,005
SOUTH CAROLINA.				
Export from Charleston—				
To foreign ports—Uplands	296,798			
" Sea Island	18,680			
Coastwise—Uplands	198,453			
Sea Island	5,771			
Burnt at Charleston	371			
Stock in Charleston, 1st September, 1855	2,085			
	522,158			
Export from Georgetown—				
To Northern Ports	4,500			
		526,658		
Deduct—				
Received from Florida	2,887			
" Savannah	7,468			
Stock in Charleston, 1st September, 1854	17,031			
		27,386		
NORTH CAROLINA.			499,272	416,754
Export—				
To foreign ports	59			
Coastwise	26,080			
		....	26,139	11,524
VIRGINIA.				
Export—				
To Foreign Ports	1,459			
Coastwise and Manufactured—taken from the ports	29,741			
Stock on hand 1st September, 1855	550			
		31,750		
Deduct—				
Stock on hand 1st September, 1854	....	750		
			31,000	21,936
Received at New York by New York & Erie Canal	....	....	377	1,182
" New York by New York & Erie R.R.	....	....	684	2,258
" Baltimore & Phila., from the South & West	....	....	6,600	8,990
Total crop of the United States	....	....	2,847,339	2,930,027

Decrease from last year .... bales..	82,688
Decrease from year before	415,543

Export to Foreign Ports, from September 1, 1854, *to August* 31, 1855.

FROM	To Great Britain.	To France.	To North of Europe.	Other Foreign Ports.	Total.
New Orleans (bales)	717,328	178,823	62,632	109,164	1,067,947
Mobile	215,248	111,090	8,257	5,716	340,311
Texas	8,926	1,570	5,664	....	16,160
Florida	28,068	5,320	1,630	....	35,018
Savannah	171,993	8,106	3,088	....	183,187
Charleston	204,102	70,656	13,700	27,020	315,478
North Carolina	59	....	....	....	59
Virginia	....	....	1,459	....	1,459
Baltimore	1,491	....	91	....	1,582
Philadelphia	300	....	....	....	300
New York	200,967	34,366	37,124	7,378	279,835
Boston	1,234	....	1,555	84	2,873
Grand total	1,549,716	409,931	135,200	149,362	2,244,209
Total last year	1,603,750	374,058	165,172	176,168	2,319,148
Decrease	54,034	....	29,972	26,806	74.939
Increase	....	35,873	....	....	....

Consumption.

Total crop of the United States, as above stated................bales.			2,847,339
Add—Stocks on hand at the commencement of the year, September 1, 1854, in the Southern ports		85,486	
" Northern ports		50.117	
			135,603
Makes a supply of			2,982,942
Deduct therefrom—The export to Foreign ports	2,244,209		
Less—Foreign included	891		
		2,243,318	
Stocks on hand September 1, 1855—			
In the Southern ports	76,644		
In the Northern ports	66,692		
		143,336	
Burnt at New York, Boston and Philadelphia		2,704	
			2,389,358
Taken for home use			593,584

ANNUAL REVIEW.

From the New Orleans Price Current, 1854—55.

The prospect of the market for our leading staple at the commencement of the past year was decidedly favorable. Although the growing crop was rather backward, compared with the large crop of 1852–53, yet, in most districts of the cotton growing region, it gave a fair promise of an abundant yield, and it was generally anticipated that the result would exhibit a considerable excess over the product of 1853–54, if it did not reach the extraordinary returns of 1852–53. At the same time, there appeared to be but one discouraging element to bear down prices, with the exception of the uncertain influences of the Eastern war ; all the ordinary causes that affect the market seem to be favorable to remunerative prices ; reduced stocks, an increased demand for consumption, the probability of abundant supples of food in Europe, and the prospects of lower freights, all combined to act favorably on the market. With regard to production, the event has proved, that although in no case could the yield have been as liberal as anticipated, yet, had it not been for the prejudicial effect of the low stage of water in the Southern and South-western rivers, it would have given a material excess over last year ; but, unfortunately, extraordinary droughts, have prevented the usual rise in the rivers, and a large amount, reaching about 145,000 bales in the Mississippi Valley, and some 100,000 bales in Alabama and Texas, has been retained in the interior, thus reducing the apparent amount of the crop, to swell to that of the present year.

The semi-weekly Price and Weekly Sales and Receipts at New York, Weekly Exports from New York and Rates of Freight to Liverpool 1st of each month, for the Crop Year ending September 1, 1855.

1854.		Price of Mid. Fair New Orleans New York Classificat'n.	Price of Mid. Fair Upland, New York Classificat'n.	Sales for week.	Receipts for week.	EXPORTS FOR THE WEEK.					Rates of Freight to Liverpool.
						To Great Britain.	To France.	North of Europe.	Other Fo'n Ports	Total Exports.	
Septem.	5..	10¾	10	4,200	2,291	5,562	1,697	187	48	7,494	⅛d.
"	8..	10½	9¾								
"	12..	10¼	9½	5,200	6,149	1,424	1,806	262	....	3,492	
"	15..	10¾	10⅛								
"	19..	10¾	10¼	5,500	5,716	2,557	....	....	....	2,557	
"	22.	10⅝	10⅛								
"	26..	10¾	10¼	3,600	7,850	2,009	1,495	1,066	....	4,570	
"	29..	10¾	10¼								
October	3..	10¾	10¼	5,000	4,853	1,963	323	689	....	2,975	⅛@5-32d
"	6..	11	10½								
"	10..	10¾	10⅜	4,300	7,865	2,880	264	142	190	3,476	
"	13..	10¾	10⅜								
"	17..	11	10½	5,300	2,473	6,328	....	216	....	6,544	
"	20..	11	10½								
"	24..	10¾	10⅜	4,500	4,771	1,743	1,126	319	....	3,188	
"	27..	10¾	10¼								
"	31..	10¾	10⅜	4,000	5,949	2,153	....	563	84	2,800	
Novem.	3..	10¾	10⅛								⅛d.
"	7..	10¾	10⅛	3,500	8,514	4,671	686	536	20	5,913	
"	10..	10⅝	10⅛								
"	14..	10¼	9½	3,000	15,569	2,444	111	400	4	2,959	
"	17..	10¼	9½								
"	21..	10½	9¾	2,400	9,288	2,711	91	100	110	3,012	
"	24..	10	9⅝								
"	28..	10	9⅜	5,500	14,656	1,972	....	253	....	2,225	
Decem.	1..	10	9¼								⅛@5-32d
"	5..	10	9¼	3,500	7,356	3,546	336	471	....	4,353	
"	8..	9¾	9¼								
"	12..	9½	9	8,100	4,747	5,333	16	616	52	6,017	
"	15..	9¼	8¾								
"	19..	9½	8¾	2,400	4,330	4,567	1,003	753	313	6,636	
"	22..	9½	9								
"	26..	9¾	9¼	5,600	9,953	3,467	704	378	326	4,875	

General Remarks.

This year opened with the prospect of an unprecedented crop, and with unfavorable foreign accounts, and stagnation here in the goods market, cotton was very quiet and a little lower; on the 13th September, advices were received by telegraph, of injury to the crop, which gave more tone to the market. Early in October, the money market became very stringent, the shipments of specie were large, a city bank suspended payments, followed in November, by the failure of numerous Western banks; some failures also occurred in Liverpool, and some here, and with favorable weather for picking, the market was very dull.

This partial stagnation continued through December, aggravated by the Russian-Turkish war in Europe, dear food, and consequent slack demand for more goods.

On the 1st of January, the stock in Liverpool was ascertained to be 45,000 bales less than the previous estimates, which, with an easier feeling in money, strengthened the market.

Decem. 29..	$9\frac{5}{8}$	$9\frac{1}{8}$								
1855.										
January 2..	$9\frac{1}{2}$	$8\frac{1}{2}$	3,000	9,418	5,862	......	671	......	6,533	5-32@$\frac{3}{16}$
" 5..	$9\frac{3}{4}$	9								
" 9..	$9\frac{3}{4}$	9	8,000	5,501	3,748	1,081	......	......	4,829	
" 12..	10	$9\frac{1}{4}$								
" 16..	10	$9\frac{1}{4}$	11,000	5,631	889	......	2,209	......	3,098	
" 19..	10	$9\frac{1}{4}$								
" 23..	$10\frac{1}{4}$	$9\frac{1}{4}$	10,000	9,171	1,514	721	1,263	......	3,498	
" 26..	$10\frac{1}{2}$	$9\frac{1}{2}$								
" 30..	$10\frac{1}{2}$	$9\frac{1}{2}$	7,000	11,251	3,600	505	197	1	4,303	
February 2..	$10\frac{1}{2}$	$9\frac{1}{2}$								$\frac{1}{8}$d.
" 6..	$10\frac{1}{2}$	$9\frac{1}{2}$	11,000	9,177	4,330	349	1,363	......	6,042	
" 9..	$10\frac{1}{2}$	$9\frac{1}{2}$								
" 13..	$10\frac{1}{2}$	$9\frac{1}{4}$	3,500	4,600	596	......	560	......	1,156	
" 16..	$10\frac{1}{2}$	$9\frac{1}{4}$								
" 20..	$10\frac{1}{2}$	$9\frac{1}{4}$	7,000	28,880	2,199	......	584	......	2,783	
" 23..	$10\frac{1}{2}$	$9\frac{3}{8}$								
" 27..	$10\frac{1}{2}$	$9\frac{3}{8}$	7,400	7,675	5,136	2,713	1,035	72	8,956	
March 2..	$10\frac{1}{2}$	$9\frac{1}{2}$								5 32@$\frac{3}{16}$
" 6..	$10\frac{1}{2}$	$9\frac{1}{2}$	10,000	13,186	891	1,785	713	......	3,389	
" 9..	$10\frac{1}{2}$	$9\frac{1}{2}$								
" 13..	$10\frac{1}{2}$	$9\frac{1}{2}$	12,500	17,104	1,504	......	1,242	......	2,746	
" 16..	$10\frac{1}{2}$	$9\frac{3}{4}$								
" 20..	$10\frac{3}{4}$	10	25,000	12,564	2,542	......	872	......	3,414	
" 23..	$10\frac{3}{4}$	10								
" 27..	$10\frac{3}{4}$	10	6,000	10,359	5,904	2,769	760	......	9,433	
" 30..	11	$9\frac{7}{8}$								
April 3..	$10\frac{3}{4}$	$9\frac{7}{8}$	8,000	13,662	6,288	2,800	2,012	......	11,100	$\frac{5}{32}$@3-16
" 6..	$10\frac{3}{4}$	10								
" 10..	$10\frac{3}{4}$	$10\frac{1}{8}$	15,000	14,208	492	149	114	......	755	
" 13..	11	10								
" 17..	11	$10\frac{1}{4}$	20,500	18,900	4,259	......	......	......	4,259	
" 20..	11	$10\frac{1}{4}$								
" 24..	11	$10\frac{1}{4}$	13,500	20,311	6 124	905	78	1,231	8,338	
" 27..	$11\frac{1}{4}$	$10\frac{1}{4}$								
May 1..	$11\frac{1}{2}$	$10\frac{3}{8}$	16,000	11,759	5,790	1,309	934	1,037	9,070	3-16@$\frac{7}{32}$
" 4..	$11\frac{1}{2}$	$10\frac{1}{2}$								
" 8..	$11\frac{1}{2}$	$10\frac{5}{8}$	14,000	13,234	3 455	3,609	2.892	875	10,831	
" 11..	$11\frac{1}{2}$	$10\frac{3}{4}$								
" 15..	$11\frac{1}{2}$	$10\frac{3}{4}$	22,000	12,223	5,833	12	......	85	5,930	
" 18	$11\frac{3}{4}$	11								

The business in February was light, owing to a change in the British Ministry; unfavorable prospects for peace there; tight money markets abroad, and English manufactories working short time.

The death of the Emperor of Russia in March, encouraged the hope for peace and there was temporarily a better feeling.

In May, the extreme low stage of the Western and South-western rivers, detained supplies in the interior, and consequently, receipts at the ports, and exports thence, largely fell off; free speculative purchases were made, both in this, and the Liverpool market, at advanced prices; in June, the same features were prominent, and it was estimated that at one time, 800,000 bales were waiting, detained in the interior.

The great drought also had an unfavorable effect upon the growing crop, and prices rapidly advanced; toward the latter part of June, timely rains swelled the rivers to some extent, and supplies came forward more freely.

In July, the appearance of the growing crop was much more favorable, and prices shaded a little, though the prevalence of the yellow fever at New Orleans in August, interfering with shipments, prevented any serious decline.

The semi-weekly Price and Weekly Sales and Receipts at New York, Weekly Exports from New York and Rates of Freight to Liverpool 1st of each Month, for the Crop Year ending September 1, 1855—Concluded.

1855.		Price of Mid. Fair New Orleans New York Classificat'n.	Price of Mid. Fair Upland. New York Classificat'n.	Sales for week.	Receipts for week.	EXPORTS FOR WEEK.					Rates of Freight to Liverpool.	GENERAL REMARKS.
						To Great Britain.	To France.	North of Europe.	Other Fo'n Ports	Total Exports.		
May	21..	12	11⅜	16,000	20,737	4,718	3,716	1,580	300	10,314		
"	25..	12½	11¾									
"	29..	12½	11¾	20,000	8,517	9,839	....	1,190	....	11,029		
June	1..	13	12⅜								⅛@5-32d	
"	5..	13	12½	28,000	15,363	4,118	262	1,086	....	5,466		
"	8..	14	13									
"	12..	14	13	26,000	14,259	5,454	1	....	....	5,455		
"	15..	14	13									
"	19..	14	12¾	12,500	10,331	4,772	....	209	....	4,981		*Exchange.*
"	22..	13½	12½									
"	26..	13	12½	8,000	9,082	2,446	....	97	....	2,543		The quotation for 60 days' bills on London ranged in September, from 9¼@10 per cent.; October the same; November, 9@9¾; in December, it fell from 9@9½ to 7½@8; in January, the lowest price was 6¾@7½, and the highest 8⅞@9½; in February the quotations fluctuated between 8¾@9⅜ and 9½@10; March, 9¼@10¼; April, 9¾@10¼; May, 9⅞@10¼; June, 9⅝@10⅛; July, 9½@10⅛; and August, 9½@10⅛; or the same as July.
"	29..	12½	11¾									
July	3..	12½	11¾	5,500	8,540	5,344	123	880	750	7,097	3-16d.	
"	6..	12	11¼									
"	10..	12	11¼	9,000	7,690	3,785	....	....	....	3,785		
"	13..	12½	11⅝									
"	17..	12½	11⅝	11,000	3,152	3,953	102	1,215	876	6,146		
"	20..	12¼	11¼									
"	24..	12¼	11¼	3,500	6,095	5,720	360	1,015	....	7,095		
"	27..	12¼	11¼									
"	30..	12	11½	6,500	10,941	4,962	150	1,800	75	6,987		
August	3..	12½	11¾								⅛d.	
"	7..	12½	12	11,000	3,981	8,943	1,123	989	....	11,055		
"	10..	12½	12									
"	14..	12½	12	7,000	7,690	2,192	26	780	600	3,598		
"	17..	12½	12									
"	21..	12¾	11¾	4,500	2,305	3,906	138	119	29	4,192		
"	24..	12¾	11¾									
"	28..	12¾	11¾	5,000	9,309	4,451	....	1,474	....	5,925		
"	31..	12½	11¾								$\frac{5}{32}$@3-16	
Average price and total sales, receipts and exports.		11.16	10.39	479,500	509,136	200,889	34,366	36,884	7,078	279,217		

LIVERPOOL STATEMENT FOR 1855.

UNITED STATES, 1854–1855.

Stock Sept. 1, 1854	134,000	Export	2,244,000
		Consumption	594,000
Crop	2,847,000	Stock Sept. 1, 1855	143,000
Bales	2,981,000	Bales.	2,981,000

Tot. Europe.	Continent.	France.	Gt. Britain.	Stock 1st Jan., 1855, in	Gt. Britain.	France.	Continent.	Tot. Europe.
				United States ... Bales.	311,000	73,000	38,000	422,000
				Brazil	48,000	....	7,000	55,000
				West Indies	4,000	1,000	4,000	9,000
				East Indies	202,000	....	11,000	213,000
				Egypt	59,000	6,000	4,000	69,000
				Bales	624,000	80,000	64,000	768,000
CONSUMPTION.				IMPORT.				
2,461,000	446.000	437,000	1,578,000	United States ... Bales.	1,623,000	422,000	428,000	2,352,000
153,000	37,000	2,000	114,000	Brazil	135.000	3,000	32,000	165,000
31,000	13,000	9,000	9,000	West Indies	9,000	12,000	9,000	31,000
470,000	193,000	....	277.000	East Indies	396.000	....	186,000	393,000
201,000	49,000	31,000	121,000	Egypt	115,000	30,000	52,000	194,000
3,316,000	738,000	479,000	2,099,000	Bales	2,278,000	467,000	707,000	3,135,000
....	....	....	317,000	Export.				
587,000	33,000	68,000	486,000	Stock, Dec. 31 ... Stock above,	624,000	80,000	64,000	768,000
3,903,000	771,000	547,000	2.902,000	Total supply, bales	2,902,000	547,000	771,000	3,903,000

COTTON AT LIVER

Week Ending.	RECEIPTS.						SALES.			
	American	E. I.	Egypt.	Brazil.	Other.	Total.	Consumption.	Speculation.	Export	Total.
Jan. 5..	14,207	9,932	85	2,083	......	26,307	34,200	820	880	35,900
" 12..	12,210	4,090	2,468	1,066	......	19,834	53,320	5,770	5,180	64,270
" 19..	1,592	......	448	......	......	2,040	39,460	10,250	6,300	56,010
" 26..	5,738	......	......	......	3	5,741	33,080	3,450	940	37,470
Feb. 2..	22,604	......	546	......	......	23,150	31,160	1,310	4,590	37,060
" 9..	11,671	.....	1,171	862	......	13,704	32,360	1,170	2,350	35,880
" 16..	3,150	2,591	506	......	......	6,247	32,740	830	1,750	35,320
" 23..	12,198	......	1,652	3,574	......	17,424	29,710	2,710	1,330	33,750
March 2..	189,828	1,750	1,398	3,544	......	196,520	31,700	1,070	3,780	36,550
" 9..	138,807	14,011	2,515	8,683	35	164,051	57,940	17,140	12,090	87,170
" 16..	52,571	......	2,377	2	......	54.950	52,290	3,690	11.230	67,210
" 23..	56,542	9.649	......	3,245	936	70,372	50,140	14,620	12,900	77,660
" 30..	3,517	14,489	......	702	......	18,708	56,510	11,200	15,290	83,000
April 5..	53,505	8,088	2,626	7,307	270	71,796	52,600	18,840	10,720	82,160
" 13..	100,904	2,382	1,642	1,356	667	106,951	49,610	15,640	7,790	73,040
" 20..	19,761	22,154	769	1,292	......	43,976	35,890	6,490	7,200	49,580
" 27..	15,601	......	......	......	......	15,601	60,970	31,980	10,040	102.990
May 4..	21,634	3,572	670	......	......	25,876	53,820	48,730	4,240	106,790
" 11..	39,638	......	5,359	5,887	......	50,884	47,050	53,570	5,280	105,900
" 18..	16,356	17,912	7,041	5,739	......	77,048	58,480	49.210	4,980	112.670
" 25..	39,446	383	2,859	4,007	.....	46,695	69,980	76,010	7,090	153,080
June 1.	26,491	70	3.906	2,928	......	33,395	65,600	81,600	5,200	152,400
" 8..	33,972	......	3,333	2,818	44	40,167	41,860	62,930	2,530	107,320
" 15..	35,208	974	1,613	1,476	......	39,271	19,650	17,060	2,000	38,710
" 22..	30,557	......	178	1,643	......	32,378	19,690	3,780	940	24,410
" 29..	52,398	......	3,475	2,921	......	58,794	23,430	17,960	1,950	43,340
July 6..	43,034	4,925	2,137	2,856	......	52,952	31,300	11,780	2,330	45,410
" 13..	19,996	1,437	3,817	1,901	30	27,181	23,260	4,120	820	28,200
" 20..	43,109	3,538	11,077	4,587	11	62,322	36,480	6,980	2,870	46,330
" 27..	23,945	454	749	1,160	......	26,308	28,160	3,690	2,820	34,670
Aug. 3..	19,537	8,817	.....	4,408	......	32,762	27,260	3,340	4,260	34,860
" 10..	9,139	......	3,112	5,991	......	18,242	31,590	3,900	3,250	38,740
" 17..	14,386	2,695	1,447	2,880	......	21,408	51,740	24,340	4,560	80,640
" 24..	15,437	6,562	1,814	2.423	104	26,340	35,630	4,710	4,610	44,950
" 31..	28,426	7,355	6,854	4.568	......	47,203	40,690	5,080	5,700	51,470
Sep. 7..	11,476	5,018	592	......	......	17,086	43,560	7,020	5,720	56,300
" 14..	14,011	......	1,074	240	......	15,325	34,500	5,930	5,900	46,330
" 21..	34,500	18,878	2,501	6,858	183	62,920	26,650	1,390	4.710	32,750
" 28..	1,865	8,805	3,712	862	.....	15,244	30,430	2,300	3,230	35.960
Oct. 5..	12,935	2,416	2,189	4,981	800	23,321	33,390	1,740	5,750	40,880
" 12..	23,464	13,789	3,578	3,649	15	44,495	26,030	1,400	3,620	31,050
" 19..	108	2.869	1,098	3.198	......	7,273	43,020	5,090	5,990	54,100
" 26..	11,102	11,409	1,575	2,016	63	26,165	30,700	1,400	6,710	38,810
Nov. 2.	10,349	16,287	885	1,053	......	28.574	50,150	7,800	7,770	65,720
" 9..	11,223	11,760	......	1,987	587	25,557	52,630	17,390	7.560	77,580
" 16..	39,228	11,275	4,079	3,203	139	57,924	33,020	9,110	6,600	48,730
" 23..	14,399	3,806	497	......	139	18,841	28,300	1.600	5,110	35,010
" 30..	....	......	214	......	......	214	40,780	2,900	5,790	49,470
Dec. 7..	18,880	25	4,424	923	......	24,252	40,040	5,920	4,840	50,800
" 14..	17.228	......	......	715	......	17,943	41,200	9,510	5,490	56,200
" 21..	26.817	14.526	1,218	1,218	440	44.219	31,050	2,660	2,420	36,130
" 28..	83,099	11,172	4,797	8,642	.....	107,710	19,700	3,860	540	24,100
" 31..	....	......	......	....	......	19,161	1,590	1,340	250	3,180
Average prices & total sales, receipts & stocks.	1,587,799	279865	110077	131,434	4,466	2,132,802	2,046,090	714,130	267780	3,028,000

POOL. YEAR 1855.

STOCKS.			PRICES.			ACTUAL EXPORT.	CON-SUMPTION.	REMARKS.
Amer'n	Other.	Total.	Mid. Up.	Mid. Orl.	Dhol.			
281,777	260,790	542,567	$4\frac{15}{16}$	5	$3\frac{5}{8}$	880	34,200	
250.607	256,194	506,801	$4\frac{15}{16}$	$5\frac{1}{8}$	$3\frac{5}{8}$	5,180	87,520	Small arrivals; easterly winds.
218,919	246,262	465,181	5	$5\frac{3}{16}$	$3\frac{5}{8}$	6,300	126,980	
198,327	239,375	437,702	5	$5\frac{3}{16}$	$3\frac{5}{8}$	940	160,060	
195,181	230,921	426.102	5	$5\frac{1}{8}$	$3\frac{5}{8}$	4,590	191,220	
181,022	224,834	405,856	5	$5\frac{1}{8}$	$3\frac{5}{8}$	2,350	223,580	
157,650	221,077	378,727	$4\frac{15}{16}$	$5\frac{1}{8}$	$3\frac{5}{8}$	1,750	256,320	
145,698	220,073	365,771	$4\frac{15}{16}$	$5\frac{1}{8}$	$3\frac{5}{8}$	1,330	286,030	
310,586	217,925	528,511	$5\frac{1}{2}$	$5\frac{3}{8}$	$3\frac{1}{2}$	3,780	317,730	Overdue ships arrived; large receipts.
405,813	228,219	634,032	5	$5\frac{3}{16}$	$3\frac{1}{2}$	12,090	375,670	
415,104	220.358	635,462	$4\frac{7}{8}$	$5\frac{1}{8}$	$3\frac{5}{8}$	11,230	427,960	
421,946	222,398	644,344	5	$5\frac{3}{16}$	$3\frac{5}{8}$	12,900	478,100	
378,657	223 341	601,998	5	$5\frac{3}{16}$	$3\frac{5}{8}$	15,290	534,610	
387,372	228,062	615,434	$5\frac{1}{16}$	$5\frac{1}{4}$	$3\frac{3}{4}$	10,720	587,210	More favorable accounts from manufacturing districts.
449,046	220,929	669,075	$5\frac{1}{8}$	$5\frac{1}{4}$	$3\frac{3}{4}$	7,790	636,820	
438,237	235,224	673,461	$5\frac{1}{8}$	$5\frac{1}{4}$	$3\frac{3}{4}$	7,200	672,710	
396,938	223,514	620,452	$5\frac{3}{16}$	$5\frac{3}{16}$	$3\frac{7}{8}$	10,040	733,680	
372,234	214,150	586,384	$5\frac{3}{16}$	$5\frac{7}{16}$	4	4.240	787,500	
374,352	210,588	584,940	$5\frac{9}{16}$	$5\frac{11}{16}$	$4\frac{1}{8}$	5.280	834.550	Large deficiency of stock at Liverpool.
373,288	225,360	598,648	$5\frac{3}{4}$	$5\frac{7}{8}$	$4\frac{1}{4}$	4,980	893,030	
359,024	210,589	569,613	6	$6\frac{1}{8}$	$4\frac{1}{2}$	7,090	963,010	Large consumption.
332,095	196,853	528,948	$6\frac{9}{16}$	$6\frac{11}{16}$	$4\frac{3}{4}$	5,200	1,028,610	
343,437	188,698	532,135	$6\frac{9}{16}$	$6\frac{11}{16}$	$4\frac{3}{4}$	2,530	1,070,470	Small receipts at United States ports.
361,865	287,571	549,436	$6\frac{9}{16}$	$6\frac{11}{16}$	$4\frac{7}{8}$	2,000	1,090,120	
373,652	181,732	555,384	$6\frac{3}{8}$	$6\frac{1}{2}$	$4\frac{3}{4}$	940	1,109,810	
405,580	182,228	587,808	$6\frac{3}{8}$	$6\frac{1}{2}$	$4\frac{3}{4}$	1.950	1,133,240	
423,134	173,996	607,130	$6\frac{1}{4}$	$6\frac{3}{8}$	$4\frac{3}{4}$	2,330	1,164,540	
420,556	185,235	605,791	$6\frac{3}{16}$	$6\frac{3}{8}$	$4\frac{3}{4}$	820	1,187,800	
434,055	195,208	629,263	$6\frac{3}{16}$	$6\frac{3}{8}$	$4\frac{3}{4}$	2,870	1,224,280	
434,160	208,631	624,791	$6\frac{1}{8}$	$6\frac{5}{8}$	$4\frac{3}{8}$	2,820	1,252,440	
431,727	197,466	629,193	6	$6\frac{3}{16}$	$4\frac{3}{8}$	4,260	1.279,700	Strikes in manufacturing districts for short hours.
414,146	197,449	611,595	$5\frac{15}{16}$	$6\frac{9}{16}$	$4\frac{3}{8}$	3.250	1,311,290	
384,522	190,481	575,003	$6\frac{3}{16}$	$6\frac{7}{16}$	$4\frac{3}{4}$	4,560	1,363,030	
372,809	190,094	562,903	$6\frac{1}{8}$	$6\frac{3}{8}$	$4\frac{3}{4}$	4,610	1,398,660	
367,075	208,681	575,756	$6\frac{3}{16}$	$6\frac{7}{16}$	$4\frac{3}{4}$	5,700	1,439,350	
345,221	190,781	536,002	$6\frac{3}{16}$	$6\frac{7}{16}$	$4\frac{3}{4}$	5,720	1,482,910	
329,972	183,705	513,677	$6\frac{1}{8}$	$6\frac{3}{8}$	$4\frac{5}{8}$	5,900	1,517,410	
343,602	201,965	545,567	6	$6\frac{1}{16}$	$4\frac{1}{2}$	4,710	1,544,060	Fall of Sebastopol, &c.
318,967	207,994	526,961	$5\frac{3}{4}$	6	$4\frac{3}{8}$	3,230	1,574,490	
302,902	209,760	512,662	$5\frac{5}{8}$	$5\frac{7}{8}$	$4\frac{1}{4}$	5,750	1,607,880	
302,636	222,661	525,297	$5\frac{3}{4}$	$5\frac{5}{8}$	$4\frac{1}{8}$	3,620	1,633,910	Favorable reports of growing crop.
269,634	216,466	486,100	$5\frac{3}{16}$	$5\frac{5}{8}$	$4\frac{1}{8}$	5,990	1,676,930	
256,126	222,179	478,305	$5\frac{1}{8}$	$5\frac{7}{16}$	$4\frac{1}{8}$	6,710	1,707,630	
225,555	228,384	453,939	$5\frac{3}{8}$	$5\frac{11}{16}$	$4\frac{1}{4}$	7,770	1,757,780	
196,818	224,848	421,666	$5\frac{3}{4}$	$5\frac{7}{8}$	$4\frac{1}{4}$	7,560	1,810,410	
209,186	231,414	440,600	$5\frac{5}{8}$	$5\frac{7}{8}$	$4\frac{1}{4}$	6,600	1.843,430	
199,315	226,187	426,502	$5\frac{1}{2}$	$5\frac{3}{4}$	$4\frac{1}{4}$	5,110	1,871,730	
166,965	216,311	383,276	$5\frac{1}{2}$	$5\frac{11}{16}$	$4\frac{1}{4}$	5,790	1,912,510	
154,875	208,963	363,838	$5\frac{1}{2}$	$5\frac{3}{4}$	$4\frac{1}{4}$	4,840	1,952.550	
135,373	197,918	333,291	$5\frac{1}{2}$	$5\frac{3}{4}$	$4\frac{3}{8}$	5,490	1,993,750	
128,850	183,092	311,942	$5\frac{1}{2}$	$5\frac{11}{16}$	$4\frac{3}{8}$	2,420	2.024,800	
191,309	191,093	382,402	$5\frac{1}{2}$	$5\frac{11}{16}$	$4\frac{3}{8}$	540	2.044 500	
......	......	428,810	......	......	......	250	2,046,290	
			5.6	5.78	4.19	267.780	39,346.08	

1856.

During this year and the one following, the production of Algerian cotton fell off in quantity, owing to a rush of cultivators into the trade, who had neither the requisite capital nor skill. (*See* years 1850 and 1855.)

Cotton mills in Great Britain, 2,210; moving power therein, 97,000 horse power; spindles, 28,000,000; power looms, 299,000; children employed, 24,684; total hands employed, 379,249, of whom 222,027 were females. (*See* year 1850.)

The Factory Commissioners of England, in their returns for this year, estimate the number of spindles at work in Great Britain at 28,010,217. (*See* year 1846.) The number in the United States was estimated at 3,950,000.

COTTON CROP OF THE UNITED STATES.

Statement and Total Amount of the Cotton Crop of the United States, for the Year ending August 31, 1856.

	Bales	Bales.	TOTAL.		
			1856.	1855.	1854.
NEW-ORLEANS.					
Export—					
To Foreign Ports	1,572,923				
Coastwise	222,100				
Burnt at New Orleans	1,200				
Stock on hand, 1st September, 1856	6,995				
		1,803,218			
Deduct—					
Received from Mobile, Montgomery, &c	73,573				
" " Florida	5,186				
" " Texas	23,601				
Stock on hand, 1st September, 1855	39,425				
		141,785			
			1,661,433	1,232,644	1,346,925
MOBILE.					
Export—					
To Foreign Ports	485,035				
Coastwise	196,286				
Consumed in Mobile	1,936				
Stock on hand, 1st September, 1856	5,005				
		688,262			
Deduct—					
Received from New Orleans	5				
Stock on hand, 1st September, 1855	28,519				
		28,524			
			659,738	454,595	538,684
TEXAS.					
Export—					
To Foreign Ports	34,002				
Coastwise	83,515				
Stock on hand, 1st September, 1856	623				
		118,140			
Deduct—					
Stock on hand, 1st September, 1855	....	2,062			
			116,078	80,737	110,325
FLORIDA.					
Export—					
To Foreign Ports—Uplands	35,858				
Coastwise—Uplands	97,738				
Sea Island	10,900				
Stock on hand, 1st September, 1856	74				
		144,570			
Deduct—					
Stock on hand, 1st September, 1855	....	166			
			144,404	136,597	155,444
GEORGIA.					
Export—					
To Foreign Ports—Uplands	177,182				
Sea Island	8,138				
Coastwise—Uplands	00,426				
Sea Island	7.346				
Stock in Savannah, 1st September, 1856	1.550				
Stock in Augusta, 1st September, 1856	1,781				
		396,423			

Statement and Total Amount of the Cotton Crop of the United States, for the Year ending August 31, 1856.—*Concluded.*

	Bales.	Bales.	TOTAL. 1856.	1855.	1854.
Deduct—					
Received from Florida—Sea Island	2,755				
Uplands	386				
Stock in Savannah, 1st September, 1855	2,130				
Stock in Augusta, 1st September, 1855	1,707				
		6,978			
			389,445	378,694	316,005
SOUTH CAROLINA.					
Export from Charleston—					
To Foreign Ports—Uplands	352,346				
Sea Island	18,765				
Coastwise—Uplands	133,451				
Sea Island	9,286				
Burnt at Charleston	751				
Stock in Charleston, 1st September, 1856	3,144				
	517,743				
Export from Georgetown, S. C.—					
To Northern Ports	2,893				
		520,636			
Deduct—					
Received from Florida—Sea Island	6,027				
Uplands	578				
Received from Savannah—Sea Island	2,689				
Uplands	13,281				
Stock in Charleston, 1st September, 1855	2,085				
		24,660			
			495,976	499,272	416,754
NORTH CAROLINA.					
Export—					
To Foreign Ports	96				
Coastwise	26,002				
		......	26.098	26,139	11,524
VIRGINIA.					
Export—					
To Foreign Ports	70				
Coastwise, and Manufactured—taken from the ports.	20,748				
Stock on hand, 1st September, 1856	842				
		21,660			
Deduct—					
Received from Mobile	652				
Stock on hand, 1st September, 1855	550				
		1,202			
			20,458	31,000	21,936
Received at New-York by New York and Erie Canal, &c	....	....	305	377	1,182
Received at New York by New York and Erie Railroad	....	....	1,781	684	2,258
Received at Baltimore and Philadelphia, from the West	....	....	12,129	6,600	8,990
Total Crop of the United States	....	....	3,527,845	2,847,339	2,930,027

Increase over crop of 1855 .... bales. 680,506

Increase over crop of 1854 597,818

Increase over crop of 1853 264,963

Export to Foreign Ports, from September 1, 1855, *to August* 31, 1856.

FROM	To Great Britain.	To France.	To North of Europe.	Other F'n Ports.	Total.
New Orleans........bales.	986,622	244,814	162,675	178,812	1,572,923
Mobile........	351,690	96,262	29,016	8,067	485,035
Texas........	19,661	5,166	9,175	....	34,002
Florida........	30,899	2,939	2,020	....	35,858
Savannah........	162,748	16,857	2,907	2,808	185,320
Charleston........	180,532	87,396	49,727	53,456	371,111
North Carolina........	96	....	....	....	96
Virginia........	70	....	....	....	70
Baltimore........	424	48	....	....	472
Philadelphia........	178	....	....	....	178
New York........	181,045	27,155	42,893	5,371	256,464
Boston........	7,421	....	5,592	64	13,077
Grand total........	1,921,386	480,637	304,005	248,578	2,954,606
Total last year........	1,549,716	409,931	135,200	149,362	2,244,209
Increase........	371,670	70,706	168,805	99,216	710,397

Consumption.

Total crop of the United States, as before stated........			3,527,845 bales.
Add--Stocks on hand at the commencement of the year, 1st September, 1855.—In the Southern ports........		76,644	
" Northern "		66,692	
			143,336
Makes a supply of........			3,671,181
Deduct therefrom—The export to Foreign ports........	2,954,606		
Less, foreign included........	835		
		2,953,771	
Stocks on hand at the close of the year, 1st September, 1856—			
In the Southern ports........	20,014		
" Northern "	44,157		
		64,171	
Burnt at New York and Boston........		500	
			3,018,442
Taken for home use........bales.			652,739

ANNUAL REVIEW.

From the New Orleans Price Current, 1855–56.

And, first we have to congratulate our planting and commercial friends upon the pleasing and profitable contrast presented by the course of business, as compared with the season which immediately preceded it—a season which was marked by extraordinary fluctuations and disasters, resulting partly from the existence of a European war, whose duration and possible complications were out of the reach of human foresight, and partly referable to an extraordinary prevalence of drought, which rendered the tributary streams unnavigable for a length of time seldom, if ever, before known, and kept back, for an unwonted period, an immense amount of produce, which had been mostly advanced upon, in the usual course of business, trusting to the receipt and sale of the produce in question to reimburse the factor. But the good faith of both planter and factor was, to an important extent, baffled by the operations of nature, and several of our prominent and respectable commission houses were compelled to suspend payment of such acceptances, as had not, to some extent at least, been provided for by the drawers. The past season, however, we are happy to state, had been favored with a fair average of natural facilities for reaching market; and, with a promptitude characteristic of the New Orleans merchants, the suspended houses, almost without an exception, availed themselves of the earliest moment of returning prosperity, to resume payment in full, with interest. We have already stated that the early receipts of the new crop were unusually abundant; and we now add that they were also of an unusually high average of quality, up to the latter part of September, when the effects of a storm that had previously occurred were prominently presented in a sudden and remarkable falling off, in the character of the receipts; and this change, aided by subsequent unfavorable circumstances, such as severe frosts, etc., has run through the remainder of the season, reducing the general average of the crop below middling. And thus it has been that for the greater part of the year the supply of middling to good middling descriptions has not been equal to the demands; and for months past, the market may almost be said to have been bare of them, as they could only occasionally be met with, in limited parcels. Indeed, for the period last mentioned, the scarcity has embraced middling fair, and several Spanish vessels, which usually take good middling to middling fair, and which arrived in the latter part of June, are yet here awaiting cargoes; these qualities having been exhausted in the old crop, after reaching the high figures of 11¾ and 12¼ cents per lb. early in July. We may note, as a prominent feature of the season's operations, their more than ordinary speculative character. The probability of continued war, or a restoration of peace, entered largely into the calculation of chances, while a wide diversity in the estimates of crops constituted another element, among the incentives to action or inaction. And thus it was that many of the accustomed operators were comparatively sparing in their purchases, while other parties, more confident, came forward and operated freely.

The semi-weekly Price and Weekly Sales and Receipts, at New York, Weekly Exports from New York and Rates of Freight to Liverpool 1st of each Month, for the Crop Year ending September 1, 1856.

1855.	Price of Middling New Orleans New York Classificat'n.	Price of Middling Upland, New York Classificati'n	Sales for week, including lots in transit.	Receipts for week, including lots in transit.	EXPORTS FOR WEEK.					Rates of Freight to Liverpool.
					To Great Britain.	To France.	North of Europe.	Other Fo'n Ports	Total Exports.	
Septem. 4..	11¼	11	6,000	1,026	3,905	297	....	....	4,202	3-16d.
" 7..	11⅛	10¾								
" 11..	11⅛	10¾	2,000	4,651	1,836	....	....	....	1,836	
" 14..	10¾	10½								
" 18..	10¾	10⅜	10,500	2,243	4,586	....	54	....	4,640	
" 21..	10⅜	10⅛								
" 25 .	10⅜	10⅛	7,000	1,420	2,181	1,086	....	1	3,268	
" 28..	10¼	10								
October 2..	10¼	9⅞	4,500	9,079	4,889	448	822	....	6,159	7-16@½d
" 5..	9¾	9¾								
" 9..	9¾	9¾	5,000	12,611	3,503	668	....	....	4,171	
" 12..	9¾	9½								
" 16..	9¾	9½	5,000	7,839	3,591	592	1,208	....	5,391	
" 19..	9½	9¼								
" 23..	9½	9¼	6,000	7,681	3,324	1,282	2,158	....	6,764	
" 26..	9½	9⅛								
" 30..	9½	9⅛	7,000	9,672	3,642	506	400	....	4,548	
Novem. 2..	9½	9⅛								¼@5-16d
" 6..	9¼	9⅞	5,000	4,532	8,430	1,333	3,770	....	13,533	
" 9..	9¼	9								
" 13..	9½	9¼	7,500	2,712	4,347	637	410	....	5,394	
" 16..	9⅞	9½								
" 20..	9⅞	9½	6,000	8,975	3,996	1,520	522	148	6,186	
" 23..	10	9¾								
" 27..	9⅞	9½	8,000	12,843	918	1,393	482	....	2,793	
" 30..	9⅞	9½								
Decem. 4..	9⅜	9	5,000	11,904	2,519	10	1,721	12	4,262	7-32@¼d
" 7..	9¼	9								
" 11..	9⅜	9⅛	12,000	10,490	1,271	1,022	1,266	128	3,687	
" 14..	9⅜	9⅛								
" 18..	9⅜	9⅛	9,000	14,012	3,848	200	463	....	4,511	
" 21..	9½	9¼								
" 25..	9½	9¼	5,700	5,799	869	169	1,939	....	2,977	

General Remarks.

The year opened with favorable crop accounts, both as regards quantity and quality. The drought of the summer had interfered with our own consumption, stopping many New England mills, while the consumption in England, notwithstanding the war, was 160,780 bales larger, from January 1 to September 1, 1855, than for the like period the year before.

On the 28th of September, accounts were received of the capture of Sebastopol, which was expected to end the war in the East; but the English money markets were now becoming very stringent, owing to the heavy drain of specie to the Crimea, and this with a continuation of the high prices of food, tended to depress the markets, both of Great Britain and this country.

The weather during the fall was very favorable to picking and field labors, in this direction, were two weeks earlier than the previous year.

In December, the suspension of intercourse of our Government with the British Minister, in connection with alleged attempts at enlistments here, for the British army, in the Crimea, caused a little flurry, which, however, soon subsided. In early

Decem. 28..	9½	9¼								
1856.										
January 1..	9½	9¼	6,000	6,866	2,437	514	128	....	3,079	7-32@¼d
" 4..	9½	9¼								
" 8..	9½	9¼	7,000	4,532	2,561	95	1,790	535	4,981	
" 11..	9½	9¼								
" 15..	9⅜	9⅛	5,000	5,060	1,451	....	231	....	1,682	
" 18..	9½	9¼								
" 21..	9½	9¼	14,000	11,551	3,397	501	685	....	4,583	
" 25..	9½	9¼								
" 29..	9½	9¼	5,000	5,857	4,138	882	280	....	5,300	
February 1..	10	9⅝								3-16@7/32
" 5..	10	9¾	20,000	14,590	778	91	870	....	1,739	
" 8..	10	9¾								
" 12..	10¼	10	16,000	14,106	2,121	....	824	....	2,945	
" 15..	10½	10¼								
" 19..	10⅝	10⅜	25,000	8,135	3,180	2,098	1,839	678	7,795	
" 22..	10⅝	10⅜								
" 26..	10⅝	10⅜	20,000	15,616	3,828	....	1,168	....	4,996	
" 28..	10⅝	10⅜								
March 4..	10½	10¼	14,000	15,946	4,029	....	237	....	4,266	¼@9-32d
" 7..	10⅜	10⅛								
" 11..	10⅛	9⅞	10,000	9,876	4,766	652	1,501	102	7,021	
" 14..	10	9¾								
" 18..	10¼	10	17,000	12,570	4,357	....	1,328	....	5,685	
" 21..	10⅜	10⅛								
" 25..	10½	10¼	22,000	18,378	4,859	474	1,485	....	6,818	
" 28..	10½	10¼								
April 1..	10⅝	10⅜	27,000	8,746	4,789	3,538	1,414	60	9,801	¼d.
" 4..	10⅝	10¾								
" 8..	10¾	10½	19,000	20,597	3,888	....	3,185	....	7,073	
" 11..	11	10¾								
" 14..	11⅛	10⅞	26,000	8,312	2,773	1,066	1,574	....	5,413	
" 18..	11⅜	11⅛								
" 22..	11⅜	11⅛	25,000	14,043	7,072	....	422	....	7,494	
" 25..	11½	11¼								
" 29..	11½	11¼	16,000	7,745	3,561	486	806	722	5,575	
May 2..	11¼	11								⅛@3-16d
" 6..	11	10¾	6,000	6,274	5,998	1,057	1,191	1,202	9,448	
" 9..	11⅛	10⅞								
" 13..	11	10¾	11,000	9,912	2,217	....	877	362	3,456	
" 16..	10¾	10½								

February, peace rumors, in connection with low water in Western rivers, and decreased receipts at the ports, affected the markets favorably, both here and in England.

In March, the receipts increased again, and with unfavorable accounts from abroad, prices yielded a little.

Peace was proclaimed in Europe, in April, and the markets on both sides the water became buoyant, with considerable speculative business at advanced figures.

Our relations with England were esteemed to be critical during May and June, owing to the Central American dispute, and there was a very unsettled feeling in commercial circles, with some shrinkage in the price of cotton; but, in July, these difficulties were amicably settled, and from thence, to the close of the year, the market was quite steady.

New York Statement for 1856.—*Concluded.*

1856.		Price of Middling New Orleans New York Classificat'n.	Price of Middling Upland, New York Classificat'n.	Sales for week, including lots in transit.	Receipts for week, including lots in transit.	EXPORTS FOR THE WEEK. To Great Britain.	To France.	North of Europe.	Other Fo'n Ports	Total Exports.	Rates of Freight to Liverpool.
May	20..	$10\frac{7}{8}$	$10\frac{5}{8}$	9,000	8,261	4,155	812	126	480	5,573	
"	23..	11	$10\frac{3}{4}$								
"	27..	11	$10\frac{3}{4}$	10,000	6,228	2,224	....	143	....	2,367	
"	30..	11	$10\frac{3}{4}$								
June	3..	11	$10\frac{3}{4}$	8,500	9,031	7,035	491	3,056	....	10,582	3-16d.
"	6..	11	$10\frac{3}{4}$								
"	10..	$11\frac{1}{8}$	$10\frac{7}{8}$	12,000	4,457	4,768	396	495	....	5,659	
"	13..	$11\frac{1}{4}$	11								
"	17..	$11\frac{1}{2}$	$11\frac{1}{4}$	12,000	8,896	5,301	175	281	....	5,757	
"	20..	$11\frac{1}{2}$	$11\frac{1}{4}$								
"	24..	$11\frac{1}{2}$	$11\frac{1}{4}$	5,000	7,854	3,708	901	567	446	5,622	
"	27..	$11\frac{1}{2}$	$11\frac{1}{4}$								
July	1..	$11\frac{1}{2}$	$11\frac{1}{4}$	4,500	9,457	5,581	787	24	475	6,867	$\frac{5}{32}$@3-16
"	4..	$11\frac{1}{2}$	$11\frac{1}{4}$								
"	8..	$11\frac{1}{2}$	$11\frac{1}{4}$	3,000	4,643	4,457	228	285	....	4,970	
"	11..	$11\frac{3}{4}$	$11\frac{1}{2}$								
"	15..	$11\frac{7}{8}$	$11\frac{5}{8}$	11,000	1,581	3,962	....	....	....	3,962	
"	18..	$11\frac{7}{8}$	$11\frac{5}{8}$								
"	21..	$11\frac{7}{8}$	$11\frac{5}{8}$	8,000	8,593	2,174	....	....	....	2,174	
"	25..	$11\frac{7}{8}$	$11\frac{5}{8}$								
"	29..	$11\frac{7}{8}$	$11\frac{5}{8}$	2,500	2,723	1,349	....	....	....	1,349	
August	1..	$11\frac{7}{8}$	$11\frac{5}{8}$								$\frac{1}{8}$@$\frac{5}{32}$d.
"	5..	$11\frac{3}{4}$	$11\frac{1}{2}$	3,000	508	2,525	68	255	20	2,868	
"	8..	$11\frac{3}{4}$	$11\frac{1}{2}$								
"	12..	$11\frac{3}{4}$	$11\frac{1}{2}$	2,500	3,016	1,310	8	....	....	1,318	
"	15..	$11\frac{3}{4}$	$11\frac{1}{2}$								
"	19..	$11\frac{3}{4}$	$11\frac{1}{2}$	7,500	1,098	1,114	....	....	....	1,114	
"	22..	$11\frac{3}{4}$	$11\frac{1}{2}$								
"	26..	$11\frac{3}{4}$	$11\frac{1}{2}$	3,000	1,067	627	154	....	....	781	
"	29..	$11\frac{5}{8}$	$11\frac{3}{8}$								
Septem.	2..	$11\frac{1}{2}$	$11\frac{1}{4}$	7,500	1,098	900	518	611	....	2,029	$\frac{1}{8}$d.
Average price and total sales, receipts and exports.		10.55	10.30	531,200	424,712	181,045	27,155	42,893	5,371	256,464	

GENERAL REMARKS.

Exchange.

Bills on London ranged through September, at $9\frac{1}{2}$@$10\frac{1}{4}$; October, $8\frac{1}{4}$@$9\frac{3}{4}$; November, $7\frac{3}{4}$@$8\frac{3}{4}$; December, 8@9; January, $7\frac{3}{4}$@$8\frac{3}{4}$; February, 8@$9\frac{1}{4}$; March, $8\frac{7}{8}$@$9\frac{3}{4}$; April, $9\frac{1}{8}$@$9\frac{7}{8}$; May, $9\frac{5}{8}$@10; June, $9\frac{1}{2}$@10; July, the same; and August the same.

LIVERPOOL STATEMENT FOR 1856.

UNITED STATES—1855–1856.

Stock Sept. 1, 1855	143,000	Export	2,954,000
		Consumption	653,000
Crop	3,528,000	Stock Sept. 1, 1856	64,000
Bales	3,671,000	Bales	3,671,000

Stock Jan. 1, 1856, in	Gt. Britain.	France.	Continent.	Tot. Europe.
United States.....Bales	236,000	58,000	19,000	313,000
Brazil	63,000	1,000	3,000	67,000
West Indies	4,000	4,000	1,000	9,000
East Indies	133,000	....	3,000	136,000
Egypt	50,000	5,000	7,000	62,000
Bales	486,000	68,000	33,000	587,000

CONSUMPTION. / IMPORT.

Tot. Europe.	Continent.	France.	Gt. Britain.		Gt. Britain.	France.	Continent.	Tot. Europe.
2,875,000	710,000	478,000	1,687,000	United States	1,758,000	464,000	729,000	2,822,000
190,000	38,000	4,000	148,000	Brazil	122,000	4,000	42,000	158,000
38,000	10,000	14,000	14,000	West Indies	11,000	10,000	11,000	31,000
499,000	216,000	2,000	281,000	East Indies	464,000	2,000	222,000	472,000
211,000	49,000	28,000	134,000	Egypt	113,000	26,000	45,000	182,000
3,813,000	1,023,000	526,000	2,264,000	Bales	2,468,000	506,000	1,049,000	3,665,000
....	....	....	358,000	Export.				
439,000	59,000	48,000	332,000	Stock Dec. 31. Stock above,	486,000	68,000	33,000	587,000
4,252,000	1,082,000	574,000	2,954,000	Total supply, bales	2,954,000	574,000	1,082,000	4,252,000

COTTON AT LIVER

Week Ending.	Receipts.						Sales.			
	Americ'n.	E. I.	Egypt.	Brazil.	Other.	Total.	Consumption.	Speculation.	Export	Total.
Jan. 4..	21,064	3,724	......	403	......	25,191	21,140	1,200	1,380	23,720
" 11..	26,954	......	530	......	......	27,484	38,380	2,750	3,890	45,020
" 18..	5,840	......	515	......	249	6,604	44,700	8,930	5,120	58,750
" 25..	52,933	6,941	3,892	3,131	924	67,821	49,550	10,400	10,270	70,220
Feb. 1..	78,438	6,545	47	6,650	212	91,892	58,030	14,380	12,090	84,500
" 8..	10,101	......	......	......	......	10,101	36,040	10,380	10.920	57,340
" 15..	70,203	2,465	2,824	3,653	......	79,145	55,100	27,530	9.440	92,070
" 22..	50,675	......	2,532	4,403	......	57,610	39,540	7,420	3,900	50,860
" 29..	23,990	......	2,551	2,318	27	28,886	49,010	13,320	3,250	65,580
Mch. 7..	31,388	......	423	1,100	613	33,524	53,560	8,540	2.110	64,210
" 14..	14,728	10,600	2,078	3,033	......	30,439	42,140	2,400	2,300	46,840
" 20..	5,963	......	2,129	......	......	8,092	27,790	2,400	1,660	31,850
" 28..	64,237	5,269	3,836	3,738	......	77,080	29,800	7,340	4,570	41.710
April 4..	68,345	5,221	4,298	5,837	......	83,701	49,800	13,980	3,120	66,900
" 11..	159,165	17,213	285	11,454	115	188,232	42,390	23,320	4,230	69,940
" 18..	44,240	15,146	1,514	4,676	365	65,941	117,230	39,930	5,010	162,170
" 25..	18,768	11,108	3,739	......	......	33,615	52,690	38,210	7,160	98,060
May 2..	48,307	12,538	3,034	1,164	......	65,043	31,850	7,070	2.760	41.680
" 9..	15,570	4,243	......	2,920	......	22,733	30,050	12,760	4,680	47,490
" 16..	131,105	2,553	4,670	3,540	42	141,910	27,850	7,770	3,210	38,830
" 23..	69,748	6,097	......	......	100	75,945	29,580	6,130	1,730	37,440
" 30..	42,833	4,069	2,313	......	5	49,220	33,030	3,400	3,980	40,410
June 6..	32,937	4,069	540	......	16	37,562	41,810	6,780	1,510	50,100
" 13..	79,034	......	2,758	1,090	211	83,093	32,680	8,560	3,290	41,530
" 20..	46,316	7,508	1,089	65	211	55,189	45,720	17,560	2,800	66,080
" 27..	46,645	5,883	2,528	4,379	......	59,435	41,820	12,240	1,700	55,760
July 4..	17,591	......	3,784	......	24	21,399	62,970	15,180	3,570	81,720
" 11..	21,137	5,901	2,459	1,195	681	31,373	28,340	2,980	3,830	35,150
" 18..	81,922	17,584	4,013	6,124	213	109,856	30,010	2,570	5,500	38,080
" 25..	36,295	1,699	1,624	1,966	58	41,642	35,830	4,860	4,580	45,270
Aug. 1..	45,666	4,062	6,406	1,309	......	57,443	49,810	9,980	9,800	69,590
" 8..	37,110	4,178	693	......	......	41,981	40,460	5,090	7,780	53,330
" 15..	38,389	16,103	9,510	3,958	......	67,960	26,750	3,140	7,260	37,150
" 22..	5,187	......	2,454	......	......	7,641	35,650	2,340	9,330	47,320
" 29..	4,008	11,776	2,676	4,821	280	23,561	31,690	8,440	8,440	48,570
Sept. 5..	27,706	......	1,868	......	3	29,577	63,560	14,650	6,200	84.410
" 12..	5,393	7,056	623	......	......	13,072	36,040	7,580	6,550	50,170
" 19..	11,064	8,224	3,137	3,024	897	26,346	35,790	6,240	6,550	48,580
" 26..	420	17,323	1,661	3,569	10	22,983	47,290	20,140	5,320	72,750
Oct. 3..	7,039	23,627	476	3,658	......	34,800	44,060	12,190	4,390	60,640
" 10..	366	6,563	3,575	1,574	......	12,078	40,150	13,010	10,570	63,730
" 17..	11,416	7,017	1,665	4,998	40	25,136	50,390	14,450	8,160	73,000
" 24..	2,050	10,399	988	1,576	344	15,357	58,580	59,160	12,150	129,890
" 31..	4,101	......	1,027	......	......	5,128	36,080	8,870	8,160	53,110
Nov. 7..	2,882	5,422	426	2,566	......	11,296	23,830	2,300	3,130	29,260
" 14..	3,210	200	633	2,677	......	6,720	33,520	5,280	2,360	41,160
" 21..	4,316	3,996	838	......	......	9,150	27,540	1,930	2,130	31,600
" 28..	7,485	27,451	2,692	2,841	......	40,469	31,010	2,180	1,070	34,260
Dec. 5..	5,982	7,237	2,777	3,992	......	19,988	48,620	9,380	1,600	59,600
" 12..	19,963	13,002	200	2,555	......	35,720	39,710	8,570	5,530	53,810
" 19..	29,447	24,621	1,674	37	......	55,779	42,840	9,690	3,380	55,910
" 24..	13,941	5,275	448	5,821	......	25,485	89,410	27,980	7,070	124,460
" 31..	....	......	......	......	......	4,324	26,450	18,820	5,330	50,600
Average prices & total sales, receipts & stocks.	1,703,613	359908	98,661	121,815	5,640	2,293,961	2,239,660	611,690	276120	3,127,470

POOL. YEAR 1856.

STOCKS.			PRICES.			ACTUAL EXPORT.	CONSUMPTION.	REMARKS.
American.	Other.	Total.	Mid. Up.	Mid. Orl.	Dhol.			
235,554	195,537	431,091	$5\frac{3}{8}$	$5\frac{5}{8}$	$4\frac{5}{16}$	1,380	21,140	
232,518	186,437	418,955	$5\frac{5}{16}$	$5\frac{7}{16}$	$4\frac{5}{16}$	3,890	59,520	
202,678	175,971	378,649	$5\frac{1}{2}$	$5\frac{3}{8}$	$4\frac{5}{16}$	5,120	104,220	
222,161	168,509	390,670	$5\frac{9}{16}$	$5\frac{11}{16}$	$4\frac{3}{8}$	10,270	153,770	
253,929	165,703	419,632	$5\frac{11}{16}$	$5\frac{13}{16}$	$4\frac{1}{2}$	12,090	211,800	Small estimates of crop.
236,260	154,771	391,031	$5\frac{11}{16}$	$5\frac{13}{16}$	$4\frac{1}{2}$	10,920	247,840	
264,383	145,853	410,236	$5\frac{7}{8}$	$5\frac{15}{16}$	$4\frac{5}{8}$	9,440	302,940	Peace news from Continent (Crimea).
283,808	135,638	419,446	$5\frac{13}{16}$	$5\frac{7}{8}$	$4\frac{5}{8}$	3,900	342,480	
270,788	125,184	395,972	$5\frac{7}{8}$	$5\frac{15}{16}$	$4\frac{3}{4}$	3,250	391,490	
257,056	114,430	371,486	$5\frac{7}{8}$	6	$4\frac{3}{4}$	2,110	445,050	
235,504	121,051	356,555	$5\frac{13}{16}$	$5\frac{15}{16}$	$4\frac{3}{4}$	2,300	487,190	
219,567	114,390	333,957	$5\frac{3}{4}$	$5\frac{7}{8}$	$4\frac{3}{4}$	1,660	514,980	
258,164	121,413	379,577	$5\frac{3}{4}$	$5\frac{7}{8}$	$4\frac{5}{8}$	4,570	544,780	
285,869	124,649	410,218	$5\frac{7}{8}$	6	$4\frac{5}{8}$	3,120	594,580	
412,934	141,466	554,400	$5\frac{15}{16}$	$6\frac{1}{16}$	$4\frac{5}{8}$	4,230	636,970	
396,344	142,257	538,601	$6\frac{1}{8}$	$6\frac{1}{4}$	$4\frac{3}{4}$	5,010	754,200	
372,722	145,524	518,246	$6\frac{1}{4}$	$6\frac{3}{8}$	$4\frac{7}{8}$	7,160	806,890	Less favorable accounts of crop.
396,469	152,330	548,799	$6\frac{3}{16}$	$6\frac{5}{16}$	$4\frac{7}{8}$	2,760	838,740	
386,599	150,473	537,072	$6\frac{3}{16}$	$6\frac{5}{16}$	$4\frac{7}{8}$	4,680	868,790	Large receipts at shipping ports.
489,164	152,578	641,742	$6\frac{3}{16}$	$6\frac{5}{16}$	$4\frac{7}{8}$	3,210	896,640	
533,752	148,045	681,797	$6\frac{1}{8}$	$6\frac{1}{4}$	$4\frac{7}{8}$	1,730	926,220	
595,385	148,212	693.597	$6\frac{1}{8}$	$6\frac{1}{4}$	$4\frac{7}{8}$	3,980	959,250	
543,972	137,818	681,790	$6\frac{1}{16}$	$6\frac{3}{16}$	$4\frac{7}{8}$	1,510	1,001,060	
596,806	130,637	727,443	$6\frac{1}{16}$	$6\frac{3}{16}$	$4\frac{7}{8}$	3,290	1,033,740	
603,982	130,059	734,041	$6\frac{1}{16}$	$6\frac{3}{16}$	$4\frac{7}{8}$	2,800	1,079,460	
616,317	132,714	749,031	$6\frac{1}{8}$	$6\frac{1}{4}$	$4\frac{7}{8}$	1,700	1,121,280	
582,398	121,542	703,940	$6\frac{3}{16}$	$6\frac{5}{16}$	$4\frac{7}{8}$	3,570	1,184,250	
583,385	123,098	706,483	$6\frac{3}{16}$	$6\frac{5}{16}$	$4\frac{7}{8}$	3,830	1,212,590	
638,697	140,522	779,219	$6\frac{3}{16}$	$6\frac{5}{16}$	$4\frac{7}{8}$	5,500	1,243,600	
645,702	135,109	780,811	$6\frac{3}{16}$	$6\frac{5}{16}$	$4\frac{7}{8}$	4,580	1,279,430	
650,688	133,086	783,774	$6\frac{3}{16}$	$6\frac{3}{8}$	5	9,800	1,329,240	Very bad weather for harvest, &c.
654,138	127,097	781,235	$6\frac{3}{16}$	$6\frac{3}{8}$	5	7,780	1,369,700	
669,157	149,998	819,155	$6\frac{1}{8}$	$6\frac{5}{16}$	5	7,260	1,397,450	
641,924	138,762	780,686	$6\frac{1}{8}$	$6\frac{1}{4}$	5	9,330	1,433,100	
619,192	148,667	767,859	$6\frac{1}{8}$	$6\frac{1}{4}$	5	8,440	1,464,790	
595,658	132,816	728,474	$6\frac{1}{4}$	$6\frac{3}{8}$	5	6,200	1,528,350	
569,231	129,345	698,576	$6\frac{1}{4}$	$6\frac{3}{8}$	$5\frac{1}{8}$	6,550	1,564,390	
548,685	131,637	680,322	$6\frac{1}{4}$	$6\frac{3}{8}$	$5\frac{1}{8}$	6,550	1,600,180	
505,055	142,320	647,375	$6\frac{3}{8}$	$6\frac{7}{16}$	$5\frac{1}{8}$	5,320	1,647,470	
473,234	155,991	629,225	$6\frac{3}{8}$	$6\frac{7}{16}$	$5\frac{1}{8}$	4,390	1,691,530	
443,670	152,013	595,683	$6\frac{7}{16}$	$6\frac{1}{2}$	$5\frac{1}{8}$	10,570	1,731,680	Bad accounts of crop from United States.
415,526	140,743	556,269	$6\frac{1}{2}$	$6\frac{9}{16}$	$5\frac{1}{8}$	8,160	1,782,070	
371,106	122,250	493,356	$6\frac{7}{8}$	$6\frac{15}{16}$	$5\frac{3}{8}$	12,150	1,840,650	
345,837	107,347	453,184	$6\frac{7}{8}$	$6\frac{15}{16}$	$5\frac{3}{8}$	8,160	1,876,730	
328,917	104.151	432,870	$6\frac{13}{16}$	$6\frac{7}{8}$	$5\frac{3}{8}$	3,130	1,900,560	
303,929	197,461	401,390	$6\frac{7}{8}$	$6\frac{15}{16}$	$5\frac{3}{8}$	2,360	1,934,080	Alarming accounts of crop failure, disturbed state of things on the Continent, &c.
285,375	90,145	375,520	$6\frac{7}{8}$	$6\frac{15}{16}$	$5\frac{3}{8}$	2,130	1,961,620	
267,100	116.489	383,589	$6\frac{13}{16}$	$6\frac{7}{8}$	$5\frac{1}{4}$	1,070	1,992,630	
235,752	116.345	352,097	$6\frac{7}{8}$	$6\frac{15}{16}$	$5\frac{1}{4}$	1,600	2,041,250	
224,865	117,612	342,477	$6\frac{7}{8}$	$6\frac{15}{16}$	$5\frac{1}{4}$	5,530	2,080,960	
222,002	129,764	351,766	$6\frac{13}{16}$	7	$5\frac{1}{4}$	3,380	2,123,800	
206,253	120,928	327,181	$7\frac{3}{16}$	$7\frac{5}{16}$	$5\frac{3}{8}$	7,070	2,213,210	
......	......	281,430	......	......	......	5,330	2,239,660	
			6.22	6.31	4.91	276,120	42,238.87	

1857.

A contrast was this year presented to the harvest year (*see* year 1845), of English cotton manufacturers. They paid this year for raw cotton, £26,000,000, receiving for yarn and manufactured cotton goods, £56,000,000, leaving a margin of £30,000,000 for machinery, fuel, dyeing, bleaching, printing, wages, interest on capital, and profit. They had done much more business than in 1845, but under less favorable circumstances.

"The cotton supply Association," of Great Britain, was established this year, the cotton manufacturers of the United Kingdom, "feeling it to be a duty to inquire whether an increased supply of cotton can be obtained from other countries, so as to lessen the dependence of Great Britain on the United States." (*See* year 1858.)

Alderman John Bayrus, in a lecture delivered at Blackburn, England, this year, gave the following statistics as to the number of spindles in use in the cotton manufactories of Austria:

Lower Austria	569,979
Upper "	83,590
Styria	25,464
Krain Gorr	30,300
Tyrol	214,094
Bohemia	449,906
Lombardy	129,046
Venice	28,464
Hungary	2,400
Total in Austria ... spindles.	1,533,243

COTTON CROP OF THE UNITED STATES.

Statement and Total Amount of the Cotton Crop of the United States, for the Year ending August 31, 1857.

	Bales.	Bales.	TOTAL. 1857.	1856.	1855.
NEW ORLEANS.					
Export—					
To Foreign Ports	1,293,717				
Coastwise	223,204				
Stock on hand 1st September, 1857	7,321				
		1,524,242			
Deduct—					
Received from Mobile	41,040				
" Montgomery, &c	18,996				
" Florida	4,708				
" Texas	17,503				
Stock on hand 1st September, 1856	6,995				
		89,242			
			1,435,000	1,661,433	1,232,644
MOBILE.					
Export—					
To Foreign Ports	314,989				
Coastwise	174,055				
Manufactured in Mobile, &c	2,246				
Burnt at Mobile	12,700				
Stock on hand 1st September, 1857	4,504				
		508,494			
Deduct—					
Received from New Orleans	10				
Shipments to Boston, returned	302				
Stock on hand 1st September, 1856	5,005				
		5,317			
			503,177	659,738	454,595
TEXAS.					
Export—					
To Foreign Ports	20,907				
Coastwise	68,636				
Stock on hand 1st September, 1857	962				
		90,505			
Deduct—					
Stock on hand 1st September, 1856	....	623			
			89,882	116,078	80,737
FLORIDA.					
Export—					
To Foreign Ports—Uplands	30,889				
Coastwise—Uplands	82,636				
Sea Island	20,365				
Burnt at Apalachicola	2,472				
Stock on hand 1st September, 1857	56				
		136,418			
Deduct—					
Stock on hand 1st September, 1856	....	74			
			136,344	144,404	136,597
GEORGIA.					
Export—					
To Foreign Ports—Uplands	152,228				
" Sea Island	6,611				
Coastwise—Uplands	158,791				
Sea Island	10,028				
Stock in Savannah, 1st September, 1857	1,926				
" Augusta, 1st September, 1857	2,747				
		332,331			

Statement and Total Amount of the Cotton Crop of the United States, for the Year ending August 31, 1857—*Concluded.*

	Bales.	Bales.	TOTAL. 1857.	1856.	1855.
Deduct—					
Received from Florida—Sea Island	6,889				
Stock in Savannah 1st September, 1856	1,550				
" Augusta, 1st September, 1856	1,781	10,220			
			322,111	389,445	378,694
SOUTH CAROLINA.					
Export from Charleston—					
To Foreign Ports—Uplands	212,604				
" Sea Island	16,581				
Coastwise—Uplands	162,541				
Sea Island	6,908				
Burnt and Manufactured at Charleston	461				
Stock in Charleston, 1st September, 1857	5,644				
	404,739				
Export from Georgetown, S. C.—					
To Coastwise Ports	9,500	414,239			
Deduct—					
Received from Florida—Sea Island	8,307				
Received from Key West and Nassau, N.P. (wrecked)—Uplands	431				
Received from Savannah—Sea Island	1,589				
" Uplands	3,437				
Stock in Charleston, 1st September, 1856	3,144	16,908			
			397,331	495,976	499,272
NORTH CAROLINA.					
Export—					
To Coastwise Ports	27,147	....	27,147	26,098	26,139
VIRGINIA.					
Export—					
To Foreign Ports	200				
Coastwise	5,454				
Manufactured (taken from the ports)	18,541				
Stock on hand, 1st September, 1857	420	24,615			
Deduct—					
Stock on hand 1st September, 1856	....	842			
			23,773	20,458	31,000
Received at New York, from Memphis, Nashville, &c., Tenn	....	....	2,022	2,086	1,061
Received at Philadelphia, from Memphis, Nashville, &c., Tenn	....	....	1,236	7,938	3,100
Received at Baltimore, from Memphis, Nashville, &c., Tenn	....	....	1,496	4,191	3,500
Total crop of the United States	....	....	2,939,519	3,527,845	2,847,339

	bales.
Decrease from crop of 1856	588,326
Increase over crop of 1855	92,180
Increase over crop of 1854	9,492

Export to Foreign Ports, from September 1, 1856, *to August* 31, 1857.

FROM	To Great Britain.	To France.	To North of Europe.	Other F'n Ports.	Total.
New Orleans........................Bales	749,485	258,163	156,450	129,619	1,293,717
Mobile....................................	211,231	84,840	16,570	2,348	314,989
Texas......................................	9,792	4,428	6,687	....	20 907
Florida....................................	29,125	....	1,764	....	30,889
Savannah................................	138,694	3,504	5,976	10,665	158,839
Charleston..............................	138,876	40,821	28,296	21,192	229,185
North Carolina........................	....	....	....	....	....
Virginia..................................	200	....	....	....	200
Baltimore...............................	....	....	....	....	....
Philadelphia...........................	820	....	....	....	820
New York................................	145,984	21,601	28,600	808	196,993
Boston....................................	4,663	....	1,455	....	6,118
Grand total........................	1,428.870	413,357	245,798	164,632	2,252,657
Total last year....................	1,921,386	480,637	304,005	248,578	2,954,606
Decrease..............................	492,516	67,280	58,207	83,946	701,949

Consumption.

Total crop of the United States, as before stated........................bales.			2,939,519
Add—Stocks on hand at the commencement of the year, September 1, 1856—			
In the Southern ports....................................		20,014	
In the Northern ports....................................		44,157	
			64,171
Makes a supply of..			3,003,690
Deduct therefrom—The Export to Foreign ports............	2,252,657		
Less Foreign included..................	1,161		
		2,251,496	
Stocks on hand, September 1, 1857—			
In the Southern ports........................	23,580		
In the Northern ports........................	25,678		
		49,258	
Burnt at New York and Baltimore...........................		798	
			2,301,552
Taken for home use...bales.			702,138

ANNUAL REVIEW.

From the New Orleans Price Current—1856–57.

In the early part of October, there were accounts of killing frosts, which had more or less influence, and at the same time supplies were interrupted by unusually low waters, even the main channel of the Mississippi being so shallow that the packets could only bring in parts of cargoes. Under these circumstances prices were run up still further, and on the 10th of October, the quotations for middling, were 12¼ to 12½ cents, but at this point the first reaction of the season took place, the demand not having fully responded to the increased supply. The month of April, witnessed quite a severe struggle between holders and purchasers, but the remarkable frosts of the 5th, 13th, and 23d of the month, put beyond all question, that the crop of 1857–8, whatever might be its ultimate extent, must be a late one, and this conviction, with a rapid falling off in the receipts and reduced rates of freight, tended to the advantage of holders and enabled them to maintain still further advanced prices, with remarkable steadiness throughout the month. This question of crop is one which we have always touched with great caution, and we can only speak in general terms of what are understood to be its present prospects. We will, however, promise that, beyond a doubt, preparations were made for the largest crop ever grown. There was every motive for such a course, and such is understood to be the fact.

LIVERPOOL STATEMENT FOR 1857.

UNITED STATES—1856–1857.			
Stock Sept. 1, 1856	64,000	Export	2,252,000
		Consumption	702,000
Crop	2,939,000	Stock Sept. 1, 1857	49,000
Bales	3,003,000	Bales	3,003,000

	Gt. Britain.	France.	Continent.	Tot. Europe.
Stock Jan. 1, 1857, in				
United States ... Bales	178,000	44,000	38,000	260,000
Brazil	27,000	1,000	7,000	35,000
West Indies	1,000	....	1,000	2,000
East Indies	99,000	....	10,000	109,000
Egypt	27,000	3,000	3,000	33,000
Bales	332,000	48,000	59,000	439,000

CONSUMPTION. / IMPORT.

Tot. Europe.	Continent.	France.	Gt. Britain.		Gt. Britain.	France.	Continent.	Tot. Europe.
2,224,000	494,000	378,000	1,352,000	United States	1,482,000	394,000	504,000	2,275,000
186,000	24,000	7,000	155,000	Brazil	168,000	8,000	19,000	190,000
29,000	8,000	14,000	7,000	West Indies	11,000	17,000	10,000	38,000
608,000	223,000	23,000	362,000	East Indies	680,000	39,000	245,000	738,000
137,000	27,000	26,000	84,000	Egypt	76,000	26,000	29,000	130,000
3,184,000	776,000	448,000	1,960,000	Bales	2,417,000	484,000	807,000	3,371,000
....	....	....	337,000	Export.				
626,000	90,000	84,000	452,000	Stock Dec. 31. Stock above,	332,000	48,000	59,000	439,000
3,810,000	866,000	532,000	2,749,000	Total supply, bales	2,749,000	532,000	866,000	3,810,000

COTTON AT LIVER

Week Ending.	Receipts.						Sales.			
	American	E. I.	Egypt.	Brazil.	Other.	Total.	Consumption.	Speculation.	Export.	Total.
Jan. 9..	36,872	6,521	722	4,778	......	48,893	59,810	18,740	8,230	86,780
" 16..	56,214	2,267	4,293	11,932	48	74,754	21,970	2,970	2,460	27,400
" 23..	34,689	......	......	11,098	34	45,821	33,960	5,520	2,240	41,720
" 30..	6,826	......	2,379	2,420	......	11,625	42,690	12,220	6,450	61,360
Feb. 6..	18,808	6,675	......	780	155	26,418	33,480	9,020	3,160	45,660
" 13..	56,420	22,811	775	13,161	413	93,580	36,210	25,100	5,440	66,750
" 20..	11,126	......	3,922	1,419	......	16,467	40,180	16,950	4,770	61,900
" 27..	53,437	2,932	332	1,744	410	58,855	31,080	6,320	6,350	43,750
Mch. 6..	37,118	10,738	735	....	......	48,591	9,770	4,700	5,070	19,540
" 13..	24,396	7,542	3,242	6,435	......	41,615	40,730	2,480	5,440	48,650
" 20..	65,413	18,253	906	9,776	......	94,348	33,450	6,220	4,260	43,930
" 27..	29,773	5,841	......	14	......	35,628	39,140	470	3,580	43,190
April 3..	64,672	22,584	2,615	3,082	......	92,953	44,500	4,160	4,370	53,030
" 9..	79,155	26,856	1,816	2,468	332	110,627	33,590	2,430	4,330	40,350
" 17..	66,763	27,175	1,194	2,365	......	97,497	34,770	2,120	5,620	42,510
" 24..	43,539	12,543	761	1,830	64	58,737	37,710	1,850	4,060	43,620
May 1..	8,189	7,241	832	2,228	2	18,492	43,440	4,750	5,180	53,370
" 8..	7,342	......	1,697	....	......	9,039	46,430	7,020	6,700	60,150
" 15..	41,619	8,610	5,574	3,334	170	59,307	38,710	6,930	5,750	51,390
" 22..	184,967	17,266	970	8,192	352	211,747	28,420	1,950	4,450	34,820
" 29..	58,798	57	2,038	3,044	186	64,123	34,370	1,420	4,820	40,610
June 5..	19,333	14,589	2,635	124	......	36,681	34,880	4,380	5,820	45,080
" 12..	25,092	23,794	2,051	7,055	406	58,398	53,900	12,780	7,960	74,640
" 19..	17,561	......	1,167	....	19	18,747	34,720	3,380	5,100	43,200
" 26..	7,637	......	1,442	....	......	9,079	55,440	7,760	7,500	60,700
July 3..	36,578	8,304	2,024	7,124	......	54,030	54,960	8,010	7,150	70,120
" 10..	34,997	2,916	1,107	4,440	70	43,530	53,550	7,110	7,640	68,300
" 17..	13,648	8,210	733	1,518	......	24,109	63,750	10,000	5,240	78,990
" 24..	15,661	10,613	1,192	590	......	28,056	58,890	11,950	4,620	75,460
" 31..	16,053	21,008	1,988	1,343	......	40,392	48,750	9,230	4,480	62,410
Aug. 7..	13,800	7,179	1,263	1,219	......	23,461	27,110	2,500	3,410	33,020
" 14..	11,597	11,646	......	1,490	......	24,733	28,710	3,630	6,310	38,650
" 21..	19,468	7,727	......	....	......	27,195	64,040	17,070	5,600	86,710
" 28..	4,632	19,404	1,254	2,000	3	27,293	73,830	33,700	3,030	110,560
Sep. 4..	1,342	3,776	1,061	1,574	80	7,833	47,360	15,480	2,250	65,090
" 11..	3,928	16,634	729	2,160	......	23,451	56,950	28,040	2,080	87,070
" 18..	12,806	33,009	1,215	2,728	540	50,298	42,370	29,500	1,710	73,580
" 25..	557	7,255	......	2,070	370	10,252	22,780	4,410	1,710	28,900
Oct. 2..	6,052	48,605	788	5,428	1,198	62,071	17,410	6,710	2,650	26,770
" 9..	987	8,758	108	37	325	10,215	43,050	13,280	2,790	59,120
" 16..	4,533	18,500	......	2,540	......	25,573	15,470	9,410	500	25,380
" 23..	7,004	9,718	663	7,899	400	25,684	17,580	2,870	790	21,240
" 30..	8,633	1,899	3,006	123	9	13,670	12,030	1,400	840	14,270
Nov. 6..	9,585	17,079	1,460	....	......	28,124	19,750	1,370	710	21,830
" 13..	11,012	2,240	1,023	627	......	14,902	9,020	1,600	310	10,930
" 20..	5,651	......	2,469	3,071	366	11,557	13,800	3,480	2,250	20,530
" 27..	14,482	4,366	3,908	12,296	......	35,052	19,960	1,590	5,070	26,620
Dec. 4..	13,770	704	7	1,290	281	16,052	31,310	10,360	4,420	46,090
" 11..	41,582	6,209	483	4,562	......	52,836	17,670	1,640	990	20,300
" 18..	23,503	34,115	413	3,688	......	61,719	24,020	3,130	1,170	28,320
" 24..	22,502	2,207	1,285	1,776	27	27,797	25,000	9,930	4,070	39,000
" 31..	....	......	......	....	......	33,752	32,550	5,470	2,170	40,190
Average prices & total sales, receipts & stocks.	1,410,122	556376	70,277	468,872	6,260	2,545,659	1,885,030	424,510	213070	2,522,610

POOL. YEAR 1857.

STOCKS.			PRICES.			ACTUAL EXPORT.	CONSUMPTION.	REMARKS.
Amer'n.	Other.	Total.	Mid. Up.	Mid. Orl.	Dhol.			
169,552	99,271	268,823	7 1/2	7 3/4	5 5/8	8,230	59,810	
209,486	109,053	318,539	7 7/16	7 5/8	5 1/2	2,460	81,780	
216,675	117,333	328,408	7 7/16	7 5/8	5 1/2	2.240	115,740	
191,531	103,792	295,323	7 1/2	7 5/8	5 1/2	6,450	158,430	
188,209	97,962	286,171	7 1/2	7 5/8	5 1/2	3,160	191,910	
220,909	121,042	341,951	7 9/16	7 3/4	5 5/8	5,440	228,120	Favorable crop accounts.
202,115	113,973	316,088	7 5/8	7 7/8	5 5/8	4,770	268,300	
233,542	106,061	339,603	7 9/16	7 13/16	5 5/8	6,350	299,380	
239,730	102,004	341,734	7 9/16	7 13/16	5 5/8	5,070	309,150	
232,496	102,663	335.159	7 5/8	7 13/16	5 5/8	5,440	349,880	
274,139	117,848	391,987	7 9/16	7 13/16	5 5/8	4,260	383,330	
276.312	108,731	385,043	7 9/16	7 13/16	5 3/4	3,580	422,470	
307,824	121,712	429,536	7 9/16	7 13/16	5 3/4	4,370	466,970	
363,179	139,534	502,713	7 9/16	7 3/4	5 3/4	4,330	500,560	
400,632	164,878	565,510	7 1/2	7 3/4	5 5/8	5,620	535,330	
418,561	162,586	581,147	7 7/16	7 11/16	5 1/2	4.060	573,040	
390,520	158,469	548,989	7 7/16	7 5/8	5 1/2	5,180	616,480	Dullness of trade in Manchester districts.
359,912	142,836	502,748	7 9/16	7 3/4	5 1/2	6,700	662,910	
368,901	143,244	512,145	7 5/8	7 13/16	5 1/2	5,750	701,620	
527,288	156,144	683,432	7 9/16	7 3/4	5 1/2	4,450	730,040	
553,916	148,159	702,075	7 5/8	7 3/4	5 1/2	4,820	764,410	
541,049	154.197	695,246	7 5/8	7 13/16	5 1/2	5,820	799,290	
526,071	167,623	693,694	7 13/16	7 15/16	5 1/2	7,960	853,190	
515,622	154,739	670,421	7 13/16	7 15/16	5 5/8	5,100	887,920	
490,689	135,191	625,880	7 7/8	8	5 5/8	7,500	943,360	
492,237	131,973	624,230	7 7/8	8	5 5/8	7,150	998,320	
483,414	120,176	603,590	8	8 1/8	5 3/4	7,640	1,051,870	Bad news from India.
451,232	104,338	555,570	8 1/16	8 3/16	5 3/4	5,240	1,115,620	
426,353	98,973	525,326	8 3/16	8 5/16	5 3/4	4,620	1,174,510	
406,786	100,642	507,628	8 1/4	8 3/8	6	4,480	1,223,260	
399,346	97,663	497,009	8 1/4	8 3/8	6	3,410	1,250,370	Uncertain character of news from India.
382,733	100,589	483,322	8 1/4	8 3/8	6 1/8	6,310	1,279,080	
360,881	88,246	449,127	8 7/16	8 9/16	6 1/4	5,600	1,343,120	
315,883	82,657	398,540	8 3/4	8 7/8	6 1/2	3.030	1,416,950	
281,565	73,918	355,483	8 13/16	8 15/16	6 1/2	2,250	1,464,310	
244,083	74.781	318,864	9	9 1/8	6 7/8	2,080	1,521,260	
221,889	103,023	324.912	9 1/8	9 1/4	6 7/8	1,710	1,563,630	Large business in Manchester and districts.
209,806	99,428	309,234	9	9 3/16	6 3/4	1.710	1,586,410	
201,958	150,187	352,145	9	9 3/16	6 3/4	2,650	1,603,820	
172,245	145,635	317,880	9 1/8	9 5/16	6 3/4	2,790	1,646,870	
166,698	159,865	326,563	9 1/16	9 1/4	6 5/8	500	1,662,340	
162,542	168,095	330,637	8 7/8	9 1/16	6 3/8	790	1,679,920	Unfavorable financial accounts from United States.
163,835	165,742	329,577	8 1/4	8 1/2	5 1/2	840	1,691,950	
161,710	175,191	336,901	7 3/4	8	5 3/4	710	1,711,700	
166,152	174,511	340,663	7 1/8	7 1/2	5 1/4	310	1,720,720	
160.123	176,537	336,660	6 1/2	6 3/4	......	2,250	1,734,520	Financial crisis in London and elsewhere.
160.953	189,462	350,415	6 7/8	6 3/8	......	5,070	1,754,480	
153,665	179,022	332,687	6 1/4	6 9/16	4 5/8	4,420	1,785,790	
182,997	183,156	366,153	6	6 1/4	4 1/2	990	1,803,460	
187,900	214,212	402,112	5 1/2	5 3/4	4 1/4	1,170	1,827,480	
190,982	210,567	401,549	5 3/4	6	4 1/4	4,070	1,852,480	
.....	....	....	......	......	......	2,170	1,885,030	
			7.73	7.93	5.51	213,070	36,250.6	

The semi-weekly Price and Weekly Sales and Receipts at New York, Weekly Exports from New York and Rates of Freight to Liverpool 1st of each month, for the Crop Year ending September 1, 1857.

1856.	Price of Middling New Orleans New York Classificat'n.	Price of Middling Upland, New York Classificat'n.	Sales for week.	Receipts for week.	EXPORTS FOR THE WEEK. To Great Britain.	To France	North of Europe.	Other Fo'n Ports	Total Exports.	Rates of Freight to Liverpool.
Septem. 5..	$11\frac{3}{4}$	$11\frac{5}{8}$								$\frac{1}{8}$d.
" 9..	12	$11\frac{3}{4}$	12,000	2,089	239	....	....	....	239	
" 12..	12	$11\frac{3}{4}$								
" 16..	12	$11\frac{3}{4}$	7,000	1,406	1,925	....	292	....	2,217	
" 19..	$12\frac{1}{8}$	$11\frac{7}{8}$								
" 23..	$12\frac{1}{8}$	$11\frac{7}{8}$	6,500	4,011	878	165	165	....	1,208	
" 26..	$12\frac{1}{2}$	$12\frac{1}{8}$								
" 30..	$12\frac{1}{2}$	$12\frac{1}{8}$	8,000	3,849	1,949	61	95	....	2,105	
October 3..	$12\frac{3}{4}$	$12\frac{1}{2}$								$\frac{1}{8}$d.
" 7..	$12\frac{3}{4}$	$12\frac{1}{2}$	16,000	7,010	1,443	18	265	....	1,726	
" 10..	13	$12\frac{3}{4}$								
" 14..	$13\frac{1}{4}$	13	11,000	6,933	1,363	418	298	....	2,079	
" 17..	$13\frac{1}{2}$	$12\frac{7}{8}$								
" 21..	$12\frac{7}{8}$	$12\frac{1}{2}$	6,500	11,027	1,921	....	....	....	1,921	
" 24..	$12\frac{3}{8}$	$12\frac{1}{4}$								
" 28..	$12\frac{3}{4}$	$12\frac{3}{8}$	6,000	8,977	1,138	98	257	....	1,493	
" 31..	$12\frac{3}{4}$	$12\frac{3}{8}$								
Novem. 4..	$12\frac{3}{4}$	$12\frac{3}{8}$	8,000	22,835	2,520	231	190	....	2,941	$\frac{1}{8}$@5-32d
" 7..	$12\frac{3}{4}$	$12\frac{3}{8}$								
" 11..	$12\frac{1}{4}$	12	7,000	7,379	1,801	267	300	....	2,368	
" 14..	$12\frac{1}{4}$	$11\frac{7}{8}$								
" 18..	12	$11\frac{3}{4}$	9,000	6,662	5,935	820	....	....	6,755	
" 21..	$12\frac{1}{4}$	$11\frac{7}{8}$								
" 25..	$12\frac{3}{8}$	12	11,500	14,064	3,555	....	49	....	3,604	
" 28..	$12\frac{3}{8}$	$12\frac{1}{8}$								
Decem. 2..	$12\frac{5}{8}$	$12\frac{1}{4}$	10,000	8,373	3,587	....	959	....	4,546	3-16d.
" 5..	$12\frac{5}{8}$	$12\frac{1}{4}$								
" 9..	$12\frac{5}{8}$	$12\frac{1}{4}$	22,000	10,755	3,539	1,044	179	....	4,762	
" 12..	$12\frac{1}{2}$	$12\frac{1}{4}$								
" 16..	$12\frac{5}{8}$	$12\frac{3}{8}$	6,500	21,419	4,901	596	280	....	5,777	
" 19..	$12\frac{7}{8}$	$12\frac{5}{8}$								
" 23..	$13\frac{1}{8}$	$12\frac{3}{4}$	15,000	8,501	4,141	480	....	....	4,621	
" 26..	$13\frac{1}{8}$	$12\frac{3}{4}$								

GENERAL REMARKS.

Owing to the falling off in the receipts at the ports, and accounts of frosts in the growing sections, prices, early in September, ruled higher here than in England, and the business in consequence was almost entirely for home use and on speculation.

Heavy shipments of specie were now made, and there was some stringency in the money market.

Throughout October and November, financial affairs, both in England and France, were much deranged, owing to the continued drain of specie thence to the East.

In November, the receipts at the ports began to increase, and this, together with the money troubles in Europe, had for a time a depressing effect; but subsequently, unfavorable reports as to the crop neutralized all other influences, and, as the year wore on, the tendency, for the most part, was steadily upward, checked only now and then by unfavorable news from Europe or occasionally a slight increase in receipts at the shipping ports.

In March again, prices were higher here than in Liverpool; in April, frost accounts gave a further stimu-

Decem. 30..	13⅜	13	17,500	7,241	3,973	......	....	....	3,973	
1857.										
January 2..	13⅜	13⅛								3/16@7/32d.
" 6..	13⅝	13¼	18,000	10,483	1,414	680	397	....	2,491	
" 9..	13⅝	13¼								
" 13..	13⅜	13	9,500	9,136	2,435	....	223	....	2,658	
" 16..	13⅛	12¾								
" 20..	13¼	13	12,000	9,791	2,016	....	....	....	2,016	
" 23..	13½	13⅛								
" 27..	13¼	12⅞	22,000	8,648	6	....	412	....	418	
" 30..	13¼	12⅞								
February 3..	13½	13¼	21,000	14,067	5,003	947	....	....	5,950	5-16d.
" 6..	13½	13¼								
" 10..	13¾	13⅜	18,000	14,747	5,683	970	1,192	....	7,845	
" 13..	13¾	13½								
" 17..	13⅞	13½	25,000	4,310	8,549	134	1,775	....	10,458	
" 20..	14	13⅝								
" 24..	14	13¾	24,000	17,976	4,843	2,003	772	....	7,618	
" 27..	14	13¾								
March 3..	14⅜	14	17,000	10,426	1,196	....	351	780	2,327	5/32@3/16d.
" 6..	14⅛	13⅞								
" 10..	14⅛	13¾	15,000	8,049	3,105	1,222	1,654	....	5,981	
" 13..	14⅛	13¾								
" 17..	14	13¾	19,000	14,158	7,872	....	2,141	....	10,013	
" 20..	14	13¾								
" 24..	14¼	13⅞	21,000	11,900	3,402	203	2,870	20	6,495	
" 27..	14⅜	14								
" 31..	14½	14⅛	19,000	8,178	3,388	1,419	740	....	5,547	
April 3..	14½	14¼								5/32@3/16d.
" 7..	14½	14¼	22,000	5,965	3,669	....	2,226	....	5,895	
" 10..	14⅝	14⅜								
" 14..	14⅝	14⅜	13,000	3,572	1,675	380	176	....	2,231	
" 17..	14½	14⅛								
" 21..	14¼	13⅞	7,000	11,189	3,096	....	53	....	3,149	
" 24..	14¼	13⅞								
" 28..	14⅜	14	10,000	13,530	1,943	....	....	....	1,943	
May 1..	14⅝	14¼								3-52@⅛d
" 5..	14¼	13⅞	9,000	9,555	4,743	390	1,956	....	7,089	
" 8..	14⅛	13¾								
" 12..	14⅛	13⅞	5,000	3,392	1,691	....	492	....	2,183	
" 15..	14	13⅝								
" 19..	14⅛	13¾	7,000	7,570	3,455	....	1,385	....	4,840	

lus to the market. April also was characterized by the same features, annulling the unfavorable position of the European market.

In May, prices fluctuated somewhat, as the accounts relative to the crop were more favorable, and the war now going on between China and England had the effect to depress the Liverpool market, the demand for manufactured goods for the former country having largely fallen off.

The weather in July was very favorable to the growing crop, but accounts came in better from abroad, and prices gradually advanced.

In August, toward the latter part of the crop year, a severe storm swept over the Southern States, inflicting much damage, and stiffened further the ruling prices.

New York Statement for Year 1857—*Concluded.*

1857.	Price of Middling New Orleans New York Classificat'n.	Price of Middling Upland, New York Classificat'n.	Sales for week.	Receipts for week.	EXPORTS FOR THE WEEK To Great Britain.	To France.	North of Europe.	Other Fo'n Ports	Total Exports.	Rates of Freight to Liverpool.
May 22..	$14\frac{1}{4}$	14								
" 26..	$14\frac{3}{8}$	14	10,000	6,155	1,943	969	676	....	3,588	
" 29..	$14\frac{3}{8}$	14								
June 2..	$14\frac{3}{8}$	14	4,000	8,454	2,022	744	123	....	2,889	$\frac{1}{8}$d.
" 5..	$14\frac{1}{4}$	14								
" 9..	$14\frac{1}{8}$	$13\frac{3}{4}$	3,500	1,528	4,062	....	864	....	4,926	
" 12..	$14\frac{1}{8}$	$13\frac{7}{8}$								
" 16..	$14\frac{1}{4}$	14	6,000	5,294	3,406	883	613	....	4,902	
" 19..	$14\frac{1}{4}$	14								
" 23..	$14\frac{1}{2}$	$14\frac{1}{4}$	6,000	4,975	953	....	233	....	1,186	
" 26..	$14\frac{5}{8}$	$14\frac{3}{8}$								
" 30..	$14\frac{7}{8}$	$14\frac{5}{8}$	9,500	712	4,324	2,113	1,260	....	7,697	
July 2..	$14\frac{7}{8}$	$14\frac{5}{8}$								$\frac{1}{8}$d.
" 7..	15	$14\frac{5}{8}$	5,000	2,796	....	968	....	8	976	
" 10..	15	$14\frac{5}{8}$								
" 14..	15	$14\frac{3}{4}$	5,000	5,041	1,409	....	....	....	1,409	
" 17..	$15\frac{1}{4}$	$14\frac{7}{8}$								
" 21..	$15\frac{1}{4}$	15	4,500	5,337	760	1,419	589	....	2,768	
" 24..	$15\frac{1}{4}$	15								
" 28..	$15\frac{1}{2}$	$15\frac{1}{4}$	4,000	5,290	4,125	....	....	....	4,125	
" 31..	$15\frac{1}{2}$	$15\frac{1}{4}$								
August 4..	$15\frac{1}{2}$	$15\frac{1}{4}$	5,500	1,378	2,309	1,449	....	....	3,758	$\frac{1}{8}$d.
" 7..	$15\frac{3}{4}$	$15\frac{3}{8}$								
" 11..	$15\frac{3}{4}$	$15\frac{3}{8}$	5,000	2,024	706	....	77	....	783	
" 14..	$15\frac{3}{4}$	$15\frac{1}{2}$								
" 18..	$15\frac{7}{8}$	$15\frac{5}{8}$	6,000	1,094	835	....	462	....	1,297	
" 21..	$15\frac{7}{8}$	$15\frac{5}{8}$								
" 25..	16	$15\frac{3}{4}$	3,500	2,102	2,355	409	200	....	2,964	
" 28..	$16\frac{1}{8}$	$15\frac{7}{8}$								
Septem. 1..	16	$15\frac{3}{4}$	3,000	1,272	737	102	966	....	1,805	$\frac{1}{8}$d.
Average prices and total sales, receipts and exports.	13.83	13.51	573,500	402,625	143,938	21,602	28,207	808	194,555	

General Remarks.

Exchange was remarkably steady throughout the crop year.

The quotations for 60 days' bills on London, in September, ranged from $9\frac{3}{8}$@10 per cent. premium; in October, $9\frac{1}{4}$@$9\frac{7}{8}$; in November, $8\frac{7}{8}$@$9\frac{3}{4}$; in December, $8\frac{5}{8}$@$9\frac{1}{4}$; in January, $8\frac{1}{4}$@$8\frac{3}{4}$; in February, 8@$8\frac{7}{8}$; in March, $7\frac{7}{8}$@$8\frac{1}{2}$; in April, 8@$9\frac{1}{2}$; in May, 9@$9\frac{5}{8}$; in June, $9\frac{1}{8}$@$9\frac{5}{8}$; in July, $9\frac{1}{8}$@$9\frac{7}{8}$; and in August, $9\frac{3}{8}$@$9\frac{3}{4}$.

The season of 1856–57, is worthy of some special remark, because it illustrates some principles of general application.

Careful observers were very early satisfied that the crop was materially short of the previous one. A short crop speculation commenced as soon as that conviction became general, but it soon broke down and was a total failure. There was no change in the crop estimate, among well informed parties, and consequently the speculation was resumed with the new year. Even that was too early, for it was only by the greatest efforts that a break down was prevented in the spring. By the most determined efforts on the part of holders, disaster was prevented and the price sustained, until even Liverpool and Manchester were convinced, and holders realized handsome profits.

The panic of 1857 did not reach Liverpool, until the latter end of October, when the old crop was almost entirely disposed of. The season had been a very profitable one; and consequently cotton recovered from the effects of the panic sooner than almost any other article. Cotton was, in fact, considered an exception to the general rottenness. This gave impetus and confidence to trade and manufacturing, while the high price stimulated production of the raw material, and was the beginning of that movement which culminated in the great crop of 1859. From 1857 to 1860, the stimulated trade and the stimulated production contended against each other.

During the three years named, the cotton crop of this country increased more than 20 per cent. per annum, and the cotton manufacturing machinery was increased with almost equal rapidity. In Great Britain the increase averaged, in 1859 and 1860, nearly 40,000 spindles per week.

It is generally believed that, when our war broke out, the whole world was over-stocked with goods.

The five years, from 1857 to 1862, was the periodical term of contraction, to which I have before alluded, but it was probably one of the mildest we have ever had; as the circulation and deposits only declined below 14 dollars, per capita, a short time, and were up again to the dangerous figure of 15 dollars, in 1859. This was, no doubt, owing to the fact that our banking system had been greatly improved, and went through the panic of 1857 triumphantly. Very few of the banks were even under the necessity of passing a dividend on account of their losses.

1858.

Dr. Livingstone, the celebrated explorer, went to Africa in March, prepared to prosecute cotton culture, having recently explored the country, and decided that the American cotton plant had there become perennial.

July 12th, Lord Palmerston, in the House of Commons, prophesied that the western coast of Africa would outstrip all other cotton districts in the world, excepting only the United States.

The "Cotton Supply Association" held its first meeting, April 9th, at Manchester. The object of this association is to "engage in gathering and distributing information respecting the capacity of various districts, and furnishing the best seed, tools, and other implements, wherever they are likely to be advantageously employed."

The "Cotton Supply Association" of Great Britain held its first anniversary meeting in April of this year (*see* year 1857) A general opinion prevailed that India was the source to look to. A project was drawn up to propose to the government an expenditure of £20,000,000 in that country during five years, in the construction of roads, bridges, railways, tram-ways, piers, landings, ships, irrigation canals and navigation facilities, the interest of the money, and, possibly, a redemption fund, to be provided for by tolls. Whether such a large demand would have been acceded to, under any circumstances, is doubtful, but the whole movement was speedily checked by the formidable mutiny.

At this time there had been, in England alone, no less than 256 patents granted, relating more or less to the cleaning, spinning, separating, scutching and batting of fibrous materials; 82, containing provisions relating to the carding, combing, drawing, doubling and roving of the materials thus prepared; and, the enormous number of 1,376, touching in a greater or less degree, the processes and apparatus for spinning, twisting, and thread-making. Not all of these related *solely to cottons*, but those which did, were *some hundreds in number*.

COTTON CROP OF THE UNITED STATES.

Statement and Total Amount of the Cotton Crop of the United States, for the Year ending August 31, 1858.

	Bales.	Bales.	TOTAL. 1858.	1857.	1856.
NEW ORLEANS.					
Export—					
To Foreign Ports	1,495,070				
Coastwise	164,637				
Stock on hand 1st September, 1858	30,230				
		1,689,937			
Deduct—					
Received from Mobile	67,451				
" Montgomery, &c	....				
" Florida	9,160				
" Texas	29,596				
Stock on hand 1st September, 1857	7,321				
		113,528			
			1,576,409	1,435,000	1,661,433
MOBILE.					
Export—					
To Foreign Ports	387,032				
Coastwise	128,013				
Manufactured in Mobile, &c	1,807				
Stock on hand 1st September, 1858	10,495				
		527,347			
Deduct—					
Received from New Orleans	479				
Stock on hand 1st September, 1857	4,504				
		4,983			
			522,364	503,177	659,738
TEXAS.					
Export—					
To Foreign Ports	50,338				
Coastwise (and burnt, 70 bales)	94,011				
Stock on hand 1st September, 1858	1,899				
		146,248			
Deduct—					
Stock on hand 1st September, 1857	....	962			
			145,286	89,882	116,078
FLORIDA.					
Export—					
To Foreign Ports—Uplands	25,737				
" Sea Island	34				
Coastwise—Uplands	70,305				
Sea Island	25,651				
Burnt at Apalachicola	600				
Stock on hand 1st September, 1858	80				
		122,407			
Deduct—					
Stock on hand, 1st September, 1857	....	56			
			122,351	136,344	144,404
GEORGIA.					
Export—					
To Foreign Ports—Uplands	159,141				
" Sea Island	8,561				
Coastwise—Uplands	117,680				
Sea Island	7,417				
Stock in Savannah, 1st September, 1858	684				
" Augusta, etc., 1st September, 1858	1,901				
		295,414			

Statement and Total Amount of the Cotton Crop of the United States, for the Year ending August 31, 1858—*Concluded.*

	Bales.	Bales.	TOTAL. 1858.	1857.	1856.
Deduct—					
Received from Florida—Sea Island.........	7,768				
Stock in Savannah, 1st September, 1857....	1,926				
" Augusta, &c., 1st September, 1857.	2,747				
		12,441			
			282,973	322,111	389,445
SOUTH CAROLINA.					
Export from Charleston—					
To foreign ports—Uplands.................	276,547				
" Sea Island..............	22,857				
Coastwise—Uplands......................	115,158				
Sea Island.....................	2,806				
Burnt and manufactured at Charleston.....	771				
Stock in Charleston, 1st September, 1858....	11,715				
	429,854				
Export from Georgetown, S. C.—					
To Coastwise Ports—Uplands..............	1,918				
		431,772			
Deduct—					
Received from Florida—Sea Island.........	7,519				
" Savannah—Sea Island......	1,575				
" " Uplands.........	10,783				
Stock in Charleston, 1st September, 1857.	5,644				
		25,521			
			406,251	397,331	495,976
NORTH CAROLINA.					
Export—					
To Coastwise ports........................	23,999				
		....	23,999	27,147	26,098
VIRGINIA.					
Export—					
To Foreign Ports..........................	495				
Coastwise.................................	8,942				
Manufactured—taken from the ports........	15,088				
Stock on hand 1st September, 1858.........	600				
		25,125			
Deduct—					
Stock on hand 1st September, 1857.........	....	420			
			24,705	23,773	20,458
Received at New York, Overland, from Tennessee, &c.	....	....	3,363	2,022	2,086
Received at Philadelphia, Overland, from Tennessee, &c.	....	....	3,275	1,236	7,938
Received at Baltimore, Overland, from Tennessee, &c.	....	....	2,986	1,496	4,191
Total crop of the United States......	....	....	3,113,962	2,939,519	3,527,845

Increase over crop of 1857 ....bales.	174,443
Decrease from crop of 1856	413,883
Increase over crop of 1855..	266,623

Export to Foreign Ports, from September 1, 1857, *to August* 31, 1858.

FROM	To Great Britain.	To France.	To North of Europe.	Other Foreign Ports.	Total.
New Orleans (bales)	1,016,716	236,596	116,304	125,454	1,495,070
Mobile	265,464	89,887	21,462	10,219	387,032
Texas	33,933	1,689	14,716	....	50,338
Florida	25,771	....	....	....	25,771
Savannah	149,346	7,376	7,680	3,300	167,702
Charleston	192,251	35,503	33,126	38,524	299,404
North Carolina	....	....	....	....	....
Virginia	495	....	....	....	495
Baltimore	164	....	....	....	164
Philadelphia	995	....	....	....	995
New York	110,721	12,951	20,308	3,841	147,821
Boston	14,110	....	1,549	4	15,663
Grand total	1,809,966	384,002	215,145	181,342	2,590,455
Total last year	1,428,870	413,357	245,798	164,632	2,252,657
Increase	381,096	....	....	16,710	337.798
Decrease	....	29,355	30,653	....	....

Consumption.

Total crop of the United States, as before stated .... bales.			3,113,962
Add—Stocks on hand at the commencement of the year, September 1, 1857, in the Southern ports		23,580	
" Northern ports		25,678	
			49,258
Makes a supply of			3,163,220
Deduct therefrom—The export to Foreign ports	2,590,455		
Less—Foreign included	723		
		2,589,732	
Stocks on hand September 1, 1858—			
In the Southern ports	57,604		
In the Northern ports	45,322		
		102,926	
Burnt at New York, Apalachicola and Galveston	711		
Burnt and Manufactured at Mobile and Charleston	2,578		
Manufactured in Virginia	15,088		
		18,377	
			2,711,035

Taken for home use north of Virginia .... bales..	452,185
Taken for home use in Virginia, and south and west of Virginia	143,377
Total consumed in the United States (including burnt), 1857-8	595,562

ANNUAL REVIEW.

From the New Orleans Price Current, 1857—58.

The year opened with great buoyancy in prices, and flattering prospects with regard to the business of the season. The crops of cotton and sugar, it was known, would not be large; and, in view of the injuries suffered from late spring frosts and subsequent unfavorable weather, it was apprehended that the former would fall short of the crop of the preceding year. But it was expected that this deficiency would be counterbalanced by a continuance of a high range of prices for that and other staples. This favorable prospect, however, was changed by the commercial and financial revulsion, which, originating at the North, spread disaster through the country, and resulted in a general change of market values and prospects. There were some weeks of gloom and depression, many losses, and some heavy failures, but the crisis here was soon passed, and trade had resumed its usual channels by the time the active business season had fairly opened. Business became settled on a more secure basis, and the feverish and excited condition of the markets, which had prevailed for some months preceding the revulsion, gave way to a healthy system of trade, prices having fallen from the stilted position which they had occupied, to a more reasonable and natural level. With a favorable autumn, the cotton crop recovered, in a measure, from the disasters of a late spring, and has proved larger than had been anticipated, exceeding that of any previous year, except 1855–6 and 1852–3. In valuation it exceeds last year's crop, $1,872,261. The opening rates for October exhibited a decline of 2 to $2\frac{1}{4}$c. Our quotations on the 3d being for low middling, new crop, $13\frac{7}{8}$ to 14c.; middling, new crop, $14\frac{1}{8}$ to $14\frac{1}{4}$c.; good middling, new crop, $14\frac{3}{8}$ to $14\frac{1}{2}$c., and the tendency continued downwards, sterling exchange falling off during the first week of the month to 2 and 4 per cent. discount for clear bills, and the best sixty days' bills on the North being usually unsalable at $4\frac{1}{2}$ to 5 per cent. discount. On the 17th, we reduced our quotations to $9\frac{1}{4}$ and $9\frac{1}{2}$c. for middling to strict middling, showing a decline within three weeks, of fully $6\frac{3}{4}$c. per lb.; exchange on London giving way to about 5 to 10 per cent. discount, and sixty days' bills on the North being entirely unsalable.

The gloom of this period was increased by advices from New York on the 15th, of the suspension of specie payment by all New York banks, producing a panic with us, and resulting in a run on our city banks, three of which, working under the free banking law, stopped specie payments the same day. A slight reaction followed, and, with some limited facilities in passing exchange, the cotton market attained a somewhat steadier and firmer position, and the rates for middling improved to $9\frac{3}{4}$ and 10c. per pound, the average prices for that grade during the month being about $10\frac{7}{8}$ to $11\frac{1}{8}$c. The reported sales for the month were 79,300 bales. Early in November, the market was further relieved, to some extent, by some arrivals of specie, and by the middle of the month, bills on London had advanced to par, having ruled below that rate during a period of six weeks, and sales being made in some instances as low as 15 to $17\frac{1}{2}$ per cent. discount. During the severity of the pressure, many planters withheld their cotton from the market; and, up to the end of October, the total receipts at this port were only about 194,500 bales, against 314,700 bales for the same period of the previous year.

The semi-weekly Price and Weekly Sales and Receipts, at New York, Weekly Exports from New York and Rates of Freight to Liverpool 1st of each Month, for the Crop Year ending September 1, 1858.

1857.	Price of Middling New Orleans New York Classificat'n.	Price of Middling Upland. New York Classificati'n	Sales for week, including lots in transit.	Receipts for week, including lots in transit.	EXPORTS FOR WEEK.					Rates of Freight to Liverpool.	GENERAL REMARKS.
					To Great Britain.	To France.	North of Europe.	Other Fo'n Ports	Total Exports.		
Septem 4..	16	$15\frac{3}{4}$								$\frac{1}{8}$d.	This crop year opened amid some gloom; early in September, a monetary panic pervaded financial circles, caused by the unexpected failure of several prominent houses; great stringency was felt, and several banks, here and elsewhere, suspended; rates of interest went up to 3@5 per cent. a month, and business was of course brought almost to a stand. In England, the revolt in India also caused some uneasiness and great caution was observed.
" 8..	16	$15\frac{3}{4}$	4,000	3,652	1,214	....	131	....	1,345		
" 11..	16	$15\frac{3}{4}$									
" 15..	$15\frac{7}{8}$	$15\frac{5}{8}$	3,500	2,203	1,724	....	....	....	1,724		
" 18..	$15\frac{7}{8}$	$15\frac{5}{8}$									
" 22..	16	$15\frac{3}{4}$	3,000	3,439	1,428	30	841	....	2,299		
" 25 .	16	$15\frac{3}{4}$									
" 29..	16	$15\frac{3}{4}$	2,000	1,828	2,731	....	67	....	2,798		
October 2..	16	$15\frac{3}{4}$								7-32@$\frac{1}{4}$d	
" 6..	Nominal.	Nominal.	1,500	1,261	4,179	....	148	....	4,327		
" 9..	"	"									
" 13..	"	"	1,600	1,145	5,113	100	....	....	5,213		
" 16..	"	"									
" 20..	"	"	1,000	2,493	1,567	236	104	....	1,907		
" 23..	"	"									The above noted unsatisfactory state of affairs continued through October, and the banks throughout the country generally suspended specie payments; many manufacturers stopped their mills and shipped their cotton to Europe; trade was entirely paralyzed, and failures were numerous.
" 27..	"	"	500	1,559	2,538	....	100	3,568	6,206		
" 30..	"	"									
Novem. 3..	13	$12\frac{1}{2}$	600	3,438	1,883	....	....	....	1,883	5-32d.	
" 6..	Nominal.	$12\frac{1}{4}$@$12\frac{1}{2}$									
" 10..	$13\frac{1}{4}$	13	3,500	3,627	1,864	228	439	....	2,531		
" 13..	$13\frac{1}{4}$	13									
" 17..	13	$12\frac{3}{4}$	1,500	4,518	2,601	60	....	....	2,661		
" 20.	13	$12\frac{1}{2}$									
" 24..	Nominal.	Nominal.	3,000	6,459	325	....	....	....	325		The panic, and financial embarrassment, now extended to Europe, money became scarce, as well on the Continent, as in England; the Bank of England advanced its rate of discount to 10 per cent., and several bank and mercantile failures took place, business there being at a stand, as well as here
" 27..	12	$11\frac{3}{4}$								5-32d.	
Decem. 1..	$12\frac{1}{4}$	$11\frac{3}{4}$	2,500	2,713	3,442	....	339	....	3,781		
" 4..	Nominal.	$11\frac{1}{4}$@$11\frac{1}{2}$									
" 8..	"	Nominal.	3,000	3,629	1,386	....	50	....	1,436		
" 11..	"	$11\frac{1}{4}$									
" 15..	"	$11\frac{1}{4}$	2,000	9,267	550	658	....	....	1,208		
" 18..	"	Nominal.									
" 22..	$10\frac{3}{8}$	$10\frac{1}{8}$	2,500	3,611	1,996	....	....	....	1,996		
" 25..	$10\frac{1}{8}$	$9\frac{3}{4}$									

Decem. 29..	9½	9	2,000	10,398	....	....	287	41	328	
1858.										
January 1..	Nominal.	9								⅛@5-32d
" 5..	8⅞	8⅞	6,000	7,800	961	....	....	....	961	
" 8..	9⅝	9⅜								
" 12..	Nominal.	9⅞	6,500	4,189	1,241	1,476	56	....	2,773	
" 15..	10½	10¼								
" 19..	11	10½	10,000	8,348	2,121	566	....	....	2,687	
" 22..	10⅝	10⅜								
" 26..	10½	10¼	4,000	7,741	2,559	115	936	....	3,610	
" 29..	10¾	10½								
February 2..	10¾	10½	11,000	3,777	1,526	1,356	425	494	3,801	3-16d.
" 5..	11⅛	11								
" 9..	11⅞	11⅝	9,000	8,977	1,869	....	414	....	2,283	
" 12..	12½	12⅛								
" 16..	12¼	12	13,500	4,267	1,377	....	....	123	1,500	
" 19..	12	11¾								
" 23..	12¼	11⅞	9,500	8,780	893	1,098	280	....	2,271	
" 26..	12⅜	12								
March 2..	12⅜	12⅛	17,000	12,993	818	....	122	....	940	⅛@5-32d
" 5..	12¼	11⅞								
" 9..	12¼	11⅞	14,500	7,367	1,118	....	892	912	2,922	
" 12..	12¼	11⅞								
" 16..	12¼	12	21,000	22,302	598	586	365	....	1,549	
" 19..	12⅛	11⅞								
" 23..	11⅞	11¾	6,800	14,541	2,843	....	1,730	....	4,573	
" 26..	11⅞	11¾								
" 30..	12¼	12	17,000	8,852	4,879	432	1,168	9	6,488	
April 2..	12⅛	11⅞								3-16d.
" 6..	12¼	12	5,500	9,096	4,708	....	1,145	....	5,853	
" 9..	12½	12¼								
" 13..	12½	12¼	14,000	20,582	1,553	....	248	268	2,069	
" 16..	12½	12¼								
" 20..	12⅝	12⅜	7,000	13,278	1,290	1,910	86	....	3,286	
" 23..	12¾	12½								
" 27..	13	12⅝	20,000	8,540	1,871	....	3,858	....	5,729	
" 30..	13	12⅝								
May 4..	13	12⅝	12,800	8,383	1,897	....	506	....	2,403	5-32d.
" 7..	13	12⅝								
" 11..	12⅞	12⅝	7,000	13,790	794	....	60	602	1,456	
" 14..	12¼	12								
" 18..	12½	12¼	7,500	14,242	1,284	205	234	....	1,723	

In December, a better feeling began to obtain, and it was evident the worst of the storm was over; some of our city banks resumed specie payments, but the accounts from Europe were still very gloomy and unassuring.

A better state of things, however, prevailed, as the new year opened; early in January, the panic in England exhausted itself; the Bank of England rapidly reduced its rate of discount, from 10, down to 3½ per cent., our mills resumed "half time," and cotton began to rise.

In February, the banks throughout the country resumed specie payments, and, though the receipts at the ports were large, the more favorable condition of our own and the foreign markets sustained prices.

Towards the latter part of April, the receipts at the ports fell off, and with an advance abroad, and telegraphic advices of danger to the growing crop by frosts in Alabama and Georgia, the market was a rising one.

Through part of May and June, the receipts were larger again, and prices here being relatively higher than in Liverpool, the market was weak and softer.

June, July and August, the situation was constantly favorable to holders, the crop accounts being discouraging, owing in part, to the overflow of the Mississippi and its tributaries.

New York Statement for 1858.—*Concluded.*

1858.		Price of Middling New Orleans New York Classificat'n.	Price of Middling Upland, New York Classificat'n.	Sales for week, including lots in transit.	Receipts for week, including lots in transit.	EXPORTS FOR THE WEEK. To Great Britain.	To France.	North of Europe.	Other Fo'n Ports	Total Exports.	Rates of Freight to Liverpool.
May	21..	12½	12¼								
"	25..	12⅜	12⅛	10,000	14,031	1,410	....	9	....	1,419	
"	28..	12⅛	11⅞								
June	1..	12⅛	11⅞	6,000	7,579	632	979	452	....	2,063	7-32d.
"	4..	12	11¾								
"	8..	12	11¾	7,000	8,260	4,404	....	348	....	4,752	
"	11..	12	11¾								
"	15..	12⅛	11⅞	5,000	8,788	2,689	....	1,245	....	3,934	
"	18..	12¼	12								
"	22..	12⅝	12⅜	11,500	2,825	4,113	....	136	....	4,249	
"	25..	12⅝	12⅜								
"	29..	12½	12¼	6,000	10,952	9,397	1,378	208	....	10,983	
July	2.	12½	12¼								⅛@3-16d
"	6..	12½	12¼	4,000	2,322	2,057	870	404	....	3,331	
"	9..	12⅝	12⅜								
"	13..	12⅞	12⅝	10,000	1,898	2,126	....	....	....	2,126	
"	16..	12⅞	12¾								
"	20..	12⅞	12¾	14,500	3,901	3,732	....	709	....	4,441	
"	23..	12⅞	12¾								
"	27..	12¾	12⅝	5,500	3,137	1,507	296	183	....	1,986	
"	30..	12¾	12⅝								
August	3.	12⅞	12⅝	8,500	2,957	258	....	358	....	616	3-16d.
"	6..	12⅞	12⅝								
"	10..	12⅞	12⅝	4,000	6,458	1,832	372	210	32	2,446	
"	13..	12¾	12½								
"	17..	13	12⅝	5,000	2,102	1,674	....	....	....	1,674	
"	20..	13	12¾								
"	24..	13	12¾	7,500	1,901	2,602	....	481	....	3,083	
"	27..	13	12¾								
"	31..	13	12¾	7,000	1,401	1,547	....	534	....	2,081	⅛@5-32d
Average price and total sales, receipts and exports.		12.58	12.23	368,800	351,597	110,721	12,951	20,308	6,049	150,029	

GENERAL REMARKS.

Exchange

was very much disturbed by the money panics here and abroad.

The quotation for 60 days' bills on London opened in September, at 8½@9¼ per cent. premium, and declined to 4@6½, October, opened at 5 per cent. premium, touched 2 per cent. discount to 2 per cent. premium, and closed at 3@7 per cent. premium; in November, the fluctuations were between 5 and 9 per cent. premium; in December, 8@9¾; January, 8¾@10; February, 9@10⅛; March, the quotation fell from 8½@9½, down to 6¼@7¼, and then advanced to 7¼@8¼; the range in April, was 7½@9; in May, 8½@10; in June, 8¾@9½; in July, 9⅛@10; and in August, 9¼@9⅞ per cent. premium.

LIVERPOOL STATEMENT FOR 1858.

UNITED STATES, 1857–1858.

Stock Sept. 1, 1857	49,000	Export	2,591,000
		Consumption	469,000
Crop	3,114,000	Stock Sept. 1, 1858	103,000
Bales	3,163,000	Bales.	3,163,000

Stock 1st Jan., 1858, in	Gt. Britain.	France.	Continent.	Tot. Europe.
United States ... Bales.	202,000	61,000	48,000	311,000
Brazil	36,000	1,000	2,000	39,000
West Indies	5,000	3,000	3,000	11,000
East Indies	191,000	16,000	32,000	239,000
Egypt	18,000	3,000	5,000	26,000
Bales	452,000	84,000	90,000	626,000

CONSUMPTION. / IMPORT.

Tot. Europe.	Continent.	France.	Gt. Britain.		Gt. Britain.	France.	Continent.	Tot. Europe.
2,677,000	594,000	444,000	1,639,000	United States ... Bales.	1,863,000	494,000	578,000	2,778,000
138,000	20,000	5,000	113,000	Brazil	106,000	7,000	23,000	126,000
39,000	12,000	16,000	11,000	West Indies	7,000	14,000	13,000	34,000
627,000	273,000	32,000	322,000	East Indies	361,000	26,000	247,000	460,000
143,000	33,000	20,000	90,000	Egypt	106,000	23,000	35,000	157,000
3,624,000	932,000	517,000	2,175,000	Bales	2,443,000	564,000	896,000	3,555,000
....	....	....	348,000	Export.				
557,000	54,000	131,000	372,000	Stock, Dec. 31 ... Stock above,	452,000	84,000	90,000	626,000
4,181,000	986,000	648,000	2,895,000	Total supply, bales	2,895,000	648,000	986,000	4,181,000

COTTON AT LIVER

Week Ending.	RECEIPTS.						SALES.			
	American	E. I.	Egypt.	Brazil.	Other.	Total.	Consumption.	Speculation.	Export	Total.
Jan. 8..	19,355	2,147	1,180	7,089	......	29,771	38,920	5,220	5,650	49,790
" 15..	51,761	......	378	4,361	506	57,006	30,250	1,310	3 840	35,400
" 22..	17,723	......	758	....	.....	18,481	47,870	15,350	4,160	67,380
" 29..	20,802	6,588	......	2,155	493	30,038	49,920	11,390	3,330	64,640
Feb. 5..	12,218	157	1,455	2,865	104	16,799	45,540	5,450	3,190	54,180
" 12..	12,269	4,154	533	....	146	17,102	60,710	10,610	7,510	78,830
" 19..	63	......	......	....	......	63	50,280	12,560	5,760	68,600
" 26..	6,904	9,402	1,092	409	255	18,062	38,430	9,760	3,100	51,290
March 5..	4,372	......	......	....	......	4,372	29,610	4,570	1,570	35,750
" 12..	102,011	......	3,026	6,779	13	111,829	16,080	2,480	2,580	21,140
" 19..	85,446	14,909	911	5,424	......	106,690	24,470	3,960	2,840	31,270
" 26..	96,865	......	72	3,050	......	99,987	51,170	4,340	9,410	64,920
April 1..	82,234	7,400	3,774	1,107	888	95,403	31,680	6,480	8,880	47,040
" 9..	18,097	7,871	1,238	....	......	27,206	54,670	13,870	8,060	76,600
" 16..	19,092	4,280	4,450	180	......	28,002	32,430	4,450	7,480	44,360
" 23..	56,937	7,326	2,250	3,385	......	69,898	65,090	14,270	8,760	88,120
" 30..	99,762	10,887	2,594	3,922	8	117,173	44,720	5,770	4,600	55,090
May 7..	86,405	3,948	....	....	......	90,353	62,470	11,480	7,310	81,260
" 14..	14,563	9,056	2,973	....	......	26,592	41,290	12,360	4,180	57,830
" 21..	126,826	11,691	4,686	2,991	130	146,324	33,820	1,620	1,640	37,080
" 28..	83,718	5,918	1,286	2,809	665	94,396	27,240	3,210	1,670	32,120
June 4..	55,402	9,016	.....	831	......	65,249	43,910	1,890	3,430	49,230
" 11..	49,202	2.744	2,480	1,556	......	55,982	42,650	3,020	3,870	49,540
" 18..	15,967	3,274	6,161	1,959	16	27,377	32,180	930	5,800	38,910
" 25..	19,419	4.818	3,706	2,337	66	30,346	35,030	3,240	7,010	45,280
July 2..	44,564	1,458	......	1,358	17	47,397	58,320	10,940	8,040	77,300
" 9..	21,592	......	1,011	1,087	......	23,690	46,270	6,650	6,690	59,610
" 16..	28,355	......	5,499	4,721	22	38,597	31,560	1,830	3,530	36,920
" 23..	60,771	4,808	3,583	12	45	69,219	38,950	1,540	4,100	44,590
" 30..	20,044	1,506	4,132	110	......	25,792	31,660	1,900	4,850	38,410
Aug. 6..	19,300	2,110	2,601	2,711	110	26,832	54,480	9,520	5,840	69,840
" 13..	30,550	11,364	2,409	425	1	44,749	38,450	1,260	5,500	45,210
" 20..	41,342	5,436	1,431	1,720	......	49,929	30,630	1,290	4,460	36,380
" 27..	43,453	8,467	1,808	1,844	......	55,572	50,700	4,700	11,000	66,400
Sep. 3..	28,820	8,467	4,942	2,075	......	44,304	49,480	1,440	3,760	54,680
" 10..	14,854	21,401	1,259	3,969	172	41,655	54,570	4,630	8,760	67,960
" 17..	3,605	3,391	......	1,550	......	8,546	56,220	17,270	9,280	82,770
" 24..	6,952	4,571	1,812	....	......	13,335	44,160	4,700	9,410	58,270
Oct. 1..	12,958	......	6,606	1,995	27	21,586	32,780	1,590	6,600	40,970
" 8..	21,786	16,274	1,301	6,056	160	45,577	39,700	2,530	3,210	45,440
" 15..	4,127	32,008	399	3,049	......	39,583	37,150	1,440	4,680	43,270
" 22..	....	7,175	2,221	1,108	549	11,053	40,910	420	6,280	47,610
" 29..	4,541	1,704	.....	5	......	6,250	26,980	480	5,250	32,710
Nov. 5..	8,326	7,968	1,374	2,279	185	20,132	36,410	2,270	8,340	47,020
" 12..	2,279	4,370	379	.....	......	7,028	37,700	3,350	7,800	48,850
" 19..	....	454	77	4,106	......	4,637	46,730	4,370	7,170	58.270
" 26..	158	3,300	3,870	....	......	7,328	36,860	490	2,540	39,890
Dec. 3..	60,460	7,613	804	3,107	......	71,984	34,510	1,040	3,620	39,170
" 10..	43,341	4,814	1,466	1,495	45	51,161	37,360	2,110	2,590	42,060
" 17..	13,767	3,161	3,229	1,847	......	22,004	58,050	6,590	2,750	67,390
" 23..	65,110	2,142	2,859	3,603	10	73,724	38,130	3,320	4,410	45,860
" 31..	....	......	......	....	......	46,331	....	790	2,950	3,740
Average prices & total sales, receipts & stocks.	1,758,468	289548	100075	103,441	4,633	2,302,496	2,119,150	268,050	279040	2,660,981

POOL. YEAR 1858.

STOCKS.			PRICES.			ACTUAL EXPORT.	CON-SUMPTION.	REMARKS.
Amer'n	Other.	Total.	Mid. Up.	Mid. Orl.	Dhol.			
190,825	198,825	389,650	6 7/16	6 5/8	4 5/8	5,650	38,920	
222,126	193,070	415,196	6 5/16	6 1/2	4 5/8	3,840	69,170	
205,449	177,778	383,227	6 9/16	6 11/16	4 3/4	4,160	117,040	
196,091	164,204	360,295	6 13/16	6 7/8	5	3,330	166,960	
175,579	154,585	330,164	6 3/4	6 7/8	5	3,190	212,500	
147,938	137,468	285,406	7 1/16	7 1/4	5 1/8	7,510	273,210	
116,171	114,878	231,049	7 3/16	7 1/2	5 7/16	5,760	323,490	Small stock of American
99,845	107,966	207,711	7 1/2	7 11/16	5 1/2	3,100	361,920	
84,327	96,356	180,683	7 11/16	7 7/8	5 1/2	1,570	391,530	
175,318	97,344	272,662	7 3/8	7 9/16	5 1/2	2,580	407,610	
242,094	110,358	353,452	6 7/8	7 1/8	5 1/4	2,840	432,080	Tremendous arrival of overdue ships.
299,169	102,630	401,799	6 3/4	7 1/16	5 1/4	9,410	483,250	
356,503	107,519	464,022	6 1/2	6 13/16	5 1/4	8,880	514,930	
330,040	104,718	434,758	6 3/4	6 15/16	5 1/4	8,060	569,600	
324,212	100,558	424,770	6 9/16	6 13/16	5 1/4	7,480	602,030	
327,339	97,679	425,018	6 7/8	7 1/16	5 1/4	8,760	667,120	
389,911	101,510	491,421	6 7/8	7 1/16	5 3/8	4,600	711,840	
426,556	86,988	513,544	7 1/8	7 5/16	5 1/2	7,310	774,310	Unsettled state of political affairs.
405,709	81,417	487,126	7 1/8	7 3/16	5 3/8	4,180	815,600	
500,225	88,955	589,180	7	7 3/16	5 3/8	1,640	849,420	
559,503	91,623	651,126	7	7 1/8	5 3/8	1,670	870,660	
574,575	90,490	665,065	6 13/16	7 1/16	5 1/2	3,430	920,570	
585,917	86,550	672,467	6 3/4	6 15/16	5 3/8	3,870	963,220	
574,224	88,160	662,384	6 11/16	6 7/8	5 3/8	5,800	995,400	
559,633	89,460	649,090	6 3/4	6 7/8	5 3/8	7,010	1,030,430	
555,837	76,560	632,397	6 7/8	7 1/8	5 3/4	8,040	1,088,750	
578,066	65,428	643,494	6 13/16	7 1/8	5 3/4	6,690	1,135,020	
574,871	66,280	641,151	6 7/8	7	5 3/4	3,530	1,166,580	
602,572	62,298	664,870	6 13/16	6 13/16	5 3/4	4,100	1,205,530	
591,636	77,088	668,724	6 3/4	6 7/8	5 7/8	4,850	1,237,190	
562,426	72,822	635,248	6 7/8	7 1/16	5 7/8	5,840	1.291,670	
558,776	77,859	636,635	6 7/8	7 1/16	5 7/8	5,500	1,330,120	
571,628	78,386	650,014	6 13/16	7 1/16	5 3/4	4,460	1,360,750	
571,061	79,5[illegible]5	650,646	6 13/16	7 1/8	5 3/8	11,000	1,411,450	Favorable news of crop from India and China.
563,271	75,392	638,663	6 7/8	7 1/16	5 1/4	3,760	1,460,930	
532,625	87,873	620,498	7	7 1/16	5 1/2	8,760	1,515,500	
482,430	77,674	560,104	7 1/8	7 3/16	5 5/8	9,280	1,571,720	
449,263	69,627	518,890	7 3/16	7 3/8	5 3/4	9,410	1,615,880	
432,271	68,315	500,586	7 3/16	7 3/16	5 3/4	6,600	1,648,660	
417,757	80,226	497,983	7 1/4	7 3/8	5 3/4	3,210	1,688,360	
388,144	103,212	491,356	7 1/4	7 3/8	5 3/4	4,680	1,725,510	
363,364	104,910	468,274	7 1/8	7 1/4	5 3/4	6,280	1,766,420	
346,119	98,175	444,294	6 15/16	7 1/16	5 3/4	5,250	1,793,400	
323,251	99,945	423,196	6 3/4	6 15/16	5 5/8	8,340	1,829,810	
288,760	95,164	383,924	6 13/16	6 15/16	5 5/8	7,800	1,867,510	Less encouraging accounts from manufacturing districts.
251,120	84,101	335,221	6 7/8	7	5 5/8	7,170	1,914,240	
220,358	79,361	299,719	6 7/8	7	5 5/8	2,540	1,951,100	
251,578	80,955	332,533	6 3/4	6 7/8	5 5/8	3,620	1,985,610	
261,199	78,305	339,504	6 11/16	6 15/16	5 5/8	2,590	2,022,970	
222,406	70,672	293,078	6 13/16	7	5 5/8	2,750	2,081,020	
253 986	71,546	325,532	6 3/4	6 15/16	5 3/8	4,410	2,119,150	
....	....	....	......	......	......	2,950	....	
			6.91	7.07	5.56	279,040	41,551.96	

1859.

The Cotton Supply Association of Great Britain (*see* years 1857 and 1858) held its second annual meeting. It had, during its second year, received two thousand communications from the Government departments, the new Indian Council, the British Consuls abroad, and societies and individuals in various parts of the world, relating to the encouragement of cotton growing in places suitable for it. Grants of cotton seed, varying from one bag to two hundred bags each, had been made and forwarded to Bombay, Madras, Calcutta, Ahmedabad, Hyderabad, Malabar, Ceylon, Singapore, Sydney, Savanilla, and Baranguilla, in South America; Honduras, Guatemala, Cuba, Jamaica, Hayti, Tunis, Lagos, Fernando Po, Sierra Leone, Cape Coast Castle, Natal, Monrovia, Macedonia, Aleppo, Jaffa, Sidon, Kaiffa, Broussa, Salonica, Constantinople, Messina, Attica, Argolis, Laconia, Arcadia, Achaia, Eubæa, and many other places. Cotton gins were forwarded to several of the above towns and countries, and cotton presses were sent to Cape Coast Castle. Medals and prizes were offered for the best samples of cotton grown in Liberia. A periodical, called *The Cotton Supply Reporter*, was established and regularly forwarded to all associations and individuals likely to be able to aid in the general object. Public trials of cotton gins were held, with the view of concentrating the attention of machinists and inventors, in the hope of ultimately obtaining machines of a more efficient character than were then in use.

The same kind of yarn which sold for 38s. per lb. in 1786, could now be sold, with a profit, at 2s. 6d. per lb.

COTTON CROP OF THE UNITED STATES.

Statement and Total Amount of the Cotton Crop of the United States, for the Year ending August 31, 1859.

	Bales	Bales.	TOTAL. 1859.	1858.	1857.
NEW-ORLEANS.					
Export—					
To Foreign Ports	1,580,581				
Coastwise	196,590				
Burnt at New Orleans	11,335				
Stock on hand, 1st September, 1859	26,022	1,814,528			
Deduct—					
Received from Mobile	59,703				
" " Montgomery, &c	13,540				
" " Florida	6,684				
" " Texas	35,097				
Stock on hand, 1st September, 1858	30,230	145,254	1,669,274	1,576,409	1,435,000
MOBILE.					
Export—					
To Foreign Ports	514,935				
Coastwise	179,854				
Manufactured in Mobile, &c	1,120				
Stock on hand, 1st September, 1859	20,106	716,015			
Deduct—					
Received from New Orleans	782				
" " Texas	154				
Stock on hand, 1st September, 1858	10,673	11,609	704,406	522,364	503,177
TEXAS.					
Export—					
To Foreign Ports—including 2,000 to Mexico	79,534				
Coastwise	111,672				
Manufactured in Galveston	100				
Stock on hand, 1st September, 1859	2,655	193,961			
Deduct—					
Stock on hand, 1st September, 1858	....	1,899	192,062	145,286	89,882
FLORIDA.					
Export—					
To Foreign Ports—Uplands	40,102				
Sea Island	750				
Coastwise—Uplands	112,873				
Sea Island	19,603				
Stock on hand, 1st September, 1859	236	173,564			
Deduct—					
Stock on hand, 1st September, 1858	....	80	173,484	122,351	136,344
GEORGIA.					
Export—					
To Foreign Ports—Uplands	253,743				
Sea Island	8,298				
Coastwise—Uplands	197,266				
Sea Island	8,493				

Statement and Total Amount of the Cotton Crop of the United States, for the Year ending August 31, 1859.—*Concluded.*

	Bales.	Bales.	TOTAL. 1859.	1858.	1857.
Stock in Savannah, 1st September, 1859....	9,320				
Stock in Augusta, &c., 1st September, 1859.	9,063				
		486,183			
Deduct—					
Received from Florida—Sea Island.........	7,349				
Uplands...........	461				
Stock in Savannah, 1st September, 1858.....	684				
Stock in Augusta, &c., 1st September, 1858.	1,901				
		10,395			
			475,788	282,973	322,111
SOUTH CAROLINA..					
Export from Charleston—					
To Foreign Ports—Uplands................	316,585				
Sea Island.............	23,339				
Coastwise (including 1,242 bales from Georgetown, S. C.)—Uplands..................	150,955				
Sea Island.............	3,680				
Burnt at Charleston......................	22				
Stock in Charleston, 1st September, 1859....	17,592				
		512,173			
Deduct—					
Received from Florida—Sea Island.........	8,733				
Uplands...........	754				
Received from Savannah—Sea Island.......	895				
Uplands.......	8,863				
Received from Savannah, per steamer Huntsville, and reshipped—Uplands............	560				
Stock in Charleston, 1st September, 1858...	11,715				
		31,520			
			480,653	406,251	397,331
NORTH CAROLINA.					
Export—					
To Coastwise Ports......................	37,482	....	37,482	23,999	27,147
VIRGINIA.					
Export—					
To Foreign Ports..........................	....				
Coastwise.....	21,537				
Manufactured—taken from the ports.	11,699				
Stock on hand, 1st September, 1859........	375				
		33,611			
Deduct—					
Stock on hand, 1st September, 1858........	....	600			
			33,011	24,705	23,773
Received at New-York, Boston, &c., from Tennessee, &c......................	....	....	47,175	3,363	2,022
Received at Philadelphia, &c., from Tennessee, &c.......	....	....	29,463	3,275	1,236
Received at Baltimore, &c., from Tennessee, &c..........................	....	....	8,683	2,986	1,496
Total Crop of the United States........	....	....	3,851,481	3,113,962	2,939,519

Increase over crop of 1858.......................................bales.	737,519
Increase over crop of 1857..	911,962
Increase over crop of 1856..	323,636

Export to Foreign Ports, from September 1, 1858, *to August* 31, 1859.

FROM	To Great Britain.	To France.	To North of Europe.	Other F'n Ports.	Total.
New Orleans........................ bales.	994,696	256,447	182,475	146,963	1,580,581
Mobile....................................	351,384	105,770	38,287	19,494	514,935
Texas.....................................	46,623	7,875	23,036	2,000	79,534
Florida...................................	40,801	....	51	....	40,852
Savannah.................................	238,402	7,815	11,264	4,560	262,041
Charleston...............................	218,047	42,284	40,590	39,003	339,924
North Carolina........................	....	....	....	....	....
Virginia..................................	....	....	....	....	....
Baltimore................................	20	....	....	84	104
Philadelphia............................	1,715	....	....	....	1,715
New York................................	122,234	30,505	31,417	9,304	193,460
Boston....................................	5,330	....	2,892	35	8,257
Grand total..............................	2,019,252	450,696	330,012	221,443	3,021,403
Total last year.........................	1,809,966	384,002	215,145	181,342	2,590,455
Increase..................................	209,286	66,694	114,867	40,101	430,948

Consumption.

Total crop of the United States, as before stated.......................3,851,481 bales

Add--Stocks on hand at the commencement of the year, 1st September, 1858.—In the Southern ports 57,604

" Northern " 45,322

———— 102,926

Makes a supply of.. 3 954,407

Deduct therefrom—The export to Foreign ports............. 3,021,403

Less, foreign included.................. 884

———— 3,020,519

Stocks on hand at the close of the year, 1st September, 1859—

In the Southern ports.................. 85,369

" Northern " 63,868

———— 149,237

Burnt at New Orleans, New York and Philadelphia.......... 11,492

Burnt and manufactured at Mobile, Charleston and Galveston 1,242

Manufactured in Virginia.................................. 11,699

———— 24,433

———— 3,194,189

Taken for home use north of Virginia...........................bales. 760,218

Taken for home use in Virginia and south and west of Virginia.......... 167,433

Total consumed in the United States (including burnt at the ports), 1858-9.... 927,651

ANNUAL REVIEW.

From the New Orleans Price Current, 1858—59.

With favorable promise for abundant crops, the world at peace, and confidence restored, the season opened with high hopes and flattering prospects, and these have, in the main, been realized. It is true that in the very flush of apparently prosperous progress, the world was startled by a few words from the lips of the Emperor of the French, the meaning of which has since been practically interpreted on the bloody fields of Italy. Sceptical as many were in regard to the actual breaking out of a war, but few, we think, were prepared for so just an adjustment of a treaty of peace. This startling episode has been, to some extent, a disturbing element in the commercial and financial world, and its introduction has not been unattended with disappointment and disaster in some communities. Our leading staple, cotton, felt a most depressing influence; but, nevertheless, the result of the season's operations in this article, taken in the aggregate, should, we think, be highly satisfactory, at least to the planting interest; for we find by our calculations that, although the crop of the year just closed has exceeded the one immediately preceding it, in the large amount of upward of 680,000 bales, yet the aggregate price obtained in value, for the total crop received at the ports, exceeds thirty-four millions of dollars. In February, the political affairs in Europe, tending to apprehensions of war in Italy, began to exercise a more marked influence on the European markets, and that influence was brought to act here at a period of heavy receipts, and an accumulation of stock beyond all precedent, the amount, including all on shipboard, having reached 534,380 bales. Our receipts at this port having reached about 1,475,000 bales, against 1,164,000 bales at the same time the year previous; and our general cotton table showed an increase in the market, at all the ports, of 1,019,000 bales. Under these circumstances, the market for all qualities gave way, but the heavy weight of stock consisted of low-running mixed grades, embracing an unusual proportion of dusky and sandy cottons, which were wholly unsaleable, and for which no quotations could be given.

LIVERPOOL STATEMENT FOR 1859.

UNITED STATES, 1858–1859.

Stock Sept. 1, 1858	103,000	Export	3,021,000
Crop	3,851,000	Consumption	784,000
		Stock Sept. 1, 1859	149,000
Bales	3,954,000	Bales	3,954,000

Stock Jan. 1, 1859, in	Gt. Britain.	France.	Continent.	Tot. Europe.
United States	269,000	111,000	32,000	412,000
Brazil	19,000	3,000	5,000	27,000
West Indies	1,000	1,000	4,000	6,000
East Indies	56,000	10,000	6,000	72,000
Egypt	27,000	6,000	7,000	40,000
Bales	372,000	131,000	54,000	557,000

CONSUMPTION. Tot. Europe.	Continent.	France.	Gt. Britain.	IMPORT.	Gt. Britain.	France.	Continent.	Tot. Europe.
3,068,000	708,000	453,000	1,907,000	United States	2,086,000	379,000	707,000	3,030,000
124,000	14,000	5,000	105,000	Brazil	125,000	3,000	10,000	130,000
32,000	9,000	17,000	6,000	West Indies	7,000	16,000	7,000	30,000
442.000	250,000	15,000	177,000	East Indies	510,000	7,000	269,000	514,000
173,000	38,000	36,000	99,000	Egypt	101,000	31,000	32,000	149,000
3,839,000	1,019,000	526,000	2,294,000	Bales	2,829,000	436,000	1,025,000	3,853,000
....	....	....	437,000	Export.				
571,000	60,000	41,000	470,000	Stock Dec. 31. Stock above,	372,000	131,000	54,000	557,000
4,410,000	1,079,000	567,000	3,201,000	Total supply, bales	3,201,000	567,000	1,079,000	4,410,000

COTTON AT LIVER

Week Ending.	Receipts.						Sales.			
	Americ'n.	E. I.	Egypt.	Brazil.	Other.	Total.	Consumption.	Speculation.	Export	Total.
Jan. 7..	21,317	......	531	2,016	14	23,878	31,607	670	2,213	34,490
" 14..	33,964	1,698	452	5,504	214	41,832	35,101	1,530	9,359	45,990
" 20..	52,246	1,185	3,406	....	......	56,837	41,208	430	2,572	44,210
" 28..	124,482	167	385	659	......	125,693	36,133	970	2,897	40,000
Feb. 4..	49,816	......	1,079	2,302	......	53,197	51,719	7,510	2.691	61,920
" 11..	16,644	9,941	1,347	3,363	105	31,400	52,330	9,630	5,750	67,710
" 18..	21,719	9,841	2,007	2,016	......	35,583	69,584	14,810	4,036	88,430
" 25..	44,865	1,804	3,372	2,400	......	52,441	35,481	3,710	2,859	42,050
Mch. 4..	36,125	310	1,263	300	63	38,061	62,005	12,490	3.395	77,890
" 11..	47,011	1,725	3,253	1,673	136	53,798	48,006	16,380	6,854	71,240
" 18..	29,882	......	2,445	1,206	113	33,646	32,201	7,670	2,289	42,160
" 25..	34,446	6,006	210	995	84	41,741	58,292	9,650	2,928	70 870
April 1..	21,925	4,186	157	977	......	27,245	51,726	9,070	2,464	63,260
" 8..	45,704	25,804	4,809	1,499	83	77,899	33,634	2,410	2,986	39,030
" 15..	60,023	10,901	2,532	4,238	......	77,694	34,726	3,190	5,404	43,320
" 21..	36,271	14,440	1,693	599	70	53,073	23,849	870	4,491	29,210
" 29..	18,812	......	5,417	....	......	24,229	20,881	600	8,539	30,0[illegible]0
May 6..	9,961	......	2,802	....	......	12,763	22,702	1,100	4,888	28,690
" 13..	220,998	15,659	5,295	3,205	302	245,459	34,535	4,580	14,225	53,340
" 20..	32,531	7,621	......	2,659	......	42,811	26,203	200	11,927	38,330
" 27..	94,513	5,514	3,139	4,922	52	108,140	43,738	1,470	6,782	51,990
June 3..	13,942	585	790	....	......	15,317	78,285	9,160	9,865	97,310
" 10..	83,764	9,840	2,486	2,639	......	98,729	24,270	460	12,960	37,690
" 17..	71,462	6,747	3,244	358	78	81,889	49,886	4,880	10,944	65,710
" 24..	141,267	14,144	1,223	2,689	......	159,323	23,118	920	9,442	33,480
July 1..	24,473	13,676	1,613	923	......	40,685	48,881	1,500	6,089	56,470
" 8..	32,512	11,857	1,778	1,032	25	47,204	56,158	5,350	8,082	69,590
" 15..	43,346	3,523	803	1,110	16	48,798	77,329	19,460	9,991	106,780
" 22..	34,908	872	2,649	2,053	......	40,482	39,140	6,580	4,620	50,340
" 29..	31,879	1,224	4,850	1,407	12	39,372	49,597	5,380	8,393	63,370
Aug. 5..	21,840	592	1,891	6,876	137	31,336	31,260	3.670	11,270	46,200
" 12..	14,735	1,663	642	1,099	6	18,145	32,481	2,760	8,069	43,310
" 19..	35,634	14,647	4,925	4,959	......	60,165	21,457	3,740	8,773	33,970
" 26..	22,330	6,577	2,337	2,347	115	33,706	31,093	2,670	5,897	39,660
Sept. 2..	17,200	28,005	2,153	3,542	......	50,900	46,595	4,070	6,495	57,160
" 9..	16,568	18,601	......	2,516	129	37,814	48,996	1,230	6,484	56,710
" 16..	8,877	6,246	3,162	5,109	159	23,553	34,423	910	9,357	44,690
" 23..	11,969	8,558	622	1,035	70	22,254	44,090	980	1,500	46,570
" 30..	14,467	29,881	2,646	5,160	......	52,154	37,353	3,680	9,237	50,270
Oct. 7..	5,390	19.866	2,766	1,573	146	29,741	41,955	3,150	9,615	54,720
" 14..	4,066	5,882	1,199	2,635	18	13,800	44,279	3,110	12,161	59,550
" 21..	5,363	26,667	1,280	....	......	33,310	54,610	2,300	1.900	58,810
" 28..	22,023	11,666	2,668	2,947	109	39,413	60,631	10,360	16,999	87,990
Nov. 4..	19,155	20,189	382	7,627	86	47,439	53,852	7,760	8,438	70,050
" 11..	17,564	10,732	1,285	2,218	132	31.931	31,675	3,120	13,855	48,650
" 18.	8,610	7,286	1,218	4,801	247	22,162	36,808	1,740	5,952	44,500
" 25..	23,598	5,461	......	7,691	1,117	37,867	31,349	1,050	5,411	37,810
Dec. 2..	22,535	7,596	1,482	1,397	171	33,181	39,276	3,440	8.734	51,450
" 9..	48,658	4,998	1,822	1,619	57	57,154	35,664	1,330	5,016	42,010
" 16..	21,633	8,659	......	....	207	30,499	36,116	10,320	16,324	62,760
" 23..	65,733	7,082	2,340	....	......	75,155	35,009	1,810	7,741	44,560
" 30..	....	......	......	....	......	84,282	34,852	5,530	5,588	45,970
Average prices & total sales, receipts & stocks.	1,958,756	430124	99,850	117,895	4,273	2,610,898	2,157,249	241,460	374751	2,773,460

POOL. YEAR 1859.

STOCKS.			PRICES.			ACTUAL EXPORT.	CONSUMPTION.	REMARKS.
American.	Other.	Total.	Mid. Up.	Mid. Orl.	Dhol.			
258,417	80,751	339,168	6	6 1/4	5 5/8	2,213	31,607	
254,741	77,129	331,870	6	5 1/4	5 3/8	9,359	66,708	Good accounts from Manchester and elsewhere.
272,417	73,430	345,847	6	6 1/4	5 1/2	2,572	107,916	
365,479	67,501	432,980	6	6 1/4	5 1/2	2,897	144,049	
362,755	60,622	423,377	6	6 1/4	5 1/2	2,691	195,768	
331,229	62,778	394,007	6	6 1/4	5 1/2	5,750	248,098	
300,288	58,752	359,040	6 1/4	6 1/2	5 5/8	4,036	317,682	
317,683	58,748	376,431	6 1/8	7 1/8	5 5/8	2,859	353.163	
302,538	50,904	353,442	7	7 3/16	5 3/4	3.395	415,168	
313,122	43,701	356,823	6 15/16	7 1/4	5 3/4	6,854	463,174	Unfavorable accounts of crop.
319,334	41,055	360,389	7 1/8	7 3/8	5 3/4	2,289	495,375	
308,240	38,600	346,840	7 1/8	7 3/8	5 3/4	2,928	553,667	
292,575	34,930	327,505	7 1/16	7 5/16	5 3/4	2,464	605,393	
311,139	58,495	369,634	7	7 1/4	5 5/8	2,986	639,027	
341,512	65,916	407,428	6 15/16	7 1/4	5 5/8	5,404	673,753	Outbreak of war on Continent.
358,323	73,698	432,021	6 5/8	6 7/8	5 1/2	4,491	697,602	
351,555	69.065	420,620	6 3/8	6 5/8	5 1/8	8,539	718,483	
336.626	66,157	402,783	6 1/2	6 13/16	5 1/8	4,888	741,185	
509,334	78,828	588,162	6 3/8	6 3/4	5 1/8	14,225	775,720	
502,995	81,918	584,913	6 3/8	6 3/8	5	11.927	801,923	
554,108	86,625	640,733	6 5/8	6 7/8	5 1/8	6,782	845,761	
492,550	73,090	565.640	6 3/8	6 7/8	5	9.865	924,046	
540,684	79,225	619,909	6 3/4	6 15/16	5	12,960	948,316	Anticipating speedy conclusion of war on Continent.
557,726	78,642	636,368	6 11/16	6 7/8	5	10,944	998,202	
671,333	84,588	755,921	6 11/16	6 7/8	5	9,442	1,021,320	
656,946	90,100	747,046	6 11/16	6 7/8	5	6,089	1,070,201	
634,508	90,462	724,970	6 13/16	6 15/16	5	8,082	1,126,359	
609,164	79,544	688,708	7 1/8	7 5/16	5 3/8	9,991	1,203,688	
620,892	72,848	693,740	7	7 1/8	5 1/4	4,620	1,242,828	
608,691	65,921	674.612	7	7 3/16	5 1/4	8,393	1,292,425	
594,061	63,817	657,878	6 15/16	7 1/8	5 1/4	11,270	1,323,685	Favorable accounts from India.
577,506	55,457	632,963	6 15/16	7 1/8	5 1/4	8,069	1,356,166	
588,950	69,158	658.108	6 7/8	7 1/16	5 1/4	8,773	1,377,623	
583,230	70,974	654,204	6 7/8	7 1/16	5 1/4	5,897	1,408,716	
556,390	96,166	652,556	6 7/8	7 1/16	5 1/4	6.495	1,455,311	
533,154	106,712	639,866	6 7/8	7 1/16	5 1/4	6,484	1,504,307	
511,250	107,122	618,372	6 3/4	7	5 1/4	9,357	1,538,730	
490,654	101,844	592,498	6 5/8	6 15/16	5 1/4	1,500	1,582,820	
475.282	124,513	599,795	6 5/8	6 7/8	5 1/8	9,237	1,620,173	Inclination of consumers to buy.
446,103	133,488	579,591	6 3/4	7	5	9,615	1,662,128	
410,221	125,989	536,210	6 3/4	7	5	12,161	1,706,407	
372,670	146,460	519,130	6 3/4	7 1/8	5	1,900	1,761,017	
337,632	136,002	473,634	6 7/8	7 1/4	5 1/8	16,999	1,821,648	
308,684	149,521	458.205	7	7 1/2	5 1/8	8,438	1,875.500	Favorable accounts from Manchester.
294,641	141,910	436,551	6 7/8	7 3/8	5 1/16	13,855	1,907,175	
269,949	144,222	414,171	6 7/8	7 3/8	5	5,952	1,943,983	
266,728	158,879	425,607	6 13/16	7 3/8	5	5,411	1,975,332	
256.287	154,517	410,804	6 3/4	7 3/8	4 7/8	8,734	2,015,608	
276,845	151,357	428,202	6 3/4	5 3/4	4 3/4	5.016	2,051,272	
262,327	138,560	400,887	6 5/8	7 1/8	4 3/4	16,324	2,087,388	
296,068	134,083	430,151	6 1/2	6 15/16	4 3/4	7,741	2,122,397	
....	....	441,710	......	......	......	5,588	2,157,249	
			6.68	6.93	5.25	374,751	42,254.79	

The semi-weekly Price and Weekly Sales and Receipts at New York, Weekly Exports from New York and Rates of Freight to Liverpool 1st of each month, for the Crop Year ending September 1, 1859.

1858.	Price of Middling New Orleans New York Classificat'n.	Price of Middling Upland, New York Classificat'n.	Sales for week, including lots in transit.	Receipts for week, including lots in transit.	EXPORTS FOR THE WEEK. To Great Britain.	To France.	North of Europe.	Other Fo'n Ports	Total Exports.	Rates of Freight to Liverpool.
Septem. 3..	13	12¾								⅛@5-32d
" 7..	13¼	13	10,500	2,105	2,182	....	....	....	2,182	
" 10..	13¼	13⅛								
" 14..	13⅜	13¼	9,800	3,327	490	....	....	2	492	
" 17..	13¼	13⅛								
" 21.	13¼	13⅛	4,000	1,279	870	....	100	25	995	
" 24..	13⅜	13¼								
" 28..	13½	13⅜	11,200	573	....	....	95	....	95	
October 1..	13⅝	13⅜								5-32d.
" 5..	13¾	13½	20,000	3,918	1,202	782	503	....	2,487	
" 8..	13⅜	13¼								
" 11..	13⅛	13	12,800	7,565	987	....	252	....	1,239	
" 15..	12¾	12½								
" 19..	12⅜	12¼	17,000	5,084	599	223	657	....	1,479	
" 22..	12½	12⅜								
" 26..	12½	12⅜	9,000	4,557	4,194	....	1,368	....	5,562	
" 29..	12¼	12⅛								
Novem. 2..	12¼	12	10,000	12,707	....	....	699	....	699	$\frac{3}{16}$@$\frac{7}{32}$d.
" 5..	12	11¾								
" 9..	11¾	11½	12,500	14,882	1,032	122	646	50	1,850	
" 12..	11½	11¼								
" 16 .	11⅝	11⅜	8,800	6,513	2,492	1,602	1,138	149	5,381	
" 19 .	11¾	11½								
" 23..	12⅛	11¾	7,500	7,941	1,933	....	1,405	1	3,339	
" 26..	12¼	12								
" 30..	12⅜	12⅛	16,000	12,004	2,188	....	....	....	2,188	
Decem. 3..	12⅜	12⅛								¼d.
" 7..	12¼	11⅞	6,000	11,263	567	1,095	578	....	2,240	
" 10..	12⅛	11¾								
" 14..	12¼	11⅞	10,000	12,101	....	....	655	....	655	
" 17..	12⅜	12								
" 21..	12⅛	12	9,500	12,741	1,563	1,061	840	600	4,064	
" 24..	12	11⅞								

General Remarks.

It was generally anticipated that at the conclusion of the war between England and France against China, there would be much more activity in cotton; but accounts were received early in September of a conclusion of a treaty of peace between those powers, and the news had but little perceptible effect, either in Europe or here.

The receipts at the ports, in October, were large; the weather for picking very fine, and with only a moderate demand for home use, and discouraging advices from abroad, prices yielded.

There was a better feeling in November, partly caused by accounts of killing frosts on the 5th, in Louisiana and Alabama; the season, however, was four weeks earlier than the previous one, and the amount of crop outstanding was comparatively small.

The receipts increased in December, and, with estimates of the crop enlarging, the market was in buyers' favor.

In January, the manufacturing industry seemed to be in a more flourishing condition, and, with easy money market, here and abroad, the staple rose.

Decem. 28..	12	11⅞	4,200	12,912	1,344	....	420	2,372	4,136	
" 31..	12	11⅞								
1859.										
January 4..	12¼	12⅛	10,100	3,486	654	....	286	....	940	$\frac{3}{16}$@$\frac{7}{32}$d.
" 7..	12⅜	12⅛								
" 11..	12⅜	12	7,500	19,757	1,571	....	....	468	2,039	
" 14..	12⅜	12								
" 18..	12⅜	12	11,800	11,207	218	....	830	....	1,048	
" 21..	12½	12¼								
" 25..	12⅝	12¼	23,000	7,203	81	....	416	....	497	
" 28..	12½	12⅛								
February 1..	12⅜	12	6,500	7,527	3,405	1,065	61	804	5,335	¼d.
" 4..	12¼	11⅞								
" 8..	12⅛	11¾	5,500	14,086	100	830	459	....	1,389	
" 11..	12	11⅝								
" 15..	12	11⅝	5,000	10,256	101	....	1,134	1,034	2,269	
" 18..	12	11¾								
" 22..	12⅛	11¾	6,600	21,619	2,121	....	683	....	2,804	
" 25..	12¼	11⅞								
March 1..	12¼	12	15,000	13,725	791	....	861	....	1,652	⅛@$\frac{3}{16}$d
" 4..	12⅜	12⅛								
" 8..	12⅝	12¼	30,000	9,181	2,015	1,297	1,433	1,054	5,798	
" 11..	12¾	12⅜								
" 15..	12¾	12½	17,000	10,992	5,878	....	609	110	6,597	
" 18..	12¾	12½								
" 22..	12¾	12½	21,500	10,327	1,350	....	1,988	548	3,886	
" 25..	12¾	12½								
" 29..	12¾	12½	49,000	20,975	5,377	....	5,106	1	10,484	
April 1..	12¾	12½								3-16d.
" 5..	12¾	12½	24,400	11,593	5,126	1,152	640	547	7,465	
" 8..	12¾	12½								
" 12..	12¾	12½	30,900	8,330	4,218	811	....	....	5,029	
" 15..	12¾	12½								
" 19..	12¾	12½	11,250	10,894	3,001	....	1,381	....	4,382	
" 22..	12½	12⅛								
" 26..	12⅜	12⅛	5,150	11,081	3,462	....	59	....	3,521	
" 29..	12½	12								
May 3..	12⅜	11⅞	4,000	7,709	5,129	1,000	960	448	7,537	$\frac{5}{32}$@$\frac{3}{16}$d.
" 6..	12⅜	11⅞								
" 10..	12¼	11⅞	5,000	5,050	2,480	1,680	....	....	4,160	
" 13..	11¾	11¼								
" 17	11⅝	11	4,800	446	2,028	....	901	....	2,929	

In February, the apprehension of war between France and Italy, against Austria, depressed the English markets in a measure, though counteracted to some extent by the increased demand in England for goods from China and India, and the falling off in the receipts at our ports. This market responded, and the business was light at lower prices.

March and April were pretty active months, the dealings being largely on speculation, owing to reduced receipts.

In May and June, the conflict between the Allies and Austria unsettled trade, the fear being generally entertained in England, that this was but a precursor to a general European war.

Toward the close of July, news was received of an armistice between the contending European powers, which was succeeded later by a treaty of peace, and a stimulus was given to trade, both at Liverpool and in this market, which continued until the close of the crop year.

New York Statement for Year 1859—*Concluded.*

1859		Price of Middling New Orleans New York Classificat'n.	Price of Middling Upland. New York Classificat'n.	Sales for week, including lots in transit.	Receipts for week, including lots in transit.	EXPORTS FOR WEEK.					Rates of Freight to Liverpool.
						To Great Britain	To France.	North of Europe.	Other Fo'n Ports	Total Exports.	
May	20..	$11\frac{3}{8}$	11								
"	24..	$11\frac{5}{8}$	$11\frac{1}{4}$	7,000	15,608	2,007	906	....	....	2,913	
"	27..	$11\frac{5}{8}$	$11\frac{1}{4}$								
"	31..	$11\frac{3}{8}$	11	5.000	1,494	1,104	....	624	473	2.201	$\frac{1}{8}$d.
June	3..	11	$10\frac{5}{8}$								
"	7..	$11\frac{1}{4}$	$10\frac{7}{8}$	5,500	3,747	1,409	2,060	....	....	3,469	
"	10..	$11\frac{1}{2}$	$11\frac{1}{8}$								
"	14..	$11\frac{3}{4}$	$11\frac{3}{8}$	11,450	4,724	1,393	1,126	....	....	2,519	
"	17..	$12\frac{1}{8}$	$11\frac{3}{4}$								
"	21..	$12\frac{1}{8}$	$11\frac{3}{4}$	5,000	5,548	4,929	2,108	83	....	7.120	
"	24..	12	$11\frac{5}{8}$								
"	28..	12	$11\frac{5}{8}$	6,000	4,535	6,808	30	657	....	7,495	
July	1..	12	$11\frac{5}{8}$								$\frac{1}{8}$d.
"	5..	12	$11\frac{5}{8}$	6,200	5,182	2,710	1,860	679	....	5,249	
"	8..	12	$11\frac{5}{8}$								
"	12..	12	$11\frac{5}{8}$	5,600	8,625	2,925	2,274	....	....	5,199	
"	15..	12	$11\frac{3}{4}$								
"	19..	$12\frac{1}{8}$	$12\frac{3}{4}$	6,500	4,753	1,725	768	20	....	2,513	
"	22..	$12\frac{3}{8}$	12								
"	26..	$12\frac{5}{8}$	$12\frac{1}{4}$	14,100	7,880	6,053	2,762	70	....	8,885	
"	29..	$12\frac{7}{8}$	$12\frac{1}{2}$								
August	2..	$12\frac{3}{4}$	$12\frac{1}{4}$	9,600	2,522	3,499	428	....	....	3,927	
"	5..	$12\frac{3}{4}$	$12\frac{1}{4}$								$\frac{5}{32}$@$\frac{3}{16}$d.
"	9..	$12\frac{1}{2}$	$12\frac{1}{8}$	2,500	5,099	5,653	1,130	409	....	7,192	
"	12..	$12\frac{3}{8}$	12								
"	16..	$12\frac{3}{8}$	$11\frac{7}{8}$	6,000	5,606	657	661	407	....	1,725	
"	19..	$12\frac{1}{4}$	$11\frac{3}{4}$								
"	23..	$12\frac{1}{4}$	$11\frac{3}{4}$	7,500	2,658	5.525	494	253	130	6,402	
"	26..	$12\frac{1}{4}$	$11\frac{3}{4}$								
"	30..	$12\frac{1}{4}$	$11\frac{3}{4}$	4,700	10,372	3,502	2,009	425	29	5,965	
Septem.	2..	$12\frac{1}{8}$	$11\frac{5}{8}$								$\frac{3}{16}$@$\frac{7}{32}$d.
Average price and total sales, receipts and exports.		12.36	12.08	575,450	435,269	120,648	31,335	30,790	8,845	191,618	

General Remarks.

Exchange

was quite steady throughout the year. The range for sixty days' bills on London, in September was $9\frac{1}{4}$@$10\frac{1}{4}$; October, $9\frac{1}{4}$@$10\frac{1}{8}$; November, $8\frac{3}{4}$@$9\frac{3}{4}$; December, 9@$9\frac{3}{4}$; January, the same; February, $9\frac{1}{4}$@$9\frac{5}{8}$; March, $9\frac{3}{8}$@$9\frac{3}{4}$; April, $9\frac{3}{8}$@$10\frac{1}{8}$; May, $9\frac{7}{8}$@$10\frac{1}{4}$; June, $9\frac{3}{4}$@$10\frac{1}{4}$; July, 10@$10\frac{3}{8}$; and in August, $9\frac{3}{4}$@$10\frac{3}{8}$.

1860.

The census taken this year reveals the following statistics touching cotton manufactures:

NEW ENGLAND STATES.

(*Maine, New Hampshire, Vermont, Massachusetts, Rhode Island and Connecticut.*)

MANUFACTURES.	No. of establishments.	Capital invested.	Cost of raw material.	No. of hands employed. Male.	No. of hands employed. Female.	Annual cost of labor.	Annual value of products.
Calico engraving......	1	$100	$500	1	1	$840	$1,400
" printing........	11	1,730,000	1,798,577	2,086	290	608,980	4,332,256
Cotton bags...........	3	92,700	83,620	113	83	40,380	177,000
" batting.........	37	274,000	473,370	266	32	91,020	747,797
" cordage	22	166,300	144,476	133	82	45,240	256,650
" gins	2	70,000	28,950	62	......	34,680	78,600
" goods	359	65,947,819	34,559,883	27,584	49,045	15,702,888	73,638,957
" lines and twine	21	201,900	211,455	128	118	52,488	344,230
" thread	5	281,000	73,548	173	253	84,864	427,148
" yarn	24	335,700	399,295	261	305	124.224	698.321
" " etc........	59	1,230,000	1,028,990	782	1,066	364,020	1,890.516
" " thread, etc.	40	730,860	696,145	451	533	215,796	1,178,281
Total...........	584	71,060,379	39,498,809	32,040	51,808	17,365.420	83,771,156

MIDDLE STATES.

(*New York, New Jersey, Pennsylvania, Delaware, Maryland and District of Columbia.*)

MANUFACTURES.	No. of establishments.	Capital invested.	Cost of raw material.	No. of hands employed. Male.	No. of hands employed. Female.	Annual cost of labor.	Annual value of products.
Cotton batting	14	$85,200	$130,899	101	34	$28,620	$207,630
" braid	1	1,500	1,550	8	4	2,160	20,000
" coverlets	17	33,475	44,020	66	18	20,628	99,675
" flannel carding..	3	6,000	23,973	29	....	4,836	54,482
" gins	1	15,000	10,200	25	....	13,500	45,000
" goods	270	17,140,719	12,507,907	11,202	15,563	5,052,836	24,031,639
" lamp wick......	2	70,000	52,909	43	30	8,700	119,124
" mosquito netting	2	53,000	32,720	58	71	25,956	138,392
" table cloths.. ..	13	23,550	18,127	68	6	15,900	40,318
" thread	1	2,500	12,500	16	30	6,000	22,000
" twine	2	1,800	2,900	4	4	1,560	5,000
" yarn	35	1,410,800	1,169,159	708	1,124	323,040	1,950,597
Total...........	361	18,843,544	14,006,864	12.328	16,884	5,503,736	26,733,857

WESTERN STATES.

(*Ohio, Indiana, Michigan, Illinois, Wisconsin, Iowa, Minnesota, Nebraska, Missouri, Kansas and Kentucky.*)

MANUFACTURES.	No. of establishments.	Capital invested.	Cost of raw material.	NO. OF HANDS EMPLOYED.		Annual cost of labor.	Annual value of products.
				Male.	Female.		
Cotton batting........	1	$3,000	$13,500	3	....	$864	$18,000
" " & wadding	2	3,200	10,530	8	....	1,980	15,987
" ginning........	1	200	612	1	....	240	1,050
" goods...	17	926,000	915,280	761	859	307,068	1,595,120
Total...........	21	932,400	939,922	773	859	310,152	1,630,157

SOUTHERN STATES.

(*Virginia, North Carolina, South Carolina, Georgia, Florida, Alabama, Mississippi, Louisiana, Texas, Arkansas, and Tennessee.*)

MANUFACTURES.	No. of establishments.	Capital invested.	Cost of raw material.	NO. OF HANDS EMPLOYED.		Annual cost of labor.	Annual value of products.
				Male.	Female.		
Calico printing........	1	$1,200	$6,400	3	....	$1,200	$9,000
Cotton ginning........	88	92,457	367,134	264	6	52,404	552,585
" gins............	54	673,225	248,338	527	2	217,980	1,028,715
" goods.........	157	9,129,221	4,683,631	3,859	6,082	1,425,770	8,072,067
" pressing.......	5	149,700	3,610	64	....	25,920	89,650
" yarn...........	2	37,000	11,600	14	11	4,428	23,000
Total...........	307	10,082,803	5,320,713	4,731	6,101	1,727,702	9,775,017

TERRITORIES.

(*Utah, New Mexico, and Washington.*)

MANUFACTURES.	No. of establishments.	Capital invested.	Cost of raw material.	NO. OF HANDS EMPLOYED.		Annual cost of labor.	Annual value of products.
				Male.	Female.		
Cotton yarn...........	1	$6,000	$6.000	4	3	$3,420	$10,000
Total.........	1	6,000	6,000	4	3	3,420	10,000

TOTALS IN THE UNITED STATES.

MANUFACTURES.	No. of establishments.	Capital invested.	Cost of raw material.	No. of hands employed.		Annual cost of labor.	Annual value of products.
				Male.	Female.		
Cotton bags	3	$92,700	$83,620	113	83	$40,380	$177,000
" batting and wad'g.	54	365,400	628,299	378	66	122,484	989,414
" braid, thread, lines, twine and yarn	191	4,239,060	3,613,142	2,549	3,451	1,182,000	6,569,093
" cordage	22	166,300	144,476	133	82	45,240	257,650
" coverlets	18	34,975	45,420	68	19	21,288	102,675
" flannel carding	3	6,000	23,973	29	......	4,836	54,482
" ginning	89	92,657	367,746	265	6	52,644	553,635
" gins	57	758,825	287,488	614	2	266,160	1,152,315
" goods	803	93,143,759	52,666,701	43,406	71,549	22,488,562	107,337,783
" lamp wick	2	70,000	52,909	43	30	8,700	119,124
" mosquito netting	2	53,000	32,720	58	71	25,956	138,392
" pressing	5	149,700	3,610	64	......	25,920	89,650
" table cloths	13	23,550	18,127	68	6	15,900	40,318
Totals	1,262	99,195,926	57,968,231	47,788	75,365	24,300,070	117,581,531

COTTON CROP OF THE UNITED STATES.

Statement and Total Amount of the Cotton Crop of the United States, for the Year ending August 31, 1860.

	Bales.	Bales.	TOTAL. 1860.	1859.	1858.
LOUISIANA.					
Export from New Orleans—					
To Foreign Ports	2,005,662				
Coastwise	208,634				
Burnt at New Orleans	5,240				
Stock on hand 1st September, 1860	73,934				
		2,293,470			
Deduct—					
Received from Mobile	34,179				
" Montgomery, &c	28,473				
" Florida	16,335				
" Texas	49,036				
Stock on hand 1st September, 1859	26,022				
		154,045	2,139,425	1,669,274	1,576,409
ALABAMA.					
Export from Mobile—					
To Foreign Ports	659,481				
Coastwise	158,332				
Burnt at Mobile	3,387				
Manufactured in Mobile	1,220				
Stock on hand 1st September, 1860	41,682				
		864,102			
Deduct—					
Received from New Orleans	984				
Stock on hand 1st September, 1859	20,106				
		21,090	843,012	704,406	522,364
TEXAS.					
Export from Galveston, &c—					
To Foreign Ports (including 1865 to Mexico)	111,967				
Coastwise	139,767				
Manufactured in Galveston	177				
Stock on hand 1st September, 1860	3,168				
		255,079			
Deduct—					
Stock on hand 1st September, 1859	....	2,655			
			252,424	192,062	145,286
FLORIDA.					
Export from Apalachicola, St. Marks, &c.—					
To Foreign Ports—Uplands	58,353				
" Sea Islands	755				
Coastwise—Uplands	117,394				
Sea Island	14,200				
Burnt at Apalachicola	1,394				
Stock on hand 1st September, 1860	864				
		192,960			
Deduct—					
Stock on hand 1st September, 1859	....	236			
			192,724	173,484	122,351
GEORGIA.					
Export from Savannah—					
To Foreign Ports—Uplands	331,159				
" Sea Island	6,596				
Coastwise—Uplands	190,937				
Sea Island	18,345				
Stock in Savannah, 1st September, 1860	4,307				
" Augusta, &c., 1st September, 1860	5,252				
		556,596			

Statement and Total Amount of the Cotton Crop of the United States, for the Year ending August 31, 1860—*Concluded.*

	Bales.	Bales.	TOTAL. 1860.	1859.	1858.
Deduct—					
Received from Florida—Sea Island	6,308				
" Uplands	686				
Stock in Savannah 1st September, 1859	9,320				
" Augusta, &c., 1st September, 1859.	9,063				
		25,377			
SOUTH CAROLINA.			531,219	475,788	282,973
Export from Charleston—					
To Foreign Ports—Uplands	365,654				
" Sea Island	21,116				
Coastwise—Uplands	153,393				
Sea Island	5,946				
Burnt at Charleston	284				
Stock in Charleston, 1st September, 1860	8,897				
Export from Georgetown, S. C.—					
To Northern Ports—Uplands	801				
		556,091			
Deduct—					
Received from Florida—Sea Island	6,844				
" Uplands	539				
Received from Savannah—Sea Island	1,411				
" Uplands	19,596				
Stock in Charleston, 1st September, 1859	17,592				
		45,982	510,109	480,653	406,251
NORTH CAROLINA.					
Export—					
To Coastwise Ports	41,194				
		....	41,194	37,482	23,999
VIRGINIA.					
Export—					
To Foreign Ports	3,259				
Coastwise	33,462				
Manufactured (taken from the ports)	17,841				
Stock on hand, 1st September, 1860	2,800				
		57,362			
Deduct—					
Stock on hand 1st September, 1859	....	375			
			56,987	33,011	24,705
TENNESSEE, &c.					
Shipments from Memphis	391,918				
" Nashville	23,000				
" Columbus & Hickman, Ky.	4,500				
Burnt and manufactured at Memphis	1,482				
Stock at Memphis, 1st September, 1860	1,709				
		422,609			
Deduct—					
Shipments to New Orleans	263,589				
" Norfolk	160				
Manufactured on the Ohio, &c	49,000				
Stock on hand 1st September, 1859	1,184				
		313,933			
			108,676	85,321	9,624
Total crop of the United States	....	....	4,675,770	3,851,481	3,113,962

Increase over crop of 1859 .... bales.	824,289
Increase over crop 1858	1,561,808
Increase over crop of 1857	1,736,251
Iucrease over crop of 1856	1,147,925

Export to Foreign Ports, from September 1, 1859, *to August* 31, 1860.

FROM	To Great Britain.	To France.	To North of Europe.	Other F'n Ports.	Total.
New Orleans........................bales.	1,426,966	313,291	136,135	129,270	2,005,662
Mobile	445,663	148,918	21,806	43,094	659,481
Galveston	83,972	5,471	19,569	2,955	111,967
Florida	52,986	1,420	2,634	2,068	59,108
Savannah	291,403	20,422	24,809	1,121	337,755
Charleston	240,151	64,895	47,056	34,668	386,770
Virginia	3,259	....	....	....	3,259
New York	121,200	35,110	39,916	6,802	203,028
Baltimore	29	60	50	18	157
Philadelphia	289	....	....	3	292
Boston	3,514	....	3,097	83	6,694
Grand total	2,669,432	589,587	295,072	220,082	3,774,173
Total last year	2,019,252	450,696	330,012	221,443	3,021,403
Increase	650,180	138,891	....	....	752,770
Decrease	....	....	34,940	1,361	....

Consumption.

Total crop of the United States, as before stated ....bales.			4,675,770
Add—Stocks on hand at the commencement of the year, September 1, 1859—			
In the Southern ports		85,369	
In the Northern ports		63,868	
			149,237
Makes a supply of			4,825,007
Deduct therefrom—The Export to Foreign ports	3,774,173		
Less Foreign included	917		
		3,773,256	
Stocks on hand, September 1, 1860—			
In the Southern ports	142,613		
In the Northern ports	85,095		
		227,708	
Burnt at New Orl., Apalachicola, Charleston & N Y...	7,415		
Burnt and manuf'd at Mobile, Galveston & Memphis...	6,266		
Manufactured in Virginia	17,841		
		31,522	
			4,032,486
Taken for home use north of Virginia........bales.			792,521
Taken for home use in Virginia and south and west of Virginia			185,522
Total consumed in the U. S. (including burnt at the ports), 1859-60			978,043

ANNUAL REVIEW.

From the New Orleans Price Current, 1859–60.

The production of cotton for the year now under review has shown a further large increase, in amount and value, and with the return of peace in Europe, and prosperous manufacturing interests at home and abroad, the crop has been disposed of at an average range of remunerative prices. An important feature in the operations of the season, was the depressing influence of large stocks of our leading staple in Europe of which the accumulation was rapid from December to June. On the 2d of September, 1859, the stock at Liverpool was estimated at 653,000 bales, including 557,000 American, from which time there was a gradual falling off, until the beginning of December, when the estimate was 411,000 bales, of which, 270,000 were American; the stock at all the ports of Great Britain being 441,000 bales, including 263,000 American. By the 20th of January, these figures had swollen, from Liverpool alone, to 606,000 bales, including 452,000 American; and, with a rapid increase during the next five months, on the 8th of June, they had reached 1,358,000 bales, of which 1,154,000 were American. At the end of June the stock was reduced to 1,333,000 bales, of which 1,038,000 bales were American. With us, the largest stock appearing in our tables, was 573,000 bales, at the beginning of March, from which time the amount fell off to 470,000 bales, on the 1st of April; 285,000, on the 2d of May; 119,000, on the 2d of June; 69,000, on the 1st of July; and 54,000, on the 1st of August. As respects the coming crop, we may remark that the planting was to an increased extent, and preparations were made for a large crop, the expectations in that regard being encouraged by a favorable spring. As the season advanced, however, the absence of rain diminished the prospects considerably, and, during the summer, intense heat prevailed, with an atmosphere almost destitute of moisture, some sections of the cotton region being without a shower of rain for upwards of three months. In the upland portions of the country, the crop suffered materially from this protracted drought, and some injury resulted in the bottom lands, though in the latter, comparatively little damage has been experienced. The plant matured rapidly, and the picking season having opened much earlier than usual, the receipts thus far have been larger than any previous year, for the same period, but the under-growth has been light, and is not expected to hold out for a late picking season, and it is now very generally conceded that a large crop cannot possibly be made.

The semi-weekly Price and Weekly Sales and Receipts at New York, Weekly Exports from New York and Rates of Freight to Liverpool 1st of each month, for the Crop Year ending September 1, 1860.

1859.	Price of Middling New Orleans New York Classificat'n.	Price of Middling Upland, New York Classificat'n.	Sales for week, including lots in Transit.	Receipts for week, including lots in Transit.	EXPORTS FOR THE WEEK. To Great Britain.	To France.	North of Europe.	Other Fo'n Ports	Total Exports.	Rates of Freight to Liverpool.
Septem. 6..	$12\frac{1}{8}$	$11\frac{5}{8}$	2,400	3,882	866	....	96	....	962	$\frac{3}{16}$@$\frac{7}{32}$d.
" 9..	$12\frac{1}{8}$	$11\frac{5}{8}$								
" 13..	$12\frac{1}{8}$	$11\frac{3}{4}$	5,800	1,775	587	471	....	2	1,060	
" 16..	$12\frac{1}{8}$	$11\frac{3}{4}$								
" 20..	$12\frac{1}{8}$	$11\frac{3}{4}$	5,300	2,689	7,373	953	1,230	....	9,556	
" 23..	$12\frac{1}{8}$	$11\frac{3}{4}$								
" 27..	12	$11\frac{5}{8}$	3,000	3,252	1,867	1,318	....	102	3,287	
" 30..	12	$11\frac{5}{8}$								
October 4..	$11\frac{7}{8}$	$11\frac{1}{2}$	3,500	4,819	1,933	....	1,546	....	3,479	$\frac{1}{4}$d.
" 7..	$11\frac{7}{8}$	$11\frac{1}{2}$								
" 11..	$11\frac{3}{4}$	$11\frac{3}{8}$	2,500	4,736	3,053	293	40	....	3,386	
" 14..	$11\frac{3}{4}$	$11\frac{3}{8}$								
" 18..	$11\frac{3}{4}$	$11\frac{3}{8}$	3,500	5,710	459	....	831	....	1,290	
" 21..	$11\frac{7}{8}$	$11\frac{1}{2}$								
" 25..	$11\frac{7}{8}$	$11\frac{1}{2}$	11,200	8,060	3,752	318	....	....	4,070	
" 28..	$11\frac{7}{8}$	$11\frac{1}{2}$								
Novem. 1..	12	$11\frac{5}{8}$	10,700	9,603	2,315	....	683	....	2,998	$\frac{5}{32}$@$\frac{3}{16}$d.
" 4..	$11\frac{7}{8}$	$11\frac{5}{8}$								
" 8..	$11\frac{7}{8}$	$11\frac{5}{8}$	16,500	9,935	1,388	777	331	....	2,496	
" 11..	$11\frac{7}{8}$	$11\frac{3}{8}$								
" 15..	$11\frac{7}{8}$	$11\frac{3}{8}$	17,000	9,252	690	527	932	....	2,149	
" 18..	$11\frac{7}{8}$	$11\frac{3}{8}$								
" 22..	$11\frac{7}{8}$	$11\frac{3}{8}$	15,000	19,875	2,883	887	522	....	4,292	
" 25..	$11\frac{5}{8}$	$11\frac{1}{4}$								
" 29..	$11\frac{1}{2}$	$11\frac{1}{8}$	8,500	16,456	2,194	1,008	560	....	3,762	
Decem. 2..	$11\frac{1}{8}$	11								7-32@$\frac{1}{4}$d
" 6..	$11\frac{1}{8}$	11	11,700	13,350	1,505	....	2,603	....	4,108	
" 9..	$11\frac{1}{8}$	11								
" 13..	$11\frac{5}{8}$	$11\frac{1}{8}$	15,500	9,422	4,826	1,160	....	....	5,986	
" 16..	$11\frac{5}{8}$	$11\frac{1}{8}$								
" 20..	$11\frac{5}{8}$	$11\frac{1}{8}$	8,600	13,006	5,059	738	1,308	....	7,105	
" 23..	$11\frac{1}{2}$	11								
" 27..	$11\frac{1}{2}$	11	16,500	15,674	385	....	616	788	1,789	

GENERAL REMARKS.

No very remarkable event occurred this crop year, and prices varied but little, its entire course. Early in September, the free receipts at the ports caused a dull, heavy feeling here and abroad, which continued through October. About the 1st of November, accounts of frost stiffened the market temporarily, but with subsequent fine and very favorable picking weather, prices soon fell off again Owing to a rise in the Southern rivers, the receipts for a time in December were larger than ever before known, but under an improved demand, prices were sustained.

In January, there was a more cheerful feeling, and prices advanced. In February and March, the large receipts and exports were adverse to the article, and it was barely steady; the stocks in Liverpool in April were double the quantity of those on hand at the same time the previous year, reaching near a million of bales of all kinds.

In May, there was a better feeling, both here and in Europe, and prices advanced. The political troubles in Sardinia and Sicily, and the apprehension that the disorders would

Decem: 30..	$11\frac{1}{2}$	11								
1860.										
January 3..	$11\frac{1}{2}$	11	5,500	6,470	4,920	4,030	825	....	9,775	$\frac{7}{32}$@$\frac{1}{4}$d.
" 6..	$11\frac{5}{8}$	$11\frac{1}{8}$								
" 10..	$11\frac{5}{8}$	$11\frac{1}{8}$	7,000	6,520	1,676	733	921	....	3,330	
" 13..	$11\frac{5}{8}$	$11\frac{1}{8}$								
" 17..	$11\frac{3}{4}$	$11\frac{1}{4}$	6,300	8,148	1,437	2,173	446	746	4,802	
" 20..	$11\frac{7}{8}$	$11\frac{3}{8}$								
" 24..	$11\frac{7}{8}$	$11\frac{3}{8}$	22,800	12,594	6,232	....	881	....	7,113	
" 27..	$11\frac{7}{8}$	$11\frac{3}{8}$								
" 31..	$11\frac{3}{4}$	$11\frac{1}{4}$	10,000	18,829	560	1,234	528	700	3,022	
February 3..	$11\frac{3}{4}$	$11\frac{1}{4}$								$\frac{7}{32}$@$\frac{1}{4}$d.
" 7..	$11\frac{3}{4}$	$11\frac{1}{4}$	8,800	10,491	2,459	1,102	1,816	....	5,377	
" 10..	$11\frac{3}{4}$	$11\frac{1}{4}$								
" 14..	$11\frac{3}{4}$	$11\frac{1}{4}$	10,500	9,716	3,433	....	1,496	....	4,929	
" 17..	$11\frac{3}{4}$	$11\frac{1}{4}$								
" 21..	$11\frac{3}{4}$	$11\frac{1}{4}$	16,500	14,977	1,501	....	....	891	2,392	
" 24..	$11\frac{3}{4}$	$11\frac{1}{4}$								
" 28..	$11\frac{5}{8}$	$11\frac{1}{4}$	8,500	18,764	3,542	....	525	....	4,067	
March 2..	$11\frac{5}{8}$	$11\frac{1}{8}$								$\frac{1}{4}$@9-32d
" 6..	$11\frac{5}{8}$	$11\frac{1}{8}$	9,200	14,900	2,525	1,425	2,684	....	6,634	
" 9..	$11\frac{5}{8}$	$11\frac{1}{8}$								
" 13..	$11\frac{5}{8}$	$11\frac{1}{8}$	17,000	14,233	3,953	....	2,323	....	6,276	
" 16..	$11\frac{5}{8}$	$11\frac{1}{8}$								
" 20..	$11\frac{5}{8}$	$11\frac{1}{8}$	5,500	10,693	4,896	....	1,932	....	6,828	
" 23..	$11\frac{5}{8}$	$11\frac{1}{8}$								
" 27..	$11\frac{5}{8}$	$11\frac{1}{8}$	6,100	8,847	3,226	550	1,708	498	5,982	
" 30..	$11\frac{5}{8}$	$11\frac{1}{8}$								
April 3..	$11\frac{5}{8}$	$11\frac{1}{8}$	6,000	11,539	5,218	659	921	....	6,798	$\frac{1}{4}$d.
" 6..	$11\frac{5}{8}$	$11\frac{1}{8}$								
" 10..	$11\frac{5}{8}$	$11\frac{1}{8}$	10,100	10,302	1,092	967	2,128	....	4,187	
" 13..	$11\frac{5}{8}$	$11\frac{1}{8}$								
" 17..	$11\frac{5}{8}$	$11\frac{1}{8}$	8,000	6,490.	7,465	921	768	....	9,154	
" 20..	$11\frac{5}{8}$	$11\frac{1}{8}$								
" 24..	$11\frac{5}{8}$	$11\frac{1}{8}$	5,000	4,804	1,812	314	766	948	3,840	
" 27..	$11\frac{5}{8}$	$11\frac{1}{8}$								
May 1..	$11\frac{5}{8}$	$11\frac{1}{8}$	5,300	5,164	3,527	140	852	....	4,519	3-16d.
" 4..	$11\frac{5}{8}$	$11\frac{1}{8}$								
" 8..	$11\frac{5}{8}$	$11\frac{1}{8}$	9,000	6,848	1,350	765	177	....	2,292	
" 11..	$11\frac{5}{8}$	$11\frac{1}{8}$								
" 15..	$11\frac{3}{4}$	$11\frac{1}{4}$	9,300	11,517	694	661	222	....	1,577	
" 18..	$11\frac{7}{8}$	$11\frac{3}{8}$								

spread to other parts of Europe, seriously interfered with trade in the Summer months, and depressed prices, assisted by the very large stock in Liverpool, and the small demand at Manchester for goods from the East.

There was at this time a great drought experienced at the South, more severe in its effects than any that had occurred before, in many years, the crop of Texas especially suffering; but these accounts had but little effect, as against the continued unfavorable position of the European markets.

The menacing aspect of affairs between Italy and Austria occasioned much concern in trade circles abroad; great caution was observed, and but little disposition to operate in any of the markets here or in Europe beyond the exigencies of the time.

New York Statement for Year 1860—*Concluded.*

1860.		Price of Middling New Orleans New York Classificat'n.	Price of Middling Upland, New York Classificat'n.	Sales for week, including lots in Transit.	Receipts for week, including lots in Transit.	Exports for the Week					Rates of Freight to Liverpool.	General Remarks.
						To Great Britain.	To France.	North of Europe.	Other Fo'n Ports	Total Exports.		
May	22..	$11\frac{7}{8}$	$11\frac{3}{8}$	9,000	8,670	1,615	376	493	40	2,524		
"	25..	$11\frac{7}{8}$	$11\frac{3}{8}$									
"	29..	$11\frac{7}{8}$	$11\frac{3}{8}$	9,500	7,273	1,701	1,945	275	....	3,921		*Exchange.*
June	1..	$11\frac{7}{8}$	$11\frac{3}{8}$								$\frac{1}{8}$d.	Sixty days' bills on London were steady in September at $9\frac{7}{8}$@$10\frac{3}{8}$ per cent. premium; in October, $9\frac{3}{4}$@$10\frac{1}{4}$; in November, the same; in December, the range was from 9@10; in January, $8\frac{1}{2}$@$9\frac{1}{4}$; in February, $8\frac{1}{4}$@$9\frac{1}{8}$; in March, $8\frac{1}{2}$@9; in April, $8\frac{5}{8}$@$9\frac{1}{2}$; in May, $9\frac{1}{4}$@$9\frac{3}{4}$; in June, $9\frac{3}{8}$@$9\frac{7}{8}$; in July, $9\frac{1}{2}$@$9\frac{7}{8}$; and in August, $9\frac{5}{8}$@10.
"	5..	$11\frac{3}{4}$	$11\frac{1}{4}$	8,500	9,132	1,278	....	859	....	2,137		
"	8..	$11\frac{3}{4}$	$11\frac{1}{4}$									
"	12..	$11\frac{3}{4}$	$11\frac{1}{4}$	7,000	6,084	342	982	737	....	2,061		
"	15..	$11\frac{1}{2}$	11									
"	19..	$11\frac{1}{2}$	11	6,600	18,628	692	76	500	....	1,268		
"	22..	$11\frac{3}{8}$	$10\frac{7}{8}$									
"	26..	$11\frac{3}{8}$	$10\frac{7}{8}$	5,000	2,337	1,184	185	1,406	....	2,775		
"	29..	$11\frac{1}{4}$	$10\frac{3}{4}$									
July	2..	$11\frac{1}{4}$	$10\frac{3}{4}$	2,400	8,265	1,937	975	44	....	2,956	5-32d.	
"	6..	$11\frac{1}{4}$	$10\frac{3}{4}$									
"	10..	$11\frac{1}{4}$	$10\frac{3}{4}$	3,200	11,428	1,360	547	263	40	2,210		
"	13..	$11\frac{1}{4}$	$10\frac{3}{4}$									
"	17..	$11\frac{1}{4}$	$10\frac{3}{4}$	6,000	3,737	741	....	1,102	59	1,902		
"	20..	11	$10\frac{1}{2}$									
"	24..	11	$10\frac{1}{2}$	6,000	3,590	1,636	1,286	....	....	2,922		
"	27..	11	$10\frac{1}{2}$									
"	31..	11	$10\frac{1}{2}$	7,500	6,073	....	1,285	....	767	2,052		
August	3..	11	$10\frac{1}{2}$								3-16d.	
"	7..	11	$10\frac{1}{2}$	5,500	3,067	35	....	980	771	1,786		
"	10..	11	$10\frac{1}{2}$									
"	14..	$11\frac{1}{8}$	$10\frac{5}{8}$	7,500	5,604	376	1,119	....	450	1,945		
"	17..	$11\frac{1}{4}$	$10\frac{3}{4}$									
"	21..	$11\frac{1}{4}$	$10\frac{3}{4}$	10,800	4,703	1,146	2,210	40	....	3,396		
"	24..	$11\frac{1}{4}$	$10\frac{3}{4}$									
"	28..	$11\frac{1}{4}$	$10\frac{3}{4}$	7,300	1,500	934	....	....	....	934		
"	31..	$11\frac{1}{4}$	$10\frac{3}{4}$								7-32@$\frac{1}{4}$d	
Average prices and total sales, receipts and exports.		11.61	11.00	445,400	463,433	117,630	35,110	39,916	6,802	199,458		

LIVERPOOL STATEMENT FOR 1860.

UNITED STATES, 1859–1860.

Stock Sept. 1, 1859	149,000	Export	3,774,000
		Consumption	823,000
Crop	4,676,000	Stock Sept. 1, 1860	228,000
Bales	4,825,000	Bales.	4,825,000

Stock 1st Jan., 1860, in	Gt. Britain.	France.	Continent.	Tot. Europe.
United States ....Bales	306,000	38,000	30,000	374,000
Brazil	31,000	1,000	1,000	33,000
West Indies	1,000	....	3,000	4,000
East Indies	116,000	2,000	26,000	144,000
Egypt	16,000	....	....	16,000
Bales	470,000	41,000	60,000	571,000

CONSUMPTION. / IMPORT.

Tot. Europe.	Continent.	France.	Gt. Britain.	IMPORT.	Gt. Britain.	France.	Continent.	Tot. Europe.
3,481,000	688,000	551,000	2,242,000	United States ....Bales	2,582,000	610,000	707,000	3,648,000
125,000	9,000	3,000	113,000	Brazil	103,000	2,000	10,000	106,000
46,000	14,000	26,000	6,000	West Indies	10,000	26,000	11,000	47,000
524,000	340,000	8,000	176,000	East Indies	563,000	12,000	344,000	573,000
145,000	16,000	33,000	96,000	Egypt	110,000	35,000	17,000	158,000
4,321,000	1,067,000	621,000	2,633,000	Bales	3,368,000	685,000	1,089,000	4,532,000
....	....	....	610,000	Export.				
782,000	82,000	105,000	595,000	Stock, Dec. 31 ....Stock above,	470,000	41,000	60,000	571,000
5,103,000	1,149,000	726,000	3,838,000	Total supply, bales	3,838,000	726,000	1,149,000	5,103,000

COTTON AT LIVER

Week Ending.	Receipts.						Sales.			
	American	E. I.	Egypt.	Brazil.	Other.	Total.	Consumption.	Speculation.	Export.	Total.
Jan. 6..	111,063	16,608	4,213	10,390	34	142,308	53,680	1,890	11,430	67,000
" 13..	103,157	8,347	1,268	4,210	296	117,278	63,515	5,010	7,517	76,042
" 20..	103,471	4,320	3,399	....	9	111,199	83,846	14,350	5,584	103,780
" 27..	42,662	145	1,276	1,005	......	45,088	58,879	15,950	8,231	83,060
Feb. 3..	61,374	10,461	926	4,502	1,295	78,558	64,072	11,490	8,718	84,280
" 10..	81,642	5,976	......	13,871	......	101,489	44,340	16.680	16.940	77,960
" 17..	47,091	......	2,414	....	252	49,757	35,817	3,510	11,043	50,370
" 24..	33,432	19,240	1,784	618	......	55.074	51,100	2,540	6,720	60,360
Mch. 2..	171,803	......	9,007	2,002	29	182,841	50,208	1,860	8,502	60,570
" 9..	118,872	1,844	35,200	173	.....	156,089	41,718	960	4.472	47,150
" 16..	20,077	7,542	1,137	....	102	28,858	44,613	3,580	12,827	61,020
" 23..	73,005	17,958	4,166	1,371	18	96,518	48,768	5,230	9,762	63,760
" 30..	39,481	23,018	5.532	1,053	23	69,107	55,019	1,140	5,841	62,000
April 5..	67,920	22,199	3,491	20,470	......	114,080	34,672	1,140	4,738	40,550
" 13..	69,110	23,648	1,503	2,018	......	96,279	34,461	1,260	9,369	45,090
" 20..	95,534	5,159	8,627	2,800	50	112,170	49,416	3,370	6,624	59,410
" 27..	68,871	9,587	2,642	1,823	......	82,923	71,830	11,740	7,980	91,550
May 4..	32,855	18,011	3,149	510	230	54,755	61,165	4,480	7,815	73,460
" 11..	67,300	4,587	4 720	2,303	85	78,995	44,476	4,380	16,894	65,750
" 18..	104,930	22,467	4,461	1,051	31	132,940	30,774	2,720	14,346	47,840
" 25..	122,307	1,588	1,353	3,596	129	128,973	32,993	1,820	8.017	42,830
June 1..	138,231	4,106	3,781	3,326	48	149,492	42,739	1,050	10.971	54,760
" 8..	101,904	4.674	3,120	2,091	8	111,797	36,648	770	10,192	47,610
" 15.	26,864	2,205	33,088	940	17	63 114	31,478	1,710	16,662	49,850
" 22..	18 424	6,795	358	186	......	25,763	52,237	2,040	9,923	64,200
" 29..	64,305	8,811	1,335	6,734	......	81.185	47,229	6 980	10,061	64,270
July 6..	18.481	3,936	719	957	......	24,093	60,345	2,170	6,625	69,140
" 13..	4.885	......	389	....	......	5,274	47,520	5,640	22,700	75,860
" 20..	101,199	3,206	699	3,863	210	109,177	47,180	4,040	10,430	61,650
" 27..	40,571	2,728	1,480	....	......	44,779	32,142	1,610	13,938	47,690
Aug. 3..	9,055	4,209	5,099	5,484	60	23,907	53,713	5,920	11,787	71,420
" 10..	12,527	9,824	576	6,647	......	29,574	54,484	10,740	17,136	82,360
" 17..	10,177	18,437	......	3	2	28,619	54,964	15,960	19,506	90,430
" 24..	4,571	27,640	1,324	1,849	20	35,404	38,091	6,860	18,129	63,080
" 31..	3,591	10,201	88	1,289	10	15,179	31,880	2,160	17,720	51,760
Sep. 7..	2,769	4,122	......	1,914	112	8,917	47,199	28,950	18,771	94,920
" 14..	706	10.493	......	1,743	......	12,942	75,218	24,720	16,282	116,220
" 21..	9,616	43.759	648	2,076	183	56,282	33,418	14,900	7,992	56,310
" 28..	2,909	6,458	364	......	......	9,731	53,512	18,790	8,218	80,520
Oct. 5..	3.836	4,773	504	759	......	9,872	68,907	32,580	9,873	111.360
" 12..	2,517	13,651	1,513	1,395	1,378	20,454	55,917	20,610	15,593	92,120
" 19..	8,570	11,416	534	1,738	.....	22,258	54,707	17,470	5,963	78,140
" 26..	12,756	2,107	1,699	315	356	17,233	43,540	5,780	4,160	53,480
Nov. 2..	1,220	......	......	1,449	......	2.669	57,340	46,970	7,160	111,470
" 9..	8.652	.....	937	....	......	9,589	31,457	41.320	12,693	85,470
" 16..	2,613	2,006	817	891	40	6,367	19,980	11,940	12.240	44,160
" 23..	13,089	4,388	1,255	3,977	1,232	23,941	22,441	3,885	6.544	32,870
" 30..	8,107	3,138	158	10	128	11,541	57,746	8.730	4,464	70 940
Dec. 7..	97,158	2,352	2,982	852	355	103,699	40,297	3,220	5.843	49,360
" 14..	38,610	8,112	2,934	4,833	196	54,685	44,387	7,450	6.373	58,510
" 21..	46,438	3,587	4,387	517	79	55,008	97,817	31,930	3,673	133,420
" 28..	41,830	.....	3,771	1,933	20	47,554	50,301	20.180	1,799	72,280
Average prices & total sales, receipts & stocks.	2,492,138	449839	174847	131,537	7,037	3,255,398	2,540,196	522,175	526791	3,589,162

POOL. YEAR 1860.

Stocks.			Prices.			Actual Export.	Consumption.	Remarks.
American.	Other.	Total.	Mid. Up.	Mid. Orl.	Dhol.			
366,329	160,401	526,730	6½	6⅞	4¾	11,430	53,680	
418,492	157,519	576,011	6 5/16	6⅝	4¾	7,517	117,195	
452,303	153,504	605,807	6 7/16	6⅞	4¾	5,584	201,041	
446,273	140,851	587,124	6½	6⅞	4¾	8,231	259,920	
449,304	145,230	594,534	6 7/16	6⅞	4⅞	8,718	323,992	
476,993	139,470	616,463	6 7/16	6⅞	4⅞	16,940	368,332	
483,647	129,370	613,017	6 7/16	6⅞	4⅞	11,043	404,149	
478,879	141,122	620,001	6 7/16	6 13/16	4¾	6,720	455,249	Large receipts at ports in United States.
609,534	140,226	749,760	6 3/16	6 13/16	4¾	8,502	505,457	
696,543	133,166	829,709	6¼	6⅜	4⅝	4,472	547,175	
674,858	126,351	801,209	6¼	6 11/16	4⅜	12,827	591,788	
702,832	136,563	839,395	6 5/16	6¾	4⅝	9,762	640,556	
691,412	155,189	846,601	6¼	6⅜	4⅝	5,841	695,575	
731,272	174,388	905,660	6 3/16	6¾	4½	4,738	730,247	
766,081	189,159	955,240	6 1/16	6½	4½	9,369	764,708	
825,295	190,611	1,015,906	6	6½	4¼	6,624	814,124	
836,099	190,780	1,026,879	6 1/16	6⅜	4¼	7,980	885,954	
818,021	198,201	1,016,222	6⅛	6⅜	4¼	7,815	947,119	
833,626	193,107	1,026,733	6⅛	6⅜	4¼	16,894	991,595	
904,542	206,325	1,110,867	6⅛	6¾	4⅜	14,346	1,022,369	
995,751	204,582	1,200,333	6⅛	6⅜	4⅜	8,017	1,055,362	Unsettled state of foreign politics.
1,088,891	206,303	1,295,194	6	6⅜	4⅜	10,971	1,098,101	
1,154,240	203,999	1,358,239	5⅞	6½	4¼	10,192	1,134,749	
1,137,197	197,434	1,334,631	5¾	6	4⅛	16,662	1,166,227	
1,114,004	193,467	1,307,471	5⅝	6¼	4⅛	9,923	1,218,464	
1,132,412	201,003	1,333,415	5½	6	4⅛	10,061	1,265,693	
1,102,530	195,563	1,298,093	5½	6	4	6,625	1,326,038	
1,049,553	178,034	1,227,587	5½	5¾	4	22,700	1,373,558	Rumors of possible failures, &c.
1,110,132	176,988	1,287,120	5 9/16	6	4	10,430	1,420,738	
1,113,751	169,080	1,282,831	5 7/16	5¾	3⅞	13,938	1,452,880	
1,076,420	164,551	1,240,971	5 7/16	5¾	3¾	11,787	1,506,593	
1,035,994	167,355	1,203,349	5½	6	3¾	17,136	1,561,077	
986,409	170,793	1,157,202	5¾	6½	3⅞	19,506	1,616,041	Reported injury done to crops by unusually hot and dry weather.
940,558	187,249	1,127,807	5⅞	6⅝	4	18,129	1,654,132	
906,403	181,383	1,087,786	5⅞	6⅛	4	17,720	1,686,012	
855,268	166,664	1,021,932	6	6¼	4⅛	18,771	1,733,211	
784,075	157,307	941,382	6⅛	6⅜	4⅜	16,282	1,808,429	
759,454	195,388	954,842	6⅛	6⅜	4⅜	7,992	1,841,847	
715,918	186,837	902,755	6⅛	6⅜	4⅜	8,218	1,895,359	
660,541	173,673	834,214	6¼	6½	4½	9,873	1,964,266	
616,396	169,656	786,052	6 3/16	6½	4⅜	15,593	2,020,183	
580,991	173,346	753,437	6½	6⅜	4¾	5,963	2,074,890	Injuring storms in Southern States.
558,361	168,399	726,760	6½	6⅝	4¾	4,160	2,118,430	
511,448	156,098	667,546	6¾	7	5	7,160	2,175,770	
477,144	140,898	618,042	6⅞	6¼	5⅛	12,693	2,207,227	
447,505	135,415	582,920	6⅜	7	5	12,240	2,227,207	
436,674	138,482	575,156	6¾	6 15/16	5	6,544	2,249,648	
390,640	134,013	524,653	6 13/16	7	5	4,464	2,307,394	Cessation of hostilities in China.
449,754	129,405	579,159	6 11/16	6⅞	5	5,843	2,347,691	
443,918	137,053	580,971	6 11/16	6⅞	5	6,373	2,392,078	
404,603	137,098	541,701	6⅞	7⅛	5¾	3,673	2,489,895	
402,891	138,595	541,486	7⅛	7⅜	5¼	1,799	2,540,196	
			5.97	6.53	4.50	526,791	48,849.92	

1861.

The population of Great Britain was 23,266,755, and the number of spindles about 33,000,000; an average of 1,418 spindles to 1,000 inhabitants. (*See* years 1840, 1850, and 1860.)

During this year the production of cotton was as follows, in the designated sections:

Greece	8,300	lbs.
Turkish Dominions	12,660	"
Egypt	7,302,160	"
Western Coast of Africa	27,780	"
Mauritius	145,760	"
Madras	3,513,640	"
Bengal	9,240	"
New Granada	27,660	"
Brazil	3,087,560	"

War with the South. (*See* year 1862.)

COTTON CROP OF THE UNITED STATES.

Statement and Total Amount of the Cotton Crop of the United States, for the Year ending August 31, 1861.

	Bales.	Bales.	TOTAL.		
			1861.	1860.	1859.
LOUISIANA.					
Export from New Orleans—					
To Foreign Ports	1,783,673				
Coastwise	132,179				
Burnt at New Orleans	3,276				
Stock on hand 1st September, 1861	10,118				
		1,929,246			
Deduct—					
Received from Mobile	48,270				
" Montgomery, &c	11,551				
" Florida	13,279				
" Texas	30,613				
Stock on hand 1st September, 1860	73,934				
		177,647			
			1,751,599	2,139,425	1,669,274
ALABAMA.					
Export from Mobile—					
To Foreign Ports	456,421				
Coastwise	127,574				
Manufactured in Mobile (estimated)	2,000				
Stock on hand 1st September, 1861	2,481				
		588,476			
Deduct—					
Stock on hand 1st September, 1860	....	41,682			
			546,794	843,012	704,406
TEXAS.					
Export from Galveston, &c.—					
To Foreign Ports	63,209				
Coastwise	84,254				
Stock on hand 1st September, 1861	452				
		147,915			
Deduct—					
Stock on hand 1st September, 1860	....	3,168			
			144,747	252,424	192,062
FLORIDA.					
Export from Apalachicola, St. Marks, &c.—					
To Foreign Ports	28,073				
Coastwise	85,953				
Burnt at St. Marks	150				
Stock on hand 1st September, 1861	7,860				
		122,036			
Deduct—					
Stock on hand, 1st September, 1860	....	864			
			121,172	192,724	173,484
GEORGIA.					
Export from Savannah—					
To Foreign Ports—Uplands	293,746				
" Sea Island	8,441				
Coastwise—Uplands	170,572				
Sea Island	11,512				
Stock in Savannah, 1st September, 1861	4,102				
" Augusta, etc., 1st August, 1861	5,991				
		494,364			

Statement and Total Amount of the Cotton Crop of the United States, for the Year ending August 31, 1861—Concluded.

	Bales.	Bales.	TOTAL. 1861.	1860.	1859.
Deduct—					
Received from Florida—Sea Island	1,033				
" " Uplands	6,188				
Stock in Savannah, 1st September, 1860	4,307				
" Augusta, &c., 1st September, 1860	5,252				
		16,780	477,584	525,219	475,788
SOUTH CAROLINA.					
Export from Charleston and Georgetown—					
To foreign ports—Uplands	199,345				
" Sea Island	15,043				
Coastwise—Uplands	121,663				
Sea Island	8,355				
Burnt at Charleston	564				
Stock in Charleston, 1st September, 1861	2,899				
		347,869			
Deduct—					
Received from Florida and Savannah— Sea Island	255				
" " Uplands	2,378				
Stock in Charleston, 1st September, 1860	8,897				
		11,530	336,339	510,109	480,653
NORTH CAROLINA.					
Export—					
To Foreign ports	195				
Coastwise	56,100				
		....	56,295	41,194	37,482
VIRGINIA.					
Export—					
To Foreign Ports	810				
Coastwise	61,129				
Manufactured—taken from the ports	16,993				
Stock on hand 1st September, 1861	2,000				
		80,932			
Deduct—					
Stock on hand 1st September, 1860	......	2,800	78,132	56,987	33,011
TENNESSEE, &c.					
Shipments from Memphis, Tenn	369,857				
" " Nashville, Tenn	16,471				
" " Columbus and Hickman, Ky.	5,500				
Stock at Memphis, 1st September, 1861	1,671				
		393,499			
Deduct—					
Shipments to New Orleans	196,366				
Manufactured on the Ohio, &c.	52,000				
Stock on hand 1st September, 1860	1,709				
		250,075			
			143,424	108,676	85,321
Total crop of the United States	....	....	3,656,086	4,669,770	3,851,481

Export to Foreign Ports, from September 1, 1860, *to August* 31, 1861.

FROM	To Great Britain.	To France.	To North of Europe.	Other Foreign Ports.	Total.
New Orleans, La. (bales)	1,159,348	388,925	122,042	113,358	1,783,673
Mobile, Ala.	340,845	96,429	6,601	12,546	456,421
Galveston, Tex	47,229	3,640	12,315	25	63,209
Florida	27,140	....	933	....	28,073
Savannah, Ga.	282,994	10,061	6,165	2,967	302,187
Charleston, S. C.	136,513	29,886	24,401	23,588	214,388
Virginia	810	....	....	....	810
North Carolina	144	....	....	51	195
New York	158,415	49,122	35,197	5,315	248,049
Baltimore	975	....	2,483	87	3,545
Philadelphia	3,793	....	....	....	3,793
Boston	17,019	....	6,113	93	23,225
Grand total	2,175,225	578,063	216,250	158,030	3,127,568
Total last year	2,669,432	589,587	295,072	220,082	3,774,173
Decrease	494,207	11,524	78,822	62,052	646,605

Consumption.

Total crop of the United States, as before stated .... bales.			3,656,086
Add—Stocks on hand at the commencement of the year, September 1, 1860, in the Southern ports		142,613	
" Northern ports		85,095	
			227,708
Makes a supply of			3,883,794
Deduct therefrom—The export to Foreign ports	3,127,568		
Less—Foreign included	701		
		3,126,867	
Stocks on hand September 1, 1861—			
In the Southern ports	37,574		
In the Northern ports	45,613		
		83,187	
Burnt at New Orleans, St. Marks, Charleston and Philadelphia,	4,390		
Manufactured in Virginia and Mobile	18,993		
		23,383	
			3,233,437
Taken for home use north of Virginia .... bales.			650,357
Taken for home use in Virginia, and south and west of Virginia			193,383
Total consumed in the United States (including burnt at the ports), 1860–61			843,740

ANNUAL REVIEW.

From the New Orleans Price Current—1860–61.

The sales in January amounted to 320,000 bales, while the receipts comprised 393,200 bales, and the nominal stock had been reduced to 247,566 bales. We say the nominal stock, because towards the close of the month heavy clearances, amounting, during the week ending on the 27th, to 242,367 bales, had been made in anticipation of the custom-house being transferred from the old Federal Government to that of the Confederate States, and a vague apprehension that there might be some delay or disadvantage in clearance, after the change. Many vessels, indeed, cleared with hardly a bale of cotton on board, but specifying as cargo the amount engaged, or on the wharf for shipment. The blockade was fully enforced on the 10th of June, since which, operations have been confined to filling small orders for Southern manufactures, or picking up trifling lots offered, to close consignments, at very low figures.

LIVERPOOL STATEMENT FOR 1861.

UNITED STATES—1860–1861.

Stock Sept. 1, 1860	228,000	Export	3,128,000
		Consumption	673,000
Crop	3,656,000	Stock Sept. 1, 1861	83,000
Bales	3,884,000	Bales	3,884,000

Stock Jan. 1, 1861, in	Gt. Britain.	France.	Continent.	Tot. Europe.
United States (Bales)	395,000	97,000	49,000	541,000
Brazil	12,000	....	2,000	14,000
West Indies	4,000	....	1,000	5,000
East Indies	157,000	6,000	30,000	193,000
Egypt	27,000	2,000	....	29,000
Bales	595,000	105,000	82,000	782,000

CONSUMPTION. / IMPORT.

Tot. Europe.	Continent.	France.	Gt. Britain.		Gt. Britain.	France.	Continent.	Tot. Europe.
2,732,000	547,000	494,000	1,691,000	United States	1,842,000	521,000	516,000	2,620,000
88,000	5,000	1,000	82,000	Brazil	99,000	1,000	4,000	102,000
40,000	4,000	22,000	14,000	West Indies	11,000	22,000	5,000	37,000
794,000	420,000	19,000	355,000	East Indies	986,000	19,000	408,000	999,000
177,000	24,000	42,000	111,000	Egypt	97,000	41,000	27,000	163,000
3,831,000	1,000,000	578,000	2,253,000	Bales	3,035,000	604,000	960,000	3,921,000
....	....	....	678,000	Export.				
872,000	42,000	131,000	699,000	Stock Dec. 31. / Stock above,	595,000	105,000	82,000	782,000
4,703,000	1,042,000	709,000	3,630,000	Total supply, bales	3,630,000	709,000	1,042,000	4,703,000

COTTON AT LIVER

Week Ending.	RECEIPTS.						SALES.			
	American	E. I.	Egypt.	Brazil.	Other	Total.	Consumption.	Speculation.	Export	Total.
Jan. 4..	9,224	......	725	....	45	9,994	18,544	1,640	2,706	22,890
" 11..	45,664	7,728	5,764	1,263	671	61,090	47,683	8,540	3,197	59,420
" 18..	58,483	......	1,047	....	209	59,739	93,428	36,590	1,532	131,550
" 25..	103,186	9,854	1,320	4,742	91	119,193	56,623	29,230	1,787	87,640
Feb. 1..	71,278	11,296	2,256	2,370	31	87,231	30,115	2,310	1,975	34,400
" 8..	95,059	2,191	2,014	....	.....	99,264	39.355	3,700	5,735	48,790
" 15..	32,978	......	1,738	....	42	34,758	22,473	2,670	6,977	32,120
" 22..	92,319	......	2,270	2,798	699	98,086	10,728	1,970	13,202	25,900
March 1..	110,824	.....	6,375	1,839	6	119,044	56,312	2,300	5,278	63,890
" 8..	69,365	14,388	3,802	1,142	346	89,043	49,581	16,030	15,949	81,560
" 15..	72,756	......	1,810	3,632	.....	78,198	33,395	9,570	15,165	58,130
" 22..	66,420	9,412	1,434	....	.....	77,266	97,938	26,210	8,772	132,920
" 29..	67,707	......	866	1,779	830	71,182	48,384	12,620	7,866	68,870
April 5..	64,246	6,690	2,695	1,553	20	75,204	21,139	6,080	19,141	46,360
" 12..	12,379	10,774	4,125	1,707	13	28,998	52,428	18,850	23,752	95,030
" 19..	21,015	22,899	2,257	800	30	47,001	43,291	15,400	9,949	68,640
" 26..	117,395	22,573	5,100	386	99	145,553	56,276	13,050	10,744	80,070
May 3..	87,877	3,794	2,245	2,604	.....	96,520	57,985	23,350	8,525	89,860
" 10..	49,102	13,792	4,285	....	10	67,189	49,451	20,950	23,529	93.930
" 17..	79,300	23,016	1,762	3,797	41	107.916	30,501	9,390	7,259	47,150
" 24..	93,674	21,582	4,380	5,287	205	125,128	43,617	18,960	17,943	80,520
" 31..	41,272	52,616	1,823	1,446	.....	97,157	45,022	10,220	12,018	67,260
June 7..	45,439	13,025	3,112	....	238	61.814	56,792	16,590	5,118	78,500
" 14..	19,413	21,905	2,980	....	8	44,306	30,780	8,450	19,290	58,520
" 21..	8,601	15,330	795	614	.....	25,340	31,387	10,400	13,033	54,820
" 28..	42,719	31,679	4,626	2,402	206	81,632	48,326	16,910	14,714	79,950
July 5..	71,545	6,588	1,594	......	.....	79,727	97,004	49,330	12,216	158,550
" 12..	42,189	6,446	179	2,912	50	51,776	52,916	13,200	14,874	80.990
" 19..	17,763	9,878	1,640	2,405	.....	31,686	77,275	22,610	13,235	113,120
" 26..	21,893	28,263	2,741	3,175	.....	56,072	85.434	37,680	21,176	144,290
Aug. 2..	34,255	40,208	1,506	12,304	21	88,294	36,056	12,200	21,254	69,510
" 9..	3,226	13,660	671	3,924	49	21,530	24.586	19,200	19,294	63,080
" 16..	....	3,706	584	2,304	37	6,631	20,390	5,580	20,220	46,190
" 23..	165	25,095	1,430	5,779	343	32,812	56.326	20,930	12,614	89,870
" 30..	....	35,324	1,236	549	15	37,124	58,434	49,770	11,896	120,100
Sep. 6..	....	46,657	1,018	695	386	48,756	42,188	24,510	10,792	77.490
" 13..	75	33,270	530	50	605	34,455	33,832	16,960	15,638	66,430
" 20..	24	15,253	464	......	138	15,879	76,540	56.850	9,860	143,250
" 27..	6	2,155	251	1,518	122	4,052	74,597	84,560	13,203	172,360
Oct. 4..	5	4,129	563	....	54	4,751	25,109	28,910	13.311	67,330
" 11..	....	17,966	241	1,293	661	20,161	40,350	63,790	16,500	120,640
" 18..	48	14,006	721	6,572	137	21,484	44,673	55,980	15,847	116,500
" 25..	....	52,317	1,571	285	228	54,401	49,342	82,090	14,378	145,810
Nov. 1..	....	11	17	....	.....	28	27,391	28,280	6,879	62,550
" 8..	408	66,627	86	675	90	67,886	17,281	36,290	6,769	60 340
" 15..	46	24,258	576	2,264	163	27,307	44,646	50,530	10,194	105,370
" 22..	588	......	......	491	.....	1,079	15,038	20,490	5,452	40,980
" 29..	789	18,355	100	2,303	106	21,653	12,671	4,520	4,639	21.830
Dec. 6..	298	32,403	505	4,886	25	38,117	12,807	9,170	6,913	28,890
" 13..	352	11,341	4,044	1,652	505	17,894	21,478	8,800	5,142	35.420
" 20..	....	......	1,919	617	12	2,548	17,981	7,360	2,989	28,330
" 27..	....	400	......	30	.....	430	13,146	9,520	8,464	31,130
" 31..	....	......	......	....	.....	3,109	28,140	11,430	4,750	44,320
Average prices & total sales, receipts & stocks.	1,711,370	823660	95.898	83,850	8,087	2,725,954	2,277,195	1,173,160	605453	4,055,808

POOL. YEAR 1861.

Stocks.			Prices.			Actual Export.	Consumption.	Remarks.
Amer'n	Other.	Total.	Mid. Up.	Mid. Orl.	Dhol.			
371,644	157,829	529,473	7 1/16	7 1/4	5 1/4	2,706	18,544	
373,504	166,752	540,256	7 1/16	7 5/16	5 1/4	3,197	66,227	
364,572	155,301	519,873	7 1/4	7 1/2	5 3/8	1,532	159,655	
431,796	163,933	595,729	7 1/4	7 1/2	5 1/2	1,787	216,278	
484,779	175,046	659,825	7 1/8	7 3/8	5 3/8	1,975	246,393	Cotton advancing in N. O.
546,948	172,486	719,434	7	7 1/4	5 1/4	5,735	285,748	Heavy receipts.
558,342	168,153	726,495	6 3/4	7	5 1/8	6,977	308,221	
628,543	162,716	791,259	6 7/16	6 13/16	5	13,202	318,949	
693,536	160,969	854,505	6 1/2	6 7/8	5	5,278	375,261	
717,972	163,357	881,329	6 5/8	6 7/8	5 1/8	15,949	424,842	
750,535	156,357	906,892	6 11/16	6 15/16	5 1/8	15,165	458,237	
738,627	156,329	894,956	7	7 1/4	5 3/8	8,772	556,175	
768,074	152,078	920,152	7 1/16	7 5/16	5 1/2	7,866	604,559	
790,571	151,814	942,385	7 1/16	7 5/16	5 3/8	19,141	625,698	
735,944	148,967	884,911	7 1/4	7 7/16	5 3/8	23,752	678.126	
712,462	161,691	874,153	7 3/8	7 9/16	5 1/2	9,949	721,417	Receipts fell off in the S.
777,457	176,535	953,992	7 7/16	7 5/8	5 1/2	10 744	777,693	Rumors of war in the United States.
818.196	172,541	990,737	7 11/16	7 7/8	5 1/2	8,525	835,688	
803,097	172,760	975,857	7 11/16	7 7/8	5 3/8	23,529	885,139	Civil war in United States commenced.
859,027	189,587	1,048,614	7 1/2	7 3/4	5 3/8	7,259	915,640	
903.640	206,919	1,110,559	7 5/8	7 7/8	5 3/8	17,943	959,257	
899,261	250,807	1,150,068	7 5/8	7 7/8	5 3/8	12,018	1,016,049	
898,199	249,405	1,147,604	7 11/16	7 7/8	5 1/4	5,118	1,046.829	
873,064	256,976	1,130,040	7 11/16	7 7/8	5 1/8	19,290	1,078,216	
843,929	261,878	1,105.807	7 11/16	7 7/8	5 1/8	13,033	1,126,542	
834,549	288,473	1,123,022	7 3/4	8	5	14.714	1,223,546	
838,054	269,209	1,107,263	7 15/16	8 1/4	5 1/4	12,216	1,276,462	
841,168	253,897	1,095,065	7 15/16	8 1/4	5 3/8	14,874	1,353,737	
798,656	254,010	1,052,666	8 1/16	8 5/16	5 1/2	13,235	1,439,171	Warlike accounts from United States.
745,776	254,256	1,000,032	8 1/4	8 1/2	5 5/8	21,176	1,475,227	
737,999	280.943	1,018,942	8 1/4	8 1/2	5 5/8	21,254	1,499,813	
709,837	279,071	988,908	8 5/16	8 5/8	5 5/8	19,294	1,520,203	
678,909	264,420	943,329	8 5/16	8 5/8	5 5/8	20,220	1,576,529	
630.741	281,226	911,967	8 1/2	8 3/4	5 5/8	12,614	1,634,963	
588,510	297,565	886,075	8 3/4	9	5 3/4	11,896	1,677,151	
553,366	332,273	885.639	8 7/8	9 1/8	5 3/4	10,792	1,722,173	
521,176	346,035	867,211	9	9 1/4	5 3/4	15,638	1,756.005	Unfavorable accounts from United States.
487,301	329,959	817,260	9 1/4	9 9/16	5 7/8	9,860	1,832,545	
444,859	305,700	750,559	9 3/4	10	6 1/4	13.203	1,907,142	
418,823	293,066	711,889	9 7/8	10 1/8	6 1/8	13,311	1,932,251	
386,453	288,817	675,270	10 1/4	10 1/2	6 3/8	16,500	1,972,601	No hope of early settlement of war.
350,696	280,881	631,577	10 3/4	10 7/8	6 5/8	15,847	2 017,274	
321,642	300,327	621,969	11 5/8	11 7/8	7 1/8	14,378	2,066,616	War news indicates long struggle.
311,428	275,290	586,718	11 1/2	11 7/8	7 1/8	6,879	2,094.007	
300,304	330,791	631,095	11 5/8	12	7 1/4	6,769	2,111,288	
281,521	324,197	605,718	11 5/8	12	7 1/2	10,194	2,155,934	
273,594	311,231	584,825	11 3/8	11 7/8	7 1/2	5,452	2.170.972	
264,117	325 091	589,208	11	11 1/2	7 1/4	4,639	2,183,643	News of the "Trent" affair caused decline.
253,677	351,095	604,772	10 1/8	10 5/8	6 3/4	6,913	2,196,450	
240,434	354,450	594,884	10 3/8	10 7/8	6 3/4	5,142	2,217,928	
230,777	348,630	579,413	10 1/2	10 7/8	6 5/8	2,989	2,235,909	
216,317	340,172	556,489	11 1/2	11 7/8	6 3/4	8,464	2,249.055	More peaceable accounts.
....	....	622,565	11 1/2	11 7/8	6 3/4	4,750	2,277,195	
			8.5	8.8	5.61	605,453	42,965.75	

The semi-weekly Price and Weekly Sales and Receipts, at New York, Weekly Exports from New York and Rates of Freight to Liverpool 1st of each Month, for the Crop Year ending September 1, 1861.

1860.	Price of Middling New Orleans New York Classificat'n.	Price of Middling Upland, New York Classificati'n	Sales for week, including lots in transit.	Receipts for week, including lots in transit.	EXPORTS FOR WEEK. To Great Britain.	To France.	North of Europe.	Other Fo'n Ports	Total Exports.	Rates of Freight to Liverpool.
Septem. 4..	11¼	10¾	5,500	4,114	....	....	....	....	....	¼d.
" 7..	11¼	10¾								
" 11..	11¼	10¾	7,300	1,802	457	....	....	....	457	
" 14..	11¼	10¾								
" 18..	11⅛	10⅝	6,700	4,256	1,378	1,527	60	134	3,099	
" 21..	11⅛	10⅝								
" 25 .	11⅛	10⅝	12,500	6,995	1,520	922	129	150	2,721	
" 28..	11⅛	10⅝								
October 2..	11¼	10¾	13,600	4,862	3,479		800	....	5,142	7-32d.
" 5..	11¼	10¾				863				
" 9..	11⅜	10⅞	20,000	7,504	2,935	....	256	....	3,191	
" 12..	11⅜	10⅞								
" 16..	11⅝	11⅛	28,000	8,923	2,452	2,945	722	121	6,240	
" 19..	11⅞	11⅜								
" 23..	12⅛	11⅝	34,000	9,779	7,173	....	351	....	7,524	
" 26..	12⅛	11⅝								
" 30..	12⅛	11⅝	14,500	23,623	2,401	....	547	....	2,948	
Novem. 2..	12⅛	11⅝								¼d.
" 6..	12⅛	11⅝	9,000	17,427	5,340	555	559	150	6,604	
" 9..	12⅛	11⅝								
" 13..	12⅛	11⅝	16,000	12,482	4,091	1,896	888	....	6,875	
" 16..	12	11½								
" 20..	11¾	11¼	12,000	20,021	5,238	1,378	1,140	....	7,756	
" 23..	11¾	11¼								
" 27..	11⅛	10⅝	9,000	17,215	2,937	....	907	....	3,844	
" 30..	11	10½								
Decem. 4..	10¾	10¼	6,500	11,411	1,964	819	1,252	1,000	5,035	¼@9-32d
" 7..	10½	10								
" 11..	10½	10	6,200	13,471	4,437	....	892	132	5,461	
" 14..	10⅝	10⅛								
" 18..	11	10½	14,000	11,052	6,042	3,013	1,360	....	10,415	
" 21..	11¼	10¾								
" 25..	11½	10⅞	9,500	6,604	6,836	....	1,411	84	8,331	

GENERAL REMARKS.

This crop year ushered in the American civil war, though it was not until towards its close that the gravity of the crisis seemed to be fairly estimated.

In September, 1860, this market remained very dull, the English food crop prospects were discouraging, business in Liverpool was very stagnant and the supply there of cotton was "enormous, being over a million of bales."

In October, the Foreign advices were more cheerful, and accounts from the Gulf States were received, noting much damage to the crop by a hurricane, which imparted a feeling of excitement to the market; a further impetus was given to the advancing prices by cold weather and light frosts at the South, as early as the 17th of October, and sustained prices until about the middle of November, when political affairs here began to assume a serious aspect. Foreign and Domestic Exchanges became deranged, great caution and stringency obtained in financial circles, and specie payments were suspended in Philadelphia, Baltimore, and some of the Western and Southern cities.

Decem. 28..	11 3/4	11 1/8									Stagnation continued throughout December, but up to this time, in Europe, very little stress had been laid upon the accounts relative to our political troubles, they being rather incredulous as to any serious results.
1861.											
January 1..	12 5/8	12	15,000	7,995	5,276	1,556	536	....	7,368	3/8d.	
" 4..	13	12 1/2									
" 8..	13 1/4	12 7/8	28,000	10,368	5,895	1,737	594	....	8,226		
" 11..	13 1/4	12 7/8									
" 15..	13 1/8	12 3/4	12,600	12,811	5,771	....	250	1,058	7,079		
" 18..	12 5/8	12 1/4									In January, the English markets began to feel the effect of the American accounts, and bought freely; the demand there being also strenghtened by the conclusion of peace between England and China; money here was now more accessible, and. with free receipts of specie, the market became buoyant, notwithstanding the Southern States were passing ordinances of secession.
" 22..	12 5/8	12 1/4	14,500	20,039	3,028	2,912	78	....	6,018		
" 25..	12 5/8	12 1/4									
" 29..	12 5/8	12 1/4	23,000	9,073	3,548	820	2,128	....	6,496		
February 1..	12 5/8	12 1/4								3/8d.	
" 5..	12 1/2	12 1/8	16,000	19,516	7,366	918	324	....	8,608		
" 8..	12 3/8	12									
" 12..	12 3/8	11 3/4	7,700	16,057	3,541	2,205	1,457	....	7,203		
" 15..	12 3/8	11 3/4									
" 19..	12 3/8	11 3/4	8,800	21,832	5,424	3,767	1,898	....	11,089		
" 22..	12 5/8	11 1/2									
" 26..	12 1/2	11 7/8	16,000	20,965	4,754	5,406	790	47	10,997		
March 1..	12 1/2	11 7/8								3/8d.	February and March, there was but little activity, the Liverpool market becoming very dull, upon the advance there in bank discounts.
" 5..	12 3/8	11 3/4	5,500	16,092	3,908	814	257	....	4,979		
" 8..	12 1/2	11 7/8									
" 12..	12 5/8	12	6,700	14,994	2,427	3,525	3,974	....	9,926		
" 15..	12 5/8	12									The bombardment and surrender of Fort Sumter in April, brought trade to a stand, and, amid much excitement, cotton was generally held out of the market, and the subsequent great rise in prices was now inaugurated.
" 19..	12 3/4	12 1/8	13,000	10,133	3,807	3,298	3,099	....	10,204		
" 22..	13	12 3/8									
" 26..	13 1/8	12 1/2	17,000	19,018	870	402	72	....	1,344		
" 29..	13 3/8	12 3/4									
April 2..	13 1/2	12 7/8	26,000	14,060	3,923	2,178	652	72	6,825	3-16d.	
" 5..	13 1/2	12 7/8									
" 9..	13 3/8	12 3/4	19,500	6,053	558	728	1,218	....	2,504		In April and May, the market continued in a feverish unsettled state, Foreign exchange was very difficult to negotiate, which interfered with exports, while money became very stringent.
" 12..	13 3/8	12 3/4									
" 16..	13 3/8	12 3/4	9,500	9,196	3,432	209	690	103	4,434		
" 19..	13 5/8	13									
" 23..	13 7/8	13 3/8	24,500	11,874	2,685	356	1,956	....	4,997		
" 26..	14 1/2	14									
" 30..	14 1/2	14	19,500	4,368	2,589	3	1,055	....	3,647		In May and June, the troubles here, caused a panic-feeling in Europe, as the demand for goods both from the Continent and England for the American markets was very light.
May 3..	14 1/4	13 3/4								7-32@1/4d	
" 7..	14 1/4	13 3/4	9,200	2,412	5,838	1,145	81	....	7,064		
" 10..	14 3/8	13 7/8									
" 14..	14 3/8	13 7/8	13,000	3,165	6,417	1,055	1,757	....	9,229		
" 17..	14 3/8	13 7/8									

New York Statement for 1861.—*Concluded.*

1861.		Price of Middling New Orleans New York Classificat'n.	Price of Middling Upland, New York Classificat'n.	Sales for week, including lots in transit.	Receipts for week, including lots in transit.	EXPORTS FOR THE WEEK. To Great Britain.	To France.	North of Europe.	Other Fo'n Ports	Total Exports.	Rates of Freight to Liverpool.
May	21..	14⅜	13⅞	8,500	371	3,957	846	303	....	5,106	
"	24..	14⅜	13⅞								
"	28..	14⅜	13⅞	9,500	343	3,621	783	....	....	4,404	
"	31..	14⅜	13⅞								
June	4..	14⅜	13⅞	5,000	878	3,978	172	53	....	4,203	5-32d.
"	7..	14⅜	13⅞								
"	11..	14⅜	13⅞	5,200	416	1,906	260	8	....	2,174	
"	14..	14½	14								
"	18..	14½	14	6,500	713	2,583	6	89	....	2,678	
"	21..	14⅝	14⅛								
"	25..	14¾	14¼	10,000	113	868	50	36	....	954	
"	28..	15	14½								
July	2.	15¼	14¾	7,200	....	460	....	274	400	1,134	5-32d.
"	5..	15¼	14¾								
"	9..	15¾	15¼	8,500	97	14	....	....	....	14	
"	12..	16¼	15¾								
"	15..	16¼	15¾	5,800	....	34	63	....	....	97	
"	19..	16¼	15¾								
"	23..	16¼	15¾	4,800	....	34	....	....	....	34	
"	26..	16¼	15¾								
"	30..	16½	16	8,000	2	165	....	....	1,003	1,168	
August	2..	16¾	16¼								5-32d.
"	6..	17½	17	11,000	....	....	....	....	....	....	
"	9..	18¼	17¾								
"	13..	18¾	18¼	13,500	....	....	....	....	800	800	
"	16..	18¾	18¼								
"	20..	18¾	18¼	4,500	213	....	....	....	....	....	
"	23..	18⅞	18⅜								
"	27..	19	18½	12,000	503	24	....	....	....	24	
"	30..	20	19½								
Septem.	3..	22½	22	17,000	50	....	50	....	....	50	Nominal
Average price and total sales, receipts and exports.		13.53	13.01	666,300	435,261	157,381	49,182	34,903	5,254	246,720	

General Remarks.

From this, until the close of the year, the business was mostly of a speculative character, prices for a time being higher here than in Liverpool, so that in August, an import of 494 bales was received from that port.

Exchange.

The occurrences noted above unsettled the market, and prices fluctuated considerably.

The range of bills on London in September was from 9¼@10 per cent. premium; in October, the rate dropped from 9⅛@9⅝, down to 8@8⅜; in November, it declined from 7¾@8⅛, to par to 4 per cent premium; in December, the range was from par to 5 per cent. premium; in January, the rate advanced from 2½ per cent. premium, to 5¾@7; in February, the range was from 5@7; in March, the extremes were 5@8 per cent. premium; April, opened at 7½@8½, and fell off to 5@6; in May, the range was from 3¾@6 per cent. premium; June, 4¾@6; in July, 5½@7¾; and in August, 6¼@7¾ per cent. premium.

1862.

During this year, the production of cotton was as follows, in the designated sections:

Greece	37,300	lbs.
Turkish dominions	824,240	"
Egypt	10,537,940	"
Western coast of Africa	67,600	"
Mauritius	353,760	"
Madras	6,708,640	"
Bengal	1,845,840	"
New Grenada	206,840	"
Brazil	4,167,680	"
China	293,900	"

War with the South. (*See* year 1861.)

The semi-weekly Price and Weekly Sales and Receipts at New York, Weekly Exports from New York and Rates of Freight to Liverpool 1st of each month, for the Crop Year ending September 1, 1862.

1861.	Price of Middling New Orleans	Price of Middling Upland.	Sales for week.	Receipts for week.	EXPORTS FOR THE WEEK. To Great Britain.	To France.	North of Europe.	Other Fo'n Ports	Total Exports.	Rates of Freight to Liverpool.
Septem. 6..	22½	22	16,600	250						
" 13..	22	21½								
" 17..	22	21½	2,500	992						
" 20.	22	21½								
" 24..	22	21½	2,000	582	95			2	97	
" 27..	22	21½								
October 1..	22	21½	1,800	418		8			8	
" 4..	22	21½								
" 8..	22	21½	4,800	1,699						
" 11..	22	21½								
" 15..	22½	22	5,500	957						
" 18..	22½	22								
" 22..	22½	22	3,300	266						
" 25..	22½	22								
" 29..	22½	22	4,200	945	200				200	
Novem. 1..	23	22½								$\frac{3}{16}$@$\frac{7}{32}$d.
" 5..	25	24½	12,500	887	300				300	
" 8..	25	24½								
" 12..	25	24½	7,000	1,010						
" 15..	25	24½								
" 19..	24½	24	2,200	1	562				562	
" 22..	24¾	24¼								
" 26..	25⅛	25	12,500	497	88				88	
" 29..	27	26½								
Decem. 3..	29	28½	11,000	931						3-16d.
" 6..	31	30½								
" 10..	31	30½	13,500	131						
" 13..	34½	34								
" 17..	38½	38	12,500	975	11				11	
" 20..	38½	38								
" 24..	37½	37	4,500	390	54				54	
" 27..	36½	36								
" 31..	36½	36	2,500	770						

GENERAL REMARKS.

This, and the succeeding three years, the absorbing question was our own domestic troubles; comparatively but little attention was given to foreign markets; and, after the suspension of our banks, and the issue of paper money by the Government, the market for this, as well as all other articles, fell into a chaotic state, values rising or falling with the great fluctuations in the gold premium, not unfrequently several cents per lb. in an hour.

In September, this market was very dull, the European accounts being unfavorable, the demand for goods for the States, as well as for India, having largely fallen off.

October witnessed a more active demand, the Liverpool market being excited by large speculative purchases; toward the latter part of November, the market was unsettled by the arrest of Messrs. Mason and Slidell, on a British mail steamship, and when the account reached Liverpool, there was great excitement in that market, in the prospect of a war between England and the United States; on the 28th December, our city banks suspended, and foreign and domestic exchanges

1862.										
January 3..	36½	36								3-16d.
" 7..	36½	36	1,600	417	......	......	......	......	......	
" 10..	36½	36								
" 14..	34½	34	1,500	1,227	......	......	......	......	......	
" 17..	33½	33								
" 21..	33½	33	900	3,297	46	......	......	......	46	
" 24..	33½	33								
" 28..	33½	33	2,800	1,993	......	......	......	......	......	
" 31..	32½	32								
February 4..	31½	31	2,000	3,320	......	......	......	......	......	
" 7..	30½	30								
" 11..	30½	30	1,300	1,728	......	......	......	......	......	
" 14..	28½	28								
" 18..	—	—	2,000	1,891	131	......	......	......	131	
" 21..	21½	20								
" 25..	21½	20	4,300	6,505	957	......	......	......	957	
" 28..	22½	21								
March 4..	24	22½	6,000	492	1,102	......	......	......	1,102	⅛@5/32d
" 7..	27	25½								
" 11..	29	27½	7,200	2,379	1,049	......	20	......	1,069	
" 14..	29	27½								
" 18..	29	27½	2,000	5,402	241	......	......	......	241	
" 21..	29	27½								
" 25..	29	27½	2,300	4,127	660	......	......	......	660	
" 28..	29	27½								
April 1..	29	27½	5,800	4,813	295	......	......	......	295	⅛@5/32d.
" 4..	29	27½								
" 8..	29	28	5,000	2,321	263	......	......	......	263	
" 11..	29	28								
" 15..	29	28	6,700	2,379	579	......	......	......	579	
" 18..	30	29								
" 22..	30½	29½	11,500	1,438	283	......	......	......	283	
" 25..	30½	29½								
" 29..	30	29	4,500	496	......	......	......	......	......	
May 2..	28½	27½								
" 6..	28½	27½	1,300	1,254	......	......	......	......	......	
" 9..	28½	27½								
" 13..	28½	27½	3,600	1,582	......	......	......	......	......	
" 16.	28½	27½								
" 20..	28½	27½	2,200	2,877	469	......	......	......	469	
" 23..	29	28								

became very much unsettled; in January, gold began to command a small premium, which soon largely advanced; the surrender of the Confederate Commissioners to England dissipated the fear of war between the two countries, and the public feeling became more calm; at this time considerable purchases of cotton were made in Liverpool, to come here.

The expectation of considerable receipts, here and at Boston, from England, caused prices, through February, to run down very rapidly.

In March, there was more activity, and prices advanced again; April was steady; the receipt of news early in May, of the capture of New Orleans, caused prices to droop, but at the close of the month, unfavorable accounts from the army in Virginia counteracted this news, and prices advanced again; June was quiet until toward the latter part, when the further issue of United States Treasury notes put up the gold premium, and cotton largely advanced with it.

The battles before Richmond, that occurred in July, with results so unfavorable to the Government, unsettled gold and exchange, and cotton was feverish, and steadily tended upward, and when, upon this, advices were received from Liverpool noting "the wildest excitement there," three-fifths of the entire stock changed hands in a few days, and large purchases to arrive; prices here rapidly advanced.

Subsequently there was a reaction,

New York Statement for Year 1862—Concluded.

1862.	Price of Middling New Orleans	Price of Middling Upland.	Sales for week.	Receipts for week.	EXPORTS FOR WEEK.					Rates of Freight to Liverpool.
					To Great Britain	To France.	North of Europe.	Other Fo'n Ports	Total Exports.	
May 27..	31½	30½	13,000	2,873	309	....	....	....	309	
" 30..	32	31								
June 3..	32	31	7,500	2,872	64	10	....	....	74	3-16d.
" 6..	32	31								
" 10..	32	31	4,500	590	30	....	....	....	30	
" 13..	32	31								
" 17..	32	31	4,000	2,401	364	....	....	....	364	
" 20..	32½	31½								
" 24..	34½	33½	10,500	2,654	236	....	....	....	236	
" 27..	38	37								
July 1..	38	37	11,000	2,699	71	....	....	....	71	3-16d.
" 4..	39½	38½								
" 8..	41½	40½	4,600	1,671	29	....	....	....	29	
" 11..	43	42								
" 15..	45½	44½	10,500	3,965	26	....	....	....	26	
" 18..	50½	49½								
" 22..	50½	50	6,300	5,273	87	....	....	....	87	
" 25..	47½	47								
" 29..	42½	42	1,700	5,496	58	....	....	....	58	
August 1..	50	49½								¼@⅜d.
" 5..	48½	48	7,800	8,354	317	....	....	....	317	
" 8..	48½	48								
" 11..	48½	48	4,500	1,671	....	....	....	....	....	
" 15..	47½	47								
" 19..	47	46½	4,300	3,924	129	....	....	....	129	
" 22..	46½	46								
" 26..	48	47½	8,000	2,909	35	....	....	....	35	
" 29..	48½	48								
Septem. 2..	52	51½	13,500	6,436	117	31	....	....	148	¼@⅜d.
Average price and total sales, receipts and exports.	32.06	31.29	305,100	115,427	9,257	49	20	2	9,328	

General Remarks.

and values fell off again; in August, the market was not active, and in prices there was considerable fluctuation.

Exchange.

Quotations for 60 days' commercial bills on London—the range in September was from 6¾@8½ per cent. premium; October, 6¼@8; November, the extremes were from 6½@7½ to 8½@9½; December, 8¼ to 10¾; in January, the lowest price was 9½@12, and the highest 13½@15; in February, the range was from 12 to 15¾; in March, the quotations dropped to 11@12; in April, the range was 11¼ to 12¾; in May, 12½@15; June opened at 13¾@14½, and steadily advanced to 19½@21; in July, the rate opened at 19½@21, and went up to 26@29; in August, the lowest quotation was 24@26 and the highest 26¾@28 per cent. premium.

LIVERPOOL STATEMENT FOR 1862.

UNITED STATES, 1861–1862.			
Stock Sept. 1, 1861	83,000	Export	unknown
Crop	unknown	Consumption	"
		Stock Sept. 1, 1862	"
Bales		Bales	

CONSUMPTION. Tot. Europe.	Continent.	France.	Gt. Britain.		Gt. Britain.	France	Continent.	Tot. Europe.
				Stock Jan. 1, 1862, in				
				United States	283,000	124,000	22,000	429,000
				Brazil	27,000	....	1,000	28,000
				West Indies	1,000	....	1,000	2,000
				East Indies	378,000	6,000	14,000	398,000
				Egypt	10,000	1,000	4,000	15,000
				Bales	699,000	131,000	42,000	872,000
				IMPORT.				
440,000	98,000	143,000	199,000	United States	72,000	24,000	84,000	94,000
135,000	24,000	8,000	103,000	Brazil	134,000	11,000	26,000	145,000
40,000	8,000	8,000	24,000	West Indies	32,000	6,000	7,000	40,000
1,131,000	326,000	95,000	710,000	East Indies	1,072,000	138,000	316,000	1,075,000
216,000	49,000	57,000	110,000	Egypt	135,000	59,000	45,000	232,000
1,962,000	505,000	311,000	1,146,000	Bales	1,445,000	238,000	478,000	1,586,000
....	....	....	564,000	Export.				
507,000	15,000	58,000	434,000	Stock Dec. 31. Stock above,	699,000	131,000	42,000	883,000
2,469,000	520,000	369,000	2,144,000	Total supply, bales	2,144,000	369,000	520,000	2,469,000

COTTON AT LIVER

Week Ending.	Receipts.						Sales.			
	American.	E. I.	Egypt.	Brazil.	Other.	Total.	Consumption.	Speculation.	Export	Total.
Jan. 10*.	436	4,247	4,685	3,636	76	13,080	67,197	89,960	11,453	168,610
" 17..	65	19,059	4,181	3,806	......	27,111	13,015	17,700	9,675	40,390
" 24..	1,246	7,094	......	5,613	800	14,753	4,414	10,540	9,376	24,330
" 31..	814	13,324	101	414	135	14,788	25,624	14,560	5,776	45,960
Feb. 7..	1,129	17,330	7,543	3,992	713	30,707	16,140	3,630	8,060	27,830
" 14..	398	1,801	3,501	140	253	6,093	32,270	16,850	4,400	53,520
" 21..	......	276	6,534	1,016	1,233	9,059	57,544	22,060	5,416	85,020
" 28..	913	......	7,979	838	......	9,730	33,686	14,640	5,924	54,250
Mch. 7..	926	......	7,251	1,001	61	9,239	16,722	6,000	8,878	31,600
" 14..	1,160	5,297	8,853	5,780	15	21,105	17,080	9,200	5,940	32,220
" 21..	266	3,538	4,314	560	53	8,731	20,239	4,230	4,881	29,350
" 28..	69	12,880	2,325	206	25	15,505	16,617	4,500	2,363	23 480
April 4..	3,603	60,246	3,288	9,741	431	77,309	25,500	5,350	1,380	32,230
" 11..	1,131	5,437	9,518	3,732	451	20,369	40,693	16,750	4,517	61,960
" 18..	79	99	2,858	1,109	......	4,145	46,398	28,820	5,162	80,380
" 25..	4,465	32,274	7,467	3,631	651	48,488	33,176	19,820	6,444	59,440
May 2..	632	......	2,379	754	......	3,765	24,605	7,210	7,905	39,720
" 9..	......	......	5,422	703	......	6,125	19,383	2,470	5,797	27,650
" 16..	2,639	34,272	3,006	3,723	142	43,782	14,063	6,460	10,097	30,620
" 23..	17	26,099	2,692	10,010	254	39,072	17,713	10,360	5,807	33,880
" 30..	1,584	18,482	5,598	3,773	105	29,042	30,479	10,920	5,031	46,430
June 6..	69	200	2,818	......	......	3,087	55,254	14,290	2,776	72,320
" 13..	494	5,894	1,103	1,232	177	8,900	54,231	22,660	6,739	83,630
" 20..	1,797	20,247	950	3,591	73	26,648	71,969	44,840	7,901	124,710
" 27..	2,173	4,928	1,222	9,294	477	18,094	77,117	69,690	12,063	158,870
July 4..	1,171	5,725	1,183	1,851	......	9,930	36,112	104,920	14,138	155,170
" 11..	......	13,600	1,749	1,421	283	17,053	2,523	38,240	28,237	69,000
" 18..	1,747	21,707	2,119	2,486	60	28,119	9,448	24,710	18,702	52,860
" 25..	395	31,308	1,730	1,711	213	35,357	3,150	9,320	8,970	21,440
Aug. 1..	462	16,996	1,116	1,560	19	20,153	18,013	20,020	13,667	51,700
" 8..	2,296	19,981	933	2,068	......	25,278	20,466	11,500	9.094	41,060
" 15..	537	100	633	314	58	1,642	23,073	13,070	13,477	49,620
" 22..	188	5,285	677	4,612	115	10,877	44,332	54,990	13,508	112,830
" 29..	777	......	47	2,944	77	3,845	17,266	65,610	8,444	91,320
Sept. 5..	......	14,357	296	......	20	14,673	1,327	49,990	11,923	63,240
" 12..	1,388	29,302	424	5,302	25	36,441	6,480	10,550	1,700	18,730
" 19..	1,427	9,787	1,463	590	539	13,806	5,377	12,160	7,063	24,600
" 26..	2,244	......	......	337	......	2,581	5,200	2,230	2,780	10,210
Oct. 3..	305	131,334	1,347	3,623	......	136,609	9,177	14,310	3,403	26,890
" 10..	794	11,417	611	3,892	......	16,714	15,796	20,280	8,494	44,570
" 17..	601	74,824	630	1,355	163	77,573	2,289	6,170	7,031	15,490
" 24..	3,549	55,237	558	......	2	59,346	9,357	10,340	9,243	28,940
" 31..	2,136	8,369	1,798	1,439	12	13,754	828	2,060	9,582	12,470
Nov. 7..	2,783	3,086	1,908	4,262	5	12,044	21,628	18,220	6,762	46,610
" 14..	3,322	10,873	1,494	733	131	16,553	10,740	2,510	1,840	15,090
" 21..	3,598	15,203	977	1,712	1,076	22,566	5,510	5,160	6,450	17,120
" 28..	1,778	2,318	1,249	174	......	5,519	21,478	14,380	2,952	38,810
Dec. 5..	2,816	17,269	2,501	2,290	......	24,876	34,176	16,960	1,574	52,710
" 12..	2,288	15,917	3,362	7,125	18	28,710	28,355	30,890	6,935	66,180
" 19..	1,258	3,442	4,105	1,448	2,215	12,468	35,345	37,020	2,975	75,340
" 24..	......	......	......	107	......	107	14,203	18,380	1,937	34,520
" 31..	......	......	......	......	......	51,320	23,191	26,920	3,569	53,680
Average prices & total sales, receipts & stocks.	62,903	783,461	139,290	136,151	11,156	1,184,281	1,250,476	1,114,250	443,411	2,808,137

* In first line of figures, from 1st to 10th January, is included.

POOL. YEAR 1862.

Stocks.			Prices.			Actual Export.	Consumption.	Remarks.
American.	Other.	Total.	Mid. Up.	Mid. Orl.	Dhol.			
248,016	317,596	565,612	......	$13\frac{5}{8}$	$8\frac{5}{8}$	11,453	67,197	
235,586	331,212	566,798	......	$13\frac{1}{8}$	$8\frac{1}{4}$	9,675	80,212	
227,817	333,828	561,645	......	$12\frac{3}{4}$	8	9,376	84,626	
216,951	329,336	546,287	......	$12\frac{7}{8}$	$8\frac{1}{8}$	5,776	110,250	
205,049	344,995	550,044	......	$12\frac{3}{4}$	$8\frac{1}{8}$	8,060	126,390	
194,554	330,243	524,797	......	$12\frac{7}{8}$	$8\frac{1}{8}$	4,400	158,660	
177,959	301,131	479,090	......	$12\frac{15}{16}$	$8\frac{3}{8}$	5,416	216,204	
170,858	281,201	452,059	......	$12\frac{15}{16}$	$8\frac{1}{2}$	5,924	249,890	
166,061	263,789	429,850	......	$12\frac{1}{2}$	$8\frac{1}{4}$	8,878	266,612	
158,095	265,400	423,495	......	$12\frac{1}{4}$	8	5,940	283,692	
150,036	253,169	403,205	......	$12\frac{3}{8}$	$8\frac{1}{8}$	4,881	303,931	
144,148	254,759	398,907	......	$12\frac{9}{16}$	$8\frac{1}{8}$	2,363	320,548	Prices steady.
142,442	313,104	455,546	......	$12\frac{5}{8}$	$8\frac{1}{8}$	1,380	346,048	
134,010	295,308	429,318	......	$12\frac{15}{16}$	$8\frac{1}{4}$	4,517	386,741	
126,690	262,491	389,181	......	$13\frac{1}{8}$	$8\frac{1}{2}$	5,162	433,139	
124,393	272,922	397,315	......	$13\frac{1}{2}$	$8\frac{5}{8}$	6,444	466,315	
118,698	248,217	366,915	......	$13\frac{1}{4}$	$8\frac{5}{8}$	7,905	490,920	
115,071	232,602	347,673	......	13	$8\frac{3}{8}$	5,797	510,303	
114,415	249,733	364,148	......	$12\frac{5}{8}$	$8\frac{1}{4}$	10,097	514,366	
108,695	268,508	377,203	......	$12\frac{1}{4}$	$8\frac{1}{8}$	5,807	532,079	
103,409	266,185	369,594	......	$12\frac{9}{16}$	$8\frac{1}{4}$	5,031	562,558	
97,374	226,561	323,935	......	$12\frac{7}{8}$	$8\frac{1}{2}$	2,776	617,312	
91,632	196,744	288,376	......	$13\frac{1}{4}$	$8\frac{3}{4}$	6,739	671,542	
82,228	175,251	259,479	......	$13\frac{3}{4}$	$9\frac{3}{8}$	7,901	743,512	
71,356	137,914	212,270	......	$14\frac{7}{8}$	$11\frac{1}{4}$	12,063	820,629	
62,024	122,158	184,182	......	17	14	14,138	856,741	Poor prospects of supplies
52,034	103,694	155,728	......	$17\frac{5}{8}$	$13\frac{1}{2}$	28,237	859,264	from United States.
47,430	107,315	154,745	......	$18\frac{1}{8}$	$13\frac{3}{4}$	18,702	868,712	
42,920	127,257	170,177	......	$18\frac{1}{8}$	13	8,970	871,862	
36,303	123,920	160,223	......	$18\frac{1}{2}$	$13\frac{3}{8}$	13,667	889,875	
33,547	124,470	157,917	......	$19\frac{1}{4}$	$13\frac{3}{8}$	9,094	910,341	
27,493	97,009	124,502	......	$19\frac{9}{16}$	$13\frac{3}{4}$	13,477	933,414	
19,588	61,583	81,171	......	$23\frac{1}{2}$	$15\frac{1}{4}$	13,508	977,746	Owing to improved state
17,501	44,661	62,162	......	$26\frac{1}{2}$	$17\frac{1}{4}$	8,444	995,012	of Manchester market.
16,082	41,260	57,342	......	29	$18\frac{1}{2}$	11,923	996,339	Small arrivals of ships;
15,401	60,117	75,518	......	$27\frac{1}{4}$	$17\frac{3}{4}$	1,700	1,002,819	east winds.
15,413	74,308	89,721	......	27	$17\frac{1}{2}$	7,063	1,008,196	
16,445	67,243	83,688	......	$25\frac{1}{2}$	$17\frac{1}{4}$	2,780	1,013,396	Large arrivals of over-
13,534	197,460	210,994	......	$27\frac{1}{2}$	$17\frac{1}{4}$	3,403	1,022,573	due Bombay ships.
12,179	197,865	210,044	......	$27\frac{7}{8}$	$17\frac{1}{2}$	8,494	1,038,379	
12,531	263,845	276,376	......	26	16	7,031	1,040,668	Conflicting accounts
15,910	298,769	314,679	......	25	$16\frac{3}{8}$	9,243	1,050,025	from United States.
17,366	293,685	311,051	......	$22\frac{1}{2}$	$15\frac{1}{2}$	9,582	1,050,853	
19,189	275,084	294,273	......	$23\frac{1}{2}$	$15\frac{1}{2}$	6,762	1,072,481	
21,203	262,969	284,172	......	$22\frac{1}{2}$	$15\frac{1}{4}$	1,840	1,083,221	
24,121	265,467	289,588	......	21	$14\frac{1}{2}$	6,450	1,088,731	
22,904	247,321	270,225	......	22	$14\frac{3}{4}$	2,952	1,110,209	Manchester market flat
23,523	238,484	262,007	......	23	$15\frac{1}{2}$	1,574	1,144,385	
23,391	227,991	251,382	......	$23\frac{1}{2}$	16	6,935	1,172,740	
21,572	207,551	229,123	......	24	17	2,975	1,208,085	
20,531	193,742	214,273	......	24	$17\frac{1}{4}$	1,937	1,227,288	
....	....	392,461	......	$24\frac{1}{2}$	$17\frac{3}{4}$	3,569	1,250,476	
			......	18,375	12.42	443,411	24,047.61	

1863. *The semi-weekly Price and Weekly Sales and Receipts at New York, Weekly Exports from New York and Rates of Freight to Liverpool 1st of each month, for the Crop-Year ending September* 1, 1863.

1862.	Price of Middling New Orleans	Price of Middling Upland.	Sales for week.	Receipts for week.	EXPORTS FOR THE WEEK. To Great Britain.	To France	North of Europe.	Other Fo'n Ports	Total Exports.	Rates of Freight to Liverpool.
Septem. 5..	53	52½								¼@⅛d.
" 9..	58½	58	16,500	4,118	236	....	....	....	236	
" 12..	57½	57								
" 16..	57	56½	7,100	6,180	917	....	414	....	1,331	
" 19..	53½	53								
" 23..	55½	55	3,200	4,380	774	....	....	....	774	
" 26..	56½	56½								
" 30..	56½	56½	6,500	5,472	1,325	....	312	....	1,637	
October 3..	56	56								5-16d.
" 7..	55	55	3,400	4,513	1,447	74	....	....	1,521	
" 10..	56½	56½								
" 14..	61	61	16,500	4,188	1,923	....	....	....	1,923	
" 17..	61	61								
" 21..	60	60	6,400	4,716	2,542	78	....	....	2,620	
" 24..	60	60								
" 28..	59½	59½	7,500	3,694	1,420	....	150	....	1,570	
Novem. 1..	60	60								¼d.
" 4..	60	60	4,500	2,270	3,784	....	....	....	3,784	
" 7..	62	62								
" 11..	62½	62½	10,600	1,215	742	....	....	....	742	
" 14..	64½	64½								
" 18..	69½	69½	17,000	1,881	....	....	....	....	....	
" 21..	67½	67½								
" 25..	66	66	3,400	1,855	71	....	....	....	71	
" 28..	66	66								
Decem. 2..	68	68	5,000	2,829	....	186	....	....	186	¼d.
" 5..	68	68								
" 9..	67	67	3,700	1,585	24	....	....	....	24	
" 12..	67	67								
" 16..	66	66	2,000	1,241	....	25	....	....	25	
" 19..	66	66								
" 23..	66	66	2,500	2,499	....	....	....	....	....	
" 26..	66½	66½								

GENERAL REMARKS.

Accounts were received here in September, from Liverpool, noting a very excited market, prices there having advanced within a week 4@4½d., and more than the entire stock there turned over; this had the effect to put prices up here, and there was some excitement.

Part of October was dull, but later, with an advance in gold and exchange, prices largely appreciated. December was quiet.

In January, the market was greatly unsettled, by the passage through Congress of the "Legal Tender" Act, everything advanced, cotton among the rest.

The market was greatly excited in February; very favorable foreign accounts were received, the English market being full of orders for goods for India; added to this, the President's "Proclamation of Emancipation," caused the greatest excitement there, with respect to future supplies, as it was supposed cotton could be only successfully cultivated here by compulsory labor; these advices, however, had less effect here than might be supposed; for this market was now no longer governed by Liverpool prices.

Decem. 30..	67	67	5,500	2,770	284	....	....	....	284	
1863.										
January 2 .	67½	67½								¼d.
" 6..	69	69	9,100	2,431	203	141	....	....	344	
" 9..	69	69								
" 13..	70	70	9,800	4,563	34	....	....	....	34	
" 16..	73	73								
" 20..	75½	75½	12,700	6,628	36	....	....	....	36	
" 23..	76	76								
" 27..	77½	77½	7,500	4,036	....	....	....	....	....	
" 30..										
February 3..	88	88	9,200	3,818	33	....	....	....	33	—@—
" 6..	86	86								
" 10..	91	91	8,800	2,834	....	....	....	....	....	
" 13..	92	92								
" 17..	91	91	10,700	4,978	8	....	....	....	8	
" 20..	89	89								
" 24..	91	91	5,200	5,239	....	....	....	....	....	
" 27..	90	90								
March 3..	88	88	5,000	6,361	15	....	....	....	15	¼d.
" 6..	84	84								
" 10..	86½	86½	5,500	3,501	....	....	....	....	....	
" 13..	85	85								
" 17..	83	83	3,700	5,079	....	....	....	....	....	
" 20..	78	78								
" 24..	73	73	2,000	5,862	....	100	....	....	100	
" 27..	62	62								
" 31..	70	70	3,300	9,517	....	....	....	....	....	
April 3..	73	73								¼d.
" 7..	70	70	3,600	4,305	10	....	....	....	10	
" 10..	64	64								
" 14..	69	69	1,800	7,017	61	....	....	....	61	
" 17..	66	66								
" 21..	64	64	2,500	6,784	74	....	....	....	74	
" 24..	66	66								
" 28..	67	67	5,100	3,310	12	....	....	....	12	
May 1..	67	67								¼d.
" 5..	65	65	1,300	7,028	....	....	....	....	....	
" 8..	65	65								
" 12..	62	62	10,500	4,957	55	....	....	5	60	
" 15..	60	60								
" 19..	55	55	2,800	3,206	46	....	....	....	46	

In March, the market was a dull and declining one, for the most part, owing to the lower range of the gold premium. April was characterized by the same features, and in May there was a large decline; prices now touching the lowest point of the year, owing to the anticipations of the opening of the Mississippi, at Vicksburg.

The market in June was much excited, and prices rapidly ran up, in consequence of the invasion of Maryland and Pennsylvania by the Confederate armies.

With the defeat of the Confederates in Pennsylvania, and the capture of Vicksburg, in July, gold dropped and cotton fell off.

In August, there was an improved spinning and speculative demand, which carried prices up again.

New York Statement for Year 1863—Concluded.

1863.	Price of Middling New Orleans	Price of Middling Upland.	Sales for week.	Receipts for week.	EXPORTS FOR THE WEEK					Rates of Freight to Liverpool.
					To Great Britain.	To France.	North of Europe.	Other Fo'n Ports	Total Exports.	
May 22..	55	55								
" 26..	51	51	4,000	3,450	445				445	
" 29..	52	52								
June 2..	54	54	3,500	1,565	1,845	352			2,197	¼d.
" 5..	56½	56½								
" 9..	57	57	9,000	4,095	2,914	50			2,964	
" 12..	57	57								
" 16..	58	58	5,000	4,932	371	20			391	
" 19..	59	59								
" 23..	61	61	9,500	2,035	1,022				1,022	
" 26..	66	66								
" 30..	75	75	21,000	5,237	1,155				1,155	
July 3..	69	69								
" 7..			3,300	3,737	494				494	¼d.
" 10..	60	60								
" 14..	63	63	2,900	1,859	77				77	
" 17..	60	60								
" 21..	61	61	2,000	3,822	201				201	
" 24..	62	62								
" 28..	62	62	5,800	3,361	231				231	
" 31..	62½	62½								
August 4..	65	65	12,000	2,146	69				69	¼d.
" 7..	68	68								
" 11..	68	68	5,200	1,607	151				151	
" 14..	68	68								
" 18..	68	68	5,500	1,078	35				35	
" 21..	67	67								
" 25..	66	66	3,700	5,542						
" 28..	65	65								
Septem 1..	67	67	4,100	2,903	55	4			59	—@—
Average prices and total sales, receipts and exports.	67.23	67.21	337,900	204,229	25,141	1,030	876	5	27,052	

General Remarks.

Exchange.

Quotations for 60 days' commercial bills on London:—September, 28@29 per cent. premium was the opening price, and gradually advanced to 33@35¼. October opened at 34½@35½, and went up to 44@47; in November, the range was not very wide, lowest, 42@43¼, highest, 45½@47. In December, the lowest was 41¾@44¾, highest, 44½@47½. The rate at the opening of January was 46@47½ per cent. premium, and rapidly went up to 69@74½. February opened at 69@77, and advanced to 78@88. The quotations rapidly yielded in March, opening at 87@89 per cent. premium, it declined to 51@54; April, it advanced again, from 56@64 to 65@72; in May, it fell off from 68@70 to 56½@58. In June, prices were comparatively steady, the extreme range being from 54@55½ to 59@61½; in July, the decline was large, the month opening at 57@58½, and the quotations running down to 35½@39; in August, the quotations fluctuated between 40@41¾ and 35½@36¼ per cent. premium, the latter being the closing figure.

LIVERPOOL STATEMENT FOR 1863.

UNITED STATES, 1862–1863.

Stock Sept. 1, 1862	unknown	Export	unknown
Crop	"	Consumption	"
		Stock Sept. 1, 1863	"
Bales		Bales	

Stock 1st Jan., 1863, in	Gt. Britain.	France.	Continent.	Tot. Europe.
United States ... Bales.	70,000	15,000	3,000	88,000
Brazil	33,000	2,000	3,000	38,000
West Indies	4,000	......	......	4,000
East Indies	301,000	38,000	7,000	346,000
Egypt	26,000	3,000	2,000	31,000
Bales	434,000	58,000	15,000	507,000

CONSUMPTION. / **IMPORT**

Tot. Europe.	Continent.	France.	Gt. Britain.		Gt. Britain.	France.	Continent.	Tot. Europe.
195,000	58,000	19,000	118,000	United States ... Bales.	132,000	8,000	55,000	149,000
178,000	52,000	14,000	112,000	Brazil	137,000	12,000	50,000	150,000
68,000	14,000	2,000	52,000	West Indies	66,000	2,000	14,000	72,000
1,518,000	432,000	191,000	895,000	East Indies	1,391,000	176,000	428,000	1,450,000
403,000	87,000	116,000	200,000	Egypt	206,000	117,000	87,000	398,000
2,362,000	643,000	342,000	1,377,000	Bales	1,932,000	315,000	634,000	2,219,000
......	......	......	662,000	Export.				
364,000	6,000	31,000	327,000	Stock, Dec. 31 ... Stock above,	434,000	58,000	15,000	507,000
2,726,000	649,000	373,000	2,366,000	Total supply, bales	2,366,000	373,000	649,000	2,726,000

COTTON AT LIVER

Week Ending.	Receipts.						Sales.			
	Amrc'n	E. I.	Egypt.	Brazil.	Other.	Total.	Consumption.	Speculation.	Export.	Total.
Jan. 9..	81	26,630	4,858	5,436	......	37,005	21,490	22,270	7,530	51,290
" 16..	35	6,994	9,256	3,320	......	19,605	22,810	26,270	5,340	54,420
" 23..	5	12,999	3,898	742	......	17,644	13,140	7,900	2,520	23,560
" 30..	1,596	3,842	13,344	8,260	161	27,203	11,800	7,480	5,130	24,410
Feb. 6..	628	966	10.645	2,258	......	14,497	16,130	7,750	6,720	30,600
" 13..	813	2,385	3,765	2,856	......	9,819	11,850	4,930	8,900	25,680
" 20..	33	25,164	8,360	136	169	33,862	11,400	1,960	4,890	18,250
" 27..	1,865	5,488	9,162	1,725	7	18,247	14,450	2,860	6,880	24,190
Mch. 6..	1,283	9,407	3,540	4,544	11	18,785	14,830	5,020	8,290	28,140
" 13..	15	17,706	4,184	1,811	......	23,716	16,140	6,190	9,750	32,080
" 20..	3,194	3,775	5,587	740	......	13,296	32,240	16,100	9,060	57,400
" 27..	247	26,673	3,198	16,123	349	46,590	22,200	4,780	8,640	35,620
April 2..	1,000	20,998	1,485	508	24	24,015	29,060	8,710	10,900	48,670
" 10..	254	1,665	5,953	3,510	......	11,382	23,590	3,130	8,680	35,400
" 17..	2,559	15,168	8,045	2,839	49	28,665	30,530	14,190	19,520	64,240
" 24..	1,178	45,522	4,607	3,347	......	54,654	26,620	11,310	18,370	56,300
May 1..	511	26,309	2,117	891	61	29,889	14,940	2,120	7,390	24,450
" 8..	379	15,943	6,538	578	11	23,449	48,600	23,160	18,780	90,540
" 15..	......	....	......	....	......	54,857	....	....	......	78,270
" 22..	3,208	28,129	6,346	4,052	36	41,771	19,820	1,880	6,510	28,210
" 29..	5	7,114	10,153	....	......	17,272	30,140	10,210	13,220	52,570
June 5..	6,790	56,338	3,357	14,283	1	80,769	16,130	10,340	6,520	32,990
" 12..	3,183	23,285	7,401	7,765	1	41,635	20,920	5,120	6,050	32,090
" 19.	6,440	6,727	5,230	2,414	......	20,811	28,740	5,590	13,330	47,660
" 26..	1,934	21,890	5,136	3,904	5	32,869	20,060	1,740	10,460	32,260
July 3..	1,678	12,747	2,470	4,321	142	21,358	22,750	3,530	7,440	33,720
" 10..	507	290	5,180	2,106	213	8,296	27,390	3,090	9,010	39,490
" 17..	616	2,639	1,225	....	60	4,540	17,750	1,570	7,700	27,020
" 24..	1,195	116	1,457	1,689	1,738	6,195	31,640	11,410	9,190	52,240
" 31..	4,083	26,400	2,078	1,064	......	33,625	33,730	11,950	2,940	48,620
Aug. 7..	7,483	60,108	1,766	5,867	37	75,261	19,560	4,100	3,140	26,800
" 14..	6,929	20,632	1,633	353	122	29,669	37,240	9,730	12,500	59,470
" 21..	4,303	26,350	735	5,774	643	37,805	42,630	16,890	19,530	79,050
" 28..	2,505	19,592	6,926	900	......	29,923	56,080	39,450	18,600	114,130
Sep. 4..	2,372	24,971	2,469	1,028	......	30,840	24,970	10,140	11,870	46,980
" 11..	3,262	27,847	1,185	4,352	725	37,371	55,690	58,040	19,190	132,920
" 18..	5,352	6,887	5,245	2	193	17,679	42,790	56,856	10,960	110,606
" 25..	4,948	63,300	2,376	560	15	71,199	30,070	24,040	10,370	64,480
Oct. 2..	4.033	17,668	2,524	928	......	25,153	25,540	13,340	9,620	48,500
" 9..	3,592	20,797	4,823	29	164	29,405	28,040	16,560	15,840	60,440
" 16..	7,878	19,326	3,060	1,435	225	31,924	45,150	62,440	24,670	132,260
" 23..	387	57,048	974	3,170	99	61,678	30,240	56,580	14,660	101,480
" 30..	742	19,780	2,864	....	......	23,386	21,740	31,530	10,550	63,820
Nov. 6..	5,156	8,527	1,119	2,287	231	17,320	16.740	7,650	5,520	29,910
" 13..	6,855	52,009	1,109	2,404	122	62,499	19,130	7,820	6,120	33,070
" 20..	4,777	82,289	1,750	4,194	31	93,041	11,370	12,230	5,160	28,760
" 27..	1,835	37,674	5,574	1,000	......	46,083	37,960	21,260	15,680	74,900
Dec. 4..	2	13,664	7,089	26	......	20,781	25,670	15,220	10,690	51,580
" 11..	5,252	10,834	9,584	1,227	15	26,912	15,690	5,160	11,370	32,220
" 18..	3,045	380	25	2,996	438	6,884	14,420	5,180	10,310	29,910
" 24..	1,695	23,558	6,057	3,844	360	35,514	12,280	2,930	6,390	21,600
" 31..	......	....	......	....	......	38,998	....	......	......	43,020
Average prices & total sales, receipts & stocks.	127610	1,087,960	227422	143,588	6,458	1,686,895	1,263,800	719,206	469100	2,573,486

POOL. YEAR 1863.

STOCKS.			PRICES.			ACTUAL EXPORT.	CON-SUMPTION.	REMARKS.
American	Other.	Total.	Mid. Up.	Mid. Orl.	Dhol.			
67,522	335,284	402,806	24	24½	17¾	7,530	21,490	
66,551	328,493	395,044	24	24½	18	5,340	44,300	
65,165	330,516	395,681	23	24	17⅝	2,520	57,440	Dull accounts from Manchester.
65,901	340,259	406,160	22	22½	17	5,130	69,240	
65,909	336,995	402,904	21½	22½	16⅞	6,720	85,370	Small Continental demand.
65,465	328,913	394,378	21	22	16¾	8,900	97,220	
61,962	353,396	415,358	20	21	16½	4,890	108,620	Flat accounts from India.
60,689	351,989	412,678	20½	21½	16½	6,880	123,070	
57,644	350,884	408,528	20	21	16	8,290	137,900	
53,613	353,897	407,510	20½	21½	16½	9,750	154,040	
53,102	325,192	378,294	20½	21½	17	9,060	186,280	
50,359	343,283	393,642	21	22	17¼	8,640	208,480	
49,073	334,052	383,125	21	22	17½	10,900	237,540	
46,800	316,721	363,521	21	22	17¼	8,680	261,130	
46,527	304,380	350,907	21½	22½	17½	19,520	291,660	
46,088	319,004	365,092	21½	22½	17½	18,370	318,280	
45,790	323,912	369,702	21	21¾	17½	7,390	333,220	
43,150	285,076	328,226	22	22½	17¾	18,780	381,820	Improved feeling in Manchester, &c.
....	....	....	......	......	......	....	....	
42,811	297,433	340,244	21½	23	18	6,510	401,640	
40,757	275,095	315,852	22	23	18¼	13,220	431,780	
45,504	327,025	372,529	21¾	22½	18¼	6,520	447,910	
47,048	339,136	386,184	21½	22½	18	6,050	488,830	
51,469	309,529	360,998	21¼	22	18	13,330	497;570	
51,388	312,983	364,371	20¾	21½	17¾	10,460	517,630	Rumors of an armistice.
49,811	303,631	353,442	20½	21	18	7,440	540,380	
45,479	273,721	319,200	20½	21	18	9,010	567,770	
42,914	254,419	297,333	20½	21	18	7,700	585,520	
39,872	222,824	262,696	21½	22	18¼	9,190	617,160	
39,935	217,767	257,702	22	22½	18½	2,940	650,890	
45;763	262,574	308,337	21¾	22½	18½	3,140	670,450	
49,038	244,748	293,786	22	22¾	18¾	12,500	707,690	
48,401	236,137	284,538	22¼	23	19	19,530	750,320	Large demand for Manchester.
43,440	205,010	248,450	23⅜	23¾	19½	18,600	806,400	
42,957	197,753	240,710	23¼	23¾	19¼	11,870	831,370	
39,612	268,999	208,611	24½	25	20	19,190	887,060	
39,807	133,273	173,080	26½	27¼	21¾	10,960	929,850	Rumors of a prolonged struggle in United States.
41,007	161,445	202,452	27	27½	22	10,370	959,920	
41,558	152,778	194,336	27	27½	22	9,620	985,460	
41,729	143,598	185,327	26¾	27¼	21¾	15,840	1,013.500	Satisfactory accounts from manufacturing districts.
43,888	119,545	163,433	28¼	28¾	23½	24,670	1,058,650	
38,396	140,241	178,637	29¼	29½	24½	14,660	1,088,890	
33,992	130,818	164,810	29¼	29¾	24½	10,550	1,110,630	
34,942	115,341	150,283	28	28½	23¾	5,520	1,127,370	Bank rate advanced from 4 to 6 per cent. in the week.
38,241	144,653	182,894	28	28½	23½	6,120	1,146,500	
40,774	216,778	257,552	27	27½	22¼	5,160	1,157,870	
37,530	222,782	260,312	27¾	28¼	23¾	15,680	1,195,830	
33,763	216,841	250,604	26½	27	22	10,690	1,221,500	Bank rate increased to 7 per cent.
36,251	218,319	254,570	27	27¼	22¾	11,370	1,213,190	
36,036	201,377	237,413	26½	27	22	10,310	1,251,610	
35,591	216,022	251,613	26½	27	22	6,390	1,263,890	
....	....	....	......	......	......	....	....	
			22.46	21.2	18.43	469,100	24,305.57	

1864.

ANNUAL REVIEW.

From the New Orleans Price Current, 1863-64.

The accounts from New York being unfavorable, and the United States Army having successfully penetrated up the Red River country as far as Alexandria, this was followed by a decided falling off, which was still more marked at a later period, the market closing at the end of the month at the following quotations, gold ruling at the same time at 165½ per dollar, showing but little variation: Ordinary, 53 to 58 cents per pound; good ordinary, 61 to 64; low middling, 65 to 67; middling, 70. The reported sales for the month amounted to 7,000 bales, while the receipts embraced 16,809 bales, and the exports 14,150. Early in April it becoming manifest that the anticipation of liberal supplies from Red River, would be disappointed, and there being a marked falling off in the receipts from other sources, prices rallied and continued to improve, from week to week, until low middling had advanced to 82 cents, which being materially above the New York market, was not fully maintained, the business of the month closing at 70 to 72 cents for ordinary, 74 to 76 for good ordinary, 78 to 79 for low middling, and 82 to 83 for middling. From this time the advance was more rapid, and by the close of June, low middling had risen to $1.50, and middling to $1.60 per pound, showing an improvement during the previous three months of 128 per cent., or an average of about 80 cents per pound, equal to $360 per bale.

This extraordinary advance must be mainly ascribed to the marked falling off of the receipts, both at this port and Memphis, and the prospect of scant supplies, but was also influenced by the course of the New York market and the rise of gold. Notwithstanding the tenor of most of the commercial correspondence from this city, and the indications in our own columns, the spinners and dealers at the North had confidently expected large receipts from Northern Louisiana, Texas, and Arkansas, and further arrivals from Mississippi and Tennessee. At last, when there could no longer be any doubt as to the facts, and it became evident that the supplies would be insufficient to meet the indispensable wants of the spinners, who would be compelled to add to their stock by imports from Liverpool, the New York market became greatly excited and bounded up with a rapidity and to an extent unprecedented in the annals of the trade.

LIVERPOOL STATEMENT FOR 1864.

UNITED STATES—1863–1864.			
Stock Sept. 1, 1863......	unknown	Export	unknown
		Consumption	"
Crop........	"	Stock Sept. 1, 1864	"
Bales	"	Bales	"

Stock Jan. 1, 1864, in........	Gt. Britain.	France.	Continent.	Tot. Europe.
United States........Bales.	38,000	4,000	....	42,000
Brazil........	9,000	....	1,000	10,000
West Indies........	1,000	....	....	1,000
East Indies........	252,000	23,000	3,000	278,000
Egypt........	27,000	4,000	2,000	33,000
Bales........	327,000	31,000	6,000	364,000

CONSUMPTION.								
Tot. Europe.	Continent.	France.	Gt. Britain.	IMPORT.	Gt. Britain.	France.	Continent.	Tot. Europe.
238,000	60,000	19,000	159,000	....United States........	198,000	15,000	61,000	220,000
251,000	67,000	28,000	156,000	....Brazil........	212,000	30,000	67,000	260,000
74,000	17,000	11,000	46,000	....West Indies........	60,000	12,000	17,000	79,000
1,619,000	476,000	184,000	959,000	....East Indies........	1,798,000	205,000	483,000	1,897,000
542,000	92,000	164,000	286,000	....Egypt........	319,000	167,000	96,000	552,000
2,724,000	712,000	406,000	1,606,000	Consumption ....Bales.	2,587,000	429,000	724,000	3,008,000
....	....	....	732,000	Export.				
648,000	18,000	54,000	576,000	Stock Dec. 31. Stock above,	327,000	31,000	6,000	364,000
3,372,000	730,000	460,000	2,914,000	Total supply, bales........	2,914,000	460,000	730,000	3,372,000

COTTON AT LIVER

Week Ending.	Receipts.						Sales.			
	Amer'n.	E. I.	Egypt.	Brazil.	Other.	Total.	Consumption.	Speculation.	Export.	Total.
Jan. 8..	262	3,165	4,542	2,783	......	10,752	26,270	13,986	2,924	43,180
" 15..	2,368	9,801	2,496	2,538	......	17,203	21,680	5,040	2,920	29,640
" 22..	2,145	27,141	5,835	2,998	446	38,565	17,910	1,343	6,437	25,690
" 29..	1,672	153	6,149	8,575	1,041	17,590	23,250	3,342	4,698	31,290
Feb. 5..	7,826	12,008	10,628	4,681	27	35,170	21,310	4,367	5,423	31,100
" 12..	949	1,327	6,505	2,092	......	10,873	33,140	7,809	5,561	46,510
" 19..	1,466	25,955	12,948	5,383	......	45,752	27,790	5,510	6,130	39,430
" 26..	606	2,506	9,418	3,326	368	16,224	24,320	2,913	7,127	34,360
March 4..	3,724	29,611	8,412	1,291	......	43,038	26,890	4,628	6,082	37,600
" 11..	6,165	22,112	4,250	9,026	53	41,606	20,050	3,190	3,920	27,160
" 18..	7,787	40,624	6,610	3,305	252	58,578	49,420	21,877	4,653	75,950
" 24..	4,460	49,290	14,545	4,135	......	72,430	20,500	6,405	5,385	32,290
April 1..	2,281	7,270	9,616	5,450	732	25,349	27,269	2,284	8,436	37,989
" 8..	7,241	38,315	9,259	3,463	41	58,319	34,340	4,450	7,270	46,060
" 15..	7,771	80,208	11,181	8,850	1	110,011	44,870	29,899	9,741	84,510
" 22..	6,853	81,365	13,919	7,853	715	110,705	54,320	20,816	8,134	83,270
" 29..	1,990	2,246	11,256	642	2,210	18,344	53,680	31,215	11,525	96,420
May 6..	9,363	30,859	5,380	13,020	2,079	60,701	56,490	28,596	13,884	98,970
" 13..	4,606	7,400	13,475	1,611	234	27,326	46,640	27,940	17,810	92,390
" 20..	1,779	54,770	10,896	5,192	67	72,404	23,920	6,920	16,270	47,110
" 27..	6,886	22,418	5,912	3,185	1,465	39,866	26,590	4,790	9,650	41,030
June 3..	2,424	11,720	13,759	....	1,554	29,457	28,830	3,730	12,570	45,130
" 10..	3,081	9,561	1,110	5,663	2,190	21,605	36,310	5,730	10,820	52,860
" 17..	11,887	24,901	9,182	16,941	2,092	65,003	45,070	8,990	17,130	71,190
" 24..	3,122	34,048	6,395	5,293	170	49,028	30,360	3,190	11,200	44,750
July 1..	2,953	39,133	2,928	6,167	3,054	54,235	49,880	14,080	22,220	86,180
" 8..	3,147	17,162	6,942	7,014	1,517	35,782	45,630	14,670	24,580	84,880
" 15..	211	1,435	5,903	....	262	7,811	53,030	35,830	23,860	112,720
" 22..	538	4,542	5,005	1,620	1,301	13,006	30,060	7,340	8,990	46,390
" 29..	11,634	68,110	1,881	8,894	1,055	91,574	23,670	2,900	7,070	33,640
Aug. 5..	5,032	49,151	5,137	3,041	1,607	63,968	24,760	3,490	5,680	33,930
" 12..	2,465	4,137	2,963	3,861	2,858	16,284	21,960	4,230	8,830	35,020
" 19..	2,276	9,878	6,497	30	986	19,667	37,400	11,770	15,470	64,640
" 26..	1,369	14,225	606	3,622	120	19,942	18,550	2,130	5,550	26,230
Sep. 2..	3,950	68,751	1,081	4,590	3,563	81,935	25,990	2,830	10,540	39,360
" 9..	1,210	100,029	3,759	10,014	176	115,188	16,100	4,210	7,810	28,120
" 16..	7,782	65,390	1,680	3,432	1,837	80,121	11,030	2,920	7,450	21,400
" 23..	4,309	16,537	943	391	910	23,090	16,100	5,440	15,480	37,020
" 30..	3,619	41,496	1,003	2,167	893	49,178	19,250	5,960	9,370	34,580
Oct. 7..	890	16,113	135	....	86	17,224	20,250	6,510	12,370	39,130
" 14..	142	10,106	1,159	....	190	11,597	12,660	3,270	8,480	24,410
" 21..	4,383	54,885	2,249	4,337	1,604	67,458	16,490	7,190	5,840	29,520
" 28..	2,825	7,869	1,042	1,680	50	13,466	34,240	25,200	16,560	76,000
Nov. 4..	922	38,529	786	960	2,405	43,542	44,120	27,650	18,000	89,770
" 11..	1,264	13,154	1,064	1,325	120	16,927	39,440	19,690	12,950	72,080
" 18..	11,305	44,264	2,744	3,585	437	62,335	49,080	24,250	11,410	84,740
" 25..	3,301	68,229	3,043	2,580	1,134	78,287	61,750	47,000	13,790	122,540
Dec. 2..	4,194	40,498	3,479	2,556	278	51,005	16,980	8,050	6,170	32,200
" 9..	1,299	28,187	6,952	8,006	2,866	47,310	37,240	25,000	6,230	68,470
" 16..	3,134	9,214	4,826	6,757	881	24,812	25,980	11,720	4,720	42,420
" 23..	144	9,974	7,349	1,330	238	19,035	35,030	33,920	8,730	77,680
" 30..	4,144	2,834	17,907	583	10	25,478	24,970	11,960	6,960	43,890
Average prices & total sales, receipts & stocks.	186,956	1,396,856	312,641	219,378	45,775	2,161,606	1,632,850	626,570	506,350	2,765,779

POOL. YEAR 1864.

STOCKS.			PRICES.			Actual Export.	Consumption.	REMARKS.
Amer'n	Other.	Total.	Mid. Up.	Mid. Orl.	Dhol.			
33,976	228,919	262,895	27½	28¼	23½	2,924	26,270	
33,023	217,255	250,278	27¼	28	23	2,920	47,950	Disturbed state of Continental politics, &c.
31,753	232,743	264,496	27	27¾	23	6,437	65,860	
29,174	224,864	254,038	27	27½	23	4,698	89,110	German troubles.
34,109	228,376	262,485	26¾	27½	22¾	5,423	110,420	
30,439	205,218	235,657	27	27½	23	5,561	143,560	
27,340	219,445	246,785	27	27½	23	6,130	171,350	
23,852	207,710	231,562	26½	27¼	22½	7,127	195,670	Unsettled European politics.
23,712	217,916	241,628	26½	27¼	22½	6,082	222,560	
26,793	238,583	256,356	26¼	26¾	22	3,920	242,610	
28,406	232,455	260,861	26¾	27	22	4,653	292,030	
29,805	277,601	307,406	26½	26¾	22	5,385	312,530	Unfavorable Indian telegrams.
27,912	269,147	297,059	26¼	26¾	21½	8,436	339,799	
29,587	280,172	309,759	26⅛	26½	21½	7,270	374,139	
29,934	336,375	366,209	26½	27	21½	9,741	419,009	
28,045	386,415	414,460	26¾	27¼	21½	8,134	473,329	Heavy import.
22,529	345,070	367,599	27¼	27¾	21½	11,525	527,009	
24,438	333,488	357,926	27½	28	22	13,884	583,499	Bank rate advanced from 7 to 9 per cent.
21,109	301,480	322,589	28	28½	22½	17,810	630,139	
17,426	334,020	351,446	28	28½	22½	16,270	654,059	
18,707	322,903	341,610	28	28½	22½	9,650	680,649	Active trade and export demand.
15,642	310,313	325,955	28	28½	22½	12,570	709,479	
11,527	285,957	297,484	28¼	28¾	22¼	10,820	745,789	
17,664	287,060	304,724	29	29½	22½	17,130	790,859	
16,384	295,167	311,551	28¾	29¼	22½	11,200	821,219	
12,756	291,079	303,835	29½	30	22¾	22,220	871,099	
8,727	272,650	281,377	30¼	30¾	23¼	24,580	916,729	
1,721	210,455	212,176	30	31½	24	23,860	969,759	
3,810	181,090	184,900	31¼	31¾	24	8,990	999,819	
12,330	255,540	237,870	31	31½	24	7,070	1,023,489	
15,070	250,910	265,980	30½	31¼	24	5,680	1,048,249	Unfavorable crop accounts
14,360	232,280	246,640	30	30¾	23¾	8,830	1,070,209	
12,850	209,060	221,910	29¾	30¾	23¾	15,740	1,107,609	
10,250	196,370	206,620	29	30	23	5,550	1,126,159	
10,050	240,840	250,890	30	31	23	10,540	1,152,149	
8,600	338,390	346,990	28¾	29½	21½	7,810	1,168,249	Severe money pressure.
11,800	381,980	393,980	27¾	28½	19½	7,450	1,179,279	
12,330	378,860	391,180	27	27¾	18½	15,480	1,195,379	
12,880	399,551	412,430	26½	27	18	9,370	1,214,629	
12,170	383,620	395,790	25½	26	17½	12,370	1,234,879	Renewed depression.
10,850	371,290	382,140	23	24	16	8,480	1,247,539	
14,300	411,210	425,510	22½	23	14½	5,840	1,264,029	Rumors of large failures.
15,020	382,790	397,810	21½	22	15	16,560	1,298,269	
13,530	376,720	390,250	23	23½	15	18,000	1,342,389	
11,170	345,860	357,030	24	24¾	16¼	12,950	1,381,829	
17,750	337,410	355,560	24¾	25¼	17	11,410	1,430,909	
15,090	345,030	360,120	26¾	27½	18½	13,790	1,492,659	Strong warlike accounts from United States.
16,310	367,490	383,800	25¾	26½	17½	6,170	1,509,639	
13,280	370,290	383,570	26	26¾	17½	6,230	1,546,879	
12,310	363,910	376,220	25¾	26½	17½	4,720	1,572,859	
8,940	346,220	355,160	26¼	27	18½	8,730	1,607,889	Rumored capture of Savannah.
10,530	342,040	352,570	26½	27	18½	6,960	1,632,859	
			27.17	27.62	20.96	506,350	31,401.13	

The semi-weekly Price and Weekly Sales and Receipts, at New York, Weekly Exports from New York and Rates of Freight to Liverpool 1st of each Month, for the Crop Year ending September 1, 1864.

1863.		Price of Middling New Orleans	Price of Middling Upland.	Sales for week.	Receipts for week.	EXPORTS FOR WEEK.					Rates of Freight to Liverpool.	GENERAL REMARKS.
						To Great Britain.	To France.	North of Europe.	Other Fo'n Ports	Total Exports.		
Septem.	4..	70	70								—@—	The fluctuations in prices, this crop year, were many, and very wide; toward its close the highest point was touched, viz.: $1.90 currency per pound.
"	8..	69½	69½	12,000	4,001	....	....	....	....	....		
"	11..	68	68									
"	15..	69	69	5,400	2,910	118	....	....	....	118		
"	18..	70	70									
"	22..	75½	75½	14,500	3,024	190	....	....	....	190		In September, there was, for the most part, an active demand from spinners and speculators, and, with very favorable European advices and advancing rates for gold, prices steadily ascended, an advance of 9 cents per pound being established in three days.
"	25..	74	74									
"	29..	83	83	16,200	1,015	126	....	....	....	126		
October	2..	84½	84½								—@—	
"	6..	87	87	8,000	3,872	96	....	....	....	96		
"	9..	90	90									
"	13..	93	93	10,000	2,879	1,218	....	....	....	1,218		
"	16..	92	92									
"	20..	86	86	5,300	3,024	....	....	....	....	....		Through October, there was also a good demand, checked for a time by account of considerable receipts at New Orleans.
"	23..	85	85									
"	27..	87½	87½	7,700	5,221	939	....	....	....	939		
"	30..	87	87									
Novem.	3..	81	81	5,400	6,089	678	....	....	....	678	—@—	Advices of considerable receipts continuing at New Orleans, unsettled the market for a time in November, and prices were fluctuating, and, towards the last of the month, free receipts and favorable war news depressed the market. The market in December, was comparatively steady; the receipts at New Orleans fell off at one time and there was some advance.
"	6..	85½	85½									
"	10..	86½	86½	8,000	6,171	470	....	....	....	470		
"	13..	86	86									
"	17..	86	86	8,200	5,819	....	....	....	....	....		
"	20..	84½	84½									
"	24..	83½	83½	6,400	8,780	9	....	....	....	9		
"	27..	80	80									
Decem.	1..	80	80	4,300	5,537	196	....	....	....	196	—@—	
"	4..	80	80									
"	8..	80	80	8,500	5,666	88	....	....	....	88		
"	11..	80	80									In January, there was, for much of the month, an active spinning and speculative demand at advanced figures, closing, however, dull and easier.
"	15..	82½	82½	9,500	6,660	9	....	....	....	9		
"	18..	81½	81½									
"	22..	80	80	6,300	7,838	45	....	....	....	45		
"	25..	80	80									

Decem. 29.. 1864.	81½	81½	8,000	3,097						
January 1..	81	81								—@—
" 5..	81½	81½	5,100	4,545						
" 8..	80½	80½								
" 12..	81½	81½	8,900	3,442						
" 15..	82½	82½								
" 19..	84	84	12,300	6,838	44				44	
" 22..	84	84								
" 26..	84	84	9,300	5,425						
" 29..	83	83								
February 2..	83½	83½	7,200	2,160						—@—
" 5..	83½	83½								
" 9..	83½	83½	5,400	7,048	64				64	
" 12..	82	82.								
" 16..	81	81	4,800	7,075	51				51	
" 19..	80½	80½								
" 23..	80	80	3,500	5,773	7	53			60	
" 26..	80	80								
March 1..	79	79	6,500	7,341	42				42	—@—
" 4..	78	78								
" 8..	77	77	5,300	12,576	664				664	
" 11..	78	78								
" 15..	77	77	6,500	8,883	2,062				2,062	
" 18..	74	74								
" 22..	73	73	5,700	9,634	689	38			727	
" 25..	75	75								
" 29..	76	76	12,100	5,462	1,810				1,810	
April 1..	76	76								3-16@$\frac{7}{32}$
" 5..	76	76	5,800	7,190	1,202	60			1,262	
" 8..	76½	75⅛								
" 12..	80	79	14,800	7,574	1,116				1,116	
" 15..	81	80								
" 19..	79	78	10,500	9,435	1,601				1,601	
" 22..	81	80								
" 26..	84	83	13,700	7,240	1,846				1,846	
" 29..	84	83								
May 3..	86	85	11,000	4,926	1,972				1,972	⅛d.
" 6..	85	84								
" 10..	83	82	8,100	5,700	2,826	34	56		2,916	
" 13..	86	85								
" 17..	88	87	12,000	6,465	1,565		1		1,566	

February witnessed a dull and drooping market, the war between Prussia and Denmark, and high rates of discount in England, unsettling trade there.

March was dull, until the latter part, when an active demand set in, and values advanced.

The market, in the early part of April, was quiet, but later there was much activity at a considerable advance.

In May, more pacific foreign accounts came to hand, and with large purchases, for export and on speculation, prices again appreciated.

Through June, the market was much excited and very feverish, the course of prices generally being rapidly upward.

July was less active, but high figures were obtained.

In August, the stock became much reduced, and the highest prices ever known before or since, were touched, though the month and crop year closed comparatively quiet, with rather lower rates ruling.

Exchange

was very much unsettled by the course of gold. The quotation for 60 days' commercial bills on London, at the opening of September, was 39½@40½ per cent. premium, and steadily advanced until 53@57½ was touched; October opened at 54½@57½ per cent. premium, and advanced to 68½@72½, then fell back to 60@61; November opened at 59½@61 per cent. premium, and

New York Statement for 1864.—*Concluded.*

1864.	Price of Middling New Orleans	Price of Middling Upland.	Sales for week.	Receipts for week.	EXPORTS FOR THE WEEK.					Rates of Freight to Liverpool.
					To Great Britain.	To France.	North of Europe.	Other Fo'n Ports	Total Exports.	
May 20..	91	90								
" 24..	98	97	14,400	4,469	2,860	45	....	....	2,905	
" 27..	1.03	1.02								
" 31..	1.08	1.07	13,700	8,848	1,344	50	....	....	1,394	
June 3..	1.08	1.07								
" 7..	1.08	1.07	6,200	5,763	1,118	68	....	....	1,186	—@—
" 10..	1.20	1.19								
" 14..	1.30	1.29	13,000	3,798	701	....	....	....	701	
" 17..	1.50	1.49								
" 21..	1.48	1.47	5,600	5,281	152	....	....	....	152	
" 24..	1.47	1.46								
" 28..	1.47	1.46	3,600	3,400	213	....	....	....	213	
July 1..	1.55	1.54								
" 5..	1.55	1.54	6,700	3,520	42	....	....	....	42	—@—
" 8..	1.66	1.65								
" 12..	1.70	1.68	7,000	4,584	....	....	....	....	....	
" 15..	1.62	1.61								
" 19..	1.62	1.61	2,800	5,237	998	....	....	....	998	
" 22..	1.62	1.61								
" 26..	1.62	1.61	2,900	5,530	41	15	....	....	56	
" 29..	1.63	1.62								
August 2..	1.66	1.65	6,700	2,296	528	....	42	....	570	—@—
" 5..	1.74	1.73								
" 9..	1.77	1.76	5,500	4,457	434	....	....	....	434	
" 12..	1.74	1.73								
" 16..	1.77	1.76	3,200	3,420	130	....	....	....	130	
" 19..	1.81	1.80								
" 23..	1.90	1.89	7,000	2,342	147	....	4	....	151	
" 26..	1.88	1.87								
" 30..	1.80	1.80	2,700	2,514	37	....	....	....	37	
Septem. 2..	1.87	1.87								—@—
Average price and total sales, receipts and exports.	1.02	1.01½	413,200	281,794	30,488	363	103	....	30,954	

GENERAL REMARKS.

the premium went up to 66½@68½; in December, the range was not wide, the lowest was 60@62½ per cent. premium, and the highest 66@67; the quotation in early January was 65½@66½, advanced to 72½@74¼, and closed at 71@72; in February, the range was from 71¼@72¼, to 74½@75; March opened at 73¼@74¾, and went up to 79½@@82; April was very fluctuating, opening at 73¼@74¾ and closing at 95@98, touching at one time 98@100 per cent. premium; May opened at 94½@97½, fell back to 84@86, and then radidly went up to 105@110 per cent. premium; June opened at 108@110 per cent. premium, and steadily advanced to 135@138. The quotation 1st of July was 170@195 per cent premium, then fell off to 158@168, then advanced to 198@210 per cent. premium, and then declined to 170@174½; August opened at 176@180 per cent. premium, the ruling rate went up to 179@183, then fell off to 152@168, and then advanced to 160@175 per cent. premium.

LIVERPOOL STATEMENT FOR 1865.

UNITED STATES, 1864–1865.

Stock Sept. 1, 1864	unknown	Export	unknown
Crop	"	Consumption	"
		Stock Sept. 1, 1865	"
Bales		Bales	

CONSUMPTION. Tot. Europe.	Continent.	France.	Gt. Britain.	Stock Jan. 1, 1865, in	Gt. Britain.	France	Continent.	Tot. Europe.
				United States	23,000	....	1,000	24,000
				Brazil	16,000	2,000	1,000	19,000
				West Indies	5,000	1,000	....	6,000
				East Indies	502,000	44,000	10,000	556,000
				Egypt	30,000	7,000	6,000	43,000
				Bales	576,000	54,000	18,000	648,000
				IMPORT.				
362,000	55,000	26,000	281,000	United States	462,000	36,000	55,000	493,000
354,000	99,000	36,000	219,000	Brazil	340,000	37,000	99,000	375,000
179,000	37,000	33,000	109,000	West Indies	131,000	33,000	38,000	187,000
1,837,000	511,000	277,000	1,049,000	East Indies	1,408,000	253,000	508,000	1,491,000
716,000	143,000	197,000	376,000	Egypt	414,000	201,000	142,000	720,000
3,448,000	845,000	569,000	2,034,000	Bales	2,755,000	560,000	842,000	3,266,000
....	....	....	891,000	Export.				
466,000	15,000	45,000	406,000	Stock, Dec. 31. Stock above,	576,000	54,000	18,000	648,000
3,914,000	860,000	614,000	3,331,000	Total supply, bales	3,331,000	614,000	860,000	3,914,000

COTTON AT LIVER

Week Ending.	Receipts.						Sales.			
	American.	E. I.	Egypt.	Brazil.	Other.	Total.	Consumption.	Speculation.	Export.	Total.
Jan. 6..	4,368	22,944	11,222	7,452	158	46,144	10,850	2,480	4,150	17,480
" 13..	7,036	16,721	9,957	4,097	2,632	40,443	20,240	6,700	5,640	32,580
" 20..	5,403	4,820	6,521	1,842	382	18,968	13,350	2,050	3,310	18,710
" 27..	2,453	3,621	12,929	4,127	91	23,221	26,460	1,900	6,860	35,220
Feb. 3..	294	2,846	21,169	3,146	40	27,495	20,370	6,540	4,680	31,590
" 10..	10,514	29,452	16,355	18,183	23	74,527	33,080	14,810	5,390	53,280
" 17..	6,924	46	18,825	7,694	7,795	41,284	17,130	3,260	2,940	23,330
" 24..	8,647	28,184	1,848	9,111	1,734	49,524	25,780	7,160	3,580	36,520
Mch. 3..	10,698	12,693	6,801	15,027	3,115	48,334	32,760	9,110	5,780	47,650
" 10..	5,103	15,897	9,967	3,955	114	35,036	30,940	4,790	5,130	40,860
" 17..	1,319	18,549	8,866	5,295	3,503	37,532	49,790	15,490	9,720	75,000
" 24..	1,962	28,530	7,818	18,779	64	57,153	20,520	2,840	5,240	28,600
" 31..	2,660	9,941	6,397	1,833	839	21,670	25,460	7,420	5,450	38,330
April 7..	15,963	22,035	8,614	16,031	500	63,143	31,760	4,780	5,940	42,480
" 13..	2,383	510	13,150	4,194	2,660	22,897	22,510	2,000	5,360	29,870
" 21..	1,703	11,129	4,940	1,228	3,652	22,652	42,180	12,920	13,010	68,110
" 28..	2,898	1,419	7,210	1,061	190	12,778	62,980	17,820	25,100	105,900
May 5..	6,938	31,326	6,365	12,535	5,231	62,395	35,900	4,800	18,890	59,590
" 12..	8,356	39,094	2,962	5,734	2,226	58,372	37,330	12,130	21,580	71,040
" 19..	3,897	10,190	5,871	4,400	6,103	30,461	46,820	9,210	32,010	88,040
" 26..	8,001	39,005	2,787	10,390	3,362	63,545	59,230	16,770	32,770	108,770
June 2..	5,715	18,112	6,470	1,017	3,834	35,148	62,810	21,740	23,790	108,340
" 9..	1,689	20,516	4,920	1,242	285	28,652	38,660	14,730	27,370	80,760
" 16..	1,944	11,873	7,681	2,665	1,859	26,022	49,160	25,120	30,970	105,250
" 23..	1,822	2	6,064	500	33	8,421	64,000	45,600	30,230	139,830
" 30..	7,419	31,784	7,707	7,598	4,769	59,276	26,510	28,030	14,360	68,900
July 7..	3,193	57,113	2,444	13,939	956	77,645	17,910	6,740	6,870	31,520
" 14..	4,451	14,792	8,003	5,727	2,740	35,713	26,970	7,580	10,680	45,230
" 21..	3,945	11,197	8,519	2,318	1,657	27,636	48,030	22,390	17,390	87,810
" 28..	7,041	15,019	7,303	2,785	55	32,203	31,970	7,310	10,200	49,480
Aug. 4..	4,847	43,932	5,552	2,000	7,661	63,992	30,220	3,100	8,790	42,110
" 11..	1,787	56,638	10,938	15,846	37	85,246	47,920	4,490	12,860	65,270
" 18..	4,179	119,743	7,401	9,099	1,961	142,383	32,070	3,440	14,200	49,710
" 25..	6,007	3,062	2,592	111	6,112	17,884	49,110	12,460	25,020	86,590
Sept. 1..	8,902	39,995	7,666	3,927	191	60,681	67,490	15.520	26,320	109,330
" 8..	3,027	16,075	9,036	2,503	2,596	33,237	63,990	21,330	28,360	113,680
" 15..	8,032	94,344	5,284	16,001	3,216	126,877	40,250	10,820	19,630	70,700
" 22..	4,193	11,527	6,003	2,541	1,293	25,557	84,720	64,430	38,820	187,970
" 29..	6,059	5,181	8,181	3,909	1,511	24,841	81,430	84,230	23,170	188,830
Oct. 6..	8,519	6,946	8,066	828	2,161	26,520	70,100	98,800	10,290	179,190
" 13..	4,179	33,345	10,401	484	1,438	49,847	55,780	54,100	12,250	122,130
" 20..	23,877	83,493	5,416	12,543	1,427	126,756	34,590	32,260	19,610	86,460
" 27..	16,230	11,077	14,459	760	2,852	45,378	15,360	17,130	11,690	44,180
Nov. 3..	19,313	14,708	4,620	13,300	3,798	55,739	22,080	15,920	13,130	51,130
" 10..	9,067	13,845	3,329	1,192	310	27,743	29,620	12,130	14,890	56,640
" 16..	17,517	3,292	6,412	1,032	1,419	29,672	28,350	8,300	9,070	45,720
" 23..	33,478	51,897	9,506	21,754	3,343	119,978	33,350	6,880	10,850	51,080
" 30..	29,343	16,752	2,444	5,548	240	54,327	58,100	18,190	19,940	96,230
Dec. 7..	18,264	7,554	3,184	7,899	3,417	40,318	45,840	25,140	28,120	99,100
" 14..	17,840	27,794	4,556	14,211	4,097	68,498	46,360	11,770	22,260	80,390
" 21..	34,586	16,787	5,080	3,333	3,119	62,905	38,110	7,730	13,080	58,920
" 28..	25,384	14,269	12,895	10,128	528	63,204	32,070	7,080	9,430	48,580
Average prices & total sales, receipts & stocks.	461,369	1,208,616	396,692	334,674	113,419	2,514,770	2,138,370	889,470	786,110	3,813,950

POOL. YEAR 1865.

Stocks.			Prices.			Actual Export.	Consumption.	Remarks.
American.	Other.	Total.	Mid. Up.	Mid. Orl.	Dhol.			
26,360	439,900	496,260	26	26¾	18	4,150	10,850	
31,350	478,940	510,290	25½	26¼	18	5,640	31,090	
34,440	474,900	509,340	24¼	24¾	17¼	3,310	44,440	Limited demand.
34,870	464,480	499,350	23¼	24¼	17	6,860	70,900	
32,930	467,110	500,040	22	22¾	16	4,680	91,270	Rumored peace negoti'ns.
39,870	496,790	536,660	22	22¾	16	5,390	124,350	Peace nego's unsuccessful.
45,490	511,480	556,970	20¼	22½	15	2,940	141,480	Important peace negoti-
50,970	506,690	577,660	18¾	19½	14	3,580	167,260	ations.
58,120	531,240	589,360	18	18½	13	5,780	200,020	
66,580	521,830	588,410	16	16½	11½	5,130	230,960	Charleston evacuated.
55,420	516,090	572,110	16¾	17	12	9,720	280,750	
55,400	647,630	603,030	16	16½	11	5,240	301,270	
48,650	531,360	580,010	14½	15½	10	5,450	326,730	
61,880	544,040	605,920	14¼	14¾	9½	5,940	358,490	
62,070	538,460	600,530	13¾	14½	9	5,360	381,000	
58,730	516,750	575,480	13	13½	9	13,010	423,180	Richmond taken.
54,760	465,390	520,150	14¼	14¾	9	25,100	486,160	Surrender of Gen. Lee—
57,440	479,570	537,010	14¼	14¾	9	18,890	522,060	Pres. Lincoln assassin'd.
57,050	484,720	541,770	14	14½	9½	21,580	559,390	Large destruction of cot-
51,480	454,940	506,420	14¼	14½	9½	32,010	606,210	ton in United States.
49,710	444,740	494,450	15¼	15½	10	32,770	665,440	Favorable Manchester reports.
46,540	393,560	440,100	16½	16¾	10¾	23,790	728,250	
41,530	364,960	406,490	16¾	17	10¾	27,370	766,910	
36,360	322,460	358,820	17¾	18	12	30,970	816,070	
29,600	247,390	276,990	19½	20	14	30,230	880,070	Better feeling — larger demand.
28,480	279,550	308,030	19¾	20¼	14½	14,360	906,550	
29,130	339,100	368,230	19¼	19½	13¼	6,870	924,490	
29,520	331,550	361,070	19	19¼	13¼	10,680	951,460	
27,210	305,490	332,700	19¾	20	14	17,390	999,490	
30,460	291,840	322,300	19½	19¾	14	10,200	1,031,460	
31,320	316,260	347,580	19	19¼	13¼	8,790	1,061,680	
26,280	347,460	373,740	19	19¼	13¼	12,860	1,109,600	
26,700	443,890	470,590	18½	18¾	12½	14,200	1,141,670	
27,250	402,810	430,060	18¼	18½	12½	25,020	1,190,780	
29,980	381,780	411,760	18¼	18¾	12½	26,320	1,258,270	Great activity in manufacturing districts.
25,850	335,290	361,140	18½	18¾	13	28,360	1,322,260	
28,410	398,980	427,590	18¼	18¾	12¾	19,630	1,362,510	
23,590	326,280	349,870	19	19¼	14	38,820	1,447,230	
20,710	240,790	261,500	21½	21¾	14¾	23,170	1,528,660	
22,690	195,270	217,960	24	24¼	17½	10,290	1,598,760	
19,230	182,740	201,970	24½	24⅜	18¼	12,250	1,654,540	
37,350	243,020	280,370	23¾	24	18¼	19,610	1,689,130	Lord Palmerston's death
50,400	252,690	303,090	22	22¼	17¾	11,690	1,704,490	announced.
64,140	258,930	323,070	20½	20¾	16½	13,130	1,726,570	
64,170	242,090	306,260	20½	21	16¼	14,890	1,756,190	Heavy receipts in United States.
73,990	220,140	294,130	19¾	20	16	9,070	1,884,540	
97,770	272,050	369,820	19½	20	15¾	10,850	1,917,890	
115,100	241,110	356,210	20¾	21	16½	19,940	1,975,970	
120,530	218,700	339,230	21¼	21½	17	28,120	2,021,830	Better feeling.
126,960	221,130	348,090	21¼	21¾	18	22,260	2,068,190	
148,170	204,580	352,750	21	21½	18	13,080	2,106,300	
163,410	205,080	368,490	20¾	20¼	17¾	9,430	2,138,370	
			19.11	20.71	13.87	786,110	41,122.5	

The semi-weekly Price and Weekly Sales and Receipts at New York, Weekly Exports from New York and Rates of Freight to Liverpool 1st of each month, for the Crop Year ending September 1, 1865.

1864.	Price of Middling New Orleans	Price of Middling Upland.	Sales for week.	Receipts for week.	EXPORTS FOR THE WEEK. To Great Britain.	To France.	North of Europe.	Other Fo'n Ports	Total Exports.	Rates of Freight to Liverpool.
Septem. 6..	1.82	1.82	3,900	788	....	....	24	....	24	—@—
" 9..	1.83	1.82								
" 13..	1.76	1.75	1,900	3,735	25	....	....	....	25	
" 16..	1.76	1.75								
" 20.	1.70	1.69	1,300	2,930	....	....	....	....	....	
" 23..	1.55	1.55								
" 27..	1.20	1.20	1,400	3,681	101	30	....	....	131	
" 31..	1.20	1.20								
October 4..	1.20	1.20	1,641	4.719	127	....	....	....	127	—@—
" 7..	1.15	1.15								
" 11..	1.05	1.05	800	4.941	47	....	....	....	47	
" 14..	1.15	1.12								
" 18..	1.20	1.18	3.600	5,752	....	....	....	....	....	
" 21..	1.17	1.15								
" 25..	1.24	1.22	2,200	1,272	49	....	....	....	49	
" 28..	1.23	1.23								
Novem. 1..	1.28	1.28	4,000	3,228	....	....	....	....	....	—@—
" 4..	1.27	1.27								
" 8..	1.35	1.35	5,600	3,172	49	....	....	....	49	
" 11..	1.41	1.41								
" 15..	1.41	1.41	6,000	4,791	....	....	....	....	....	
" 18..	1.30	1.30								
" 22..	1.31	1.31	3,500	2.376	199	....	....	....	199	
" 25..	1.30	1.30								
" 29..	1.30	1.29	3,500	5,679	....	....	....	....	....	—@—
Decem. 2..	1.30	1.29								
" 6..	1.28	1.27	2,700	6,174	....	....	....	....	....	
" 9..	1.33	1.32								
" 13..	1.33	1.32	7,800	3,955	14	....	....	....	14	
" 16..	1.33	1.32								
" 20..	1.25	1.24	4,200	6,816	874	....	....	....	874	
" 23..	1.25	1.24								
" 27..	1.15	1.14	2,100	2.643	....	....	....	....	....	

GENERAL REMARKS.

This crop year witnessed an immense fall, as calculated in currency, of the staple, prices running down from 183 cents per lb. to 37 cents, though subsequently rallying to 44 cents.

Through September, the market was very dull, with prices rapidly receding, the fall, at one period, within three days, being 35 cents per lb.

October was dull and very irregular. An active speculative demand prevailed through part of November, which rallied the market. December was dull, and prices steadily fell off.

The occupation of Savannah by the Government forces, early in January, had a most depressing effect, and prices dropped 35 cents per lb.

The receipt of about 20,000 bales from Savannah, in February, and the occupation of Charleston by the Federal army, still further reduced prices; in March, the market was very unsettled, and the fall of Richmond, with the accompanying circumstances, in April, put values down to their lowest ebb for the crop year; at the low rates now ruling, there set in a good spinning and speculative demand, in May,

Decem. 30..	1.20	1.19								
1865.										
January 3..	1.21	1.20	3,800	7,607	....	....	....	....	....	—@—
" 6..	1.20	1.19								
" 10..	1.16	1.15	3,600	3,480	9	....	....	....	9	
" 13..	1.11	1.10								
" 17..	1.02	1.01	2,400	6,248	....	....	....	....	....	
" 20..	1.00	98								
" 24..	85	84	3,400	4,244	540	....	....	....	540	
" 27..	90	89								
" 31.	86	85	5,000	7,221	1,263	....	....	....	1,263	
February 3..	80	78								—@—
" 7..	88	87	5,500	4,273	916	....	....	....	916	
" 10..	88	87								
" 14..	82	81	4,300	2,921	3,147	....	89	....	3,236	
" 17..	85	84								
" 21..	84	83	4,100	12,094	1,000	....	....	....	1,000	
" 24..	84	83								
" 28..	84	83	5,600	10,442	610	....	....	....	610	
March 3..	83	82								⅛@¼d.
" 7..	81	80	3,700	14,864	533	324	....	....	857	
" 10..	73	72								
" 14..	71	70	2,800	7,100	434	....	76	....	510	
" 17..	60	58								
" 21..	55	53	2,000	8,575	1,066	....	....	....	1,066	
" 24..	47	45								
" 28..	50	50	4,000	10,197	987	....	....	....	987	
" 31..	48	47								
April 4..	37	35	2,600	8,306	618	66	62	....	746	⅛@¼d.
" 7..	37	35								
" 11..	37	35	6,200	10,909	2,626	....	15	....	2,641	
" 14..	37	35								
" 18..	37	35	8,300	4,703	978	....	78	....	1,056	
" 21..	41	40								
" 25..	51	50	9,000	4,302	762	....	....	....	762	
" 28..	51	50								
May 2..	45	45	11,400	1,669	149	250	126	....	525	5-32d.
" 5..	49	48								
" 9..	57	56	23,500	1,203	113	....	....	....	113	
" 12..	54	53								
" 16	50	49	8,100	4,300	203	....	....	....	203	
" 19..	55	54								

and during part of the month there was much activity at better prices.
Through the most of June and July, the market was active—under favorable foreign advices the export, spinning, and speculative demand being large.
In August, there was less activity, and prices yielded.

New York Statement for Year 1865—*Concluded.*

1865.		Price of Middling New Orleans	Price of Middling Upland.	Sales for week.	Receipts for week.	EXPORTS FOR WEEK. To Great Britain.	To France.	North of Europe.	Other Fo'n Ports	Total Exports.	Rates of Freight to Liverpool.
May	23..	55	54	17,300	9,449	219	....	....	....	219	
"	26..	49	49								
"	30..	48	48	4,800	4,772	28	....	1	....	29	
June	2..	46	46								3-16@¼d
"	6..	43	42	14,300	8,550	....	....	....	....	....	
"	9..	41	41								
"	13..	43	42	6,100	8,062	96	....	....	....	96	
"	16..	42	41								
"	20..	40	40	6,800	7,669	1,242	....	....	....	1,242	
"	23..	42	41								
"	27..	46	46	18,100	6,760	971	100	101	....	1,172	
"	30..	44	44								
July	4..	50	50	27,500	11,176	....	....	....	....	....	3-16@¼d
"	7..	50	50								
"	11..	53	52	10,100	9,941	6,172	....	....	....	6,172	
"	14..	52	51								
"	18..	51	49	11,400	16,387	1,980	....	....	....	1,980	
"	21..	48	48								
"	25..	47	46	15,700	12,790	1,215	....	....	....	1,215	
"	28..	48	48								
August	1..	48	48	19,300	21,689	3,366	....	....	....	3,366	⅛@3-16d
"	4..	47	47								
"	8..	45	45	10,000	14,782	4,573	124	....	....	4,697	
"	11..	45	45								
"	15..	44	43	14,600	17,428	6,379	....	....	....	6,379	
"	18..	44	44								
"	22..	46	45	16,500	22,046	3,958	95	30	....	4,083	
"	25..	45	44								
"	29..	44	43	12,400	18,824	4,874	....	100	....	4,974	
Septem.	1..	44	43								⅛@3-16d
Average price and total sales, receipts and exports.		84.22	83.38	380,341	391,635	52,512	989	702	..	54,203	

General Remarks.

Exchange.

The quotation for 60 days' commercial bills on London, at the opening of September, was 160@175 per cent. premium, but rapidly went down until at the close it was 108@110 premium; in October, the lowest quotation was 105@108 per cent. premium, and the highest, 118@130; in November, the quotation was put upon a gold basis, and henceforward the market was comparatively steady the rate of bills fluctuating between 8¾@10 per cent. premium, gold; December, steady, at 9¼@9¾ per cent. premium, gold; January, 9½@9¾; in February, the quotation declined from 9½@9⅝ to 6½@7½ premium; in March, the range was from 6¾@9 per cent premium; April, 7½@9½; May, 8¼@9⅞; in June, the lowest quotation was 8@9 per cent. premium, the highest, 9¼@10; in July, the extremes were 7½ up to 9 premium; and in August, the range was from 7¼@9 per cent. premium, gold.

1866.

COTTON CROP OF THE UNITED STATES.

Statement and Total Amount of the Cotton Crop of the United States, for the Year ending August 31, 1866.

	Bales.	Bales.	TOTAL. 1866.	1861.	1860.
LOUISIANA.					
Export from New Orleans—					
To Foreign Ports	516,188				
Coastwise	252,355				
Stock on hand 1st September, 1866	102,082				
		870,625			
Deduct—					
Received from Mobile	26,483				
" Montgomery, Ala	4,378				
" Florida	12,785				
" Texas	32,111				
Stock on hand 1st September, 1865	83,239				
		158,996			
			711,629	1,751,599	2,139,425
ALABAMA.					
Export from Mobile—					
To Foreign Ports	270,934				
Coastwise	142,764				
Burnt and lost	6,307				
Stock on hand 1st September, 1866	29,009				
		449,014			
Deduct—					
Stock on hand 1st September, 1865	....	24,290			
			424,724	546,794	843,012
TEXAS.					
Export from Galveston, &c—					
To Foreign Ports	64,308				
Coastwise	116,023				
Stock on hand 1st September, 1866	8,511				
		188,842			
Deduct—					
Stock on hand 1st September, 1865	....	13,857			
			174,985	144,747	252,424
FLORIDA.					
Export from Apalachicola, St. Marks, &c.—					
To Foreign Ports	37,977				
Coastwise	123,650				
Stock on hand 1st September, 1866	162				
		161,789			
Deduct—					
Stock on hand 1st September, 1865	....	12,650			
			149,139	121,172	192,724
GEORGIA.					
Export from Savannah—					
To Foreign Ports—Uplands	90,425				
" Sea Island	4,937				
Coastwise—Uplands	162,267				
Sea Island	6,020				
Stock in Savannah, 1st September, 1866	3,240				
	266,889				
Export from Darien, Ga.—					
To New York	489				
		267,378			
Deduct—					
Stock in Savannah 1st September, 1865	....	4,005			
			263,373	477,584	525,219

Statement and Total Amount of the Cotton Crop of the United States, for the Year ending August 31, 1866—*Concluded.*

	Bales.	Bales.	TOTAL.		
			1866.	1861.	1860.
SOUTH CAROLINA.					
Export from Charleston, S. C.—					
To Foreign Ports—Uplands and Sea Island.	53,807				
Coastwise—Uplands and Sea Island........	54,147				
(Total export of Sea Island, 5,630 bales)					
Stock in Charleston, 1st September, 1866 ..	5,535				
	113,489				
Exp't from Georget'n & Port Royal, S.C.—					
To New York 1,645					
Boston 56					
	1,701				
		115,190			
Deduct—					
Stock in Charleston, 1st September, 1865 ..	....	1,972			
			113,218	336,339	510,109
NORTH CAROLINA.					
Export—					
To Foreign Ports........................	21				
Coastwise...............................	64,538				
		....	64,559	56,295	41,194
VIRGINIA.					
Export—					
To Coastwise Ports......................	27,732				
Manufactured (taken from the ports)	6,333				
Stock on hand, 1st September, 1866	3,466				
		....	37,531	78,132	56,987
TENNESSEE, &c.					
Shipments from Memphis, Tenn............	218,504				
" other places in Tennessee ..	35,000				
" Kentucky	5,000				
" Illinois, Indiana, &c	30,000				
Stock at Memphis, 1st September, 1866	10,831				
		299,335			
Deduct—					
Shipments to New Orleans	40,000				
Manufactured on the Ohio, &c..............	35,000				
Stock on hand 1st September, 1865	12,450				
		87,450			
			*211,885	143,424	108,676
Total crop of the United States........	....	....	2,151,043	3,656,086	4,669,770
Receipts at all the ports from close of the war to September 1, 1865	....	....	420,000		
Total Receipts at the ports since the close of the war—say from May 1, 1865, to September 1, 1866 (16 months).............	....	....	2,571,043		

* Being the amount received at New York, Philadelphia, Baltimore and Boston, overland—say New York 136,517 bales, Philadelphia 51,002, Baltimore 3,300, and Boston 21,066—total 211,885 bales.

Export to Foreign Ports, from September 1, 1865, *to August* 31, 1866.

FROM	To Great Britain.	To France.	To North of Europe.	Other F'n Ports.	Total.
New Orleans, La .... bales.	358,878	134,510	5,422	17,378	516,188
Mobile, Ala	229,171	40,184	270	1,369	270,934
Galveston, Texas	59,435	1,739	3,014	120	64,308
Apalachicola, Fla	37,977	....	....	....	37,977
Savannah, Ga	93,870	1,492	....	....	95,362
Charleston, S. C	46,935	6,050	....	822	53,807
Virginia	....	....	....	....	....
Wilmington, N. C	21	....	....	....	21
New York	415,481	36,675	39,695	3,458	495,309
Baltimore	6,709	....	....	....	6,709
Philadelphia	2,035	....	....	....	2,035
Boston	11,759	....	246	9	12,014
Grand total 1865-66	1,262,271	220,650	48,647	23,096	1,554,664
Total 1860-61	2,175,225	578,063	216,250	158,030	3,027,568
Decrease	912,954	357,413	167,603	134,934	1,572,904

Consumption.

Total crop of the United States, as before stated .... bales.			2,151,043
Add--Stocks on hand at the commencement of the year, 1st September, 1865.—In the Southern ports		152,463	
" Northern "		95,662	
			248,125
Makes a supply of			2,399,168
Deduct therefrom—The export to Foreign ports	1,554,664		
Less, foreign included	7,763		
		1,546,901	
Stocks on hand, September 1, 1866—			
In the Southern ports	162,836		
" Northern "	120,856		
		283,692	
Burnt at New York and Mobile	21,590		
Manufactured in Virginia	6,333		
		27,923	
			1,858,516
Taken for home use north of Virginia .... bales.			540,652
Taken for home use in Virginia and south and west of Virginia			126,640
Total consumed in the United States (including burnt at the ports), 1865-66...			667,292

ANNUAL REVIEW.

From the New Orleans Price Current—1865-66.

The accounts from the interior in regard to the maturing crop, indicate that it will not exceed 400,000 to 500,000 bales. A much larger amount was left on hand from previous years, but, from the best sources of information at the time, it was supposed that it did not exceed 1,000,000 bales, while many authorites put it down at 200,000 bales less. Had these estimates been correct, the entire supply, from the close of the war up to the present date, would have been from 1,400,000, to 1,700,000 bales, or about 850,000 less than the actual supply. This extraordinary discrepancy will be regarded as less remarkable, when it is considered that planters, country merchants, newspapers published in the interior, officials employed by the two governments, and traveling speculators overestimated the amount destroyed by the casualties of war and exposure to the weather, and equally underestimated that which had been concealed from observation by the prudence of its owners. Up to the middle of October, the course of the Liverpool market had been well calculated to stimulate and strengthen our own. Under a serious reduction in stock, a limited amount at sea, and an active demand for goods, the trade entertained grave apprehensions of a cotton famine. Up to the beginning of the month the prevailing estimates of the stock in this country, and the incoming crop did not exceed one and a half millions of bales, at least half of which it was expected would be required in the United States, leaving but seven hundred and fifty thousand bales for export. It is not surprising that, under such circumstances, the Liverpool market should have been excited, with a strong tendency to a speculative paroxysm. This spirit, however, was checked by the bank of England having suddenly raised its rate of discount, during the first week of October, from 4½ to 7 per cent.; but still the movement continued with little abatement, showing the great confidence of the trade in the future course of the market. Towards the latter part of the month, however, unusually heavy arrivals—the heaviest since the commencement of the war—which, in consequence of their paying large profits, were offered freely, caused a sharp turn in favor of buyers, and the market lost much of its previous buoyancy. The downward tendency was increased by the growing conviction that the estimates of the American supply had been much too low; this change of opinion being predicated on the unexpectedly liberal receipts at the Southern ports, and the views of the Northern correspondents who variously predicated a supply up to the present date, of, from two and a half millions, to three millions of bales. February opened with a nominal stock of 178,480 bales, and with rather more on sale than had been offered during the previous month.

With no increase in the demand, nor other modifying influences, prices would have consequently ruled in favor of buyers, without, perhaps any material decline; but the downward tendency was increased by a heavy reduction in foreign exchange, on the rates for which all foreign orders were predicated. The large sum realized from the sale of American securities on the Continent and Great Britain, had gone far to liquidate the amount due from this country for imports; the demand for sterling at New York fell off, and the market was completely glutted with cotton bills, which fell to two or three per cent. below the real par, while even at this decline the supply exceeded the demand. Extending its influence to our own market, commercial sterling declined to 143½ to 145½, against 147 to 148, in January. The month of July opened under the depressing influences of a net decline at Liverpool of ¾d., and prices gave way 1 to 2c. per pound; the falling off being the greatest in the lower grades, and as factors met the demand pretty freely at the reduction, the business during the first week was, to a fair extent, on the basis of 30 to 32c. for low middling; but the next Liverpool steamer bringing an account of an advance of ½d. and anticipations of a termination of the Continental war, the market rallied, and factors were enabled to establish an advance of 2c. per pound, putting low middling at 32 to 34c.

The semi-weekly Price and Weekly Sales and Receipts at New York, Weekly Exports from New York and Rates of Freight to Liverpool 1st of each month, for the Crop Year ending September 1, 1866.

1865.	Price of Middling New Orleans	Price of Middling Upland.	Sales for week.	Receipts for week.	EXPORTS FOR THE WEEK.					Rates of Freight to Liverpool.	GENERAL REMARKS.
					To Great Britain.	To France.	North of Europe.	Other Fo'n Ports	Total Exports.		
Septem. 5..	44	43	13,600	23,173	2,615	....	....	....	2,615	3-16d.	
" 8..	44½	43½									
" 12..	45½	44½	24,200	26,102	3,985	....	....	....	3,985		The perturbations in the market were less serious in this, than in the previous crop year, though still very considerable.
" 15..	45½	44									
" 19..	46	44½.	17,500	23,198	6,025	500	....	....	6,525		
" 22..	46	44½									
" 26..	45½	44	17,100	22,697	6,695	....	137	....	6,832		In September, there was a good demand, at higher prices than those ruling the previous month.
" 29..	46	44½									
October 3..	45	44	30,000	24,630	5,582	....	126	....	5,708	⅜@7-16d	
" 6..	52	50@51									October was very active, the demand being heavy from all buyers, spinners, exporters, and speculators, the foreign accounts being very exciting, the deliveries at Liverpool being "enormous," and the stock there greatly reduced; prices touched their highest point for this crop year this month.
" 10..	58@59	57	52,000	26,807	11,667	....	171	....	11,838		
" 13..	61	59@60									
" 17..	61	59@60	44,000	26,396	8,624	....	249	....	8,873		
" 20..	60	58									
" 24..	59@60	57	24,000	34,074	14,133	....	153	....	14,286		
" 27..	60	58									
" 31..	58@59	56	29,800	19,483	13,784	....	430	....	14,214		
Novem. 3..	58@59	56								5-16@⅜d	
" 7..	56	53@54	15,300	32,575	15,093	1,178	814	....	17,085		In November, the business fell off, and prices yielded; in December, there was a better demand again, and the downward tendency was, for a time, arrested.
" 10..	53	51									
" 14..	52	50	19,500	15,260	12,670	602	785	....	14,057		
" 17..	53	51									
" 21..	54	52	24,000	22,258	8,871	....	1,620	....	10,491		
" 24..	54	52									During a portion of January, the market was dull and unsettled, but afterward there was considerable activity, with some advance in prices.
" 28..	54	52	19,000	24,105	5,742	387	614	169	6,912		
Decem. 1..	52	50								5-16d.	
" 5..	52	50	19,500	31,330	12,665	....	142	....	12,807		
" 8..	51	49									
" 12..	49	47½	12,000	30,342	6,038	2,090	633	....	8,761		In February, the month, as a whole, was dull, and prices softened.
" 15..	51	49									
" 19..	52	50	35,800	19,975	13,530	....	499	....	14,029		March was irregular, though for the most of the month there was considerable activity, the general course
" 22..	53	51									
" 26..	53	51	20,700	18,983	11,064	518	844	440	12,866		

Decem. 29..	$53\frac{1}{2}$	$51\frac{1}{2}$								
1866.										
January 2..	$53\frac{1}{2}$	$51\frac{1}{2}$	22,300	30,206	5,531	2,162	883	....	8,576	$\frac{3}{8}$@7-16d
" 5..	$53\frac{1}{2}$	$51\frac{1}{2}$								
" 9..	51	49	25,300	16,676	7.306	431	875	....	8,612	
" 12..	51	49								
" 16..	53	51	29,000	18,093	11,325	771	424	....	12,520	
" 19..	53	51								
" 23..	52	$49\frac{1}{2}$	15,800	34,889	12,213	787	792	....	13,792	
" 26..	50	$48\frac{1}{2}$								
" 30..	51	49	22,000	17,084	7,014	1,074	1,664	42	9,794	
February 2..	50	48								$\frac{3}{8}$d.
" 6..	49	$47\frac{1}{2}$	13,700	25,952	17,990	2,667	1,030	....	21,687	
" 9..	48	46								
" 13..	47	45	17,200	12,081	6,969	....	938	....	7,907	
" 16..	47	45								
" 20..	47	45	17,500	19,208	6,281	735	2,242	....	9,258	
" 23..	47	45								
" 27..	46	$44\frac{1}{2}$	14,600	17.078	5,411	1,735	2,010	....	9,156	
March 2..	$45\frac{1}{2}$	44								5-16@$\frac{3}{8}$d
" 6..	45	$43\frac{1}{2}$	19,700	8,429	15,210	529	2,568	....	18,307	
" 9..	43	41								
" 13..	43	41	15,000	12,304	5,411	2,372	1,246	....	9,029	
" 16..	43	41								
" 20..	$41\frac{1}{2}$	$39\frac{1}{2}$	19,800	20,108	14,172	3,117	2,507	....	19,796	
" 23..	43	41								
" 27..	42	40	21,500	16,160	14,134	2,151	1,162	....	17,447	
" 30..	43	41								
April 3..	$41\frac{1}{2}$	$39\frac{1}{2}$	20,700	21,468	13,784	....	863	225	14,872	5-16@$\frac{3}{8}$d
" 6..	$40\frac{1}{2}$	$38\frac{1}{2}$								
" 10..	$39\frac{1}{2}$	$37\frac{1}{2}$	18,000	14,533	15,679	2,416	1,132	....	19,227	
" 13..	$39\frac{1}{2}$	$37\frac{1}{2}$								
" 17..	$39\frac{1}{2}$	$37\frac{1}{2}$	22,000	11,151	19,049	485	1,688	145	21,367	
" 20..	$38\frac{1}{2}$	$36\frac{1}{2}$								
" 24..	37	35	10,000	12,551	15,580	2,295	3,880	747	22,502	
" 27..	$34\frac{1}{2}$	$32\frac{1}{2}$								
May 1..	36	34	5,000	7,990	10,140	1,895	2,371	....	14,406	$\frac{1}{2}$@9-32d
" 4..	36	34								
" 8..	36	34	9,800	9,262	9,675	213	1,751	....	11,639	
" 11..	36	34								
" 15..	$36\frac{1}{2}$	35	14,500	3,297	6,422	1,005	636	....	8,063	
" 18..	38	36								

of prices, however, being in buyers' favor.

In April, the market was unsettled, by the apprehension of a war between Prussia and Austria, afterward realized; and with a panic in the Liverpool market, prices here rapidly fell off.

Through the early part of May, the Continental war affected business very unfavorably; the financial panic now prevailing in London, in consequence of the prominent failures there, depressed the Liverpool market; but toward the close of the month, these discouraging accounts were more than neutralized, by the unfavorable crop accounts from the South, the weather there being very wet, and rivers overflowing.

In June, prices fell off again, the Liverpool accounts continuing of an unfavorable character.

July witnessed the close of the Austro-Prussian war, and there was a better feeling, though not accompanied by any very marked advance.

Through part of August, there was more activity, but subsequently the foreign accounts were less assuring, and the reports relative to the growing crop being more favorable, prices receded. This month witnessed the successful laying of the Atlantic Telegraph.

35

New York Statement for Year 1866—*Concluded.*

1866.		Price of Middling New Orleans	Price of Middling Upland.	Sales for week.	Receipts for week.	EXPORTS FOR THE WEEK					Rates of Freight to Liverpool.
						To Great Britain.	To France.	North of Europe.	Other Fo'n Ports	Total Exports.	
May	22..	40	38	19,700	4,443	2,774	383	619	350	4,126	
"	25..	42½	40@40½								
"	29..	41	39	24,000	11,099	94	....	300	....	394	
June	1..	40½	38½								⅛d.
"	5..	40	38	5,800	7,299	37	327	71	....	435	
"	8..	40	38								
"	12..	41	39½	20,800	14,020	474	....	....	....	474	
"	15..	40½	38								
"	19..	40½	38	8,500	9,507	937	560	25	384	1,906	
"	22..	39	37								
"	26..	39	37	6,500	11,038	732	....	....	112	844	
"	29..	38	36								
July	3..	38	36	4,250	5,477	381	....	....	20	401	⅛d.
"	6..	38	36								
"	10..	37	35	3,900	10,468	626	46	....	....	672	
"	13..	37	35								
"	17..	37	35	6,400	7,993	804	34	....	35	873	
"	20..	37½	36								
"	24..	37½	36	15,000	7,837	2,808	....	....	....	2,808	
"	27..	37½	36								
"	31..	37½	36	8,300	5,987	3,451	75	....	....	3,526	
August	3..	37½	36								⅛d.
"	7..	37	35½	6,500	7,657	3,199	460	....	....	3,659	
"	10..	36	34½								
"	14..	35	33	7,800	5,646	4,415	483	....	....	4,898	
"	17..	36	34								
"	21..	35½	33½	12,000	6,462	3,350	....	483	548	4,381	
"	24..	35½	33½								
"	28..	35½	33½	12,000	4,707	13,320	1,533	318	241	15,412	
"	31..	35	33								5-32d.
Average prices and total sales, receipts and exports.		45.06	43.20	932,850	869,548	415,481	36,016	39,695	3,458	494,650	

General Remarks.

Exchange.

The quotation for 60 days' commercial bills on London ranged in September from 7½@9½ per cent. gold premium; October, 7@9½; November, 7@8⅞; December, 8¼@9¼; January, 7@9; February, 6@8¼; March, 5¾@8½; in April, the extreme range was from 5½@6¼ up to 7¼@8; May, 7½@9½; June, 7@9½; July, 7@8; August opened at 6½@7¼, and declined to 4¾@5½ per cent. premium, gold.

LIVERPOOL STATEMENT FOR 1866.

UNITED STATES, 1865–1866.

Stock Sept. 1, 1865	229,000	Export	1,555,000
		Consumption	544,000
Crop	2,154,000	Stock Sept. 1, 1866	284,000
Bales	2,383,000	Bales	2,383,000

	Gt. Britain.	France.	Continent.	Tot. Europe.
Stock 1st Jan., 1866, in				
United States, Bales	144,000	10,000	1,000	155,000
Brazil	36,000	3,000	1,000	40,000
West Indies	12,000	1,000	1,000	14,000
East Indies	183,000	20,000	7,000	210,000
Egypt	31,000	11,000	5,000	47,000
Bales	406,000	45,000	15,000	466,000

CONSUMPTION.								
Tot. Europe.	Continent.	France.	Gt. Britain.	IMPORT.	Gt. Britain.	France.	Continent.	Tot. Europe.
1,345,000	209,000	204,000	932,000	United States, Bales	1,163,000	241,000	215,000	1,411,000
488,000	135,000	63,000	290,000	Brazil	408,000	73,000	141,000	510,000
163,000	45,000	30,000	88,000	West Indies	111,000	36,000	47,000	173,000
1,776,000	656,000	187,000	933,000	East Indies	1,866,000	210,000	658,000	1,952,000
426,000	101,000	130,000	195,000	Egypt	201,000	128,000	101,000	416,000
4,198,000	1,146,000	614,000	2,438,000	Consumption, Bales	3,749,000	688,000	1,162,000	4,462,000
....	....	....	1,137,000	Export.				
730,000	31,000	119,000	580,000	Stock, Dec. 31 .... Stock above,	406,000	45,000	15,000	466,000
4,928,000	1,177,000	733,000	4,155,000	Total supply, bales	4,155,000	733,000	1,177,000	4,928,000

COTTON AT LIVER

Week Ending.	RECEIPTS.						SALES.			
	American	E. I.	Egypt.	Brazil.	Other.	Total.	Con-sumption.	Specu-lation.	Export.	Total.
Jan. 4..	43,163	13,835	4,727	5,017	569	67,311	35,110	6,820	13,440	55,370
" 11..	23,411	9,311	9,670	1,570	1,836	45,798	32,700	5,740	11,760	50,200
" 18..	34,329	17,900	2,916	11,335	3,711	70,191	39,300	3,340	14,250	56,890
" 25..	22,886	19,618	5,636	17,317	390	66,247	39,700	4,790	13,860	58,350
Feb. 1..	17,315	11,494	8,298	2,470	5,532	40,109	35,740	3,920	11,780	51,440
" 8..	43,358	12,871	8,450	12,193	951	77,823	43,350	9,260	16,970	69,580
" 15..	28,072	13,080	3,321	13,847	1,870	60,190	36,230	7,380	12,580	56,190
" 22..	21,141	340	4,251	2,028	2,865	30,625	44,160	3,440	11,000	58,600
Mch. 1..	3,003	....	4,206	11,551	92	18,852	52,060	9,760	14,180	76,000
" 8..	19,118	5,699	10,460	9,977	283	45,537	54,270	6,590	12,900	73,760
" 15..	23,426	13,137	2,222	4,047	231	43,063	58,260	16,310	16,720	91,290
" 22..	47,376	56,852	10,507	19,971	2,055	136,761	45,580	14,370	10,830	70,780
" 28..	59,170	67,018	6,817	16,503	2,143	151,651	27,120	3,270	8,100	38,490
April 5..	53,575	51,637	7,350	25,223	1,257	139,042	28,360	3,360	9,780	41,500
" 12..	25,116	12,338	15,790	12,053	1,894	67,191	32,400	3,560	13,140	49,100
" 19..	45,048	108,307	10,287	22,792	5,493	191,414	39,850	8,710	19,280	67,840
" 26..	33,893	16,854	4,318	8,732	12,063	75,860	60,710	10,370	24,470	95,550
May 3..	16,502	33	32	......	......	16,567	35,430	4,650	9,550	49,630
" 10..	65,124	57,088	3,786	20,812	4,141	150,951	38,300	3,520	4,010	45,830
" 17..	77,239	70,474	624	24,848	608	173,793	49,950	5,140	4,980	60,070
" 24..	15,293	58,584	1,246	3,293	1,838	80,254	36,210	1,520	6,110	43,840
" 31..	16,377	16,268	594	9,161	256	42,656	63,970	12,710	9,480	86,160
June 7..	23,145	52,257	286	9,902	3,158	88,748	55,200	6,710	9,580	71,490
" 14..	60,690	36,282	2,632	12,898	1,581	114,083	56,600	4,710	10,400	71,710
" 21..	32,574	15,265	841	5,744	2,469	56,893	40,070	3,110	5,320	48,500
" 28..	13,226	30,054	711	3,950	165	48,106	62,270	5,360	10,450	78,080
July 5..	26,659	34,927	1,772	10,216	175	73,749	64,850	8,630	14,360	87,840
" 12..	8,104	19,123	190	2,785	1,836	32,038	64,520	10,370	23,180	98,070
" 19..	5,797	10,146	2,418	....	2,679	21,040	45,440	5,030	20,420	70,890
" 26..	6,801	61,679	....	8,269	2,029	78,778	56,160	13,420	32,930	102,510
Aug. 2..	10,585	61,707	2,881	9,766	1,392	86,331	44,750	5,070	24,780	74,600
" 9..	12,272	28,844	1,717	5,321	3,173	51,327	29,370	1,750	20,570	51,690
" 16..	8,416	54,580	1,571	1,927	889	67,383	57,230	3,790	24,200	85,220
" 23..	4,402	40,536	1,578	1,750	269	48,535	37,570	1,840	21,100	60,510
" 30..	9,197	58,678	2,032	4,496	1,043	75,444	40,720	2,310	21,650	64,680
Sep. 6..	14,774	6,089	1,296	4,354	821	27,334	34,090	2,880	12,600	49,570
" 13..	4,108	41,482	2,970	4,230	536	53,326	51,530	2,690	15,780	70,000
" 20..	13,140	105,744	1,593	8,353	1,541	130,371	68,230	6,800	29,490	104,520
" 27..	13,219	15,444	2	1,519	1,430	31,614	69,500	17,960	44,370	131,830
Oct. 4..	2,323	19,045	1,555	....	147	23,070	60,540	31,830	34,200	126,570
" 11..	4,355	22,388	975	7,158	1,463	36,339	49,420	18,480	27,580	95,480
" 18..	5,502	13,970	2,200	3,537	910	26,119	46,320	32,030	25,460	103,810
" 25..	10,120	99,146	2,010	5,717	791	117,784	34,890	9,600	14,660	59,150
Nov. 1..	7,702	3,412	1,089	4,357	3,151	19,711	36,150	6,920	12,280	55,350
" 8..	9,033	10,747	3,284	6,494	903	30,461	48,920	9,500	17,140	75,560
" 15..	13,927	4,293	5,351	4,328	284	28,183	28,980	940	10,990	40,910
" 22..	12,742	15,302	6,119	3,882	3,108	41,153	42,790	6,960	14,930	64,680
" 29..	6,508	20,941	2,376	7,168	1,191	38,184	51,540	7,370	13,260	72,170
Dec. 6..	11,766	2,219	1,231	2,871	568	18,655	46,660	4,800	13,900	65,360
" 13..	11,695	6,301	9,038	1,798	74	28,906	69,050	19,460	18,650	107,160
" 20..	18,986	87	11,662	2,808	925	34,468	62,390	26,190	22,660	111,240
" 27..	13,429	19,233	3,789	2,316	1,495	40,262	45,010	13,470	12,500	70,980
Average prices & total sales, receipts & stocks.	1,102,828	1,515,769	200,608	383,654	90,774	3,293,633	2,429,520	448,670	845,580	3,723,770

POOL. YEAR 1866.

STOCKS.			PRICES.			ACTUAL EXPORT.	CONSUMPTION.	REMARKS.
American.	Other.	Total.	Mid. Up.	Mid. Orl.	Dhol.			
176,140	226,490	402,630	20½	21	17¾	13,440	35,110	
182,290	209,910	392,200	19¾	20	17¾	11,760	67,810	
196,630	221,080	417,710	19¾	20	17½	14,250	107,110	
196,890	227,570	424,460	19⅛	19½	17	13,860	146,810	
194,980	213,220	418,200	18½	18¾	16	11,780	182,550	Stringent money market.
207,800	221,510	429,310	19¼	19½	16	16,970	225,900	
215,160	227,810	442,970	19	19⅜	16	12,580	262,130	Large rec'ts at U.S. ports.
214,650	204,030	418,680	18½	18⅞	15¾	11,000	306,290	" "
218,400	182,450	400,850	18⅞	19¼	16¼	14,180	358,350	Reduced receipts "
203,850	102,900	366,750	18⅞	19¼	16½	12,900	412,620	
197,230	143,020	340,250	19⅜	19¾	17	16,720	470,880	
221,580	198,890	420,470	19½	20	16¾	10,830	516,460	Large import.
267,285	285,498	552,783	18¾	19¼	15½	8,100	543,580	
307,430	356,410	663,840	18	18½	15	97,800	571,940	
306,920	369,300	676,220	15⅝	16	13½	13,140	604,340	Market depressed, conti-
330,060	489,740	819,800	14½	14¾	11	19,280	644,190	nental difficulties.
334,320	487,630	821,950	15¼	15¾	11	24,470	704,900	
328,430	456,200	784,630	13¾	14½	10	9,550	740,330	Stoppage of Barned's B'k.
368,200	501,730	869,930	12¾	13½	9	4,010	778,630	
415,190	555,310	970,500	12¾	13½	8¼	4,980	828,580	
411,360	593,970	1,005,330	12	12½	8	6,110	864,790	Much anxiety, Continent.
399,800	575,230	975,030	13½	14	8¾	9,480	928,760	
419,280	579,490	998,770	13	13½	8¾	9,580	983,960	Failure Overend, Gurney
440,790	599,160	1,039,950	14	14½	8¾	10,400	1,040,560	& Co.
454,470	588,190	1,042,660	12½	13¼	8½	5,320	1,080,630	
444,950	513,720	958,670	13⅛	14	8½	10,450	1,142,900	Suspension Agra. & M.
429,700	537,870	967,570	14	14½	9¼	14,360	1,207,750	Bank.
411,750	511,440	923,190	14	14½	9¾	23,180	1,272,270	Outbreak of hostilities in
397,790	480,380	878,170	13¾	14¼	9½	20,420	1,317,710	Germany.
401,350	481,480	882,830	14	14½	10½	32,930	1,373,870	1st message by Atl. Cable.
389,350	512,880	902,330	14	14½	10½	24,780	1,418,620	Preston Bank suspended.
383,160	515,340	898,500	13½	14¼	10¼	20,570	1,447,990	Peace declared on Cont'nt.
343,600	438,740	882,340	13¾	14½	10¼	24,200	1,505,220	Bank rate 10 per cent.
326,190	539,550	865,740	13¾	14¼	10¼	21,100	1,542,790	" reduced 8 "
313,190	567,020	880,210	13½	14	10	21,650	1,583,510	" " 7 "
309,610	540,440	850,050	13	13½	9½	12,600	1,617,600	" " 6 "
285,630	541,940	827,570	13	13½	9¾	15,780	1,669,130	" " 5 "
266,700	605,260	871,960	13⅜	14	10⅛	29,490	1,737,360	Manchester accounts un-
268,990	505,490	774,480	14¼	14¾	10¾	44,370	1,806,860	favorable.
253,020	484,020	737,040	14⅝	15	11½	34,200	1,867,400	Unfavorable crop accounts
234,990	727,732	692,720	13½	13¾	11½	27,580	1,916,820	from United States.
218,550	428,450	647,000	15¼	15¾	12	25,460	1,963,140	
209,440	493,060	702,500	15	15¼	12	14,660	1,998,030	
190,730	474,430	665,160	15	15¼	11½	12,280	2,034,180	
189,740	451,250	630,990	14¾	15¼	11¼	17,140	2,083,100	Unfavorable Manchester
191,670	427,020	618,670	14	14½	10¾	10,990	2,112,080	accounts.
188,180	416,530	604,710	14	14½	10¾	14,930	2,154,870	
176,930	401,250	578,180	14	14¼	10¾	13,260	2,206,410	
172,830	366,180	539,010	13⅞	14½	10¾	13,900	2,253,070	
159,860	324,460	484,320	14	14¼	11¼	18,650	2,322,120	
154,840	287,390	442,230	14½	14¾	12	22,660	2,384,510	Improving.
150,700	276,050	426,750	15	15¼	12½	12,500	2,429,520	
			15.3	15.76	11.98	845,580	46,721.53	

The season ending on August 31, 1866, was, in some respects, one of the most remarkable on record. The war ended in the Spring of 1865, leaving Southern society totally disorganized. The railroads were torn up, or deficient in rolling stock, the planters were poorly supplied with draught animals; and besides, being doubtful of the policy of the General Government, many persons were in no hurry to expose any cotton they might possess. In the meantime, a four years' blockade left the South destitute of not only the luxuries, but the common necessaries of civilization. Four years of paper money *ad libitum* had upset all ideas of real value. The people bought eagerly at any price. This caused great excitement among the merchants at the North, who sent their agents in hot haste to Europe to make purchases. The great object was to get the goods and have them here as soon as possible. Price was a secondary consideration. In their turn, the manufacturers bought cotton with a somewhat similar feeling. In the meantime, receipts at the ports were small, and there was a cry of cotton famine. The result was an advance of 8d. in Liverpool and 20c. in New York, in an incredibly short time. It is a curious fact that at the very beginning of a peace which opened up to the world the best of all cotton producing countries, it should also have started a cotton famine panic; yet such was the case. Throughout the whole winter, Liverpool held up to about 18d.; a price which was amply sufficient to draw to that market every spare bale in the world—no, not quite! for people in this country were as much deluded as in Liverpool. Large stocks were retained in our ports, and a good deal on the plantations in the interior. Some of that old cotton never was sold until Liverpool went down to the neighborhood of 7d.

The crisis was reached in April, 1866, when the market declined 6d. in a very short time. During the distribution of the small crop of 1866, the decline was arrested; but, with the more favorable season of 1867, the decline was resumed, and continued until Liverpool touched 7d., about Christmas, 1867. The revulsion of 1866 was not confined to cotton. In fact, the catastrophe in cotton was merely one of the symptoms of a wide-spread disease, consequent to our civil war. It is probable that the losses in the cotton trade, in 1866, were three or four times greater than were ever before known in any revulsion in its history.

1867.

COTTON CROP OF THE UNITED STATES.

Statement and Total Amount of the Cotton Crop of the United States, for the Year ending August 31, 1867.

	Bales.	Bales.	TOTAL. 1867.	1866.	1861.
LOUISIANA.					
Export from New Orleans—					
To Foreign Ports	618,940				
Coastwise	248,376				
Stock on hand, 1st September, 1867	15,256				
		882,572			
Deduct—					
Received from Mobile	36,676				
" " Montgomery	10,792				
" " Florida	11,810				
" " Texas	19,081				
Stock on hand, 1st September, 1866	102,082				
		180,441			
			702,131	711,629	1,751,599
ALABAMA.					
Export from Mobile—					
To Foreign Ports	153,424				
Coastwise	108,950				
Burnt at Mobile,	2,437				
Stock on hand, 1st September, 1867	3,714				
		268,525			
Deduct—					
Stock on hand, 1st September, 1866	....	29,009			
			239,516	429,102	546,794
TEXAS.					
Export from Galveston, &c.—					
To Foreign Ports—including 6,470 to Mexico	76,918				
Coastwise	113,936				
Stock in Galveston, 1st September, 1867	2,654				
		193,508			
Deduct—					
Stock in Galveston, 1st September, 1866	....	7,589			
			185,919	174,985	144,747
FLORIDA.					
Export from Apalachicola, St. Marks, &c.—					
To Foreign Ports	3,019				
Coastwise—Uplands	42,875				
Sea Island	11,521				
Burnt at Apalachicola, &c.	1,189				
Stock in Apalachicola and St. Marks, 1st September, 1867	9				
Deduct—		58,613			
Stock in Apalachicola and St. Marks, 1st September, 1866	....	264			
			58,349	149,139	121,172
GEORGIA.					
Export from Savannah—					
To Foreign Ports—Uplands	106,449				
Sea Island	8,053				
Coastwise—Uplands	142,142				
Sea Island	7,058				
Burnt at Savannah	51				
Stock in Savannah, 1st September, 1867	633				
Export from Darien, Ga.—	264,386				
To New York	5				
		264,391			

Statement and Total Amount of the Cotton Crop of the United States, for the Year ending August 31, 1867.—Concluded.

	Bales.	Bales.	TOTAL. 1867.	1866.	1861.
Deduct—					
Received from Florida—Uplands	190				
Sea Island	4,996				
Stock in Savannah, 1st September, 1866	3,240				
		8,426			
SOUTH CAROLINA.			255,965	263,373	477,584
Export from Charleston—					
To Foreign Ports—Uplands	72,909				
Sea Island	7,987				
Coastwise—Uplands	80,942				
Sea Island	8,766				
Burnt at Beaufort and Hilton Head, S. C.—Sea Island	45				
Stock in Charleston, 1st September, 1867	1,228				
Export from Georgetown, Port Royal, &c.—	171,877				
To Northern Ports Uplands 915					
Sea Island 637					
	1,552				
Deduct—		173,429			
Received from Florida—Uplands	258				
Sea Island	5,389				
Stock in Charleston, 1st September, 1866	5,535				
		11,182			
NORTH CAROLINA.			162,247	112,273	336,339
Export—					
To Foreign Ports	534				
Coastwise	37,988				
		....	38,522	64,559	56,295
VIRGINIA.					
Export—					
To Foreign Ports	11,900				
Coastwise	96,693				
Manufactured—taken from the ports	15,000				
Burnt at Norfolk	2,500				
Stock in Norfolk and Petersburg, 1st September, 1867	1,000				
		127,093			
Deduct—					
Stock on hand, 1st September, 1866	....	3,466			
			123,627	37,531	78,132
TENNESSEE, &c.					
Shipments from Memphis, Tenn	227,377				
Nashville, Tenn	55,548				
other places in Tennessee, Kentucky, &c.	66,531				
Crop of Illinois, Indiana and Missouri	20,000				
Stock in Memphis and Nashville, 1st September, 1867	1,602				
Deduct—		371,058			
Shipments to New Orleans, from Memphis and Nashville	49,615				
Manufactured on the Ohio	49,000				
" in Pennsylvania, New York, &c.	75,000				
Stock in Memphis & Nashville, 1st Sept., 1866	11,731				
		185,346			
			*185,712	211,885	143,424
Total Crop of the United States 1866–67	....	....	1,951,988	2,154,476	3,656,086

*Being the amount received at New York, Philadelphia, Baltimore and Boston, Overland, from Tennessee, &c.

Export to Foreign Ports, from September 1, 1866, *to August* 31, 1867.

FROM	To Great Britain.	To France.	To North of Europe.	Other Foreign Ports.	Total.
New Orleans, La. (bales)	403,521	160,852	22,217	32,350	618,940
Mobile, Ala.	145,566	4,352	630	2,876	153,424
Galveston, Tex	60,751	....	9,697	6,470	76,918
Apalachicola, Florida	3,019	....	....	....	3,019
Savannah, Ga.	111,993	959	....	....	112,952
Charleston, S. C.	75,547	3,524	....	1,825	80,896
Norfolk, Va.	11,900	....	....	....	11,900
Wilmington, N. C.	534	....	....	....	534
New York	376,101	28,460	62,519	3,516	470,596
Baltimore	7,820	....	155	....	7,975
Philadelphia	2,650	....	....	....	2,650
Boston & Portland (Portland, 236 to Gt. Bt'n)	16,860	....	124	266	17,250
Grand total 1866-7	1,216,262	198,147	95,342	47,303	1,557,054
Total 1865-66	1,262,271	220,650	48,647	23,096	1,554,664
Increase	....	....	46,695	24,207	2,390
Decrease	46,009	22,503	....	....	....

Consumption.

Total crop of the United States, as before stated ....bales.			1,951,988
Add—Stocks on hand at the commencement of the year, September 1, 1866, in the Southern ports		162,836	
" Northern ports		120,856	
			283,692
Makes a supply of			2,235,680
Deduct therefrom—The export to Foreign ports	1,557,054		
Less—Foreign included	3,709		
		1,553,345	
Stocks on hand September 1, 1867—			
In the Southern ports	24,574		
In the Northern ports	55,722		
		80,296	
Burnt at New York, Mobile, Apalachicola, Norfolk, &c.	13,672		
Manufactured in Virginia	15,000		
		28,672	
			1,662,313
Taken for home use north of Virginia ....bales.			573,367
Taken for home use in Virginia, and elsewhere throughout the United States			280,672
Total consumed in the United States (including burnt at the ports), 1866-67			854,039

ANNUAL REVIEW.

From the New Orleans Price Current, 1866—67.

The severe drought which had succeeded the copious rains in June, along the southern belt of the cotton-growing section, had extended to the northern districts; there had been a general complaint that the plants had shed their forms, and had been checked in their growth, and caterpillars were doing more or less injury. Such was the prospect on the 1st of September, and the accounts which came to hand during the ensuing four or five weeks were well calculated to confirm all its unfavorable features. The reports of the ravages of the worm were more alarming, as well as more general, and parties not addicted to extravagant views materially reduced their previous estimates of the yield, as stated above. At the same time, the trade generally under-estimated the amount of old crops remaining in the interior towns and in planters' hands. While such was the prospect on this side, it was manifest, that, unless there should be a much larger supply of American, than could be relied on, the diminution in the stock of long staple kinds at Liverpool would cause a further advance in that market. The movement in January was quite animated, but with a general downward tendency in prices, resulting in a decline of fully 1 cent per pound, caused by the unfavorable tenor of the accounts from Liverpool, a continuance of the previous liberal receipts, and a steadily accumulating stock. Up to this period there had been few important changes in the Liverpool market, after the advance established towards the close of September, by the prospect of small supplies. Prices had ranged from 14 to 15, for middling uplands. The controlling elements were the state of trade at Manchester, and the probable extent of future receipts. The highest point was reached when the accounts with regard to our crop were the most gloomy. The same causes had their influence on this side, low middling ranging in our market from 28 to 29 cents up to 39 and 40 cents. But, from this time, the tendency of prices was steadily downward, falling off with few or no interruptions, and causing heavy losses to all parties engaged in the trade.

Thus documentary bills purchased by our banks, with a wide and apparently safe margin, came back dishonored, and the proceeds of the shipments largely failed to cover the face of the exchange. Not only those who were proverbial for their prudence, but others, who were equally cautious, and moreover fully advised with regard to the marked prospect abroad, were sufferers by this unexpected revulsion.

COTTON AT LIVER

Week Ending.	Receipts.						Sales.			
	American	E. I.	Egypt.	Brazil.	Other.	Total.	Consumption.	Speculation.	Export.	Total.
Jan. 3..	22,428	5,344	7,437	7,591	3,043	45,843	45,710	9,790	8,350	63,850
" 10..	7,456	6,779	18,467	9,281	......	41,983	28,440	4,910	7,570	40,920
" 17..	9,118	46	6,065	1,720	36	16,985	27,260	6,810	7,560	41,630
" 24..	5,039	1,204	9,874	344	......	16,461	32,190	2,230	7,790	42,210
" 31..	65,369	21,369	10,977	16,212	1,320	115,247	48,190	5,930	14,560	68,680
Feb. 7..	47,665	8,569	10,057	5,054	10,057	81,402	31,180	1,520	10,260	42,960
" 14..	24,294	5,064	9,964	4,967	3,409	47,698	42,820	2,760	15,650	61,230
" 21..	25,064	2,572	9,933	6,495	120	44,184	38,890	3,800	14,100	56,790
" 28..	47,681	1,913	5,741	6,356	2,350	64,241	38,290	3,370	11,070	52,730
March 7..	12,085	927	280	7,129	5,045	25,466	42,000	1,950	11,270	55,220
" 14..	7,614	2,364	3,302	4,857	1,139	19,276	46,100	3,840	12,760	62,700
" 21..	6,938	6,424	10,238	5,816	1,046	30,462	47,400	4,810	15,370	67,580
" 28..	124,065	36,038	5,037	29,912	9,638	204,690	38,140	2,240	10,000	50,380
April 4..	60,782	1,700	146	5,080	2,153	69,861	46,300	780	12,860	59,940
" 11..	41,486	19,214	7,912	26,507	2,202	97,321	41,860	3,410	12,670	57,940
" 17..	79,878	15,758	4,952	25,926	4,952	131,466	34,240	2,370	13,940	50,550
" 25..	43,746	16,687	6,011	11,790	234	78,468	39,230	3,280	13,050	55,560
May 2..	39,234	14,705	1,958	15,516	5,087	76,500	68,620	10,150	19,910	98,680
" 9..	37,839	6,445	1,025	7,757	297	53,363	51,600	2,300	19,250	73,150
" 16..	32,068	30,639	1,653	17,817	4,413	86,590	58,590	8,300	41,530	108,420
" 23..	19,202	697	1,380	1,750	846	23,875	47,990	1,830	19,640	69,460
" 30..	45,251	45,866	685	25,542	6,295	123,639	55,240	4,400	22,220	81,860
June 6..	68,433	56,270	1,597	12,466	743	139,509	71,170	5,400	18,900	95,470
" 13..	21,314	41,229	613	10,721	3,688	77,565	42,280	3,750	14,930	60,960
" 20..	17,159	7,756	679	4,050	2,801	32,445	47,000	2,120	16,090	65,210
" 27..	8,611	18,935	722	2,483	......	30,751	43,860	3,190	13,770	60,820
July 4..	13,889	6,351	3,442	2,982	3,706	30,370	40,840	1,680	11,210	53,730
" 11..	25,959	31,911	1,605	9,672	5,399	74,546	43,180	1,030	12,140	56,350
" 18..	34,179	11,787	52	9,106	254	55,378	59,220	1,860	15,660	76,740
" 25..	8,428	14,910	2,087	3,109	2,588	31,122	51,620	3,850	17,000	72,470
Aug. 1..	10,355	22,132	725	578	864	34,654	40,680	2,580	13,430	56,690
" 8..	7,290	97,967	296	11,655	418	117,626	43,610	1,810	20,370	65,790
" 15..	14,941	19,132	859	321	457	35,710	70,910	9,080	16,710	96,700
" 22..	24,616	100,537	755	11,485	5,540	142,933	38,130	1,160	20,050	59,340
" 29..	6,809	70,937	221	2,835	2,072	82,874	35,680	1,550	14,890	52,120
Sep. 5..	7,727	45,759	962	4,262	3,923	62,633	49,520	2,300	18,090	69,910
" 12..	5,648	82,642	1,012	7,397	587	97,286	45,310	4,510	16,780	66,600
" 19..	3,124	72,447	37	14,237	1,696	91,541	54,210	4,570	16,460	75,240
" 26..	2,087	41,849	446	3,765	884	49,031	44,200	1,690	20,010	65,900
Oct. 3..	2,699	29,887	357	6,161	3,851	42,955	51,060	8,460	16,550	76,070
" 10..	2,222	3,701	954	793	854	8,524	47,660	3,520	18,590	69,770
" 17..	3,909	60,498	404	3,126	1,467	69,404	66,160	6,380	22,710	95,250
" 24..	2,099	36,896	1,907	8,277	55	49,234	75,730	11,800	28,050	115,580
" 31..	4,749	5,995	191	......	3,067	14,002	44,830	5,980	18,340	69,150
Nov. 7..	3,336	10,335	3,294	......	......	16,965	53,590	5,190	13,040	71,820
" 14..	4,231	9,749	2,755	1,399	587	18,721	48,540	2,240	9,590	60,370
" 21..	8,910	5,383	......	3,704	1,194	19,191	49,060	1,770	9,780	60,610
" 28..	5,786	20,764	4,160	13,528	565	44,803	47,680	2,450	7,320	57,450
Dec. 5..	11,002	3,385	4,994	9,396	711	29,488	57,340	4,330	11,810	73,480
" 12..	21,603	30,792	7,965	10,238	4,038	74,636	52,570	3,650	11,570	67,790
" 19..	27,221	36,814	5,421	17,989	839	88,084	55,460	1,670	11,400	68,530
" 24..	28,240	2,787	10,458	9,606	841	51,932	29,450	820	10,370	40,640
" 31..	10,057	13,299	5,724	1,186	......	30,266	28,550	530	5,910	34,990
Average prices & total sales, receipts & stocks.	1,221,722	1,357,170	198,801	446,786	118,671	3,343,150	2,523,060	206,500	792,300	

POOL. YEAR 1867.

STOCKS.			PRICES.			ACTUAL EXPORT.	CONSUMPTION.	REMARKS.
Amer'n	Other.	Total.	Mid. Up.	Mid. Orl.	Dhol.			
172,030	345,030	517,060	15¼	15⅝	12¾	8,350	45,710	
167,270	462,740	520,010	14⅞	15¼	12½	7,570	74,150	Considerable demand, but freely supplied.
163,840	337,480	501,320	14¾	15⅛	12½	7,560	101,410	
153,330	324,970	478,300	14¾	15⅛	12¼	7,790	133,600	
199,950	338,930	538,880	14¾	15⅛	12¼	14,560	181,790	
235,070	335,700	570,770	14½	14⅝	12	10,260	212,970	Heavy receipts; lower American quotations.
241,420	324,080	565,500	14	14¼	12	15,650	255,790	
248,060	308,880	556,940	13¾	14¼	12¼	14,100	294,680	
274,240	276,290	570,630	13⅜	13⅞	11¾	11,070	332,970	
262,270	271,560	533,830	16	16½	11¾	11,270	374,970	
245,830	251,390	497,220	13¾	13¾	11½	12,760	421,070	
227,800	239,900	467,700	13⅜	13¾	11½	15,370	468,470	
332,240	312,070	644,310	13⅛	13½	11¼	10,000	506,610	Heavy imports; depressed market.
369,890	298,090	667,980	12⅝	13	11	12,860	552,910	
385,200	327,140	712,340	12	12¼	10½	12,670	594,770	Unfavorable Manchester reports.
440,790	335,030	795,820	11⅝	11⅞	10	13,940	629,010	
454,720	371,390	826,110	10½	10¾	9	13,050	668,240	
455,480	367,590	823,070	11½	11¾	9¼	19,910	736,860	
459,450	353,600	813,050	11	11¼	9	19,250	788,460	
457,010	379,110	826,120	11⅜	11⅝	9	41,530	846,990	
439,500	334,340	773,840	11	11¼	8¾	19,640	894,980	
444,240	366,380	810,620	11⅛	11⅜	8¾	22,220	966,150	
463,550	389,300	852,950	11⅜	11⅝	8¾	18,900	1,008,430	
451,490	410,000	861,490	11⅜	11⅝	8¾	14,930	1,055,430	
438,780	385,670	824,450	11¼	11⅜	8¾	16,090	1,099,290	
389,760	358,910	748,670	11	11¼	8½	13,770	1,140,130	
386,090	352,110	738,200	10½	11	8¼	11,210	1,183,310	
384,320	364,380	748,700	10¼	10½	8	12,140	1,242,530	
383,490	346,320	729,810	10¼	10½	8	15,660	1,294,130	
362,660	334,730	697,390	10¼	10½	8	17,000	1,334,810	
349,200	326,600	675,800	10¼	10⅝	8	13,430	1,378,420	
332,240	401,450	733,690	10¼	10⅜	8	20,370	1,449,330	Large E. I. imports depress markets.
315,540	364,560	680,100	10⅞	11⅛	8	16,710	1,487,460	
323,230	443,410	766,640	10⅝	11	7¾	20,050	1,524,140	Very heavy import.
313,550	480,010	793,560	10⅛	10½	7½	14,890	1,573,660	
302,420	483,810	786,230	10	10¼	7¼	18,090	1,616,970	
288,790	628,790	817,580	9½	9¾	6¾	16,780	1,671,180	
270,060	566,590	836,650	9⅜	9⅝	6¾	16,460	1,715,380	
253,360	567,160	820,520	8¾	9	6½	20,010	1,766,440	
233,550	562,130	795,680	8½	8¾	6¼	16,550	1,814,100	Suspension of Royal Bank.
215,580	521,420	737,000	8⅛	8½	6	18,590	1,880,260	
193,110	524,890	718,000	8⅜	8⅝	6¼	22,710	1,955,990	
165,050	513,150	678,200	8¾	9	6½	28,050	2,000,820	Strong desire to sell.
150,860	476,690	627,550	8¾	9	6¼	18,340	2,054,410	
133,190	438,610	571,800	8⅝	9	6¼	13,540	2,102,950	
117,410	410,630	528,040	9¼	10¾	6¼	9,590	2,152,010	
107,220	376,190	483,410	8¼	8½	6	9,780	2,299,690	
91,110	377,710	468,820	7⅝	7⅞	5¾	7,320	2,357,030	
78,050	353,640	431,690	7⅜	7⅞	5⅝	11,810	2,409,600	
79,480	363,980	443,460	9½	9½	5½	11,570	2,465,060	Increased desire to sell.
82,060	387,310	469,370	7⅜	7⅝	5½	11,400	2,494,510	
96,610	376,000	472,610	7¾	8½	5¼	10,370	2,523,060	
103,420	361,580	465,000	7⅛	7⅜	5¼	5,910	....	
			10.98	11.12	8.63	792,300	47,604.9	

The semi-weekly Price and Weekly Sales and Receipts, at New York, Weekly Exports from New York and Rates of Freight to Liverpool 1st of each Month, for the Crop Year ending September 1, 1867.

1866.	Price of Middling New Orleans	Price of Middling Upland.	Sales for week.	Receipts for week.	EXPORTS FOR WEEK.					Rates of Freight to Liverpool	GENERAL REMARKS.
					To Great Britain.	To France.	North of Europe.	Other Fo'n Ports	Total Exports.		
Septem. 4..	35	33	9,800	6,161	....	....	....	....	....	5-32d.	Early in September, 1866, the market was very dull, owing to unfavorable foreign accounts; the low rates of exchange, considerable specie coming this way, and advices of shipments of cotton from Liverpool hither; later, however, the crop accounts were gloomy, and, with some improvement in Liverpool, prices here advanced.
" 7..	35	33									
" 11..	35½	33½	16,700	5,891	7,333	289	524	....	8,146		
" 14..	35½	33½									
" 18..	36½	34½	20,200	4,490	3,959	344	....	8	4,311		
" 21..	37½	35½									
" 25..	39½	37½	21,000	4,682	5,174	49	251	771	6,245		
" 28..	40½	38½									
October 2..	44	42	29,000	7,731	2,537	144	75	16	2,772	5-32d.	
" 5..	42	40									
" 9..	39½	38	8,000	5,799	2,299	....	....	....	2,299		In October, the market was unsettled; there were two or three spasms of activity, but, on the whole, prices yielded, closing at 3 cents decline from those current at its opening.
" 12..	41½	40									
" 16..	44	42	21,400	8,686	677	521	....	....	1,198		
" 19..	43	41									
" 23..	42	40	13,300	12,465	4,653	100	133	....	4,886		
" 26..	41	39									In November, the market was unsettled by accounts of frost in the growing sections, night of October 31, but the foreign advices were still unfavorable, not responding to these frost reports, and with improved weather again, the market yielded.
" 30..	41	39	14,700	17,477	4,242	74	....	....	4,316		
Novem. 2..	41	39								3-16d.	
" 6..	41	39	13,000	14,929	8,073	199	740	....	9,012		
" 9..	39	37									
" 13..	37½	35½	6,000	17,450	3,797	....	692	....	4,489		
" 16..	36	34									
" 20..	36	34½	11,500	24,000	8,344	305	781	....	9,430		
" 23..	35½	34									December was comparatively a steady month, with a fair business.
" 27..	36½	35	11,200	15,804	8,578	....	740	....	9,318		
" 30..	35	33½									Through January, there was considerable excitement and speculation, but subsequently the stock at Liverpool being found to be 90,200 bales in excess of the previous estimates, and larger receipts at our ports, the buoyancy was lost.
Decem. 4..	35	33½	12,500	17,578	7,898	80	921	....	8,899	3-16@¼d	
" 7..	35	33½									
" 11..	35	33¾	14,700	20,609	6,101	5	617	....	6,723		
" 14..	36	34¾									
" 18..	36	34½	19,300	15,830	8,028	362	1,153	65	9,608		
" 21..	36	34½									
" 25..	35½	34	13,200	15,643	7,567	284	825	....	8,676		

Decem. 28..	35½	34								
1867.										
January 1..	36½	35½	4,500	11,820	11,166	316	829	....	12,311	¼@5-16d
" 4..	36½	35½								
" 8..	36½	35½	25,700	29,845	5,500	....	1,559	....	7,059	
" 11..	35½	34½								
" 15..	35½	34½	9,300	16,027	14,844	491	1,254	....	16,589	
" 18..	36	35								
" 22..	35	34	9,800	18,220	6,264	270	1,434	....	7,968	
" 25..	34½	33¾								
" 29..	35	34	11,000	2,637	10,466	472	767	....	11,705	
February 2..	34	33								¼@5-16d
" 5..	33½	32½	4,000	19,407	9,954	....	1,060	....	11,014	
" 8..	34	33								
" 12..	34	33	10,200	30,915	8,697	397	2,703	....	11,797	
" 14..	34	33								
" 19..	33½	32½	13,950	25,696	8,142	790	620	....	9,552	
" 22..	33	32								
" 26..	32	31	12,400	23,213	11,530	65	1,880	....	13,475	
March 1..	33	32								5-16@⅜d
" 5..	33	32	24,000	19,734	12,858	2,403	1,303	....	16,564	
" 8..	31½	30								
" 12..	31½	30	15,800	18,124	21,027	650	2,160	....	23,837	
" 15..	33½	32								
" 19..	33½	32	29,500	15,897	8,091	2,630	3,135	....	13,856	
" 22..	32½	31								
" 26..	32	31	11,500	9,284	16,554	485	2,427	....	19,466	
" 29..	32	30½								
April 2..	31½	30	12,800	20,454	14,131	578	3,354	800	18,863	¼d.
" 5..	30	28½								
" 9..	29½	28	13,600	14,518	15,550	2,070	3,020	....	20,640	
" 12..	28½	27								
" 16..	28¾	27¾	15,200	11,689	6,887	924	2,623	....	10,434	
" 19..	26	25								
" 23..	26½	25½	10,000	5,857	10,516	4,028	4,319	....	18,863	
" 26..	27½	26½								
" 30..	30	28½	18,200	6,332	9,805	2,197	723	....	12,725	
May 3..	29	27½								$\frac{5}{32}$@3-16
" 7..	29	27½	9,900	8,118	12,245	392	1,794	....	14,431	
" 10..	29	27½								
" 14..	29½	28	11,500	9,529	7,335	567	536	....	8,438	
" 17..	29½	28½								

In February, there was some speculative demand in consequence of falling off in the receipts at the ports, with a shrinkage in the crop estimates to 1,750,000 bales, but the Liverpool accounts were unfavorable, the Manchester mills reducing their time, &c., and prices yielded.

March was for the most part dull, though for a few days in the middle of the month, there was a speculative movement.

The market in April was unsettled by the drooping prices in Liverpool, consequent upon the threatening aspect of political affairs on the Continent.

The political skies of Europe became clearer in May, and, with more activity, prices advanced again; the feeling in June was less cheerful, and the course of the market was in buyers' favor.

Early in July, the market was dull and weak, but later there was quite an active speculative movement, which, with a small stock, carried prices up.

The market was considerably agitated in August, by the unfavorable crop accounts; the army worm was reported as doing much mischief, in addition to which the Mississippi overflowed, and the culture generally was said to be unsatisfactory; these statements carried prices up for a time, but foreign markets remained immovable, and the business was wholly for home use or on speculation.

New York Statement for 1867.—Concluded.

1867.		Price of Middling New Orleans	Price of Middling Upland.	Sales for week.	Receipts for week.	EXPORTS FOR THE WEEK. To Great Britain.	To France.	North of Europe.	Other Fo'n Ports	Total Exports.	Rates of Freight to Liverpool.
May	21..	28½	27½	13,600	5,637	6,955	1,450	1,080	152	9,637	
"	24..	28½	27½								
"	28..	28½	27½	11,000	14,173	2,263	711	516	....	3,490	
"	31..	28½	27½								
June	4..	29	27¾	12,400	12,210	4,100	1,593	1,072	....	6,765	5-32d.
"	7..	28	27								
"	11..	27¾	26¾	12,000	5,110	4,575	915	520	....	6,010	
"	14..	28	27								
"	18..	28	27	15,700	10,140	11,737	104	1,149	....	12,990	
"	21..	28	26¾								
"	25..	27¾	26½	7,600	5,159	5,933	194	2,175	....	8,302	
"	28..	27¾	26½								
July	2.	27¾	26½	9,300	10,814	7,982	274	3,662	991	12,909	3-16@¼d
"	5..	27¼	26¼								
"	9..	27¼	26½	6,700	5,758	4,322	371	1,192	....	5,885	
"	12..	27¼	26½								
"	16..	27¼	26½	9,900	7,061	5,101	742	2,447	....	8,290	
"	19..	27½	26½								
"	23..	28	27	16,200	5,495	4,774	....	1,023	....	5,797	
"	26..	28	27								
"	30..	28½	27¾	10,200	5,723	2,180	30	484	....	2,694	
August	2..	29	28								⅛@5-32d
"	6..	30	29	14,300	5,523	1,715	....	657	322	2,694	
"	9..	29¼	28¼								
"	13..	30	28¼	7,000	5,708	1,852	....	390	30	2,272	
"	16..	29¼	28¼								
"	20..	29½	28½	9,100	9,193	789	....	66	105	960	
"	23..	29	28								
"	27..	28	27	6,300	5,207	930	116	495	197	1,738	
"	30..	27½	26½								
Septem.	3..	27½	26½	5,700	2,759	864	....	118	59	1,041	⅛d.
Average price and total sales, receipts and exports.		32.86	31.59	695,350	648,211	376,894	28,981	61,998	3,516	471,389	

GENERAL REMARKS.

Exchange.

The range in September for 60 days' bills on London, was 4¾@5½ per cent. gold, at the opening, and 7@7¾ at the close; in October, the range was from 5¾@6¼, up to 8½@9; November, steady at 8½@9; December, 8⅛@9¾; January, 8@9; February, 7¾@8⅝; March, 7¼@8¾; April, 7½@9¼; May, 8@9⅝; June, 9¼@9¾; July, 9¼@9⅞; and August, 8¾@9¾.

1868.

COTTON CROP OF THE UNITED STATES.

Statement and Total Amount of the Cotton Crop of the United States, for the Year ending August 31, 1868.

	Bales.	Bales.	TOTAL. 1868.	1867.	1866.
LOUISIANA.					
Export from New Orleans—					
To Foreign Ports	581,477				
Coastwise	100,215				
Stock on hand 1st September, 1868	1,959				
		683,651			
Deduct—					
Received from Mobile	67,043				
" Montgomery, Ala	8,659				
" Florida	5,770				
" Texas	7,692				
Stock on hand 1st September, 1867	15,256				
		104,420			
			579,231	702,131	711,629
ALABAMA.					
Export from Mobile—					
To Foreign Ports	236,511				
Coastwise	130,893				
Burnt at Mobile	342				
Stock on hand 1st September, 1868	2,161				
		369,907			
Deduct—					
Stock on hand 1st September, 1867	....	3,714			
			366,193	239,516	429,102
TEXAS.					
Export from Galveston, &c.—					
To Foreign Ports (including 6,168 to Mexico)	68,595				
Coastwise	49,138				
Stock in Galveston, 1st September, 1868	166				
		117,899			
Deduct—					
Stock in Galveston, 1st September, 1867	....	3,233			
			114,666	185,919	174,985
FLORIDA.					
Export from Apalachicola, St. Marks, &c.—					
To Foreign Ports	9				
Coastwise—Uplands and Sea Island	34,241				
Burnt at St. Marks	398				
Stock in Apalachicola and St. Marks, 1st September, 1868	....				
		34,648			
Deduct—					
Stock in Apalachicola and St. Marks, 1st September, 1867	....	9			
			34,639	58,349	149,139
GEORGIA.					
Export from Savannah—					
To Foreign Ports—Uplands	253,556				
" Sea Island	6,048				
Coastwise—Uplands	235,708				
Sea Island	5,245				
Burnt and manufactured at Savannah	98				
Stock in Savannah, 1st September, 1868	696				
		501,351			

36

Statement and Total Amount of the Cotton Crop of the United States, for the Year ending August 31, 1868—*Concluded.*

	Bales.	Bales.	TOTAL. 1868.	1867.	1866.
Deduct—					
Received from Florida—Uplands	666				
" " Sea Island	4,997				
From Mobile, by railroad	50				
Stock in Savannah, 1st September, 1867	633				
		6,346			
			495,005	255,965	263,373
SOUTH CAROLINA.					
Export from Charleston, S. C.—					
To foreign ports—Uplands	99,847				
" Sea Island	5,966				
Coastwise—Uplands	135,031				
Sea Island	3,328				
Stock in Charleston, 1st September, 1868	1,945				
	246,117				
Export from Georgetown, Port Royal, &c.					
To Northern ports—Uplands and Sea Island	133				
		246,250			
Deduct—					
Received from Florida—Uplands	180				
Sea Island	4,617				
Stock in Charleston, 1st September, 1867	1,228				
		6,025			
			240,225	162,247	112,273
NORTH CAROLINA.					
Export—					
To Coastwise ports	......	......	38,587	38,522	64,559
VIRGINIA.					
Export—					
To Foreign Ports	8,215				
Coastwise	160,111				
Manufactured—taken from the ports	20,000				
Stock in Norfolk and Petersburg, 1st September, 1868	161				
		188,487			
Deduct—					
Stock on hand 1st September, 1867	......	1,000			
			187,487	123,627	37,531
TENNESSEE, &c.					
Shipments from Memphis, Tenn.	254,240				
" " Nashville, Tenn.	93,126				
" " other places in Tennessee, Kentucky, &c.	90,844				
Crop of Illinois, Indiana, Missouri, &c.	10,000				
Stock in Memphis and Nashville, 1st September, 1868	107				
		448,317			
Deduct—					
Shipments to New Orleans, from Memphis, Nashville, &c	71,855				
Stock in Memphis and Nashville, 1st September, 1867	1,602				
		73,457			
			*374,860	185,712	211,885
Total crop of the United States, 1867–8.	......	......	2,430,893	1,951,988	2,154,476

* Of which received at New York, Philadelphia, Baltimore, Portland, and Boston, overland from Tennessee, &c. .... bales. 204,337

Export to Foreign Ports, from September 1, 1867, *to August* 31, 1868.

FROM	To Great Britain.	To France.	To North of Europe.	Other F'n Ports.	Total.
New Orleans........................bales.	327,689	147,120	50,235	56,433	581,477
Mobile	211,154	10,432	7,794	7,131	236,511
Galveston	40,782	1,625	20,020	6,168	68,595
Apalachicola	9	....	....	....	9
Savannah	240,505	9,904	9,195	....	259,604
Charleston	89,651	2,936	3,710	9,516	105,813
Norfolk	8,215	....	....	....	8,215
Wilmington	....	....	....	....	....
New York	291,663	25,498	50,935	5,414	373,510
Baltimore	13,388	....	2,921	....	16,309
Philadelphia	1,440	....	....	....	1,440
Boston and Portland (Portland, 2,892 to Great Britain)	4,100	....	232	1	4,333
Grand total 1867–8	1,228,596	197,515	145,042	84,663	1,655,816
Total 1866–7	1,216,262	198,147	95,342	47,303	1,557,054
Increase	12,334	....	49,700	37,360	98,762
Decrease	....	632	....	....	....

Consumption.

Total crop of the United States, as before stated ....bales.			2,430,893
Add—Stocks on hand at the commencement of the year, September 1, 1867—			
In the Southern ports		24,574	
In the Northern ports		55,722	
			80,296
Makes a supply of			2,511,189
Deduct therefrom—The Export to Foreign ports	1,655,816		
Less Foreign included	4,190		
		1,651,626	
Stocks on hand, 1st September, 1868.			
In the Southern ports	7,195		
In the Northern ports	30,203		
		37,398	
Burnt at New York, Mobile, St. Marks, Savannah, Baltimore, &c	2,348		
Manufactured in Virginia	20,000		
		22,348	
			1,711,372
Taken for home use north of the Potomac and Ohio Rivers.........bales.			799,817
Taken for home use south of the Potomac and Ohio Rivers, and burnt....			168,348
Total consumed in the U. S. (including burnt at the ports), 1867–68.......			968,165

COTTON AT LIVER

Week Ending.	Receipts.						Sales.			
	American	E. I.	Egypt.	Brazil.	Other.	Total.	Con-sumption.	Specu-lation.	Export.	Total.
Jan. 9..	23,009	18,901	8,864	8,205	2,826	61,805	90,130	6,780	17,220	114,130
" 16..	47,178	26,500	6,881	14,279	112	94,950	71,210	9,130	17,450	97,790
" 23..	33,555	2,197	6,458	13,752	191	56,153	70,870	9,870	20,150	100,890
" 30..	62,916	4,845	7.994	9,345	557	85.657	80,990	12,270	19,390	112,650
Feb. 6..	20,526	5,703	7,002	7,004	624	40,859	77,860	9.280	18,070	105,210
" 13..	24,346	2,403	10.017	1,403	778	38,947	69,680	25,060	19,440	114,180
" 20..	61,810	2,097	4,274	1,773	378	70,332	79,540	40,940	24,860	145,340
" 27..	50,956	3,584	5,917	9,507	1.263	71,227	35.010	9.530	6,870	51,410
Mch. 5..	53,832	7,017	4,585	32,954	2,507	100,895	52,260	12,530	12,350	77,140
" 12..	39,368	4,818	6,548	6,630	4,738	62,102	51,630	19,310	13,240	84.180
" 19..	69,264	2.243	3,374	10,522	550	85,953	52,100	12,640	14,580	79,320
" 26..	33,268	800	9,261	13,941	627	57,897	52,260	10,380	11,370	73,010
April 2..	15,473	7,513	1,897	11,338	129	36,350	100,780	48,640	13,710	163,130
" 8..	60,454	25,978	3,495	27,660	1,935	119,522	70,230	29,380	8,320	107,930
" 16..	56,373	7,484	8,840	12,168	3,594	88,459	32,680	17,200	5,900	55,780
" 23..	111,015	1,832	2,495	16,868	420	132,630	49,370	24,580	9,470	83,420
" 30..	46,317	21,291	1,878	23,127	4,077	96.690	52,010	23,320	11,140	86,470
May 7..	26,123	17,511	4,182	9,017	2,257	59,090	34,990	6,980	5,670	47,640
" 14..	40,412	13,520	4,949	30.761	1,409	91,051	32,310	8,170	6,590	47,070
" 21..	21,144	6,736	3,557	16,772	400	48,609	37,500	4,070	5,000	46,570
" 28..	27,358	13,702	3,383	14,939	5,647	65,029	27,240	3.920	6,210	37,370
June 4..	32,577	25,132	1,174	14,483	4,758	78,124	43,980	4,320	5,500	53,800
" 11..	30,563	6,856	1,325	6,587	860	46,191	37,080	1.960	6,810	45,850
" 18..	11,217	6,119	615	8,671	1,893	28,515	62,230	11,210	15,330	88,770
" 25..	27,420	690	842	16,944	2,894	48,790	45,400	8,560	14,330	68,290
July 2..	7,405	15,231	430	22,560	4,985	50,611	44,400	4,150	7,320	55,870
" 9..	12,182	13,311	1,127	10,479	4,181	41,280	51,620	8,900	10,330	70,850
" 16..	3,767	8,114	246	......	983	13,110	36,120	2,410	8,860	47,420
" 23..	10,188	50,871	279	25,423	3,740	90,501	30,700	3.480	10,370	44,550
" 30..	440	19,740	641	16,501	930	38,252	44,330	7,250	11,480	63,060
Aug. 6..	6,207	59,015	549	10,447	744	76,962	49,730	7,900	10,820	68,450
" 13..	2,369	59,796	543	12,624	2,478	77,810	68,440	5,520	15,100	89,060
" 20..	663	7,068	746	6,911	3,793	19,181	72,620	13,570	22,840	109,030
" 27..	5,854	23,813	169	5,145	2,706	37,787	71,630	14,640	23,620	109,890
Sept. 3..	731	86,743	295	24,717	1,653	114,339	45,740	5,160	18,290	69,190
" 10..	4,056	11,937	89	17,463	3,137	36,682	41,660	4,080	15,770	61,510
" 17..	1,893	34,011	508	3,222	1,050	40,684	54,050	10,070	19,510	83,630
" 24..	776	24,257	574	7,208	884	33,699	47,710	5,890	13,820	67,420
Oct. 1..	4,399	69,202	463	9,332	1,348	84,744	68,850	8,820	17,180	94,850
" 8..	741	64,246	542	7,073	257	72,859	54,140	23,280	18,590	96,010
" 15..	819	40,557	906	8,309	4,149	54,740	57,110	11,460	17,820	86,390
" 22..	1,573	75,320	2,961	4,093	1,017	84,964	59,570	7,440	17,830	84,840
" 29..	3,083	108,459	1,284	10,160	......	122,986	68,010	18,150	32,830	118,999
Nov. 5..	3,945	27,664	1,255	4,921	2,198	39.983	68,360	26,400	26,680	121,440
" 12..	6,296	16,768	7,292	11,128	1.296	42,780	41,770	7,680	14,760	64,210
" 19..	8,453	11,892	2,547	3,825	415	27,132	32,140	2,640	7,800	42,580
" 26..	10,069	20,209	6,093	10,002	310	46,683	74,520	12,610	21,950	109,080
Dec. 3..	31,545	32,366	9,709	21,191	2,002	96,813	48,950	13,770	22,730	85,450
" 10..	20,543	19,487	7,016	8,298	315	55,659	45,700	3.400	14,630	63,730
" 17..	19,659	10,879	6,080	15,235	245	52,098	46,820	2,200	7,860	56,880
" 23..	26,114	613	9,606	11,995	168	48,496	53,820	10,660	16,260	80,740
" 30..	42,016	7,730	5,914	1,350	2,853	59,863	58,770	16,430	12,110	87,310
Average prices & total sales, receipts & stocks.	1,262,778	1,144,871	187,471	624,262	93,361	3,312,743	2,836,220	669,770	746,430	4,252,420

POOL. YEAR 1868.

STOCKS.			PRICES.			ACTUAL EXPORT.	CONSUMPTION.	REMARKS.
American.	Other.	Total.	Mid. Up.	Mid. Orl.	Fair Dhol.			
93,940	325,260	419,200	$7\frac{1}{8}$	$7\frac{1}{2}$	$5\frac{5}{8}$	17,220	90,130	January 9th, market opened
111,890	322,400	434,290	$7\frac{1}{2}$	$7\frac{3}{4}$	$5\frac{7}{8}$	17,450	161,340	with animation, but freely
117,400	288,740	406,140	$7\frac{3}{4}$	8	$6\frac{1}{8}$	20,150	232,210	supplied; 16th, reduced re-
146,700	249,970	396,670	$7\frac{7}{8}$	$8\frac{1}{8}$	$6\frac{1}{2}$	19,390	313,200	ceipts, active demand; 23d,
134,290	205,860	340,150	$7\frac{15}{16}$	$8\frac{3}{16}$	$6\frac{7}{8}$	18,070	391,660	trade buying freely, large ar-
121,670	166,330	288,000	$8\frac{5}{8}$	$8\frac{7}{8}$	$7\frac{5}{8}$	1,944	460,740	rival business; 30th, quieter
142,770	124,010	266,780	$10\frac{1}{8}$	$10\frac{3}{8}$	$8\frac{5}{8}$	2,486	540,280	tone, prices firm.
168,830	122,920	291,750	$9\frac{3}{8}$	$9\frac{9}{16}$	$8\frac{3}{8}$	6,870	575,290	February 6th, market steady;
186,310	138,680	326,990	$9\frac{3}{4}$	10	$8\frac{1}{2}$	12,350	627,550	13th, great excitement at $\frac{3}{4}$@
192,970	151,500	344,470	$10\frac{1}{8}$	$10\frac{3}{8}$	$8\frac{3}{4}$	13,240	679,180	1d. per pound advance; 20th,
228,860	142,170	370,030	$10\frac{1}{8}$	$10\frac{3}{8}$	$8\frac{3}{4}$	14,580	739,280	apprehension of insufficient
228,210	163,320	391,530	$10\frac{1}{8}$	$10\frac{3}{8}$	$8\frac{7}{8}$	11,370	783,540	supplies, $1\frac{1}{2}$d. per pound ad-
189,180	122,860	312,040	$11\frac{5}{8}$	$11\frac{7}{8}$	$10\frac{1}{2}$	13,710	884,320	vance; 27th, heavier receipts,
196,550	160,000	356,550	$11\frac{7}{8}$	$12\frac{1}{8}$	$10\frac{3}{4}$	8,320	954,550	decline of $\frac{3}{4}$d. per pound.
235,010	174,860	409,870	$12\frac{1}{8}$	$12\frac{3}{8}$	$10\frac{5}{8}$	5,900	987,230	March 5th, market opened dull,
320,790	165,510	486,300	$12\frac{3}{8}$	$12\frac{5}{8}$	$10\frac{5}{8}$	9,470	1,036,600	but recovered; 12th, good de-
341,340	184,890	526,230	$12\frac{5}{8}$	$12\frac{7}{8}$	$10\frac{7}{8}$	11,140	1,088,610	mand; 26th, steady.
342,150	200,250	542,400	$12\frac{3}{8}$	$12\frac{5}{8}$	$10\frac{1}{2}$	5,670	1,123,600	April 2d, unusual excitement,
358,040	229,400	587,440	12	$12\frac{1}{4}$	$10\frac{1}{4}$	6,590	1,155,980	through fear of short crops;
354,080	236,490	590,570	$11\frac{1}{2}$	$11\frac{3}{4}$	$9\frac{7}{8}$	5,000	1,183,410	8th, gradual taming down to-
363,220	257,030	620,250	$11\frac{1}{4}$	$11\frac{1}{2}$	$9\frac{1}{2}$	6,210	1,210,250	wards the end of the week;
371,170	277,650	648,820	$11\frac{1}{2}$	$11\frac{3}{4}$	$9\frac{3}{8}$	5,500	1,254,230	16th, stock declared, 900 bales
381,390	271,350	652,740	11	$11\frac{3}{8}$	9	6,810	1,291,310	below estimate; 23d, fair de-
364,390	247,590	611,980	$11\frac{1}{4}$	$11\frac{1}{2}$	9	15,330	1,353,940	mand; 30th, animated.
369,400	237,930	607,330	$11\frac{1}{4}$	$11\frac{1}{2}$	9	14,330	1,398,940	May 7th, closes irregular; 14th,
353,300	249,880	603,180	$11\frac{1}{8}$	$11\frac{3}{8}$	$8\frac{7}{8}$	7,320	1,443,340	dull; 21st, declining; 28th,
341,360	240,510	581,870	$11\frac{1}{4}$	$11\frac{5}{8}$	9	10,330	1,494,960	limited request.
329,450	223,550	553,000	$11\frac{1}{8}$	$11\frac{3}{8}$	$8\frac{7}{8}$	8,860	1,531,030	June 4th, fair demand; 11th,
323,710	278,790	602,500	$10\frac{3}{8}$	$10\frac{5}{8}$	$8\frac{1}{2}$	10,370	1,561,780	heavy tone; 18th, increased
278,300	274,770	553,070	$9\frac{5}{8}$	$9\frac{7}{8}$	$7\frac{3}{4}$	11,480	1,606,110	inquiry; 25th, no change.
266,290	314,340	580,630	$9\frac{5}{8}$	$9\frac{7}{8}$	$7\frac{1}{2}$	10,820	1,655,840	July 16th, slight decline; 23d,
240,160	337,420	577,580	10	$10\frac{1}{4}$	$7\frac{3}{8}$	15,100	1,724,280	pressure to sell; 30th, heavy
215,090	295,120	510,210	$10\frac{1}{2}$	$10\frac{3}{4}$	$7\frac{3}{4}$	22,840	1,796,900	and declining market.
193,080	268,280	461,360	11	$11\frac{1}{4}$	$8\frac{1}{8}$	23,620	1,868,530	August 6th, actual stock 28,980
175,350	336,480	511,830	$10\frac{7}{8}$	$11\frac{1}{8}$	$8\frac{1}{8}$	18,290	1,914,270	bales below estimate; 27th,
162,080	317,710	479,790	$10\frac{1}{2}$	$10\frac{3}{4}$	$7\frac{7}{8}$	15,770	1,955,930	more animation, hardening
142,990	309,360	452,350	$10\frac{1}{8}$	$10\frac{3}{8}$	$7\frac{3}{4}$	19,510	2,009,980	tendency.
125,930	296,210	422,140	10	$10\frac{1}{4}$	$7\frac{5}{8}$	13,820	2,057,690	September 3d, little or no change;
106,760	317,420	424,180	$10\frac{1}{4}$	$10\frac{1}{2}$	$7\frac{3}{4}$	17,180	2,126,540	10th, dull and inanimate; 17th,
88,870	338,230	427,100	$10\frac{3}{4}$	$10\frac{7}{8}$	$7\frac{7}{8}$	18,590	2,180,680	depression continued; 24th,
67,700	339,300	407,000	$10\frac{3}{4}$	11	8	17,820	2,237,790	further decline.
47,110	360,980	408,090	$10\frac{3}{4}$	11	8	17,830	2,296,360	October 1st, good demand at ad-
31,700	413,690	445,390	$11\frac{1}{8}$	$11\frac{1}{4}$	$8\frac{1}{4}$	32,830	2,365,370	vancing prices; 8th, extensive
47,870	378,940	426,810	$11\frac{1}{8}$	$11\frac{3}{8}$	$8\frac{1}{2}$	26,680	2,433,730	business; 22d, steady; 29th
44,110	361,350	405,460	$10\frac{7}{8}$	$11\frac{1}{8}$	$8\frac{1}{4}$	14,760	2,475,500	general advance of $\frac{1}{4}$d. per lb.
39,960	340,070	380,030	$10\frac{3}{4}$	11	$8\frac{1}{4}$	7,800	2,507,640	November 12th, dull in tone;
27,640	302,840	330,480	$11\frac{3}{8}$	$11\frac{1}{2}$	$8\frac{1}{2}$	21,950	2,582,160	26th, good trade demand.
44,200	323,160	367,360	$11\frac{1}{4}$	$11\frac{1}{2}$	$8\frac{1}{2}$	22,730	2,631,110	December 3d, quotations barely
48,730	316,270	365,000	$10\frac{7}{8}$	11	$8\frac{3}{8}$	14,630	2,676,810	maintained: 10th, heavy and
50,780	303,500	354,280	$10\frac{5}{8}$	$10\frac{3}{4}$	$8\frac{1}{4}$	7,860	2,723,630	declining; 17th, quiet; 30th,
58,100	277,750	335,850	$10\frac{5}{8}$	$10\frac{3}{4}$	$8\frac{3}{8}$	16,260	2,777,450	trade operating freely.
82,360	269,980	352,340	$10\frac{7}{8}$	11	$8\frac{5}{8}$	12,110	2,836,220	
			10.52	10.76	8.48	746,430	54,542.68	

The semi-weekly Price and Weekly Sales and Receipts at New York, Weekly Exports from New York and Rates of Freight to Liverpool 1st of each month, for the Crop Year ending September 1, 1868.

1867.		Price of Middling New Orleans	Price of Middling Upland.	Sales for week.	Receipts for week.	EXPORTS FOR THE WEEK.					Rates of Freight to Liverpool.
						To Great Britain.	To France.	North of Europe.	Other Fo'n Ports	Total Exports.	
Septem.	6..	$27\frac{1}{2}$	$26\frac{1}{2}$								$\frac{1}{8}$d.
"	10..	$27\frac{1}{2}$	26	5,829	1,965	....	....	....	....	....	
"	13..	$26\frac{1}{2}$	25								
"	17..	$26\frac{1}{2}$	25	6,114	3,331	2,513	....	361	....	2,874	
"	20.	$25\frac{1}{2}$	$24\frac{1}{2}$								
"	24..	$24\frac{1}{2}$	$23\frac{1}{2}$	5,084	4,597	1,411	22	....	....	1,433	
"	27..	23	22								
October	1..	$22\frac{1}{2}$	$21\frac{1}{2}$	6,721	4,129	920	120	374	....	1,414	3-16d.
"	4..	21	20								
"	8..	21	20	9,938	6,392	2,199	32	213	....	2,444	
"	11..	20	19								
"	15..	$20\frac{1}{2}$	$19\frac{1}{2}$	14,705	8,890	3,250	....	648	....	3,898	
"	18..	20	19								
"	22..	$20\frac{1}{2}$	$19\frac{1}{2}$	13,393	10,864	3,759	132	688	....	4,579	
"	25..	$21\frac{1}{2}$	20								
"	29..	$21\frac{1}{2}$	20	17,766	10,103	5,739	526	1,148	....	7,413	
Novem.	1..	$20\frac{1}{2}$	19								5-16@$\frac{3}{8}$d
"	5..	$19\frac{1}{2}$	$18\frac{1}{4}$	11,585	20,459	9,054	135	1,811	850	11,850	
"	8..	$20\frac{1}{4}$	19								
"	12..	20	$18\frac{1}{4}$	17,313	11,399	5,803	....	1,546	....	7,349	
"	15..	$19\frac{1}{2}$	$18\frac{1}{4}$								
"	19..	$19\frac{1}{4}$	$18\frac{1}{8}$	19,022	24,380	7,541	1,353	1,675	....	10,569	
"	22..	$18\frac{1}{2}$	$17\frac{1}{2}$								
"	26..	18	$16\frac{3}{4}$@17	15,520	17,872	9,883	....	4,093	652	14,628	
"	29..	$17\frac{1}{4}$	16								
Decem.	3..	$16\frac{3}{4}$	$15\frac{3}{4}$	19,667	24,958	10,192	339	2,503	....	13,034	$\frac{1}{4}$@5-16d
"	6..	$18\frac{1}{4}$	$17\frac{1}{4}$								
"	10..	17	16	21,923	25,165	11,776	2,078	1,395	1,014	16,263	
"	13..	$16\frac{3}{4}$	$15\frac{3}{4}$								
"	17..	$16\frac{3}{4}$	$15\frac{3}{4}$	15,068	18,244	5,026	2,383	2,458	....	9,867	
"	20..	$16\frac{3}{4}$	$15\frac{3}{4}$								
"	24..	$16\frac{1}{2}$	$15\frac{1}{2}$	14,692	18,719	11,895	....	2,660	499	15,054	
"	27..	$16\frac{1}{2}$	$15\frac{1}{4}$@$15\frac{1}{2}$								

GENERAL REMARKS.

The fluctuations in prices this crop year were numerous and with, the market being very sensitive as to the crop reports.

In September, the estimates of the crop then being picked were reduced to $2\frac{1}{4}$ million bales, and the market, though quiet, was firm; the goods market, both here and abroad, was very dull, and though the weather at the South was very wet and unfavorable, the foreign markets were stagnant, and prices at length gave way here.

Early in October there was a dull, weak feeling, the supply of Surats in the Liverpool market was large, which caused Americans to be neglected; later, however, there was more activity at the reduced prices, and better tone, with some advance.

In November, the market was, for the most part, dull and drooping.

The repeal of the internal revenue tax on cotton having passed Congress in December, there was considerable activity at better prices, (the opinion hitherto being that the act would take immediate effect, but the bill only exempted cotton grown after 1867,) but subsequently the foreign accounts were discouraging and the improvement was lost.

Decem. 31..	17	16	20,438	18,312	8,260	2,513	2,798	....	13,571	
1868.										
January 3..	17¾	16¾								5-16@⅜d
" 7..	17¼	16¼	30,487	22,232	4,149	....	1,641	....	5,790	
" 10..	17¼	16⅜								
" 14..	17½	16½	28,979	23,478	8,348	1,590	1,958	....	11,896	
" 17..	18	17¼								
" 21..	18	17	34,270	16,655	10,062	2,140	2,911	18	15,131	
" 24..	18½	17¾								
" 28..	19	18¼	40,734	18,240	6,925	780	1,942	....	9,647	
" 31..	20¼	19½								
February 4..	20	19	35,838	19,155	12,298	2,227	2,751	....	17,276	⅜@½d.
" 7..	21	20								
" 11..	21¾	20¾	47,048	21,539	11,680	430	1,840	32	13,982	
" 14..	21½	20½								
" 18..	25	24	59,491	21,841	8,375	1,943	903	....	11,221	
" 21..	24½	23½								
" 25..	23½	22¾	36,057	23,185	8,189	224	655	550	9,618	
" 28..	23	22								
March 3..	24½	23½	30,546	27,405	12,692	1,538	1,939	....	16,169	Steam,
" 6..	25¾	25								7-16@½d
" 10...	25¾	25	49,048	20,799	5,052	61	1,097	590	6,800	Sail,
" 13..	25¾	25								@⅜d
" 17..	25¾	25	23,347	21,827	10,200	....	789	500	11,489	
" 20..	25¾	25								
" 24..	25½	24½	12,559	16,833	4,848	99	670	....	5,617	
" 27..	26¾	26								
" 31..	27¾	27	68,119	18,094	12,153	....	1,195	....	13,348	
April 3..	29¾	29								Steam,
" 7..	29¼	28½	56,014	12,544	12,603	2,095	916	109	15,723	7-16@½d
" 10..	30¾	30								Sail.
" 14..	32½	31½	41,666	10,672	8,235	....	759	....	8,994	5-16@⅜d
" 17..	31½	30½								
" 21..	32	31	13,052	8,770	14,126	....	520	....	14,646	
" 24..	33½	32½								
" 28..	34	33	31,996	4,791	13,179	928	779	....	14,886	
May 1..	33½	32½								Steam,
" 5..	33¾	32¾	13,450	4,424	3,613	531	150	....	4,294	3-16d.
" 8..	33	32								Sail,
" 12..	31½	30½	10,769	5,524	8,486	....	173	....	8,659	⅛@3-16d
" 15	32½	31½								
" 19..	32½	31½	18,302	4,277	3,192	820	815	....	4,827	

The light receipts at the ports, and low prices ruling in January, stimulated a speculative feeling, and prices advanced.

In February there was much activity, consequent upon the continued light receipts at the ports, and stocks, both at Liverpool and Havre, being much reduced, the market was buoyant, though at the close, the demand subsided, and prices fell back.

In March, favorable foreign accounts were received, which, together with light receipts, caused an active demand at advanced prices.

The favorable position noted in March continued pretty well through April, and prices advanced very considerably, with a very large business.

In May, there was an abatement in the speculative fever, and prices having been run up above those current in Liverpool, and the prospects for the growing crop good, prices receded.

During the early part of June, the market was comparatively quiet, though the receipts at the ports were still light; but, later, the foreign advices were more encouraging, and there was more activity at advanced figures.

The market in July, for the most part, was dull; the growing crop looked well, and, upon the report that purchases had been made in Liverpool to come here, prices receded.

In August, reports of injury to the growing crop, by rains and worms,

New York Statement for Year 1868—*Concluded.*

1868.		Price of Middling New Orleans	Price of Middling Upland.	Sales for week.	Receipts for week.	EXPORTS FOR WEEK.					Rates of Freight to Liverpool.
						To Great Britain.	To France.	North of Europe.	Other Fo'n Ports	Total Exports.	
May	22..	$31\frac{1}{2}$	$30\frac{1}{2}$								
"	26..	$31\frac{1}{2}$	$30\frac{1}{2}$	3,217	6,334	4,658	....	641	....	5,299	
"	29..	32	31								
June	2..	32	31	9,287	3,537	2,013	51	149	600	2,813	Steam, $\frac{1}{8}$d.
"	5..	$31\frac{1}{2}$	$30\frac{1}{2}$								
"	9..	31	30	6,662	6,161	287	....	72	....	359	Sail, —@—
"	12..	$30\frac{1}{2}$	$29\frac{1}{2}$								
"	16..	30	29	9,173	2,564	471	115	310	....	896	
"	19..	32	31								
"	23..	31	30	12,797	5,393	815	....	543	....	1,358	
"	26..	$32\frac{1}{2}$	$31\frac{1}{2}$								
"	30..	$32\frac{1}{2}$	$31\frac{1}{2}$	15,628	4,770	567	31	100	....	698	
July	3..	33	$32\frac{1}{4}$								Steam, 3-16d.
"	7..	$33\frac{1}{4}$	$32\frac{3}{4}$	16,691	4,631	871	....	....	....	871	
"	10..	33	$32\frac{1}{2}$								Sail, —@—
"	14..	$32\frac{1}{2}$	32	9,204	6,526	555	....	116	....	671	
"	17..	32	$31\frac{1}{2}$								
"	21..	$31\frac{1}{2}$	31	7,462	4,556	124	....	41	....	165	
"	24..	$31\frac{1}{2}$	31								
"	28..	30	$29\frac{1}{2}$	5,180	3,765	60	....	188	....	248	
"	31..	31	$30\frac{1}{2}$								
August	4..	$30\frac{1}{2}$	30	12,373	5,113	116	....	....	....	116	Steam, 3-16d.
"	7..	$29\frac{1}{2}$	29								
"	11..	30	$29\frac{1}{2}$	6,637	2,301	107	....	....	....	107	Sail, —@—
"	14..	30	$29\frac{1}{2}$								
"	18..	$30\frac{1}{2}$	30	11,236	1,021	31	....	....	....	31	
"	21..	$30\frac{1}{2}$	30								
"	25..	$30\frac{1}{2}$	30	9,418	2,366	1,333	232	....	....	1,565	
"	28..	31	$30\frac{1}{2}$								
Septem.	1..	31	$30\frac{1}{2}$	11,791	1,987	2,039	30	1	....	2,070	Steam, $\frac{5}{32}$@$\frac{3}{16}$d.
Average price and total sales, receipts and exports.		25.57	24.85	1,063,306	613,089	291,673	25,498	50,935	5,414	373,520	Sail, 5-32d.

General Remarks.

were received, and there was more activity at better prices, stocks being light.

Exchange.

The quotations for bills on London, in September, ranged from $8\frac{1}{4}$ to $9\frac{1}{2}$ per cent. premium, gold; in October, 8@$9\frac{1}{4}$; in November, 8@9; in December, $8\frac{1}{2}$@10; in January, $8\frac{1}{2}$@10; in February, 9@$9\frac{5}{8}$; in March, 9@$9\frac{3}{8}$; in April, 9@$9\frac{3}{4}$; in May, $9\frac{1}{4}$@10; in June, $9\frac{1}{2}$@$9\frac{7}{8}$; in July, $9\frac{5}{8}$@10, and in August, $8\frac{1}{2}$@$9\frac{3}{4}$ per cent., gold.

ANNUAL REVIEW.

From the New Orleans Price Current, 1867—68.

During the month of September, the heavy losses on the year's operations, unabated dullness in the manufacturing districts, and the views entertained with regard to future supplies, kept down prices at Liverpool, and caused a further decline in our own market. Whatever doubts prevailed toward the close of January, with regard to the course of prices, they were entirely dispelled at the commencement of March. The supply question had assumed features that made the markets extremely sensitive, and ready to advance under the slightest improvement in the demand. Under this prospect, middling advanced at Liverpool $\frac{3}{4}$d. by the 6th of the month, $\frac{3}{8}$d. and $\frac{1}{2}$d. in the latter part, and $\frac{3}{4}$d. and 1d. at the close.

1869.

COTTON CROP OF THE UNITED STATES.

Statement and Total Amount of the Cotton Crop of the United States, for the Year ending August 31, 1869.

	Bales.	Bales.	TOTAL. 1869.	TOTAL. 1868.	TOTAL. 1867.
LOUISIANA.					
Export from New Orleans—					
To Foreign Ports	619,534				
Coastwise	222,871				
Stock on hand 1st September, 1869	770				
		843,175			
Deduct—					
Received from Mobile	36,515				
" Montgomery, Ala.	2,373				
" Florida	747				
" Texas	7,376				
Stock on hand 1st September, 1868	1,959				
		48,970			
ALABAMA.			794,205	579,231	702,131
Export from Mobile—					
To Foreign Ports	163,154				
Coastwise	84,194				
Stock on hand 1st September, 1869	1,169				
		248,517			
Deduct—					
Received from New Orleans	15,630				
Stock on hand 1st September, 1868	2,161				
		17,791			
TEXAS.			230,726	366,193	239,516
Export from Galveston, &c—					
To Foreign Ports (including 3,165 to Mexico)	83,376				
Coastwise	64,505				
Stock in Galveston, 1st September, 1869	202				
		148,083			
Deduct—					
Received from New Orleans	100				
Stock in Galveston, 1st September, 1868	166				
		266			
FLORIDA.			147,817	114,666	185,919
Export from Apal'cola, Jacksonville, &c.—					
To Foreign Ports	810				
Coastwise—Uplands	5,816				
Sea Island	6,748				
Stock in Fernandina, 1st September, 1869	18				
		13,392			
Deduct—					
Stock on hand 1st September, 1868	....	....			
			13,392	34,639	58,349
GEORGIA.					
Export from Savannah—					
To Foreign Ports—Uplands	161,516				
" Sea Island	6,021				
Coastwise—Uplands	189,989				
Sea Island	5,174				
Stock in Savannah, 1st September, 1869	313				
		363,013			
Deduct—					
Received from Florida—Uplands	240				
" Sea Island	4,824				
Stock in Savannah 1st September, 1868	696				
		5,760			
			357,253	495,005	255,965

Statement and Total Amount of the Cotton Crop of the United States, for the Year ending August 31, 1869—*Concluded.*

	Bales.	Bales.	TOTAL. 1869.	1868.	1867.
SOUTH CAROLINA.					
Export from Charleston, S. C.—					
To Foreign Ports—Uplands	52,814				
" Sea Island	3,995				
Coastwise—Uplands	142,024				
Sea Island	3,313				
Stock in Charleston, 1st September, 1869	250				
	202,396				
Export from Georgetown, S. C.—					
To Northern Ports—Uplands & Sea Island	348				
		202,744			
Deduct—					
Received from Florida—Uplands	156				
" Sea Island	1,700				
Stock in Charleston, 1st September, 1868	1,945				
		3,801			
			198,943	240,225	162,247
NORTH CAROLINA.					
Export—					
To Coastwise Ports	....	....	35,912	38,587	38,522
VIRGINIA.					
Export—					
To Foreign Ports	6,258				
Coastwise	134,276				
Manufactured (taken from the ports)	20,000				
Stock in Petersburg, 1st September, 1869	50				
		160,579			
Deduct—					
Stock on hand 1st September, 1868	....	161			
			160,418	187,487	123,627
TENNESSEE, &c.					
Shipments from Memphis, Tenn	247,651				
" Nashville, Tenn	65,825				
" other places in Tenn., K'y, &c.	75,304				
Stock in Memphis & Nashville, 1st Sept, 1869	94				
		388,874			
Deduct—					
Shipments to New Orleans, from Memphis and Nashville	30,767				
Shipments to Norfolk, from Memphis and Nashville	34,707				
Received from New Orleans	1,402				
Stock in Memphis & Nashville, 1st Sept., 1868	107				
		66,983			
			321,891	374,860	185,712
Total crop of the United States 1868–9	....	....	2,260,557	2,430,893	1,951,988
Decrease from Crop of 1867–8	....	....	....	....	170,336
Increase over Crop of 1866–7	....	....	....	....	308,569

Export to Foreign Ports, from September 1, 1868, *to August* 31, 1869.

FROM	To Great Britain.	To France.	To North of Europe.	Other F'n Ports.	Total.
New Orleans, La .... bales.	342,249	165,282	73,743	38,260	619,534
Mobile, Ala..............................	137,484	16,133	2,981	6,556	163,154
Galveston, Texas.........................	57,582	....	22,629	3,165	83,376
Jacksonville, Fla........................	....	810	....	....	810
Savannah, Ga.............................	133,678	20,869	12,990	....	167,537
Charleston, S. C	53,753	....	652	2,404	56,809
Norfolk, Va..............................	6,253	....	....	....	6,253
Wilmington, N. C.........................	....	....	....	....	....
New York.................................	246,311	21,433	54,093	5,863	327,700
Baltimore................................	9,091	....	10,094	....	19,185
Philadelphia.............................	98	....	....	....	98
Boston & Portland (Port'd, 1,695 to Gt. B'n.)	3,001	....	....	186	3,187
Grand total 1868–69......................	989,500	224,527	177,182	56,434	1,447,643
Total 1867–68	1,228,596	197,515	145,042	84,663	1,655,816
Increase.................................	....	27,012	32,140	....	....
Decrease.................................	239,096	....	....	28,229	208,173

Consumption.

Total crop of the United States, as before stated ....bales.			2,260,557
Add--Stocks on hand at the commencement of the year, 1st September, 1868.—In the Southern ports		7,195	
" Northern "		30,203	
			37,398
Makes a supply of...			2,297,955
Deduct therefrom—The export to Foreign ports.............	1,447,643		
Less, foreign included..................	2,975		
		1,444,668	
Stocks on hand, September 1, 1869—			
In the Southern ports............................	2,772		
" Northern "	8,388		
		11,160	
Burnt in transit from Cedar Keys to Fernandina, Fla........	203		
Manufactured in Virginia..................................	20,000		
		20,203	
			1,476,031
Taken for home use north of the Potomac and Ohio rivers ....bales.			821,924
Taken for home use south of the Potomac and Ohio rivers, and burnt....			,173,203
Total consumed in the United States (including burnt at the ports), 1868–69...			995,127

ANNUAL REVIEW.

From the New Orleans Price Current, 1868–69.

The rainy and unusually cold weather in the spring threw the planting of the crop back about three weeks later than usual. With the advance in the season, however, the weather became more genial, and latterly it has been so hot and generally propitious as to compensate to a great extent for the lateness of the planting. The receipts at this port, the coming season, are expected materially to exceed those of the one just closed, but the picking season has but just commenced, and the crop is yet liable to many contingencies. Experience has taught us the impropriety of making crop estimates at this early period of the season, but we may say, nevertheless, that the figures of those best acquainted with the subject, range from 2,750,000 to 3,000,000 bales for the crop of the United States, the prevalent belief being that the larger estimate cannot be reached, however, without considerable additions to the present picking force, and without, also, a very favorable and protracted picking season.

COTTON AT LIVER

WEEK ENDING.	RECEIPTS.						SALES.			
	American	E. I.	Egypt.	Brazil.	Other.	Total.	Con-sumption.	Specu-lation.	Export.	Total.
Jan. 7..	28,197	5,628	4,662	11,161	4,050	53,698	49,460	13,410	19,610	82,480
" 14..	16,120	5,949	7,842	7,508	3,665	41,084	84,100	51,950	20,380	156,430
" 21..	25,757	17,072	7,456	7,667	819	58,771	55,620	18,310	13,020	86,950
" 28..	18,110	....	3,539	13,671	553	35,873	46,470	18,980	11,940	77,390
Feb. 4..	33,753	6,870	6,613	7,080	1,586	55,902	61,930	62,530	15,230	139,690
" 11..	27,398	3,243	5,840	16,367	933	53,781	42,710	22,210	3,860	68,780
" 18..	93,973	5,214	8,873	16,046	3,036	127,142	32,480	7,860	4,270	44,610
" 25..	29,331	7,587	3,230	5,401	2,372	47,921	39,710	9,260	7,950	56,920
Mch. 4..	33,520	11,899	7,622	13,316	3,151	69,508	38,710	5,590	7,640	51,940
" 11..	11,354	4,260	2,686	8,989	181	27,470	47,880	9,230	10,580	67,690
" 18..	16,536	5,068	2,573	6,594	767	31,538	44,270	5,020	4,490	53,780
" 24..	28,723	4,763	1,168	6,561	3,589	44,804	45,090	9,790	8,300	63,180
April 1..	13,145	3,191	1,388	6,776	2,543	27,043	45,590	6,530	9,080	61,200
" 8..	51,824	36,643	5,312	25,105	4,947	123,831	35,520	3,330	8,500	47,350
" 15..	22,980	14,450	3,656	12,256	3,330	56,672	54,940	11,010	8,310	74,260
" 22..	46,315	25,083	2,083	16,258	1,633	91,372	37,970	3,900	8,160	50,030
" 29..	18,526	....	....	3,089	9,687	31,302	39,020	5,870	8,960	53,850
May 6..	12,010	17,123	8,687	16,489	1,904	56,213	36,840	2,650	7,160	46,650
" 13..	38,197	17,156	1,230	3,644	2,576	62,803	42,150	2,980	6,380	51,910
" 20..	22,569	13,498	2,192	7,253	6,182	51,694	34,700	1,240	6,310	42,250
" 27..	67,252	6,283	741	16,091	2,418	92,785	47,450	4,660	10,140	62,250
June 3..	20,908	7,013	2,667	5,384	4,146	40,118	64,000	11,090	15,060	90,150
" 10..	53,576	25,844	4,410	24,268	1,356	109,454	56,480	9,680	12,120	78,280
" 17.	10,778	34,795	1,495	12,151	4,249	63,468	54,860	13,340	17,680	85,880
" 24..	8,972	5,113	3,254	6,895	3,549	27,783	57,000	12,370	15,690	85,060
July 1..	5,993	14,094	3,861	802	1,411	26,161	72,230	11,330	18,640	102,200
" 8..	38,761	33,518	1,411	17,256	4,330	95,276	46,120	7,410	9,830	63,360
" 15..	26,952	39,915	1,689	9,008	3,519	81,083	53,760	7,810	11,730	73,300
" 22..	544	13,137	1,954	70	5,912	21,617	35,320	6,820	11,790	53,930
" 29..	4,724	13,555	3,549	16,295	1,172	39,295	67,370	17,750	16,020	101,140
Aug. 5..	9,394	12,369	1,622	3,176	3,800	30,361	55,800	14,680	16,380	86,860
" 12..	5,177	15,272	1,999	685	535	23,668	49,280	9,990	16,220	75,490
" 19..	3,316	28,478	1,707	4,019	2,342	39,862	62,710	30,470	18,010	111,190
" 26..	3,217	42,059	1,481	8,074	3,580	58,411	22,870	6,170	9,270	38,310
Sep. 2..	10,447	192,276	1,672	6,090	3,652	214,137	31,810	14,740	11,070	57,620
" 9..	4,623	50,665	2,038	15,390	1,272	73,988	29,950	8,620	11,450	50,020
" 16..	1,801	30,359	654	18,683	1,520	53,017	24,800	13,260	17,530	55,590
" 23..	2,026	25,462	1,101	1,125	466	30,180	28,090	11,040	17,100	56,230
" 30..	360	41,288	1,205	10,364	518	53,735	37,550	13,040	12,590	63,180
Oct. 7..	3,233	68,490	369	7,080	1,469	75,641	44,110	10,270	10,410	64,790
" 14..	6,832	7,072	757	7,005	417	22,083	40,210	8,130	7,590	55,930
" 21..	8,037	56,210	2,263	2,702	1,401	70,613	53,650	21,300	17,010	91,960
" 28..	7,244	33,624	....	13,014	2,545	56,427	49,860	7,780	9,410	67,050
Nov. 4..	16,136	17,058	1,876	6,971	1,709	43,750	62,940	23,630	19,920	106,490
" 11..	13,404	53,015	6,648	12,770	274	86,111	38,020	4,220	7,470	49,710
" 18..	18,455	17,458	3,988	12,646	443	52,990	56,200	7,460	10,880	74,540
" 25..	10,384	1,973	4,926	10,663	5,223	33,169	50,940	4,910	9,970	65,820
Dec. 2..	12,810	17,832	5,112	2,829	1,393	39,976	82,580	28,250	16,990	127,820
" 9..	16,267	11,433	5,852	660	288	34,500	38,190	5,060	10,280	53,530
" 16..	61,950	8,628	9,143	17,423	3,603	100,747	60,530	21,260	15,980	97,770
" 22..	32,880	15,201	15,167	11,305	1,143	75,696	44,030	7,530	6,830	58,390
" 30..	34,327	2,157	4,961	7,226	2,718	51,389	48,310	8,670	11,880	68,860
Average prices & total sales, receipts & stocks.	1,109,018	1,144,583	189,124	499,358	129,987	3,072,070	2,592,170	754,290	621,070	3,967,530

POOL. YEAR 1869.

STOCKS.			PRICES.			ACTUAL EXPORT.	CONSUMPTION.	REMARKS.
American.	Other.	Total.	Mid. Up.	Mid. Orl.	Fair Dhol.			
93,590	258,500	352,090	11	11 1/8	8 3/4	19,610	49,460	
78,650	221,890	300,540	11 1/2	11 3/4	9 3/8	20,380	133,560	Market opened strong and
81,880	212,850	294,730	11 1/2	11 3/4	9 1/2	13,020	189,180	became very excited.
76,770	192,470	269,240	11 1/2	11 11/16	9 1/2	11,940	235,650	
84,090	172,470	256,560	12 1/4	12 7/16	10 1/4	15,230	297,580	February 4th, excitement,
93,210	167,180	260,390	12 5/16	12 1/2	10 3/8	3,860	340,290	large demand from all
96,360	181,170	277,530	11 7/8	12 1/8	10	4,270	372,770	classes of buyers; 11th,
107,610	174,470	282,080	11 7/8	12 1/8	10	7,950	412,480	firm market; 18th, mar-
120,910	189,060	309,970	11 7/8	12 1/8	10	7,640	451,180	ket heavy and declining;
108,680	174,860	283,540	12	12 5/16	10 1/8	10,580	499,060	25th, quiet but steady.
105,350	158,320	263,670	12	12 1/4	10	4,490	543,330	
115,830	142,420	258,250	12 1/8	12 3/8	10 1/8	8,300	588,420	Improved tone.
108,260	127,870	236,130	12 3/8	12 5/8	10 3/8	9,080	634,010	April 1st, at first animated,
145,100	173,860	318,960	12 1/8	12 3/8	10 1/4	8,500	669,530	then quiet, bank rate
141,750	172,830	314,580	12 1/4	12 1/2	10 3/8	8,310	724,470	raised to 4 per cent.; 8th,
167,330	195,650	362,980	12	12 1/4	10 1/8	8,160	762,440	large import, market
168,450	183,090	351,540	11 7/8	12 1/8	10	8,960	801,460	dull; 15th, better feel-
165,080	196,660	361,740	11 3/4	12	10	7,160	838,300	ing: 22d, limited de-
181,990	191,920	373,910	11 1/2	11 7/8	9 7/8	6,380	880,450	mand.
185,140	197,810	382,950	11 3/8	11 3/4	9 3/4	6,310	915,150	
231,380	191,600	422,980	11 1/4	11 1/2	9 5/8	10,140	962,600	
220,950	171,180	392,130	11 1/2	11 7/8	9 7/8	15,060	1,026,600	Large business at har-
248,500	186,940	435,440	11 3/4	12	10	12,120	1,083,080	dening prices.
229,280	299,180	428,460	12 1/16	12 5/16	10	17,680	1,137,940	Very active.
212,150	176,330	388,480	12 1/4	12 1/2	10 1/8	15,690	1,194,940	
188,340	144,080	332,420	12 1/2	12 3/4	10 1/4	18,640	1,267,170	
207,110	158,690	365,800	12 1/2	12 3/4	10 1/4	9,830	1,313,290	Steady prices.
210,310	171,300	381,610	12 5/8	12 7/8	10 3/8	11,730	1,367,050	Bank rate reduced to 3 per cent.
193,090	159,840	352,930	12 1/2	12 3/4	10 1/4	11,790	1,402,370	Quiet tone.
167,500	147,960	315,460	12 3/4	12 15/16	10 3/8	16,020	1,479,740	Good general demand.
150,430	126,500	276,930	12 7/8	13 1/16	10 1/2	16,380	1,535,540	
123,270	127,030	250,300	13	13 1/4	10 9/16	16,220	1,584,820	August 12th, stock declar-
105,470	122,120	227,590	13 3/4	14	10 7/8	18,010	1,647,530	ed, proving 15,930 above
95,430	152,200	247,630	13 11/16	13 7/8	10 3/4	9,270	1,670,400	estimate; 19th, large de-
93,870	325,510	419,380	13 11/16	13 13/16	10 5/8	11,070	1,702,210	mand and bare supply,
87,030	362,510	449,540	13 3/8	13 1/2	10 1/4	11,450	1,732,160	bank rate reduced to 2 1/2
79,520	380,450	459,970	13	13 1/4	10 1/16	17,530	1,756,960	per cent.; 26th, without
69,210	372,820	442,030	12 1/4	12 1/2	9 5/8	17,100	1,785,050	animation.
56,510	385,500	442,010	12 1/4	12 1/2	9 5/8	12,590	1,822,600	September 16th, market
46,460	412,790	459,250	12 3/8	12 5/8	9 3/8	10,410	1,866,710	depressed and irregular;
41,840	483,590	425,430	12 1/8	12 1/2	9	7,590	1,906,920	23d, further depression.
36,370	397,300	433,670	12	12 1/4	9 1/8	17,010	1,960,570	October 7th, firm for Am-
29,080	399,160	428,240	12	12 1/4	8 7/8	9,410	2,010,430	erican, but Surats dull
30,680	367,820	398,500	12 1/8	12 1/4	9 1/8	19,920	2,048,450	of sale; 14th, dull in
30,870	403,930	434,800	11 1/2	11 3/4	8 7/8	7,470	2,104,650	tone; 21st, fair inquiry.
29,900	389,950	419.850	11 5/8	11 7/8	8 3/4	10,880	2,155,590	
24,600	363,460	388,060	11 1/2	11 7/8	8 3/4	9,970	2,238,170	Quiet, but steady.
18,630	316,400	335,030	12 1/8	12 3/8	9 3/16	16,990	2,276,360	Market very active.
21,230	297,990	319,220	11 3/4	12	9	10,280	2,336,890	Cotton freely offered.
64,090	285,010	349,100	11 3/4	12	9 5/16	15,980	2,380,920	
81,970	288,440	370,410	11 1/2	11 3/4	9 3/8	6,830	2,429,230	
76,900	260,860	337,760	11 1/2	11 3/4	9 7/16	11,880	2,592,170	Fair demand, prices firm.
			12.12	12.29	9.875	621,070	49,849.88	

The semi-weekly Price and Weekly Sales and Receipts at New York, Weekly Exports from New York and Rates of Freight to Liverpool 1st of each month, for the Crop Year ending September 1, 1869.

1868.	Price of Middling New Orleans.	Price of Middling Upland.	Sales for week.	Receipts for week.	EXPORTS FOR THE WEEK. To Great Britain.	To France.	North of Europe.	Other Fo'n Ports	Total Exports.	Rates of Freight to Liverpool.
Septem. 4..	30	29½	11,682	2,225	2,004	1	296	....	2,301	Steam, $\frac{5}{32}$@$\frac{3}{16}$d. Sail, 5-32d.
" 8..	29	28½								
" 11..	27	26½								
" 15..	25½	25	7,034	2,620	1,667	....	....	....	1,667	
" 18..	27	26½								
" 22..	26½	26	10,789	5,528	1,864	400	146	....	2,410	
" 25..	26	25½								
" 29..	26½	26	9,814	7,964	1,495	....	16	328	1,839	Steam, 3-16d. Sail, 3-16d.
October 2..	27½	27								
" 6..	27½	27	15,538	12,054	1,070	1,099	111	....	2,280	
" 9..	26½	26								
" 13..	26¾	26¼	15,039	12,791	793	....	100	....	893	
" 16..	26	25½								
" 20..	26	25½	16,306	23,731	3,081	691	456	....	4,228	
" 23..	25¾	25¼								
" 27..	26	25½	23,707	18,134	4,346	1,386	882	....	6,614	
" 30..	26	25½								
Novem. 3..	26	25½	22,644	20,113	7,070	759	2,190	....	10,019	Steam, ⅜@½d. Sail, ¼d.
" 6..	25¾	25¼								
" 10..	25	24½	16,887	26,908	7,775	....	2,795	....	10,570	
" 13..	24¾	24¼								
" 17..	24¾	24¼	19,987	20,304	8,486	2,705	2,701	....	13,892	
" 20..	25	24¾								
" 24..	25¾	25¼	28,364	18,584	9,478	....	4,004	....	13,482	
" 27..	25¾	25¼								
Decem. 1..	25¾	25¼	23,106	24,265	10,244	2,902	4,413	972	18,531	Steam, ⅝@¾d. Sail, 5-16@⅜d
" 4..	25½	25								
" 8..	25¼	24⅝	15,437	21,377	10,941	....	3,288	....	14,229	
" 11..	25¾	25¼								
" 15..	25¾	25¼	21,215	20,067	10,318	2,472	4,083	....	16,873	
" 18..	25¾	25¼								
" 22..	25¾	25¼	20,262	19,014	7,981	....	1,856	926	10,763	
" 25..	25½	25⅛								

GENERAL REMARKS.

Early in September, there was a good demand at full prices, but later, the prompt arrival at the ports of new cotton, together with a dull market in Liverpool, induced more desire to sell, and prices declined.

October opened with a better demand from spinners and some speculative inquiry, and prices advanced; at the rise, there was some pressure to realize on speculative lots, and the market fell back.

Larger receipts at the ports caused a dull and declining market in November, though toward the latter part there was more activity, and prices rallied.

In December, there was a fair demand, and prices, for the most part, were steady, though occasionally affected by the contracting for future delivery, which now began to be a feature in the market, at prices 1@2c. below those current for spot lots.

January was a very active month, the business being largely on

Decem. 29..	$25\frac{3}{4}$	$25\frac{3}{8}$	16,865		18,359	4,654	387	3,922	....	8,943		
1869.			Spot, &c.	Contracts							Steam,	
January 1	$26\frac{1}{2}$	26									$\frac{3}{8}$@7-16d	
" 5..	28	$27\frac{1}{2}$	31,611	1,850	16,931	3,180	....	1,387	....	4,567	Sail,	
" 8..	$28\frac{1}{4}$	$27\frac{3}{4}$									$\frac{1}{4}$@5-16d	
" 12..	$29\frac{1}{4}$	$28\frac{1}{4}$	31,967	4,600	20,276	8,681	725	1,250	....	10,656		
" 15..	$30\frac{1}{4}$	$29\frac{3}{4}$										
" 19..	$29\frac{1}{2}$	29	40,728	4,150	28,497	6,036	....	1,722	....	7,758		
" 22..	30	$29\frac{1}{2}$										
" 26..	29	$28\frac{1}{2}$	20,123	3,100	23,761	2,720	184	651	600	4,155		
" 29..	$29\frac{1}{2}$	29									Steam,	
February 2..	$30\frac{1}{4}$	$29\frac{3}{4}$	32,765	3,150	18,942	1,953	....	140	....	2,093	$\frac{1}{4}$@5-16d	
" 5..	$30\frac{3}{4}$	$30\frac{1}{4}$									Sail,	
" 9..	$30\frac{3}{4}$	$30\frac{1}{4}$	42,826	2,800	18,022	5,661	....	365	....	6,026	3-16d.	
" 12..	$30\frac{1}{2}$	30										
" 16..	30	$29\frac{1}{2}$	14,291	2,000	15,329	6,812	....	714	....	7,526		
" 19..	$29\frac{1}{4}$	$28\frac{3}{4}$										
" 23..	$30\frac{1}{4}$	$29\frac{3}{4}$	16,949	1,000	18,169	11,313	346	103	....	11,762		
" 26..	$29\frac{3}{4}$	$29\frac{1}{4}$									Steam,	
March 2..	$29\frac{3}{4}$	$29\frac{1}{4}$	7,057	2,600	20,701	4,457	....	951	....	5,408	$\frac{1}{8}$@3-16d	
" 5..	$29\frac{3}{4}$	$29\frac{1}{4}$									Sail, $\frac{1}{8}$d.	
" 9..	$29\frac{3}{4}$	$29\frac{1}{4}$	12,661	3,500	19,218	2,006	209	314	....	2,529		
" 12..	29	$28\frac{1}{2}$										
" 16..	29	$28\frac{1}{2}$	10,435	2,800	13,535	4,759	....	143	....	4,902		
" 19..	29	$28\frac{1}{2}$										
" 23..	29	$28\frac{1}{2}$	16,514	3,000	10,619	5,647	509	120	....	6,276		
" 26..	$29\frac{1}{4}$	$28\frac{3}{4}$										
" 30..	$29\frac{1}{2}$	29	18,251	750	10,092	7,642	....	137	....	7,779	Steam,	
April 2..	$29\frac{1}{4}$	$28\frac{3}{4}$									3-16@$\frac{1}{4}$d	
" 6..	29	$28\frac{1}{2}$	13,725	900	9,127	9,907	291	1,378	....	11,576	Sail,	
" 9..	29	$28\frac{1}{2}$									—@—	
" 13..	$29\frac{1}{4}$	$28\frac{3}{4}$	23,619	500	8,334	5,625	....	1,644	....	7,269		
" 16..	$29\frac{1}{2}$	29										
" 20..	$29\frac{1}{2}$	$28\frac{3}{4}$	18,677	....	6,039	8,665	1,536	1,198	....	11,399		
" 23..	$29\frac{1}{2}$	$28\frac{3}{4}$										
" 27..	$29\frac{1}{2}$	$28\frac{3}{4}$	10,851	....	4,991	12,466	....	1,444	862	14,772		
" 30..	$29\frac{1}{2}$	$28\frac{3}{4}$									Steam,	
May 4..	$29\frac{1}{2}$	$28\frac{3}{4}$	16,885	600	7,455	7,726	244	2,120	....	10,090	5-32d.	
" 7..	$29\frac{1}{2}$	$28\frac{3}{4}$									Sail,	
" 11..	$29\frac{1}{2}$	$28\frac{5}{8}$	14,192	600	3,600	6,195	....	1,943	824	8,962	—@—	
" 14..	$29\frac{1}{2}$	$28\frac{3}{4}$										
" 18..	$29\frac{1}{2}$	$28\frac{3}{4}$	18,031	....	7,823	5,290	676	1,499	....	7,465		

speculation and to cover contracts, the market being less influenced by Liverpool and Manchester accounts than before in several years.

Early in February, the market was excited, the stock in Liverpool having run down to 260,000 bales of all kinds, and the purchases by speculators and exporters were large; spinners here, however, hesitated to go on at the ruling rates, and subsequently the market became dull and prices fell off.

The market in March was, for the most part, dull and declining.

In April, there was a better feeling, as the receipts at the ports fell off, and the accounts from the India crop were unfavorable; but the English manufacturers reduced their consumption, and our own pursued a very cautious policy, goods selling relatively lower than cotton.

May was, on the whole, comparatively quiet, but prices were steady, being strengthened to some extent by the cool weather at the South, which retarded the growth of the plant.

In June, the market was active and buoyant, with some advance in prices.

The activity continued through the early part of July, the supply on both sides the water being light, but later, with more

New York Statement for Year 1869—*Concluded.*

1869.		Price of Middling New Orleans.	Price of Middling Upland.	SALES FOR WEEK.		Receipts for week.	EXPORTS FOR THE WEEK					Rates of Freight to Liverpool.
				Spot, &c.	Contracts		To Great Britain.	To France.	North of Europe.	Other Fo'n Ports	Total Exports.	
May	21..	29$\frac{1}{2}$	28$\frac{3}{4}$									
"	25..	29$\frac{1}{2}$	28$\frac{3}{4}$	10,510	300	3,416	5,909	930	933	....	7,772	
"	28..	29$\frac{1}{2}$	28$\frac{3}{4}$									Steam,
June	1..	30$\frac{1}{2}$	29$\frac{3}{4}$	17,155	....	5,669	4,083	802	495	1,351	6,731	$\frac{1}{4}$d.
"	4..	31	31$\frac{1}{4}$									Sail,
"	8..	32	31$\frac{1}{4}$	16,998	1,000	4,966	1,204	....	1,151	....	2,355	5-32d.
"	11..	32$\frac{1}{4}$	31$\frac{1}{2}$									
"	15..	33$\frac{1}{4}$	32$\frac{3}{4}$	20,669	1,700	7,449	538	482	648	....	1,668	
"	18..	34	33$\frac{1}{2}$									
"	22..	33$\frac{1}{2}$	33	11,740	1,750	8,515	1,206	....	200	....	1.406	
"	25..	33$\frac{3}{4}$	33									
"	29..	34$\frac{3}{4}$	34$\frac{1}{4}$	18,432	3,400	10,270	132	20	....	....	152	Steam,
July	2..	35	34$\frac{1}{2}$									$\frac{1}{4}$d.
"	6..	35	34$\frac{1}{2}$	12,866	2,170	7,330	778	202	....	....	980	Sail,
"	9..	35	34$\frac{1}{2}$									5-32d.
"	13..	35	34$\frac{1}{2}$	7,313	5,650	4,623	147	10	....	....	157	
"	16..	35	34$\frac{1}{2}$									
"	20..	35	34$\frac{1}{2}$	5,622	3,350	4,368	1,036	....	10	....	1,046	
"	23..	34$\frac{1}{2}$	34									
"	27..	34$\frac{1}{4}$	33$\frac{3}{4}$	4,860	2,900	4,555	143	7	....	150	300	
"	30..	34	33$\frac{1}{2}$									Steam,
August	3..	34	33$\frac{1}{2}$	6,762	6,320	2,460	1,164	....	....	....	1,164	$\frac{1}{4}$d.
"	6..	34	33$\frac{1}{2}$									Sail,
"	10..	34	33$\frac{1}{2}$	8,451	11,150	2,130	1,879	524	174	....	2,577	5-32d.
"	13..	33$\frac{3}{4}$	33$\frac{1}{4}$									
"	17..	33$\frac{3}{4}$	33$\frac{1}{4}$	11,175	11,700	2,079	4,129	182	....	....	4,311	
"	20..	35$\frac{3}{4}$	35									
"	24..	35$\frac{1}{2}$	35	8,760	7,175	1,217	2,719	752	....	....	3,471	
"	27..	35$\frac{1}{4}$	34$\frac{3}{4}$									Steam,
"	31..	35$\frac{1}{2}$	35	5,416	5,200	2,290	1,256	....	....	....	1,256	$\frac{1}{4}$d.
Average prices and total sales, receipts and exports.		29.48	29.01	873,563	101,665	626,836	246,311	21,433	54,093	6,013	327,850	Sail, 3-16d.

General Remarks.

cheerful crop accounts, prices receded.

The crop estimates began to widen in August and settled down to 2$\frac{3}{4}$@3 millions of bales, and the market for future delivery was weak; spot and transitu lots, however, were very firm, stocks everywhere being at a low ebb, and a large advance was only prevented by a general reduction of the consumption.

Exchange.

The quotation in September for 60 days' bills on London ranged from 8@8$\frac{7}{8}$ per cent. premium; in October, 8@9$\frac{5}{8}$; in November, the same; in December, 8$\frac{3}{4}$@9$\frac{3}{8}$; in January, the same; in February, 8@9$\frac{1}{4}$; in March, 7@8$\frac{5}{8}$; in April, the quotation advanced from 6$\frac{3}{4}$@7$\frac{1}{4}$ to 8@8$\frac{5}{8}$; May, steady, 8@9; June, 8$\frac{1}{4}$@9$\frac{1}{2}$; July, 9@9$\frac{3}{4}$; and in August the range was 9@9$\frac{3}{4}$ per cent. premium, gold.

1870.

COTTON CROP OF THE UNITED STATES.

Statement and Total Amount of the Cotton Crop of the United States for the Year ending August 31, 1870.

	Bales.	Bales.	TOTAL. 1870.	1869.	1868.
LOUISIANA.					
Export from New Orleans—					
To Foreign Ports	1,005,530				
Coastwise	179,520				
Burnt and Manufactured	2,357				
Stock on hand, 1st September, 1870	20,696				
		1,208,103			
Deduct—					
Received from Mobile	49,890				
" " Florida	3,477				
" " Texas	11,869				
Stock on hand, 1st September, 1869	770				
		66,006			
			1,142,097	794,205	579,231
ALABAMA.					
Export from Mobile—					
To Foreign Ports	200,838				
Coastwise	97,685				
Stock on hand, 1st September, 1870	9,743				
		308,266			
Deduct—					
Received from New Orleans	1,141				
Stock on hand, 1st September, 1869	1,169				
		2,310			
TEXAS.			305,956	230,726	366,193
Export from Galveston, &c.—					
To Foreign Ports—including 5,522 to Mexico	152,559				
Coastwise	89,132				
Stock in Galveston, 1st September, 1870	4,795				
		246,486			
Deduct—					
Stock in Galveston, 1st September, 1869	....	202			
			246,284	147,817	114,666
FLORIDA.					
Export from Apalachicola, Jacksonville, &c.					
To Foreign Ports—Uplands	240				
Sea Island	16				
Coastwise—Uplands	12,852				
Sea Island	10,097				
Stock in Apalachicola, 1st September, 1870	7				
		23,212			
Deduct—					
Stock on hand, 1st September, 1869	....	18			
			23,194	13,392	34,639
GEORGIA.					
Export from Savannah—					
To Foreign Ports—Uplands	259,102				
Sea Island	6,529				
Coastwise—Uplands	214,188				
Sea Island	9,606				
Burnt at Savannah (300 Sea Island)	540				
Stock in Savannah, 1st September, 1870	2,833				
		492,798			
Deduct—					
Received from Florida—Uplands	417				
Sea Island	6,377				

Statement and Total Amount of the Cotton Crop of the United States, for the Year ending August 31, 1870.—*Concluded.*

	Bales.	Bales.	TOTAL. 1870.	1869.	1868.
Received from Beaufort, S. C.	317				
Stock in Savannah, 1st September, 1869	313				
		7,424			
SOUTH CAROLINA.			485,734	357,253	495,005
Export from Charleston—					
To Foreign Ports—Uplands	89,851				
Sea Island	7,258				
Coastwise—Uplands	146,760				
Sea Island	5,686				
Stock in Charleston, 1st September, 1870	1,399				
	250,954				
Export from Georgetown, S. C.—					
To New York—Uplands	472				
From Beaufort, S. C., to Savannah	317				
		251,743			
Deduct—					
Received from Florida—Uplands	1,462				
Sea Island	3,438				
Stock in Charleston, 1st September, 1869	250				
		5,150			
NORTH CAROLINA.			246,593	198,943	240,225
Export—					
To Foreign Ports	50				
Coastwise	58,834				
		....	58,884	35,912	38,587
VIRGINIA.					
Export—					
To Foreign Ports	9,689				
Coastwise	193,357				
Stock in Norfolk and Petersburg, 1st September, 1870	985				
		204,031			
Deduct—					
Stock on hand, 1st September, 1869	....	50			
			203,981	160,418	187,487
TENNESSEE, &c.					
Shipments from Memphis, Tenn	286,457				
Nashville, Tenn	58,263				
other places in Tennessee, Kentucky, &c.	108,768				
Stock in Memphis and Nashville, 1st September, 1870	6,481				
		459,969			
Deduct—					
Shipments to New Orleans, from Memphis and Nashville	45,462				
Shipments to Norfolk and Charleston, from Memphis and Nashville	92,027				
Stock in Memphis and Nashville, 1st September, 1869	94				
		137,583			
			322,386	321,891	374,860
Manufactured at the South	....	....	79,843		
Total Crop of the United States 1869–70	....	....	3,114,592	2,260,557	2,430,893

Increase over crop of 1868–9 .... bales. 854,035
Increase over crop of 1867–8 683,699
Increase over crop of 1866–7 1,162,604

Export to Foreign Ports, from September 1, 1869, *to August* 31, 1870.

FROM	To Great Britain.	To France.	To North of Europe.	Other Foreign Ports.	Total.
New Orleans, La....................bales.	549,603	259,223	124,049	72,655	1,005,530
Mobile, Ala..................................	165,989	15,910	10,413	8,526	200,838
Galveston, Tex..............................	122,106	7,939	16,992	5,522	152,559
Jacksonville, Florida......................	....	256	....	....	256
Savannah, Ga................................	204,570	43,796	17,265	....	265,631
Charleston, S. C............................	87,287	1,825	1,338	6,659	97,109
Norfolk, Va..................................	9,689	....	....	....	9,689
Wilmington, N. C...........................	50	....	....	....	50
New York	324,421	17,757	60,516	2,446	405,140
Baltimore	7,494	....	24,668	10	32,172
Philadelphia	....	....	....	....	....
Boston & Portland (Portland, 2941 to G. Bt'n)	4,235	....	74	276	4,585
Grand total 1869–70.................	1,475,444	346,706	255,315	96,094	2,173,559
Total 1868–9	989,500	224,527	177,182	56,434	1,447,643
Increase over 1868–9	485,944	122,179	78,133	39,660	725,916

Consumption.

Total crop of the United States, as before stated................bales.			3,114,592
Add—Stocks on hand at the commencement of the year, September 1, 1869, in the Southern ports..		2,772	
" Northern ports..........		8,388	
			11,160
Makes a supply for the year ending August 31, 1870, of			3,125,752
Deduct therefrom—The export to Foreign ports.............	2,173,559		
Less—Foreign included..................................	3,301		
		2,170,258	
Stocks on hand September 1, 1870—			
In the Southern ports................................	46,939		
In the Northern ports................................	18,386		
		65,325	
Burnt at New York, Baltimore, &c..........................	5,422		
Manufactured at the South................................	79,843		
Shipped to Canada	27,563		
		112,828	
			2,348,411
Taken for home use north of the Potomac and Ohio Rivers...............bales.			777,341
Taken for home use south of the Potomac and Ohio Rivers, and Burnt..........			85,265
Total consumed in the United States (including burnt at the ports), 1869–70.....			862,606

COTTON AT LIVER

Week Ending.	RECEIPTS.						SALES.			
	American	E. I.	Egypt.	Brazil.	Other.	Total.	Consumption.	Speculation.	Export.	Total.
Jan. 6..	31,272	8,782	7,131	12,819	1,309	61,313	48,110	8,120	9,390	65,620
" 13..	43,767	3,338	5,530	11,394	3,328	67,347	45,910	7,850	10,360	64,120
" 20..	55,240	......	2,349	2,405	1,752	61,746	73,100	17,360	13,990	104,450
" 27..	7,595	7,370	6,776	5,123	1,016	27,880	59,290	24,750	13,390	97,430
Feb. 3..	59,849	3,442	4,357	19,820	3,504	90,972	35,110	10,190	5,770	51,070
" 10..	30,150	18,197	6,487	8,810	1,138	64,782	50,120	8,920	6,100	65,140
" 17..	13,068	......	......	1,195	1,312	15,575	46,760	8,190	6,660	61,610
" 24..	16,030	3,648	9,762	7,442	2,023	38,905	42,310	6,260	5,080	53,650
March 3..	9,046	1,000	2,172	5,348	3,966	21,532	38,230	4,730	5,140	48,100
" 10..	9,612	8,106	5,101	1,256	......	24,075	38,440	2,980	4,900	46,320
" 17..	32,289	8,032	7,801	12,931	566	61,619	48,930	5,100	4,590	58,620
" 24..	95,834	7,536	2,257	13,631	3,510	122,768	47,970	3,610	7,100	58,680
" 31..	103,233	18,597	3,942	8,395	2,118	136,285	48,510	5,700	7,270	61,480
April 7..	53,616	10,125	4,117	5,515	2,473	75,846	63,320	16,540	14,610	94,470
" 13..	48,805	11,141	913	9,920	959	71,738	44,610	6,500	8,300	59,410
" 21..	19,965	29,517	1,930	400	1,218	53,030	39,600	7,060	3,840	50,500
" 28..	95,078	21,246	1,261	14,735	2,001	135,321	50,260	7,600	8,730	66,590
May 5..	8,319	16,271	6,071	......	2,189	32,850	48,540	6,970	5,250	60,760
" 12..	23,189	9,632	5,232	17,182	521	55,756	58,490	16,830	7,390	82,710
" 19..	95,207	1,385	2,062	17,755	4,289	120,698	46,000	6,220	6,580	58,800
" 26..	25,659	29,843	2,992	7,803	2,223	68,520	46,170	4,320	5,410	55,900
June 2..	21,325	25,357	1,158	2,651	2,580	53,071	42,470	4,690	4,400	51,560
" 9..	56,259	6,420	771	5,391	578	69,419	39,600	2,080	5,040	46,720
" 16..	12,930	12,062	3,069	3,753	2,817	34,631	35,950	3,580	3,140	42,670
" 23..	36,342	15,792	3,187	5,870	2,004	63,195	44,580	4,480	5,920	54,980
" 30..	12,412	14,831	1,322	......	2,111	30,676	51,050	2,010	6,160	59,220
July 7..	39,618	25,340	968	5,316	650	71,892	52,750	5,610	6,990	65,350
" 14..	36,744	31,307	654	3,270	267	72,242	47,330	4,290	6,840	58,460
" 21..	25,492	12,774	319	16,960	1,760	57,305	39,050	2,970	1,030	43,050
" 28..	18,158	8,735	1,358	2,267	6,960	37,478	50,190	2,370	4,960	57,520
Aug. 4..	17,886	5,337	141	6,773	565	30,702	49,520	7,200	8,150	64,870
" 11..	21,802	37,426	803	12,632	1,752	73,415	65,850	9,560	21,410	96,820
" 18..	4,095	8,568	154	8,133	685	21,635	52,340	4,600	14,390	71,330
" 25..	8,801	35,331	1,554	5,209	6,394	57,289	48,760	4,330	11,570	64,660
Sep. 1..	6,835	18,323	1,883	6,048	1,877	34,966	50,020	4,210	9,020	63,250
" 8..	23,595	42,223	55	4,229	392	70,494	55,190	4,200	12,020	71,410
" 15..	9,517	39,127	1,947	7,263	4,131	61,985	58,200	6,760	11,660	76,620
" 22..	18,163	23,157	1,934	8,787	920	52,961	43,680	3,050	11,290	58,020
" 29..	17,079	64,533	1,080	1,531	4,758	88,981	43,590	1,730	13,160	58,480
Oct. 6..	38,664	40,427	2,300	6,720	2,092	90,203	62,160	2,610	13,570	78,340
" 13..	7,516	65,434	2,804	6,122	7,123	88,999	48,370	5,140	15,360	68,870
" 20..	24,858	28,022	925	10,194	439	64,438	73,780	9,450	19,250	102,480
" 27..	14,319	17,393	4,230	11,185	2,691	49,818	66,310	8,820	23,640	98,770
Nov 3..	21,557	13,722	4,358	8,264	545	48,446	74,010	10,700	25,460	110,170
" 10..	14,771	16,935	7,949	1,881	3,532	45,068	47,960	5,180	14,150	67,290
" 17..	18,958	16,253	9,598	1,245	633	46,687	58,860	5,210	12,160	76,230
" 24..	22,723	17,869	672	6,203	2,181	49,648	67,360	12,410	18,050	97,820
Dec. 1..	48,783	15,193	13,068	6,595	3,015	86,654	53,440	4,650	14,290	72,380
" 8..	26,262	3,842	9,378	3,769	169	43,420	51,800	3,470	15,800	71,070
" 15..	48,284	7,114	10.538	13,322	402	79,660	50,490	3,120	11,200	64,810
" 22..	52,927	1,246	6,177	7.683	4,760	72,793	68,940	4.330	16,420	89,690
" 29..	75,224	7,788	9,707	1,419	1,251	95,389	42,610	2,060	9,060	53,730
Average prices & total sales, receipts & stocks.	1,678,722	898,089	202,034	374,804	111,449	3,265,098	2,645,990	348,650	525,410	3,520,050

POOL. YEAR 1870.

STOCKS.			PRICES.			ACTUAL EXPORT.	CON-SUMPTION.	REMARKS.
Amer'n	Other.	Total.	Mid. Up.	Mid. Orl.	Fair Dhol.			
90,930	261,000	351,930	11 1/2	11 3/4	9 1/2	9,390	48,110	
114,020	252,010	366,030	11 3/8	11 5/8	9 3/8	10,360	94,020	
135,150	206,120	341,270	11 5/8	11 7/8	9 5/8	13,990	167,120	Market active.
113,670	188,680	302,350	11 3/4	11 15/16	9 7/8	13,390	226,410	Abundant supply; quota-
155,190	193,260	348,450	11 1/2	11 11/16	9 3/4	5,770	261,520	tions reduced.
157,510	194,920	352,430	11 1/2	11 3/4	9 5/8	6,100	301,640	
146,630	169,550	316,180	11 3/8	11 11/16	9 5/8	6,660	348,400	
139,200	171,740	310,940	11 5/16	11 9/16	9 1/2	5,080	390,710	
127,700	164,060	291,760	11 1/8	11 7/16	9 3/8	5,140	428,940	
115,190	156,020	271,210	11	11 5/16	9 3/16	4,900	467,380	
111,070	183,900	294,970	11 1/8	11 7/16	9 1/4	4,590	516,310	
183,110	187,670	370,780	11 1/8	11 7/16	9 3/8	7,100	564,280	
256,160	195,710	451,870	10 7/8	11 3/16	9 1/4	7,270	612,790	Increased demand at hard-
273,820	181,730	455,550	11 1/4	11 1/2	9 1/2	14,610	676,110	ening prices.
297,290	179,710	477,000	11 1/8	11 7/16	9 1/2	8,300	720,720	Quieter tone
291,630	192,100	483,730	11 1/8	11 7/16	9 1/2	3,840	760,320	
357,990	204,800	562,790	11	11 5/16	9 1/2	8,730	810,580	
334,890	205,000	339,990	10 7/8	11 3/16	9 7/16	5,250	859,120	
323,850	203,050	526,900	11 1/8	11 7/16	9 1/2	7,390	917,610	Better feeling.
388,020	201.980	590,000	10 15/16	11 1/4	9 7/16	6,580	963,610	Heavy import; market in-
384,370	222,010	606,380	10 13/16	11 3/16	9 3/8	5,410	1,009,780	active.
377,140	232,030	609,170	10 11/16	11	9 1/4	4,400	1,052,250	
397,090	230,930	628,020	10 3/8	10 5/8	8 7/8	3,040	1,091,850	
366,400	236,860	603,260	10 5/8	10 15/16	9 1/4	3,140	1,127,800	
402,830	239,520	642,350	10 1/8	10 3/8	8 3/4	5,920	1,172,380	
310,130	244,500	554,630	9 7/8	10 1/8	8 1/2	6,160	1,223,430	July 21st, declaration of
314,640	259,130	573,770	9 3/4	10	8 1/8	6,990	1,276,180	war between France and
320,270	272,340	592,610	9 5/8	9 7/8	7 7/8	6,840	1,325,510	Germany; 23d, bank
318,420	288,340	606,760	8 3/8	8 5/8	6 7/8	1,030	1,362,560	rate raised to '4 per
310,070	278,960	589,030	7 3/4	8	6 1/4	2,960	1,412,750	cent., and on 28th to 5
286,890	278,460	565,350	7 7/8	8 1/8	6 1/8	8,150	1,462,270	per cent.
272,230	296,490	568,720	8 7/8	9 1/8	6 7/8	2,410	1,528,120	Hopes of an early peace;
246,200	282,730	528,930	8 7/8	9 1/8	7 1/4	14,390	1,580,460	considerable advance in
221.570	302,740	524,310	9	9 1/4	7 3/8	11,520	1,629.220	prices.
195.420	298,930	494,350	9	9 1/4	7 1/4	9,020	1,679,240	
186,400	307,500	493,900	9 1/4	9 1/2	7 3/8	12,020	1,734,430	Capitulation of Sedan.
168,450	321,090	489,540	9 1/2	9 3/4	7 3/8	11,660	1,792,630	
161,060	325,710	486,770	9 1/4	9 1/2	7 3/16	11,290	1,836,310	
150,600	370,230	520,830	8 3/4	9	6 5/8	13,160	1,879,900	Market irregular.
153,350	384,360	537,710	8 3/4	9	6 1/2	13,510	1,942,060	
130,240	434,460	564,700	8 1/2	8 3/4	6 1/4	15,360	1,990,430	
116,610	422,740	539,330	8 5/8	8 7/8	6 1/2	19,250	2,064,210	
75,660	433,810	509,470	8 13/16	9 1/16	6 5/8	2,364	2,130,520	
64,300	399,200	463,500	9 1/8	9 7/16	7	25,460	2,204,530	Hardening tendency.
51,770	387,050	438,820	9	9 3/16	6 11/16	14,150	2,252,490	
43.100	365,020	408,120	9 3/16	9 7/16	6 9/16	12,160	2,311,350	
38,250	335,670	373,920	9 1/4	9 7/16	6 3/4	18.050	2,378.710	
61,800	331.830	393,630	9	9 3/8	6 11/16	14,290	2,432,150	
61,720	309,030	370,750	8 9/16	8 7/8	6 9/16	15,800	2,483,950	Dull market.
81,560	303,380	384,940	8 1/4	8 1/2	6 5/16	11,200	2,534,440	
98,980	274,750	373,730	8 3/8	8 5/8	6 1/2	16,240	2,603,380	
109,710	269,010	378,720	8 1/4	8 1/2	6 1/2	9,060	2,645,990	
			9.89	10.2	8.1	525,410	50,884.42	

The semi-weekly Price and Weekly Sales and Receipts, at New York, Weekly Exports from New York and Rates of Freight to Liverpool 1st of each Month, for the Crop Year ending September 1, 1870.

1869.	Price of Middling New Orleans	Price of Middling Upland	SALES FOR WEEK.		Receipts for week.	EXPORTS FOR WEEK.					Rates of Freight to Liverpool
			Spots, &c.	Contracts		To Great Britain.	To France.	North of Europe.	Other Fo'n Ports	Total Exports.	
Septem. 3..	35½	35									Steam,
" 7..	35¼	34¾	3,753	4,050	1,288	1,382	766	....	....	2,148	¼d.
" 10..	34½	34									Sail,
" 14..	33	32	4,129	5,200	1,613	479	....	....	....	479	3-16d.
" 17..	31¼	30¼									
" 21..	29¼	28¼	11,985	9,731	12,722	400	1,537	100	....	2,037	
" 24..	29¼	28¾									
" 28..	29¼	28¾	10,735	3,700	12,942	5,296	559	703	....	6,558	Steam,
October 1..	28	27½									⅛d.
" 5..	28¼	27¾	13,393	6,150	14,058	5,093	461	1,580	....	7,134	Sail, ¼d.
" 8..	28	27⅜									
" 12..	27	26½	17,927	5,250	17,842	7,127	....	794	....	7,921	
" 15..	27	26⅛									
" 19..	27¼	26¾	18,012	7,200	29,213	7,876	393	3,164	....	11,433	
" 22..	26¾	26¼									
" 26..	26¾	26¼	13,944	6,700	22,184	10,971	....	2,129	....	13,100	
" 29..	26½	26⅛									Steam,
Novem. 2..	26⅝	26¼	14,913	8,075	12,238	8,832	744	1,441	....	11,017	½@9-16d
" 5..	26¼	25⅞									Sail, ¼d.
" 9..	25⅞	25⅛	14,057	7,800	17,987	13,756	....	1,745	....	15,501	
" 12..	25½	25⅛									
" 16..	25¾	25⅜	15,667	9,975	23,703	6,480	664	781	....	7,925	
" 19..	25¾	25⅜									
" 23..	25½	25⅛	11,318	7,700	19,606	9,903	....	3,186	....	13,089	
" 26..	25½	25⅛									
" 30..	25⅝	25⅛	19,783	11,425	19,452	7,151	440	756	1,654	10,001	Steam.
Decem. 3..	25⅝	25⅛									¼@5-16d
" 7..	25⅝	25⅛	15,426	6,000	17,191	11,049	....	2,572	104	13,725	Sail,
" 10..	25½	25									3-16d.
" 14..	26	25½	21,219	4,000	24,931	9,532	332	1,240	....	11,104	
" 17..	26	25½									
" 21..	25¾	25¼	13,532	5,500	25,303	11,219	160	3,296	....	14,675	
" 24..	25⅝	25⅛									

GENERAL REMARKS.

The market throughout this crop year, was a declining one, relieved occasionally by a short rally, but only to fall back to lower figures. The elements of depression being too strong to prevent the downward tendency.

These elements consisting of good present, and prospective crops, taken as a whole; the unremunerative returns for Manchester goods and the European war existing toward its close between France and Germany.

September opened very quiet, the stock being light, and buyers waiting for the new crop, which, toward the latter part of the month, began to come forward in quantity, and prices dropped.

There was but little cheerfulness in October, Liverpool accounts were unfavorable; great distress was reported in the manufacturing districts of England; and, though the stock of Americans was small, prices there remained drooping. The great flurry that occurred about this time in gold, unsettled exchange and deterred shipping, so that prices steadily receded.

In November, there was an improved demand, for a portion of the month,

Decem. 28..	25⅝	25⅛	9,037	7,350	17,421	7,461	150	394	....	8,005		under which prices rallied a little,
" 30..	25⅝	25⅛										but, subsequently with accumula-
1870.												ting supplies at the ports, and un-
January 4..	25⅞	25⅜	10,724	8,350	30,556	4,426	....	409	....	4,835	Steam,	favorable foreign advices, the mar-
" 7..	25¾	25¼									¾d.	ket again r ceded.
" 11..	25⅜	25⅛	12,128	8,400	24,902	5,902	512	4,743	....	11,157	Sail. ¼d.	Prices in December, were quite irreg-
" 14..	25¾	25¼										ular, there were occasional spurts
" 18..	25⅞	25⅜	17,940	12,475	16,607	4,132	....	99	....	4,231		of activity, with some speculative
" 21..	25⅞	25⅜										feeling, and values were better, but
" 25..	25⅞	25⅜	20,119	10,900	16,636	4,680	1,295	1,805	....	7,780		later the market fell back to about
" 28..	26½	25⅞										the same figures as those current
February 1..	26	25½	12,785	13,475	21,590	8,856	....	3,028	....	11,884	Steam,	at the opening of the month.
" 4..	25⅞	25⅜									—@—	The receipts now in January, were
" 7..	25⅞	25⅜	12,047	12,600	12,760	4,548	601	1,147	51	6,347	Sail,	about 10 per cent larger than those
" 11..	25¾	25¼									5/32@3-16	of the previous year, at the same
" 15..	25½	25⅛	11,307	14,375	22,960	4,425	....	489	....	4,914		time, and prices fell off a little, the
" 18..	24⅞	24⅜										crop estimates widening to two and
" 22..	24⅝	24⅛	11,969	17,450	10,295	4,748	692	1,365	....	6,805		three-quarter millions bales, after-
" 25..	24	23½										wards there was a better export de-
March 1..	23¾	23¼	25,659	25,850	16,604	6,980	....	2,373	....	9,353	Steam,	mand, which carried figures up
" 4..	23⅛	22⅝									—@—	higher than before in three months.
" 8..	21¾	21¼	26,182	27,650	11,446	11,344	2,073	3,243	....	16,660	Sail,	The improvement was since lost, in
" 11..	22	21½									5/32@3-16	February, the market being unable
" 15..	22¼	21¾	22,135	26,750	9,978	10,237	....	1,924	....	12,161		to stand up against heavy receipts
" 18..	24	23½										and depressing foreign accounts.
" 22..	23⅜	22⅞	11,807	24,375	15,108	11,304	569	2,320	....	14,193		The market was unsettled through
" 25..	23¼	22¾										March; at one time, there was much
" 29..	23¼	22¾	13,159	9,900	9,371	7,957	1,366	4,037	....	13,360	Steam,	pressure to sell, and prices rapidly
April 1..	23¼	22¾									¼d	declined; afterwards, with some
" 5..	23¾	23¼	9,121	11,780	13,477	6,667	1,561	567	....	8,795	Sail, ⅛d.	falling off in the receipts, there was
" 8..	24	23½										a reaction, which, however, proved
" 12..	23⅞	23⅜	11,516	10,800	12,566	6,515	....	1,903	....	8,418		but temporary.
" 15..	23¾	23½										The advices from abroad were, in
" 19..	23⅝	23⅛	6,004	4,200	13,165	7,759	779	1,513	....	10,051		April, of a more favorable charac-
" 22..	24	23½										ter, and there was considerable ac-
" 26..	24	23½	16,478	9,550	12,479	5,967	....	432	....	6,399		tivity at rather better prices.
" 29..	23⅞	23⅛									Steam,	The market in May was unsettled; at
May 3..	23½	23	8,896	8,700	16,529	5,750	1,714	143	....	7,607	¼d	one time, with diminished receipts,
" 6..	23¼	22⅞									Sail,	there was a better feeling, but ac-
" 10..	23¾	23¼	14,392	15,350	14,258	6,061	....	368	....	6,429	⅛@3-16d	counts from abroad were less en-
" 13..	24	23½										couraging, and at the close prices
" 17..	23½	23	13,688	27,300	10,845	6,203	138	829	....	7,170		gave way.

New York Statement for 1870.—*Concluded.*

1870.		Price of Middling New Orleans	Price of Middling Upland	Sales for Week. Spots, &c.	Sales for Week. Contracts	Receipts for week.	Exports for the Week. To Great Britain.	To France.	North of Europe.	Other Fo'n Ports	Total Exports.	Rates of Freight to Liverpool.
May	20..	23⅝	23⅛									
"	24..	23½	23	7,116	10,850	15,437	6,473	....	499	....	6,972	
"	27..	23⅝	23⅛									
"	31..	23	22½	11,349	11,000	9,746	4,797	.....	548	....	5,345	
June	3..	23	22½									Steam,
"	7..	22⅝	22⅛	8,640	10,400	7,390	6,792	59	249	....	7,100	¼d.
"	10..	22½	22									Sail,
"	14..	22⅜	21⅞	7,176	10,550	19,480	4,623	58	441	....	5,122	5-32d
"	17..	22	21½									
"	21..	21⅞	21⅜	6,572	15,800	7,865	5,541	....	1,010	....	6,551	
"	24..	21½	21									
"	28..	21½	21	4,780	18,750	10,467	5,757	21	280	....	6,058	
July	1.	21	20½									Steam,
"	5..	20¾	20¼	2,869	12,850	8,407	6,750	99	50	....	6,899	¼d.
"	8..	20¾	20¼									Sail,
"	12..	20½	20	7,061	9,700	7,785	6,510	14	631	....	7,155	⅛@5-32d
"	15..	20⅝	20⅛									
"	19..	20½	20	8,132	11,150	7,389	6,649	....	190	....	6,839	
"	22..	20½	20									
"	26..	21	20½	6,813	12,800	8,227	4,383	....	....	....	4,383	
"	29..	21	20									Steam,
August	2..	21	20	5,401	11,950	11,317	2,647	....	....	600	3,247	¼d.
"	5..	20⅛	19⅝									Sail,
"	9..	20	19½	3,599	9,900	5,933	214	....	....	....	214	—@—
"	12..	20	19½									
"	16..	19¾	19¼	5,092	5,850	1,747	1,300	....	....	....	1,300	
"	19..	20	19½									
"	23..	20⅜	19⅞	9,081	12,500	9,496	1,000	....	....	37	1,037	
"	26..	20¼	19¾									Steam,
"	30..	20¼	19¾	5,843	11,500	6,718	1,599	....	....	....	1,599	¼d
Septem.	2..	20⅜	19⅞									Sail,
Average price and total sales, receipts and exports.		24.41	23.98	616,410	591,586	739,230	323,503	17,757	60,516	2,446	404,222	—@—

General Remarks.

June witnessed a steadily declining market, the crop estimates had enlarged to over three millions of bales, and with good prospects for the growing plant, and dull foreign markets, there was scarcely an attempt made to resist the downward tendency

In the early part of July, there was much depression, consequent upon fears of war between France and Prussia, but later, heavy rains, &c., in some sections of the South, caused some speculative and spinning demand, and prices recovered.

The market in August became very dull, shippers were without margins, prices here being higher than abroad, so that the business for the most part was for spinning and on speculation.

Exchange.

The range of bills on London in September, was from 8@9½ per cent. premium, gold, down to 6@7¼; in October, 7@9; in November, 8@8⅞; in December, 7¾@8½; in January, 8¼@9; in February, 8@8⅞; in March, 7½@8½; in April, 7¾@8¾; in May, 8½@9½; in June, 9@9⅝; in July, 9¼@9⅞; and in August, 8½@9¼ per cent. premium, gold.

ANNUAL REVIEW.

From the New Orleans Price Current—1869–70.

On July 11th, the telegrams from London and Paris assumed an unexpected and warlike complexion. In our own cotton circles they were variously construed, some regarding them as a merely transitory excitement, and scouting the idea of war; but others, viewing them in a much more serious light, and as well calculated to make all prudent men adopt a more cautious policy; on the assumption of the probability, at least, of the pending complications resulting in actual hostilities. While buyers were reluctant to come forward, or had their orders cancelled, factors evinced more disposition to realize, and prices gave way ½c. in low middling, with a greater reduction in the lower grades, while stained and dusky lots were entirely nominal. On the ensuing day the dispatches were of a more serious character, bringing further cancellation of orders, reporting prices at Liverpool unsettled and declining, and discouraging shipments. On the 13th there was some revival of confidence, which continued on the 14th, but, although prices recovered ¼c. at New York, the only response in our market was more steadiness in the better qualities, while the lower grades, especially when stained and dusky, were more than ever demoralized. On the 15th, all hopes of a pacific solution of the European complications were dispelled, more orders were cancelled, factors evinced greater anxiety to realize, and prices gave way ¼c. in the better qualities and ½c. in the lower grades, at which, moreover, the business was of very limited extent. After this period, notwithstanding the fluctuations and general advance in gold and foreign exchange, prices have continued to decline, giving way ½c. during the third week, when low middling ruled at 16 to 16½, against 17¾ to 18¼c., at the commencement of the month, showing a total decline of 1¾c.

1871.

COTTON CROP OF THE UNITED STATES.

Statement and Total Amount of the Cotton Crop of the United States, for the Year ending August 31, 1871.

	Bales.	Bales.	TOTAL. 1871.	1870.	1869.
LOUISIANA.					
Export from New Orleans—					
To Foreign Ports	1,302,535				
Coastwise	238,824				
Burnt and manufactured	2,150				
Stock on hand 1st September, 1871	25,323				
		1,568,832			
Deduct—					
Received from Mobile	76,581				
" Florida	2,694				
" Texas	22,371				
Stock on hand 1st September, 1870	20,696				
		122,342			
			1,446,490	1,142,097	794,205
ALABAMA.					
Export from Mobile—					
To Foreign Ports	287,074				
Coastwise	130,429				
Burnt and lost	502				
Stock on hand 1st September, 1871	5,466				
		423,471			
Deduct—					
Received from New Orleans	9,055				
Stock on hand 1st September, 1870	9,743				
		18,798			
			404,673	305,956	230,726
TEXAS.					
Export from Galveston, &c.—					
To Foreign Ports (including 39 to Mexico)	213,922				
Coastwise	94,867				
Stock in Galveston, 1st September, 1871	10,490				
		319,279			
Deduct—					
Stock in Galveston, 1st September, 1870	....	4,795			
			314,484	246,284	147,817
FLORIDA.					
Export from Apalachicola, Pensacola, &c.—					
To Foreign Ports—Uplands	98				
Coastwise—Uplands	7,931				
Sea Island	8,666				
Stock in Apalachicola, 1st September, 1871	....				
		16,695			
Deduct—					
Stock on hand, 1st September, 1870	....	7			
			16,688	23,194	13,392
GEORGIA.					
Export from Savannah—					
To Foreign Ports—Uplands	460,676				
" Sea Island	2,835				
Coastwise—Uplands	260,529				
Sea Island	6,839				
Stock in Savannah, 1st September, 1871	3,215				
		734,094			
Deduct—					
Received from Florida—Uplands	1,300				
" " " Sea Island	3,968				

Statement and Total Amount of the Cotton Crop of the United States, for the Year ending August 31, 1871—*Concluded.*

	Bales.	Bales.	TOTAL.		
			1871.	1870.	1869.
Received from Beaufort, &c., Sea Island....	465				
Stock in Savannah, 1st September, 1870	2,833				
		8,566			
			725,528	485,374	357,253
SOUTH CAROLINA.					
Export from Charleston, S. C.—					
To foreign ports—Uplands................	170,543				
" Sea Island	5,107				
Coastwise—Uplands	172,359				
Sea Island	6,562				
Stock in Charleston, 1st September, 1871....	3,443				
	358,014				
Export from Georgetown, S. C.—					
To New York and Boston.................	397				
From Beaufort, S. C.—Sea Island..........	465				
		358,876			
Deduct—					
Received from Florida—Uplands...........	1,593				
Sea Island.........	4,698				
Received from Savannah—Sea Island.......	494				
Stock in Charleston, 1st September, 1870....	1,399				
		8,184			
			350,692	246,593	198,943
NORTH CAROLINA.					
Export—					
To foreign ports.........................	70				
Coastwise	77,153				
		....	77,223	58,884	35,912
VIRGINIA.					
Export—					
To Foreign Ports	5,417				
Coastwise................................	333,906				
Stock in Norfolk and Petersburg, 1st September, 1871........................	837				
		340,160			
Deduct—					
Stock on hand 1st September, 1870.........	....	985			
			339,175	203,981	160,418
TENNESSEE, KENTUCKY, &c.					
Shipments from Memphis, Tenn............	513,536				
" " Nashville, Tenn...........	114,829				
" " Louisville...............	320,964				
" " other places in Tennessee, Kentucky, &c..........	93,604				
Stock in Memphis, Nashville, &c., 1st September, 1871..............................	4,628				
		1,047,561			
Deduct—					
Shipments to Norfolk and New Orleans.....	193,137				
Received at Louisville from Memphis and Nashville................................	267,130				
Stock in Memphis and Nashville, 1st September, 1870.............................	6,481				
		466,748			
			580,813	322,386	321,891
Manufactured at the South...............	....	....	91,240	79,843	....
Total crop of the United States, 1870–1.	....	....	4,347,006	3,114,592	2,260,557

Increase over crop of 1869-70 ....bales.	1,232,414
Increase over crop of 1868-69	2,086,449
Increase over crop of 1867-68	1,916,113
Increase over crop of 1866-67	2,395,018

Export to Foreign Ports, from September 1, 1870, *to August* 31, 1871.

FROM	To Great Britain.	To France.	To North of Europe.	Other F'n Ports.	Total.
New Orleans....bales.	823,032	19,171	242,981	117,351	1,302,535
Mobile	240,660	....	32,558	13,856	287,074
Galveston	179,916	5,637	28,330	39	213,922
Pensacola....	98	....	....	....	98
Savannah....	350,546	7,532	105,433	....	463,511
Charleston....	135,144	....	22,895	17,611	175,650
Norfolk....	5,417	....	....	....	5,417
Wilmington....	70	....	....	....	70
New York	585,390	6,529	48,445	3,778	644,142
Baltimore	21,977	....	15,690	....	37,667
Philadelphia	806	....	....	....	806
Boston and Portland (Portland, 275 to Great Britain)....	2,942	....	45	293	3,280
Grand total 1870-1....	2,345,998	138,869	496,377	152,928	3,134,172
Total 1869-70....	1,475,444	346,706	255,315	96,094	2,173,559
Increase over 1869-70....	870,554	....	241,062	56,834	960,613
Decrease from 1868-9....	....	207,837	....	....	....

Consumption.

Total crop of the United States, as before stated ....bales.			4,347,006
Add—Stocks on hand at the commencement of the year, September 1, 1870—			
In the Southern ports....		46,939	
In the Northern ports....		18,386	
			65,325
Makes a supply for the year ending August 31, 1871, of....			4,412,331
Deduct therefrom—The Export to Foreign ports....	3,134,172		
Less Foreign included....	3,594		
		3,130,578	
Stocks on hand, 1st September, 1871—			
In the Southern ports	53,402		
In the Northern ports	55,265		
		108,667	
Burnt at New York, New Orleans and Mobile....	3,302		
Manufactured at the South....	91,240		
Shipped to Canada....	6,118		
		100,660	
			3,339,905
Taken for home use north of the Potomac and Ohio Rivers....bales.			1,072,426
Taken for home use south of the Potomac and Ohio Rivers, and burnt....			94,542
Total consumed in the U. S. (including burnt at the ports), 1870-71....			1,166,968

The semi-weekly Price and Weekly Sales and Receipts at New York, Weekly Exports from New York and Rates of Freight to Liverpool 1st of each month, for the Crop Year ending September 1, 1871.

1870.	Price of M'dling New Orleans	Price of M'dling Upland	SALES FOR WEEK.		Receipts for week.	EXPORTS FOR THE WEEK.					Rates of Freight to Liverpool.
			Spot, &c.	Contracts		To Great Britain.	To France.	North of Europe.	Other Fo'n Ports	Total Exports.	
Septem. 6..	$20\frac{3}{4}$	20	8,790	13,800	5,167	....	....	....	....	....	Steam,
" 9..	$20\frac{5}{8}$	$19\frac{7}{8}$									$\frac{1}{4}$d.
" 13..	$20\frac{1}{2}$	$19\frac{1}{4}$	7,013	12,350	6,593	3,305	....	....	....	3,305	Sail,
" 16..	$20\frac{1}{4}$	$19\frac{1}{2}$									—@—
" 20.	$19\frac{1}{4}$	$18\frac{1}{2}$	10,344	17,350	12,759	4,708	....	....	....	4,708	
" 23..	$18\frac{3}{4}$	18									
" 27..	$18\frac{1}{4}$	$17\frac{1}{2}$	10,743	32,500	12,574	4,767	....	....	....	4,767	
" 30..	—	—									Steam,
October 4..	17	$16\frac{1}{4}$	14,252	36,550	23,805	10,630	....	....	....	10,630	$\frac{3}{8}$@7-16d
" 7..	$17\frac{1}{8}$	$16\frac{3}{8}$									Sail,
" 11..	$16\frac{3}{4}$	16	14,994	15,050	15,833	11,759	....	97	....	11,856	—@—
" 14..	$16\frac{1}{8}$	$15\frac{3}{8}$									
" 18..	$16\frac{7}{8}$	$16\frac{1}{8}$	21,550	44,450	29,345	11,097	....	363	....	11,460	
" 21..	17	$16\frac{1}{4}$									
" 25..	18	$17\frac{1}{4}$	20,405	75,550	28,773	13,898	....	753	....	14,651	
" 28..	$17\frac{1}{2}$	$16\frac{3}{4}$									Steam,
Novem. 1..	$17\frac{3}{4}$	17	21,527	42,750	29,332	14,776	....	2,171	....	16,947	$\frac{1}{2}$@9-16d
" 4..	$17\frac{7}{8}$	$17\frac{1}{8}$									Sail,
" 8..	$17\frac{1}{8}$	$16\frac{3}{8}$	23,182	48,200	27,976	17,557	....	....	1,015	18,572	$\frac{1}{4}$@5-16d
" 11..	$16\frac{7}{8}$	$16\frac{1}{8}$									
" 15..	$17\frac{1}{4}$	$16\frac{1}{2}$	23,830	40,100	34,912	13,516	....	103	....	13,619	
" 18..	$16\frac{1}{4}$	16									
" 22..	$17\frac{1}{4}$	$16\frac{1}{2}$	38,230	50,850	26,356	15,577	....	789	....	16,366	
" 25..	$17\frac{1}{8}$	$16\frac{3}{8}$									
" 29..	17	$16\frac{1}{4}$	24,888	28,350	39,411	16,250	....	1,030	....	17,280	Steam,
Decem. 2..	$16\frac{5}{8}$	$15\frac{7}{8}$									9-16d.
" 6..	$16\frac{1}{2}$	$15\frac{1}{2}$	23,717	43,706	33,926	19,848	....	1,265	2	21,115	Sail,
" 9..	$16\frac{1}{4}$	$15\frac{1}{2}$									$\frac{1}{4}$@5-16d
" 13..	16	$15\frac{1}{4}$	21,886	45,450	32,324	22,717	....	1,173	....	23.890	
" 16..	$15\frac{3}{4}$	15									
" 20..	$16\frac{1}{4}$	$15\frac{1}{2}$	28,821	62,025	38,427	23,655	....	4,017	1,200	28,872	
" 23..	$16\frac{1}{4}$	$15\frac{1}{2}$									
" 27..	$16\frac{1}{8}$	$15\frac{3}{8}$	20,278	41,550	30,408	21,125	....	....	....	21,125	

GENERAL REMARKS.

The events of this year are too fresh to need dwelling upon at length, the chief interest centering in the growing crop prospects. The past year having been an exceptionally favorable one, the yield large, and the returns remunerative; it was given out, early this year, that the planting would be for a very much smaller crop; added to this, an unfavorable, wet season, and the present outlook favors a short crop —one unquestionably below that of last year; but the extent of the decrease cannot now be closely estimated, as much depends upon the appearance of early or late frosts in the growing regions, and the weather during the picking season. The full crop prospects, and weak foreign markets in September, carried prices down.

The early part of October was quiet, but later peace rumors strengthened holders, and there was considerable activity. In November, there was much fluctuation in prices; the unfavorable attitude of Russia to Turkey caused apprehensions of further European conflicts, and there was considerable hesitation.

New York Statement for Year 1871—Concluded.

1871.	Price of M'dling New Orleans	Price of M'dling Upland	SALES FOR WEEK.		Receipts for week.	EXPORTS FOR WEEK.					Rates of Freight to Liverpool.	GENERAL REMARKS.
			Spot, &c.	Contracts		To Great Britain	To France.	North of Europe.	Other Fo'n Ports	Total Exports.		
Decem. 30..	16	15¼										December was, for the most part, dull and drooping.
1871.												
January 3..	16	15¼	12,558	42,500	20,545	16,097	....	896	....	16,993	Steam, ½@9 16d	In January, there was, at one time, a little better tone, in consequence of the probability of a cessation of the war between France and Germany, and prices on the whole were a little better, notwithstanding the receipts continued large.
" 6..	15¾	15									Sail,	
" 10..	16⅛	15⅜	26,156	69,350	34,087	16,235	....	1,067	....	17,302	5-16@⅜d	
" 13..	16	15¼										
" 17..	16	15¼	26,533	43,750	30,645	14,944	....	1,076	....	16,020		
" 20..	16⅛	15⅜										
" 24..	16¼	15½	12,913	36,600	26,139	13,942	....	549	....	14,491		
" 27..	16⅜	15⅝										The market, throughout the most of February, was dull, prices of good qualities being higher here than in Liverpool, while the crop estimates widened to four million bales; prices, however, were pretty well supported, in the expectation that peace in Europe, now assured, would quicken the demand.
" 31 .	16⅜	15⅝	22,767	56,650	27,606	13,501	....	759	....	14,260	Steam,	
February 3..	16¼	15½									9/32@11/32d.	
" 7..	16⅛	15⅜	18,390	61,300	31,142	15,034	....	150	....	15,184	Sail,	
" 10..	16⅛	15⅜									⅜@7-16d	
" 14..	16	15¼	20,217	56,600	28,695	12,323	....	2,446	40	14,809		
" 17..	16	15¼										
" 21..	16⅛	15⅜	27,977	73,250	32,482	11,695	....	1,302	60	13,057		
" 24..	16⅛	15⅜										
" 28..	16	15¼	19,727	49,700	25,070	17,406	....	1,049	248	18,703	Steam,	In March, the foreign advices were construed more favorably, and with now and then some little decrease in the receipts at the ports, and, so far, very unfavorable weather for planting, at the close the market was in sellers' favor.
March 3..	15½	14¾									¼@5 16d	
" 7..	15½	14¾	23,224	141,200	28,533	23,163	....	2,999	561	26,723	Sail,	
" 10..	15⅜	14¾									9/32@11/32d.	
" 14..	15½	14⅞	20,404	105,600	29,930	10,198	....	3,538	141	13,877		
" 17..	15¾	15⅛										
" 21..	15⅝	15	24,407	99,700	28,629	19,615	666	4,487	....	24,768		
" 24..	15¾	15⅛										The equilibrium in prices between here and abroad was not yet restored, and in April the market was generally drooping.
" 28..	15⅞	15¼	21,166	76,000	25,564	18,731	2,058	3,121	....	23,910		
" 31..	15¾	15⅛									Steam,	
April 4..	15¾	15⅛	21,192	46,300	18,171	14,509	....	4,210	....	18,719	5-16@⅜d	
" 7..	15⅝	15									Sail.	In May, the unfavorable crop reports began to outweigh all adverse influences, and prices steadily advanced.
" 11..	15⅝	15	14,096	13,900	19,897	17,355	480	928	....	18,763	¼@5-16d	
" 14..	15½	14⅞										
" 18..	15⅜	14¾	21,576	33,850	18,717	19,027	....	2,872	411	22,310		
" 21..	15⅜	14¾										

April	25..	15⅜	15⅜	18,450	41,140	23,501	17,736	439	915	....	19.090	
"	28..	15⅛	15⅛									Steam, ¼d.
May	2..	15½	14⅞	16.782	56.904	16,283	10,508	1,125	61	....	11,694	Sail, ¼d.
"	5..	15⅞	15¼									
"	9..	15⅞	15¼	22,251	68,117	10.178	16,745	205	2,851	....	19.801	
"	12..	16¼	15¾									
"	16	16⅜	15⅞	21,092	106,412	10,588	10,144	450	529	....	11,123	
"	19..	16⅜	15⅞									
"	23..	16⅝	16⅛	16.567	48,400	13,932	7,909	702	8	....	8,619	
"	26..	17¼	16⅝									
"	30..	18¼	17⅝	31.004	100.550	13,331	5,531	204	99	....	5.834	Steam, ¼d.
June	2..	18½	17⅞									
"	7..	22⅞	22⅛	12,553	42,000	8,583	9,485	59	325	....	9,869	Sail, $\frac{5}{32}$@$\frac{3}{16}$d.
"	9..	20	19⅜									
"	13..	20¾	20⅛	18,280	64,500	15,732	1,836	....	....	....	1,836	
"	16..	21¼	20⅝									
"	20..	21½	20⅞	17,426	83,200	9,147	2,286	....	143	....	2,429	
"	23..	20⅞	20¼									
"	27..	20½	19⅞	4,967	76,950	12,833	607	....	....	100	707	
"	30..	21⅜	20¾									Steam, ¼d.
July	4..	—	—	12,846	41,250	5,516	70	....	....	....	70	
"	7..	21¾	21⅛									Sail, 5-32d.
"	11..	21⅝	21	19,880	78,350	11,277	544	....	....	....	544	
"	14..	21⅝	21									
"	18..	21½	20⅞	8.456	74,800	9,015	937	....	....	....	937	
"	21..	21½	20⅞									
"	25..	21⅜	20¾	16,242	53,900	7,275	1,116	....	....	....	1,116	
"	28..	20⅞	20¼									Steam, ¼d
August	1..	20⅛	19¾	9,097	81,800	6,159	1.725	....	....	....	1,725	
"	4..	20⅛	19¾									Sail,
"	8..	19¾	19⅜	8,638	53,300	7,980	2,100	....	....	....	2,100	$\frac{5}{32}$@$\frac{3}{16}$d.
"	11..	19⅜	19									
"	15..	18⅞	18½	13,836	66,750	4,711	2,162	200	....	....	2,362	
"	18..	19⅛	18¾									
"	22..	19¼	18⅞	13,098	69,300	5,235	9,556	....	....	....	9,556	
"	25..	19⅜	19									
"	29..	19¾	19⅜	5,769	98,350	4,649	6,496	....	....	....	6,496	Steam, ¼d.
Septem.	1..	20	19⅝									
Average price and total sales, receipts and exports.		17.60	16.95	954,987	2,962,804	1,049,698	586,453	6,588	48,141	3,778	644,960	Sail, 3-16d.

Throughout the most of June, the business was chiefly for home use and on speculation, prices being too high to admit of shipments; and though prices continued to be affected more or less by unfavorable crop reports, the general tendency was downward

In July, the crop accounts were "much mixed" and conflicting, and with only a moderate demand, prices yielded, the decline for the month being about ⅞ cts. per lb.

August was dull, with prices in buyers' favor, until later in the month, when, with small offerings and further unfavorable crop reports, values reacted a little, the prices at the close being only ⅛ cent lower than at the commencement of the month.

Exchange.

The range for 60 days' commercial bills on London, in September, was from 8½ to 9½ per cent. premium, gold; in October, 7½@9; in November, 8¼@9; in December, 8@9; in January, 8¼@9; in February, 8¾@9⅜; in March, 8¾@9⅝; in April, 9¼@9⅞; in May, 9½@10¼; in June, 9¼@10⅜; in July, 9⅜@10⅜, and August, 8¼@9¾.

COTTON AT LIVER

Week Ending.	Receipts.						Sales.			
	American	E. I.	Egypt.	Brazil.	Other.	Total.	Consumption.	Speculation.	Export.	Total.
Jan. 5..	129,716	35,694	9,259	5,080	2,834	182,583	41,630	1,300	6,720	49,650
" 12.	63,191	7,589	15,078	7,456	1,793	95,107	64,690	4,010	10,640	79,340
" 19..	73,025	2,019	1,807	12,354	2,268	91,473	69,090	6,320	13,110	88,520
" 26..	34,751	12,527	16,487	9,114	424	73,303	68,080	7,920	18,950	94,950
Feb. 2..	42,828	6,559	9,290	8,634	2,063	69,374	44,880	4,790	16,460	66,130
" 9..	61,137	11,585	6,037	11,905	1,586	92,250	44,210	1,900	13,050	59,160
" 16..	161,212	14,619	8,249	4,189	7,315	195,584	47,930	3,130	10,530	61,590
" 23..	52,750	4,428	14,527	20,703	1,126	93,534	62,360	4,830	17,390	84,580
Mch. 2..	18,515	....	1,862	1,110	986	22,473	52,370	2,620	14,630	69,620
" 9..	100,861	10,924	4,973	14,451	1,028	132,237	49,030	4,890	18,410	72,330
" 16..	106,040	11,142	9,066	1,829	191	128,268	57,580	6,700	14,510	78,790
" 23..	54,270	10,115	2,658	8,685	2,491	78,219	64,310	9,010	11,300	84,620
" 30..	17,223	18,742	797	9,869	3,437	50,068	57,520	9,020	15,120	81,660
April 5..	46,080	2	7,822	10,112	2,795	66,811	40,670	2,100	8,600	51,370
" 13..	101,798	14,799	4,820	4,294	39	125,750	38,190	2,040	7,160	47,390
" 20..	135,554	14,835	1,446	23,463	1,943	177,241	54,930	3,870	12,340	71,140
" 27..	43,708	31,525	5,479	14,300	4,780	99,792	57,830	8,180	15,690	81,700
May 4..	91,045	7,237	4,910	4,120	1,709	109,021	54,310	10,800	17,520	82,630
" 11..	32,501	8,849	....	5,609	1,346	48,305	54,680	10,190	13,430	78,300
" 18..	24,973	16,406	8,745	8,894	756	59,774	64,970	22,190	19,420	106,580
" 25..	45,233	36	3,938	4,474	114	53,795	62,290	17,160	19,710	99,160
June 1..	93,500	13,619	2,531	8,199	548	118,397	45,700	17,660	19,300	82,660
" 8..	17,170	7,163	791	16,700	337	42,161	77,490	46,960	19,700	144,150
" 15..	93,546	1,929	6,760	20,657	1,900	124,792	71,110	36,590	25,930	133,630
" 22..	38,457	13,515	2,361	8,595	1,089	64,017	79,070	38,570	19,820	137,460
" 29..	15,970	107	2,362	7,868	2,051	28,358	71,380	15,010	11,620	98,010
July 6..	31,354	26,760	6,010	10,250	2,169	76,543	74,810	33,620	14,110	122,540
" 13..	10,888	3,777	1,780	6,258	2,709	25,412	80,960	38,250	7,770	126,980
" 20..	25,249	13,510	3,371	9,683	1,212	53,025	43,690	15,890	9,980	69,560
" 27..	28,926	49,686	912	13,122	3,241	95,887	54,330	15,500	7,820	77,650
Aug. 3.	16,336	11,805	758	2,383	1,250	32,532	51,580	7,330	5,150	64,060
" 10..	4,861	7,024	1,413	5,009	894	19,201	42,660	4,880	5,710	53,250
" 17..	1,762	1,214	558	812	2,941	7,287	49,660	9,170	5,830	64,660
" 24..	14,837	46,325	962	14,808	245	77,177	64,190	27,320	10,500	102,010
" 31..	6,600	22,737	237	8,332	9,347	47,253	82,240	31,520	10,700	124,460
Sept. 7..	26,158	6,378	1,181	14,259	2,459	50,435	82,470	25,810	10,660	118,940
" 14..	14,233	10,194	176	3,328	5,667	33,598	52,900	14,480	9,430	76,810
" 21..	8,664	6,691	1,197	5,265	265	22,082	45,500	10,940	8,820	65,260
" 28..	8,496	2,682	415	4,028	422	16,043	52,310	7,480	6,050	65,840
Oct. 5..	25,347	140,921	1,034	22,943	6,382	196,627	118,260	49,490	15,450	183,200
" 12..	16,114	12,000	1,546	5,231	6,043	40,934	61,370	24,720	10,740	96,830
" 19..	10,998	15,297	2,329	9,179	580	68,383	56,540	17,740	10,270	84,550
" 26..	21,485	46,933	4,224	5,883	2,484	81,009	36,020	6,050	10,620	52,690
Nov. 2..	12,764	21,911	3,802	18,615	6,271	63,363	52,570	14,970	11,850	79,390
" 9..	18,211	....	3,109	11,126	2,058	34,504	45,000	5,420	9,770	60,190
" 16..	15,828	59,982	18,918	19,342	6,474	120,544	71,930	12,900	13,820	98,650
" 23..	23,946	36,107	9,586	6,027	177	75,843	79,270	19,810	12,580	111,660
" 30..	18,436	9,612	2,772	13,685	1,688	46,193	78,060	22,980	22,380	123,420
Dec. 7..	20,844	3,994	11,419	11,285	744	48,286	75,960	19,380	20,330	115,670
" 14..	63,542	23,616	19,681	7,691	3,334	117,864	89,490	23,360	19,480	132,330
" 21..	36,323	3,393	2,915	5,676	433	48,740	58,300	7,700	7,860	73,860
" 28..	56,738	6,844	12,520	13,582	3,660	93,344	38,900	5,540	8,640	53,080
Average prices & total sales, receipts & stocks.	2,222,024	885,978	264,480	509,467	120,898	4,003,027	3,135,270	763,700	676,580	4,575,550

POOL. YEAR 1871.

STOCKS.			PRICES.			ACTUAL EXPORT.	CONSUMPTION.	REMARKS.
American.	Other.	Total.	Mid. Up.	Mid. Orl.	Fair Dhol.			
212,440	307,240	519,680	7 15/16	8 3/16	6 1/2	6,720	41,630	January 5th, heavy import, market dull; 26th, aspect of foreign politics more favorable.
234,460	315,630	550,090	7 13/16	8 1/8	6 3/8	10,640	106,320	
264,160	308,330	572,490	7 7/8	8 1/8	6 3/8	13,110	175,410	
247,120	318,760	565,880	8 1/16	8 5/16	6 1/2	18,950	243,490	
255,320	323,480	578,800	7 13/16	8 1/16	6 3/8	16,460	288,370	February 2d, market heavy, prices declining; 23d, better tone, but checked by large receipts at American ports.
276,730	333,660	610,390	7 5/8	7 7/8	6 3/8	13,050	332,580	
400,350	350,470	750,820	7 1/2	7 13/16	6 1/4	10,530	380,510	
400,660	363,890	764,550	7 11/16	7 15/16	6 1/4	17,390	442,870	
369,740	340,560	710,300	7 7/16	7 11/16	6 1/8	14,630	495,240	Market dull and freely supplied.
421,630	346,570	768,200	7 3/16	7 7/16	6	18,410	544,270	
465,920	338,580	804,500	7 3/8	7 5/8	5 15/16	14,510	601,850	Increased demand, slight improvement.
452,460	327,890	780,350	7 1/2	7 3/4	6	11,300	666,160	
415,200	335,040	750,240	7 9/16	7 3/4	6 1/8	15,120	723,680	
417,470	333,760	751,230	7 1/2	7 11/16	6 1/8	8,600	764,350	Market quiet.
479,150	334,280	813,430	7 1/2	7 11/16	6 1/8	7,160	802,540	
569,850	353,740	923,590	7 1/4	7 1/2	6	12,340	857,470	Depressed in tone.
562,680	388,940	951,620	7 1/4	7 1/2	5 7/8	15,960	915,300	Market steady.
607,300	386,990	994,290	7 5/16	7 9/16	5 3/4	17,520	969,610	
592,240	373,860	966,100	7 3/8	7 5/8	5 3/4	13,430	1,024,290	
562,150	375,100	937,250	7 1/2	7 3/4	6	19,420	1,089,260	Strong feeling, prices hardening.
559,960	353,490	913,450	7 11/16	7 7/8	6 1/16	19,710	1,151,550	
611,560	355,930	967,490	7 7/8	8 1/16	6 3/16	19,300	1,197,250	Advancing tendency.
578,980	334,990	913,970	8 1/8	8 1/4	6 7/16	19,700	1,274,740	
615,500	329,740	945,240	8 5/16	8 1/2	6 5/8	25,930	1,345,550	
595,200	314,480	909,680	8 1/2	8 11/16	6 7/8	19,820	1,424,920	Large inquiry.
437,700	275,620	713,320	8 9/16	8 3/4	6 7/8	11,620	1,496,300	
423,590	291,720	715,310	9	9 3/16	7 1/4	14,110	1,571,110	Stock taken, proving 135,760 bales below estimates, market excited
392,090	268,550	660,640	9	9 1/4	7 7/16	7,770	1,652,070	
385,260	275,240	660,500	9	9 1/4	7 3/8	9,980	1,695,760	
379,720	319,470	699,190	9 1/16	9 5/16	7 5/16	7,820	1,755,090	
362,420	310,530	672,950	8 15/16	9 1/4	7 1/8	5,150	1,801,670	
340,610	303,220	643,830	8 7/8	9 1/8	7	5,710	1,844,330	
309,800	286,400	596,200	8 7/8	9 1/8	6 7/8	5,830	1,893,990	
284,220	319,740	603,960	9 1/16	9 5/16	7	10,500	1,958,180	Good general demand.
248,920	312,660	561,580	9 5/16	9 9/16	7 1/8	10,700	2,040,420	
231,100	290,330	521,430	9 1/2	9 3/4	7 1/4	10,660	2,122,890	
221,670	276,470	498,140	9 1/2	9 3/4	7 5/16	9,430	2,175,790	Market steady.
207,800	259,910	467,710	9 7/16	9 11/16	7 1/4	8,820	2,221,290	
184,710	239,170	423,880	9 7/16	9 11/16	7 1/4	6,050	2,273,600	
156,120	343,430	499,550	9 7/8	10 1/16	7 1/2	15,450	2,391,860	Market excited, prices advancing.
144,040	327,370	471,410	9 3/4	10	7 7/16	10,740	2,453,230	Bank rate raised to 5 per cent.
131,110	342,530	473,640	9 11/16	9 15/16	7 3/8	10,270	2,509,770	
134,120	384,730	518,850	9 1/2	9 3/4	7 1/8	10,620	2,545,790	Tone heavy, supply abun ant.
117,520	399,540	517,060	9 1/2	9 3/4	7 1/8	11,850	2,598,360	
111,030	386,730	497,760	9 3/8	9 11/16	7 7/16	9,770	2,643,360	
88,370	450,580	538,950	9 1/2	9 13/16	7 1/8	13,820	2,715,290	Firm market.
74,350	451,750	526,100	9 9/16	9 7/8	7 1/4	12,580	2,794.560	Hardening tendency.
56,960	429,300	486,260	9 5/8	9 15/16	7 3/8	22,380	2,872,620	
41,230	403,990	445,220	9 3/4	10 1/16	7 7/16	20,330	2,948,580	December 14th, active market, bank rate reduced to 3 per cent; 21st, quieter tone; 28th, Stock taken, proving 84,630 bales above estimate.
69,610	388,860	458,470	10	10 5/16	7 9/16	19,480	3,038,070	
77,190	360,800	437,990	9 15/16	10 1/4	7 1/2	7,860	3,096,370	
168,800	398,100	566,900	9 15/16	10 1/4	7 1/2	8,640	3,135,270	
			8.55	8.8	6.73	676,580	60,294	

1872. *The semi-weekly Price and Weekly Sales and Receipts at New York, Weekly Exports from New York and Rates of Freight to Liverpool 1st of each month, for the Crop Year ending September 1, 1872.*

1871.	Price of Middling New Orleans	Price of Middling Upland.	Sales for week.	Receipts for week.	EXPORTS FOR THE WEEK. To Great Britain.	To France	North of Europe.	Other Fo'n Ports	Total Exports.	Rates of Freight to Liverpool.	GENERAL REMARKS.
Septem. 5..	20⅛@21	19¾@20⅝									
" 8..	21¼@—	20⅞@—	12,877	1,032	7,924	....	....	....	7,924	Steam,	
" 12..	21¼@21½	20⅞@21⅛								¼d.	
" 15..	21½@21⅜	21⅛@21	9,512	1,024	3,644	....	....	....	3,644	Sail,	
" 19..	21⅛@20½	20¾@20⅛								3-16d.	
" 22..	20⅜@20⅛	20 @19¾	8,348	910	5,009	....	....	....	5,009		
" 26..	20⅛@20¼	19¾@19⅝									
" 29..	20¼@—	19¾@—	12,828	1,219	7,221	....	....	....	7,221		
October 3..	20¼@20⅞	19¾@20¼									
" 6..	20⅞@20⅝	20¼@20	16,491	1,157	11,405	....	279	....	11,684	Steam,	
" 10..	20½@—	19⅞@—								¼d.	
" 13..	20⅝@20¾	20 @20⅛	16,849	1,758	14,271	....	57	....	14,328	Sail,	
" 17..	20¾@20¼	20⅛@19⅝								3-16d.	
" 20..	20⅛@19½	19½@18⅞	10,764	2,495	14,906	24	....	....	14,930		
" 24..	19¼@—	18½@—									
" 27..	19¼@—	18½@18⅜	11,765	3,439	13,873	....	49	....	13,922	Steam,	
" 31..	19½@19⅝	18⅝@18¾								½d.	
Novem. 3..	19½@19⅜	18⅜@18½	14,952	4,168	11,267	....	1,090	....	12,357	Sail,	
" 7..	19⅜@—	18½@—								5-16d.	
" 10..	19⅜@—	18½@—	14,387	4,984	13,104	....	....	....	13,104		
" 14..	19⅜@—	18½@18⅝									
" 17..	19⅜@—	18⅝@—	21,431	4,552	12,241	29	443	....	12,713		
" 21..	19½@—	18¾@—									
" 24..	19⅝@19⅞	18⅞@19⅛	17,514	4,964	12,943	....	1,311	....	14,254	Steam,	
" 28..	19⅞@19¾	19⅛@19								⅜d.	
Decem. 1..	19¾@19⅞	19 @19⅛	12,598	4,345	11,401	....	618	....	12,019	Sail,	
" 5..	19⅞@20	19⅛@19¼								5-16d.	
" 8..	20 @20⅜	19¼@19⅝	22,673	4,330	7,478	....	573	....	8,051		
" 12..	20¾@20⅞	20 @20⅛									
" 15..	20⅞@20⅝	20⅛@19⅞	17,800	4,171	11,662	....	709	....	12,371		
" 19..	20½@20⅞	19¾@20⅛									
" 22..	21 @—	20¼@—	12,657	3,936	7,072	....	200	....	7,272		
" 26..	21 @20⅞	20¼@20⅛									

Decem.	29..	20¾@—	20 @—	4,359	4,146	3,646	160	200	....	4,006	Steam, ¼d.
1872.											
January	2..	20⅞@21⅜	20⅛@20⅝								
"	5..	21⅜@21⅞	20⅝@21⅛	14,629	3,401	4,694	48	25	1,196	5,963	Sail, 7-32d.
"	9..	22⅜@22⅛	21⅝@21⅜								
"	12..	22¼@22⅞	21½@22⅛	17,392	5,051	10,111	61	505	....	10,677	
"	16..	22⅝@22½	21⅞@21¾								
"	19..	22⅝@22⅞	21⅞@22⅛	15,757	4,334	5,806	71	23	....	5,900	
"	23..	22¾@22½	22⅛@—								
"	26..	22½@22¼	21⅞@22⅛	10,971	4,131	10,475	79	122	....	10,676	
"	30..	22½@23	22⅛@22⅝								Steam,
February	2..	23 @23⅛	22⅝@22¾	23,715	4,818	7,780	19	75	....	7,874	¼d.
"	6..	23½@23⅞	23 @23½								Sail,
"	9..	23¾@23½	23⅜@23	20,525	3,185	7,831	....	67	....	7,898	3-16d.
"	13..	23½@23¼	23⅛@22⅞								
"	16..	23¼@23¾	22⅞@23⅜	9,414	4,410	11,551	9	229	....	11,789	
"	20..	23¾@23½	23⅜@23⅛								
"	23..	23¼@23⅛	22⅞@22¾	11,329	3,519	4,066	....	155	....	4,221	
"	27..	23¼@23	22⅞@22⅝								
March	1..	22¾@22⅝	22⅜@22¼	9,680	4,112	9,738	500	17	....	10,255	Steam,
"	5..	23 @23¼	22⅝@22⅞								¼d.
"	8..	23 @22⅞	22⅝@22½	14,839	3,966	9,068	....	30	....	9,098	Sail,
"	12..	22⅞@22⅝	22½@22¼								3-16d.
"	15..	22¾@23⅛	22⅜@22¾	8,555	3,083	7,654	....	19	....	7,673	
"	19..	23 @22⅞	22⅝@22½								
"	22..	22⅞@23	22½@22⅝	18,160	3,613	7,421	....	....	....	7,421	
"	26..	23 @23⅝	22⅝@23¼								
"	29..	23⅝@23¾	23¼@23⅜	22,822	4,092	11,100	....	26	....	11,126	Steam,
April	2..	24 @23⅞	23⅝@23½								¼d.
"	5..	23⅞@—	23½@—	13,436	3,043	5,844	....	40	....	5,884	Sail, ¼d.
"	9..	23⅞@—	23½@—								
"	12..	23⅞@—	23½@—	10,074	2,922	10,622	....	....	....	10,622	
"	16..	24 @—	23⅝@—								
"	19..	24 @—	23⅝@—	12,186	2,678	6,513	....	....	....	6,513	
"	23..	23⅞@—	23½@—								
"	26..	23⅞@24	23½@23⅝	6,335	2,135	3,996	....	....	....	3,996	Steam,
"	30..	24¼@24⅛	23⅞@23¾								¼d.
May	3..	24⅛@—	23¾@—	3,275	781	984	....	....	....	984	Sail,
"	7..	24⅛@24	23¾@23⅝								3-16d.
"	10..	24⅛@24¼	23¾@23⅞	2,902	326	734	....	....	....	734	
"	14..	24¼@24⅜	23⅞@24								
"	17..	24⅝@—	24¼@—	3,047	689	2,657				2,657	

New York Statement for Year 1872—*Concluded.*

1872.		Price of Middling New Orleans	Price of Middling Upland.	Sales for week.	Receipts for week.	EXPORTS FOR THE WEEK					Rates of Freight to Liverpool.	GENERAL REMARKS.
						To Great Britain.	To France.	North of Europe.	Other Fo'nPorts	Total Exports.		
May	21..	24⅝@25¼	24¼@24⅞									
"	24..	25½@26	25⅛@26	10,862	1,043	459	......	8	......	467		
"	28..	26 @26⅜	26 @26⅜									
"	31..	26⅜@—	26⅜@—	12,332	466	2,076	......	......	......	2,076	Steam, ¼d.	
June	4..	27⅜@27	27⅜@27									
"	7..	26¾@—	26¾@—	6,528	1,065	2,354	......	......	......	2,354	Sail, 3-16d.	
"	11..	26½@26¼	26½@26¼									
"	14..	26¼@26½	26¼@26½	3,753	1,829	6,651	......	......	......	6,651		
"	18..	26½@—	26¼@—									
"	21..	26½@—	26¼@—	8,681	1,647	1,032	......	......	......	1,032		
"	25..	26½@—	26¼@—									
"	28..	26½@26⅜	26¼@26⅛	5,755	1,051	1,439	......	......	......	1,439		
July	2..	26 @25¾	25¾@25½									
"	5..	—@—	—@—	4,305	901	1,646	......	......	......	1,646	Steam, ¼d.	
"	9..	25¼@24¼	25 @24									
"	12..	24½@24¼	23¾@24½	3,800	369	789	......	......	......	789	Sail, 3-16d.	
"	16..	24 @23¼	23¾@23									
"	19..	22⅝@22⅜	22⅜@22	9,754	382	1,551	65	......	......	1,616		
"	23..	22¾@22¼	22½@22									
"	26..	22¼@—	22 @—	6,468	294	1,840	......	......	......	1,840		
"	30..	21¾@21½	21½@21¼									
August	2..	21⅝@21¾	21⅜@21½	14,590	351	6,950	......	......	......	6,950	Steam, ¼d.	
"	6..	21¾@22	21½@21⅝									
"	9..	22 @22⅛	21⅝@21¾	11,962	104	9,068	......	......	......	9,068	Sail, 3-16d.	
"	13..	22⅛@—	21¾@—									
"	16..	22⅛@—	21¾@—	8,266	190	7,688	......	......	......	7,688		
"	20..	22⅛@22¼	21¾@—									
"	23..	22⅜@22½	21⅞@22	11,558	297	5,570	......	......	......	5,570		
"	27..	22½@—	22 @—									
"	30	22¼@22⅜	21¾@21⅞	11,493	219	4,907	......	......	......	4,907		
Average prices and total sales, receipts and exports,				616,955	127,127	361,732	1,065	6,870	1,196	370,863		

CONSUMPTION OF COTTON IN EUROPE, INCLUDING CONTINENT.

I bring down to 1867, Stolterfoht & Co.'s statement of the consumption of cotton in Europe; for the period commencing with the year ending October 1, 1867, I present M. Ott-Trumpler's statement, which affords a comparative view of the five years ending Oct. 1, 1871:

THE DELIVERIES IN EUROPE DURING THE YEAR FROM OCTOBER 1, 1870, TO SEPTEMBER 30, 1871, IN BALES.

GREAT BRITAIN.

	American.	Indian.	Brazil.	Egypt.	Sundry.	Total.
Stock in the ports, Oct. 1, 1870.	146,000	303,000	54,000	20,000	36,000	559,000
Imports during the season (49,000 bales from Continent)....	2.299,000	977,000	469,000	253,000	135,000	4,133,000
Total..................	2,445,000	1,280,000	523,000	273,000	171,000	4,692,000
Exports to the Continent (962,000 bales, American 1,000 bales)......................	343,000	533,000	58,000	11,000	18,000	963,000
	2,102,000	747,000	465,000	262,000	153,000	3,729,000
Total in the ports, September 30, 1871.............	177,000	189,000	86,000	21,000	34,000	507,000
Consumption..................	1,925,000	558,000	379,000	241,000	119,000	3.222,000
CONTINENT.						
Stock in the ports, Oct. 1, 1870.	100,000	55,000	18,000	2,000	25,000	200,000
Imports of the season direct from countries of production at Havre, Marseilles, Bordeaux, Nantes, Antwerp, Rotterdam, Amsterdam, Bremen, Hamburg, Trieste, Genoa, Venice, and Naples..................	617,000	296,000	95,000	87,000	151,000	1.186,000
Export from England to the Continent, deduction being made for 49,000 bales re-exportation from England, and 1,000 bales to America..................	314,000	522,000	48,000	11,000	18,000	913,000
Total..................	1,031,000	813,000	161,000	100,000	194,000	2,999,000
Stocks in the ports, Sept. 30, 1871	112,000	80,000	21.000	4,000	36,000	253,000
Consumption..................	919,000	733,000	140.000	96,000	158,000	2,046,000

The receipts at the ports of Spain, Sweden, and Russia, from American and other countries, and the consumption in Italy of native cotton, are not included in these tables of consumption.

ENGLISH CONSUMPTION.

	American.	Indian.	Brazil.	Egypt.	Sundry.	Total.
1870-71............. ....bales.	1,925,000	558,000	379,000	241,000	119,000	3,222.000
1869-70.....................	1,304,000	834,000	361,000	168,000	93,000	2,760.000
1868-69.....................	877,000	913,000	493,000	175,000	129.000	2,587.000
1867-68.....................	1,197,000	799,000	533,000	182,000	111,600	2.822,000
1866-67.....................	1,016,000	815,000	298,000	160,000	125,000	2,414,000
CONSUMPTION OF CONTINENT.						
1870-71............... bales.	919,000	733,000	140,000	96,000	158,000	2,046,000
1869-70.....................	608,000	623,000	165,000	58,000	173,000	1,627.000
1868-69.....................	545,000	850,000	191,000	61,000	269,000	1.916.000
1867-68.....................	538,000	723,000	175.000	69,000	277,000	1,782.000
1866-67.....................	532,000	777,000	152,000	55,000	217,000	1,733 000

CONSUMPTION OF EUROPE.

	American	Indian.	Brazil.	Egypt.	Sundry.	Total.
1870-71............... bales.	2,844,000	1,291,000	519,000	337,000	277,000	5,268.000
1869-70.....................	1,912,000	1,457,000	526,000	226,000	266,000	4.387,000
1868-69.....................	1,422,000	1,763 000	684.000	236,000	398,000	4,503.000
1867-68.....................	1,735,000	1,522,000	708,000	251,000	388,000	4.604,000
1866-67.....................	1,548,000	1,592,000	450,000	215,000	342,000	4,147,000

STOCK IN ENGLAND, SEPTEMBER 30.

1871....................... bales.	507,000	1868..............................	513,000
1870............................	559,000	1867..............................	911,000
1869............................	489,000	1866..............................	945.000

DETAILS OF STOCK IN THE PORTS OF THE CONTINENT, SEPT. 30, 1871.

Havre..... .... bales.	102,000	Bremen..........................	35,000
Bordeaux and Nantes............	4,000	Hamburg..........................	12,000
Marseilles......................	20,000	Trieste..........................	13,000
Anvers.........................	15,000	Genoa	8,000
Rotterdam..........	5,000		———
Amsterdam......	39 000	Total..........	253,000

The following is a recapitulation of some of the principal figures :—

IMPORTATIONS INTO EUROPE FOR YEAR

	American.	Indian.	Sundry.	Total.
1866-67........................... bales.	1,495,000	1.524,000	1,077.000	4,096.000
1867-68.................................	1,572,000	1,307.000	1,247.000	4,426,000
1868-69.................................	1,362,000	1,856.000	1.254,000	4,472.000
1869-70........	2,084,000	1 419,000	1,060,000	4,563.000
1870-71.................................	2,887,000	1,203,000	1,180.000	5,270,000
STOCK IN EUROPE, SEPTEMBER 30.				
1867................. ....bales.	297,000	518,000	277,000	1,032,000
1868...........	134,000	303,000	177,000	614,000
1869....................................	74,000	396,000	113.000	583.000
1870....................................	246,000	358,000	155,000	759,000
1871....................................	189,000	269,000	202,000	769,000

CONSUMPTION IN EUROPE.

Av. w't bales.		American.	Indian.	Sundry	Total.	Total weight in lbs.
371 lbs.	1866-67...... bales.	1,548,000	1,592,000	1,007,000	4,147,000	1,539 millions
364 lbs.	1867-68............	1,735,000	1,522,000	1,347,000	4,604,000	1,676 "
355 lbs.	1868-69	1,422,000	1,763,000	1,318,000	4,503,000	1,599 "
374 lbs.	1869-70............	1,912,000	1,457,000	1,018,000	4,387,000	1,640 "
386 lbs.	1870-71............	2,844,000	1,291,000	1,133,000	5,268,000	2,033 "

The importations into Europe, as compared with the preceding season, are as follows:

	American.	Indian.	Brazil.	Egypt.	Sundry.	Total.
1870-71............... bales.	2,887,000	1,203,000	564,000	330,000	286,000	5,270,000
1869-70....................	2,084,000	1,419,000	532,000	226,000	302,000	4,563,000
Increase	803,000	....	32,000	104,000	....	707,000
Decrease................	....	216,000	....	....	16,000	....

RECAPITULATION.

Stock October 1, 1870.....bales.	759,000	Consumption.............bales.	5,268,000
Arrivals	5,270,000	Exported	1,000
		Stock, September 30, 1871.......	760,000
Total......................	6,029,000	Total......................	6,029,000

The consumption of Europe, estimated in pounds (see above), this season shows an increase over last season of 24 per cent., and over 1859-60 of 16 per cent.

Below is the proportion by weight which the different sources of production have yielded for European consumption:

America..	60 per cent.
India...	23 "
Sundry...	17 "
	100 per cent.

CONSUMPTION PER WEEK.

	England.	Continent.	Europe.
1866-67 .... bales.	46,423	33,327	79,750
1867-68	54,269	34,270	88,539
1868-69....................................	49,750	36,846	86,596
1869-70	53,077	31,288	84,365
1870-71....................................	61,961	30,347	101,308

THE QUESTION OF ACREAGE.

A friend, who has been traveling nearly all the past summer on business among the planters of Mississippi, Tennessee, and North Alabama, took pains to ascertain from every planter, with whom he came in contact, how many acres he planted in cotton this year, and how many last year; each time entering the facts in his pocket memorandum. The following is the result:

In 23 counties in Tennessee, 358 small planters show an increase this year over last of 75 per cent. In 21 counties in North and West Alabama, 198 planters, small and large, show an increase of 36 per cent. In Mississippi, 1,287 planters, in every county in the State, show an increase of 42 per cent.

I have long known that the returns of the Agricultural Bureau were unreliable. The increase and decrease in acreage, with the rise and fall in price, has been much greater than it has been estimated. I suppose that the increase or diminution is much greater among small planters than large ones. I annex particulars of the 358 planters in Tennessee, with the names of the counties. The facts are important, as illustrating the elasticity of the new system of labor, and how largely the production depends upon the price; or rather the relative prices of breadstuffs and provisions. No crop estimate in the future should have any weight that is not based upon a thorough understanding of the question of acreage:

Crop Statement of Tennessee for 1871 *and* 1872.

Planters.	County.	Bales, 1871.	Bales, 1872.	Planters.	County.	Bales, 1871.	Bales, 1872.
1	Benton County	3	7	26	Werkley Co., continued..	15	20
2	"	5	8	27	"	10	12
3	"	6	10	28	"	15	20
4	"	4	5	29	"	10	15
5	"	8	15	30	"	13	20
6	"	10	12	31	"	14	20
7	"	6	8	32	"	9	12
8	"	3	5	33	"	6	15
9	"	4	5	34	"	4	10
10	Werkley County..........	10	12	35	"	10	15
11	"	6	8	36	Gibson County..........	15	30
12	"	20	25	37	"	15	22
13	"	24	30	38	"	10	25
14	"	10	13	39	"	30	30
15	"	15	20	40	"	25	25
16	"	......	15	41	"	10	30
17	"	......	20	42	"	5	20
18	"	......	25	43	"	12	15
19	"	5	25	44	"	7	15
20	"	6	15	45	"	9	10
21	"	10	28	46	"	10	15
22	"	8	10	47	"	10	20
23	"	10	12	48	"	11	15
24	"	15	10				
25	"	18	20		Amount carried forward..	481	794

Crop Statement of Tennessee for 1871 *and* 1872.—*Continued.*

Planters.	County.	Bales, 1871.	Bales, 1872.	Planters	County.	Bales, 1871.	Bales, 1872.
	Amount brought forward.	481	794	110	Tipton County, coutinued	12	15
49	Gibson County, continued	13	20	111	"	20	25
50	"	16	22	112	"	30	35
51	"	17	20	113	"	20	30
52	"	15	30	114	Lauderdale County	10	17
53	"	12	15	115	"	11	18
54	"	23	30	116	"	9	12
55	"	18	25	117	"	15	18
56	Lake County	......	20	118	"	16	20
57	"	......	15	119	"	14	20
58	"	6	20	120	"	20	15
59	"	......	25	121	"	10	12
60	"	......	20	122	"	15	12
61	"	......	18	123	"	12	10
62	"	......	17	124	"	8	10
63	"	......	10	125	"	9	15
64	"	......	15	126	"	10	20
65	"	......	10	127	"	......	20
66	"	......	12	128	"	......	15
67	"	5	10	129	"	......	10
68	"	3	15	130	"	......	12
69	"	4	10	131	Haywood County	20	25
70	Oliver County	6	10	132	"	30	35
71	"	5	8	133	"	40	40
72	"	3	5	134	"	25	30
73	"	3	8	135	"	27	35
74	"	10	15	136	"	22	30
75	"	......	10	137	"	18	25
76	"	......	12	138	"	15	22
77	"	......	15	139	"	16	27
78	"	......	10	140	"	30	20
79	"	......	6	141	"	22	20
80	"	......	5	142	"	18	25
81	"	......	8	143	"	19	15
82	"	3	10	144	"	30	37
83	"	4	10	145	"	32	40
84	"	5	15	146	"	20	22
85	"	6	18	147	"	75	100
86	"	7	10	148	"	25	50
87	"	5	10	149	"	30	75
88	"	6	8	150	"	40	80
89	"	3	8	151	"	27	35
90	"	4	10	152	"	20	15
91	"	5	14	153	"	22	20
92	"	5	5	154	"	32	35
93	"	10	15	155	"	38	45
94	"	12	18	156	"	42	50
95	"	20	25	157	Fayette County	27	30
96	"	10	15	158	"	25	35
97	"	11	20	159	"	30	25
98	Tipton County	15	20	160	"	44	50
99	"	15	30	161	"	42	62
100	"	12	25	162	"	26	30
101	"	18	30	163	"	22	20
102	"	9	15	164	"	17	25
103	"	20	20	165	"	23	30
104	"	10	12	166	"	28	35
105	"	11	15	167	"	32	40
106	"	20	25	168	"	23	38
107	"	18	25	169	"	27	42
108	"	17	30				
109	"	15	20		Amount carried forward.	2,370	3,617

Crop Statement of Tennessee for 1871 *and* 1872.—*Continued.*

Planters.	County.	Bales, 1871.	Bales, 1872	Planters.	County.	Bales, 1871.	Bales, 1872.
	Amount brought forward.	2,370	3,617	231	Henry County, continued.	7	15
170	Hardman County	30	45	232	"	8	10
171	"	33	55	233	"	......	20
172	"	37	60	234	"	......	15
173	"	22	30	235	"	......	22
174	"	28	40	236	"	......	18
175	"	15	25	237	"	......	5
176	"	10	20	238	"	......	10
177	"	5	15	239	"	5	15
178	"	15	10	240	"	6	10
179	"	20	35	241	"	4	10
180	"	25	30	242	"	7	15
181	"	18	35	243	"	8	10
182	Madison County.........	22	30	244	"	5	10
183	"	17	25	245	"	10	20
184	"	18	30	246	"	20	20
185	"	25	60	247	"	5	10
186	"	22	30	248	"	10	15
187	"	18	35	249	"	6	10
188	"	15	20	250	"	4	10
189	"	10	15	251	"	5	8
190	"	10	10	252	Perry County...........	5	12
191	"	12	17	253	"	10	10
192	"	8	18	254	"	7	15
193	"	40	75	255	"	8	10
194	"	30	50	256	"	10	15
195	"	10	20	257	"	5	15
196	"	15	25	258	"	5	10
197	Henderson County	6	10	259	"	......	10
198	"	4	15	260	"	......	5
199	"	10	20	261	"	......	10
200	"	15	25	262	Dickson County.........	3	10
201	"	5	10	263	"	7	10
202	"	......	20	264	"	5	12
203	"	......	15	265	"	5	8
204	"	......	10	266	"	6	10
205	"	......	5	267	"	4	10
206	"	10	30	268	"	5	10
207	"	20	66	269	"	10	15
208	"	5	15	270	"	15	25
209	"	7	15	271	"	20	20
210	"	8	15	272	"	18	25
211	"	9	20	273	"	22	35
212	"	11	20	274	"	5	8
213	"	13	15	275	"	10	12
214	Carroll County..........	7	10	276	"	12	20
215	"	5	20	277	"	8	15
216	"	20	25	278	"	......	10
217	"	22	30	279	"	......	15
218	"	18	25	280	"	10	15
219	"	5	10	281	"	5	10
220	"	10	20	282	"	6	10
221	"	6	12	283	Wayne County..........	4	12
222	"	4	8	284	"	5	8
223	"	......	10	285	"	10	15
224	"	22	35	286	"	5	20
225	"	18	30	287	"	4	10
226	"	10	......	288	"	6	15
227	"	5	......	289	"	10	10
228	"	10	20	290	"	10	18
229	"	5	15				
230	Henry County	5	5		Amount carried forward..	3,585	5,871

Crop Statement of Tennessee for 1871 *and* 1872.—*Concluded*

Planters.	County.	Bales, 1871.	Bales, 1872.	Planters.	County.	Bales, 1871.	Bales, 1872.
	Amount brought forward.	3,585	5,871	326	Harden County, continued	4	25
291	Wayne County, continued	12	12	327	"	6	15
292	"	8	10	328	"	10	20
293	"	5	15	329	"	5	15
294	"	5	15	330	"	8	16
295	"	4	20	331	"	7	24
296	"	6	15	332	"	5	10
297	"	7	10	333	"	12	25
298	"	8	18	334	"	8	15
299	"	5	12	335	"	5	8
300	"	......	10	336	"	5	7
301	"	......	15	337	McNairy County	4	10
302	"	......	20	338	"	6	15
303	"	......	15	339	"	11	20
304	"	......	15	340	"	9	25
305	"	......	18	341	"	10	22
306	"	......	17	342	"	10	18
307	"	......	20	343	"	10	10
308	Harden County	5	10	344	"	5	12
309	"	8	15	345	"	5	8
310	"	7	10	346	"	10	15
311	"	10	20	347	"	6	10
312	"	......	10	348	"	4	15
313	"	......	20	349	"	5	10
314	"	......	15	350	"	10	20
315	"	8	20	351	"	10	18
316	"	7	15	352	"	20	35
317	"	5	10	353	"	15	30
318	"	6	15	354	"	22	50
319	"	4	10	355	"	18	20
320	"	3	10	356	"	10	15
321	"	7	15	357	"	12	25
322	"	4	10	358	"	8	15
323	"	6	15				
324	"	5	10		Total	3,943	6,907
325	"	10	25				

PRICES OF BROWN SHEETINGS

(Standard weight) in New York at the first and middle of each month for twenty-five years.

YEARS.	SEPT.		OCT.		NOV.		DEC.		JAN.		FEB.		MARCH.		APRIL.		MAY.		JUNE.		JULY.		AUGUST.		AV. FOR THE YEAR.
	First	Mid.	First	Mid.	First	Mid.	First	Mid.	First	Mid.	First	Mid.	First	Mid.	First	Mid.	First	Mid.	First	Mid.	First	Mid.	First	Mid.	
1847–8	8	8¼	8½	7⅞	7⅞	7¾	7½	7¾	7½	7¼	7¼	7	6¾	6¾	7	6¾	7	7	7	7	7	7	6¾	6	7.23
1848–9	6½	6½	6¾	6½	6½	6¼	6½	6¼	6⅜	6½	6⅝	6¾	6¾	6¾	6⅝	6½	6¾	6⅝	6¾	6¾	6½	6¾	6¾	7	6.66
1849–50	7¼	7¼	7¼	7½	7½	7½	7½	7½	7½	7¾	8½	8½	8½	8¼	8¼	8	7⅝	7⅝	7½	7½	7¾	7¾	7¾	7½	7.74
1850–1	7½	7⅜	7⅜	7½	8½	8½	8½	8½	8¼	8¼	8¼	8¼	8	8	7½	7¼	7¼	7¼	6¾	6½	6½	6⅝	6½	6½	7.56
1851–2	6⅜	6½	6½	6½	6⅝	6⅝	6⅝	6⅝	6⅝	6¾	6¾	6¾	6⅝	6¾	6¾	6¾	6¾	6¾	7⅛	7	7⅛	7⅛	7	7⅛	6¾
1852–3	7	7¼	7¼	7¼	7¼	7¼	7	7⅛	7⅜	7⅝	7⅞	7¾	7⅞	7⅞	7⅞	7⅞	8	8	8	8	8	7⅞	8	8	7.40
1853–4	8	8	8	8	7⅞	8	8	8	8	8	8	8	7⅞	8	8	8	8	8	8	8	7⅞	8	7⅞	8	7.98
1854–5	8	7⅞	8	8	7⅞	8	8	8	7¾	7¾	7¾	7¾	7¾	7¾	7¾	7¾	7½	7½	7½	7½	7½	7¾	7¾	7½	7.76
1855–6	7¾	7⅞	7½	7½	7⅞	7½	7½	7½	7½	7½	7½	7¾	7½	7½	7½	7½	7¾	7¾	7¾	7¾	7¾	7¾	7¾	7¾	7.63
1856–7	7½	7½	7½	8¼	8½	8½	8½	8½	8½	8¾	8¾	8¾	9	9⅛	9¼	9⅛	9⅛	9¼	9	9	9	8¾	9¼	9	8.60
1857–8	9	9	9¼	9	9	9¼	8⅜	8½	8½	7¾	7¾	7⅝	8	8	8¼	8½	8½	8½	8½	8½	8⅜	8½	8⅜	8⅜	8.47
1858–9	8¼	8¼	8¼	8¼	8¼	8¼	8¼	8⅜	8¼	8¼	8½	8⅝	8¾	9	8⅝	8¾	8¾	8½	8½	8¾	8¾	8¼	8½	8⅝	8.48
1859–60	8½	8½	8½	8¼	8¼	8¼	8¼	8¼	8½	8½	8¾	8¾	8¾	8¾	8¾	8¾	8¾	8¾	8¾	8¾	8¾	8¾	8¾	8¾	8.59
1860–1	8¾	8¾	8¾	8¾	8¾	8¾	8¾	8¾	8¾	8¾	8¾	8¾	8¾	8¾	8¾	8¾	8¾	8¾	8¾	8¾	8¾	9	9½	9½	8.82
1861–2	11½	11½	11½	11½	12½	13	13¼	13¼	14	16	16	16	14	14	14	14	14	14	14	14	16	18	....	....	......
1862–3	25	25	25	25	25	25	25	25	30	35	37½	42½	45	45	37½	35	33	31	25	27½	35	30	33	33½	29.44
1863–4	33½	33½	37½	40	39½	39½	40	41	42½	42½	42½	42½	41½	41½	40	41	42	42	47½	55	65	70	67½	72½	45.83
1864–5	....	....	57½	55	55	62½	60	60	60	60	53½	53	52½	50	39	30	37½	38	30	27½	32½	34	32½	31	......
1865–6	36½	37½	33½	37	33	33	31	32	33½	33½	30	30	25	25	25	25	22	20½	22½	24	23	24	22½	22	28.37
1866–7	22½	22	22	23	22½	22	21	21	22	22	21	21½	22	20	20½	20½	20	17½	17	17¾	17½	18	18	17¾	20.36
1867–8	17½	16½	16½	15½	15	15	15	15	15	15	15	17	19	18	19	19	18½	17½	17	17½	17½	17½	17	17	16.77
1868–9	16½	15	15½	16	16	16	15	16	16	16	16½	17	17	16	16	15½	15	15½	16	17	17	17	17	17	16.14
1869–70	17	17	17	15½	15	15	15	15½	16	16	16	16½	16½	15	15	15	15	15	15	15	14	14	14	14	15½
1870–1	14	14	14	14	13	13	13	13	13	13	13	13	13	13	12½	12½	12½	12	12	12½	13	13	13½	13½	13
1871–2	13½	13½	13½	13½	13	13	13	13½	13½	14	14	14½	15	15	15	15	15	15	14½	14	14½	14½	14½	14½	14¾

PRICES OF PRINTING CLOTHS

(Standard count) in New York at the first and middle of each month for twenty-five years.

YEARS.	SEPT.		OCT.		NOV.		DEC.		JAN.		FEB.		MARCH.		APRIL.		MAY.		JUNE.		JULY.		AUGUST.		AV. FOR THE YEAR.
	First	Mid.	First	Mid.	First	Mid.	First	Mid.	First	Mid.	First	Mid	First	Mid.	First	Mid.	First	Mid.	First	Mid.	First	Mid.	First	Mid.	
1847–8	5⅝	5½	5¾	5¾	5⅝	5½	5⅝	5⅜	5¼	5⅛	4⅞	4¾	4⅝	4⅝	4½	4½	4½	4½	4½	4½	4½	4¼	4	4	4.82
1848–9	4	4	4	4	3⅞	3⅞	3⅞	3⅞	4	4¼	4⅜	4½	4⅝	4¾	4½	4⅜	4¼	4¼	4⅛	4¼	4⅛	4⅜	4⅜	4¾	4.15
1849–50	4⅞	5	5	5	5	5	5	5	5¼	5¼	5¾	5¾	5½	5½	5¼	5	5	5	5	5	5	5	5⅛	5⅛	5.14
1850–1	5⅛	5	5	5	5¼	5¼	5¼	5¼	5¼	5¼	5¼	5¼	5	5	4¾	4¾	4¾	4⅝	4½	4¼	4⅜	4⅜	4¼	4¼	4.46
1851–2	4¼	4¼	4¼	4¼	4¼	4¼	4⅜	4⅜	4⅜	4⅜	4⅜	4¼	4¼	4¼	4¼	4¼	4⅜	4⅜	4½	4⅝	4¾	4¾	4⅞	5	4.41
1852–3	5	5	5	5	5	5⅛	5¼	5¾	6	6	6	6	6	6	6	6	6⅛	6¼	6¼	6¼	6¼	6⅜	6⅜	6⅜	5.81
1853–4	6¼	6⅛	6⅛	6⅛	6⅛	6⅛	6⅛	6⅛	6¼	6	6	6	6	6	6	6	6	6	5¾	5¾	5¾	5¾	5¾	5¾	6.
1854–5	5¾	5½	5½	5½	5⅝	5⅝	5⅝	5¾	4⅞	4¾	4¾	5	5	5	5	5	5	5	5⅛	5½	5½	5½	5½	5½	5.70
1855–6	5¼	5¼	5¼	5¼	4⅞	4⅞	5	5	5	5	5¼	5¼	5¼	5¼	5¼	5¼	5¼	5¼	5⅜	5⅜	5⅜	5⅜	5⅜	5⅜	5.21
1856–7	5⅜	5½	5½	5½	5⅝	5⅝	5⅝	5¾	5¾	6	5⅞	6	6	6	6⅛	6⅛	6	6⅛	5⅞	6	5⅞	6	6	5⅞	5.84
1857–8	5⅞	6	6	6	6	6	6	6	6	6	6	6	6	6	5¾	5½	5	5	5⅛	5¼	5½	5⅝	5¾	5¾	5¾
1858–9	5¾	5½	5½	5½	5½	5½	5½	5½	5⅝	5⅝	5⅝	5⅝	5¾	5¾	5¾	5¾	5¾	5¾	5¾	5⅞	5¾	5⅞	5⅞	5¾	5.67
1859–60	5⅝	5½	5½	5½	5½	5½	5⅝	5⅝	5⅝	5⅝	5½	5½	5⅝	5½	5½	5⅜	5⅜	5¾	5¾	5⅝	5⅝	5½	5⅝	5⅝	5.54
1860–1	5⅝	5⅝	5⅛	5⅜	5	4⅞	4¾	4¾	4¾	4¾	4½	4½	4¼	4⅜	4⅜	4⅜	4⅜	4⅜	4⅜	4⅜	4½	4½	4¾	5	4.74
1861–2	5¾	5¾	6	6	7	7¼	9	9	9	9	8½	8	7	7½	7¼	7½	7½	7½	7¾	7½	8⅝	9½	10	10⅛	7¾
1862–3	10⅛	11½	12⅜	12⅞	14¼	14½	14	13½	14¼	15¼	18⅝	18⅞	19	18	14½	13½	12½	12¼	10¾	12⅝	14½	13½	13¾	14½	14.17
1863–4	14	14½	15¾	17¼	16½	16⅞	16⅞	16¾	17	17¼	17¼	17	16⅜	16⅞	16¼	16⅞	17½	18	21	24	29½	33	33	33½	19.67
1864–5	38½	....	....	...	25½	28½	26	29	27	25	22	20½	18	15	10	11	14½	17½	16½	18	19	25	23½	23	......
1865–6	25¼	27¼	27¼	23½	22½	17¾	17¼	19½	19½	19	17½	16½	15	12½	13¼	13	11	12	14¼	15	14½	13½	14	13¼	17.04
1866–7	13½	13½	14	14½	13¼	12	12¼	12	12	12	11½	11½	10⅝	10⅝	10¼	9⅞	9½	9¼	9	8⅞	9	8⅞	9	9⅛	11.08
1867–8	8⅞	8⅛	7½	6⅞	6¾	6⅝	6⅝	6⅝	6½	6⅝	7⅜	7⅞	8⅝	9	9¼	9½	9⅝	9	8¼	8¼	8⅞	9	8¾	8⅜	8.14
1868–9	8	7¼	7⅜	7⅜	7⅜	7¼	7½	8⅞	9	9½	9¼	9	8½	8⅛	7⅞	7¾	7¾	7⅛	8¼	8⅞	8¾	8¾	8⅜	8⅜	8.18
1869–70	8½	8¼	7⅞	7¾	7¾	7⅞	7⅝	8¼	8⅛	8	8	7¾	7	6⅞	6¾	6⅞	6⅞	6⅞	6⅞	6¾	6½	6½	7	7⅜	7.39
1870–1	7⅝	7⅞	7⅜	6⅞	6½	6¾	7⅛	7⅛	7¼	7½	7½	7	6⅝	6⅛	6⅝	6¾	7⅛	7	7	7¼	7½	7½	7¾	7½	7.14
1871–2	7⅝	7¾	8	8⅛	7¾	7½	7⅞	....	7⅜	7½	....	7⅞	....	....	....	8½	....	7⅞	8¼	...	7⅛	....	....	8	......

THE PRICE OF GOLD

On each Friday, from the suspension of specie payments to the present time, except when Friday fell on a holiday, when the quotation of the previous day is given.

1862.

January 17	101⅞@102
“ 24	103⅜@103½
“ 31	101½@103⅝
February 7	103⅜@103⅝
“ 14	104⅜@104⅝
“ 21	103⅛@103⅛
“ 28	102⅛@102¼
March 7	102 @102⅛
“ 14	101½@101⅝
“ 21	101¼@101½
“ 28	101¼@101⅜
April 4	101⅜@101⅞
“ 11	101⅞@102
“ 18	101½@101½
“ 25	101½@101½
May 2	102¼@102½
“ 9	103⅛@103¼
“ 16	103 @103½
“ 23	103⅜@103½
“ 30	103⅜@103⅝
June 6	104 @104¼
“ 13	105¼@105⅛
“ 20	106⅞@106⅝
“ 27	109⅜@109½
July 3	109⅛@109½
“ 11	114¾@116
“ 18	119 @119¼
“ 25	114½@116½
August 1	115½@115¾
“ 8	112⅝@114
“ 15	114¾@115½
“ 22	115½@116¼
“ 29	116 @116
September 5	118¼@119⅛
“ 12	118½@118¾
“ 19	116⅞@117
“ 26	120¼@120½
October 3	122⅜@122⅝
“ 10	127¼@129
“ 17	132 @132½
“ 24	131 @132
“ 31	130¼@130¾
November 7	131½@132
“ 14	131 @132
“ 21	130¼@130⅝
“ 28	129¼@129½
December 5	131½@132½
“ 12	131½@131¾
“ 19	132¼@132¾
“ 26	131⅞@132

1863.

January 2	133⅝@133⅞
“ 9	138 @138½
“ 16	145½@145¾
“ 23	147 @148
“ 30	153 @158¼
February 6	157¼@158[illegible]
“ 13	155⅛@156
“ 20	162½@163[illegible]
“ 27	169⅝@171
March 6	150 @154
“ 13	159 @161¼
“ 20	154½@155
“ 27	140 @140⅞
April 3	153 @153½
“ 10	146½@149
“ 17	153½@153¾
“ 24	151½@152
May 1	150¾@151½
“ 8	154¼@154½
“ 15	149¾@150
“ 22	148½@149¾
“ 29	144½@145½
June 5	146 @146¼
“ 12	141¼@141⅝
“ 19	143 @143¾
“ 26	144¾@145
July 3	144 @144¼
“ 10	132¼@132½
“ 17	125¾@126
“ 24	126½@126¼
“ 31	128½@129
August 7	127 @127¾
“ 14	125¼@126½
“ 21	125¼@125⅜
“ 28	124 @124⅜
September 4	133½@134½
“ 11	129⅜@129¾
“ 18	133 @133¼
“ 25	138 @138½
October 2	142½@143¼
“ 9	143¾@147
“ 16	154⅛@154¼
“ 23	145¾@146½
“ 30	146 @146⅜
November 6	148 @148¾
“ 13	147 @147⅜
“ 20	152½@152¾
“ 27	143 @145½
December 4	152½@152¾
“ 11	151 @151½
“ 18	151⅝@152½
“ 24	151⅜@151¾
“ 31	151⅞@151⅞

1864.

January 8	151¾@152¼
“ 15	155 @155½
“ 22	156¾@157⅝
“ 29	156¾@157⅝
February 5	157¾@158¼
“ 12	159 @159¼
“ 19	158 @158⅝
“ 26	157½@158⅜
March 4	160⅛@161⅛
March 11	164⅜@164¾
“ 18	163 @ —
“ 24	166½@ —
April 1	166½@168¼
“ 8	169½@ —
“ 15	173½@173¾
“ 22	173½@173¾
“ 28	177 @180¼
May 6	174 @176¾
“ 13	170 @173⅞
“ 20	181 @ —
“ 27	186 @186¼
June 3	— @ —
“ 10	198½@198¾
“ 17	196⅝@196⅞
“ 24	213 @217
July 1	222 @250
“ 8	266¾@276½
“ 15	244 @256
“ 22	250½@257¾
“ 29	250 @253½
August 5	257½@261¼
“ 12	255⅛@257⅛
“ 19	257 @257⅞
“ 26	253⅜@256
September 2	248½@254½
“ 9	234⅛@256
“ 16	224⅛@228
“ 23	211 @217
“ 30	— @ --
October 7	198 @204
“ 14	208 @217¼
“ 21	207⅜@209
“ 28	215⅛@217¾
November 4	231⅝@238⅝
“ 11	236⅛@244½
“ 18	210 @219
“ 25	216¾@221¼
December 2	230¼@233¾
“ 9	239½@242⅜
“ 16	233¾@234¾
“ 23	220¼@222¾
“ 30	226 @229½

1865.

January 6	227 @228¼
“ 13	218½@222
“ 20	204¾@207¾
“ 27	208¼@215
February 3	205¼@209¾
“ 10	210½@211⅞
“ 17	203½@204⅝
“ 24	198⅞@200¼
March 3	198¼@199
“ 10	186⅝@191⅛
“ 17	160 @169
“ 24	148½@152½
“ 31	151¼@151⅞
April 7	147¼@150½

The Price of Gold—Continued.

April 13	246 @147¼
" 21	147¼@149⅛
" 28	146¾@148
May 5	142½@143⅞
" 12	130⅜@133¼
" 19	130⅜@131½
" 26	135¾@138
June 2	137½@138⅜
" 9	137⅛@138
" 16	143¼@145¼
" 23	141⅜@142¼
" 30	139 @141¼
July 7	139½@139¾
" 14	142½@143⅜
" 21	142⅛@143¾
" 28	144⅝@146¼
August 4	143⅜@144¼
" 11	140¼@141⅞
" 18	142½@143¾
" 25	143½@144
September 1	144⅜@145
" 8	144½@144¾
" 15	142⅞@143¼
" 22	143¾@143⅞
" 29	143⅛@144⅛
October 6	146⅛@149
" 13	144⅜@144⅞
" 20	146 @146¼
" 27	145⅜@145⅞
November 3	146⅝@147
" 10	146½@146⅝
" 17	146¾@147
" 24	146¾@146⅞
December 1	148 @148⅛
" 8	145¾@146½
" 15	146¼@146¾
" 22	145⅝@146⅛
" 29	145⅛@145⅜

1866.

January 5	142⅜@143½
" 12	138¼@139¼
" 19	137⅞@138¾
" 26	139 @139⅛
February 2	139⅞@140½
" 9	139¼@140⅝
" 16	137¼@137¾
" 23	136⅜@137⅜
March 2	135⅛@136⅛
" 9	130¼@131½
" 16	130⅛@131
" 23	126¾@128¼
" 29	127½@128¼
April 6	127 @128⅛
" 13	126⅛@127
" 20	126½@127¼
" 27	128⅜@129⅜
May 4	127⅛@127⅞
" 11	128⅞@129½
" 18	129¾@130⅞
" 25	139⅛@141½
June 1	140½@141
" 8	138⅜@141¾
June 15	147[illegible]@149⅞
" 22	148⅛@149[illegible]
" 29	153½@156
July 6	153¾@154¾
" 13	152¼@153[illegible]
" 20	149¾@150[illegible]
" 27	149⅞@150⅜
August 3	147¾@148⅛
" 10	148⅛@148⅜
" 17	150⅛@151⅜
" 24	148⅛@150¾
" 31	147⅜@148
September 7	145¾@146¼
" 14	144¾@145½
" 21	143⅛@144
" 28	144⅞@145½
October 5	148⅜@149⅜
" 12	150½@153⅝
" 19	147 @149½
" 26	147 @148¼
November 2	146¾@147½
" 9	146 @146⅜
" 16	141⅛@143⅜
" 23	138⅛@139⅛
" 30	140¼@141⅝
December 7	138¼@139
" 14	137⅜@138¼
" 21	133⅝@134⅛
" 28	132 @133⅜

1867.

January 4	133⅝@134¼
" 11	132½@134
" 18	136⅜@137⅛
" 25	133⅛@136⅛
February 1	135⅛@135¾
" 8	137½@138⅛
" 15	136½@136¾
" 21	137½@138⅛
March 1	138⅝@140⅜
" 8	133⅝@134⅜
" 15	134 @134⅜
" 22	134¼@134⅛
" 29	134⅛@134¼
April 5	132⅞@133½
" 12	136⅜@137⅜
" 18	135¼@137¾
" 26	138⅛@139⅛
May 3	135¾@136⅜
" 10	136⅞@137½
" 17	136⅞@137½
" 24	137½@138⅝
" 31	136¾@137⅛
June 7	136[illegible]@136⅞
" 14	137 @137⅛
" 21	137[illegible]@137⅞
" 28	137¾@138½
July 5	138⅞@139¼
" 12	139 @139⅛
" 19	139¾@140
" 26	139[illegible]@139[illegible]
August 2	139⅞@140¼
" 9	140 @140⅛
August 16	140⅜@140⅝
" 23	140¾@140⅞
" 30	141⅛@142⅛
September 6	142⅜@142¾
" 13	144⅜@145⅛
" 20	142⅞@143½
" 27	143¼@143½
October 4	144¾@145⅛
" 11	143⅝@143⅞
" 18	144⅛@144⅜
" 25	141⅝@142½
November 2	140⅝@140⅞
" 9	138⅞@139¼
" 16	140⅞@141⅛
" 23	138⅞@139¼
" 30	139 @139⅛
December 6	137½@137⅞
" 13	133½@133¾
" 20	133¾@134
" 27	133⅞@134⅛

1868.

January 3	133⅝@134
" 10	137⅛@137⅜
" 17	138¼@130
" 24	140 @140¾
" 31	140⅛@140⅝
February 7	141⅞@142½
" 14	139¾@140⅜
" 21	140½@140¾
" 28	141¼@141½
March 6	141⅛@141¼
" 13	139⅛@140
" 20	138¼@138¾
" 27	138⅛@138⅝
April 3	137¾@138
" 9	138¾@138¾
" 17	138¼@138⅝
" 24	139 @140
May 1	139⅜@139⅝
" 8	139½@139⅞
" 15	139⅝@139⅞
" 22	139⅞@140
" 29	139¼@139¾
June 5	139¾@140
" 12	139⅞@140⅛
" 19	140½@140⅞
" 26	140 @140¼
July 3	140¼@140½
" 10	140⅝@140⅞
" 17	142⅝@143¼
" 24	143¼@143½
" 31	144⅝@144¼
August 7	147¾@148½
" 14	146⅞@148
" 21	143⅞@144¾
" 28	144¾@145⅛
September 4	143¾@144⅛
" 11	143⅞@144⅛
" 18	144⅜@143¾
" 25	141⅞@142½
October 2	139¼@140½
" 9	138⅞@139¾

The Price of Gold—Continued.

Date	Price
October 16	137¼@137¾
" 23	135 @136
" 30	134 @134½
November 6	132 @132¼
" 13	133⅜@134¾
" 20	134¼@134⅞
" 27	135 @135⅞
December 4	135⅛@135¾
" 11	135⅝@136¼
" 18	134⅝@135¼
" 24	134⅝@135⅛
" 31	134⅝@135

1869.

Date	Price
January 8	134⅞@135⅛
" 15	136¼@136½
" 22	135½@135⅞
" 29	136⅞@136⅝
February 5	135 @135¼
" 12	135⅛@135⅜
" 19	133⅜@134¾
" 26	131⅞@132¼
March 5	131 @131⅝
" 12	131 @133⅜
" 19	130⅜@131¼
" 25	131 @131¼
April 2	131½@132
" 9	132⅛@133⅛
" 16	132⅝@133⅜
" 23	133⅜@133½
" 30	134⅛@134¼
May 7	136⅝@137¾
" 14	138½@138¾
" 21	141¾@144⅛
" 28	139¼@139¾
June 4	137¾@138¼
" 11	138⅞@139⅝
" 18	136¾@137⅞
" 25	137 @137¾
July 2	136½@137⅜
" 9	135¾@136¼
" 16	135¾@136¾
" 23	135¼@135¾
" 30	136¼@136⅝
August 6	136⅛@136⅜
" 13	134⅛@134¾
" 20	132⅝@133
" 27	132⅜@134¼
September 3	135⅜@136
" 10	135 @135¼
" 17	136⅜@136¾
" 24	133 @162½
October 1	130 @130½
" 8	130⅝@131⅝
" 15	130 @130⅛
" 22	130⅞@131¾
" 29	128⅝@129¼
November 5	126⅜@127⅛
" 12	126⅝@126⅞
" 19	126½@126¾
" 26	124½@124⅞
December 3	122⅛@122½
" 10	122⅞@123¼
" 17	120½@120⅞
" 24	120⅝@120¾
" 31	119⅝@120⅜

1870.

Date	Price
January 7	121⅛@122⅛
" 14	121½@121⅞
" 21	120½@121
" 28	121½@121¾
February 4	120⅜@120¾
" 11	119⅞@120⅛
" 19	119 @119⅜
" 26	116⅛@117⅛
March 4	112⅜@114
" 11	112½@113⅞
" 18	111⅞@112¼
" 25	111¾@112⅜
April 1	111½@111¾
" 8	112⅜@112¾
" 15	112⅝@113⅛
" 22	112¾@113
" 29	114⅜@115⅜
May 6	114⅝@114⅞
" 13	114¾@115⅛
" 20	114½@114¾
" 27	114¾@115⅛
June 3	114⅜@114½
" 10	113⅜@113⅝
" 17	112¼@113⅛
" 24	111¼@111¾
July 1	111⅛@112⅜
" 8	111¾@112¼
" 15	114 @115½
" 22	118½@119⅞
" 29	120⅝@121¾
August 5	121¼@121½
" 12	117⅛@118
" 19	116 @116¼
" 26	116¼@116⅞
September 2	116⅜@116⅝
" 9	113¾@114⅛
" 16	114 @114⅛
" 23	113 @113¾
" 29	113⅝@114
October 7	113 @113⅛
" 14	113⅜@113¾
" 21	112½@112⅞
" 28	111⅜@111⅝
November 5	110 @110½
" 12	110⅞@111⅛
" 19	112½@113¾
" 26	111¾@112⅛
December 2	110⅝@111⅛
" 9	110½@110¾
" 16	110⅝@110⅞
" 23	110½@110⅝
" 30	110⅝@110⅞

1871.

Closing prices:

Date	Price
January 6	110¾
" 13	110¾
January 20	110⅝
" 27	110⅞
February 3	111⅜
" 10	111¾
" 17	111⅜
" 24	111⅜
March 3	111
" 10	111¼
" 17	111¼
" 24	110⅞
" 31	110⅜
April 6	110⅜
" 14	110¾
" 21	111⅛
" 28	111⅛
May 5	111⅛
" 12	111⅜
" 19	112
" 26	111⅝
June 2	112¼
" 9	112⅛
" 16	112⅜
" 23	112½
" 30	113
July 7	113⅛
" 14	112⅛
" 21	112¼
" 28	112
August 4	112¼
" 11	112¼
" 18	112⅞
" 25	112½
September 1	112⅞
" 8	113⅛
" 15	114⅛
" 22	114⅝
" 29	114¾
October 6	114⅝
" 13	114¼
" 20	112¼
" 27	111⅛
November 3	111¾
" 10	111⅛
" 17	111⅛
" 24	110⅞
December 1	110⅜
" 8	110⅛
" 15	109¼
" 22	108⅜

1872.

Date	Price
January 5	109⅛
" 12	108⅞
" 19	109
" 26	109¼
February 2	109¾
" 9	110⅝
" 16	110¼
" 23	110⅛
March 1	110¼
" 8	110¼
" 15	110¼
" 22	109⅞

The Price of Gold—Concluded.

March 28	110¼	May 31	114⅛	July 26	114¾
April 5	110¼	June 7	114	August 2	115⅜
" 12	110⅝	" 14	113⅞	" 9	115½
" 19	111¼	" 21	113⅝	" 16	115¼
" 26	112⅞	" 28	113½	" 23	113⅝
May 3	112⅞	July 5	113¾	" 30	112⅞
" 10	114⅜	" 12	114	September 6	112⅞
" 17	113⅞	" 19	114¼	" 13	112⅞
" 24	113¾				

RATE OF DISCOUNT

for unexceptionable bills, at the Bank of England, for a series of years.

Month.	Rate.	Month.	Rate.	Month.	Rate.	Month.	Rate.
1847.		**1850.**		**1853.**		**1856.**	
January	4	January	2½	January	3	Jauuary	6
February	4	February	2½	February	3	February	6
March	4	March	2½	March	3	March	6
April	5	April	2½	April	3	April	6
May	5	May	2½	May	3	May	5
June	5	June	2½	June	3½	June	4½
July	5	July	2½	July	3½	July	4½
August	6½	August	2½	August	3½	August	4½
Sebtember	5½	September	2½	September	5	September	4½
October	8	October	2½	October	5	October	6
November	7	November	2½	November	5	November	6½
December	5	December	3	December	5	December	6½
1848.		**1851.**		**1854.**		**1857.**	
January	4	January	3	January	5	January	6
February	4	February	3	February	5	February	6
March	4	March	3	March	5	March	6
April	4	April	3	April	5	April	6½
May	4	May	3	May	5½	May	6½
June	3½	June	3	June	5½	June	6½
July	3½	July	3	July	5½	July	6½
August	3½	August	3	August	5	August	5½
September	3½	September	3	September	5	September	5½
October	3½	October	3	October	5	October	8
November	3	November	3	November	5	November	10
December	3	December	3	December	5	December	8
1849.		**1852.**		**1855.**		**1858.**	
January	3	January	2½	January	5	January	4
February	3	February	2½	February	5	February	3
March	3	March	2½	March	5	March	3
April	3	April	2	April	4½	April	3
May	3	May	2	May	4	May	3
June	3	June	2	June	3½	June	3
July	3	July	2	July	3½	July	3
August	3	August	2	August	3½	August	3
September	2½	September	2	September	5	September	3
October	2½	October	2	October	6	October	3
November	2½	November	2	November	6	November	3
December	2½	December	2½	December	6	December	2½

Rate of Discount—Concluded.

Month.	Rate.
1859.	
January	2½
February	2½
March	2½
April	3½
May	4½
June	3
July	2½
August	2½
September	2½
October	2½
November	2½
December	2½
1860.	
January	3
February	4
March	4
April	5
May	4½
June	4
July	4
August	4
September	4
October	4
November	5
December	5@6
1861.	
January	7
February	8
March	7
April	5
May	6
June	6
July	6
August	4
September	3
October	3½
November	3
December	3
1862.	
January	2½
February	2½
March	2½
April	2½
May	3
June	2½
July	2
August	2
September	2
October	3
November	3
December	3
1863.	
January	5
February	4
March	4
April	3
May	3½
June	4
July	4
August	4
September	4
October	4
November	6
December	7
1864.	
January	8
February	6
March	6
April	9
May	7
June	6
July	7
August	8
September	9
October	9
November	7
December	6
1865.	
January	5
February	5
March	4
April	4
May	4
June	3
July	3½
August	4
September	4½
October	7
November	6
December	7
1866.	
January	8
February	7
March	6
April	6
May	10
Jun	10
July	10
August	6
September	4½
October	4½
November	4
December	3½
1867.	
January	3½
February	3
March	3
April	3
May	2½
June	2½
July	2
August	2
September	2
October	2
November	2
December	2
1868.	
January	2
February	2
March	2
April	2
May	2
June	2
July	2
August	2
September	2
October	2
November	2½
December	3
1869.	
January	3
February	3
March	3
April	4
May	4½
June	3½
July	3
August	2½
September	2½
October	2½
November	3
December	3
1870.	
January	3
February	3
March	3
April	3
May	3
June	3
July	5
August	4
September	2½
October	2½
November	2½
December	2½
1871.	
January	2½
February	2½
March	3
April	2½
May	2½
June	2½
July	2
August	2
September	4
October	5
November	4
December	3
1872.	
January	3
February	3
March	3
April	4
May	4
June	3
July	3½
August	3½

VALUE OF EXPORTS

(in pounds sterling) of cotton cloth and cotton yarn from Great Britain for a series of calendar years.

	Cotton Cloth.	Cotton Yarn.	Total.
1856	£29,632,713	£8,652,056	£38,284,769
1857	29,912,726	9,200,183	39,112,909
1858	32,876,683	10,098,901	42,975,584
1859	38,079,498	10,128,946	48,208,444
1860	41,397,533	10,615,949	51,013,482
1861	36,969,204	9,867,545	46,836,749
1862	29,931,871	6,840,757	36,772,628
1863	38,682,428	8,761,536	47,443,964
1864	44,973,583	9,883,906	54,867,489
1865	46,150,358	11,104,487	57,254,845
1866	59,796,025	14,769,401	74,565,426
1867	54,858,272	15,985,420	70,843,692
1868	51,718,974	15,822,317	67,541,291
1869	51,857,525	15,307,439	67,158,964
1870	55,521,915	15,888,216	71,410,131

VALUE OF EXPORTS

of raw cotton and manufactures of cotton from the United States for fifteen fiscal years, ending June 30 (in gold).

	Raw Cotton.	Manuf. of Cotton.	Total Value.
1857	$131,575,859	$6,115,177	$137,691,036
1858	131,386,661	5,651,504	137,038,165
1859	161,434,923	8,316,222	169,751,145
1860	191,806,555	10,934,796	202,741,351
1861	34,051,483	7,957,038	42,008,521
1862	1,156,973	2,888,690	4,045,663
1863	4,855,770	2,121,468	6,977,238
1864	6,343,496	933,911	7,277,407
1865	3,384,356	1,708,693	5,093,049
1866	199,563,988	1,262,535	200,826,523
1867	142,886,828	3,268,252	146,155,080
1868	109,143,383	3,479,321	112,622,704
1869	122,280,490	4,416,708	126,697,198
1870	194,595,106	3,246,242	197,841,348
1871	191,036,220	3,113,349	194,149,569

NOTE.—The above are from official figures, and do not, of course, include the cotton which escaped the blockade of the Southern ports, between 1861 and 1865, inclusive.

CONTRACTS FOR FUTURE DELIVERY.

This business commenced during our civil war. Some of our manufacturers entered into large contracts with the Government, and covered themselves by purchasing the raw material for delivery at some future time. The price, within certain wide limits, was then of much less importance than the certainty of getting the cotton when needed.

The great convenience this system offered to speculators drew them into it, and the business grew steadily during the war, and still more after its termination. It soon became apparent that the business could not be conducted without special rules and authority to enforce them. To this end, in the summer of 1870, an association of merchants and brokers was completed, and a convenient room leased to serve the purpose of an exchange. Afterward a charter was obtained from the Legislature of the State, and, the success of the association being fully assured, a building was bought, altered and enlarged at a total cost of about $160,000.

I append the charter, by-laws and rules of the association, as they existed early this year. Some important changes in the rules are now in contemplation, and many more may be made from time to time. The business was entirely new, and almost every rule was an experiment. It is hoped that in time they will be made as perfect as man's work can be expected to be

CHARTER.

CHAPTER 365.

AN ACT

To Incorporate the New York Cotton Exchange,

PASSED APRIL 8, 1871.

The People of the State of New York, represented in Senate and Assembly, do enact as follows:

SECTION 1. The members of the Association known as the "New York Cotton Exchange," and all other persons who may hereafter become asssociated with them under the provisions of this act, are hereby created a body corporate, by the name of the "New York Cotton Exchange," with perpetual succession and power to use a common seal, and alter the same at pleasure, to sue and be sued, to take and hold, by grant, purchase and devise, subject to the provisions of law relating to devises and bequests by last will and testament, real and personal property to an amount not exceeding three hundred thousand dollars, for the purpose of such Association, and to sell, convey, lease and mortgage the same or any part thereof.

SEC. 2. The property, affairs, business and concerns of the corporation hereby created shall be managed by a President, Vice-President, Treasurer and fifteen managers, who, together, shall constitute a Board of Managers, to be elected annually, at such time and place as may be provided by the by-laws ; and the present officers and managers of the said Association, as now constituted, shall be the officers and managers of the said corporation until their present term of office shall expire, and until others, under the provisions of this act, shall be elected in their place. All vacancies which may occur in said board by death, resignation or otherwise, shall be filled by the said board. A majority of the members of such board shall constitute a quorum for the transaction of business.

SEC. 3. The purposes of said corporation shall be to provide, regulate and maintain a suitable building, room or rooms, for a Cotton Exchange, in the city of New York, to adjust controversies between its members, to establish just and equitable principles in the trade, to maintain uniformity in its rules, regulations and usages, to adopt standards of classification, to acquire, preserve and disseminate useful information connected with the cotton interest throughout all markets, to decrease the local risks attendant upon the business, and generally to promote the cotton trade of the city of New York, increase its amount and augment the facilities with which it may be conducted. The said corporation shall have power to make all proper and needful by-laws, not contrary to the constitution and laws of the State of New York or of the United States.

SEC. 4. The said corporation shall have power to admit new members and expel any member, in such manner as may be provided by the by-laws.

SEC. 5. The board of managers shall annually elect, by ballot, five members of the Association, who shall not be members of the board, as a committee, to be known and styled the Adjudication Committee of the New York Cotton Exchange. The Board of Managers may, at any time, fill any vacancy or vacancies that may occur in said committee for the remainder of the term in which the same shall happen. It shall be the duty of said Adjudication Committee to hear and decide any controversy which may arise between the members of the said Association, or any person claiming by, through or under them, and as may be voluntarily submitted to said committee for arbitration ; and such members and persons may, by an instrument, in writing, signed by them and attested by a subscribing witness, agree to submit to the decision of such committee any such controversy which might be the subject of an action at law or in equity, except claims of title to real estate or to any interest therein, and that a judgment of the Supreme Court shall be rendered upon the award made pursuant to such submission.

SEC. 6. Such Adjudication Committee, or a majority of them, shall have power to appoint a time and place of hearing of any such controversy, and adjourn the same, from time to time, as may be necessary, not beyond the day fixed in the submission for rendering their award, except by consent of parties, to issue subpœnas for the attendance of witnesses residing or being in the metropolitan police district. Witnesses so subpœnaed, as aforesaid, shall be entitled to the fees prescribed by law for witnesses in the courts of justices of the peace.

SEC. 7. Any number, not less than a majority of all the members of the Adjudication Committee, shall be competent to meet together and hear the proofs and allegations of the parties, and an award by a majority of those who shall have been present at the hearing of the proofs and allegations, shall be deemed the award of the Adjudication Committee, and shall be valid and binding on the parties thereto. Such award shall be made in writing, subscribed by the members of the committee concurring therein, and attested by a subscribing witness. Upon filing the submission and award in the office of the clerk of the Supreme Court of the city and county of New York, both duly acknowledged or proved in the same manner as deeds are required to be acknowledged or proved in order to be recorded, a judgment may be entered therein according to the award, and shall be docketed, transcripts filed, and executions issued thereon, the same as authorized by law in regard to judgments in the Supreme Court. Judgments entered in conformity with such award shall not be subject to be removed, reversed, modified, or in any manner appealed from by the parties thereto, except for fraud, collusions, or corruption of said Adjudication Committee, or some member thereof.

SEC. 8. This act shall take effect immediately.

STATE OF NEW YORK,
OFFICE OF THE SECRETARY OF STATE. } ss.

I have compared the preceding with the original law on file in this office, and do hereby certify that the same is a correct transcript therefrom, and of the whole of said original law.

[L. S.] Given under my hand and seal of office, at the city of Albany, this twelfth day of April, in the year one thousand eight hundred and seventy-one.

DIEDRICH WILLERS, JR.,
Dep. Secretary of State.

NEW YORK COTTON EXCHANGE
BY-LAWS.

ARTICLE I.

TITLE.

The title of this Association shall be the "NEW YORK COTTON EXCHANGE."

ARTICLE II.

MEMBERS.

All persons, who, as principals, are permanently engaged and of good standing in the Cotton Trade of the City of New York, and also, any person who has the exclusive control and management of the Cotton business of any house in good standing in the Cotton Trade of the City of New York, may be elected members of this Association.

ARTICLE III.

APPLICATION FOR MEMBERSHIP AND RIGHTS AND DUTIES OF MEMBERS.

SECTION 1. All applications for membership must first be made to the Committee on Membership, and if approved by that committee, must then be referred to the Board of Managers, and if recommended by the board by a two-third (2-3d) vote, shall, after having been posted on the bulletin of the Exchange for at least five (5) days (notice to state the time of balloting), be voted on by the members of the Exchange. The balloting shall be at the general Exchange Rooms, and on Mondays only. The Polls shall be opened at 12 M., and closed at 2 P. M. Each elector shall cast one ballot; if in favor of the candidate, the word "yes," if against, the word "no," written or printed thereon. Three-fourths (3-4ths) of the whole number of votes cast shall be required in favor of the applicant to entitle him to membership. The Secretary of the Exchange or his Assistant shall act as teller, and register the names of voters as polled. Two members, either of the Board of Managers or of one of the sub-committees, shall be present at the counting of the votes cast, and they shall promptly post the result on the bulletin of the Exchange over their own signatures. No name after being rejected shall be again proposed within six months after such rejection. Each member shall, within ten days after receiving notice of his election, subscribe to the Charter and By-Laws, and pay to the Treasurer the initiation fee and the annual dues.

Sec. 2. The Initiation fee for membership of the Exchange shall be five hundred dollars ($500), for which a certificate shall be issued, which may be transferred by any member to any other member or member elect, upon payment to the Treasurer of the Exchange of the sum of one hundred dollars ($100), or the Exchange will, unless otherwise ordered, accept the transfer of the certificate, paying therefor such sum as the Board of Managers may from time to time, in their discretion, direct. The legal representatives of any deceased member, may transfer such membership, as herein provided.

Sec. 3. Any member failing to pay his dues and assessments within ten days after the same shall become due, shall be notified by the Treasurer, in writing, of such failure, and be deprived of the privileges of the Exchange until the same are paid. One dollar per day shall be added to the amount of said dues and assessments for each day not exceeding thirty they remain unpaid, after such notice, and at the expiration of one month from the notice, if they still remain unpaid, the Board of Managers may order the rights of membership of such member to be sold, and after deducting the amount in arrears, together with all penalties and charges accrued thereon, place the balance of the proceeds to the credit of such member, payable to him or his legal representative, on demand, without interest.

Sec. 4. When a sale of the rights of membership of any member is ordered, it shall be made by the Secretary of the Exchange, to the highest bidder, at open outcry, at the Exchange Room, after 10 days notice posted on the bulletin in said room. Any member may purchase the said rights of membership of any other member, which shall give him the right to sell the same to any other member or member elect.

Sec. 5. The Secretary of the Exchange may bid for and purchase for account of the Exchange, any right of membership sold by him, at a price not exceeding limits which shall have been fixed by the Board of Managers, and when so purchased, the member sold out, shall, if otherwise in good standing, have the right of redemption, within six months from date of sale, by the payment to the Exchange, of all arrearages, penalties and charges, and 10 per cent. on amount of sale.

Sec. 6. The payment of annual dues must be made by every member admitted to the Exchange, whether by the payment of the regular initiation fee, or by transfer of rights of membership, and such payment shall only cover the current official year of the Exchange, or unexpired portion of same. Any member purchasing rights of membership, shall pay the same annual dues for each right so held by him, as if each right was held by a different member, except that such purchased right shall be exempt for the current year in which it is purchased, while held by a member who has paid the annual dues for the current year.

SEC. 7. Every transfer of rights of membership, whether direct, by one member to another, or under sale by the Secretary, shall be recorded by the Secretary of the Exchange, in a book to be kept by him, open to examination by any member.

Every member upon signing the Charter and By-laws pledges himself to abide by the same, and also by all By-Laws, Rules and Regulations, which may hereafter be adopted.

ARTICLE IV.

ELECTION OF OFFICERS.

SECTION 1. There shall be an annual election for President, Vice-President, Treasurer, and fifteen (15) Managers of this Association, to be held at the Exchange, on the first Monday in June; at which time there shall also be elected three Inspectors of the next ensuing election. The polls shall be opened at 10 A. M., and closed at 2 P. M.

SEC. 2. Each member not suspended shall be entitled to one vote, in person, by ballot, and a plurality of votes cast shall elect.

SEC. 3. The Board of Managers shall enter upon the duties of their office on the first Thursday succeeding their election, and shall continue in office until the first Thursday following the election of their successors.

At their first meeting the Board shall elect a Secretary from their own number.

ARTICLE V.

MEETINGS.

SECTION 1. An annual meeting of the Exchange shall be held at the general meeting room on the Tuesday preceding the election of Officers and Managers, at which meeting shall be presented by the Board of Managers a general statement of the affairs and finances of the Exchange, together with an estimate of the expenses for the next ensuing year, and the amount of annual dues which the Board recommend to be collected from each member for the next said ensuing year, which amount may be increased or diminished, and shall be fixed by a vote of the Exchange at said meeting; and the sum so fixed shall be due on the Tuesday succeeding the election.

The proceedings of said annual meeting shall be recorded in the book of minutes of the Board of Managers.

SEC. 2. The regular meetings of the Board of Managers shall be held on the first Monday of each month, and the President may call special meetings when deemed necessary, and shall do so on the written request of three members of the Board.

SEC. 3. The President may call a meeting of the Exchange whenever he shall deem it necessary, and shall do so when ordered by the Board of Managers, or, upon the written request of twenty-five members of the Exchange (which request must state the object of the call), and, when present, shall preside at all meetings of the Exchange and of the Board.

In the absence of the President, the Vice-President shall perform the duties of the President, and in the absence of both, a chairman *pro tem.* shall be appointed, upon whom shall devolve all the duties of the President.

SEC. 4. No member shall speak more than twice on any question under discussion at any meeting of the Exchange, unless by the consent of a majority of the members present.

SEC. 5. No notice shall be taken at any meeting of the Exchange of any resolution unless submitted in writing. Not less than forty members shall be a quorum of the Exchange.

ARTICLE VI.

BOARD OF MANAGERS.

The Board of Managers shall have general and entire management of the property and business of the Association not inconsistent with the Charter and By-Laws; shall adopt such rules and regulations as they may deem best to carry out the purposes of the Association; have power to hire or lease, and fit up such rooms as may be required for the purposes of the Exchange, and to appoint such subordinate officers or employés as they may deem necessary.

They shall appoint the following Committees at the first meeting of the Board after their election; and when by them deemed expedient, they may, by a majority vote of members present at any meeting, change in whole or in part, the composition of any of the Committees (Adjudication and Board of Appeals alone excepted), viz.:—

1. Executive Committee.
2. Finance Committee.
3. Supervisory Committee.
4. Adjudication Committee.
5. Board of Appeals.
6. Committee on Classification.
7. Committee on Membership.
8. Committee on Information and Statistics.
9. Committee on Trade.
10. Quotation Committee.
11. Arbitration Committee.

And from time to time, such other Committees as they may deem advisable.

ARTICLE VII.

DUTIES OF COMMITTEES.

EXECUTIVE COMMITTEE.

SECTION 1. The Executive Committee shall be appointed from the Board of Managers, and shall, unless otherwise ordered, have the general supervision of the property, business, and affairs of the Exchange, and make such reports and recommendations as in their judgment will best promote its interests ; shall have direction and supervision of all appointed officers and employés in the discharge of their respective duties, and shall see that all rooms provided are suitably furnished and kept in good order for the accommodation of the Exchange, the Board of Managers and Committees.

FINANCE COMMITTEE.

SEC. 2. The Finance Committee shall be appointed from the Board of Managers, and shall audit all bills, exercise a general supervision over the financial affairs of the Exchange, and audit the account of the Treasurer and Secretary of the Exchange monthly ; and shall also audit the annual accounts of Treasurer.

SUPERVISORY COMMITTEE.

SEC. 3. The Supervisory Committee shall consist of three members, who shall be appointed from the Board of Managers. They shall, unless otherwise ordered, have general supervision of members under Articles XI and XII of the By-Laws, and shall enforce said Articles in all cases.

ADJUDICATION COMMITTEE.

SEC. 4. As soon as practicable after its election, the Adjudication Committee shall proceed to organize, and appoint a Clerk, not of their own body, who shall act as Clerk of the Committee. Before entering upon the duties of their office, the members shall be sworn faithfully and fairly to hear and examine the matters in controversy, which may come before them during their tenure of office, and to make a just award according to the best of their understanding. The Committee shall have power to adjourn the hearing from time to time as circumstances may require.

All persons who may desire the services of said Committee, shall file with the Clerk an agreement in writing, to submit their case to the Committee, and to be bound by its decision, which agreement shall be signed by the parties thereto, and attested by a subscribing witness.

On such agreement being filed, the Clerk shall call a meeting of the Committee, to be held as soon thereafter as may be convenient to the parties concerned, to hear and decide such controversy ; and the Committee shall render its award in conformity with Section 5, 6, and 7 of the Charter.

The Committee shall be entitled to the sum of twenty-five dollars and the Clerk five dollars for each sitting ; said sums to be paid by the party against whom the decision may be made, except in such cases as the Committee in their discretion shall otherwise decide.

The proceedings of said Committee shall be recorded in a book to be kept for that purpose, in which shall be entered a summary of each controversy had before them, the award made thereon, and at the discretion of the Committee, the grounds for such award. Said book shall be the property of the Exchange, and subject to the inspection of its members.

BOARD OF APPEALS.

SEC. 5. The Board of Appeals shall consist of seven members, who shall hear and decide finally all controversies (except as provided in Article VII, Section 7 of these By-Laws), upon which an award has been previously made by an Arbitration Committee, and from which an appeal has been taken by one of the parties, as provided in Article XI of these By-Laws.

Any number not less than a majority of the Board, shall be competent to meet together and hear the proofs and allegations of the parties, and an award by a majority of those present and serving, shall be deemed to be the award of the Board, and shall be binding on the parties thereto.

Such award shall be in writing, subscribed to by the members concurring therein, and within twenty-four hours from the time of the decision shall be given to the Secretary of the Exchange, who shall record the same in the book of decisions of the Board of Appeals, and serve copies on both parties to the controversy.

There shall be paid to the said Board, as compensation for their services in each case, the sum of thirty-five dollars, to be equally divided between the members present and serving, and the Board shall state in their award which of the parties shall pay the same.

COMMITTEE ON CLASSIFICATION.

SEC. 6. The Committee on Classification shall, unless otherwise ordered, have charge of all questions of Classification, grade, quality, and condition of cotton; they shall hear and decide finally all controversies relating to the classification and value of cotton (in connection with the classification), upon which a decision has previously been made by an Arbitration Committee, and from whose decision an appeal is taken by one of the parties in accordance with Article XI of these By-Laws. They shall also inquire and report as to the standards of classification of all principal markets, and provide and keep on exhibition samples of the standards of such markets. They shall also procure information as to the practicability of making the standard of classification of American cotton uniform in all principal markets, and report, from time to time, such recommendations on the subject as they may deem for the interest of the Exchange. They shall also examine the references of such samplers and weighers of cotton as may apply for a license, and report to the Board of Managers the names of such applicants as they recommend shall be licensed.

COMMITTEE ON MEMBERSHIP.

SEC. 7. The Committee on Membership shall, unless otherwise ordered, have charge of all applications for membership, and for powers of Attorney ; they shall decide finally upon all such applications within one week from the date of their reception, and within twenty-four hours after any decision, they shall, in every case, report such decision to the Secretary of the Board of Managers, whether approving or disapproving.

COMMITTEE ON INFORMATION AND STATISTICS.

SEC. 8. The Committee on Information and Statistics shall, unless otherwise ordered, have charge of all matters pertaining to the supply of newspapers, market reports, telegraphic dispatches and statistical information for the use of the Exchange ; and it shall be the duty of said Committee to organize plans for obtaining early, reliable, and regular information affecting the price of cotton from all cotton producing, and all cotton consuming sections.

COMMITTEE ON TRADE.

SEC. 9. The Committee on Trade shall, unless otherwise ordered, consider and report such rules and regulations for the purchase, sale, transfer, transportation, and custody of cotton, as will best promote the interests of all parties interested in the cotton trade of the city of New York.

QUOTATION COMMITTEE.

SEC. 10. The Quotation Committee shall, unless otherwise ordered, meet twice daily, at the Exchange when the same is opened for business (except Saturdays, from and including the second (2d) Saturday of June, to and including the third (3d) Saturday of September, meeting but once upon those days), to confer upon, and, by a majority of the members present, establish the market quotations for the time being of the recognized grades of cotton, which quotations they shall immediately post upon the bulletin board of the Exchange.

ARBITRATION COMMITTEE.

SEC. 11. The Arbitration Committee shall consist of five (5) members, from which the two arbitrators, chosen by disputing parties, shall, in all cases, choose the third, as provided in Article XI. It shall be the duty of this Committee to study the Rules and Laws of the Association, and to keep themselves informed of all decisions made by Arbitration Committees, and to note particularly by an examination of "The Reports of Committees of Arbitration" upon file at the Exchange, whether the same case, in whole or in part, has or has not been adjudicated within thirty (30) days.

THE SECRETARY OF THE BOARD OF MANAGERS.

SEC. 12. The Secretary of the Board of Managers shall keep accurate minutes of the meetings both of the Exchange and of the Board ; shall cause notices to be posted on the bulletins of all the meetings of the Exchange, and give notice in writing of all the meetings of the Board.

THE TREASURER.

Sec. 13. The Treasurer shall receive all funds belonging or payable to the Exchange, and deposit or invest the same as Treasurer in such manner as the Finance Committee shall direct. He shall pay all bills against the Exchange when certified by the Finance Committee, or when authorized by the Board of Managers. He shall have custody of the corporate seal, shall keep an account of all receipts and disbursements in a book to be kept for that purpose, subject at all times to the examination of the Finance Committee and the Board of Managers; shall render a report at each regular meeting of the Board, and a general annual report at the close of each year, and at the expiration of his term of office shall transfer to his successor all funds, books, papers and other property of the Exchange that may be in his possession

DUTY OF SECRETARY OF THE EXCHANGE.

Sec. 14. The Secretary of the Exchange shall keep the accounts, except those of the Treasurer; act as the General Superintendent, and perform such other services as the Board of Managers may require.

LICENSES.

Sec. 15. Licenses may be granted by the Board of Managers to Cotton Samplers and Weighers on the recommendation of the Committee on Classification. The Board may also at their discretion, suspend or cancel such licenses.

ARTICLE VIII.

POWER OF ATTORNEY.

Any member may be represented by one Attorney (except where two or more members of one commercial firm are members of the Exchange, there shall be but one Attorney for such members), who shall be his *bona fide* recognized partner, or clerk, on the following conditions, to wit:—On the annual payment into the treasury of the Exchange, of the same dues that are assessed against the members, and filing a written agreement with the Secretary binding the member for all transactions of his Attorney. Said Attorney (who must be approved both by the Committee on Membership and the Board of Managers), shall be subject to all the rules of order which apply to members, and may be required to withdraw from the privileges of the Exchange, for cause, at the written request of five members, addressed to and approved by the presiding officer. Attorneys are not allowed to vote or to attend executive sessions of the Exchange.

ARTICLE IX.

REPORTING SALES.

Any member who shall be convicted of intentionally reporting false sales shall be expelled.

All transactions in contracts for future delivery and free on board must be reported promptly to the Secretary of the Exchange, giving exact time and terms, and name of party making the report, only.

Transactions between parties, neither of whom are brokers, must be reported by the seller ; if between parties only one of whom is a broker, by the broker ; if between two brokers, by the selling broker.

The party whose duty it is to report any transaction, shall furnish a "stamped contract" for same, and shall be responsible for any expense incurred by his neglect therein.

All exchanges giving one time of delivery for another, and all transactions covering more times of delivery than one, and made at an average price, shall be reported strictly as they occurred, but shall not be used by the Secretary in making up the daily averages. It shall be the duty of the Secretary to keep the reports on file, and without charge, to stamp "Reported" on all contracts which have been reported within twenty minutes of the time when the transaction was made.

No transaction not duly reported shall have any right under the By-Laws and Rules of the Exchange ; provided, that any buyer holding a contract which has not been reported, may have it brought under the By-Laws and Rules of the Exchange on application to the Secretary within ten days from date of the transaction, and on payment of five (5) cents per bale to the Exchange, the receipt of which shall be stamped on the contract, and the sum thus paid shall be repaid to him by the seller.

ARTICLE X.

DUTIES OF MEMBERS FAILING TO MEET THEIR OBLIGATIONS

Any member finding himself unable to meet his liabilities or obligations at maturity, shall immediately notify the members of the Exchange by letter addressed to the Secretary, who shall promptly post the same on the bulletin of the Exchange, where it shall remain five days ; such notice shall be deemed sufficient for all the members of the Exchange, and shall render it obligatory upon them to close all contracts with such failing member, by settlement, at the average price for that day for like deliveries; provided, notice is given before 3 o'clock P. M., otherwise, on average quotation of next day; and no transfer of contracts, or settlement, otherwise than is herein provided, shall be accepted by any member of the Exchange after such notice of failure.

Provided, That in any case where such notice is given, any member having a claim or claims against such failing member, may demand an investigation of the affairs of such member, by the Supervisory Committee, and if the Committee shall report that, in their opinion, such member is able to pay all his obligations, that he shall be debarred from the privilege of settlement under this By-Law.

All privileges to call, or to deliver Cotton in favor of or against the failing member, shall be settled upon such basis as shall be fixed by a Committee of Arbitration.

Any member failing to meet his obligations at maturity shall thereby forfeit all the privileges of membership, except the right of Arbitration on all claims resulting from business transactions made prior to failure, and the right to sell and transfer his membership to a member or member elect:

Provided, that if he shall have duly notified the members of the Exchange in accordance with the By-Laws, and shall have made honorable settlement with his creditors, or has offered to pay them *pro rata* to the full extent of his ability, he may, within one year from the date of his failure, on application to the Supervisory Committee, who shall report the same within one week to the Board of Managers, be fully reinstated to the rights and privileges of membership by a two-third vote of the Board at any meeting subsequent to that at which the Supervisory Committee may report upon the same.

If any member fails to meet his obligations, and does not give notice to the members of the Exchange in accordance with the By-Laws, he shall not be eligible to reinstatement, nor to be elected, nor to act as Attorney for a member. And it shall be the duty of the Supervisory Committee, upon complaint and evidence that a member has so failed to meet his obligations and to give notice as herein provided, to notify the members of the Exchange, by notice posted on the bulletin not less than five days.

ARTICLE XI.

CLAIMS OF MEMBERS.

All claims of one member against another (unless arising from transactions negotiated through brokers who are not members of this Exchange), shall be subject to arbitration by a Committee of three members, one to be chosen by each disputant, those two selecting a third (from the Arbitration Committee. See Article VII, Section 11). The award of a majority of the arbitrators to be binding, subject only to the right of appeal to the Board of Appeals (except such cases as are provided for in Article VII, Section 4). In the event, however, of one of the disputing parties appointing an arbitrator, and the other refusing or neglecting to do so for three days after notice in writing of the appointment, or in case the arbitrators appointed shall not within seven days after their appointment, make report of their award, then the President, or officer presiding for the day, shall, at the request in writing of either party, appoint an arbitrator or arbitrators to act in the case. Each arbitrator acting on such committee shall be paid five dollars ($5) for each case, and the arbitrators in their award shall decide by whom the expenses of the arbitration shall be paid.

All claims or complaints of one member against another for cause, must be preferred, in writing, within thirty (30) days from date of discovery of cause of action, and if arbitration is desired, notice of same, in writing, must be served upon the opposite party within said period of thirty (30) days, otherwise they shall be null and void.

ARTICLE XII.

COMPLAINTS OF MEMBERS.

SECTION 1. If complaint of improper conduct is made against a member of the Exchange, it must be in writing, and addressed to the Supervisory Committee, specifying the particular act complained of, together with all the documentary evidence bearing on the case that the complainant can furnish, together with a list of the witnesses he desires may be examined on said complaint. The chairman of the Committee shall cause copies of the complaint and evidence to be served on the member complained of, and it shall be the duty of the said member within five days of the receipt of the same, to return said documents to the Committee with his written answer to the complaint, and all documents bearing on the case, in his possession, with the names of witnesses he desires shall be summoned. The Chairman shall, as soon as expedient after the receipt of said answer, or if no answer is received after the expiration of five days from service of the copies of complaint, notify both parties and witnesses named of the time and place at which the Committee will meet to act on said complaint. If, after examination, the Committee deem the charges substantiated, they shall so report to the Board of Managers, and deliver to the said Board all documents relating thereto.

SEC. 2. The Board of Managers may, at their discretion, act upon the documentary evidence alone, or may call the parties or such witnesses as they may desire to examine in the case before them ; but one party shall not be called before the Board of Managers, without giving the other party an opportunity of being present at the same time.

SEC. 3. Such accused member or members may, upon a substantiation of the charges, be suspended or expelled by a vote of two-thirds (2-3ds) of the Board of Managers present, and voting at the meeting, when the same shall be acted upon.

SEC. 4. Any suspended member may be reinstated by a three-fourth (3-4th) vote of the Board of Managers present, and voting at the meeting, when the same shall be acted upon.

ARTICLE XIII.

SAMPLES OF COTTON.

No samples of cotton drawn in this City shall be arbitrated upon under the By-Laws of this Exchange, unless drawn by samplers licensed by the Board of Managers as already provided for. It shall be the duty of members to employ licensed samplers only. The party employing a sampler in violation of this article shall become responsible for any loss sustained thereby.

ARTICLE XIV.

WEIGHER'S RETURNS.

No weigher's return of cotton weighed in this City shall be arbitrated upon under the By-Laws of this Exchange, unless the same is signed by a weigh-master licensed by the Board of Managers, as already provided for.

ARTICLE XV.

VACANCIES.

All vacancies that may occur in the Board by death, resignation, or otherwise, may be filled by the Board at any regular meeting. The Board shall also have power to fill any vacancies of Inspectors of Elections.

ARTICLE XVI.

HOLIDAYS.

SECTION 1. All contracts falling due on Sundays, legal holidays, or such holidays as are established by the Exchange, shall be completed on the preceding day, and when notice of five (5) days is required, it shall be given five (5) days previous to day of delivery.

SEC. 2. The Exchange may, by a two-third (2-3d) vote, at any meeting called on one day's notice and held at least six (6) days previous to the date of the holiday, order a holiday which shall be binding upon all members so far as regards any business upon contracts for future delivery.

SEC. 3. All Saturdays from and including the second (2d) Saturday of June to and including the third (3d) Saturday of September, shall be observed as holidays so far as concerns the issuing of and transferring of notices and orders upon contracts for future delivery, and the succeeding Mondays shall be substituted for such purposes.

SEC. 4. When the fifth day previous to delivery falls on a Sunday or a Holiday, the next previous business day must be substituted for the issue of the transferable order, and for the issue and transfer of notices, except as provided in section 3d.

When the day of delivery happens on the day following a Sunday or a Holiday the next previous business day must be substituted for the termination of the circulation of the transferable and warehouse orders, except as provided in section 3d.

ARTICLE XVII.

POSTING RULES.

All rules adopted by the Board of Managers, shall, after having been posted on the bulletin of the Exchange ten (10) days, be in force and binding on the members, and shall govern all cases to which they are applicable.

Any alteration in the rules relating to contracts, made under the provisions of this Article, shall be binding on all contracts entered into before as well as after its adoption, provided said alteration does not affect the amount of money to be paid, or the quality of the cotton to be received under such contracts.

ARTICLE XVIII.

PROHIBITED APPROPRIATIONS.

Section 1. There shall be no appropriation of money voted either by the Board of Managers or the Exchange, except for the strictly legitimate business of the Association.

ARTICLE XIX.

PROHIBITED NOTICES.

Section 1. There shall be no notices posted upon the bulletin boards, or any other portion of the Exchange building, except such as relate strictly to the legitimate business of the Association. No notice of any kind shall be posted except by proper authority.

ARTICLE XX.

SUSPENDING BY-LAWS OR RULES.

No By-Law or Rule of the Exchange, Board of Managers, or of any of the standing Committees, shall be suspended at any meeting, except by the unanimous vote of the members present.

ARTICLE XXI.

VISITORS.

Complimentary cards of admission to the floor of the Exchange, not transferable, and extending not over two consecutive weeks, may be granted by the Secretary to non-residents of the City, on application of any member ; but such card of admission shall not be renewed to any individual within three months of the time of its original issue, unless by vote of the Board of Managers.

It is distinctly understood that parties receiving such cards are not to engage in any transaction on the floor, except through members of the Exchange.

ARTICLE XXII.

Any proposed alteration of the By-Laws must first be approved by a majority of the Board of Managers, and after such approval, a copy of the proposed alteration with notice of the time of meeting to vote on the same shall be posted for ten (10) days at the general meeting room of the Exchange, and to be finally adopted must have the affirmative vote of two-thirds (2-3ds) of the members of the Exchange, present at such meeting.

RULES.

FORM OF CONTRACT.

Rule 1.—The following shall be the Contract in all cases, where no other form of Contract is specified at the time of sale. Verbal contracts, when proven to the satisfaction of arbitrators, shall have the same standing as if written.

[CONTRACT A].

Office of

New York, 187

SOLD for M____________________

To M____________________

45,000 lbs. in about *one hundred* Square bales Upland Cotton, deliverable from dock or store in said City, between the *first* and *last* days of inclusive. The delivery within such time to be at seller's option in lots of not less than fifty bales, upon five days' notice to buyer. The Cotton to be of any grade, from Good Ordinary to Good Middling, inclusive, at the price of cents per pound for Low Middling, with additions or deductions for other grades, according to the rates of the New York Cotton Exchange, at the time of delivery.

Either party to have the right to call for a margin, as the variations of the market for like deliveries may warrant, and which margin shall be kept good.

This Contract is made in view of, and in all respects subject to, the rules and conditions established by the New York Cotton Exchange, and in full accordance with Article XVII, of the By-Laws.

Respectfully,

Cotton Brokers.

DEPOSITS ON CONTRACTS.

Rule 2.—Upon all contracts, either party, on signing a contract, shall have the right to call for a margin of one-half ($\frac{1}{2}$) of one cent per pound to be deposited as per Rule 6 of the Rules of the Exchange, for the security of the buyer, and the same sum per pound for the security of the seller.

CONTRACT BINDING.

Rule 3.—No transaction made by a broker shall be binding on the opposite party until the name of the principal is given, unless by stipulation at the time, in which case it shall be the duty of the broker to give a satisfactory principal, or make himself principal, on the same day, before 5 P. M. If a broker makes himself principal in a transaction, he must pay the opposite party a transfer fee, unless he gives himself as principal at the time of making the trade. All contracts must be delivered by the maker at the office of the opposite party before 5 P. M. of the first day after the transaction, under the penalty of the payment of a transfer fee by the defaulter to the opposite party, and no contract shall be valid unless delivered before 5 P. M. of the second day after the transaction, the defaulting party being held liable to the other for all damages arising from such default.

CANCELING CONTRACT.

Rule 4.—Either party to a contract may close or cancel the same by giving notice in writing to the opposite party, any day before notice of delivery has been given. The party to whom notice is given has the option either to make settlement, or to receive a satisfactory contract made equal to that held by him. Any party holding a contract against another, corresponding in all respects (except as to price) with one held by the other party against him, may close, or cancel both, by giving notice in writing (any time before notice of delivery) to the opposite party; or where a "ring" settlement may be formed, all parties thereto shall be compelled to settle upon the terms as hereinafter provided. The settlement to be made on the day after notice. All parties paying differences shall be allowed a discount from the date of payment to the 15th day of the last month, covered by the contract, at the rate of seven per cent. per annum. Any party who, at the request of another, transfers a contract, or substitutes another for one held by him, shall receive five (5) cents per bale as compensation for making the change. All brokerages for making settlements, transfers and changes in contracts shall be paid by the party employing the broker.

MARGIN.

Rule 5.—When a contract provides for requiring a margin, or for payment of a variation in the market price, and notice in writing requiring the same is given on any day before 12 M., the margin or difference must be paid before three o'clock P. M. of the same day. When the notice is given after 12 M., the margin or difference must be paid before 12 M. of the next day. If a party fails to deposit any margin called for and due, in accordance with contract (see Rules 1st and 2d), the other party may, at his option, close the contract by giving written notice of his decision to do so. The party in default shall then account to the other party for the difference between the contract price and the average quotations of the Cotton Exchange for that day, for like deliveries, with an allowance against the defaulting party of one-quarter cent per pound.

The passing of the transferable order shall not prevent the calling of margins on contracts any time before the day of delivery.

DISPOSAL OF MARGIN.

RULE 6.—Margins shall be deposited in a Trust Company by the Broker, or by a person designated by the parties to a contract, he receiving a certificate of deposit made payable to the order of the buyer or seller, as the Broker or person depositing may direct. The certificate shall be deposited with the Secretary of the Exchange, who shall deliver it only upon the order of the Broker or person depositing, and of the two parties to the contract. In case the two parties do not agree as to the delivery, the matter may, by either of them, be referred to an Arbitration Committee for decision, with the right of either party to appeal from their decision to the Board of Appeals. On the decision of an Arbitration Committee, if not appealed from, or that of the Board of Appeals, if appealed to, the Secretary shall deliver the certificate to the Broker or person depositing it, who shall promptly indorse to each party the amount to which each is entitled by such decision.

NOTICES.

RULE 7.—Where notice of delivery on part of seller, or demand of Cotton by a buyer, who has option so to do, is required by a Contract, it shall be given by the party furnishing the Cotton in the one case, and by the buyer in the other case, to the party requiring said notice, before 10 A. M. of the fifth day prior to the delivery. The party receiving the notice may transfer the same by indorsement to subsequent parties, and it may be given from one transferee to another. Every transfer must be promptly made, and every person receiving the notice shall indorse upon it the time he receives it. Any party failing to forward such notice promptly shall be liable to a penalty of one-eighth cent per pound to the party with whom the notice shall lodge at the close of the day. All transfers shall be made previous to 4 o'clock P. M. Should the office of the party, to whom notice is to be given, be closed, it shall be a good service to give the notice to the Secretary of the Exchange. He shall indorse thereon the day and time of its receipt, and post notice thereof on the bulletin of the Exchange. When no notice is given, the delivery shall be made on the last day stipulated by the contract. All notices must be for 45,000 or 22,500 pounds of Cotton, and must be a notice only and not connected with an order. Every party, who issues an original notice of intention to deliver *before* the last day covered by the contract, must either deliver to the party so notified, a transferable order, as in Rule 12, *or* an order on warehouse or place of delivery, before 12 M. of the day before the Cotton is due—*or*, in default thereof, shall pay to the party notified one-quarter ($\frac{1}{4}$) cent per pound on the quantity of Cotton designated by such notice. The contract to remain in full force.

SETTLEMENTS.

RULE 8.—When the settlement of a contract depends upon the market price on a particular day, the average quotations of the New York Cotton Exchange on that day for like deliveries, as declared by the official record of the Secretary of the

Exchange, is to be taken as the basis of settlement. In case of failure to deliver any portion of the Cotton named in the contract, when due, no obstacle being made by buyer, the basis of settlement of Cotton due on such contract for default in delivery, shall be one-quarter (¼) of one cent per pound above the average quotations for spot Cotton of the day of delivery. And if failure to receive any portion of the Cotton named in contract shall prove to be the fault of the buyer, the settlement shall be made at the average quotations for spot Cotton of the day following, with the addition of one-quarter (¼) of one cent per pound in favor of seller. Provided, however, that no seller shall be entitled to receive penalty, unless he has given the stipulated notice of intention to deliver ; and no buyer (who has not received notice), unless demand is made by him two days before expiration of the contract. Provided, also, that no defaulting party can claim settlement under this rule, except upon evidence that the default was unintentional and not premeditated. Nothing, however, in this rule shall be construed to prevent a settlement by mutual agreement.

QUOTATIONS FOR SALE REPORTS.

RULE 9.—The reported sales for each month, as made each day from the opening of the Exchange until 3 P. M., shall be taken as the full day's report, and shall govern the Secretary in making the average quotations of that day for time contracts. When no sales for any month are reported at the Exchange, the Committee on Quotations shall, if required, fix the quotations for such month.

TERMS OF DELIVERIES ON CONTRACTS.

RULE 10.—On all contracts calling for deliveries in lots of not less than 50 bales, the deliveries must be made in lots of not less than 50 bales on the same day, and at one place, provided, that when a delivery has been made on account of a contract, the balance to finish the contract shall be taken as a proper delivery, though it be less than 50 bales ; and two orders given at the same time for delivery on the same day, making together 100 bales from two places, shall be taken as a proper delivery. No one order for Cotton to be delivered on contract shall be for a larger quantity than 45,000 pounds. All orders on warehouse or other place of delivery must be delivered to the first receiver before 12 M. of the day before the Cotton is due, but such orders may be transferred until 5 P. M. of that day, provided the first transfer is made before 12½ P. M. of that day, and every transfer must be promptly made, every party noting on the order the time he receives it. Any party may reject an order which shows from the date thereon that it has been unduly delayed in transfer, and the last transferee of an order which has been so delayed shall be entitled to collect, from the party causing the delay, one-quarter cent per pound on the amount of Cotton called for by such order, as liquidated damages. *And all contracts* on which orders shall not be delivered in accordance with these stipulations, *and all transferable orders* (under Rule 12), if presented, on which orders on warehouse or place of delivery shall not be delivered to the receiver, as stipulated above, *shall be settled* at the average quotations for spot Cotton of the day the Cotton is due, with the addition of one-quarter cent per pound against the defaulting party.

TIME FOR PAYMENT OF COTTON DELIVERED ON CONTRACTS.

RULE 11.—The contract price of Cotton delivered on time contracts and the approximate value of any other Cotton delivered shall be payable on the day following delivery, provided, weigher's return and bill are presented before 12 M., of said day, following delivery of Cotton, except where payment on delivery is required. Cotton shall not be at the risk of the buyer, until it has passed the scale in course of delivery, unless otherwise expressly agreed upon.

TRANSFERABLE ORDER.

RULE 12.—An order for 45,000 pounds of Cotton, in about 100 square bales, at a price not varying more than one-quarter of one cent per pound from the average price for like deliveries of the day previous to its issue, in the following form:

Messrs. ————————

On the deliver to the order of , on account of our contract sale to them dated at *cents* per pound, 45,000 pounds of Cotton, in about 100 square bales, to be any grade from Good Ordinary to Good Middling, inclusive, which Cotton is to be received and held by the last indorser hereon as custodian for us (insured for whom it may concern), and subject to our order until we are paid at the rate of *cents* per pound.

————————

Drawn and accepted by a member of the New York Cotton Exchange, and indorsed as follows:

In consideration of one (1) dollar paid to each of the subscribers by , receipt of which is hereby acknowledged, it is agreed that the last subscriber hereon will, on or before 12 M. on the day preceding the , present the within order to , receiving the Cotton named therein, and hold the same as custodians and agents for said (insured for whom it may concern), and subject to their order as the true owners of the same, until they are paid the full amount of cents per pound, and to settle with them on the basis of Low Middling, with allowance for variation of grade in accordance with quotations of the New York Cotton Exchange on day of delivery. It is further agreed that each subscriber hereon shall continue his or their liability to each other for the fulfillment of the contracts referred to, and that the sum paid to shall be accepted by each as payment on account of their mutual obligations.

————————

Agreement Stamp.

Transferred in the following form:

We accept the within order from , with all the conditions and obligations thereof on account of our contract purchase from them, dated at cents per pound, we paying them dollars, to make price equal to *cents* per pound the contract price with

————————

If tendered by the drawer before 4 P. M. on the fifth day before delivery of Cotton is due, or if tendered by transfer before 11 A. M. of the day before the delivery of Cotton is due, shall be accepted by any member of the Exchange to whom Cotton is due under any contract; and the price shall be made equal to the price of the contract on which it shall be tendered, provided, that it is otherwise in accordance with such contract; and the amount of Cotton delivered on such order shall be accepted as so much received on account of the contracts specified in the order and transfers. Any party holding such order and failing to present the same before 12 M. of the day before the one specified in such order, shall settle for such Cotton on the average price of that day, for the like delivery, with one-quarter cent per pound in favor of the seller; and each party issuing or transferring the same shall collect that amount from the party to whom they issued or transferred such order in settlement.

Any party delivering Cotton on such order, and failing to receive payment for the same according to the terms thereof, shall give a written notice to the party to whom he issued within 48 hours from the close of such delivery, and said written notice must be promptly transferred by every party receiving such notice to the party to whom he transferred such order.

SEEDY AND FRAUDULENTLY PACKED COTTON.

Rule 13.—Seedy or fraudulently packed Cotton shall be rejected and not be deliverable; and where mixed-packed Cotton is in the bale, the whole bale shall be deemed of the grade of the poorest quality, and if that is below the lowest grade called for in the contract, or shown in the samples at the time of sale, it shall not be deliverable.

CLASSIFICATION OF COTTON ON DELIVERY, &c.

Rule 14.—In determining the classification of Cotton delivered, the sand and dirt shall be taken into account. If Cotton of lower grade than is specified in contract is delivered, settlement shall be made the same as if such grades had been included in the contract; but nothing in this rule shall be construed as requiring any one to receive other grades than those embraced in the contract. All Cotton shall be classed within 24 hours after delivery, and in delivery of F. O. B. Cotton the samples shall be classed and passed upon within 12 hours after the samples are opened. When sold on a basis, the additions or deductions for other grades shall be according to the rates of the New York Cotton Exchange on the day on which the Bills of Lading are dated, except when otherwise agreed upon. If notice for arbitration is not given within 24 hours after receipt of the Certificate of Classification, the settlement shall be made in accordance with such certificate.

BANDS AND ROPES.

Rule 15.—Six iron bands or ropes, not exceeding in weight 12 pounds in the aggregate, shall be considered sufficient for each bale of Cotton. Any excess shall, at the option of the buyer, be removed from the bale, or be deducted from its gross weight.

WEIGHT.

Rule 16.—No bale of Cotton weighing less than 300 pounds shall be deemed merchantable, and any buyer may refuse to receive the same

HOURS FOR DELIVERY OF COTTON.

Rule 17.—The day on which Cotton is due the delivery shall not be required to commence before 8 A. M., nor to continue later than sunset. Cotton shall be delivered or received at the rate of not less than 20 bales per hour, from each scale, when required by buyer or seller. When the seller requires the buyer to receive Cotton from two scales, at one place of delivery, on one order, it must be noted on the delivery order.

WEATHER SUITABLE FOR DELIVERY.

Rule 18.—If the weather be deemed unsuitable for the delivery of Cotton, by any party interested in a delivery, on any day, the Secretary of the Exchange shall, at their request, obtain the opinion thereon of three Members of the Exchange (not interested in deliveries on that day), and if a majority decide that the weather be unsuitable for the delivery of Cotton, the Secretary shall post their certificate on the Bulletin of the Exchange, dating the time of posting, which shall remain posted until a majority of said three members shall decide the weather to be suitable, when it shall be taken from the Bulletin and filed, noting the time of removal. During the time such Certificate is posted on the Bulletin, all deliveries of Cotton may be suspended, at the option of either party to any delivery; and any delivery suspended under this rule shall be entitled to an extension of time—two hours more than the time such Certificate was posted. The Secretary shall give a certified copy of said Certificate to any member requiring it, on the receipt of 50 cents, and such copy shall be a sufficient authority for the suspension and resumption of delivery of any lot of Cotton by the parties to the delivery.

TIME ALLOWED FOR TAKING DELIVERY.

Rule 19.—All Cotton shall be received within ten days of date of purchase.

COMPENSATION FOR GRADES ON AVERAGE PRICE.

Rule 20.—When Cotton of various grades is sold by sample, at an average price, and, for proper causes, rejections are made, and such rejected bales are from the grades better than the average, the seller shall make good to buyer the difference in value; and if the rejections are from grades below the average, the buyer shall make good to the seller the difference

SALES OF COTTON TO ARRIVE.

Rule 21.—The following shall be the contract in all cases, for cotton to arrive, where no other form of contract is specified at the time of sale.

[FORM OF CONTRACT].

NEW YORK, 187

SOLD for M_______________

To M________ ________

Bales Cotton to arrive, to be taken from dock or docks, depot or depots, by the buyer, after having been weighed by seller, upon any day (Sundays and legal holidays excepted), if notified by the seller before three o'clock P. M., provided the whole lot, or as much as twenty-five (25) bales are ready for delivery. It is mutually agreed by both parties that if the Cotton covered by this contract should arrive in more than one parcel, each shall constitute a separate lot, for which a distinct bill shall be rendered, and if desired, an independent arbitration.

In the event of the loss, from any cause, of the Cotton named in the contract, said contract shall be canceled, without prejudice to either party; but if the loss is only partial, the contract shall remain in full force and effect, to the extent of such portion as may be saved, and arrive consigned to and under the control of the seller; provided it arrives in merchantable order, or may be put into such condition, in the opinion of three competent and disinterested parties, without an undue sacrifice upon the part of the seller.

Nothing in this Rule shall be construed to apply to Cotton sold to arrive within a stipulated period of time, provided such period of time has expired.

STAINED COTTON.

RULE 22.—Stained Cotton not exceeding 20 per cent. of the whole amount called for on a contract, may be delivered, and, when so delivered, shall be classed down one grade from what such Cotton would class if not stained.

CLAIMS FOR IRREGULARITIES.

RULE 23.—After Cotton has been examined, received and passed upon by the Broker or agent of the buyer, no claim shall be made against the seller, except for fraudulent packing. When claims are made they shall be in writing, stating the particulars of the fraudulent packing, and the same shall be verified by oath or affirmation. The claim shall then be deemed *prima facie* valid in favor of the claimant, and it can only be defeated by the decision of an Arbitration Committee or of the Board of Appeals. All claims for fraudulent packing, giving the marks by which the Cotton was sold, and all other legible marks and numbers shall be good against all sellers for nine months from date of sale, and without limit of time against the original packer.

TRANSIT AND FREE ON BOARD COTTON.

Rule 24.—Transit Cotton, when offered for sale, should be accompanied by a memorandum giving the name of the vessel, ports of loading and destination, rate of freight, where insured, and rate of insurance, marks and quantity of each mark, and total weight when known.

If the amount of insurance is not stated, the buyer shall consider, and the seller shall guarantee, the cotton insured for not less than the price at which it may have been sold. The seller shall cause additional insurance to be made when so requested in writing, and in the event of loss or damage before the time named in the contract for delivery of the documents, the Cotton shall be considered the property of the buyer, who must promptly pay for the same on presentation of shipping documents as specified in the contract, the seller then transferring the insurance to the buyer, who from time of purchase shall accept the responsibility of the Underwriters.

If the buyer does not pay for the cotton at the time named in the contract, the seller shall have the option to make other disposition of the same holding the buyer responsible for any loss he may sustain

In case of the arrival of the Cotton at the port of destination before the seller shall have delivered the shipping documents to the buyer, the landing expenses incurred at said port are to be borne by the purchaser, but it shall be the duty of the seller, at the time of sale, to state (when within his knowledge) if the Cotton has sailed from port or ports of loading, or arrived at port or ports of destination, or if any accident has occurred to vessel or cargo; failing to do so, he shall become responsible for all such proper expenses as may be incurred by reason of such neglect on his part; and for neglecting to give such information, the buyer shall have the option of canceling the contract before the documents are delivered.

The seller shall guarantee the correctness of the samples shown, and it shall be the duty of the broker promptly to seal them, and they are only to be opened for comparison with samples to be drawn at the port of destination, under the inspection of referees, whose names shall be specified in the contract, and whose decision shall be final; all claims for difference shall be void if the samples are opened without the consent of seller or seller's referees.

Unsealed samples shall not be removed from the office of the broker in the original sale, for re-sale through another broker or otherwise, without consent of the seller.

All claims for difference in samples, if accompanied by proper vouchers, must be paid by seller within three days from presentation.

A bill or invoice for the cotton must be rendered to the buyer as soon as practicable after sale, accompanied by a sworn or licensed weigher's certificate of weights, and shall be paid for within seven days after sale, on delivery of shipping documents.

Free on board sales shall be subject to the same terms and conditions of marine insurance and guaranty of samples which attach to Cotton sold in transitu.

Settlements for difference in grades shall be determined by the quotations of the Cotton Exchange, on the day the bill of lading is dated, which must be mailed not later than two days after its date; payment shall be made on presentation of the shipping documents, upon receiving one day's notice, unless otherwise agreed upon at the time of the sale.

When sales are made for a specified grade, or by type samples, the buyer shall have the right to reject the Cotton tendered, if the quality falls off more than a half grade, but in exercising this privilege he must release the seller from the fulfillment of the contract, unless the buyer is willing to receive it at a proper allowance, but no allowance shall be made to the seller if the Cotton proves to be better than the grade or type stipulated, unless otherwise agreed upon at the time of the sale.

When claims are presented for so-called "country damage" the certificate of the weigher or shipping clerk at time of shipment, corroborated by the bills of lading, shall be taken as "*prima facie*" evidence of the condition of the cotton. Failure to obtain remedy from the vessel for delivery at port of destination in improper condition as to exterior of the package, shall not invalidate the original evidence.

MODE OF SELECTING ARBITRATORS.

RULE 25.—When more than two parties are apparently interested in any difference arising between members, an arbitration may be called by any two members interested, and every party they may require to be included in such arbitration, shall be considered parties thereto, and shall be notified of the call for arbitration; and each party shall nominate one arbitrator, and if any neglect to nominate, one shall be nominated for each party neglecting to nominate, and the Secretary of the Exchange shall draw from the whole number of names two, who shall choose from the Arbitration Committee (see Article XI of the By-Laws) a third person to act with them, and the three so chosen shall decide as to the liabilities of each party, and as to the portion of the expenses to be borne by each; and either of the parties may appeal from such decision, in which event all the parties included in the arbitration shall be included in the appeal for decision, as to the liability of each party and the portion of expenses to be paid by each.

DECISIONS BY ARBITRATORS AND APPEALS.

RULE 26.—All decisions of the Board of Appeals, and of Arbitration Committees, shall be in writing, and shall be given to the Secretary of the Exchange, who shall record the decisions as given and send copies to each party to the arbitration. The party or parties to pay the fees of the arbitrators, or of the Board of Appeals, shall, on receipt of such decisions, pay over the amount of such fees to the Secretary, who shall pay same to the arbitrators. All appeals from decisions of Arbitration Committees must be made within 24 hours of the date of notice of a decision. In case no appeal is made in such specified time, the decisions of Arbitration Committees shall be final.

RATES OF BROKERAGE.

RULE 27.—The following rates of brokerage, as proposed by the Cotton Brokers, are expected to govern all transactions made by Members of this Exchange :—For Cotton on the spot or to arrive, 25 cents per bale shall be paid by each party to a contract, which shall cover the broker's services in attending to receipt and delivery. For transit Cotton, or free on board contracts, 25 cents per bale shall be paid by each party. For contracts for future delivery, 12½ cents per bale shall be paid by each party. For settlements of contracts, 6¼ cents per bale shall be paid by the party employing the broker to make the same. When Cotton is delivered on contracts, 12½ cents per bale more shall be paid by the party delivering the Cotton, and 12½ cents per bale by the party receiving it. Brokerages are due, and may be collected, as soon as the contract for forward delivery is accepted, and after payment on all other contracts.

APPENDIX.

STATISTICS OF THE MANUFACTURES OF COTTON IN THE UNITED STATES.

While this work was in press, statistics of the Manufactures of Cotton in the United States, based upon the returns of the census of 1870, were made public. They are semi-official in their character, and are vouched for by so competent an authority as Mr. B. F. Nourse, of Boston, as substantially correct. There has been barely sufficient time to compile them in a comparative table with the returns of the census of 1860, and in that relation they are printed below. I have neither time nor room for comment, their obvious deductions must be left to the intelligent consideration of the reader.

Statistics of the Manufactures of Cotton in the United States.

STATES.	1860.		1870.	
	No. Mills.	Capital.	No. Mills.	Capital.
Maine	19	$6,018,325	23	$9,839,685
New Hampshire	47	12,586,880	36	13,332,710
Vermont	8	271,200	8	670,000
Massachusetts	217	33,704,674	191	44,714.375
Rhode Island	153	10,052,200	139	18,836,300
Connecticut	129	6,627,000	111	12,710,700
Total New England States	570	$69,260,279	508	$100,103,770
New York	79	$5,583,470	81	$8,511,336
Pennsylvania	185	9,203,940	138	12,550,720
New Jersey	44	1,320,550	27	2,762,000
Delaware	11	582,500	6	1,165,000
Maryland	20	2,254,500	22	2,734,250
District of Columbia	1	45,000	....	....
Total Middle States	340	$18,789,069	274	$27,723,306
Ohio	8	$265,000	7	$555,700
Indiana	2	251,000	4	551,250
Illinois	3	4,700	5	151,000
Iowa	....	....	1	1,500
Utah	1	6.000	3	42,000
Missouri	2	169,000	3	489,200
Kentucky	6	244,000	5	405,000
Total Western States	22	$939,700	28	$2,195,650
Virginia	16	$1,367,543	11	$1,128,000
North Carolina	39	1,272,750	33	1,030,900
South Carolina	17	801,825	12	1,337,000
Georgia	33	2,126,103	34	3,433,265
Florida	1	30,000	....	....
Alabama	14	1,316,000	13	931,000
Louisiana	2	1,000,000	4	592,000
Texas	1	450,000	4	496.000
Mississippi	4	230,000	5	750,500
Arkansas	2	37,000	2	13,000
Tennessee	30	965,000	28	970,650
Total Southern States	159	$9,596,221	146	$10,682,315
Total United States	1,091	$98.585,269	956	$140,706,041

Statistics of the Manufactures of Cotton in the United States.

STATES.	1860. RAW MATERIAL USED.		1870. RAW MATERIAL USED.	
	Pounds.	Value.	Pounds.	Value.
Maine	23,733,165	$3,319,335	25,887,771	$6,746,780
New Hampshire	51,002,324	7,128,196	41,469,719	12,318,867
Vermont	1,447,250	181,030	1,235,652	292,269
Massachusetts	134,012,759	17,214,592	130,654,040	37,371,599
Rhode Island	41,614,797	5,799,223	44,630,787	13,268,315
Connecticut	31,891,011	4,028,406	31,747,309	8,818,651
Total New England States	283,701,306	$37,670,782	275,625,278	$78,816,482
New York	23,945,627	$3,061,105	24,783,351	$6,990,626
Pennsylvania	37,496,203	7,386,213	32,953,318	10,724,052
New Jersey	9,094,649	1,165,435	7,920,035	1,964,758
Delaware	3,403,000	570,102	2,587,615	704,733
Maryland	12,880,119	1,698,413	12,693,647	3,409,426
District of Columbia	294,117	47,403	....	....
Total Middle States	87,113,715	$13,928,671	80,937,966	$23,793,595
Ohio	3,192,500	$374,100	2,226,400	$493,740
Indiana	1,813,944	229,925	2,070,318	542,875
Illinois	95,000	11,930	857,000	177,525
Iowa	....	....	20,000	4,950
Utah	12,000	6,600	23,500	7,051
Missouri	990,000	110,000	2,196,600	481,745
Kentucky	1,826,000	214,755	1,584,625	375,048
Total Western States	7,929,444	$946,710	8,908,443	$2,082,934
Virginia	7,544,297	$811,187	4,255,383	$937,820
North Carolina	5,540,738	622,363	4,238,276	963,809
South Carolina	3,978,061	431,525	4,756,823	761,469
Georgia	13,907,904	1,466,375	10,921,176	2,504,758
Florida	200,000	23,000	....	....
Alabama	5,246,800	617,633	3,249,523	764,965
Louisiana	1,995,700	226,000	748,525	161,485
Texas	588,000	64,140	1,079,118	216,519
Mississippi	638,800	79,800	580,764	123,568
Arkansas	187,500	11,600	66,400	13,780
Tennessee	4,072,710	384,548	2,872,582	595,789
Total Southern States	43,960,510	$4,739,371	32,768,570	$7,042,962
Total United States	422,704,975	$57,285,534	398,248,257	$111,735,973

Statistics of the Manufactures of Cotton in the United States.

STATES.	1860.		1870.	
	No. Spindles.	No. Looms.	No. Spindles.	No. Looms.
Maine	281,056	6,877	459,772	9,902
New Hampshire	636,788	17,336	749,843	19,091
Vermont	17,600	352	28,768	628
Massachusetts	1,673,498	42,779	2,116,521	55,343
Rhode Island	814,554	17,315	1,043,242	18,975
Connecticut	435,406	8,675	597,142	11,943
Total New England States	3,858,962	93,344	4,995,288	115,882
New York	348,584	7,887	492,573	17,218
Pennsylvania	476,979	12,994	434,246	12,862
New Jersey	123,548	1,567	200,580	2,170
Delaware	38,974	986	29,534	771
Maryland	51,835	1,670	89,112	1,947
District of Columbia	2,560	82	....	....
Total Middle States	1,042,480	25,185	1,246,045	34,968
Ohio	19,664	540	23,240	208
Indiana	11,000	375	11,736	448
Illinois	....	....	1,856	16
Iowa	....	....	....	....
Utah	70	....	1,020	11
Missouri	5,000	89	16,715	415
Kentucky	8,192	76	3,526	152
Total Western States	43,926	1,071	58,093	1,250
Virginia	49,440	2,160	77,116	1,310
North Carolina	41,884	761	39,897	618
South Carolina	30,890	525	34,940	745
Georgia	85,186	2,041	87,602	1,887
Florida	1,600	20	....	....
Alabama	35,740	623	28,056	632
Louisiana	6,725	150	13,082	292
Texas	2,700	100	8,878	235
Mississippi	6,344	90	3,526	152
Arkansas	....	....	1,125	....
Tennessee	29,850	243	27,923	313
Total Southern States	290,359	6,713	322,145	6,184
Total United States	5,235,727	126,313	6,621,571	1,582,804

Statistics of the Manufactures of Cotton in the United States.

STATES.	1860.			1870.			
	No. of Hands.		Cost of Labor.	No. of Hands.			Cost of Labor.
	Male.	Female		Male.	Female	Child'n	
Maine	1,828	4,936	$1,368,888	2,606	6,246	587	$2,565,197
New Hampshire	3,829	8,901	2,883,804	3,752	7,490	1,300	3,989,853
Vermont	157	222	78,468	125	242	74	125,000
Massachusetts	13,691	24,760	7,798,476	13,694	24,065	5,753	13,589,305
Rhode Island	6,353	7,724	2,847,804	5,583	8,028	3,134	5,224,650
Connecticut	4,028	4,974	1,743,480	4,443	4,734	3,909	3,246,783
Total New England States	29,886	51,517	$16,720,920	30,203	50,805	13,757	$28,740,788
New York	3,107	4,552	$1,405,292	2,608	4,546	1,990	$2,626,131
Pennsylvania	6,412	8,582	2,768,340	8,359	6,097	2,774	3,496,986
New Jersey	1,010	1,524	468,336	1,086	1,745	683	1,009,351
Delaware	520	589	218,352	225	286	215	190,069
Maryland	1,093	1,594	582,780	686	1,452	720	671,933
District of Columbia	70	25	19,800	......	......	......	....
Total Middle States	12,212	16,866	$5,462,900	12,964	14,126	6,382	$7,994,470
Ohio	372	468	$151,164	216	147	99	$113,520
Indiana	177	190	84,888	119	179	206	113,200
Illinois	10	1	2,640	26	31	41	25,500
Iowa	......	......	....	3	3	......	2,275
Utah	4	3	3,420	10	2	4	6,300
Missouri	85	85	30,600	107	154	100	120,300
Kentucky	130	116	41,280	77	71	121	57,951
Total Western States	778	863	$313,992	558	587	571	$439,046
Virginia	694	747	$260,856	921	507	313	$229,750
North Carolina	449	1,315	189,744	258	916	279	182,951
South Carolina	342	549	123,300	289	508	326	257,680
Georgia	1,131	1,682	415,332	1,147	1,080	619	611,868
Florida	40	25	7,872	......	......	......	....
Alabama	543	769	198,408	303	445	289	216,679
Louisiana	220	140	49,440	123	57	66	60,600
Texas	130	......	15,600	184	52	55	68,211
Mississippi	106	109	36,264	78	88	99	61,833
Arkansas	14	11	4,428	8	3	6	4,100
Tennessee	323	576	139,180	252	463	175	178,156
Total Southern States	3,983	5,923	$1,440,424	3,563	4,119	2,222	$1,871,828
Total United States	46,859	75,169	$23,940,108	47,288	69,637	22,942	$39,046,132

Statistics of the Manufactures of Cotton in the United States.

STATES.	1850.	1860.	1870.
	Value of Product.	Value of Product.	Value of Product.
Maine	$2,630,616	$6,235,623	$11,844,181
New Hampshire	8,861,749	13,699,994	16,999,672
Vermont	280,300	357,450	546,510
Massachusetts	21,394,401	38,004,255	59,493,153
Rhode Island	6,495,972	12,151,191	22,049,203
Connecticut	4,122,952	8,911,387	14,026,334
Total New England States	$43,785,990	$79,359,900	$124,759,053
New York	$5,019,323	$6,676,878	$11,178,211
Pennsylvania	5,812,126	13,650,114	17,490,080
New Jersey	1,289,648	2,217,728	4,015,768
Delaware	538,439	941,703	1,060,898
Maryland	2,021,396	2,973,877	4,832,808
District of Columbia	100,000	74,400	....
Total Middle States	$14,780,932	$26,534,700	$38,587,765
Ohio	$594,204	$723,500	$681,335
Indiana	86,660	344,350	778,047
Illinois	....	18,987	279,000
Iowa	....	....	7,000
Utah	....	10,000	16,803
Missouri	142,900	230,000	798,850
Kentucky	445,639	315,270	251,550
Total Western States	$1,269,403	$1,642,107	$2,792,585
Virginia	$1,446,109	$1,489,971	$1,435,800
North Carolina	985,411	1,046,047	1,345,052
South Carolina	842,410	713,050	1,529,930
Georgia	1,395,056	2,371,207	3,648,973
Florida	49,920	40,000	....
Alabama	398,585	1,040,147	1,088,767
Louisiana	....	466,500	251,550
Texas	....	80,695	374,598
Mississippi	22,000	176,328	234,445
Arkansas	17.360	23,000	22,362
Tennessee	508,481	698,122	941,542
Total Southern States	$5,665,362	$8,145,067	$10,873.019
Total United States	$65,501,687	$115,681,774	$177,022,422

INDEX.

www.ingramcontent.com/pod-product-compliance
Lightning Source LLC
LaVergne TN
LVHW021051110826
845150LV00001B/42

* 9 7 8 1 4 2 5 5 6 7 6 9 9 *